ACPL ITEM
DISCARDED

621.36
Fundamentals of fibre optics
in telecommunication and

**DO NOT REMOVE
CARDS FROM POCKET**

ALLEN COUNTY PUBLIC LIBRARY
FORT WAYNE, INDIANA 46802

You may return this book to any agency, branch,
or bookmobile of the Allen County Public Library.

DEMCO

FUNDAMENTALS OF FIBRE OPTICS
IN TELECOMMUNICATION AND SENSOR SYSTEMS

Edited by
BISHNU P. PAL

JOHN WILEY & SONS
NEW YORK • CHICHESTER • BRISBANE • TORONTO • SINGAPORE

First Published in 1992 by
WILEY EASTERN LIMITED
4835/24 Ansari Road, Daryaganj
New Delhi 110 002, India

Distributors:

Australia and New Zealand :
JACARANDA WILEY LIMITED
PO Box 1226, Milton Old 4046, Australia

Canada :
JOHN WILEY & SONS CANADA LIMITED
22 Worcester Road, Rexdale, Ontario, Canada

Europe and Africa :
JOHN WILEY & SONS LIMITED
Baffins Lane, Chichester, West Sussex, England

South East Asia :
JOHN WILEY & SONS (PTE) LIMITED
05-04, Block B, Union Industrial Building
37 Jalan Pemimpin, Singapore 2057

Africa and South Asia :
WILEY EASTERN LIMITED
4835/24 Ansari Road, Daryaganj
New Delhi 110 002, India

North and South America and rest of the world :
JOHN WILEY & SONS INC.
605, Third Avenue, New York, NY 10158, USA

Copyright © 1992 WILEY EASTERN LIMITED
New Delhi, India

Library of Congress Cataloging-in-Publication Data

ISBN 0-470-22051-1 John Wiley & Sons Inc.
ISBN 81-224-0469-3 Wiley Eastern Limited

Typeset by Jay Compusoft, New Delhi and printed at Taj Press, New Delhi, India

I have great pleasure in dedicating my contribution in bringing out this book to the fond memories of my parents for their unwavering encouragement and support in my academic pursuit.

Bishnu Pal
Editor

Preface

Fibre optics is a very important constituent of modern information technology. In the western world, where long distance communication is taken so much for granted, fibre optics is a common concept. One major economic benefit offered by fibre optics is very high information transmission rate at low cost per circuit-km. It also offers immunity from electromagnetic interference and cross-talk, large repeater spacings and small size. The first fibre optic telephone link went public in 1977. Since then, the industrially advanced nations around the world have been striving to deploy fibre optics in all sectors of communication. In less than 10 years since the first fibre optic public telephone link, R&D community around the world was already talking about fourth generation fibre optic communication systems. Rarely, since the discovery of the transistor, have we noticed such a fantastic growth rate of a new technology. Development of high purity optoelectronic materials and composites, which have taken place in the last two decades, allowed manufacture of high performance LEDs, laser diodes, low-loss optical fibres, high performance detectors, optical and optoelectronic integrated circuits (OIC and OEIC). They have all contributed to this rapid growth.

As an important by-product of this phenomenal progress, a new class of ultra-sensitive optical sensors based on fibre optics has taken shape. Generically called fibre optic sensors, these are immune to electromagnetic interference and they are compatible with low-loss optical fibre telemetry (and, hence, they permit remote sensing and measurements). A variety of fibre optic sensors are presently being developed for large scale use in the industrial and biomedical sectors.

Optics has long been considered an exclusive domain of Physicists; it rarely formed a component of the electronic or electrical engineering curricula until the laser was developed in the early sixties. Since then, we have seen a progressive penetration of optics into the realm of electronics. This has led to the emergence of the interdisciplinary field of R & D which we now call PHOTONICS, in which flow of photons rather than electrons become the carrier of information. Fibre optics constitutes a major constituent of this field of study. This book provides semi-tutorial presentation of the fundamentals of this emerging technology as applied to telecommunication and sensor development. The book conceived after the simultaneous publication in 1986 of the joint special issues on optoelectronics and optical communication of the two journals: JIETE and IETE Technical Review, published by Institution of Electronics and Telecommunication Engineers (India), which I had Guest Edited. It was thought that a book could be based on the reprints/updated versions of some of the articles that appeared in these special issues along with new articles so as to enable its use as a reference source for class room use at senior level UG and PG courses in Engineering and Applied Physics. With this aim each chapter is appended with a large number of references to the original publications. In this respect the book should prove useful also to researchers and R & D engineers who want a tutorial introduction to the technologies of fibre optic telecommunication and sensors. Readers having a basic knowledge of electromagnetic theory, lasers, semiconductor Physics, and electronics at the senior UG or PG level and having an interest in fibre optics should find the book useful. Most of the chapters begin with a tutorial introduction to the topic.

The book is broadly divided into three parts. The first part is devoted to fundamentals of optical waveguides. Chapter 1 and 2 of this part deal with historical evolution and current status of fibre optics and related technologies. Chapters 3-6 that follow describe electromagnetics of fibre and integrated optical waveguides and the physics behind their propagation characteristics. Nonlinearities in optical waveguides have lately attracted a great deal of attention and is a subject of great contemporary interest from an academic and device point of view. Nonlinear processes in optical waveguides are discussed in chapters 7 and 8. Chapters 9-12 in this part are devoted to fabrication and characterisation of optical fibres and cables.

It is quite natural to appreciate that optical fibres and cables or integrated optical waveguides, in isolation, would remain an academic curiosity unless telecommunication systems are built around them. The second part of this book is concerned with optical sources, detectors and communication systems designs. Principles of semiconductor laser sources and detectors/receivers for optical communication are discussed in chapters 14-17. Chapter 18 is devoted to fundamentals of digital communication, which is the most well suited modulation scheme for optical communication. Chapters 19 and 20 are concerned with optical fibre system designs for direct detection and coherent optical fibre transmissions, respectively. Several guided wave optical devices are discussed in chapter 21 on integrated optical devices.

The last part of the book is devoted to industrial applications of fibre optic devices and sensors. Several fundamental topics in this area are discussed. Chapters 23 and 24 describe principles behind intensity and phase modulated fibre sensors. Fused fibre couplers and several other in-line fibre components form the subject matter of chapters 25 and 26 while chapters 27 and 28 are concerned with signal processing in single-mode optical fibre sensors and multiplexed operation of fibre sensors, respectively. The final chapter of the book deals with application of optical fibres in power industries.

I am grateful to the authors, all of whom have made rich research contributions at various stages during the growth of this field. I would like to thank all my colleagues and students of our Fibre Optics Group at IIT Delhi for their encouragement and who are to be credited for the sustained cooperative team spirit and academically stimulating environment that we enjoy in our Group. I am indebted to Prof. A.K. Ghatak and Prof. K. Thyagarajan for their constructive criticisms and support throughout the preparation of this book. Prof. Thyagarajan had kindly agreed to write the Epilogue for chapter 8 for which I thank him and also Dr. M.R. Shenoy for a careful reading of this chapter at the final proof reading stage. I would like to thank Prof. Bharathi Bhat and Prof. S.C. Dutta Roy, the then members of the Editorial Board of IETE for their keen interest and support to bring out a special issue of JIETE on optoelectronics and optical communication. This book, as already stated, is essentially a spin-off of that special issue published in 1986. During the final stages of preparation of this book, I was spending my sabbatical year at the National Institute of Standards and Technology with the Optical Electronic Metrology Group at Boulder, Colorado, USA. I am immensely grateful to my host Dr. R.L. Gallawa for his constant help and useful suggestions during my stay there. Thanks are due to Mrs. Lata Mathur for her flawless typing and to Mr. N.S. Gupta, and Mr. R. Kapoor for their help in drawing of the figures during preparation of the manuscript. I thank my colleague Mr. Varghese Paulose for his help in preparing some of the illustrations. Finally I am grateful to my wife Subrata and daughter Parama for their patience and support.

New Delhi *Bishnu Pal*

Contributors

Govind P. Agrawal, Institute of Optics, University of Rochester, Rochester, New York. USA.

Kjell Bløtekjær, Division of Physical Electronics, Norwegian Institute of Technology, Trondheim, NORWAY.

Sujatha Chandramohan, Department of Applied Mechanics, IIT, Madras, INDIA.

M. Chown, Standard Telecommunication Laboratories, Harlow, Essex, UK.

B. Culshaw, Department of Electronic and Electrical Engineering, University of Strathclyde, Glasgow, UK.

J. Das, Flat A/2, 164/78 Lake Gardens, Calcutta, INDIA.

M.P. De Micheli, University of Nice, Nice, FRANCE.

Niloy K. Dutta, AT & T Bell Laboratories, Murray Hill, New Jersey USA.

Ajoy Ghatak, Physics Department, IIT, New Delhi, INDIA.

T.G. Hodgkinson, BT Laboratories, Martlesham, Heath, Ipswich, UK.

A. Hordvik, Optoplan a.s., Moholtan, Trondheim, NORWAY.

J.D.C. Jones, Department of Physics, Heriot-Watt University, Edinburgh, UK.

H. Karstensen, Siemens Research Laboratories, Muenchen, GERMANY.

Raman Kashyap, BT Laboratories, Martlesham, Heath, Ipswich, UK.

Arun Kumar, Physics Department, IIT, New Delhi, INDIA.

Chinlon Lin, Bell Communications Research, Red Bank, New Jersey, USA.

D.J. Malyon, BT Laboratories, Martlesham, Heath, Ipswich, UK.

B.P. Pal, Physics Department, IIT, New Delhi, INDIA.

F.P. Payne, Department of Engineering, University of Cambridge, Cambridge, UK.

J.P. Raina, Department of Electrical Engineering, IIT, Madras, INDIA.

CONTRIBUTORS

A.J. Rogers, Department of Electronic and Electrical Engineering, King's College, Strand, London, UK.

Arnab Sarkar, Orion Laboratories, 550 Via Alondra. Camarilo, CA 93012, USA.

D.W. Smith, BT Laboratories, Martlesham, Heath, Ipswich, UK.

K. Thyagarajan, Physics Department, IIT, New Delhi, INDIA.

G. Umesh, Physics Department, IIT, New Delhi, INDIA.

D. Uttamchandani, Department of Electronic and Electrical Engineering, University of Strathyclyde, Glasgow, UK.

R. Wyatt, BT Laboratories, Martlesham, Heath, Ipswich, UK.

Contents

Preface   v

Contributors   ix

PART I
FIBRE AND INTEGRATED OPTICAL WAVEGUIDES : FUNDAMENTAL PRINCIPLES

1   **EVOLUTION OF FIBRE OPTICS — A BRIEF SURVEY**   3
B.P. Pal

2   **OPTO-ELECTRONICS AND THE INFORMATION AGE: A PERSPECTIVE**   11
Chinlon Lin
1. Introduction   11
2. Optical Fibres for Information Transmission   12
    2.1 Multimode vs single-mode fibres   12
    2.2 Dispersion-shifting and dispersion-tailoring in single-mode fibres   13
    2.3 Future directions of single-mode fibre research   14
3. Semiconductor Lasers   14
    3.1. Uniqueness and applications in information technology   14
    3.2 GaAlAs lasers (0.8 μm) and InGaAsP lasers (1.3 to 1.6 μm)   15
    3.3 Ideal semiconductor laser characteristics   16
    3.4 Multi-longitudinal-mode vs single-longitudinal-mode semiconductor lasers   16
    3.5 The need for single-longitudinal-mode semiconductor lasers   16
    3.6 Coupled-cavity lasers and DFB/DBR lasers   16
    3.7 Chirping in single-frequency semiconductor lasers and implications   17

3.8 High-frequency modulation and high-speed
   semiconductor laser pulses   18
3.9 High-power semiconductor lasers and nonlinear
   optical effects in fibres   19
4. Future Directions of Technological Advancements   19
   4.1 High-speed optical transmission, multi-optical-
       channel high-density wavelength-division-multiplexing
       (HD-WDM) and optical amplifiers   19
   4.2 Coherent optical transmission systems, optical frequency-
       division-multiplexing (OFDM) and tunable single-
       frequency lasers   20
   4.3 Information technology network: wideband information
       transmission, storage and retrieval   20
   4.4 OEIC (Opto-electronic integrated circuits) and
       micro opto-electronics   20
5. Conclusion   21

3   ELECTROMAGNETICS OF INTEGRATED OPTICAL WAVEGUIDES   24
Ajoy Ghatak and K. Thyagarajan

1. Introduction   24
2. Modes of a Planar Waveguide   25
3. A Simple Asymmetric Planar Waveguide : An Example of a
   Single Polarization Single Mode (SPSM) Waveguide   27
4. Propagation Characteristics for the Symmetric Structure   28
5. Propagation Characteristics for the Asymmetric Structure   31
6. Metal Clad Waveguides   32
7. Anisotropic Polarizers   35
8. Leaky Modes in a Planar Structure   35
9. Periodic Waveguides   39
   9.1 Coupled Mode Analysis   40
10. Rectangular Core Waveguides   44
11. Conclusions   48

4   TRANSMISSION CHARACTERISTICS OF TELECOMMUNICATION OPTICAL FIBRES   52
B.P. Pal

1. Introduction   52
2. Geometry of an Optical Fibre and Concept of
   Numerical Aperture   53
3. Framework of the Electromagnetic Theory of Propagation in
   Optical Fibres   54
4. Mode Cut-off   60

	5. Temporal Pulse Dispersion in Optical Fibres	64
	5.1 Graded Index Fibres	64
	5.2 Pulse dispersion in single-mode fibres	74
	6. Signal Attenuation in Optical Fibres	79
	6.1 Absorption Loss	81
	6.2 Radiative Loss	83
	6.3 Source fibre interconnection loss	87
	6.4 Fibre-fibre splice and joint loss	89
	7. Some Important Characteristics of Single-mode Fibres	93
	7.1 Mode Field Profile and Mode Spot Size	94
	7.2 Effective Cut-off Wavelength (λ_{ce})	102
	7.3 Advanced Single-mode Fibre Designs	103
	8. Future Developments	106
	9. Conclusion	107

5   SINGLE-MODE FIBRE DESIGNS FOR TELECOMMUNICATIONS   111
Arnab Sarkar

	1. Introduction	111
	2. Single-mode Fibre Design Considerations	112
	2.1 Effective cut-off wavelength	113
	2.2 Mode field radius	114
	2.3 Loss considerations	115
	2.4 Dispersion considerations	119
	2.5 Summary of design considerations	119
	3. Present Status of Nominally Step-index Single-mode Fibre Designs	121
	4. Trends in Single-mode Fibre Design	122
	4.1 Silica-core fluorosilicate-clad fibre	122
	4.2 Dispersion-shifted fibres	123
	4.3 Dispersion-flattened fibres	124
	5. Trends in Single-mode Systems	124
	6. Conclusion	124

6   POLARISATION MAINTAINING FIBRES AND THEIR APPLICATIONS   126
Arun Kumar

	1. Introduction	126
	2. Different Types of Polarisation Maintaining Fibres	127
	2.1 High birefringent fibres	127
	2.2 Single polarisation single-mode (SPSM) fibres	138
	3. Applications	142
	4. Summary	143

7 NONLINEAR PROCESSES IN OPTICAL FIBRES: PHASE-MATCHING FOR SECOND HARMONIC GENERATION IN OPTICAL WAVEGUIDES — 146
Raman Kashyap

1. Introduction — 146
   1.1 Historical development — 147
   1.2 Polarisation response of transparent media — 150
2. Phase-matching in Waveguides — 152
   2.1 Introduction — 152
   2.2 Phase matching criteria — 153
   2.3 Phase-matching via self-written gratings in optical fibres — 166
   2.4 Electric field induced SHG (EFISH) in fibres — 169
3. Phase-matched Harmonic Generation in Waveguides: Non-periodic Structures — 170
4. Review of SHG Using Periodic Structures — 171
5. Concluding Remarks — 186
6. Epilogue — 195

8 NON-LINEAR INTEGRATED OPTICS — 198
M.P. De Micheli

1. Introduction — 198
2. Second Harmonic Generation in Integrated Optics — 199
   2.1 Bulk-waveguide comparison — 199
   2.2 Choice of a medium — 201
3. SHG in Integrated Optics — 201
   3.1 Theory — 201
   3.2 SHG in $LiNbO_3$ waveguides — 205
4. "Cerenkov" Configuration SHG — 210
   4.1 Introduction — 210
   4.2 Theory — 211
   4.3 SHG in Cerenkov configuration in PE waveguides — 217
5. Conclusion — 220
6. Epilogue : Quasi Phase Matching for SHG — 221

9 FABRICATION TECHNIQUES OF OPTICAL FIBRES — 223
H. Karstensen

1. Introduction — 223
2. Fibre Drawing — 224
3. Multicomponent Technology — 227
   3.1 Rod-in-tube method — 227
   3.2 Double crucible method — 228
4. Vapour-deposition Methods — 229
   4.1 Inside-vapour-deposition (IVD) method — 231

    4.2 Outside-vapour-deposition (OVD) method   231
    4.3 Chemical vapour deposition (CVD) method   233
    4.4 Modified chemical vapour deposition (MCVD) method   233
    4.5 Plasma enhanced MCVD (PMCVD) method   239
    4.6 Plasma activated CVD (PCVD) method   240
    4.7 Plasma outside deposition (POD) method   241
    4.8 Vapour phase axial deposition (VAD) method   242
  5. Comparison of Performance   244

10 CHARACTERISATION OF OPTICAL FIBRES FOR TELECOMMUNICATION AND SENSORS — PART I : MULTIMODE FIBRES   249

B.P. Pal, K. Thyagarajan and A. Kumar

  1. Introduction   249
  2. Key Experimental Factors Common to Fibre Characterisation Techniques   250
    2.1 Fibre-end preparation   250
    2.2 Launching light into fibres   252
    2.3 Cladding-mode stripping   254
    2.4 Coupling from fibre to detector   254
  3. Measurement Techniques   254
    3.1 Refractive index profile (RIP)   254
    3.2 Geometrical measurements   262
    3.3 Numerical aperture   263
    3.4 Total attenuation   264
    3.5 Scattering loss and differential mode loss   269
    3.6 Non-destructive loss measurements (OTDR)   271
    3.7 Transmission bandwidth and dispersion   272
    3.8 Bandwidth of jointed fibres   274
    3.9 Differential mode delay (DMD)   276
  4. Conclusion   276

11 CHARACTERISATION OF OPTICAL FIBRES — PART II: SINGLE-MODE FIBRES   280

K. Thyagarajan, B.P. Pal and A. Kumar

  1. Introduction   280
  2. Measurement Techniques   280
    2.1 Attenuation   280
    2.2 Refractive index profile (RIP)   281
    2.3 Mode field diameter   282
    2.4 Equivalent step-index (ESI) profile   286
    2.5 Mode cut-off wavelength and the single-mode operating regime   287

	2.6 Dispersion	294
	2.7 Birefringence measurement	296
	2.8 Measurement of the propagation constant of the fibre mode	304
	3. Conclusion	304

12 OPTICAL FIBRE CABLES — 309
A. Hordvik

1. Introduction — 309
2. Loss Mechanisms in Fibre Optic Cables — 310
   2.1 Macro- and microbending losses — 310
   2.2 The effect of hydrogen on fibre losses — 315
3. Mechanical Strength of Optical Fibres — 316
4. Cable Core Structures — 317
   4.1 Loose tube structures — 318
   4.2 Tight jacket secondary coating — 322
5. Varieties of Fibre Optic Cables — 322

PART II
OPTICAL SOURCES, DETECTORS AND COMMUNICATION SYSTEMS

13 OPTICAL SOURCES, DETECTORS AND ENGINEERED SYSTEMS—INTRODUCTORY REMARKS — 331
B.P. Pal

14 SEMICONDUCTOR LASERS FOR OPTICAL FIBRE COMMUNICATIONS — 333
Govind P. Agrawal

1. Introduction — 333
2. The p-n Junction — 333
3. Fabry-perot Cavity — 336
4. Heterostructure Semiconductor Lasers — 337
5. Semiconductor Materials — 340
6. Emission Characteristics — 341
7. Single-frequency Semiconductor Lasers — 344
8. Semiconductor Lasers for Coherent Systems — 347
9. Concluding Remarks — 349

15 DISTRIBUTED FEEDBACK InGaAsP LASERS — 351
Govind P. Agrawal and Niloy K. Dutta

1. Introduction — 351
2. Theory — 353
   2.1 Coupled-wave equations — 353

	2.2 Longitudinal modes of a DFB laser	355
	2.3 Operating characteristics	359
3.	Device Structure and Fabrication	361
4.	Device Performance	363
5.	Concluding Remarks	370

16. SOME FUNDAMENTALS OF PHOTODETECTORS FOR FIBRE OPTICS — 372
B.P. Pal and G. Umesh

1. Introduction — 372
2. Reverse Bias Photodetectors — 374
   - 2.1 Dark current — 375
   - 2.2 Quantum efficiency — 375
   - 2.3 Signal to noise ratio (SNR) — 376
   - 2.4 Noise equivalent power (NEP) — 378
   - 2.5 Speed of response and bandwidth — 378
   - 2.6 Linearity and dynamic range — 378
3. Types of Detectors — 379
4. Conclusions — 380

17. RECEIVERS FOR DIGITAL FIBRE OPTIC COMMUNICATION SYSTEMS — 382
Sujatha Chandramohan and J.P. Raina

1. Introduction — 382
2. Basic Components of Digital Fibre Optic Communication Receiver — 383
3. Detectors for Digital Fibre Optic Receivers — 383
   - 3.1 The PIN diode — 383
   - 3.2 The avalanche photo diode (APD) — 386
   - 3.3 The PIN diode vs APD — 388
   - 3.4 Noise and signal modelling of a photodiode — 388
4. Front Ends for Digital Fibre Optic Receivers — 390
   - 4.1 Straight terminated front end — 390
   - 4.2 High input impedance type front end amplifier — 391
   - 4.3 Transimpedance front-end — 397
5. Equalizers for Optical Communication Receivers — 402
6. PIN-FET Receivers for Longer Wavelength Communication Systems — 406
7. Typical Practical Receivers — 408
8. Conclusion — 411

18 FUNDAMENTALS OF DIGITAL COMMUNICATION 415
J. Das

1. Introduction 415
2. Analog Modulation [2] 416
   2.1 Amplitude modulation (AM) 416
   2.2 Frequency modulation (FM) 417
   2.3 Phase modulation (PM) 420
3. Discrete Representation of Analog Signals [3, 4] 421
   3.1 Pulse modulation schemes 421
   3.2 Analog-to-digital conversion (ADC) [4] 424
   3.3 Pulse code modulation (PCM) 425
4. TDM-MUX [5] 431
5. Digital Signalling [3] 433
   5.1 Error rates 436
   5.2 Carrier and clock synchronization [5] 438
6. Digital Transmission [5] 439
   6.1 Line codes [6] 443
   6.2 ISI and equalizers [5] 446
7. Data Transmission Systems [5] 448

19 PURPOSE DESIGNING OF OPTICAL FIBRE SYSTEMS 452
M. Chown

1. Introduction 452
2. System Configuration and Power Budget 452
   2.1 Path loss 454
   2.2 Working margins 454
   2.3 Power budget diagram 455
   2.4 Power budget examples 456
3. Dispersion 460
   3.1 Multimode fibre dispersion 460
   3.2 Single-mode dispersion 461
4. System Trade-offs 463
5. Spectral Penalty 465
6. Other Component Considerations 467
7. Conclusion 468

20 COHERENT OPTICAL FIBRE TRANSMISSION SYSTEMS 470
T.G. Hodgkinson, D.W. Smith, R. Wyatt and D.J. Malyon

1. Introduction 470
2. Direct Detection Performance 471
3. Coherent Detection Principles 472
4. Coherent System Performance 474
5. Comparison Of Direct And Coherent Detection Performance 475

	6. Practical Coherent System Constraints	476
	6.1 Effects of laser phase noise	476
	6.2 Polarisation penalty	478
	7. Coherent Optical Communication Experiments	478
	8. Homodyne System Experiment Using Gas Lasers	479
	9. Heterodyne System Experiment Using External Cavity Lasers	483
	10. Heterodyne System Experiment Using A DFB Laser Transmitter	486
	11. Frequency Multiplexed Coherent System Experiment	489
	12. Conclusion	493
	13. Appendix: Overview of Coherent Systems Research (1985-88)	493
	13.1 Coherent optical fibre transmission system studies	493
	13.2 Single-mode fibre/cable studies	496
	13.3 Polarisation insensitive coherent detection	497
	13.4 Phase diversity systems	499
	13.5 Multichannel systems	500
21	**INTEGRATED OPTIC DEVICES**	**512**
	K. Thyagarajan and AJoy Ghatak	
	1. Introduction	512
	2. The Electrooptic Effect	512
	3. Phase Modulator	516
	4. Polarization Modulators and Wavelength Filter	521
	5. The Mach Zehnder Interferometric Modulator	523
	5.1 Logic operations	526
	6. The Optical Directional Coupler	527
	6.1 Directional coupler wavelength filter	533
	6.2 Polarization problem	535
	6.3 Polarization splitting directional coupler	536
	7. Polarizers	537
	7.1 Leaky mode polarizers	538
	7.2 Metal clad polarizers	539
	7.3 Thin metal clad polarizers	541
	8. Conclusions	542

PART III
OPTICAL FIBRE SENSORS AND DEVICES

22	**OPTICAL FIBRE SENSORS AND DEVICES**	**547**
	B.P. Pal	
23	**INTENSITY MODULATED OPTICAL FIBRE SENSORS**	**551**
	B.P. Pal	
	1. Introduction	551
	2. General Feature	551
	2.1 Intensity modulation through light interruption	552

	2.2 Shutter/Schlieren multimode fibre optic sensors	554
	2.3 Reflective fibre optic sensors	557
	2.4 Evanescent-wave fibre sensors	559
	2.5 Microbend optical fibre sensors	563
	2.6 Fibre optic refractometers	571
	2.7 Intensity modulated fibre optic thermometers	575
	2.8 Chemical analysis	577
	2.9 Distributed sensing with fibre optics	579
3.	Conclusion	581

24 INTERFEROMETRIC OPTICAL FIBRE SENSORS — 584
B. Culshaw

1. Introduction — 584
2. Basic Principles of Interferometric Optical Fibre Sensors — 585
3. Applications of Interferometric Optical Fibre Sensors — 592
4. Components for Interferometric Sensors — 594
5. Future Trends in Interferometric Sensors — 598
6. Concluding Comments — 599

25 FUSED SINGLE-MODE OPTICAL FIBRE COUPLERS — 602
F.P. Payne

1. Introduction — 602
2. Physical Principles — 604
   2.1 Coupling co-efficient — 604
   2.2 Polarisation effects — 607
3. Experimental Properties — 608
   3.1 Wavelength dependence [2] — 608
   3.2 Polarisation effects [8-10] — 609
   3.3 Dependence on external refractive index [2,14] — 611
4. Theoretical Modelling [15,19] — 612
   4.1 Qualitative behaviour — 612
   4.2 A first approximation — 613
   4.3 A second approximation — 614
5. Comparison with Experiment — 615
   5.1 Wavelength dependence — 615
   5.2 Polarisation effects — 615
   5.3 Dependence on external refractive index — 616
6. Conclusion — 616

26 SINGLE-MODE ALL FIBRE COMPONENTS — 619
D. Uttamchandani

1. Introduction — 619
2. Directional Couplers — 620

2.1 Fused single-mode couplers	620
2.2 Polished single-mode couplers	622
3. Polarizers	627
3.1 Fibre polarizers based on polished coupler blocks	627
3.2 Polarisers fabricated in continuous fibre length	630
4. Polarization Splitters	632
4.1 Polarization effects in fused tapered couplers	632
4.2 Polarization effects in polished couplers	634
5. Polarization Controllers	635
6. Optical Isolators	637
7. Single-mode Fibre Filters	639
8. Wavelength Multiplexers and Demultiplexers	645
9. Switches and Intensity Modulators	648
10. Phase Modulators	650
11. Frequency Modulators	651
12. Conclusions	653

27 SIGNAL PROCESSING IN MONOMODE FIBRE OPTIC SENSOR SYSTEMS — 657

J.D.C. Jones

1. Introduction	657
2. Transduction Mechanisms	658
2.1 Sensor transfer function	658
2.2 Phase modulated sensors	659
2.3 Polarisation modulated sensors	660
3. Optical Processing	662
3.1 Two beam interferometers	662
3.2 Multiple beam interferometers	667
3.3 Polarimetric techniques	670
3.4 Spectral techniques	675
4. Modulators and Components	677
4.1 Phase modulation	678
4.2 Polarisation modulation	679
4.3 Frequency modulation	682
4.4 Directional couplers	684
5. Electronic Processing	687
5.1 Active homodyne processing	688
5.2 Passive homodyne processing	690
5.3 Heterodyne processing	694
5.4 Noise considerations	695
6. Conclusions	700

28 FIBRE OPTIC SENSOR MULTIPLEXING — 706
Kjell Bløtekjær

1. Introduction — 706
2. Generic Topological Configuration — 706
3. Incoherent Detection — 709
   - 3.1 Time division multiplexing — 709
   - 3.2 Improving the duty cycle — 711
   - 3.3 Wavelength division multiplexing — 715
   - 3.4 Mode division multiplexing — 715
   - 3.5 Differential detection — 718
4. Coherent Detection — 720
   - 4.1 Homodyne detection — 721
   - 4.2 Heterodyne detection — 721
   - 4.3 Coherence requirement — 722
   - 4.4 Frequency division multiplexing — 724
   - 4.5 Pseudoheterodyne — 726
   - 4.6 Two-pulse systems — 727
   - 4.7 Coherence multiplexing — 730
5. Alternative Topological Configuration — 731
   - 5.1 Recursive array — 731
   - 5.2 Reflective arrays — 733
   - 5.3 Forward transmission — 734
   - 5.4 Series topology — 736
6. Birefringent and Two-mode Fibres — 737
7. Conclusion — 737

29 OPTICAL FIBRES FOR POWER SYSTEMS — 741
A.J. Rogers

1. Introduction — 741
2. Elements of Propagation in Optical Fibres — 743
3. Optical Fibre Communication — 746
4. Optical Fibre Communication Systems in the ESI — 749
5. Measurement Applications — 753
6. The Future — 766

INDEX — 771

Part I
**Fibre and Integrated Optical Waveguides:
Fundamental Principles**

Part I

1

Evolution of Fibre Optics — A Brief Survey

B.P. Pal[*]

Glass optical fibres have attracted a great deal of interest from the point of view of telecommunication in the contemporary world as the optical transmission media of the eighties. Although hand signals, smokes, fires etc. could be cited as examples of earliest forms of optical communication, modern idea of optical communication can be traced to the photophone experiment (see Fig. 1(a)) of Graham Bell carried out in USA in 1880 almost immediately after the invention of the telephone. However though this experiment worked well upto a distance of ~ 200 m, optical telephony remained discarded in favour of electrical telephony till lasers appeared on the scene in the sixties. Until the discovery of laser, there was no optical frequency coherent source which could be modulated for practical telecommunication. Optical frequencies ($10^{14} - 10^{15}$ Hz) of the electromagnetic spectrum are preferred as the carrier waves for telecommunication due to their huge channel bandwidth (i.e. high information carrying capacity). It is well known that by use of such high carrier frequencies, one can attain a very high density of signal/information transmission in time as compared to its lower frequency counterparts namely radio, UHF/VHF and microwaves (see Chapter 18 by J. Das). Initial experiments for communication with lasers involved use of a 'radio system' e.g. transmission of modulated laser beams through open air but it was soon realised that too many extraneous factors like atmospheric dusts, rain etc. would significantly attenuate laser beams through absorption and/ scattering within a relatively short distance. We are all too familiar with low visibility in rainy and foggy weather! Further, a finite sized electromagnetic beam like a laser suffers inherent diffraction

---

[*] Physics Department, Indian Institute of Technology Delhi, New Delhi - 110016, India.

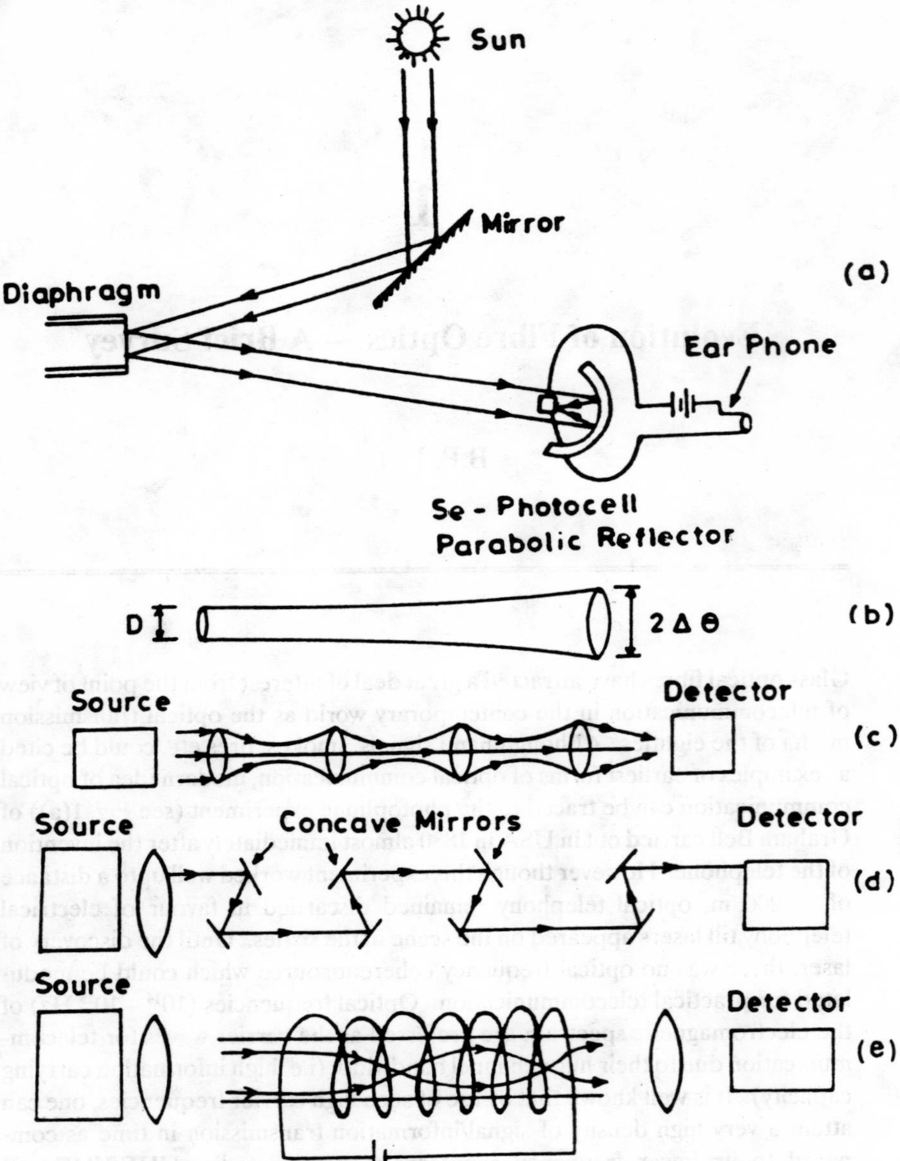

Fig. 1. (a) The Photophone experiment of Graham Bell (adapted from Midwinter [1] by permission of John Wiley & Sons, Inc. Copyright © 1979). Sunlight after deflection by a mirror was directed onto another mirror mounted on the diaphragm of a microphone, wherefrom after reflection the beam falls on a selenium photocell. Variation of the diaphragm in accordance with the speech acoustic signals results in a differential curvature of the reflecting surface, which, in turn, leads to intensity modulation of the reflected beam. Demodulation takes place at the photocell receiver (b) Diffraction divergence of an optical beam with propagation (c) Lens guide (d) Mirror guide (e) Gas lens—where electric heating of the gas inside a metallic tube leads to a graded refractive index distribution of the gas thereby resulting in lens action of the gas for the propagating beam.

divergence with propagation as shown in Fig. 1(b), the angular width of divergence being given by

$$\Delta\theta \sim \lambda/D \tag{1}$$

with λ as its wavelength and D its diameter (assuming a uniform sized beam). These drawbacks of open air optical communication (though they have their importance in tactical defence communication) led to proliferation of a variety of schemes to isolate and contain the signal carrying laser beam by guidance through either a series of lenses or mirrors or through so called 'gas lenses' as shown in Figs. 1(c)-1(e). In a gas lense, a radial temperature gradient is generated in a metal tube through which a laminar flow of gas is maintained. This temperature gradient created a corresponding radial gradient in the refractive index of the gas thereby leading to focusing of the propagating laser beam. However all these schemes were found to be unwieldy in practice and it was left to Kao and Hockham [2] from England and Wertz [3] from France to propose independently in 1966 cladded glass fibres as the most suitable optical transmission medium similar to electrical transmission media like coaxial cables. However, the major problem at that time was that the conventional quality glasses exhibited prohibitively high transmission loss of $\sim 10^5$ dB/km (e.g. window glass) while even high quality glasses that would find use in precision optics like lenses etc. were characterized by losses $\sim$ 500 dB/km. These authors speculated that the major source of loss in glass was the presence of impurities in glass. It became evident that unless the transmission loss in glass fibres could be brought down to a level of less than 20 dB/km at the operating laser wavelength, glass fibres cannot compete economically with best quality conventional metallic transmission lines for broad band transmission. An intensive R & D effort made at Corning Glass Works in USA over the following four year period (1966–70) eventually succeeded in breaking this economic barrier target of 20 dB/km set by Kao and Hockham for telecommunication by producing a fibre that exhibited a loss of 17 dB/km at 0.6328 μm wavelength of He-Ne laser [4]. This achievement immediately triggered a world-wide interest on optical fibres as the transmission line for optical telecommunication. Ever since, losses in telecommunication grade fibres have steadily decreased with attainment of loss as low as 0.2 dB/km (at $\lambda \sim 1.55 \mu$m) within the last decade itself in 1979 [5]. This figure has been brought down to a still lower figure of 0.154 dB/km in 1986, which is almost the theoretical limit [6]. Figure 2(a) depicts yearwise reduction in fibre transmission loss to demonstrate the phenomenal progress in fibre optic technology that took place within a decade of realising the first low-loss fibre in 1971.

Simultaneously, remarkable progress has been made in improving the reliability of semiconductor lasers in terms of their mean life under continuous operation at room temperatures beginning with a few hours around 1970 to one million hours (under accelerated ageing tests) towards the end of the seventies

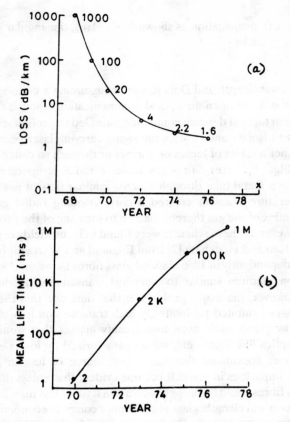

Fig. 2. (a) Loss vs year showing gradual decrease in loss (at $\lambda \sim 0.85\,\mu m$) with year in silica optical fibres; here x represents the lowest loss which was achieved at $\lambda = 1.55\,\mu m$ in the late 1970s (b) Life (MTBF) vs. year depicting increase in the life of GaAlAs semiconductor lasers for room temperature cw operation with year [both (a) and (b) adapted from [7]. Copyright © 1978 IEEE].

[*cf.* Fig. 2(b)]. The development of these two key technologies was primarily responsible for the impressive current state-of-art in optical communication [7].

The essential elements of a fibre optic communication system are shown in Fig. 3. As is common with any telecommunication system, the three major optical components are the laser diode/LED in the transmitter, the transmission medium i.e. the fibre and the photodetector within the receiver system. Electric analog signals from the telephone get coded into a stream of digital electric pulses which, after amplification, drive the electro-optic source (Laser/LED) to yield equivalent optical pulses which get launched into the fibre through a connector (could be a lensed connector). These pulsed optical signals after having been launched into the fibre travel down long lengths of concatenated (spliced and jointed) fibres till they reach the receiving end

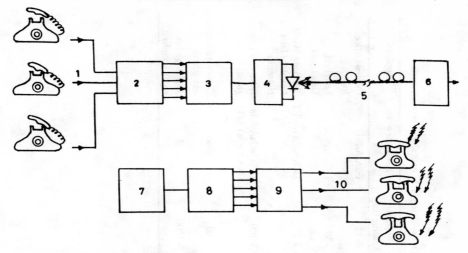

Fig. 3. Essential components of a fibre optical telecommunication system: 1. Analog telephone signals; 2. A/D converter + coder (multiplexing) for digital electrical signals; 3. modulation bias and control circuits; 4. E-O conversion through LED/laser diode; 5. Optical fibre cable—the transmission medium with/without intermediate splice/joints through which light wave propagates as a stream of optical pulses; 6. O-E conversion: photodiode (PIN/APD); 7. Bias, control, amplification of the detector output; 8. Demodulation and decoding; 9. Reconversion to analog signals; 10. Telephone receiver.

where they are detected by a photodiode (PIN/avalanche diode). The photodiode's electrical output after amplification is demodulated/decoded to retrieve the original telephone signals.

It so happened that at the early stages of the development of fibre optical communication, the efficiencies of the commercially available suitable optical sources and the photo-detectors both peaked at a wavelength $\sim 0.8\,\mu$m and added to it was the fact that the transmission loss in fibres in this wavelength region is relatively low. So it is only natural that initial fibre optic communication systems evolved around the working wavelength, $\lambda \sim 0.8\,\mu$m. Hence, these systems are now referred to as first-generation systems. The pattern of technology development in fibre optics has been so rapid in terms of practical system realisation that fourth generation systems could be realised within a span of only eight years since the first commercial telecommunication system that became operational in 1977. In Table 1, a summary of the typical characteristics of different generations of optical fibre communication is displayed. As is evident from the table, initial systems operated at relatively low bit rates and were essentially based on multimode fibres. Till the beginning of eighties, single-mode fibres were quite expensive compared to multimode fibres. However, since the commercial introduction of single-mode systems in public telecommunication networks in 1983, there has been an ever growing penetration of single-mode fibre systems into all segments of telecommunication

TABLE 1

Typical characteristics of different generation of fibre optic transmission systems that have evolved/are evolving in the optical transmission market

Generation	Bit rate	Operating wavelength (μm)	Type of Fiber	Loss (dB/km)	Repeater Spacing (km)	Source/detector	Application
First (1978-82)	34-45 Mb/s	0.85	Multimode (Graded Core)	~3	8-15	GaAlAs/Si	Central Office trunking
Second (1980-84)	45-90 Mb/s	1.3	Multimode (Graded Core)	~1	~30	InGaAsP/Ge	Longhaul, Data Links
Third (1982-87)	144-560 Mb/s	1.3	Single-mode	~0.4	≳40	InGaAsP/PINFET	Trunking-WDM Longhaul Data Networks
Fourth (1984-90)	≳1 Gb/s	1.55	Single-mode (SEGCOR/Disp. shifted)	<0.3	≳100	DFB InGaAsP/InGaAsP (Ge) APD	Advanced Longhaul
Fifth (1990-1995)	≳5 Gb/s	1.55	Single-mode	<0.25	~100	-do-	Coherent Systems and soliton
Sixth (1992-2000)?	?	>2	Infrared	<0.01	≳1000	?	Superhigh bandwidth Longhaul

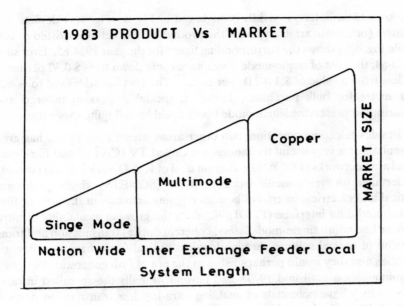

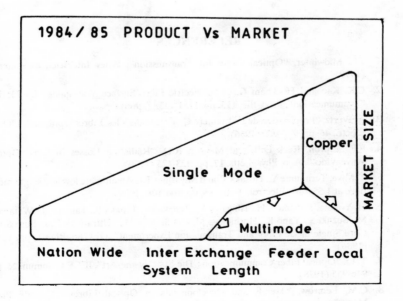

Fig. 4. Comparison of fibre optical and copper cables product vs. market for the year 1983 and 1984-85; the latter clearly demonstrates the tremendous growth of the fibre optic market, in particular, the single-mode fibre since mid-1980s (courtesy of Corning Incorporated [8]).

market which, in fact, is vividly brought out in Fig. 4. Fig. 4(a) depicts on a relative (otherwise arbitrary) scale the product vs. market as it existed in 1983 while Fig. 4(b) shows the corresponding figure for the year 1984-85. Ever since 1983-84, the cost of single-mode fibres have come down to ~ $ 0.30 per metre or less from a value of $ 1.0-2.00 per metre. The cost has stabilised to ~ $0.1 per metre for bulk purchase. However, special dispersion tailored and/ polarisation preserving single-mode fibres could be still quite expensive.

Besides telecommunication, two other areas where fibre optics has great potentials as a very useful technology are cabled TV (CATV) and fibre optic local area networks (FOLANS). A great deal of R & D work is being currently undertaken on synchronous optical networks (SONETS) which synchronise optical and electrical interfaces besides ongoing activities in the area of fibre distributed data interface (FDDI)[9]. With the growing availability of ultra-narrow line-width single-mode lasers, coherent optical transmission and transmission of optical solitons through low-loss optical fibres have now become realities and they would perhaps usher in the era of fifth generation of optical communication until mid 1990's and beyond. Finally the so called infrared fibres which have potentials of enabling ultra-low loss transmission down to < 0.01 dB/km in the mid-infrared wavelength range appear to be the course along which near-future technology development in fibre optics will take place.

REFERENCES

1. J.E. Midwinter, "Optical Fibres for Transmission", Wiley Interscience, Somerset (1979).
2. C.K. Kao and Hockham G.A., "Dielectric Fibre Surface Waveguides for Optical Communication", Proc. IEE **113**, pp. 1151-1158 (1966).
3. A. Wertz, "Propagation de La Lumiere Coherente dans les Fibres Optiques", L'Onde Electr., **46**, pp. 977-980 (1966).
4. F.P. Kapron, Keck D.B. and Maurer, R.D. "Radiation Losses in Glass Optical Waveguides", App. Phys. Letts. **17**, pp. 423-425 (1970).
5. T. Miya, Terunume Y., Hosaka T. and Miyashita T., "An ultimate low loss single-mode fibre at 1.55 μm", Electron. Letts. **15**, pp. 106-108 (1979).
6. H. Yokota, Kanamori H., Ishigawa Y., Tanaka G., Tanaka S., Tanada H., Watanabe M., Suzuki S., Yano K., Hoshikawa M. and Shimba H., "Ultra-Low-Loss Pure Silica Core Single-mode Fibre and Transmission Experiment", OFC'86, PD3-1, pp. 11-18 (1986).
7. T. Li, "Optical Fiber Communication : The-State-of-the-Art", IEEE Commun. **26**, pp. 946-955 (1978).
8. C.M. Lemrow, "Trends and Developments in Optical Fibres", Outside Plant Magazine, Practical Communications Inc., Corning Glass, 1985.
9. F. Ross, "FDDI — A Tutorial", IEEE Commun. Mag. **24**, pp. 10-17 (1986); see also IEEE Commun. Mag. April (1988).

2

Opto-electronics and the Information Age: A Perspective*

CHINLON LIN**

1. INTRODUCTION

Ever since the invention of transistors, the progress made in micro-electronics based on semiconductor electronic devices and integrated circuits has been phenomenal. The impact of semiconductor-based micro-electronics on the information society is evident in the widespread use of microprocessor-based products in the industrial, commercial, scientific, defense, and consumer-related systems. The technology of micro-electronics has progressed to such an extent in the information society, that even the kindergarten children now take personal computers, digital watches and calculators for granted, as if these microprocessor-based products existed centuries ago, just like wooden toys, bicycles, and things of that kind.

The next era of information age, as I would like to call it, is the era where micro-electronics becomes commonplace, but opto-electronic devices and systems become the center of attention. We can call it "the opto-electronics era" in analogy to the micro-electronics era.

In the micro-electronics era, one's comfort of living and level of 'civilizedness', if we can define such a status, perhaps could be measured by the number of microprocessor-based products (calculators, digital watches, personal computers, microprocessor-controlled microwave ovens and

---

* Updated reprint from J.I.E.T.E., Vol. 32 (Special issue on Optoelectronics and Optical Communication), July-August (1986).

** BELL Communication Research, 331 Newman Springs Road, Red Bank, NJ 07701-7040, USA.

refrigerators, programmable climate and energy controls, remote-controlled VCRs and TVs, keyless digital entry systems, etc.) in the house/office. In contrast, in the opto-electronics era of the information age, perhaps one can measure the same by the number of semiconductor-laser or LED-based products in the house/office. For example, laser printers and image-scanners for the personal computer, laser-based digital audio-disk (compact disk) players, video-disk players, optical data storage system, fibre-optic communication systems for two-way video, data, and audio transmission, optical sensors for automobiles, robotics, etc., all use semiconductor lasers, LEDs and photodetectors as their key components.

Therefore, the opto-electronics era is the era of high-capacity/high-speed information transmission, storage and processing, based on the emerging technology of optical fibres, semiconductor lasers and related devices, and optoelectronic components and circuits. This chapter presents a brief discussion on the past advances and the future prospects of some important optoelectronic technologies, in the hope that such an overall perspective may help some general readers gain a better understanding of the significance of optoelectronics for the coming information age.

2. OPTICAL FIBRES FOR INFORMATION TRANSMISSION

2.1 Multimode vs single-mode fibres

The farsight of Charles K. Kao on the use of glass optical fibres for long distance optical information transmission [1] has proved to be one of the most remarkable events in the history of modern science and technology. His ideas and predictions have essentially all been embodied in the present optical fibre communication systems. Historically, after a period of studying step-index single-mode fibres (the first low-loss fibre made, as reported by the Corning group in 1970, was a single-mode fibre), the attention turned to larger core size graded-index multimode fibres, because of the uncertainty about the manufacturing and the practical use of single-mode fibres whose cores were about 5 times smaller. The richness of theoretical problems associated with analyses of graded-index multimode fibres had attracted many early researchers to work on the multimode fibre waveguides [2, 3]. However, the difficulty in realizing the theoretically-possible bandwidth in the actual manufacturing and the uncertainty in predicting the effective bandwidth of the concatenated multimode optical fibres soon discouraged any significant further effort on graded-index multimode fibres. The discovery of the modal noise problem made it clear that multimode fibres have severe intrinsic limitations for high-capacity long-distance transmission, although for some short haul computer local area network (LAN) applications, coupling loss and branching loss considerations favour the large numerical aperture (NA) multimode fibres of reasonable bandwidth (>1 GHz-km).

It became obvious that single-mode optical fibres were the right answer for the optical fibre communications systems to realize their promised potential benefits. The fibre attenuation as well as the splice and connector losses were all found to be quite small if designed properly; reproducible results were obtained by many laboratories. In the same time frame, the fact that there is a wavelength region near 1.3 μm where the material dispersion of the silica glass becomes minimum and thus essentially negligible (the so-called zero-dispersion wavelength) was pointed out [3]. Soon after, a near-infrared-fibre-Raman-laser-based high-speed pulse-time-delay measurement system was proposed and implemented to measure such zero-dispersion wavelengths in km-long different silica single-mode fibres with various dopants [3]. A few years later, the lightwave technology research and development turned the attention to all aspects of single-mode fibres including manufacturing and actual system implementation. Starting in 1980, single-mode fibre technology and related components became the dominant system technology for high-bandwidth long-distance optical communications. The undersea optical fibre communication system, proposed a few years earlier, took the center stage in pushing the single-mode fibre optic technology forward with its system demands [4]. Following that, the terrestrial single-mode fibre systems were designed and installed. Starting in the early 1980's as all of a sudden, the whole lightwave industry wanted only single-mode fibres, the demand for such fibres became high. In 1988, the TAT-8 (trans-atlantic transmission) undersea single-mode optical fibre communication system was put in service; on land, Gb/s single-mode fibre systems began their widespread use.

2.2 Dispersion-shifting and dispersion-tailoring in single-mode fibres

After the research community learned more and more about the single-mode fibre dispersion characteristics and how such characteristics are affected by material dispersion of the silica glass and the dopants and doping level, and the waveguide dispersion as a function of core-cladding dimension and index profiles, various dispersion-shifting and tailoring schemes were proposed and demonstrated [3, 5]. The goal was twofold: to achieve minimum dispersion in the lowest loss wavelength region of 1.55 μm, and to have low dispersion over a wide spectral region. With step-index profiles, a single-mode fibre with a small core diameter and a large index difference compared with regular single-mode fibre can have the minimum chromatic dispersion in the 1.55 μm region, but the penalty is increased fibre loss due to increased dopant level. To reduce such an effect, graded-index profiled single-mode fibres were proposed [6] and in particular triangular profile (a special case of graded-index profile) single-mode fibres [6,7] were shown to result in minimum chromatic dispersion in the 1.55 μm spectral region with just a small amount of increase in loss. Based on similar ideas, low-loss dispersion-shifted single-mode fibres are now commercially available. System experiments using dispersion-shifted fibres have already been reported [8], and one can expect an increase in actual system installation in the near future based on 1.55 μm semiconductor lasers and dispersion-shifted single-mode fibres for very long-distance high-bandwidth transmission, especially for undersea applications where the minimization of the number of undersea repeaters is desirable.

Still on the horizon is the dispersion-flattened single-mode fibres which have low-dispersion over a wide spectral region, e.g. from 1.3 to 1.6 μm. Such fibres can be made by using quadruptly-clad index profiles. Such fibres offer the potential of high-bandwidth transmission over a wide spectral range suitable for wave-length-division-multiplexed (WDM) systems; these type of broadband low-dispersion single-mode fibres, when fully available in productions, may become the system designer's choice for WDM based systems.

2.3 Future directions of single-mode fibre research

As far as telecommunication-grade low-loss low-dispersion silica glass single-mode fibres are concerned, the research and development of the past decade or so have been so successful that it has now reached the maturing stage. Except for the low-loss broadband low-dispersion fibres, the main task remaining seems to be the improvement in fabrication and manufacturing technology in terms of yield, speed, and cost. It seems that future research and development of single-mode fibres will be in low-loss single-polarization and polarization-preserving fibres [9] and much longer wavelength infrared (2 to 10 μm) glass fibres for ultralow loss transmission. Polarization-preserving single-mode fibres are of immediate interest to fibre sensors [9] and nonlinear optical device applications. The progress from such research and development efforts will undoubtedly have a significant impact on future automation and robotic sensing technology, which are important for the information age.

Another new fibre research area of great significance is active single-mode fibres such as rare-earth-doped fibre lasers and amplifiers [10]. An optically pumped fibre laser amplifier can provide broadband or tunable amplification (with 20-40 dB gain per stage) for multi-channel optical signals in longhaul telecommunications [10].

3. SEMICONDUCTOR LASERS

3.1. Uniqueness and applications in information technology

Semiconductor diode lasers are unique and different from other lasers in that they are miniature in size, that they are based on semiconductor technology which is not entirely different from the micro-electronics technology, and that they can be directly driven and modulated by electric currents. Furthermore, by using various material systems and compositions, semiconductor laser sources covering a wide spectral range (visible to infrared) can be obtained. The range of available powers (a few mW to a few W) using different structures is another attractive feature. Moreover, due to the large gain spectral width of even a single semiconductor laser, tunability over several 100 Angstrom can be achieved with appropriately-designed dispersive cavity. With all these attributes, it is no wonder that semiconductor lasers are fast becoming the most important active devices for various information technology. More specifically, semiconductor diode lasers are useful:

(i) for information transmission

optical fibre communication systems and free-space communication systems,

(ii) for information storage

optical disk players for audio and video and optical disk data storage systems and

(iii) for information collection and processing

laser printers, laser bar-code scanners and image-scanning systems, and laser sensing and measurement systems.

3.2 GaAlAs lasers (0.8 μm) and InGaAsP lasers (1.3 to 1.6 μm)

After the first demonstration of room temperature cw operation in 1970 [11] of a GaAlAs semiconductor diode laser near 0.8 μm, strong interest in developing semiconductor lasers into pratical laser light sources followed. Mainly stimulated by the need to a reliable source for optical fibre transmission systems, scientists and engineers devoted tremendous efforts to help develop the knowledge base and the technology of semiconductor diode lasers [12]. Since GaAlAs lasers were the main material system of interest, early optical fibre communication systems were designed for operation in the 0.8 μm spectral region using GaAlAs semiconductor lasers as the optical sources [3, 12].

Soon the progress made in low-loss fibres (less than 1dB/km loss) made it clear that the infrared absorption as well as the Rayleigh scattering loss are lowest at a much longer wavelength. Interest in semiconductor laser sources for wavelengths longer than 1 μm surged; various material systems were proposed and demonstrated [12, 13]. Among these, the InGaAsP/InP system for obtaining semiconductor diode lasers in the 1.2 to 1.6 μm spectral region [13] became the winner. Worldwide research and development activities on the InGaAsP/InP semiconductor lasers (so-called long-wavelength lasers) have since dominated every major conference in semiconductor lasers and lightwave technology. Likewise, actual optical communication systems designed with various types of semiconductor lasers and detectors operating in the 1.3 μm spectral region have since become the most important optical fibre system for longhaul (undersea and terrestrial) transmission.

3.3 Ideal semiconductor laser characteristics

An ideal semiconductor laser for one application may not necessarily be suitable for a different application. Therefore, one has to know the specific applications to tailor the semiconductor laser design for optimized performance. In general, however, semiconductor lasers with a low threshold current, high differential quantum efficiency, high linearity and thermal stability (large T_0 and capable of high temperature operation), high power in a single transverse mode and high modulation bandwidth, are desirable. In addition, the semiconductor laser structure should be easy to fabricate with high yield and reproducibility in terms of the electro-optical characteristics.

3.4 Multi-longitudinal-mode vs single-longitudinal-mode semiconductor lasers

Specific applications require specific characteristics from semiconductor lasers. Here, we look at one interesting example. To reduce modal noise in a multimode optical fibre transmission system, a semiconductor laser with a large spectral width, highly multi-longitudinal-mode spectra (corresponding to low coherence length) is preferred over a single-longitudinal-mode semiconductor laser. On the other hand, for long-distance high-bandwidth single-mode fibre transmission systems, a semiconductor laser with just a few longitudinal modes (a small spectral width) or even a single-longitudinal-mode laser is needed to minimize the dispersion-related system penalty. Thus, both multi- and single- longitudinal-mode lasers are needed, depending on system requirements.

3.5 The need for single-longitudinal-mode semiconductor lasers

Single-longitudinal-mode lasers oscillate in a single spectral mode. It is essentially a single-frequency laser source when operating in the cw condition. Such single-frequency semiconductor lasers are useful for applications where (*i*) narrow spectral width, (*ii*) fixed wavelength and (*iii*) minimum mode-partition noise are required. For example, as already mentioned, in long-distance high-bit-rate single-mode optical fibre transmission, such a source is much preferred over the multi-frequency lasers. In optical systems involving dispersive, wavelength-selective optics such as gratings, prisms, and filters (for example, in laser printers, laser-based spectroscopic sensing systems and wavelength-division-multiplexers) single-frequency lasers are also desired. Furthermore, optical information systems which require an optical source with low intensity noise may find single-frequency lasers the right type of sources. Thus, single-longitudinal-mode semiconductor lasers in all spectral ranges may be needed for a large variety of opto-electronic information systems.

3.6 Coupled-cavity lasers and DFB/DBR lasers

There are several approaches to designing semiconductor lasers with single-longitudinal-mode selection capabilities [14]: the coupled-cavity approach, and the integrated grating structure approach.

The couple-cavity approach, based on the two composite cavity modulating the cavity resonance [14], is easy to implement in fabrication, but difficult to use and control due to their limited range of stability. On the other hand, the integrated grating approach, with distributed feedback (DFB) and distributed Bragg-reflector (DBR) structures being the most well known example [14, 15], is more complicated to implement into the manufacturing process, but the resultant laser is much easier to use because of the large side-mode suppression and the wide stability range. It is no wonder that those concerned with systems prefer to use the single-longitudinal-mode semiconductor lasers made by the latter approach because in such a case, the burden is on the laser maker rather than the user.

DFB semiconductor lasers of various designs (including the phase-shifted grating type) have been successfully made in research laboratories as is evident from many publications and conference reports. The large-scale manufacturing and system applications are now on the horizon. These are useful for wavelength-division-multiplexing applications as well as for high-bit-rate long-distance single-mode optical fibre transmission systems to relax the fibre-dispersion/laser-wavelength matching requirements. Mode-partition-noise will be greatly reduced in DFB lasers, with 30-40 dB side-mode-suppression under modulation thereby allowing high speed long distance transmissions. Moreover, the resultant intensity noise level is also greatly reduced, thereby making DFB lasers also useful for high-quality (analog) video disk and optical data disk optical information storage systems, if the DFB laser cost could be made low enough to justify their use is such applications. Today, DFB lasers in both the infrared and the visible wavelength ranges are available in research labs; commercial Gb/s optical fibre transmission systems use almost exclusively DFB lasers in the 1.3 - 1.55 μm region. These long-wavelength DFB lasers are also being used in advanced analog (AM and FM) multi-channel CATV-type video transmission systems using single-mode fibres.

3.7 Chirping in single-frequency semiconductor lasers and implications

Under high-speed direct modulation, even the single-frequency lasers show a time-dependent wavelength-shift of the single-longitudinal-mode, called chirping [16], resulting in a time-dependent spectral sweep and apparent spectral line broadening, which limit the transmission bandwidth of the single-mode fibre in the dispersive regime. In the low dispersion region near the dispersion minimum, the chirping does not present any serious problem unless the bit-rate is very high ($\sim$ 10-20 Gb/s modulation). However, with single-frequency lasers such as DFB lasers, system designers may use the lasers even in the dispersive region away from the dispersion minimum of the single-mode fibres. Thus, it is necessary to either reduce the degree of chirping by biasing the laser above threshold (and tolerate some loss of on/off ratio) or control the chirping sign, so that some form of dispersion-compensation might be obtained [17]. Alternatively, special low-chirp lasers or external modulators with much reduced chirping could be used. This approach has been experimentally demonstrated but has not been implemented in practical systems, because, so far, for practical systems upto 2.5 Gb/s chirping has not been the limit. As practical systems move to high Gb/s rates (eg, 10 Gb/s systems) in the future, chirping limitations in single-frequency lasers such as DFB or DBR lasers even in the low dispersion region of the fibre will have to be taken into consideration in system designs. In the very near future, the 2.5 Gb/s SONET systems are to be deployed. If the transmission fibre is long (especially with optical amplifiers based longhaul systems) and the DFB laser is far enough away from the minimum chromatic dispersion wavelength of the transmission fibre, chirping could already be significant enough even at this low Gb/s rate, that system designers will need to take it into account in determining the optimum system operating parameters.

3.8 High-frequency modulation and high-speed semiconductor laser pulses

For high-bandwidth information transmission, whether in analog or in digital modulation format, a high-bandwidth semiconductor laser [18, 19] is needed. A 3-dB modulation frequency of up to 15 GHz for GaAlAs lasers and up to 18 GHz for InGaAsP lasers (at room temperature) have been reported in research laboratories. Such lasers are being considered for very high frequency (10 to 20 GHz) microwave fibre optic links in which analog microwave signals are effectively translated to the optical domain for low-loss fibre transmission and recovered at the end of the fibre link. Such systems may find applications in radars and satellite entrance links for analog microwave signal transmission (including FM CATV signals). For optical fibre transmission of digital signals, such lasers can be used for very high bit-rates [19]. In fact, large-signal intensity modulation at a bit rate of 16 Gb/s NRZ (non-return-to-zero format) has been demonstrated [20]. Thus, it appears that for a long time to come, the speed of optical information transmission system will be limited by the capability of electronics circuits rather than the speed of semiconductor lasers.

A related subject is the generation of high-speed picosecond optical pulses from semiconductor lasers. These can be generated by mode-locking, Q-switching or gain-switching [21] techniques. The technique of gain-switching appears to be the simplest and most versatile, although it does not necessarily produce the shortest pulses. Gain switched 11 GHz, 30 ps optical pulses from a short-cavity InGaAsP laser [22] have been obtained. Higher frequency (> 200 GHz) mode locking has also been achieved [23]. Such high-repetition-rate picosecond and subpicosecond pulses could be useful for measuring fibre dispersion, detector speed, response of fast optical modulators or optical gates, as well as for time-domain signal processing applications.

3.9 High-power semiconductor lasers and nonlinear optical effects in fibres

There is a strong interest in high-power semiconductor lasers for Nd: YAG laser pumping, optical fiber amplifier pumping, long-range space communication, optical data disk writing, laser printing, and a variety of special sensing applications. Both single-element high-power devices and multi-element, diode array devices [24] have been developed. With such high power lasers, unfortunately, there are power limitations in optical fibre transmission due to the power-dependent loss and distortion of the input signals when the input power is high enough to induce various nonlinear optical effects in the silica glass fibres [25]. This is also true of fibre sensor systems using high-power laser diodes as sources.

On the other hand, high power semiconductor diode lasers provide the opportunity to construct really compact nonlinear optical frequency conversion devices such as fibre Raman amplifiers and erbium-doped fiber amplifiers for a variety of system or measurement applications [10, 26]. For example, an in-line optical fibre amplifier can be used to amplify the weak optical signals coming out of a fibre transmission line or a fibre sensor system. Practical significance of such a high-power semiconductor-laser-pumped optical amplifier device modules for longhaul undersea fiber systems and terrestrial system has been clearly demonstrated [10, 35].

4. FUTURE DIRECTIONS OF TECHNOLOGICAL ADVANCEMENTS

In the following, a brief outline on the directions of future advancements is attempted.

4.1 High-speed ($\geq$ Gb/s) optical transmission, multi-optical-channel high-density wavelength-division-multiplexing (HD-WDM) and optical amplifiers

First, the demand for higher bandwidth means higher speed optical fibre transmission systems. Analog microwave fibre-optic links operating in the 10 to 20 GHz frequency range, using semiconductor lasers and detectors of very high bandwidth, have been developed and used in a variety of microwave signal transmission applications including multi-channel FM/FDM video transmission. Digital systems operating at 10 Gb/s bit rates, based on DFB lasers and long-length of dispersion-shifted single-mode fibres have been demonstrated in the laboratory with discrete or integrated electronic circuit [27, 35]. Future practical systems operating in the 10-20 Gb/s range can be expected. In addition, multi-channel high-density wavelength-division-multiplexing (HD-WDM) will be implemented to increase or upgrade the transmission capacity, and provide broadband network flexibility [28, 35]. Combining high-speed and HD-WDM technologies, for example, with 50 DFB lasers in the 1300-1550 nm wavelength range, and 4 nm channel spacing, each laser operating at 20 Gb/s rates, Terabit/sec transmission could become a reality. Optical amplifiers (semiconductor-laser-based or fibre-based) have been developed to help meet the distribution and transmission power budget requirements [10, 29, 35].

4.2 Coherent optical transmission systems, optical frequency-division-multiplexing (OFDM) and tunable single-frequency lasers

In parallel to the surging interest in high-speed, direct-detection optical transmission systems, coherent optical transmission systems using stable, narrow-linewidth semiconductor lasers and heterodyne detection techniques have also generated a great deal of interest [30, 31]. Coherent systems promise not only increased repeater spacing (thus ideal for island-hopping undersea systems) due to increased receiver sensitivity, but more importantly very narrow channel spacing for realising optical frequency-division-multiplexing (OFDM) [31] based on stable and tunable single-frequency semiconductor lasers [32]. However, it is also clear that the demand in information transmission bandwidth has not yet caught up with even the present single-mode fibre transmission system's capacity. The need has to be created by having an information-centered design around the home, not just in the offices or industrial environments. This requires the widespread use of the information-based products (as mentioned above) at home, and the establishment of an interactive broadband information transmission environment.

4.3 Information technology network: wideband information transmission, storage and retrieval

It seems that the opto-electronic information storage and information-collection and processing systems based on semiconductor diode lasers will have to be widely introduced to stimulate the full utilization of the bandwidth potential offered by the optical communication systems. The current trend in using laser printers instead of impact printers for higher speed information output, in using laser bar-code scanners and laser image scanners rather than keyboard entries for automated information input-processing, in using optical data disks for high-capacity, fast access information storage, in using audio, video disks and cable TV channels for information retrieval video-on-demand and entertainment, etc. will undoubtedly create the future need of *broadband opto-eletronic technology-based information transmission* [33] *and switching networks (such as Gigabit networks and Fiber-to-the-Home networks [35]).*

4.4 OEIC (Opto-electronic integrated circuits) and micro opto-electronics

In all the opto-electronic information technology systems described above, semiconductor lasers, optical fibres, and associated micro-optic component technology will continue to play the dominant roles. The trend in the device and component technology is toward integration. Looking into the future, OEIC, opto-electronic integrated circuits containing many opto-electronic devices (lasers, modulators, amplifiers, multiplexers, filters, detectors, signal processors, etc) on a chip [34], will probably be developed for a variety of complex functions as required by the whole spectrum of information technology, such as information transmission, storage, processing, measurements, sensing, etc. The next era of the information age, thus, can also be possibly called "the era of micro-opto-electronics".

5. CONCLUSION

The future looks bright for the era of micro-opto-electronics. Semiconductor lasers, LEDs, photodetectors, optical amplifiers, multi-wavelength laser and detector arrays, tunable optical filters, optical gates and switches, waveguide couplers and modulators, etc., may be combined with semiconductor electronic circuits into OEICs. Together with high bandwidth transmission fibres, micro-optic fibre devices and fibre sensors, these opto-electronic information technologies will definitely be at the center stage of the next information age in the 1990's [35, 36].

REFERENCES

1. C.K. Kao and Hockham, G.A., "Dielectric-fibre surface waveguides for optical frequencies", Proc. IEE, Vol. 133, p 1151 (1966).
2. See, for example, M.J. Adams, Introduction to Optical Waveguides, New York, Wiley (1981).
3. See, for example, P.J.B. Clarricoats (ed.), Progress in Optical Communications, IEE Reprint Series 3, Peter Peregrinus, UK (1980).
4. P.K. Runge and Trischitta P.R., "Undersea lightwave communications", IEEE/OSA J. Lightwave Tech., Vol. LT-2, p 742 (1984).
5. L.G. Cohen, Chinlon Lin and French W.G., "Tailoring zero chromatic dispersion into the 1.5-1.6 μm low loss spectral region of single-mode fibres", Electron. Lett., Vol. 15, p 334 (1979).
   L.G. Cohen, Mammel W.L. and Jang S.J., "Low loss quadrupleclad single-mode lightguides with dispersion below 2 ps/nm-km over the 1.28-1.64 μm wavelength range", Electron. Lett., Vol. 18, p 1023 (1982).
6. U.C. Paek, Peterson G.E. and Carnevale A., "Dispersionless single mode lightguides with alpha index profiles", Bell System. Tech. J., Vol. 60, p 583 (1981).
7. M.A. Saifi et al., "Triangular-profile single-mode fibre", Opt. Lett., Vol. 7, p 43 (1982).
   B.L. Ainslie et al., "Monomode fibre with ultra-low loss and minimum dispersion at 1.55 μm", Electron. Lett. Vol. 18, p 842 (1982).
   V.A. Bhagavatula et al., "Segmented-core single-mode fibres with low loss and low dispersion", Electron. Lett., Vol. 19, p 713 (1983).
8. R. Goodfellow et al., "Practical demonstration of 1.3 Gb/s over 107 km of dispersion-shifted monomode fibre using a 1.55 μm multimode laser", in Tech Digest OFC'85, San Diego, CA, Paper TuB2, p 28 (1985).
9. T. Okoshi, "Review of polarization-maintaining single-mode fibres", I00C'83, Tokyo, Japan, 28A4-1 (1983).
   T.G. Giallorenzi et al., "Optical fibre sensor technology", IEEE J. Quantum Electron., Vol. QE-18, p 626 (1982).

10. D.N. Payne and Reekie L., "Rare earth doped fibre lasers and amplifiers", ECOC'88, Brighton, U.K., p. 49 (1988); also see Technical Digest, OFC'90-'92.
11. See for example, G.H.B. Thompson, Physics of Semiconductor Laser Devices, New York, Wiley, Chap 1 (1980).
12. See, Technical Digest of OFC (Optical Fibre Communications) meetings, 1975 through 1992.
13. J.J. Hsieh and Shen C.C, "Room temperatuer cw operation of buried-stripe heterostructure GaInAsP/InP diode lasers", Appl. Phys. Lett., Vol. 30, p 429 (1977).
14. T.E. Bell, "Single-frequency semiconductor lasers", IEEE Spectrum, Vol. 30 p 429 (1977).
15. Y. Suematsu, Arai S. and Kishino K., "Dynamic single-mode semiconductor lasers with a distributed reflector", IEEE/OSA J. Lightwave Tech., Vol. LT-1, p 161 (1983).
16. Chinlon Lin, Lee T.P. and Burrus C.A, "Picosecond frequency chirping and dynamic line broadening in InGaAsP injection lasers under fast excitation", Appl. Phys. Lett., Vol. 42, p 141 (1983).
17. Chinlon Lin, "Controlled chirped single-frequency laser pulse transmission in dispersion-shifted single-mode fibres for WDM Gb/s optical communication", ECOC'84, paper 10B6, p 244, Stuttgart, Germany (1984).
18. C.B. Su et al., "12.5 GHz direct modulation bandwidth of vapour phase regrown 1.3μm InGaAsP BH lasers", Appl. Phys. Lett., Vol. 46, p 344 (1985).
19. J.E. Bowers et al., "High-frequency constricted mesa lasers", Appl. Phys. Lett., Vol. 47, p 78 (1985).
20. Chinlon Lin and Bowers J.E., "High-speed large-signal digital modulation of a 1.3μm InGaAsP constricted mesa laser at a simulated bit rate of 16 Gb/s", Electron. Lett., Vol. 21, p 906 (1985).
    A.H. Gnauck and Bowers J.E., "16 Gb/s direct modulation of an InGaAsP laser", Electron, Lett,, Vol. 23, p. 801 (1987).
21. Chinlon Lin et al., "11.2 GHz picosecond optical pulse generation in gain-switched short-cavity InGaAsP injection lasers by high-frequency direct modulation", Appl. Phys. Lett., Vol. 20, p 238 (1984).
22. G. Eisenstein et al., "Active mode-locking characteristics of InGaAsP-single-mode fibre composite cavity laser", IEEE J. Quantum Electron., Vol. QE-22, p 142 (1986).
23. M.C. Wu et al, Tech. Digest OFC'91 and OFC'92.
24. K. Kobayashi and Mito I., "High light output-power single-longitudinal-mode semiconductor laser diodes", IEEE/OSA J. Lightwave Tech., Vol. LT-3, p 1202 (1985).
    D. Botez, "Laser diodes are power-packed", IEEE Spectrum, Vol. 22, p 43 (1985).
25. R.H. Stolen, "Nonlinearity in fibre transmission", Proc. IEEE, Vol. 68, p 1232 (1980).
26. Chinlon Lin, "Designing optical fibres for frequency conversion and optical amplification by stimulated Raman scattering and phase-matched four-photon mixing", J. Opt. Commun., Vol. 4 p 2 (1983).
    Chinlon Lin, "Nonlinear optics in fibres for fibre measurements and special device function," IEEE/OSA J. Lightwave Tech., Vol. LT-4, p 1103 (1986).
27. S. Fujita et al., "A 10 Gb/s-80 km optical fibre transmission experiment using a directly modulated DFB-LD and a high-speed InGaAs-APD", OFC'88, Post-deadline Paper PD16, New Orleans (1988).
28. Chinlon Lin, Kobrinski H., Frenkel A. and Brackett C.A, "A wavelength-tunable 16 optical channel transmission experiment at 2 Gb/s and 600 Mb/s for broadband subscriber distribution", Electron. Lett., Vol. 24, p. 1215 (1988).
29. G. Coquin, et al., "Simultaneous amplification of 20-channel centered at 1.54μm in a multiwavelength distribution system", ECOC'88, Post-deadline Paper, Brighton, 1988.

30. D.W. Smith and Stanley I.W., "The worldwide status of coherent optical fibre transmission systems", Tech Digest ECOC'83, Geneva, Switzerland, p 263 (1983).
    T. Okoshi, "Recent progress in heterodyne coherent optical fibre communications", IEEE/OSA J. Lightwave Tech., Vol LT-2, p 341 (1984).
31. R.E. Wagner, Cheung N.K. and Kaiser P., "Coherent lightwave systems for interoffice and loop-feeder applications", IEEE/OSA J. Lightwave Tech., Vol. LT-5, p 429 (1987).
    K. Nosu and Iwashita K., "A consideration of factors affecting future coherent lightwave communication systems", IEEE/OSA J. Lighwave Tech., Vol. LT-6, p 686 (1988).
32. K. Kobayashi and Mito H. "Single-frequency and tunable laser diodes", IEEE/OSA J. Lightwave Tech., Vol. 6, p 1623 (1988).
33. C. Baack and Heuer P. "Architecture of broadband communication", IEEE J. Selected Areas Commun., Vol. SAC-4, p 542 (1986).
34. S.R. Forrest, "Monolithic optoelectronics integration: a new component technology for lightwave communications", IEEE/OSA J. Lightwave Tech, Vol. LT-3, p 1248 (1985).
    A Suzuki, Kasahara K. and Shikada M., "InGaAsP/InP long-wavelength OEIC's for high-speed optical fibre communication systems", IEEE/OSA J. Lightwave Tech, Vol. LT-5, p 1479 (1987).
35. For most recent advances in optoelectronic technology and lightwave systems, readers should consult conference proceedings of the conferences : OFC's, ECOC's, and CLEO's, and the recent issues of Electronic Letters, IEEE Photonic Technology Letters, and IEEE/OSA J. Lightwave Technology.
36. Chinlon Lin (ed.), *Optoelectronic Technolgy and Lightwave Communication Systems*, New York, Van Nostrand, Reinhold, 1989.

3

Electromagnetics of Integrated Optical Waveguides

AJOY GHATAK[*] AND K. THYAGARAJAN[*]

1. INTRODUCTION

The field of integrated optics is primarily based on the fact that light can be guided and confined in very thin films (with dimensions ~ wavelength of light) of transparent materials on suitable substrates. By a proper choice of substrates and films and a proper configuration of the waveguides, one can perform a wide range of operations like modulation, switching, multiplexing, filtering or generation of optical waves. Due to the miniature size of these components, it is possible to obtain a high density of optical components in space unlike the case in bulk optics. In addition, the confinement of light energy in small regions of space leads to an efficient interaction of the optical energy with an applied electric field or an acoustic wave, thus leading to much more efficient electro-optic and acousto-optic modulators requiring very low drive powers.

The basic component of an integrated optic device is the optical waveguide which can be either planar (in which the confinement is only along one transverse dimension) or strip (in which the confinement is in both transverse dimensions) — see Figs. 1 and 2.

In this chapter, we give the electromagnetic analysis of many optical waveguides used in integrated optical devices. The attempt in each case will be to consider a simple refractive index distribution and bring out most of the important propagation characteristics of the waveguide. Some of the results obtained will be quite general in nature and will be applicable to optical fibres also.

---

[*] Department of Physics, Indian Institute of Technology Delhi, New Delhi - 110016, India.

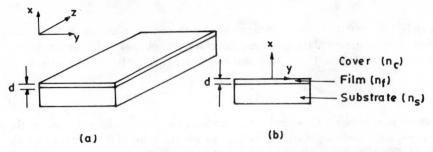

Fig. 1. A planar waveguide consisting of a film of refractive index n_f deposited on a substrate of refractive index n_s and with a cover of refractive index n_c. The refractive index varies only along the x-direction and is independent of y and z. Light confinement is only along the x-direction; light can diffract in the y-z plane.

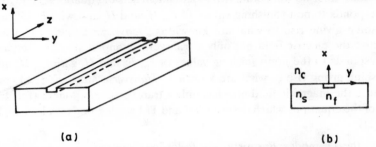

Fig. 2. A strip waveguide consisting of a high index region of refractive index n_f surrounded by media of lower refractive indices. The light confinement is along both x and y directions; and direction of propagation is the z-direction.

We will first define the TE and TM modes in planar waveguides and then give the results for a simple asymmetric waveguide. The concept of single polarization single mode (usually abbreviated as SPSM) waveguide will then be introduced. In the subsequent two sections we discuss the electromagnetics of metal clad and anisotropic waveguides which act as integrated optic polarizers; such polarizers form important components in many integrated optic devices. In the next section we consider a simple leaky structure and explicitly show how the power inside the core leaks into the cladding. Based on this a matrix method is discussed which allows easy numerical determination of propagation characteristics of graded/leaky/absorbing structures. In the last two sections we have presented an analysis to treat periodic waveguides and a perturbation method for the analysis of rectangular core waveguides.

2. MODES OF A PLANAR WAVEGUIDE

We assume that the refractive index depends only on the Cartesian coordinate x, i.e.

$$n^2 = n^2(x). \tag{1}$$

For such a configuration (which is said to describe a planar waveguide) the y and z dependences of the electric and magnetic fields will be of the form

$\exp[-i(\gamma y + \beta z)]$. However, we can *always* choose the z-axis along the direction of propagation of the wave and therefore, without any loss of generality we may put $\gamma = 0$. Thus the electric and magnetic fields can be written in the form

$$\left. \begin{array}{l} \mathcal{E}_j = E_j(x) e^{i(\omega t - \beta z)} \\ \mathcal{H}_j = H_j(x) e^{i(\omega t - \beta z)} \end{array} \right\} \quad j = x, y, z, \qquad (2)$$

where ω represents the angular frequency of the wave and β is known as the propagation constant. Corresponding to a specific value of β, there is a specific field distribution described by $\mathbf{E}(x)$ and $\mathbf{H}(x)$ which remains unchanged with propagation along the waveguide (i.e., along the z-axis); such distributions are referred to as *modes* of the waveguide. If we substitute Eq. (2) in Maxwell's equations, then we will obtain two independent sets of equations. The first set corresponds to non-vanishing values of E_y, H_x and H_z and with E_x, E_z and H_y vanishing giving rise to what are known as *Transverse Electric* (TE) modes because the electric field has only a transverse component. The second set corresponds to the non-vanishing values of E_x, E_z and H_y with E_y, H_x and H_z vanishing giving rise to what are known as *Transverse Magnetic* (TM) modes because the magnetic field now has only a transverse component (for derivation of the equations which describe TE and TM modes, one may look up Refs. [1], [2]).

For the TE modes, E_y satisfies the following equation:

$$\frac{d^2 E_y}{dx^2} + \left[k_0^2 n^2(x) - \beta^2\right] E_y(x) = 0, \qquad \text{(TE)} \qquad (3)$$

where

$$k_0 = \omega \sqrt{\varepsilon_0 \mu_0} = \omega/c \qquad (4)$$

represents the free space wave number, c being the speed of light in free space. Once E_y is known H_x and H_z can be determined from the following equations:

$$\left. \begin{array}{l} H_x = -\dfrac{\beta}{\omega \mu_0} E_y \\ H_z = \dfrac{i}{\omega \mu_0} \dfrac{\partial E_y}{\partial x} \end{array} \right\} \quad \text{(TE)} \qquad (5)$$

From Eq. (3) it is obvious that at a discontinuity of the refractive index profile

$$E_y \text{ and } \frac{dE_y}{dx} \text{ should be continuous;} \qquad (6)$$

this also follows from the fact that E_y and H_z are tangential components which are continuous across an interface.

For the TM modes $H_y(x)$ satisfies the equation

$$\frac{d^2 H_y}{dx^2} - \left(\frac{1}{n^2} \frac{dn^2}{dx}\right) \frac{dH_y}{dx} + \left[k_0^2 n^2(x) - \beta^2\right] H_y(x) = 0, \qquad (7)$$

with

$$H_y \text{ and } \frac{1}{n^2}\frac{dH_y}{dx} \quad \text{continuous everywhere} \tag{8}$$

Once H_y is known, E_x and E_z can be determined from the following equations:

$$E_x = \frac{\beta}{\omega\varepsilon_0 n^2(x)} H_y(x),$$

$$E_z = \frac{1}{i\omega\varepsilon_0 n^2(x)} \frac{dH_y}{dx}.$$

3. A SIMPLE ASYMMETRIC PLANAR WAVEGUIDE: AN EXAMPLE OF A SINGLE POLARIZATION SINGLE MODE (SPSM) WAVEGUIDE

We consider an asymmetric planar waveguide (see Fig. 3) for which

$$\begin{aligned} n(x) &= n_c \text{ for } x > 0 \text{ (cover)} \\ &= n_f \text{ for } -d < x < 0 \text{ (film)} \\ &= n_s \text{ for } x < -d \text{ (substrate)}. \end{aligned} \tag{9}$$

The symmetric waveguide corresponds to $n_c = n_s$ and is discussed in many books on optical waveguides [1, 2]. The cover is usually air and therefore without any loss of generality we will assume $n_s > n_c$. The classification of modes is as follows:

$$n_s^2 < \frac{\beta^2}{k_0^2} < n_f^2, \Leftrightarrow \text{Discrete guided modes} \tag{10}$$

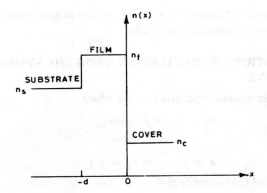

Fig. 3. The refractive index distribution of an asymmetric planar waveguide [see Eq. (9)].

$$\frac{\beta^2}{k_0^2} < n_s^2 \Leftrightarrow \text{Continuum radiation modes}$$

The discrete guided modes correspond to exponentially decaying solutions in the cover *and* substrate. The radiation modes correspond to oscillatory fields either in the substrate or in both substrate and cover. Now, for the refractive index profile given by Eq. (9), the solutions of Eqs. (3) and (7) are straightforward and application of appropriate boundary conditions lead to the following transcendental equations [2]:

$$\tan\left[V(1-b)^{\frac{1}{2}}\right] = \frac{\left(\frac{b}{1-b}\right)^{\frac{1}{2}} + \left(\frac{b+a}{1-b}\right)^{\frac{1}{2}}}{1 - \frac{[b(b+a)]^{\frac{1}{2}}}{(1-b)}} \qquad \text{TE modes} \qquad (11)$$

$$= \frac{\frac{1}{\gamma_1}\left(\frac{b}{1-b}\right)^{\frac{1}{2}} + \frac{1}{\gamma_2}\left(\frac{b+a}{1-b}\right)^{\frac{1}{2}}}{1 - \frac{1}{\gamma_1\gamma_2}\frac{[b(b+a)]^{\frac{1}{2}}}{(1-b)}}, \qquad \text{TM modes} \qquad (12)$$

where

$$b = \frac{\left(\frac{\beta}{k_0}\right)^2 - n_s^2}{n_f^2 - n_s^2}, \qquad \text{(normalized propagation constant)} \qquad (13)$$

$$V = k_0 d \left(n_f^2 - n_s^2\right)^{1/2} \qquad \text{(normalized waveguide parameter)} \qquad (14)$$

$$a = \frac{n_s^2 - n_c^2}{n_f^2 - n_s^2}, \qquad \text{(asymmetry parameter)} \qquad (15)$$

and

$$\gamma_1 = \left(\frac{n_s}{n_f}\right)^2, \quad \gamma_2 = \left(\frac{n_c}{n_f}\right)^2 \qquad (16)$$

Using the above equations, we now discuss the propagation characteristics of symmetric and asymmetric waveguides.

4. PROPAGATION CHARACTERISTICS FOR THE SYMMETRIC STRUCTURE

We first consider a symmetric structure for which

$$n_s = n_c = n_2 \text{ (say)}.$$

Thus

$$a = 0, \gamma_1 = \gamma_2 = \gamma \text{ (say)},$$

and Eqs. (11) and (12) simplify to

$$\tan\left[V(1-b)^{\frac{1}{2}}\right] = \frac{2\left(\frac{b}{1-b}\right)^{\frac{1}{2}}}{1 - \frac{b}{1-b}} \qquad \text{(TE modes)} \qquad (17a)$$

$$= \frac{\frac{2}{\gamma}\left(\frac{b}{1-b}\right)^{\frac{1}{2}}}{1 - \frac{1}{\gamma^2}\left(\frac{b}{1-b}\right)} . \quad \text{(TM modes)} \qquad (17b)$$

Typical variations of b with V are shown in Fig. 4 and as can be seen easily from the above two equations the difference between the propagation constants of the TE and TM modes become small as $\gamma \to 1$. Thus when the core and cladding refractive indices become nearly equal, the TE and TM modes become nearly degenerate. The cut-off condition ($\beta/k_0 = n_2$, i.e. $b = 0$) immediately gives the following expression for cut-off frequencies:

$$V_c = m\pi, m = 0, 1, 2, ... \qquad \text{(for TE and TM modes)} \qquad (18)$$

Now, if we operate in the domain $0 < V < \pi$ then although it is usually referred to as a single mode structure, there are actually two independent modes (the TE_0 and TM_0 modes) with slightly different propagation constants: we denote these propagation constants by β_1 and β_2. If the incident beam is y-polarized, it will excite the TE mode of the waveguide with $\beta = \beta_1$. Similarly, if the incident beam is x-polarized, it will excite the TM mode of the waveguide with $\beta = \beta_2$. In each case (provided there are no imperfections in the waveguide) the state of polarization will not change as the beam propagates

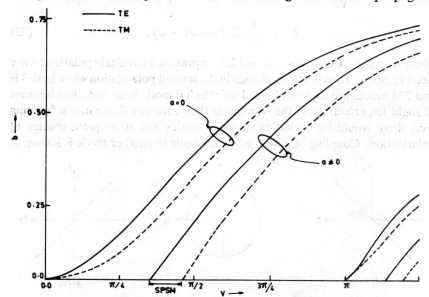

Fig. 4. Qualitative variation of the normalized propagation constant b with V for the TE and TM modes of a symmetric ($a = 0$) and an asymmetric ($a \neq 0$) planar waveguide. The region shown by a horizontal arrow corresponds to a SPSM waveguide. The solid and dashed curves correspond to the TE and TM modes, respectively.

along the axis of the waveguide. Let us next consider the incidence of a linearly polarized wave with $\mathscr{E}$ making an angle of 45° with x and y axes [see Fig. 5(a)]; thus the field at $z = 0$ can be written as

$$\left. \begin{array}{l} \mathscr{E}_x = E_0 \cos \dfrac{\pi}{4} \cos \omega t \\ \mathscr{E}_y = E_0 \sin \dfrac{\pi}{4} \cos \omega t \end{array} \right\} \quad \text{at } z = 0. \tag{19}$$

Now $\mathscr{E}_x$ will excite the TM mode and $\mathscr{E}_y$ will excite the TE mode; thus the fields at $z > 0$ will be given by

$$\mathscr{E}_x = \frac{1}{\sqrt{2}} E_0 \cos(\omega t - \beta_2 z), \quad \text{(TM mode)} \tag{20}$$

$$\mathscr{E}_y = \frac{1}{\sqrt{2}} E_0 \cos(\omega t - \beta_1 z). \quad \text{(TE mode)} \tag{21}$$

At

$$z = \frac{\pi}{2(\beta_1 - \beta_2)} = z_0.$$

we have

$$\mathscr{E}_x = \frac{1}{\sqrt{2}} E_0 \cos\left(\omega t - \phi + \frac{\pi}{2}\right) = -\frac{1}{\sqrt{2}} E_0 \sin(\omega t - \phi) \tag{22}$$

$$\mathscr{E}_y = \frac{1}{\sqrt{2}} E_0 \cos(\omega t - \phi), \tag{23}$$

where $\phi = \beta_1 z_0$. Equations (22) and (23) represent a circularly polarized wave [see Fig. 5(b)]. Thus there is a change in the state of polarization when both TE and TM modes are excited. Even if only the TE mode is excited, then because of slight imperfections in the waveguide (like core size fluctuations, bending etc.) there would be excitation of other modes and subsequent change of polarization. Coupling of power from one mode to another mode is known as

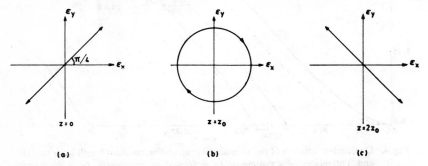

Fig. 5 (a) A linearly polarized wave is incident at $z = 0$ with the electric vector making an angle of 45° with x and y axis, (b) After propagating through a distance of z_0 [see Eq. (21)] the wave becomes circularly polarized, (c) At $z = 2z_0$ the wave again becomes linearly polarized.

mode conversion and normally in devices involving optical waveguides, such mode conversion should be avoided. On the other hand, one can make use of mode conversion in devices such as TE ←→ TM converters, wavelength filters etc. (see Chapter 21). A waveguide which supports only one particular state of polarization is known as SPSM (single polarization single mode) waveguide and in recent years there has been considerable interest in the study and fabrication of such waveguides. We will discuss examples of such waveguides in later sections.

5. PROPAGATION CHARACTERISTICS FOR THE ASYMMETRIC STRUCTURE

We next consider the asymmetric waveguide in which $n_c \neq n_s$ i.e., $a \neq 0$. If we substitute the cut-off condition

$$\frac{\beta}{k_0} = n_s \quad \Rightarrow \quad b = 0 \tag{24}$$

in Eqs. (11) and (12) we would get the following expression for the cut-off frequencies:

$$V_c = \tan^{-1}\sqrt{a} + m\pi \quad \text{(TE modes)}$$
$$m = 0, 1, 2, ... \tag{25a}$$

$$= \tan^{-1}\left(\sqrt{a}/\gamma_2\right) + m\pi. \quad \text{(TM modes)} \tag{25b}$$

Since $n_f > n_s > n_c$, we have

$$\gamma_2 < 1 \tag{26}$$

and therefore

$$V_c^{TM} > V_c^{TE} \tag{27}$$

i.e., the cut-off frequencies of TM modes are greater than the corresponding cut-off frequencies of TE modes (see Fig. 4). Considering the fundamental mode ($m = 0$), when

$$\tan^{-1}\left(\sqrt{a}/\gamma_2\right) > V > \tan^{-1}(\sqrt{a}), \tag{28}$$

we will only have the TE mode and the waveguide is usually referred to as absolutely single polarization single moded. This is one of the many examples of SPSM waveguides. It may be noted that for $V < V_c^{TE}$ there are *no* guided modes of the system.

In the above we have discussed the exact solutions for an asymmetric refractive index structure with a step profile. Exact solutions for many profiles are given in the literature (see, e.g., Ref. [1] - [4]).

6. METAL CLAD WAVEGUIDES

Metal clad optical waveguides are of considerable current interest as they have been shown to possess an extremely large differential attenuation between the TE and TM polarizations and thus can be designed to act as efficient integrated optic polarizers [5 - 12]; such a polarizer forms an important component in many integrated optical devices used in rotation sensors, directional couplers for OTDR measurements etc. A metal cladding is also usually used for applying electric or magnetic fields on optical waveguides for electro-optic or magneto-optic modulation [13]. In the former application one requires a large differential attenuation between TE and TM modes while in the latter application, it is required that there is very little additional attenuation for the TE and TM modes because of metal cladding.

We consider a structure shown in Fig. 6(a). By using the appropriate boundary conditions, one obtains the following transcendental equation which determines the propagation constants of the waveguide:

$$2a u_3 = m\pi + \tan^{-1}\left(\gamma_{34}\frac{u_4}{-iu_3}\right)$$
$$+ \tan^{-1}\left[\gamma_{32}\frac{u_2}{u_3}\tan\left\{\tan^{-1}\left(\gamma_{21}\frac{u_1}{-iu_2}\right) - u_2 b\right\}\right], \qquad (29)$$

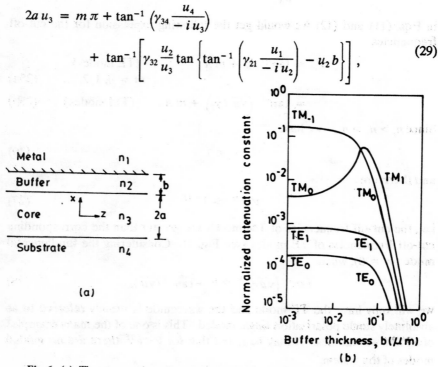

Fig. 6 (a) The cross section of a metal clad waveguide. Typically for $\lambda = 6328$ Å, $n_1 = 1.2 - i\,7.0$ (Aluminum), $n_2 = 1.457$ (SiO$_2$), $n_3 = 1.758$ and $n_4 = 1.544$ (Corning glass). (b) The attenuation constant of the metal clad waveguide as a function of the buffer thickness b. (Adapted from Y. Yamamoto, T. Kamiya and H. Yanai, Characteristics of optical guided modes in multilayer metal clad planar optical guide with low index dielectric buffer layer, IEEE J. Quantum Electron. QE-11 (1975) 729–© 1975, IEEE)

where

$$\gamma_{pq} \begin{cases} = 1 & \text{for TE mode} \\ = \left(\dfrac{n_p}{n_q}\right)^2 & \text{for TM mode} \end{cases}, \qquad (30)$$

$$u_l = k_0 \left[n_l^2 - \left(\dfrac{\beta}{k_0}\right)^2\right]^{\frac{1}{2}},$$

$$p, q, l = 1, 2, 3, 4,$$

$$\beta = \beta_r - i\beta_i,$$

u_l is the transverse wave number in each region and the refractive index in each region is indicated in Fig. 6(a); the subscripts 1, 2, 3 and 4 refer to metal cladding, buffer layer, core and substrate respectively. The dimensions a and b are shown in Fig. 6(a).

Figure 6(b) shows the attenuation characteristics of the structure shown in Fig. 6(a) as a function of the buffer thickness b. It is seen that the attenuation of the TM modes passes through a resonance peak if a dielectric buffer of proper refractive index and thickness is introduced between the waveguide and the metal cladding while the TE mode attenuation reduces monotonically as the buffer thickness is increased. This absorption peak is caused by the resonant coupling between the guided TM mode and the lossy surface mode (also known as the plasmon mode—see below). Because of the large differential attenuation between TE and TM modes, the device acts as an efficient integrated optic polarizer. Since in many cases, the waveguiding layer is graded, numerical methods for determining the propagation characteristics of graded index metal clad waveguides have also been developed [14].

It may be of interest to mention here that an interface of a dielectric and a metal can also have a waveguiding action for TM modes. We assume

$$\begin{aligned} n^2 &= n_1^2 & ; x > 0 \\ &= n_2^2 < 0 & ; x < 0 \end{aligned} \right\} \qquad (31)$$

i.e., the region $x < 0$ has a negative dielectric constant (see Fig. 7). In general, n_2^2 is complex for a metal but usually the real part is negative and large compared to the imaginary part. For example, for silver at $\lambda_0 = 0.6328\,\mu$m

$$n^2 \approx -16.4 - i\,0.54 = -\varepsilon_r - i\varepsilon_i \approx -\varepsilon_r, \qquad (32)$$

where $-\varepsilon_r$ and $-\varepsilon_i$ represent the real and imaginary parts of the dielectric constant $\varepsilon\ (= n^2)$. It is assumed that

$$\varepsilon_r > 0 \text{ and } \varepsilon_r \gg \varepsilon_i \qquad (33)$$

as a consequence of which we have neglected the imaginary part of ε. Considering TM modes, the equations satisfied by H_y would be

$$\dfrac{d^2 H_y}{dx^2} + \left(\varepsilon_1 k_0^2 - \beta^2\right) H_y = 0; \qquad x > 0 \qquad (34a)$$

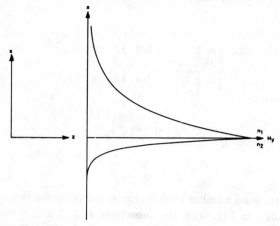

Fig. 7. The surface plasmon mode propagating along the interface of a dielectric and a metal.

$$\frac{d^2 H_y}{dx^2} - \left(\varepsilon_r k_0^2 + \beta^2\right) H_y = 0; \quad x < 0 \tag{34b}$$

where $\varepsilon_1 = n_1^2$. If $\beta^2 > \varepsilon_1 k_0^2$ then in both the media, H_y will decay exponentially. Thus

$$H_y = C e^{-\mu x}; \quad x > 0$$
$$= C e^{\nu x}; \quad x < 0 \tag{35}$$

where the constant C is the same because H_y should be continuous at $x = 0$ and

$$\mu = \left[\beta^2 - \varepsilon_1 k_0^2\right]^{1/2},$$
$$\nu = \left[\beta^2 + \varepsilon_r k_0^2\right]^{1/2} \tag{36}$$

Continuity of

$$\frac{1}{n^2} \frac{dH_y}{dx}$$

gives us

$$-\frac{\mu}{\varepsilon_1} = -\frac{\nu}{\varepsilon_r}$$

or,

$$\frac{\beta^2 - \varepsilon_1 k_0^2}{\varepsilon_1^2} = \frac{\beta^2 + \varepsilon_r k_0^2}{\varepsilon_r^2}$$

giving

$$\frac{\beta}{k_0} = \left(\frac{\varepsilon_1 \varepsilon_r}{\varepsilon_r - \varepsilon_1}\right)^{1/2} = \left(\frac{n_1^2 n_2^2}{n_2^2 + n_1^2}\right)^{1/2} \tag{37}$$

Obviously, the quantity β/k_0 would be real if

$$n_2^2 + n_1^2 < 0 \Rightarrow n_2^2 < -n_1^2. \tag{38}$$

Thus if the above condition is satisfied, the interface of a metal and a dielectric acts as a waveguide and supports one TM mode. This mode is usually referred to as the *surface plasmon mode*. For complex values of n_2^2, β would also become complex. Now, for TE modes also, $E_y(x)$ satisfies the same equation as (34a) and (34b); however, continuity of dE_y/dx leads to a condition which can never be satisfied. Thus the structure cannot support a TE mode.

7. ANISOTROPIC POLARIZERS

In the previous section we discussed metal clad optical waveguides which exhibit large differential attenuation between the TE and the TM modes so that they can be used as integrated optic polarizers. However, a metal cladding will attenuate the TE modes also. In this section we will discuss single mode anisotropic polarizers [15, 16], a typical refractive index distribution of which is shown in Fig. 8. The device uses a birefringent superstrate which is such that for one of the polarizations the superstrate refractive index is larger than that of the guiding layer [Fig. 8(a)] and for the other polarization it is smaller than the guiding layer [Fig. 8(b)]. For waveguides fabricated in LiNbO$_3$, one may use another LiNbO$_3$ crystal as the birefringent superstrate. The superstrate can be oriented in such a way that a TE - pass or TM - pass polarizer is obtained [15]. The attenuation coefficient for the mode which leaks out can be calculated by using the method given in Ref. [17] or by using the technique developed in Sec. 8. Although the final result is the same, the main advantage of the latter method is that it can be extended to graded structures.

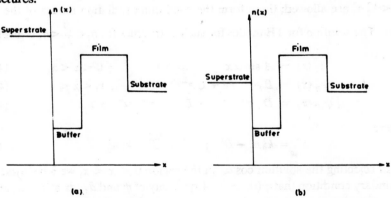

Fig. 8. The refractive index distribution of a typical anisotropic polarizer in which there is a birefringent superstrate such that for one of the polarizations, the superstrate refractive index is larger than that of the guiding layer (a) and for the other polarization it is smaller than the guiding layer (b).

8. LEAKY MODES IN A PLANAR STRUCTURE

Leaky modes in optical waveguides is a subject of considerable interest; not only are leaky modes technologically important for our understanding of the losses in waveguides, but they are also very interesting from the mathematical point of view. In this section we will carry out the electromagnetic analysis of a planar leaky structure which would bring out all the physics associated with leaky modes in a waveguide; we will closely follow Ref. [18].

We consider the TE modes for the refractive index distribution given by [see Fig. 9(a)]:

$$n(x) = n_1 \quad ; \quad 0 < x < x_1, \text{ and } x > x_2$$
$$= n_2 \ (<n_1) \quad ; \quad x_1 < x < x_2 . \tag{39}$$

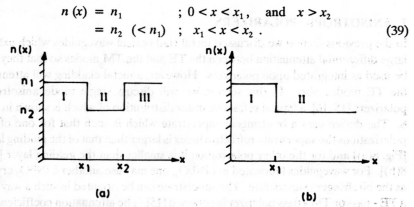

Fig. 9. The refractive index profile for (a) a leaky structure and (b) the corresponding guiding structure. The fields are assumed to vanish at $x = 0$.

At $x = 0$, we assume a metallic boundary which makes the field vanish there.* For such a structure there exists no guided mode, and all values of $\beta^2 < k_0^2 n_1^2$ are allowed; these form the continuum radiation modes of the system. The solution for TE modes for such a structure for $n_2 < \frac{\beta}{k_0} < n_1$ is given by

$$\psi_I(x) = A \sin \alpha x \qquad ; \quad 0 < x < x_1 \tag{40a}$$
$$\psi_{II}(x) = B e^{\gamma(x-x_1)} + C e^{-\gamma(x-x_1)} \quad ; \quad x_1 < x < x_2 \tag{40b}$$
$$\psi_{III}(x) = D_+ e^{i\alpha(x-x_2)} + D_- e^{-i\alpha(x-x_2)} \quad ; \quad x > x_2 \tag{40c}$$

where

$$\alpha^2 = k_0^2 n_1^2 - \beta^2 \quad ; \quad \gamma^2 = \beta^2 - k_0^2 n_2^2 . \tag{41}$$

In rejecting the solution $\cos \alpha x$ in the region $0 < x < x_1$ we have used the boundary condition that $\psi(0) = 0$. Continuity of ψ and $d\psi/dx$ at $x = x_1$ and $x = x_2$ gives us

$$B = \frac{1}{2} A \left(\sin \alpha x_1 + \frac{\alpha}{\gamma} \cos \alpha x_1 \right), \tag{42a}$$

---

* By assuming the field to vanish at $x = 0$, we are essentially considering the antisymmetric modes of the structure:
$$n(x) = n_1 \ ; \ \text{for } 0 < |x| < x_1 \text{ and } |x| > x_2$$
$$= n_2 \ ; \ \text{for } x_1 < |x| < x_2 .$$

$$C = \frac{1}{2}A\left(\sin\alpha x_1 - \frac{\alpha}{\gamma}\cos\alpha x_1\right), \qquad (42b)$$

$$D_-^* = D_+ = \frac{1}{2}B\left(1 + \frac{\gamma}{i\alpha}\right)e^{\gamma(x_2 - x_1)}$$

$$+ \frac{1}{2}C\left(1 - \frac{\gamma}{i\alpha}\right)e^{-\gamma(x_2 - x_1)} \qquad (42c)$$

Equations (40) – (42) correspond to the continuum radiation modes of the system. The normalization condition

$$\int_0^\infty \psi_{\beta'}^*(x)\,\psi_\beta(x)\,dx = \delta\left(\beta - \beta'\right) \qquad (43)$$

gives us

$$|D_\pm| = \sqrt{\beta/2\pi\alpha}. \qquad (44)$$

If we assume that $\gamma(x_2 - x_1) \gg 1$, then the exponentially decaying term in Eq. (40b) can be neglected and in this approximation

$$D_-^* = D_+ \approx \frac{L}{4}A\left(1 + \frac{\gamma}{i\alpha}\right)e^{\gamma(x_2 - x_1)}, \qquad (45)$$

where

$$L = \sin\alpha x_1 + \frac{\alpha}{\gamma}\cos\alpha x_1. \qquad (46)$$

It is readily seen that since $\gamma(x_2 - x_1) \gg 1$, unless $L \approx 0$, $\left|\frac{D_+}{A}\right|$ and $\left|\frac{D_-}{A}\right|$ are very large quantities and the amplitude of the field is large in the region $x > x_2$. However, when $L = 0$, or when

$$\cot\alpha x_1 = -\frac{\gamma}{\alpha}, \qquad (47)$$

$B = 0$ and therefore we have only the exponentially decaying solution in the region $x_1 < x < x_2$ (see Eq. 40b) and the oscillatory field in the region $x > x_2$ is much weaker (see Fig. 10(a)). Thus when $L = 0 = B$ the field has the properties of a guided mode in the region of the core although it eventually becomes oscillatory in the cladding (like a radiating field). These fields with $B = 0$ are referred to as 'quasi-modes'.

Equation (47) represents the transcendental equation for determining the guided modes of the structure shown in Fig. 9(b). For such a profile, the field corresponding to a guided mode is given by

$$\psi_r(x) = A_r \sin\alpha_r x, \qquad 0 < x < x_1$$

$$= A_r \sin\alpha_r x\, e^{-\gamma_r(x - x_1)}, \quad x > x_1 \qquad (48)$$

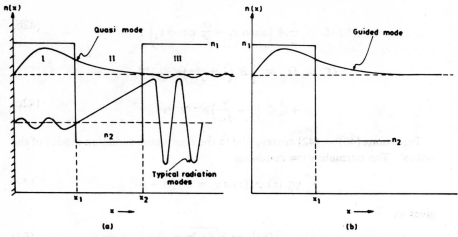

Fig. 10. (a) For the quasi modes we have only the exponentially decaying solution in the region $x_1 < x < x_2$, and the oscillatory field in the region $x > x_2$ is weak. Far away from the resonance condition the amplitude of the field is very large in the region $x > x_2$. (b) In the region $0 < x < x_2$, the fields corresponding to the guided modes of Fig. 9(b) are the same as quasi-modes of Fig. 9(a).

where α_r and γ_r satisfy Eq. (37), the subscript r referring to the *resonance condition*. Using the normalization condition

$$\int_0^\infty \left| \psi_r(x) \right|^2 dx = 1 \tag{49}$$

we get

$$A_r = \left[\frac{2\gamma_r}{1 + \gamma_r x_1} \right]^{\frac{1}{2}} \tag{50}$$

In Ref. [18] we have studied the propagation of a beam through the structure shown in Fig. 9(a) whose transverse field distribution at $z = 0$ was assumed to be a guided mode of the structure shown in Fig. 9(b), i.e.,

$$\psi(x, z = 0) = \psi_r(x) . \tag{51}$$

Such a field would excite a packet of radiation modes, i.e.,

$$\psi(x, z = 0) = \psi_r(x) = \int \phi(\beta) \psi_\beta(x) \, d\beta . \tag{52}$$

For $z > 0$, the field would be given by

$$\psi(x, z) = \int \phi(\beta) \psi_\beta(x) e^{-i\beta z} d\beta . \tag{53}$$

Using the orthonormality condition [Eq. (43)] we get

$$\phi(\beta) = \int_0^\infty \psi_\beta^*(x) \psi_r(x) \, dx . \tag{54}$$

Since only those values of β which are near $\beta = \beta_r$ are of importance, in regions I and II we may write

$$\psi_\beta(x) \approx \frac{A}{A_r} \psi_r(x) . \tag{55}$$

Since in region III (i.e., when $x > x_2$), $\psi_r(x)$ has a negligible value, we may neglect the contribution from region III to the integral in Eq. (54) to write

$$\phi(\beta) \approx \frac{A}{A_r} \int_0^\infty |\psi_r(x)|^2 dx = \frac{A}{A_r}. \tag{56}$$

Using an analysis similar to that given in Ref. [18], we can evaluate A to write [2]:

$$|\phi(\beta)|^2 = \frac{1}{\pi} \frac{\Gamma}{(\beta - \beta'_r)^2 + \Gamma^2}. \tag{57}$$

where

$$\Gamma = \frac{4\alpha_r^3 \gamma_r^3}{\beta_r \delta^4} \frac{e^{-2\gamma_r(x_2 - x_1)}}{(1 + \gamma_r x_1)}, \tag{58a}$$

$$\beta'_r = \beta_r + \Delta\beta, \tag{58b}$$

$$\Delta\beta = -\frac{\Gamma(\alpha_r^2 - \gamma_r^2)}{2\alpha_r \gamma_r}. \tag{58c}$$

The fractional power $W(z)$ that remains inside the core at z is approximately given by

$$W(z) \simeq \left| \int_0^\infty \psi^*(x, 0) \psi(x, z) dx \right|^2, \tag{59}$$

which can be evaluated to give [2]:

$$W(z) = e^{-2\Gamma z}. \tag{60}$$

The above equation shows how the power inside the core 'leaks' into the cladding. Equation (58a) gives an analytical expression for the attenuation coefficient of the 'quasi-modes'.

The above analysis suggests that if we are able to calculate $|\phi(\beta)|^2$ then by fitting it to a Lorentzian [see Eq. (57)], we should be able to calculate the real and imaginary parts of the propagation constant. This concept has been used in a series of papers to develop a method to obtain the propagation constant, leakage/absorption losses, bending losses, modal field profiles of arbitrarily graded refractive index profiles [19-22]. The technique has also been extended to the analysis of nonlinear planar waveguide structures [23].

9. PERIODIC WAVEGUIDES

Periodic waveguides in which either the thickness, refractive index or both vary periodically along the direction of propagation are used in the realization of many important functions in integrated optics. These include mode converters, polarization transformers, wavelength filters, distributed Bragg and distributed feedback laser structures, input-output couplers etc.

One of the most important characteristics of periodic coupling is that the periodic perturbation couples power mainly among two modes which satisfy a quasi phase matching condition. Thus if β_1 and β_2 are the propagation constants of two modes, then a periodic perturbation with a period Λ would induce coupling between these modes if

$$\beta_1 - \beta_2 = \pm K, \tag{61}$$

where

$$K = \frac{2\pi}{\Lambda}, \tag{62}$$

and Λ represents the spatial period of the perturbation. Because of the above requirement, periodic coupling is selective and this leads to its various applications. A typical periodic structure is shown in Fig. 11 wherein the thickness of the waveguide varies sinusoidally along the propagation direction.

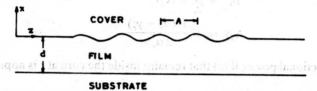

Fig. 11. A periodic waveguide whose thickness varies periodically along the z-direction with a period Λ.

In this section we shall give a brief outline of periodic coupling. It is interesting to note that the problem of periodic coupling is very similar to the quantum mechanical coupling of two energy eigenstates by the application of a time varying harmonic perturbation. Indeed, the coupled mode theory to be described here is very similar to the time dependent perturbation theory used in quantum mechanics.

9.1 Coupled Mode Analysis

To keep the analysis simple we consider the coupling among TE modes in a planar waveguide. Let $n_p^2(x, z)$ be the refractive index profile of a perturbed planar waveguide; here z-is assumed to be along the direction of propagation. We assume

$$n_p^2(x, z) = n^2(x) + \Delta n^2(x, z), \tag{63}$$

where $n^2(x)$ is the refractive index profile of the unperturbed waveguide and $\Delta n^2(x, z)$, the perturbation. The electric field E of the TE mode must satisfy the following wave equation:

$$\frac{\partial^2 E}{\partial x^2} + \frac{\partial^2 E}{\partial z^2} + k_0^2 n_p^2(x, z) E = 0. \tag{64}$$

In the coupled mode analysis, we assume that the z- dependent perturbation leads to coupling of power among the modes of the unperturbed waveguide. In view of the condition given by Eq. (61) we also assume that power gets coupled only between two modes. Thus we write

$$E(x, z) = A(z) E_1(x) e^{-i\beta_1 z} + B(z) E_2(x) e^{-i\beta_2 z}, \tag{65}$$

where $E_1(x)$ and $E_2(x)$ are the modal field profiles of two modes of the profile $n^2(x)$ and β_1 and β_2 their corresponding propagation constants. The constants $A(z)$ and $B(z)$ are z-dependent to account for coupling among the modes.

Substituting Eq. (65) in Eq. (64) and neglecting $\frac{\partial^2 A}{\partial z^2}$, $\frac{\partial^2 B}{\partial z^2}$ (which is known as the slowly varying approximation) and using the fact that $E_1(x)$ and $E_2(x)$ are modal field profiles, we arrive at the following set of coupled mode equations:

$$\frac{dA}{dz} = -i \tilde{\kappa}_{12} B e^{i\Gamma z}, \tag{66}$$

$$\frac{dB}{dz} = -i \tilde{\kappa}_{21} A e^{-i\Gamma z}, \tag{67}$$

where we have assumed

$$\Delta n^2(x, z) = \Delta n^2(x) \cos Kz \tag{68}$$

and have neglected nonsynchronous terms such as terms proportional to exp $[\pm i (\Delta\beta + K) z]$, exp $[\pm i Kz]$ etc. This is called the synchronous approximation and leads to simple analytical solutions which are also reasonably accurate. (In quantum mechanics parlance, the corresponding approximation in called the rotating wave approximation). In Eqs. (66) and (67),

$$\tilde{\kappa}_{12} = \frac{k_0^2}{\beta_1} \frac{\int E_1^* \Delta n^2 E_2 \, dx}{\int E_1^* E_1 \, dx}, \tag{69}$$

$$\tilde{\kappa}_{21} = \frac{k_0^2}{\beta_2} \frac{\int E_2^* \Delta n^2 E_1 \, dx}{\int E_2^* E_2 \, dx} \tag{70}$$

are called the coupling coefficients. One can show that

$$\tilde{\kappa}_{21} = \tilde{\kappa}_{12}^* = \kappa. \tag{71}$$

Also

$$\Gamma = \beta_1 - \beta_2 - K. \tag{72}$$

Equations (66) and (67) give the z-variation of the amplitudes of modes E_1 and E_2. These equations can be solved analytically and if we assume the initial conditions

$$A(z = 0) = 1, B(z = 0) = 0 \tag{73}$$

i.e., at $z = 0$, mode $E_1(x)$ is incident with unit amplitude, then one obtains for the power in each mode at any z as

$$P_1(z) = \cos^2 \gamma z + \frac{\Gamma^2}{4\gamma^2} \sin^2 \gamma z, \qquad (74)$$

$$P_2(z) = \frac{\kappa^2}{\gamma^2} \sin^2 \gamma z, \qquad (75)$$

where

$$\gamma^2 = \kappa^2 + \frac{\Gamma^2}{4} \qquad (76)$$

Figure 12 shows the variation of $P_2(z)$ with z for $\Gamma = 0$ and $\Gamma = 2\sqrt{3}\,\kappa$. Notice that complete power exchange from mode E_1 to mode E_2 is possible only if $\Gamma = 0$, i.e.,

$$\beta_1 - \beta_2 = K. \qquad (77)$$

If $\Gamma = 0$, then the length required for complete power conversion is

$$L_c = \frac{\pi}{2\kappa}. \qquad (78)$$

which is called the coupling length.

Condition given by Eq. (77) implies that if there is a periodic perturbation with spatial frequency K, then strong coupling takes place only between modes satisfying $\Delta\beta \simeq K$. Since $\Delta\beta$ is wavelength dependent, this coupling process can be used to build wavelength filters. Also if an anisotropic perturbation is considered, it can lead to polarization conversion which is the basis of polarization converters (see chapter 21).

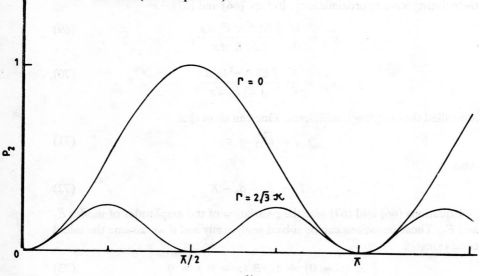

Fig. 12 Variation of power P_2 in the coupled mode with κz for $\Gamma = 0$ (i.e., $\Delta\beta = K$) and for $\Gamma = 2\sqrt{3}\kappa$ when unit power is incident in mode 1. Notice that for complete transfer of power from mode 1 to mode 2, one must have $\beta_1 - \beta_2 = K$.

In the above we have considered codirectional coupling, i.e., coupling between two modes propagating along the same direction. One can use the same periodic perturbation to couple two oppositely propagating modes. Since the z-dependence of phase of the backward propagating mode would be $e^{i\beta_2 z}$, Eq. (77) gets modified to

$$\beta_1 + \beta_2 = K. \tag{79}$$

Indeed, if the coupling is into the same spatial mode propagating in the $-z$-direction, then $\beta_2 = \beta_1$ and Eq. (79) becomes

$$\beta_1 = \frac{K}{2}. \tag{80}$$

If n_{eff} is the effective index of the mode, then $\beta_1 = \frac{2\pi}{\lambda_0} n_{\text{eff}}$ and Eq. (80) gives

$$\Lambda = \frac{\lambda_0}{2 n_{\text{eff}}}, \tag{81}$$

where $\Lambda = 2\pi/K$ is the spatial period of the perturbation. Thus a periodic perturbation with a period given by Eq. (81), can be used for coupling from a forward propagating mode to a backward propagating mode. This concept is used in distributed Bragg reflector and distributed feedback lasers.

Figure 13 shows some waveguide devices based on periodic coupling.

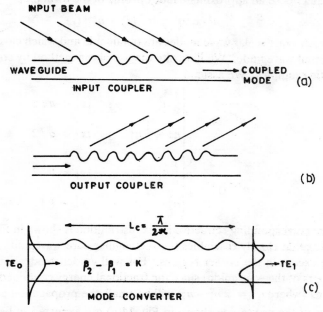

Fig. 13. Some integrated optic devices based on periodic waveguides (a) input coupler, (b) output coupler, (c) mode converter.

10. RECTANGULAR CORE WAVEGUIDES

In the planar waveguides discussed above, no lateral confinement of the light wave exists, i.e., the light wave is confined near the surface of the substrate and cover but can diffract in the plane parallel to these surfaces. The waveguides used in integrated optics such as rib waveguides and channel waveguides consist of rectangular or nearly rectangular cores, which provide a lateral confinement also. Lateral confinement becomes essential for realising integrated optical circuits and efficient modulators, directional couplers etc. Thus, it is quite important to understand the modal properties of such rectangular core waveguides which have been a subject of considerable interest in the recent past [24-29]. Achieving an accurate analytical solution of such waveguides is difficult because of the presence of corners. Considerable amount of work has been done based on analytical [25, 26, 29, 30] and numerical methods [24, 27, 28] to obtain approximate solutions corresponding to rectangular-core waveguides. Amongst the analytical methods, the effective index method [26] is the most efficient and simple. Inspite of the fact that the effective index method has been widely applied to highly sophisticated structures, it fails to give satisfactory results for some simple waveguides, especially at low frequencies [31]. Further, it gives ambiguous results for waveguides with square cross-sections [32].

In this section we show that for homogeneous core rectangular waveguides, one can choose an approximate index profile of the form [32]:

$$n^2(x,y) = n'^2(x) + n''^2(y), \qquad (82)$$

for which exact scalar wave modes are obtainable and which closely resembles the actual waveguide. We illustrate the use of the method by considering two specific examples. We assume

$$n'^2(x) = \begin{cases} \dfrac{1}{2}n_1^2 & , \quad |x| < a/2 \\ n_2^2 - \dfrac{1}{2}n_1^2 & ; \quad |x| > a/2 \end{cases} \qquad (83)$$

$$n''^2(y) = \begin{cases} \dfrac{1}{2}n_1^2 & , \quad |y| < b/2 \\ n_2^2 - \dfrac{1}{2}n_1^2 & ; \quad |y| > b/2 \end{cases} \qquad (84)$$

The corresponding refractive index distribution is shown in Fig. 14(a); such a waveguide closely resembles the rectangular core waveguide shown in Fig. 14(b) except in the corner regions. For well guided modes (i.e., for large V numbers of the waveguide) since the fractional energy contained in the corner regions (where $n^2 = 2n_2^2 - n_1^2$) will be small, the propagation constant of the modes of the waveguide shown in Fig. 14(a) are expected to be very close to those shown in Fig. 14(b). Further, in most practical cases, $n_1 \approx n_2$, and the

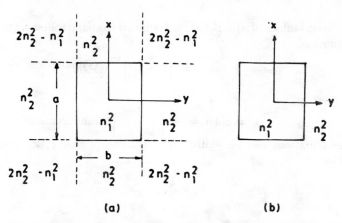

Fig. 14. A rectangular core waveguide described by Eqs. (82)-(84). This waveguide closely resembles the waveguide shown in (b) when $n_2 \simeq n_1$. Also for well guided modes the fraction of the energy present in the corner regions (where $n^2 = 2n_2^2 - n_1^2$) will be very small and the propagation constants of the waveguide shown in (a) are expected to closely match those shown in (b).

effect of the difference in the refractive index can be accurately incorporated by using first order perturbation theory (see Ref. [32]).

We next solve the scalar wave equation

$$\frac{\partial^2 \psi}{\partial x^2} + \frac{\partial^2 \psi}{\partial y^2} + \left[k_0^2 n^2(x,y) - \beta^2\right]\psi = 0, \qquad (85)$$

where k_0 is the free space wavenumber and β is the propagation constant. For $n^2(x,y)$ given by Eqs. (82)-(84), Eq. (85) can be solved by using the method of separation of variables. The field patterns can be written as $\psi(x,y) = X(x) Y(y)$, where $X(x)$ and $Y(y)$ can be symmetric or antisymmetric functions of x and y respectively. For example, for a mode which is symmetric (or antisymmetric) in x, we have

$$\begin{aligned} X(x) &= A\cos\mu\xi \quad (\text{or } A\sin\mu\xi) \quad, \quad |\xi| < 1 \\ &= B\exp\left[-\left(V_1^2 - \mu^2\right)^{1/2} |\xi|\right] \quad, \quad |\xi| > 1 \end{aligned} \qquad (86)$$

where

$$\xi = \frac{2x}{a}, \quad V_1 = \frac{1}{2}k_0 a \left(n_1^2 - n_2^2\right)^{\frac{1}{2}},$$

$$\mu = a\left(\frac{1}{2}k_0^2 n_1^2 - \beta_1^2\right)^{\frac{1}{2}},$$

and the allowed values of μ (or β_1) are determined from the following transcendental equations:

$$\mu \tan\mu = \left[V_1^2 - \mu^2\right]^{1/2} \quad \text{(symmetric in } x\text{)} \tag{87}$$

or

$$\mu \cot\mu = -\left[V_1^2 - \mu^2\right]^{\frac{1}{2}} \quad \text{(antisymmetric in } x\text{)}.$$

Similar equations can be written for $Y(y)$ except that ξ, V_1, and μ are replaced by

$$\eta = \frac{2y}{b}, \quad V_2 = \frac{1}{2} k_0 b \left(n_1^2 - n_2^2\right)^{1/2}$$

and

$$\nu = b \left(\frac{1}{2} k_0^2 n_1^2 - \beta^2\right)^{\frac{1}{2}}, \tag{88}$$

respectively. The normalised propagation constant is then given by

$$P^2 = \frac{\beta_1^2 + \beta_2^2 - k_0^2 n_2^2}{k_0^2 \left(n_1^2 - n_2^2\right)} = 1 - \frac{\mu^2}{V_1^2} - \frac{\nu^2}{V_2^2}. \tag{89}$$

In a similar manner, we can construct fields which are symmetric along x antisymmetric along y; antisymmetric along x, symmetric along y, and antisymmetric along both x and y.

It is of interest to mention that using the modal fields corresponding to the configuration shown in Fig. 14(a) it is possible to get an accurate expression for the propagation constant for a profile given in Fig. 14(b) by using perturbation method. The results are shown in Fig. 15. For a comparison, we have also plotted the propagation constant calculated using more accurate numerical calculations [24]. it can be seen that the approximate analysis becomes more and more accurate as the V value becomes farther away from the cut-off V value of the mode.

In a similar manner, for the embedded strip waveguide shown in Fig. 15, we assume

$$n'^2(x) = \begin{cases} n_0^2 - \frac{1}{2} n_1^2, & x > 0 \\ \frac{1}{2} n_1^2, & 0 > x > -a \\ n_2^2 - \frac{1}{2} n_1^2, & x < -a \end{cases} \tag{90}$$

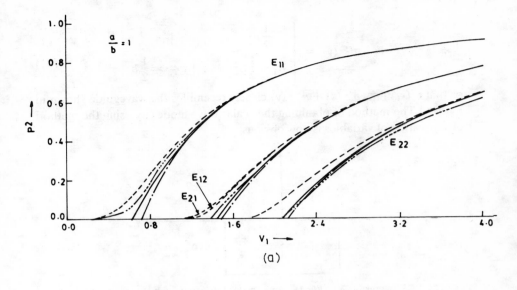

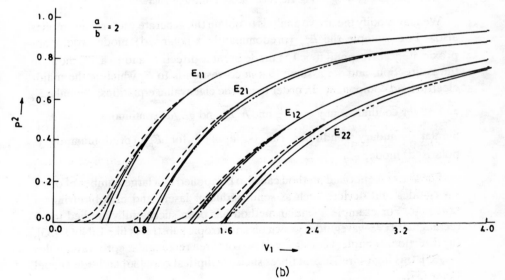

Fig. 15. Variation of the normalized propagation constant P^2 vs. the waveguide parameter V_1 for (a) a square cross section waveguide with $a = b$ and (b) a rectangular cross section guide with $a = 2b$. In both the curves, we have assumed $n_2 \simeq n_1$. The solid curves correspond to the calculations using Eq. (89), the dotted curves to the results after applying first order perturbation theory (see Ref. 32) and the dashed curves to the numerical calcuiations [24]. (Adapted from A. Kumar, K Thyagarajan and A.K. Ghatak, Analysis of rectangular core dielectric waveguides: an accurate perturbation approach, Optics Letts. **8** (1983) 63 with permission of the Optical Society of America).

$$n''^2(y) = \begin{cases} \frac{1}{2}n_1^2 & , \; |y| < \frac{1}{2}b \\ n_2^2 - \frac{1}{2}n_1^2 & , \; |y| > \frac{1}{2}b \end{cases} \quad (91)$$

so that $n^2(x,y) \equiv n'^2(x) + n''^2(y)$ closely resembles the waveguide shown in Fig. 16. The method of obtaining the scalar wave modes by using the method of separation of variables is now obvious.

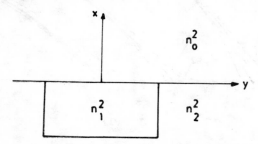

Fig. 16. An embedded strip waveguide.

We may modify the above analysis to obtain the accurate vector-wave modes also. For example the E_{pq}^x (predominantly x-polarised) mode would approximately correspond to a TM mode in the x direction and to a TE mode in the y-direction, and we assume that ψ corresponds to E_x which is the major electric field component. In order to get the eigenvalue equations, we make E_x and $\partial E_x/\partial y$ continuous at $y = \pm \frac{b}{2}$ and $n^2 E_x$ and $\partial E_x/\partial x$ continuous at $x = \pm \frac{a}{2}$. Similar boundary conditions can be applied for E_{pq}^y (predominantly y-polarised) mode.

The above mentioned method can also be applied to a large number of other waveguides and devices such as semiconductor lasers and high-birefringent fibres etc. For example a similar method has been successfully applied to (i) rectangular core waveguides grown on anisotropic substrates like LiNbO$_3$ [33], (ii) directional couplers consisting of two or three rectangular core waveguides [34, 35], (iii) high-birefringent fibres such as elliptical core [36] and side-tunnel fibres [37].

11. CONCLUSIONS

In order to understand the operation of integrated optical devices it is very necessary to have a proper understanding of the propagation characteristics of waveguides which are used in such devices. The purpose of this paper has been to highlight the important propagation characteristics of several types of planar and rectangular core waveguides. Starting from the definition of TE and TM modes we have tried to explain the working of anisotropic and metal clad waveguides which act as polarizers. We have also explained the principle of absolutely SPSM (single polarization single moded) waveguides.

The propagation of a leaky mode as a superposition of a packet of radiation modes has been discussed. The analysis explicitly shows how the power leaks into the cladding. The analysis also forms the basis of the recently developed matrix method which can be used for the analysis of planar and circularly symmetric waveguides which could be absorbing as well as leaky. The matrix method is simple and straightforward and can also be used for quantitative understanding of prism film couplers, directional couplers etc. Finally an approximate analysis of rectangular core waveguides has also been given.

ACKNOWLEDGEMENTS

The authors wish to thank Professor B.P. Pal for his invitation to write this chapter and ch. 21. They also wish to thank Dr. Arun Kumar for his numerous suggestions. Thanks are also due to Dr. M.R. Shenoy, Dr. A.N. Kaul and Ds.(Ms.) S. Diggavi for their help during the preparation of the manuscript.

REFERENCES

1. M.S. Sodha, and Ghatak A.K., *Inhomogeneous Optical Waveguides,* Plenum Press, New York, 1977.
2. A.K. Ghatak and Thyagarajan K., *Optical Electronics,* Cambridge University Press, Cambridge, UK (1989); Foundation Books (India) 1991.
3. M.J. Adams, *An Introduction to Optical Waveguides,* J. Wiley, Chichester, UK, 1981.
4. J.D. Love and Ghatak A.K., Exact Solutions for TM modes in graded index slab waveguides, *IEEE J. Quantum Electron,* vol **QE-15**, pp 14-16, 1979.
5. A. Reisinger, Characterization of optical guided modes in lossy waveguides, *Appl. Opt.,* vol **12**, pp 1015-1025, 1973.
6. J.N. Polky and Mitchell G.L., Metal-clad planar dielectric waveguide for integrated optics, *J. Opt. Soc. Am.,* vol **64**, pp 274-279, 1974.
7. I.P. Kaminow, Mammel W.L. and Weber H.P., Metal-clad optical waveguides ; analytical and experimental study, *Appl. Opt.,* vol **13**, pp 396-405, 1974.
8. Y. Yamamoto, Kamiya T. and Yanai H, Characteristics of optical guided modes in multilayer metal-clad planar optical guide with low-index dielectric buffer layer, *IEEE J. Quantum Electron,* vol **QE-11**, pp 729-736, 1975.
9. M. Masuda and Koyama J., Effects of a buffer layer on TM modes in a metal-clad optical waveguide using Ti-diffused $LiNbO_3$ c-plate, *Appl. Opt.,* vol **16**, pp 2994-3000, 1977.
10. J. Ctyroky, Janta J. and Schrafel J., Thin-film polarizer for optical waveguides, *Proc Tenth European Conf. on Optical Commun.,* Stuttgart (FRG), pp 44-45, 1984.
11. J.P.J. Bristow, Nutt A.C.C, and Laybourn P.J.R, Novel integrated optical polarizers using surface plasma waves and ion milled grooves in lithium niobate, *Electron Lett.,* vol **20**, pp 1047-1048, 1984.
12. K. Thyagarajan, Bourbin Y, Enard A., Vatoux S. and Papuchon M., Experimental demonstration of TM mode-attenuation resonance in planar metal-clad optical waveguides, *Opt. Lett.,* vol **10**, pp 288-290, 1985.

13. Y. Suematsu, Hakuta M, Faruya K., Chiba K. and Hasumi R., Fundamental transverse electric field (TE) mode selection for thin-film asymmetric light guides, Appl. Phys. Lett., vol **21**, pp 291-293, 1972.
14. K. Thyagarajan, Kaul A.N. and Hosain S.I., Attenuation Characteristics of single-mode metal-clad graded index waveguides with a dielectric buffer : A simple and accurate numerical method, Opt. Lett., vol **11**, pp 479-481 (1986).
15. T. Findakly, Chen B. and Booher D., Single-mode integrated-optical polarizers in $LiNbO_3$ and glass waveguides, ibid., vol **8**, pp 641-643, 1983.
16. R.A. Bergh, Lefevre H.C. and Shaw H.J. "Single-mode fibre optic polarizer", *Opt. Lett.*, vol **5**, pp 479-481.
17. P.K. Tien and Ulrich R., Theory of prism-film coupler and thin-film light guides. *J. Opt. Soc. Am.*, vol **60**, pp 1325-1337, 1970.
18. A.K. Ghatak, Leaky modes in optical waveguides, *Opt. Quantum Electron.*, vol **17**, pp 311-321, 1985.
19. A.K. Ghatak, Thyagarajan K. and Shenoy M.R., Numerical analysis of planar optical waveguides using matrix approach, *J. Lightwave Technol.*, vol **5**, pp 660-667, 1987.
20. K. Thyagarajan, Shenoy M.R. and Ghatak A.K., Accurate numerical method for the calculation of bending loss in optical waveguides using a matrix approach, *Opt. Lett.*, vol **12**, pp 296-298, 1987; Erratum, *Optics Letters* April 1989.
21. K. Thyagarajan, Diggavi S. and Ghatak A.K., Analytical investigations of leaky and absorbing planar structures, *Opt. Quantum Electron.*, vol **19** pp 131-137, 1987.
22. M.R. Ramadas, Garmire E., Ghatak A.K., Thyagarajan K. and Shenoy M.R., Analysis of absorbing and leaky planar waveguides: a novel method, *Opt. Lett.*, vol **14**, pp 376-378, 1989.
23. M.R. Ramadas, Varshney R.K., Thyagarajan K. and Ghatak A.K., A matrix approach to study the propagation characteristics of a general nonlinear planar waveguide, *J. Lightwave Technol.*, vol **7**, pp 1901-1905, 1989.
24. J.E. Goell, A Circular-harmonic computer analysis of rectangular dielectric waveguides, *Bell. Syst. Tech. J.*, vol **48**, pp 2133-2160, 1969.
25. E.A.J. Marcatili, Dielectric rectangular waveguide and directional coupler for integrated optics, *Bell Syst Tech. J*, vol **48**, pp 2071-2102, 1969.
26. G.B. Hocker, and Burns W.K., Mode dispersion in diffused channel waveguides by the effective index method, *Appl., Opt.*, vol **16**, pp 113-118, 1977.
27. C. Yeh, Ha K, Dang S.B., and Brown W.P. Single-mode optical waveguides, *Appl. Opt.*, vol **18**, pp 1490-1504, 1979.
28. R. Mittra, Hou Y.L. and Samejal V, Analysis of open dielectric waveguides using mode-matching technique and variational methods, *IEEE Trans.*, vol **MTT-28**, pp 36-43, 1980.
29. V.V. Cherny, Juravley G.A., and Whinnery J.R., A multilayer fibre guide with rectangular core, *IEEE Trans.* vol **MTT-28**, pp 401-404, 1980.
30. P.K. Mishra, Sharma A., Labroo S. and Ghatak A.K., Scalar variational analysis of single-mode waveguides with rectangular cross-section, ibid, vol **MTT-33**, pp 282-286, 1985.
31. F.P. Payne, A new theory of rectangular optical waveguides, *Opt. Quantum Electron.*, vol **14**, pp 525-537, 1982.
32. A. Kumar, Thyagarajan K and Ghatak A.K., Analysis of rectangular core dielectric waveguides : an accurate perturbation approach, *Opt. Lett.*, vol **8**, pp 63-65, 1983.
33. A. Kumar, Shenoy M.R. and Thyagarajan K., Modes in anisotropic rectangular waveguides : an accurate and simple perturbation approach, *IEEE Trans.*, vol **MTT-32**, pp 1415-1418, 1984.

34. A. Kumar, Kaul A.N. and Ghatak A.K., Prediction of coupling length in a rectangular-core directional coupler : an accurate analysis, *Opt. Lett.*, vol **10**, pp 86-88, 1985.
35. A.N. Kaul, Thyagarajan K. and Arun Kumar, Coupling characteristics of three channel waveguide directional coupler, *Optics Comm.*, vol **56**, pp 95-99, 1985.
36. A. Kumar, Varshney R.K. and Thyagarajan K., Birefringence calculations in elliptical-core fibres, *Electron Lett.*, vol **20**, pp 112-113, 1984.
37. R.K. Varshney, and Kumar A., Birefringence calculations in side-tunnel optical fibres: A rectangular-core waveguide model. *Opt. Lett.*, vol **11**, pp 45-47, 1986.

4

Transmission Characteristics of Telecommunication Optical Fibres

B.P. PAL[*]

1. INTRODUCTION

Glass optical fibres have emerged as the most important broadband transmission-media of the eighties for telecommunication. Besides offering very large information transmission bandwidth, these optical fibres are characterised by extremely low-loss (~0.2 dB/km in the premium grade fibres): both these features enable one to attain extremely large repeater spacings in optical communication systems. The other key advantages of optical fibres over conventional metallic transmission media like coaxial cables/twisted wire pairs etc. lie in their being much smaller in size and lighter in weight for the same transmission bandwidth, thereby requiring much less duct space and transportation cost (and also for potential on-board application in aircrafts and ships). Being dielectric, they are immune from EMI, EMP and short circuits/ground loops and are almost free from cross-talks all of which have obvious impact from the view point of secure communication in defence communication networks. These fibres are more tolerant to hostile temperature environments and they are definitely very much cost effective as compared to metallic transmission media wherever volume of information traffic is large. Price trend shows a continuous decline with time in cost of the fibres as the technology was getting established in contrast to, say, copper wires where the cost is increasing almost every year.

In Ch. 3, the principles of wave guidance in optical waveguides of planar and strip geometry were discussed. In this chapter we describe the fundamental

---

[*] Physics Department, Indian Institute of Technology Delhi, New Delhi - 110016, India

characteristics of optical waveguides of circular geometry i.e., the optical fibres which determine propagation effects in them. The great importance of the topic stems from the fact, which we shall see in the following sections, that design improvements in certain propagation characteristics of the fibre had essentially motivated and dictated technology developments of associated optoelectronic components like sources and detectors which eventually led to the emergence of newer generation of optical communication systems.

2. GEOMETRY OF AN OPTICAL FIBRE AND CONCEPT OF NUMERICAL APERTURE

Figure 1(a) shows the geometry of an optical fibre. It is a cylindrically symmetric structure consisting of a 'core' glass of diameter $\sim 4-100\ \mu m$ and is surrounded by a 'cladding' glass of slightly lower refractive index. Depending on applications, the core/cladding diameters could be $50 \pm 3/125 \pm 3\ \mu m$, $62.5 \pm 3/125 \pm 3\ \mu m$, $85 \pm 5/125 \pm 3\ \mu m$, $100 \pm 5/140 \pm 5\ \mu m$ or $4-10/125 \pm 3\mu m$. The non-circularity of core and cladding are $\leq 6\%$ and $< 2\%$, respectively while core/cladding concentricity error is supposed to be $\leq 6\%$ in standard fibres. To preserve the pristine strength of the fibres, they are normally coated with a protective soft plastic coating of about $250 \pm 15\ \mu m$ diameter as a

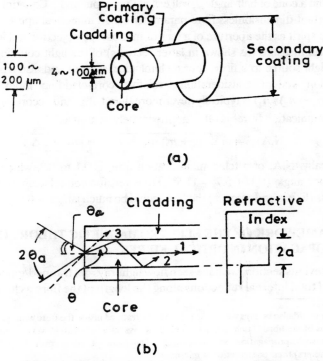

Fig. 1. (a) Geometry of an optical fibre; (b) Zig-zag path of rays in a step-index fibre and concept of N.A./acceptance cone.

primary coating followed by another harder and tighter secondary coating. For telecommunication applications, standard overall diameter of the fibre is 125 ± 3μm; for applications in local area networks it may be 140μm also. If a light ray injected through the air-core interface of the fibre is incident at the core-cladding interface at an angle greater than the critical angle for that interface then it will suffer total internal reflection there. Due to the cylindrical symmetry of the fibre, this ray would suffer repeated total internal reflections at the upper and lower interfaces [See Fig. 1(b)] and would, thus, get trapped within the fibre as a guided ray to emerge at the exit end of the fibre into air again. From Snell's law, it is easy to conclude that for total internal reflection to take place, the angle θ must be such that

$$\theta < \frac{\pi}{2} - \theta_{cr}, \qquad (1)$$

where $\theta_{cr} = \sin^{-1}(n_2/n_1)$, represents critical angle for total reflection, n_1 and n_2 are the refractive indices of the core and the cladding, respectively. Thus, applying Snell's law at the air-core interface, one gets

$$n_{\text{air}} \sin \theta_a = \left(n_1^2 - n_2^2\right)^{1/2} \qquad (2)$$

From Fig. 1(b), it is easy to see that all rays which are incident at the fibre input face within a cone of half-angle θ_a will exit at the output end.* Equation (2) defines an important dimensionless fibre parameter called numerical aperture (N.A.) in the same spirit as the aperture of a lens or microscope objective and it is in fact, a very useful measure (as shown in latter sections) of the light coupling efficiency from a light source to a fibre. For technological reasons and also from telecommunication system considerations, relative core-cladding index difference ($\Delta = (n_1 - n_2)/n_1$) is typically never more than 1-2%, and accordingly, N.A. for telecommunication fibres is often approximately written as

$$\text{N.A.} = \sqrt{(n_1 + n_2)(n_1 - n_2)} \approx n_1 \sqrt{2\Delta}. \qquad (3)$$

Typically, N.A. of a telecommunication fibre is 0.1 to 0.2, which implies an acceptance angle $(\theta_a) \approx 5.7° - 11.5°$. However, for non-telecommunication applications (like in endoscopes) the N.A. could be much larger $\gtrsim 0.5$ (i.e. $\theta_a \gtrsim 30°$).

3. FRAMEWORK OF THE ELECTROMAGNETIC THEORY OF PROPAGATION IN OPTICAL FIBRES

In the previous section, we have seen how light rays get trapped/guided through repeated total internal reflections along the length of the fibre as long as the ray

---

* There could also be rays which can get partially trapped inside the fibre without ever crossing the axis of the fibre. Such rays are called skew rays, most of them however continuously lose power with propagation and accordingly for most purposes over long fibre lengths (kilometers) their contribution is minimum.

is incident within the acceptance cone semi-angle of the fibre, which, in turn, is dictated by the core and cladding refractive indices. However, the model of light propagation through a bounded structure like optical fibre in terms of geometrical rays is only an approximate description of propagation effects in them. This approach works well as long as (i) the characteristic dimension of the fibre cross-section e.g. core diameter ($2a$, where a is the core radius) is large compared to the wavelength (λ) of the propagating light wave and (ii) the relative core-cladding index difference is not too small. In fact both a and Δ can be combined together with λ to yield a composite parameter called normalised frequency/V-number of the fibre defined through

$$V = (2\pi/\lambda)\, a n_1 \sqrt{2\Delta}. \tag{4}$$

If the V-number of the fibre is $\gg 10$, geometrical-optics results based on ray trajectories yield precise results on many of the propagation effects in optical fibres [1-4]. For $V \lesssim 10$, geometrical optics cannot explain propagation effects in fibres and one is required to make an electromagnetic analysis based on wave optics to investigate propagation effects in them. Thus, for developing a general framework for that purpose which would be applicable to any fibre waveguide of arbitray V-number, one starts out with Maxwell's equations and forms the so called vector wave equations [2, 3] satisfied by electric (E) and magnetic (H) field vectors of the light wave

$$\nabla^2 E + \nabla\left[(\nabla \varepsilon/\varepsilon).E\right] = \mu_0 \varepsilon \frac{\partial^2 E}{\partial t^2}, \tag{5}$$

$$\nabla^2 H + \left(\frac{\nabla \varepsilon}{\varepsilon}\right) \times (\nabla \times H) = \mu_0 \varepsilon \frac{\partial^2 H}{\partial t^2}, \tag{6}$$

where $\varepsilon\ (= \varepsilon_0 n^2, \varepsilon_0$ being ε for free space while 'n' is the refractive index of the fibre) is the dielectric permittivity of the fibre, and μ_0 is the free-space magnetic permeability which is same as that of the fibre, the fibre being assumed to be non-magnetic. The earliest form of refractive index distribution* proposed for an optical fibre consisted of a uniform refractive index say, n_1 across the core (of diameter $2a$) surrounded by a cladding of lower refractive index n_2 as shown in Fig. 1(b) so that one may represent algebraically the RIP through

$$n(r) = n_1, \quad r < a$$
$$ = n_2, \quad r \gtrsim a. \tag{7}$$

Fibres with RIP as Eq. (7) are known as step-index fibres. For such a homogeneous medium $\nabla \varepsilon$ term will be identically zero in both core and cladding and in each of these regions, each cartesian component of the electric and magnetic fields will satisfy

---

* Henceforth will be referred to as RIP, standing for Refractive Index Profile.

$$\nabla^2 \Psi - \mu_0 \varepsilon \frac{\partial^2 \Psi}{\partial t^2} = 0. \tag{8}$$

This is known as the scalar wave equation with Ψ as representing any of the cartesian components of the E and H fields. Since n is independent of z, the solution to Eq. (8) can, in general, be written as

$$\Psi(r, \varphi, z, t) = \psi(r, \varphi) e^{i(\omega t - \beta z)}, \tag{9}$$

where the direction of propagation is chosen along z and β is the propagation constant, and ω represents angular frequency of the wave. Equation (9) admits two kinds of solutions to Eq. (8) - one whose field decreases exponentially with r for $r \gtrsim a$ and is oscillatory inside the core ($r < a$) while the second one leads to oscillatory solutions at all values of r. We shall shortly see that the former type of solution yields discrete values of β known as guided modes of the fibre while the latter ones represent what are known as radiation modes characterised by a continuum of β's. Formally, a guided mode is defined as a specific field distribution which propagates through the waveguide with a definite group velocity ($= 1/d\beta/d\omega$) and without any change in the nature of its distribution. Depending on its geometry and physical characteristics, a fibre can support a number of modes or just one mode-the former case is called a MULTIMODE FIBRE while the latter is called a SINGLE-MODE or MONOMODE FIBRE. In fact, an arbitrary field incident at the input end of a fibre can always be expanded as [4]

$$\Psi(r, \psi, z, t) = \sum_p a_p \psi_p(r, \phi) e^{i(\omega t - \beta_p z)}$$
$$+ \int a(\beta) \psi(\beta, r, \phi) e^{i(\omega t - \beta z)} d\beta. \tag{10}$$

In Eq. (10), $\sum_p$ represents summation over the discrete guided modes while the integral is over the unbound continuum of radiation modes. Actual values of β_p will be yielded by the boundary conditions.

We may recall that in telecommunication grade fibres, relative core-cladding index difference is usually never* made more than 1-2%. Such fibres having $\Delta \ll 1$, are known as weakly guiding fibres [5]. A byproduct of this condition (which is a practical necessity) is that modes in such fibres can be shown to be almost linearly polarized [5] having the transverse field lying almost entirely along y or x with a very small longitudinal component. Further since the refractive index difference is assumed to be small ψ and $\partial \psi / \partial r$ can be asumed to be continuous across $r = a$ [2].

---

* It will become clear in latter sections that to contain pulse dispersion and scatter loss to acceptably low levels, Δ is purposely kept small while fabricating telecommunication fibres.

For a step-index fibre, 'n' is dependent on 'r' only i.e. it is cylindrically symmetric and hence one can write Eq. (8) in cylindrical system of co-ordinates as

$$\frac{\partial^2 \psi}{\partial r^2} + \frac{1}{r} \frac{\partial \psi}{\partial r} + \frac{1}{r^2} \frac{\partial^2 \psi}{\partial \phi^2} + \left[k_0^2 n^2(r) - \beta^2 \right] \psi = 0, \quad (11)$$

where $k_0 = \omega \sqrt{\varepsilon_0 \mu_0}$ ($= \omega/c$) is the free-space wave number. By adopting the method of separation of variables i.e. writing

$$\psi(r, \phi) = R(r) \Phi(\phi) \quad (12)$$

Eq. (11) can be solved separately for its radial and azimuthal parts. The azimuthal variation would be given by

$$\Phi(\phi) \sim e^{\pm il\phi} \quad (13)$$

with $l = 0, 1, 2, 3,...$ and the radial part of ψ will satisfy the following equations

$$\frac{d^2 R}{dr^2} + \frac{1}{r} \frac{dR}{dr} + \left[k_0^2 n_1^2 - \beta^2 - \frac{l^2}{r^2} \right] R = 0 \quad r < a \quad (14)$$

and

$$\frac{d^2 R}{dr^2} + \frac{1}{r} \frac{dR}{dr} - \left[\beta^2 - k_0^2 n_2^2 + \frac{l^2}{r^2} \right] R = 0. \quad r \gtrsim a \quad (15)$$

Equations (14) and (15) are standard form of Bessel's equations which admits four different types of cylindrical functions: $J_l(r)$, $Y_l(r)$ and $K_l(r)$, $I_l(r)$ respectively. However, for modal fields to be finite and bound within the core and exponentially decaying in the cladding (guided modes), one has to choose Bessel function $J_l(r)$ as solution to Eq. (14) inside the core and modified Bessel function $K_l(r)$ as solution to Eq. (15) inside the cladding. Accordingly, solution to Eqs. (14) and (15) can be written as

$$\psi(r, \phi) = E_y(r, \phi) = \frac{A_l}{J_l(u)} J_l(ur/a) e^{\pm il\phi}, \quad r < a$$

$$= \frac{A_l}{K_l(w)} K_l(wr/a) e^{\pm il\phi}, \quad r \gtrsim a \quad (16)$$

where $u = a\sqrt{(k_0^2 n_1^2 - \beta^2)}$ and $w = a\sqrt{\beta^2 - k_0^2 n_2^2}$

$$\Rightarrow \quad u^2 + w^2 = a^2 k_0^2 (n_1^2 - n_2^2) = V^2 \quad [\text{see Eq. (4)}]. \quad (17)$$

In writing Eq. (16), continuity of ψ has been used and E_y has been chosen to be the dominant transverse component of the electric field, since for $\Delta \ll 1$, modes are almost linearly polarized [2, 4]. For large real values of the argument, $J_l(ur/a)$ behaves as standing waves (of decreasing amplitude) while $K_l(wr/a)$ decreases monotonically. Thus, these functions precisely satisfy the requirements of Eq. (16) to represent the guided modes of the fibre. Thus both u and w must be real and positive for a guided mode thereby implying that for a mode to be guided, its β must satisfy the condition

$$k_0 n_1 \gtrsim \beta \gtrsim k_0 n_2. \tag{18}$$

Now, as already stated for $\Delta \ll 1$, transverse field component ψ will lie almost entirely along y (or x), thus the only non-zero field components for the modal solution Eq. (16) will be E_y, E_z, H_x, H_z of which longitudinal components E_z and H_z can be shown to be much smaller than the transverse components E_y and H_x as long as Δ is small [2, 3, 5]. If E_x was chosen as the dominant transverse component of the field, then the non-zero field components which will form the modal field will be E_x, E_z, H_z and H_y and the field would have experienced the same propagation constant β, thereby implying that these modes are two-fold degenerate. Accordingly, the modes in weakly guiding structures are known to be linearly polarized and are designated as LP_{lm}-modes. From the continuity of $\partial E_y / \partial r$ at $r = a$, one gets

$$u \frac{J_l'(u)}{J_l(u)} = w \frac{K_l'(w)}{k_l(w)}, \tag{19(a)}$$

where primes denote differentiation of the cylinder functions with respect to their argument. By use of appropriate recurrence relations governing Bessel and modified Bessel functions [2, 3, 4, 5], Eq. [19(a)] can be shown to reduce to

$$\frac{u J_{l-1}(u)}{J_l(u)} = -w \frac{K_{l-1}(w)}{k_l(w)}. \tag{19(b)}$$

Equation 19(b) is a transcendental equation whose solutions within the range specified by the condition [Eq. (18)] will yield the discrete propagation constants for different guided modes.

In practice, for calculating the β's, one expresses β in a normalised form as:

$$b = \frac{(\beta^2 / k_0^2 - n_2^2)}{n_1^2 - n_2^2} = \frac{w^2}{V^2}. \tag{20(a)}$$

Since β differs very little from $n_1 k_0$ in weakly guiding fibres Eq. 20(a) can be simplified to:

$$b \approx \frac{\beta/k_0 - n_2}{n_1 - n_2} \Rightarrow \beta \approx k_0 n_2 (1 + b\Delta). \tag{20(b)}$$

For a guided mode, it is obvious from Eq. (18) that

$$0 \lesssim b \lesssim 1. \tag{21}$$

At the lower limit $b = 0 \Rightarrow \beta = k_0 n_2$ which is nothing but the plane wave propagation constant in an infinite homogeneous medium of refractive index n_2. In terms of b and V, Eq. 19(b) can be rewritten as [4]

$$V\sqrt{1-b} \frac{J_{l-1}(V\sqrt{1-b})}{J_l(V\sqrt{1-b})} = -V\sqrt{b} \frac{K_{l-1}(V\sqrt{b})}{K_l(V\sqrt{b})} \quad \text{for } l \gtrsim 1 \tag{22(a)}$$

and

$$V\sqrt{1-b}\,\frac{J_1(V\sqrt{1-b})}{J_0(V\sqrt{1-b})} = V\sqrt{b}\,\frac{K_1(V\sqrt{b})}{K_0(V\sqrt{b})} \quad \text{for} \quad l = 0. \qquad [22(b)]$$

From a plot of the L.H.S. and R.H.S. of Eq. (22) as a function of b ($0 < b < 1$) for different values of V on the same graph and the intersections between them one can obtain b and hence, β for different modes as solutions of the transcendental Eq. (22). For a given V, the number of intersections yields the possible number of guided modes (see, for example, [4]).

A plot of b vs V is shown in Fig. 2, which is a universal plot and it does not depend explicitly on other fibre parameters and therein lies the advantage of expressing β in a normalised form. Further, in practice, β is often required to be known with very high accuracy (e.g., in dispersion calculations) upto about sixth place of decimals and for which expressing it in terms of b is always very convenient because b is a more sensitive parameter than β with respect to variations in V.

It may be mentioned here that a more rigorous approach [2, 3] would be to solve Eq. (8) in cylindrical polar co-ordinates for ψ ($= E_z$) and obtain E_r (and

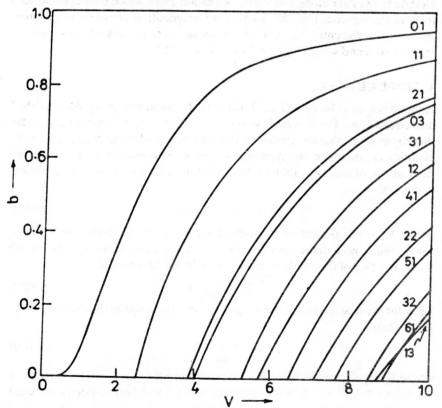

Fig. 2. Normalised propagation constant b vs. V for various LP_{lm}-modes. Here $b = \left(\beta^2/k_0^2 - n_2^2\right)/\left(n_1^2 - n_2^2\right)$ and $V = ak_0\left(n_1^2 - n_2^2\right)^{1/2}$ (reprinted from [5] with permission. Copyright 1971, Optical Society of America).

H_r) and E_φ (and H_φ) in terms of E_z and H_z from Maxwell's curl equations by writing them in terms of components in cylindrical coordinates. Thereafter, imposing the continuity of E_z (and H_z) and E_φ (and H_φ) which are the tangential components, would, in place of Eq. 19(a), result in the following transcendental equation for β [2, 3]

$$\left[\left(\frac{n_1^2}{n_2^2}\right)\left(\frac{w^2}{u}\right)\frac{J_l'(u)}{J_l(u)} + \frac{w K_l'(w)}{K_l(w)}\right]\left[\frac{w^2}{u}\frac{J_l'(u)}{J_l(u)} + \frac{w K_l'(w)}{K_l(w)}\right]$$
$$= \left[l\left(\frac{n_1^2}{n_2^2} - 1\right)\beta k_0 n_2 \left(\frac{a}{u}\right)^2\right]^2, \qquad (23)$$

where primes denote differentiation with respect to their argument. Such a derivation of Eq. (23) does not involve any approximation whatsoever. However, if the weakly guiding conditions namely, $\Delta \ll 1$ and $n_1 \sim n_2$ are applied then Eq. (23) simplifies [3] (after making use of recurrence relations like $J_l'(u) = J_{l-1}(u) - \left(\frac{l}{u}\right) J_l(u)$) to Eq. 19(b) thereby justifying our earlier assertion that modes in a weakly guiding fibre are almost linearly polarized with electric field either along y or x-axis. Equation 19(b) which is an approximate form of the rigorous Eq. (23) for yielding propagation constants of different modes under the condition $\Delta \ll 1$ has been shown to be accurate to within 1% for $\Delta < 0.01$ and within 10% for $0.01 < \Delta < 0.25$ [1].

4. MODE CUT-OFF

A mode is said to be cut-off i.e. it will cease to propagate as a guided mode if its β equals $k_0 n_2$. For $\beta = k_0 n_2$, $w = 0$ and for $\beta < k_0 n_2$, w becomes imaginary thereby implying that the guided modal field instead of being evanescent in the cladding i.e. decaying exponentially at large r will transform to an oscillatory field at all values of r, thereby becoming a radiation mode. The limiting condition:

$$\beta = k_0 n_2 \Rightarrow w = 0 \qquad [24(a)]$$

is thus known as the cut-off condition for a mode. In the limit $w \to 0$ for the lowest order mode (corresponding to $l = 0$), Eq. 22(b) shows that the cut-off frequency (V_c) of this mode would be given by the first root of

$$J_1(V_c) = 0 \qquad [24(b)]$$

while for the next mode, cut-off frequency would be given by the first root of (cf Eq. [(22a)])

$$J_0(V_c) = 0, \qquad [24(c)]$$

where V_c represents the value of V at mode cut-off ($\because w = 0$ for a mode at cut-off, its $u = V = V_c$). Since the zeros of $J_1(x)$ and $J_0(x)$ respectively, occur at $V_c = 0, 3.8317, 7.0456,....$ and at $V_c = 0, 2.4048, 5.5201, 8.6537,...$ modes having $V_c = 0, 2.4048, 3.8317,...$ are respectively designated as LP_{01}, LP_{11}, LP_{02},.... modes. The nomenclature LP_{lm} follows from the fact that these modes are linearly polarized [5]. The subscript l stands for l^{th} order Bessel function that defines the cut-off condition for corresponding mode order and it is related to the azimuthal periodicity while m which is also an integer defines the successive roots of the corresponding Bessel function. Basically l represents number of azimuthal antinodes over a semicircle while m represents number of radial antinodes in a mode's field pattern*. As an example, mode patterns of two relatively high order LP_{lm}-modes have been schematically drawn in Fig. 3 as they would have appeared on a photograph. It may be mentioned here that in practice it is extremely difficult to experimentally launch a relatively high order mode in isolation in a multimode fibre and maintain its popagation over long fibre lengths. This is because any small perturbation along the length of the fibre in the form of geometrical imperfections, inhomogeneities etc. is likely to transfer power form this mode to other modes with propagation. As a result, when a multimode fibre is excited, e.g., with a He-Ne laser what one observes at the fibre exit end is essentially a superposition of the various mode patterns. Only if the fibre is so designated that effectively its V-value lies in the range $0 < V < 2.4048$ will then one be able to maintain propagation of single fundamental mode namely, LP_{01} mode in that fibre. This is so because at $V < 2.4048$ no other mode except the LP_{01} mode can be supported by a fibre. In fact LP_{01}-mode is never cut-off! It can propagate even if core diameter or refractive index difference Δ are made aribitrarily small (i.e. arbitrarily low V-value) although we will shortly see that at very low V-values, power of LP_{01}-mode confined within the core is very little and most of it would effectively be propagating in the cladding. Fibres which can support only the LP_{01}-mode

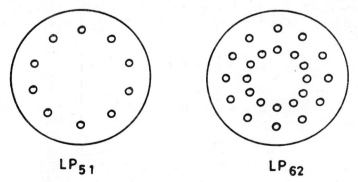

Fig. 3. Schematic representation of mode intensity patterns for the modes; (a) LP_{51} and (b) LP_{62}.

---

* Except for the LP_{0m}-modes, for which there will be no radial anti-node at the centre.

are known as single-mode fibres. Thus for purely single-mode operation, a fibre's V-value must lie in the range:
$$0 < V < 2.4048. \qquad [25(a)]$$

This condition can be used to obtain design guidelines e.g. choice of a and Δ for obtaining single mode operation at a particular λ. Since to contain scatter loss from the dopants in a fibre to acceptably low values, Δ is typically kept no more than 0.3%, in order to satisfy condition [25(a)] for single-mode operation, the core diameter ($2a$) would typically be 4-6 μm at the 1st generation wavelength $\sim 0.8\,\mu$m and 8-10 μm at the 2nd and 3rd generation wavelength $\sim 1.3\,\mu$m. The condition [25(a)] is also often alternatively expressed in terms of cut-off wavelength defined through
$$\lambda_c = \frac{2\pi a n_1 \cdot \sqrt{2\Delta}}{2.4048}. \qquad [25(b)]$$

At any $\lambda > \lambda_c$ (of a particular fibre), only LP_{01}-mode can be supported because the next higher order mode namely, LP_{11}-mode and all subsequent higher order modes will be cut-off i.e. absent in this fibre. In this context, the concept of λ_c is very important because the choice of λ_c which is actually dicated by considerations like low transmission loss and high transmission bandwidth (BW) wavelength window, wavelength at which source and detector peak efficiencies match and so on, ultimately decides the value of a and Δ. Due to inadvertent perturbations like bending etc. experienced by a single-mode fibre in its lay over long lengths, a fibre is actually single moded at an effective cut-off wavelength (λ_{ce}), which is shorter than λ_c (see latter sections and Ch. 10).

In order to obtain net amount of power carried by different modes, one is essentially required to calculate the z-component of the Poynting vector associated with each mode and integrate the same across the fibre cross-section. By simple algebra, it is possible to show [3] that in a weakly-guiding fibre, the fractional power carried by a guided mode in the core and in the cladding, respectively, are
$$\eta_{core} = \frac{P_{core}}{P_{total}} = 1 - (1-\kappa)\frac{u^2}{V^2} \qquad [26(a)]$$

and
$$\eta_{clad} = \frac{P_{clad}}{P_{total}} = (1-\kappa)\frac{u^2}{V^2}, \qquad [26(b)]$$

where
$$\kappa = \frac{K_l^2(w)}{K_{l+1}(w)K_{l-1}(w)} \simeq 1 - \frac{1}{\sqrt{1+l^2+w^2}}. \qquad [26(c)]$$

These equations clearly show that far from cut-off, since w will be, relatively, a large number, most of the guided power will reside in the core. On the other hand near cut-off, $w \ll 1$ and hence
$$\eta_{core} \simeq 1 - \frac{1}{\sqrt{1+l^2}}, \qquad [27(a)]$$

$$\eta_{\text{clad}} \simeq \frac{1}{\sqrt{1 + l^2}}. \qquad [27(b)]$$

Thus for $l = 0$ modes, most of the power would flow through the cladding which is not true for $l > 1$ modes. The former result leads one to conclude that especially in the case of single-mode fibres supporting only LP_{01}-mode, the cladding like the core must be made of very low-loss materials in order to avoid excessive loss over long fibre lengths. Figure 4 shows fractional modal power in the cladding $P_{\text{clad}}/P_{\text{total}}$ as well as several other propagation quantities in a

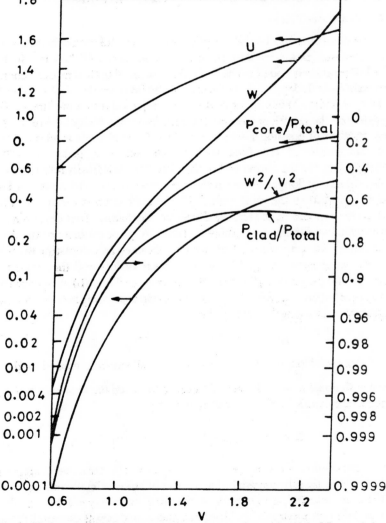

Fig. 4. Various characteristics propagation parameters/quantities of the LP_{01}-mode in a step-index fibre as a function of V-number (reprinted with permission from D. Gloge, "Propagation effects in optical fibres", IEEE Trans. Microwave Theory and Techniques, Vol. MTT-23, pp. 106-120, 1975. Copyright © 1975 IEEE).

step idex fibre as a function of V-value for the LP_{01}-mode. In passing, it is worthwhile to note here that since the modal field in the cladding varies as $K_l(wr/a)$ which asymptotically behaves as $\exp(-wr/a)$, this penetration of guided mode power into the cladding can be gainfully exploited to fabricate evanescent coupling devices like directional couplers, power dividers, switches etc. which find extensive applications in all-fibre type fibre optic sensors as well as wavelength multiplexers, filters etc. in optical telecommunication.

5. TEMPORAL PULSE DISPERSION IN OPTICAL FIBRES

5.1 Graded Index Fibres

For several reasons, most suitable optical carrier modulation scheme for optical telecommunication involves pulse-code modulation (PCM) in which the signal/information is coded in the form of a stream of optical pulses generated by directly modulating the drive current of the laser diode/LED sources (see Ch. 18 by J. Das). When an optical pulse is injected into a multimode fibre, depending on the launch conditions, typically its energy will get distributed into a large number of possible modes of the fibre. Each of these modes will travel down the length of the fibre with its characteristic group velocity, v_g ($= 1/(d\beta_{lm}/d\omega)$). At the output end of the fibre, they will finally fuse to produce a pulse usually of longer duration than the input pulse and hence of larger temporal width than the input pulse. This phenomenon is called intermodal pulse dispersion and is, in fact, the dominant mechanism that determines the transmission bandwidth of a multimode fibre. A quick estimate of the dispersion in a step-index multimode fibre may be obtained by calculating the transit time difference between the fastest (i.e. lowest order) and the slowest (i.e. highest order) propagating modes. By referring to Fig. 1(b), it is easy to see that the time taken by a typical ray (e.g. ray 2) corresponding to a specific mode to travel through a length L of the fibre is

$$\tau = n_1 L / (c \cos \theta). \qquad (28)$$

Since θ can vary from 0 to θ_c $\left(= \dfrac{\pi}{2} - \theta_{cr} \right)$, if all possible guided modes are excited with equal power, the time difference between the shortest ($\equiv \theta = 0_c$) and the longest path ($\equiv \theta = \theta_c$) will be given by

$$\Delta \tau = \tau_{\text{slowest}} - \tau_{\text{fastest}} \simeq \left(\frac{n_1 L}{c} \right) \Delta . \qquad (29)$$

In a typical fibre, if one assumes $n_1 \approx 1.46$, $\Delta = 1\%$, then over a length of 1 km, $\Delta\tau \sim 48$ ns thereby implying that a pulse of infinitesimally small temporal width (i.e. an impulse) if injected into the fibre its energy will be distributed between different modes of the fibre and these modes will eventually fuse to form the output pulse of duration 48 ns i.e. approximately 50 ns after travelling one kilometre through the fibre. This would imply that the pulse transmission rate through this fibre cannot be more than $\sim(1/50) \times 10^9 = 20$ Mb/s for distortion-free reception after 1 km at the receiving end. This is not a very attractive figure from the point of view of large information carrying capacity that an optical communciation system is supposed to offer. If, however, the core RIP of the fibre is made to be a graded profile, it leads to a substantial increase in the transmission bandwidth of the fibre. The most popular form of RIP that has been chosen to represent a general graded profile is known as power-law profile which is analytically expressed as [6]

$$n^2(r) = n_1^2 \left[1 - 2\Delta \left(\frac{r}{a}\right)^q\right] \quad , \quad r < a$$
$$= n_1^2 [1 - 2\Delta] \quad , \quad r \gtrsim a \quad (30)$$

where n_1 is the axial refractive index, a the core radius, q defines the shape of the RIP and Δ represents the relative core cladding index difference

$$\Delta = \frac{n_1^2 - n_2^2}{2 n_1^2} \simeq \frac{n_1 - n_2}{n_1}.$$

(for weakly guiding fibres)

A wide variety of RIPs can be represented through Eq. (30) simply by varying q, e.g., $q = 1$ would correspond to a triangular profile, $q = 2$ a parabolic, while $q = \infty$ would make the core RIP a step function [see Fig. 5(a)]. In the case of a $q = 2$ fibre, dispersion [cf. Eq. (29)] $\Delta \tau$ is given by [7]

$$\Delta \tau \simeq \left(\frac{n_1 L}{2c}\right) \Delta^2, \quad (31)$$

which clearly shows that $\Delta \tau$ would be much less in this case as compared to a $q = \infty$ fibre for the same n_1 and Δ. For a fibre having $n_1 = 1.46$, $\Delta \simeq 0.01$ and $L = 1$ km, $\Delta \tau$ would be $\simeq 48$ ns in a $q = \infty$ fibre while in a $q = 2$ fibre, it will be 0.24 ns which is 200-times less as compared to that in a $q = \infty$ fibre. Physically, in a graded core fibre because of the succeedingly lower refractive indices that a ray encounters as it travels towards the core cladding interface away from the fibre axis, it suffers continuous refraction as shown in Fig. 5(b). Thus, if we consider optical paths of two typical rays numbered 1 and 2 (see Fig. 5(b)), ray 2 will effectively propagate through a different refractive index region (less than the axial region) this eventually leads to a kind of equalisation between the transit times of different rays/modes and hence, the dispersion will be low. The transmission bandwidth (BW $\approx 1/\Delta \tau$) of a graded-core multimode fibre could be anything from 300 MHz-km to as large a figure as 18 GHz-km (in the so called optimum profiled fibres discussed later) while a figure around 500-600 MHz-km is more typical.

In order to make an electromagnetic analysis of propagation in graded core fibres, we note that refractive index varies continuously with r ($r < a$) in contrast

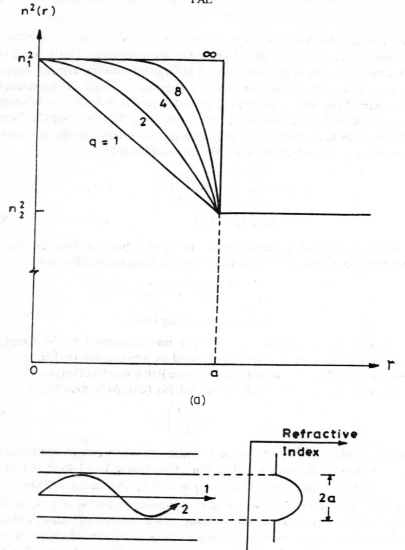

Fig. 5. (a) Power law class of refractive index profiles; q defines profile shape and a is the core radius; (b) Path of rays in a graded-core fibre.

to a step index fibre where $n(r) = n_1$ for $0 < r < a$. The scalar wave equation [Eq. (8)] can still be used[*]. For graded-core fibres since $n = n(r)$ we re-write Eq. (14) for radial part of ψ by replacing constant refractive index n_1 with $n(r)$ as

---

[*] Even though n is a function of r in graded-core fibres, as long as $\nabla \varepsilon / \varepsilon$ is small in distances of the order of wavelength, this term may be neglected in vector wave Eqs. (5) and (6); weakly guiding fibres do satisfy this condition [2].

$$\frac{d^2 R}{dr^2} + \frac{1}{r}\frac{dR}{dr} + \left[k_0^2 n^2(r) - \beta^2 - \frac{l^2}{r^2}\right] R = 0. \tag{32}$$

However, there exists no general analytically exact solution to Eq.(32) for a cladded arbitrary core RIP except for the parabolic ($q = 2$) (for which R's are Weber functions [2,4]) and the already obtained step index ($q = \infty$) core RIPs. Several approximate techniques like perturbation theory, WKB, variational, finite element analyses, and so on have been reported in the literature to study graded core fibres. Out of these, in terms of popularity, the WKB method[*] of solving (32) has been found to be most convenient. The applicability of the WKB method demands that the following condition be satisfied:

$$\left|\frac{1}{n}\frac{dn}{dr}\right| \gg \lambda^{-1}, \tag{33}$$

which stated in words would imply that the refractive index is a very slowly varying function of the spatial co-ordinate, being almost a constant in distances of the order of wavelength. In weakly-guiding multimode graded-core fibres, Eq. (33) is generally obeyed and one can apply standard WKB results directly by transforming Eq. (32) to an one dimensional wave equation form, through the following substitution [3]:

$$f(r) = \frac{R(r)}{r^{1/2}}. \tag{34}$$

In term of this new function $f(r)$, Eq. (32) gets transformed to

$$\frac{d^2 f(r)}{dr^2} + q^2(r) f(r) = 0, \tag{35(a)}$$

where
$$q^2(r) = k_0^2 n^2(r) - \beta^2 - \frac{\left(l^2 - \frac{1}{4}\right)}{r^2}. \tag{35(b)}$$

Using quantum mechanical analogy of bound energy states under WKB approximation, one can show that for bound modes in a fibre, one must have

$$\int_{r_1}^{r_2} q(r) dr = (2m - 1)\frac{\pi}{2}, \tag{36}$$

where $m = 1, 2,...$ will represent the radial mode number. The limits of integration: r_1 and r_2 are yielded by the solution of $q(r) = 0$ such that $q(r)$ is real between r_1 and r_2 which implies that the field is oscillatory within this region while outside this range of r, $q(r)$ is imaginary thereby leading to exponentially decaying fields in these regions.

---

[*] Method was originally developed by Wentzel, Krammers and Brillouin and it finds extensive applications in Quantum Mechanical calculations.

Equation (36) represents the eigen value equation under WKB approximation for graded-core fibres and solution of Eq. (36) for different combinations of l and m will yield the corresponding modal propagation constant β_{lm}. However, as long as we do not discuss individually very low order modes, in order to obtain a general view of the mode structure in graded-core fibres, one can rewrite Eq. (36) approximately as [3]

$$\int_{r_1}^{r_2} \left[k_0^2 n^2(r) - \beta^2 - \frac{l^2}{r^2} \right]^{1/2} \approx m\pi \qquad (37)$$

and take Eq. (37) as defining the characteristic or eigen value equation for obtaining the propagation constant β_{lm} in general graded-core fibres as long as Eq. (33) is satisfied. Making use of Eq. (37), one can also make a count of the number of modes as was shown in [3, 6]

$$M(\beta') = a^2 k_0^2 n_1^2 \Delta \left(\frac{q}{q+2} \right) \left(\frac{k_0^2 n_1^2 - \beta'^2}{2 \Delta k_0^2 n_1^2} \right)^{\frac{q+2}{q}} \qquad (38)$$

and
$$N \simeq \frac{1}{2} \left(\frac{q}{q+2} \right) V^2, \qquad (39)$$

where $M(\beta')$ represents number of modes having propagation constant β greater than β' and N corresponds to total number of possible guided modes. For a step index fibre since $q = \infty$

$$N \approx V^2 / 2, \qquad [40(a)]$$

while for a parabolic core fibre for which $q = 2$

$$N \approx V^2 / 4. \qquad [40(b)]$$

Thus for a given V, total number of guided modes in a step index fibre is twice of that in an equivalent (in the sense, both having same Δ and a) parabolic index fibre. Typically the V-value of a multimode fibre is ~ 24 at $\lambda = 1.3\,\mu$m, then in a $q = \infty$ fibre N will be $\simeq 288$ while in a $q = 2$ fibre it will be $\simeq 144$. As already stated, in a PCM optical transmission system since the signal is transmitted in the form of a series of optical pulses, these pulses get broadened with propagation inside the fibre mainly due to intermode dispersion and a measure of dispersion will determine permissible information/data transmission rate through the fibre. In reality, the dispersion or pulse broadening in a fibre can be shown to originate from three mechanisms: (i) intermodal, (ii) material and (iii) waveguide dispersion — the latter two comprise what is known as intramodal or chromatic dispersion. Even if through appropriate grading one is able to minimise intermode dispersion, the finite spectral width of practically available optical sources as well as the Fourier spectrum of the pulsed signal lead to intramodal dispersion. Material dispersion which is one component of it, owes its origin to wavelength dependence of the refractive index of the bulk material that make up the fibre composition. On the other hand, waveguide dispersion arises due to the explicit dependence of a mode's propagation constant on wavelength and it is essentially dependent on the ratio: a/λ, which is a characteristic of the fibre at the operating λ.

If we take, as an example, the propagation of a Gaussian temporal pulse:

$$f(t) = e^{-\frac{1}{2}(t/T)^2 + i\omega_c t}, \qquad (41)$$

where T represents half-width of the pulse at $1/e$ intensity point and ω_c is the carrier frequency, then it is possible to show from the Fourier spectrum of $f(t)$ that after propagating through a length L of the fibre via its pth mode, the intensity distribution of pth mode will be given by [4, 8]

$$\left|p(t)\right|^2 = \left[1 + \left(\frac{\gamma_{2,p}L}{T^2}\right)^2\right]^{-1/2} e^{-\frac{(t-\gamma_{1,p}L)^2}{T^2 + (\gamma_{2,p}L)^2/T^2}} \qquad (42)$$

Here p-stands for composite mode number (introduced for convenience only) in place of the two indices l, m normally used to represent an arbitrary mode,

$$\gamma_{1,p} = \left.\frac{\partial \beta_p}{\partial \omega}\right|_{\omega = \omega_c},$$

and

$$\gamma_{2,p} = \left.\frac{\partial^2 \beta_p}{\partial \omega^2}\right|_{\omega = \omega_c}$$

Equation (42) shows that like the input pulse, the output pulse is also Gaussian in shape and the half-width of the output pulse corresponding to p^{th} mode is greater than that of the input pulse. In contrast to T it is

$$T_2 = T\left[1 + \left(\frac{\gamma_{2,p}L}{T^2}\right)^2\right]^{1/2} \qquad (43)$$

Typically, since a multimode fibre supports a large number of modes, overall intensity that will be recorded by a square-law receiver at its output end will be given by [2, 3]

$$\sum_p \left|A_p\right|^2 \left|p(t)\right|^2, \qquad (44)$$

where A_p is the amplitude co-efficient of the pth mode. The important thing to be noted here is that transit time delay or group delay (τ_g) associated with each mode, as expected, is given by $Ld\beta_p/d\omega$ and hence intermode dispersion will result from the difference in transit times between different modes. Further, distortion in the pulse shape is primarily determined by the second derivative: $\partial^2\beta_p/\partial\omega^2$.

By making use of Eqs. (37) and (38) it can be shown that [3, 6]

$$\beta_M^2 = k_0^2 n_1^2 \left[1 - 2\Delta \left\{ \left(\frac{q+2}{q}\right) \frac{M}{a^2 k_0^2 n_1^2 \Delta} \right\}^{\frac{q}{q+2}} \right] \quad [45(a)]$$

$$\Rightarrow \quad \beta_\alpha^2 = k_0^2 n_1^2 (1 - 2\alpha), \quad [45(b)]$$

where $\quad \alpha = \Delta \left(\dfrac{M}{N}\right)^{\frac{q}{q+2}}$

Simple algebraic manipulation leads one to [3]

$$\tau_g = L \frac{d\beta_M}{d\omega} = \frac{LN_1}{c} \frac{d\beta_M}{d(k_0 n_1)}$$

$$= \frac{LN_1}{c} \left[1 + \left(\frac{q-2-\varepsilon}{q+2}\right)\alpha + \left(\frac{3q-2-2\varepsilon}{q+2}\right) \frac{\alpha^2}{2} + O(\Delta^3) \right] \quad (46)$$

where $N_1 \left(= n_1 - \lambda_0 \dfrac{dn_1}{d\lambda_0} \right)$ is called the group index, λ_0 being the free-space wavelength and ε, known as profile dispersion parameter, is given by [3]

$$\varepsilon = \frac{2k_0 n_1}{\Delta} \cdot \frac{d\Delta}{d(k_0 n_1)} = -\frac{2n_1 \lambda_0}{N_1} \cdot \frac{\Delta'}{\Delta}, \quad (47)$$

Δ' being $d\Delta / d\lambda_0$. It is apparent from Eq. (46) that for

$$q = 2 + \varepsilon \equiv q_0 \quad (48)$$

the delay difference amongst the modes will vanish to first order in Δ, since all the modes will be effectively taking the same transit time ($= LN_1/c$). In Fig. (6) a plot of q_0 vs. wavelength is shown to depict this variation for different core glass dopant materials commonly used in fabricating telecommunication grade fibres. It clearly shows that for achieving wide bandwidth optical communication simultaneously at multiple wavelengths (for wavelength multiplexed systems), one should choose ternary comound materials like $P_2O_5 - GeO_2 - SiO_2$ for the core [9].

If we neglect profile dispersion i.e. $\varepsilon \to 0$, the transit time difference between the fastest and the slowest mode in such power-law profiled fibres will reducce to Eqs. (29) and (31) for $q = \infty$ and $q = 2$, respectively, provided we disregard material dispersion [3]. These results confirm the equivalence between ray optics and the zeroeth order WKB analysis for multimode fibres. Sometimes it is convenient to represent group delay in terms of a normalised group delay (t_α), such that the time delay that is common between the modes is eliminated where t_α is defined as

$$t_\alpha \equiv \frac{\tau_g}{(LN_1/c)} - 1 = \frac{q-2}{q+2}\alpha + \frac{3q-2}{q+2}\frac{\alpha^2}{2} \quad (49)$$

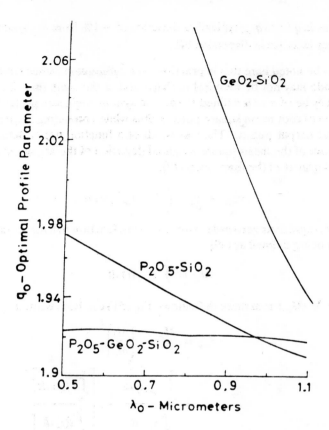

Fig. 6. q_0 vs. λ for different core material compositions (reprinted with permission of McGraw-Hill, Inc., New York from A.H. Cherin, "An Introduction to Optical Fibres", Fig. 6.14, 1983).

Fibres having $q > 2$ are called undercompensated while fibres with $q \ll 2$ are called overcompensated index profiled fibres because for $q_{opt} \equiv 2 - 2\Delta$, highest and lowest order modes would take the same transit time to reach the fibre output end (if we neglect term of $O(\Delta^2)$ under the assumption $\Delta \ll 1$. The fastest mode in such optimum profile fibres corresponds to $\alpha_{min} \simeq \Delta/2$, which implies that the dispersion in such fibres is

$$\Delta t_\alpha \equiv t_{\alpha = 0, \Lambda} - t_{\alpha_{min}} \simeq \frac{\Delta^2}{8} \qquad [50(a)]$$

$$\Rightarrow \qquad \Delta \tau_g = \frac{LN_1}{c} \cdot \frac{\Delta^2}{8}, \qquad [50(b)]$$

which is an improvement by a factor of 1/4 as compared to that in a parabolic core fibre. It may be mentioned that pulse dispersion is quite sensitive to deviations in q from q_{opt}; typically a deviation of $\simeq 1\%$ from q_{opt} leads to almost four times increase in dispersion [4].

It may be noted here that in practice since subsequent pulses emitted by the laser diode may not be identical in shape and at the same time they may not necessarily be of a well defined shape, the system engineers prefer to use the definition of root mean square pulse widths while computing dispersion from input and output pulses. The rms width of a function $f(t)$ is defined as the square root of the mean square weighted deviation of the argument t from the centre of gravity of the function, t_0 [10].

$$\sigma_{rms}^2 = \frac{1}{M_0} \int_{-\infty}^{\infty} (t - t_0)^2 f(t)\, dt, \tag{51}$$

where M_0 represents zero order moment of the function $f(t)$, nth moment of the function being defined as [10]

$$M_n = \int_{-\infty}^{\infty} t^n f(t)\, dt \tag{52}$$

and $t_0 = M_1/M_0$. From these definitions, Eq. (51) can be rewritten as [10]

$$\sigma_{rms}^2 = \frac{M_2}{M_0} - \left(\frac{M_1}{M_0}\right)^2$$

$$= \frac{\int_{-\infty}^{\infty} t^2 f(t)\, dt}{\int_{-\infty}^{\infty} f(t)\, dt} - \left[\frac{\int_{-\infty}^{\infty} t f(t)\, dt}{\int_{-\infty}^{\infty} f(t)\, dt}\right]^2 \tag{53}$$

For example, if we consider $f(t)$ as a Gaussian function of $1/e$ half-width as ω_0 i.e. $f(t) \sim \exp(-t^2/\omega_0^2)$, its rms half-width will be $\omega_0/\sqrt{2}$ (from Eq. (53)).

If these definitions are applied to pulse propagation problem, M_0 would represent total energy in the pulse while t_0 would imply mean pulse arrival time [11]. With these definitions Olshansky and Keck [12] have shown that

$$\sigma_{intermodal} = \left[\langle \tau_g^2 \rangle - \langle \tau_g \rangle^2\right]^{1/2} = (LN_1/2c)\frac{q}{q+2}\left[\frac{q+2}{3q+2}\right]^{1/2}$$

$$\times \left[C_1^2 + 4C_1 C_2 \frac{(q+1)\Delta}{(2q+1)} + 16\Delta^2 C_2^2 \frac{(q+1)^2}{(5q+2)(3q+2)}\right]^{1/2} \tag{54}$$

where $\quad C_1 = \dfrac{q - 2 - \varepsilon}{q + 2}\quad$ and $\quad C_2 = \dfrac{3q - 2 - 2\varepsilon}{2q + 4}.\quad$ (55)

Minimisation of intermodal rms width yields optimum value of q in place of Eq. (48) as

$$q_{opt} = 2 + \varepsilon - \Delta \cdot \frac{(4 + \varepsilon)(3 + \varepsilon)}{5 + 2\varepsilon}. \tag{56}$$

In the absence of profile dispersion ($\varepsilon = 0$) Eq. (56) reduces to $2-(12/5)\Delta$, which was 2 if optimum profile was derived by minimisation of mode transit times alone.

If one defines total pulse broadening in a multimode fibre as the sum of rms widths due to both intermodal and intramodal dispersion then,

$$\sigma_{total} = (\sigma^2_{intermodal} + \sigma^2_{intramodal})^{1/2}, \tag{57}$$

where [13]

$$\sigma_{intramodal} \simeq \left(\frac{L}{c}\right)\left(\frac{\sigma_s}{\lambda_0}\right)\left[\left(-\lambda_0^2 \frac{d^2 n_1}{d\lambda^2}\right)^2 - 2\lambda_0^2 N_1 \Delta \left(\frac{q-2-\varepsilon}{q+2}\right)\right.$$
$$\left. \times \frac{q}{q+1} \frac{d^2 n_1}{d\lambda^2} + (N_1 \Delta)^2 \left(\frac{q-2-\varepsilon}{q+2}\right)^2 \left(\frac{2q}{3q+2}\right)\right]^{1/2}, \tag{58}$$

where $\sigma_s^2 = \int_{-\infty}^{\infty} (\lambda - \lambda_0)^2 S(\lambda) \, d\lambda$ is the rms width of the source spectral distribution, $\lambda_0 \equiv \int \lambda S(\lambda) \, d\lambda$ is the mean wavelength. A plot of σ_{total} vs. q for three different common types of sources around $\lambda_0 \simeq 0.9 \, \mu$m but each of different spectral widths is shown in Fig. 7. It can be seen from Eq. (54) that in the absence of material ($n \neq f(\lambda)$) and profile ($\varepsilon = 0$) dispersions [14]

$$\sigma_{q=\infty}/\sigma_{opt} \approx 10/\Delta. \tag{59}$$

Typically for $\Delta = 1\%$, rms pulse spreading in a step index fibre will be about 14 ns/km while in an optimum index profiled fibre for the same Δ, it will be ~ 0.014 ns/km, i.e. thousand times less [14].

It may be noted here that due to manufacturing tolerances, it is rarely possible to attain a perfectly optimum profile and even a small perturbation in an otherwise near optimum profiled fibre will lead to an increase in pulse dispersion [15-17]. For example, it is well known that the most commonly used fibre manufacturing process namely the MCVD* method usually leads to a characteristic dip in the index profile around the fibre axis due to evaporation of some of the dopants from the innermost layers of the fibre preform in the so called collapsing step [18]. It has been shown by Khullar et. al. [17] that typically in parabolically graded-core fibre having $n_1 = 1.476, n_2 = 1.458$ and $V = 30$, the transit time difference between the slowest and the fastest travelling modes increases from 0.37 ns/km (in the absence of dip) to 1.73 ns/km due to the presence of a 10% gaussian axial dip of relative width: $b/a = 0.08$, b being the half-width of the dip and a core radius.

---

* Modified Chemical Vapour Deposition [see Ch. 9 by H. Karstensen].

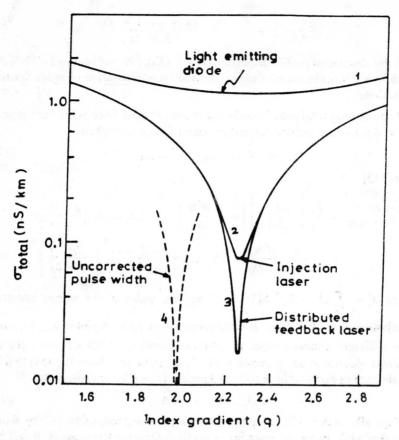

Fig. 7. RMS pulse width vs. profile shape parameter q for various sources (reprinted from [12] with permission. Copyright 1976, Optical Society of America).

5.2 Pulse dispersion in single-mode fibres

We may recall from Sec. 4 that if the operating V-number of a step index fibre satisfies condition [25(a)], then the fibre will support only one guided mode namely, the LP_{01}-mode. Obviously intermode dispersion will be absent in single-mode fibres. However, as already stated in the previous section pulse transmission rate in such fibres would be limited by intramodal or chromatic dispersion which originates due to finite wavelength spread in the emission pattern of practical sources (see Table 1) if the inherent spectral width (i.e. Fourier components) of the pulse is negligible compared to the source spectral width. If $\Delta\lambda$ represents this source spectral width, then the signal pulse may be taken as being carried by a large number of individual carrier wavelengths spread over $\Delta\lambda$ and accordingly when they eventually fuse at the exit end to form the output pulse, the pulse shows distortion and broadening leading to

TABLE 1

Typical spectral widths of various optical communication sources (adapted from: Y, Suematsu, Long wavelength optical fibre communication, Proc. IEEE, Vol. 71, pp. 692-721, June 1983, copyright© 1983, IEEE)

Source			Spectral wavelength (μm)	Spectra width (nm)		Suitability of coupling to fibre	
				DC	rapidly modulated	MM	SM
Light emitting diode (LED)			0.83	30	30	Yes	No
			1.3	120	120		
Conventional Laser diode (LD)	MM		0.85	2-3	2-3	Yes	Yes
			1.3	8-10	8-10		
			1.55	10	10		
	SM		0.85	SM	2-3	Yes	Yes
			1.3	SM	8-10		
			1.55	SM	10		
DSM		LD	1.55	SM	0.3-0.4	Not applicable	Yes

MM: Multi-frequency (mode)
DSM: Dynamic single mode
SM: Single-frequency (mode)

temporal dispersion. Thus the increase in pulse width in a single-mode fibre due to the above mentioned mechanism will be [3]

$$\Delta \tau = \frac{d\tau_g}{d\lambda_0} \Delta \lambda = \frac{d\tau_g}{dk_0} \frac{dk_0}{d\lambda_0} \Delta \lambda = D.L.\Delta \lambda, \quad (60)$$

where

$$D = -\frac{k_0}{c\lambda_0} \frac{d^2\beta}{dk_0^2} \quad (61)$$

is known as dispersion coefficient, which is usually expressed in units of ps/nm-km and L is the length of fibre. Equation (61) shows that $\Delta \tau$ is essentially proportional to $d^2\beta/dk_0^2$. In order to make an estimate of the magnitude of D in a step index single-mode fibre, we notice from Eq. (20) that for weakly-guiding fibres, i.e., for $\Delta \ll 1$

$$\beta \simeq k_0 n_2 (1 + \Delta b). \quad (62)$$

Equation (62) shows that the propagation constant β of a mode essentially has two components: one is related to the purely material property of the fibre ($k_0 n_2$) and the other is related to waveguide/mode parameter b. For the sake of simplicity if we consider one in the absence of the other, then by considering only the second term (which corresponds to waveguide dispersion) on the R.H.S. of Eq. (62) we get [3].

$$\frac{d\beta}{dk_0} \simeq n_2 \Delta \frac{d(Vb)}{dV}. \quad (63)$$

Therefore,

$$\frac{d^2\beta}{dk_0^2} = \left(\frac{n_2\Delta}{k_0}\right) V \frac{d^2(Vb)}{dV^2}. \tag{64}$$

Inserting Eq. (64) in Eq. (61) we get

$$D_{wg} = -\left[\frac{n_2\Delta}{c\lambda_0} V \frac{d^2(Vb)}{dV^2}\right], \tag{65}$$

where the suffix wg stands for waveguide to indicate that Eq. (65) represents contribution to temporal dispersion in a single-mode fibre by the waveguide property of the fibre[*]. Accordingly, $D_{wg} \cdot L$ is called the waveguide dispersion. Though rigorously speaking, to compute D_{wg} one is required to solve Eq. 19(b) for $l = 0$ (corresponding to LP_{01}-mode) to obtain β and hence $V d^2(Vb)/dV^2$ at a particular V within the single-mode region defined by Eq. (25), one may also use the following empirical relation [19] for calculating $V d^2(Vb)/dV^2$:

$$V \frac{d^2(Vb)}{dV^2} \simeq 0.80 + 0.549(2.834 - V)^2. \tag{66}$$

Equation (66) is accurate to within 5% for $1.3 < V < 2.6$. Following the same procedure by neglecting momentarily the second term in Eq. (62), one can show that the contribution to mode dispersion from material effect in single-mode fibres is given by [3]

$$D_m = -\left(\frac{\lambda_0}{c}\frac{d^2 n_2}{d\lambda_0^2}\right), \tag{57}$$

where D_m is called the material dispersion coefficient. Interestingly, in telecommunication optical fibres which are usually based on pure fused SiO_2 as the cladding, D_m goes through a zero at $\lambda_0 \simeq 1.27 \mu m$ ($\equiv \lambda_m^0$) [20] and it gets shifted to a longer wavelength (but shorter λ, if doped with F or B) in doped SiO_2 as shown in Fig. 8 (dotted curve, for which $\lambda_m^0 \simeq 1.285 \mu m$) [21]. In pure SiO_2 at $\lambda_0 \simeq 0.85 \mu m$, D_m is $\simeq 90$ ps/nm-km while at $\lambda_0 \simeq 1.3 \mu m$ it is < 0.1 ps/nm-km increasing to 20 ps/nm-km again at $\lambda_0 \simeq 1.55 \mu m$. Waveguide dispersion coefficient (D_{wg}) is also plotted on the same figure as dashed-dotted curve corresponding to a single-mode fibre having diameter of 7.0 μm and containing 3.0 mole % GeO_2-doped SiO_2 with fused SiO_2 forming the cladding. Total dispersion coefficient, D ($= D_m + D_{wg}$) is shown as solid curve in the same figure. It is apparent that at a λ_0 ($\equiv \lambda_T^0$) $\simeq 1.325 \mu m$ which is longer than 1.28 μm, total intramodal dispersion passes through a zero. This wavelength λ_T^0 at which total dispersion is zero is known as the wavelength of zero disper-

---

[*] It may be noted here, in passing, that waveguide dispersion can not be explained by ray or geometrical optics. This fact shows why wave optics/modal analysis is necessary to explain guidance in single-mode fibres.

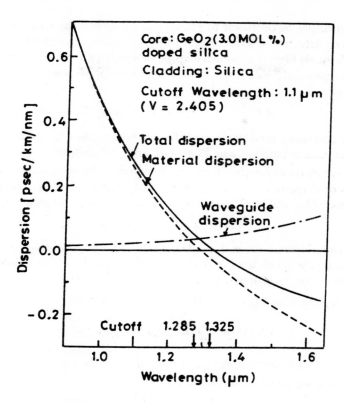

Fig. 8. Dispersion (in a single-mode fibre) vs wavelength (after [21]). Material dispersion curve corresponds to the doped SiO_2-core (3.0 mole % GeO_2) while waveguide dispersion curve has been computed by subtracting material dispersion from total dispersion. Here D is plotted by neglecting $-$ve sign in its definition (cf. Eq. (61)).

sion in the literature. By varying dopant concentration and waveguide parameters, e.g. core diameter it is possible to tailor [22] λ_T^0 to fall anywhere within the lowest loss wavelength window: 1.3 - 1.6 μm of silica fibres, in which the second and third generation systems operate. Thus, if a fibre is so designed that its λ_T^0 coincides with the lowest loss wavelength, one can achieve an enormously large repeater spacings of more than 100 km at as high pulse transmission rate as 4.2 G bit/s [23]. We may note here that the assumption that waveguide and material dispersions are additive as separate effects is a good approximation unless very high precision is required in any particular situation [24, 25].

It may be worthwhile to point out that the term 'zero dispersion' is a misnomer to certain extent because dispersion is zero only to 1st order at this wavelength. If $d^2\beta/dk_0^2$ term vanishes then terms like $d^3\beta/dk_0^3$ will determine residual dispersion to second-order [26, 27]. This second order dispersion can also, in principle, be overcome by phase compensation at the receiver through heterodyne detection scheme. Accordingly, maximum theoretical bandwidth of a single-mode fibre at minimum dispersion wavelength will be yielded by the fourth derivative of β with respect to k_0 and which is approximately given at 1.3 μm by [28]

$$BW\bigg|_{max} L^{1/4} = 3\,THz - (km)^{1/4}. \tag{68}$$

In the commonly used single-mode fibres of today, the core index profile is a step/quasi-step function and $\lambda_T^0 \simeq 1.31\,\mu$m. In addition to λ_T^0, a related quantity which is extensively referred in the literature for system calculations is called zero dispersion slope, S_0. In practice, dispersion charateristics of a single-mode fibre can be estimated by fitting an empirical relation [10]

$$\tau_g(\lambda) = A\lambda^{-2} + B + C\lambda^2 \tag{69}$$

to the experimental datas for $\tau_g(\lambda)$ generated by measuring at various wavelengths; the coefficients A, B, C being determined by a least square fitting of Eq. (69) to the measured datas for $\tau_g(\lambda)$.

Accordingly, from Eq. (69), $D(\lambda)$ will be given by

$$D(\lambda) = \frac{d\tau_g}{d\lambda} = 2C\lambda - 2A\lambda^{-3}. \tag{70}$$

The zero dispersion wavelength λ_T^0 can, therefore, be obtained by equating Eq. (70) to zero, which yields [10, 3]

$$\lambda_T^0 = (A/C)^{1/4}. \tag{71}$$

Differentiation of Eq. (69) once more with respect to λ and substitution of Eq. (71) in the resulting expression yields [3]

$$S_0 = 8C \tag{72}$$

$$\Rightarrow \qquad D(\lambda) = \frac{S_0}{4}\left(\lambda - \frac{\lambda_T^{0\,4}}{\lambda^3}\right). \tag{73}$$

Since, in practice it may not always be possible to obtain a source operating exactly at λ_T^0 for a given fibre, knowledge of S_0 and λ_T^0 enables an estimate of the dispersion penalty involved in operating the fibre at a wavelength away from its λ_T^0. The value of S_0 in practical fibres is ~ 0.085 ps/nm^2-km.

5.2.2 Dispersion in graded-core single-mode fibres

In the previous section, entire discussion was based on ideal step idex core RIP. If the index profile deviates from this like having an axial index dip [29] or graded-core* [30], the predicted total dispersion would change and so would

---

* In practice, it is hard to obtain a perfectly-stepped RIP and some grading around the core-cladding boundary inevitably occurs during the fabrication process.

λ_T^0. Gambling et al [30] have studied dispersion in graded-core single-mode fibres by considering index profile of a single-mode fibre to be given by

$$n^2(r) = n_1^2 \left[1 - 2\Delta (r/a)^q\right], \quad 0 \leq r \leq a$$
$$= n_1^2 \left[1 - 2\Delta\right] = n_2^2 \qquad r > a \qquad (74)$$

where $\Delta \simeq (n_1 - n_2)/n_2$. After a considerable algebra, the intramodal dispersion in such fibres was shown to be yielded by [30]

$$\Delta \tau = (D_{cmd} + D_{wd} + D_{cpd}) \cdot L \cdot \Delta \lambda = D_T \cdot L \cdot \Delta \lambda \qquad (75)$$

with D_{cmd} (= composite material dispersion coefficient)

$$= \frac{\lambda_0}{c} \left[A(V) \frac{d^2 n_1}{d\lambda^2} + \{1 - A(V)\} \frac{d^2 n_2}{d\lambda^2}\right], \qquad [76(a)]$$

D_{wd} (= waveguide dispersion coefficient)

$$= \left(\frac{n_2 \Delta}{c\lambda_0}\right) B(n) V \frac{d^2 (V \cdot b)}{dV^2}, \qquad [76(b)]$$

D_{cpd} (= composite profile dispersion coefficient)

$$= -\frac{n_2}{c} C(n) D(V) \frac{d\Delta}{d\lambda}, \qquad [76(c)]$$

where $A(V) = \frac{1}{2} \left[b + \frac{d(V \cdot b)}{dV}\right], \qquad [77(a)]$

$$B(n) = \left[1 - \frac{\lambda_0}{n_2} \frac{dn_2}{d\lambda}\right]^2, \qquad [77(b)]$$

$$C(n) = 1 - \frac{\lambda_0}{n_2} \frac{dn_2}{d\lambda} - \frac{\lambda_0}{4\Delta} \frac{d\Delta}{d\lambda}, \qquad [77(c)]$$

$$D(V) = V \frac{d^2(Vb)}{dV^2} + \frac{d(Vb)}{dV} - b. \qquad [77(d)]$$

By considering a fibre composed of 11.1 mole % GeO_2-doped SiO_2 and pure SiO_2 as the cladding, Gambling et. al. [30] had reported computations of Eq. (75) for a parabolic core [$q = 2$ in Eq. (74)] and a step index ($q = \infty$) fibre as being two extreme cases of Eq. (74); their results are shown in Fig. 9. Figure 9(a) corresponds to step index fibres with three different core diameters while Fig. 9(b) is representative of parabolic core single-mode fibres of different core diameters. These figures clearly show that for a given core cladding combination (i.e., a given Δ) the shape of the core RIP may have a considerable effect on $\Delta \tau$.

6. SIGNAL ATTENUATION IN OPTICAL FIBRES

As is true with any transmission medium, an optical fibre also suffers from transmission loss as a signal propagates through it. In fact, attenuation and dispersion together determine repeater spacings in a fibre optical communica-

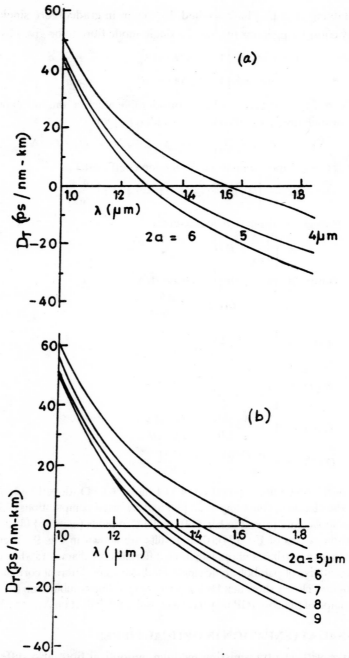

Fig. 9. (a) Total dispersion coefficient in a step-index single-mode fibre for different core diameters; (b) Effect of grading (a parabolic core fibre) on total dispersion is shown for comparison with (a) (both (a) and (b) are reproduced from [30] with permission. Copyright 1979 IEE, Stevenage).

tion system; e.g. if a single-mode fibre link is operated at its zero dispersion wavelength, the system would be essentially loss limited in terms of repeater spacings. If $P(0)$ represents the power launched at the input of a fibre of length 'L', then power at the end of the fibre will be given by Bouger's law

$$P(L) = P(0)e^{-\alpha' L}, \qquad (78)$$

where α' is the attenuation coefficient in nepers per unit length. In practice attenuation is expressed in dB/km and it is defined as

$$\alpha = 10 \log_{10}\left[P(0)/P(L)\right]/L \qquad [79(a)]$$

with 'L' expressed in kilometers. From Eqs. (78) and [79(a)] it is easy to relate α to α' as (because $\ln(x) = 2.3026 \times \log_{10}(x)$)

$$\alpha = 4.34\,\alpha'. \qquad [79(b)]$$

The principal sources of attenuation in a fibre waveguide could be broadly classified into two groups: absorptive and radiative. Acordingly, α (in dB/km) is usually expressed in terms of its components as [3]

$$\alpha = B + C(\lambda) + E(\lambda) + \frac{A}{\lambda^4}. \qquad (80)$$

Typical loss spectrum of an optical fibre alongwith its associated components that contribute to total loss at a particular wavelength are shown in Fig. 10. This figure also shows the low-loss wavelength windows for optical communication and the corresponding sources and detector that are used in exploiting these low-loss wavelength windows. In the following, we discuss each component of the loss spectrum in Eq. (80).

6.1 Absorption Loss

The absorption loss may again be subdivided into intrinsic and extrinsic losses. Intrinsic loss may be caused by interaction of the propagating light wave with one or more major components of the fibre material—an interaction which may eventually lead to quantum transitions between different electronic as well as vibrational energy levels of the fibre materials; the second term in Eq. (80), namely, $C(\lambda)$ represents this loss. Since band gap energy in pure fused silica[*] (SiO_2) is ~ 8.9 eV, peak absorption of light due to electronic transition occurs at a wavelength $\sim 0.14\,\mu m$ in the UV-region. On the other hand, fundamental vibrational absorption band in SiO_2 is centered at $\lambda \sim 9.2\,\mu m$ in the IR-region with the possibility of weaker absorption bands occurring at $\lambda \simeq 3.2, 3.8$ and $4.4\,\mu m$ due to anharmonic vibrations of the Si-O bond. These absorption bands in the UV- and IR-regions decay exponentially resulting in what are known as absorption tails that spill over into the neighbouring wavelengths. However, in the wavelength ranges (0.8-0.9 μm and 1.2-1.5 μm) of interest to optical communication, these tails are of no serious concern. For the current and near

---

[*] Which constitutes major fibre material component.

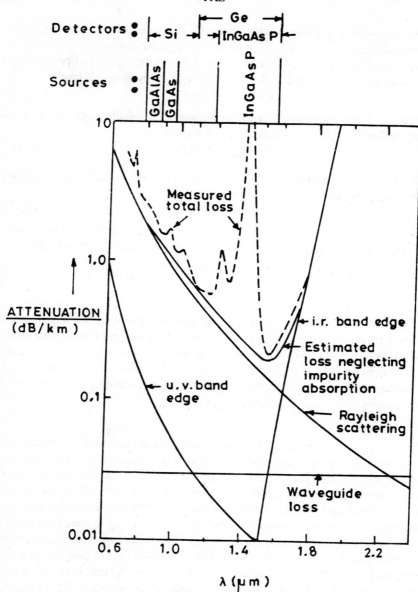

Fig. 10. Measured total loss spectrum of a very low loss 2.2 km long single-mode fibre of $\Delta = 0.0019$ shown as dashed curve. Estimated loss contributions from various components to the loss spectrum are also shown in the figure (adapted with permission from: T. Miya, Y. Terume, T. Hosaka, and T. Miyashita, Electronics Letters, Vol. 15, pp. 106-108, 1979. Copyright 1979 IEE, Stevenage).

future optical comunication systems, IR absorption could have some effect beyond $1.5\,\mu$m but in any case at $1.5\,\mu$m its contribution would be less than 0.05 dB/km. Thus, in low-loss fibres, $C(\lambda)$ can be neglected.

On the other hand, extrinsic absorption (represented by $E(\lambda)$ in Eq. (80)) has been found to be caused by presence of even minute traces (in parts per million) of transition metal ions like copper, manganese, iron, vanadium etc. as also presence of water (in the form of OH⁻ ions) dissolved in glass as shown in Table 2. However, the state-of-art technologies of low-loss fibre fabrication are all essentially based on vapour phase reactions, a process which inherently purifies the basic materials (due to differential vapour pressures) forming the fibre from the presence of these impurities except water. The OH⁻ ion vibrates at a fundamental frequency corresponding to IR wavelength of 2.7 μm. However, due to slight anharmonicity present in the O-H bond, overtone absorption peaks may appear at approximate wavelengths of 0.72, 0.95 and 1.38 μm. Further one or more combinational absorption peaks may also occur at 0.88, 1.13 and 1.24 μm. However, fortuitously, these OH-absorption peaks are sufficiently narrow (in contrast to IR absorption bands) to yield ultra low-loss fibres around wavelengths of current interest namely 1.3 μm and 1.55 μm. As such $E(\lambda)$ can also be neglected in Eq. (80) for practical purposes in low-loss fibres of today. Sometimes, long term increase in attenuation is observed in cabled fibres due to diffusion of hydrogen from the cladding and/or silica support jacket into the core. This can be prevented by selecting a synthetic cladding layer, minimising P_2O_5 as the dopant and using UV-curable resine with nylon [31].

TABLE 2
Absorption loss in SiO_2-glass due to presence of trace amounts of different metals and hydroxyl ion as impurities

Impurities	Loss (dB/km) due to one part in 10^9	Absorption peak wavelength (μm)
V^{4+}	2.7	0.725
Cu^{2+}	1.1	0.85
Fe^{2+}	0.68	1.1
	Loss (dB/km) due to 10 parts in 10^9	
OH-	0.01	0.95
OH-	0.015	1.24
OH-	0.35	1.38

6.2 Radiative Loss

6.2.1 Rayleigh scatter loss

Radiative loss is said to occur if a part or whole of the guided optical energy is lost through radiation from the fibre. Most dominant source of inherent radiation loss in a fibre is caused by Rayleigh scattering [the last term in Eq. (80)] and it is caused by small scale (small compared to wavelength of the propagating light) compositional and density fluctuations that gets frozen into glass lattice at the glass softening point during melting and subsequent cooling.

The resulting inhomogeneities lead to attenuation that varies with wavelength as λ^{-4}. Thus by operating a system at relatively longer wavelengths, one can minimise the contribution of Rayleigh scatter loss in an optical fibre. Theory predicts a Rayleigh scatter loss of ~ 0.14 dB/km in fused SiO_2 at 1.55 μm, a figure which may slightly increase if refractive index modifiers like GeO_2, P_2O_5, B_2O_3 etc. are mixed with SiO_2. In fact the coefficient A in Eq. (80) is approximately given by [32]

$$A = A_1 + A_2 \Delta^n , \tag{81}$$

where A_1 = 0.7 dB-μm^4/km in fused SiO_2 and $A_2 \simeq$ 0.4 - 1 dB-μm^4/km for Δ in % while n = 0.7-1 depending on the dopant and Δ. Typically, A $\simeq$ 0.8 dB-μm^4/km in modern single-mode fibres [32]. Assuming Δ to be ~ 0.3% in such fibres, it implies that $A_2 \simeq$ 0.4 dB-μm^4/km and n ~1 in such fibres. Relatively losses in single-mode fibres are much less than in multimode fibres because Δ is relatively large in the latter. Radiative loss may also be caused by intrinsic fibre imperfections/deformations like core-cladding interface irregularities and diameter fluctuations and so on and this explains the presence of the term B in Eq. (80). However, with the sophistications/skills that now-a-days go into the manufacturing process of an optical fibre, chances of intrinsic irregularities are practically negligible.

6.2.2 Macrobend loss

Extrinsic perturbation like bending a fibre may arise during cabling and/laying of the cable. A simple experiment that involves launching a visible laser light (e.g. from a He-Ne laser) into a fibre that is once laid straight and once bent into an arc of a circle will immediately reveal that the fibre suffers radiation loss at bends/curves along its path. Physically, it can be accounted for as follows: the fraction of the mode field in a bent fibre that travels along the periphery of the circular arc in the cladding may at some stage, be required to travel at a rate faster than the local plane wave velocity in order to maintain equiphasefronts at radial planes (see Fig. 11). This being physically disallowed, that part of the modal field dissociates itself from the fibre and it gets lost through radiation from the sides [3, 7]. The loss in a bent fibre is approximately given by [3, 11, 33]

$$\alpha_B \simeq K \exp\left[-\frac{2}{3} n_1 k_0 R \left(\frac{\gamma^2(0)}{n_1^2 k_0^2} - \frac{2a}{R}\right)^{3/2}\right], \tag{82}$$

where
$$\gamma(0) = \sqrt{\beta^2 - k_0^2 n_2^2},$$

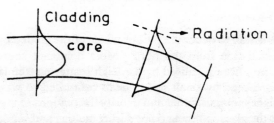

Fig. 11. Schematic representation of the radiation loss of a mode at a fibre bend.

R is the radius of curvature, and K is a constant for a particular fibre though it is relatively unimportant in determining the magnitude of α_B. It can be shown [11, 33], that the number of modes in a bent multimode fibre will be given by

$$N_R \simeq N_{R=\infty} \left[1 - \frac{q+2}{2q\Delta} \left\{ \frac{2a}{R} + \left(\frac{3}{2n_1 k_0 R} \right)^{2/3} \right\} \right]. \quad (83)$$

where $N_{R=\infty}$ represents total number of guided modes in a straight fibre (cf. Eq. (39)). If all the guided modes in a fibre are excited with equal power, then

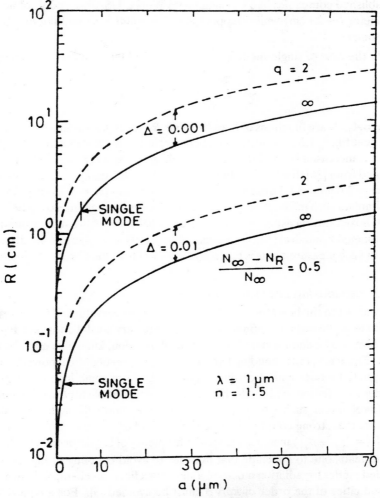

Fig. 12. Radius of curvature (R) of a bent fibre that costs 50% of the guided modes (of a straight fibre) as a function of core radius for a step-index (full curve) and a parabolic index (dashed curve) core fibre having $\Delta = 0.01$ and $\Delta = 0.001$; the vertical bar denotes single-mode limit (reprinted with permission of Academic Press, Orlando, Fl from "Guided properties of fibres", by D. Marcuse, D. Gloge and E.A.J. Marcatilli in [11]. copyright 1977).

the power guided in a straight and a bent fibre would be proportional to $N_{R=\infty}$ and N_R, respectively. Figure 12 shows a plot of radius of curvature ($R_{50\%}$) (with which if a fibre is bent would have caused loss of 50% of the total modes guided by the same fibre when laid straight) versus core radii for two parabolic and step index fibres of different N.A.'s. This figure shows that fibres with less core diameter as well as fibres having larger Δ are more tolerant to sharp bends in terms of fractional power lost due to bending. The above results roughly provide guidelines for designing fibre spooling reels as also precaution to be taken during cable laying steps in order to avoid excessive bend loss. Well established companies like Corning Glass Works, USA use spools of nominal diameter ($\sim$ 20 cm) while shipping multikilometer long multimode fibres to customers.

In the case of single-mode fibres, macro-bend induced loss is given by [34, 35]

$$\alpha_{\text{Bend}}^{\text{SM}} = \frac{A_3}{R^{1/2}} \exp(-A_4 R). \tag{84}$$

where A_3, A_4 are functions of V and u, while R represents radius of curvature of the bent fibre. Loss sensitivity of a bent single-mode fibre increases as operating λ is increased beyond its cut-off wavelength. As a typical result, it may be quoted from [36] that bend-induced loss in a fibre having $\lambda_c = 1.1\,\mu$m increases from $\sim$ 2 dB/m at 1.3 μm to $\sim$ 100 dB/m at 1.55 μm ($R = 1.25$ cm). In order to simulate such inadvertent bends that may occur during fibre laying or in fibre joint boxes, one measures excess loss due to 100 turns of the fibre around a 7.5 cm diameter mandrel [32]; excess loss upto 0.1 dB at 1.3 μm and 1.0 dB at 1.55 μm in such measurements is taken as acceptability of the fibre for field deployments.

6.2.3 Radiative loss due to microbending

In contrast to the bend loss due to a constant fibre curvature, if the fibre lay is made to go through a continuous succession of very small bends (see Fig. 13), the fibre may exhibit a significant rise in attenuation, known as microbending loss. Physically, microbending leads to redistribution of optical power amongst the guided modes and also coupling of power from some high order guided modes to radiation modes-which eventually is responsible for the loss exhibited by fibres under such circumstances. It can be shown through some simple algebra that strong coupling between the pth and qth mode of a fibre will occur if $\Delta\beta (= |\beta_p - \beta_q|)$ matches the spatial frequency of the deformer [3, 37]. This result leads one to conclude that in order to avoid coupling of highest order guided modes to radiation modes in a step index fibre, mechanical deformation periodicities of the order of $\pi a/\sqrt{\Delta}$ must be avoided [3]. For a typical SiO$_2$-based fibre of core diameter 62.5 μm, N.A. $\sim$ 0.3, the required periodicity for coupling core modes to radiation modes will be $\sim$ 0.69 mm. Since microbending essentially gets transmitted to fibre during the jacketing of the fibre for

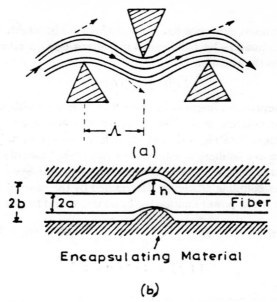

Fig. 13 (a) Geometry of a microbend fibre (b) Model for calculating loss from a fibre microbend due to a bump in the cable (after J.E. Goell, "Optical Fibre Cable", in [39]).

cabling, one should ensure avoidance of any periodic deformations of above magnitude to avoid excessive microbend loss in this fibre. This can be made possible by choosing right type of coating and fibre packaging so as to insulate the fibre from irregular external pressures.

According to a model proposed by Olshansky [38], microbend induced loss in a cabled step index multimode fibre can be expressed in dB as

$$\alpha_{mb} \simeq 0.9 \nu \frac{h^2}{b^6} (E_e/E_f)^{3/2} \frac{a^4}{\Delta^3}. \tag{85}$$

Here ν stands for number of bumps/unit length, h effective rms bump height, $2b$ overall fibre diameter, a the core radius, E_e refers to modulus of encapsulating material and E_f modulus of fibre core material. Typically for a fibre with $a = 25\,\mu m$, $b = 62.5\,\mu m$, $\Delta = 0.01$, $E_e = 7 \times 10^7$ Nt/mm^2 and $E_f = 7 \times 10^{10}$ Nt/mm^2, α_{mb} would be $\simeq 0.018$ dB for every $10\,\mu m$ bump. Thus, on an average if there are 100 such $10\,\mu m$ bumps over one kilometer length of a step index fibre due to cabling, then the cabling induced excess loss over and above the fibre loss would be 1.8 dB. With the state-of-art cabling process, the excess loss in cabled fibres is well within 0.1 dB. Microbend induced loss in single-mode fibres can be described in terms of mode (LP_{01}-mode) spot size (see Sec. 7.1).

6.3 Source fibre interconnection loss

In addition to the above sources of loss, there are two more equally important sources of loss in any fibre optic system. These are source-to-fibre coupling loss and fibre interconnection loss both of which are inevitable in any telecommunication system. The light emission pattern from a source can be approximated by the equation

$$I(\theta) = I_0 (\cos\theta)^m, \tag{86}$$

where $I(\theta)$ represents intensity in a direction θ defined with respect to a line normal to the emitting area. Here m represents directionality of the source radiation pattern (see Fig. 14). For $m = 1$, the source would correspond to what is known as a lambertian source while higher the value of m, narrower will be the emission pattern. If a source is directly butted against the fibre, the optical power coupling efficiency determined by the ratio of power coupled into the fibre to the power emitted by the source would be given by [39]

$$\eta = P_{\text{fibre}}/P_{\text{source}} = \left(\frac{q}{q+2}\right)\left(\frac{m+1}{2}\right) (\text{N. A.})^2. \tag{87}$$

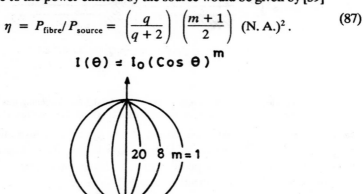

Fig. 14. Typical emission pattern $I(\theta)$ of a fibre optical communication source.

Equation (87) shows that for the same N.A. and m, the power coupling efficiency in a graded-core fibre would be less by a factor of $[q/(q + 2)]$ as compared to a step index fibre. This result is also consistent with the observation made earlier that the number of guided modes in a step index fibre is twice that in a parabolic core fibre and since the power coupled would be approximately proportional to number of guided modes excited, the above result follows automatically. If an intervening optical element like a lens is introduced between the source and the fibre to increase power coupling efficiency*, then it will typically increase over that of the direct butt coupling case by a factor M which is given by the ratio of fibre area to source area [39]

---

* An intervening optics can increase coupling efficiency from a lambertian source only if the source emission area times its emission solid angle is less than the corresponding product for the fibre core. If this product for the source > that for the fibre, most efficient coupling would be a direct butting of the source to the fibre [40].

$$M = A_{\text{fibre}} / A_{\text{source}} = (d_{\text{fibre}} / d_{\text{source}})^2, \qquad (88)$$

where d_{fibre} and d_{source} represent the diameters of the fibre and the source, respectively. Similarly, use of an intervening taper would also enhance coupling efficiency. As compared to 3.2% coupling efficiency obtainable by direct butt-coupling of a lambertian source to a 0.18 N.A. fibre, by use of a taper an increase in coupling efficiency upto ~ 53% achieved experimentally has been reported [41]. If a non-lambertian radiator like a double-heterojunction laser diode is substituted for the lambertian source in the above experiment then direct butt-coupling yields for the coupling efficiency a figure of ~ 30% which can be increased to 97% with a 4.3 mm long intervening taper of taper ratio ($= d_{\text{thick end}}/d_{\text{thin end}}$) 3.4 [42]. Experiments have also been carried out by forming (through heating/etching) a self-aligning micro ball lens at the LED surface [43]. Such systems can yield a maximum coupling efficiency of $(d_{\text{fibre}}/d_{\text{LED}})^2 \times$ (fibre N.A.)^2. These results are important from the point of view of designing source-to-fibre coupling connectors. For developing connectors, design principles also require an estimate of tolerances in mechanical alignments of the source and fibre axis on the input coupling efficiency. To give an idea of the order of magnitude of coupling loss due to various misalignments, we may quote a typical result from Barnosky [39]: in the case of coupling from a LED to a 50 μm core step index fibre of N.A. 0.14, (i) a transverse misalignment of the fibre axis of $\pm$ 20 μm with respect to a 50 μm active area LED will result in an additional loss of about 1 dB in the coupling; (ii) a longitudinal separation as large as 150 μm will lead to additional loss no more than 1 dB while (iii) angular misalignment upto 10° would make additional loss well within 0.25 dB. These results show that very tight tolerance is required to align fibre axis with the LED with respect to transverse misalignments.

In the case of coupling laser diodes* to single-mode fibres one usually employs a lens in between. The results of a study made on the LP_{01}-mode excitation efficiency in single-mode graded-core fibres, by considering a plane wave segment being focused through a lens as a function of various misalignments/mismatches between the lens and the fibre N.A. and axes are reproduced in Table 3 from [44]. This table shows that in case of single-mode fibres as in the case of multimode fibres, source to fibre coupling is highly tolerant to longitudinal separations while transverse alignment is quite critical.

6.4 Fibre-fibre splice and joint loss

Another inevitable source of loss is interconnection/splicing of fibres. The first step in any fibre splicing process requires preparation of smooth and square

---

* LEDs are usually considered to be unsuitable for injecting light into a single-mode fibre in view of the fact LEDs emit light over much wider angles than the acceptance cone semi-angle (N.A.) of the fibre.

TABLE 3

LP$_{01}$-mode excitation efficiency in graded-core single-mode fibres when operated at V (near their corresponding V_c) for various lens-fibre misalignments/mismatches
(After Tewari and Pal [44])

q (cf. Eq. (74))	V_c	Maximum launched power for Z=D=0 %	(1) θ_L/θ_a	(2) r_A/a	(3) D_h	(4) $Z_{90\%}$
1.0	4.38	77.4	0.70	1.248	0.604	15.3
2.0	3.5	78.1	0.78	1.401	0.675	15.5
4.0	3.0	78.4	0.83	1.555	0.749	16.3
8.0	2.7	78.4	0.84	1.686	0.813	17.3
10.0	2.65	78.4	0.84	1.718	0.827	17.6
20.0	2.5	78.4	0.84	1.821	0.876	18.7
∞	2.4	78.6	0.85	1.875	0.902	19.0

Z = z/a: Longitudinal offset;
D = d/a: transverse off-set;
(1) Value of the lens half angle for maximum launched power at $\theta_a = \sin^{-1}(0.07)$ and D = Z = 0.
(2) r_A: First Airy disc radius corresponding to θ_L of the preceding column
(3) Value of D at which the launched power falls to half of its maximum value shown in colum 3.
(4) Value of Z at which power falls to 90% of its value shown in column 3.

end faces (end angles less than 1°). Several commercial fibre splicing tools/machines are now available to obtain such mirror-like end face by means of controlled fracture under tension. Once the fibre ends are prepared, one is required to align them before joining them mechanically or through fusion in an electric arc or flame. There could be three different kinds of misalignments at a fibre splice/joint: the axes of two fibres could be offset in the transverse direction, there could be longitudinal separation and/angular misalignment between them.

In Fig. 15(a), we have plotted loss vs. misalignment obtained experimentally in our Fibre Optics laboratory at IIT Delhi with two pieces of a 50 μm graded core ($q \sim 2.1$) fibre having N.A. 0.21. This figure clearly shows that coupling loss is most sensitive to transverse offset. For maintaining a loss < 0.1 dB, transverse off-set between the fibre axes must be well below 4.5 μm in such fibres. Figure 15(b) depicts constant loss curves for different misalignments. Such curves are very important as they provide valuable guidelines in the designs of low-loss fibre connectors and splices. An additional source of loss at a fibre to fibre connector could be Fresnel reflection loss (at most $\sim$ 0.2 dB) due to refractive index mismatch between the fibre and the air gap*, which however can be substantially overcome by incorporating index matching liquids at the joint.

---

* Analogous to impedance mismatch in electrical cables.

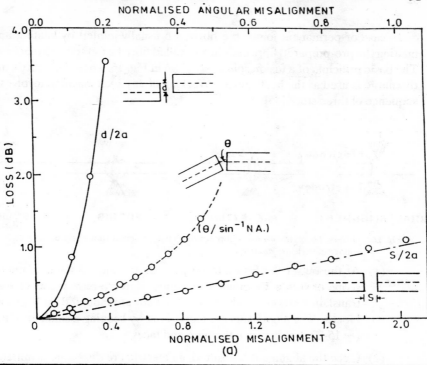

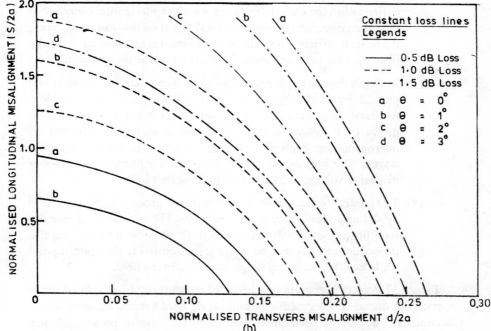

Fig. 15. (a) Loss measured in our Fibre Optics laboratory at IIT Delhi in dB due to various misalignments between two 50/125 μm multimode fibres separated by a small air-gap (∼ 1-2 μm); (b) Constant loss curves generated from datas of (a) above.

In case of permanent joints, the fibres are usually jointed by heating and melting the pre-prepared fibre ends in a so called fibre fusion splicing machine. The basic principle of a fusion splicer is shown in Fig. 16, where an electric arc discharge is used as the heat source. Working of any such machine involves a sequence of three steps [45]:

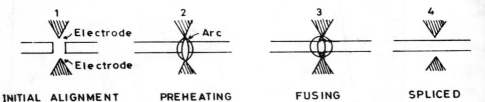

Fig. 16. Different steps involved in a fibre fusion splicing machine that works under a high voltage electric arc discharge.

(1) At the outset the fibres, to be jointed, are aligned under a microscope or with a TV monitor by means of micro-movements of x-y-z translation stages on which the fibre ends are mounted on V-groove blocks. The alignments are carried out by keeping a small axial gap ($\sim$ 15-20 μm) between the fibre end faces.

(2) Once the alignment is ensured, an electric arc discharge is initiated to fire-polish the ends of the fibres, which results in little rounding of the fibre edges, which helps in achieving a good low-loss joint during the next step. In this step, the arc also removes any dust etc. from the fibre ends for a smooth joint. Step 2 is known as prefusion step.

(3) In step 3, the two so prepared fibre ends are brought into contact and pushed longitudinally against each other by an additional longitudinal movement of $\sim$ 10-15 μm of the micropositioner. This distance is called stroke length and it can be controlled by microprocessor controlled stepper motor drives of the translation stages. This butt joint is then fused by the arc discharge at a higher discharge voltage and current to fuse the butt joint.

(4) The jointed ends is finally lifted from the V-grooves and is protected by sliding a plastic sleeve over the joint. The ends of the sleeve are then heated separately and cured to fix securely on the fibre; this procedure ensures that no stress is transmitted to the spliced point while handling or laying the jointed fibre in the field.

Figure 17 depicts histogram of a series of measurements on spliced fibres carried out during development of a multimode fibre splicing machine in our laboratory. It may be mentioned that there are several parameters like electrode gap, prefusion time, fusion time, stroke length, fusion current etc., which are very important in optimising any such splicing machine [45]. For the

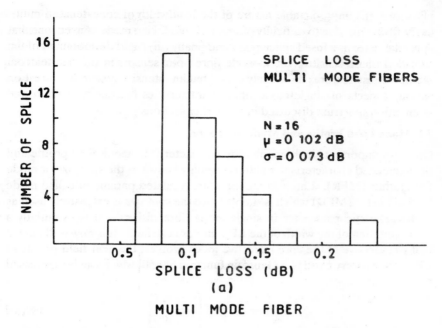

Fig. 17. Histogram depicting splice loss vs number of splices obtained in a multimode fibre fusion splicing machine [45].

machine which yielded results for Fig. 17, the optimum electrode gap and fusion time were 1.0 to 1.2 mm and 1.5 sec, respectively. Depending on shape of the electrodes, these numbers may vary to a certain extent from machine to machine. Geometrical alignments are much more stringent in the case of single-mode fibres. Typically a transverse offset of $\sim 2\,\mu$m between two $9\,\mu$m core single-mode fibres may result in as much as ~ 0.9 dB loss. A few other splicing techniques have also been proposed and tried with success like elastometric splice [46] and loose tube splice [47]. A complete book on splices and connectors is now available, which the readers may refer to for further details [48]. It may be mentioned here that it is very important that spliced fibres exhibit strength comparable to the unspliced fibres and in any splicing machine design, one should incorporate features to fulfill this requirement [45].

7. SOME IMPORTANT CHARACTERISTICS OF SINGLE-MODE FIBRES

Since early eighties, most of the R & D interest in fibre optics has shifted to single-mode fibres primarily due to three factors: (i) attainment of extremely low-loss (~ 0.15 dB/km at $1.55\,\mu$m, which is almost the intrinsic loss limit set by Rayleigh scattering) along with development of high quality single-mode fibre splicing machines that yield loss < 0.1 dB/splice for telecommunication applications, (ii) unpredictable nature of the bandwidth of concatenated multi-mode fibre links due to difficulty of precisely modelling mode conversions that may take place at a fused splice point and finally (iii) rapid development of the so called phase sensitive single-mode fibre optic sensors in the non-telecom sector. These features have lately generated an intensive research interest on various aspects of single-mode fibre characteristics besides dispersion and attenuation spectrum discussed in the earlier sections.

7.1 Mode Field Profile and Mode Spot Size

One very important single-mode fibre parameter is its mode field profile and the associated characteristic radius commonly known as the spot size or mode field radius (MFR). Lately in the literature a related parameter called mode field diameter (MFD) which is simply twice the spot size is extensively used as a characteristic parameter in single-mode fibre disigns[*]. It is essentially a representation of the width of the LP_{01}-modes near field. It is now well known that irrespective of the core refractive index profiles, the near field profile of LP_{01}-mode is very close to a Gaussain function [49-50] which can be expressed as

$$E(r) = A \exp\left[-\frac{1}{2}\frac{r^2}{\omega_g^2}\right] \quad [89(a)]$$

⇒

Near field intensity:

$$P(r) = \left|E(r)\right|^2 = A^2 \exp\left[-\frac{r^2}{\omega_g^2}\right] \quad [89(b)]$$

Here ω_g is normally referred to as the spot size of the LP_{01}-mode's near field intensity[**] and physically, it simply represents the half-width of the modal intensity at $1/e$ point such that at $r = \omega_g$, intensity of the LP_{01}-mode has dropped to $1/e$ of its maximum (which occurs at $r = 0$). Accordingly, MFD would be 2 ω_g. In Fig. 18, near field intensity distribution for a typical step index single-mode fibre having Δ as 0.3%, $n_2 = 1.444$ and core diameter ($2a$) of 8.5 μm (⇒ $\lambda_c = 1245$ nm), is shown with operating λ (⇒ V) as the labelling parameter. It is evident from the figure that as λ is moved further away from λ_c (⇒ $V < V_c$) the field spreads more and more into the cladding. Thus if a single-mode fibre designed for operation at 1.3 μm is operated at the longer wavelength of 1.55 μm, the modal field will spread so much into the cladding that the fibre will be highly loss sensitive to external perturbations like macrobending and microbending. In view of this fact, the fundamental mode spot size (or alternatively, MFD), has a major influence on the losses that arise at single-mode fibre

---

[*] It is now well known that dimensionally a single mode fibre is characterised in terms of its spot size or MFD and core diameter is of little interest in such fibres.

[**] In contrast, the spot size of the modal near field would be $\sqrt{2}\,\omega_g$ [51].

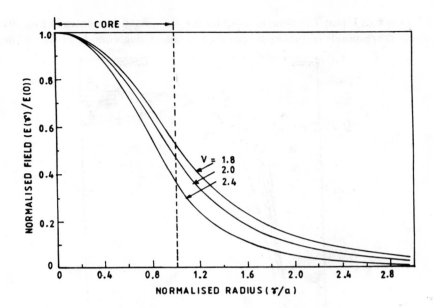

Fig. 18. Variation of LP_{01}-mode near field in a step-index fibre optimised for operation at 1.3 μm; different curves are labelled with operating λ as variable. As operating V-number decreases ($\Rightarrow \lambda$ increases), the field spreads more and more into the cladding.

splices, connectors, source to single-mode fibre coupling as well as at bends, microbends and so on. For example, joint losses in two different single-mode fibres (both having approximately Gaussian mode intensity profile but having different V_c's and N.A.'s) would be small as along as their spot sizes are designed to be the same. In fact, for system engineers refractive index profile and core diameter are only of theoretical significance and manufacturers rarely include these information in their data sheets. It is the spot size or MFD which determines coupling losses at various interconnections, e.g., source to fibre, fibre to fibre and even waveguide dispersion [36] in single-mode fibres.

In Sec. 3, we have seen that the fundamental mode (LP_{01}) field of a step index fibre is exactly described by J_0 (ur/a) in the core and K_0 (wr/a) in the cladding. However, in practice, fabrication difficulties to yield a perfect step index core results in a quasi-step index profile besides an axial dip in the core refractive index profile. But unfortunately there are no closed form analytical solutions to the LP_{01}-mode field in a single-mode fibre of an arbitrary core refractive index profile. However, as already mentioned, it turns out [49] that the LP_{01}-mode field of practical fibres having a smooth refractive index profile closely resembles that of a Gaussian field. In fact, it can be shown [49] from calculations of the power transmission coefficient (*T*) between an optimally adjusted free space Gaussian laser beam and a step index single-mode fibre as a function of V that T is close to unity for $2.1 \lesssim V \lesssim 3.0$ (as shown in Fig. 19) which corresponds to practical operating range of single-mode fibres. Now

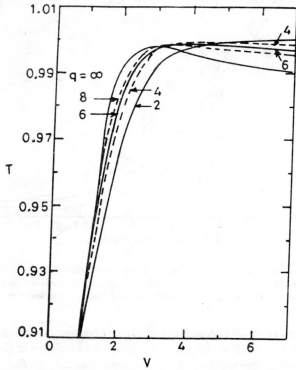

Fig. 19. Normalised power transmission between a free-space Gaussian beam and graded core fibres supporting only the LP_{01}-mode; core profile-shape parameter 'q' labels different curves (adapted with permission from [49]. Copyright 1978 Optical Society of America).

power transmission coeffcient/coupling efficiency is essentially given by the overlap between the free space Gaussian field distribution and the LP_{01}-mode field of the fibre. Thus the above result obviously points to the fact that atleast over the above range of V-values the description of the LP_{01}-mode field in a step index fibre can be described by a Gaussian distribution to a good approximation. These observations, namely, that the LP_{01}-mode field distribution in a step index fibre as well as graded-core fibres closely resembles a Gaussian distribution led Matsumura *et al*, [52] to suggest an interesting technique to characterise a graded-core single-mode fibre by deriving an equivalent step index (ESI) fibre; equivalent in the sense that the so derived hypothetical step index fibre yields almost an identical LP_{01}-mode intensity profile and propagation constant as that of the original graded-core fibre. Since the mode parameters like u, w and so on can be calculated exactly for a given step index fibre, a large number of propagation characteristics like losses at bends, microbends, splice, source to fibre coupling etc., can be determined for that fibre at any wavelength. In this context, the concept of ESI profile is quite convenient because it enables one to estimate most of the propagation characteristics[*] of a graded-core fibre to a reasonable accuracy without recourse to extensive numerical simulations. Miller [53] was first to suggest a direct method for determining the ESI profile of a single-mode fibre without necessitating any knowledge of its actual index profile. The method is based on the empirical relation for the mode spot size of a step index fibre originally suggested by Marcuse[**] [50]:

$$\omega_s = a_s f(V_s), \tag{90}$$

where
$$\sqrt{2} f(V_s) = 0.65 + \frac{1.619}{V_s^{3/2}} + \frac{2.879}{V_s^6}. \tag{91}$$

Here the suffix s stands for step index fibre for which these equations are valid. Differentiation of Eq. (90) with respect to wavelength (λ) yields

$$\frac{d\omega_s}{d\lambda} = a_s \frac{df(V_s)}{dV_s} \cdot \frac{dV_s}{d\lambda}$$

$$= \frac{a_s V_s}{\lambda} \cdot \frac{1}{\sqrt{2}} \cdot \left[\frac{2.4285}{V_s^{5/2}} + \frac{17.274}{V_s^7}\right]$$

$$\Rightarrow \quad \sqrt{2}\lambda \frac{d\omega_s}{d\lambda} = a_s \left[2.4285 V_s^{-3/2} + 17.274 V_s^{-6}\right] \tag{92}$$

After substituting for $V_s^{-3/2}$ from Eq. (91) in Eq. (92) we will get

$$\sqrt{2}\lambda \frac{d\omega_s}{d\lambda} = \omega_s \left[2.1213 - \frac{0.975 a_s}{\omega_s}\left(1 - 13.2877 V_s^{-6}\right)\right] \tag{93}$$

Simple algebraic manipulations of Eq. (93) yield

$$a_s = \frac{2.1757\omega_s - 1.45047\lambda \frac{d\omega_s}{d\lambda}}{1 - 13.2877 V_s^{-6}}. \tag{94}$$

If one makes a direct measurement of the mode spot size ($1/e$ intensity width) of a single-mode fibre, experimentally as a function of λ, the spot size would decrease almost linearly with decrease in λ. As λ is decreased beyond the wavelength of LP_{11}-mode cut-off (λ_c) the plot of $\omega_s(\lambda)$ would exhibit a sudden kink at $\lambda = \lambda_c$ due to onset of the second, i.e. LP_{11}-mode, thereby leading to a

---

[*] In the case of calculation of dispersion/bandwidth, possibly the accuracy of ESI approach is questionable.

[**] Here ω_s represents half-width (spot size defined with respect to intensity of the LP_{01}-mode) of the LP_{01}-mode intensity distribution at its $1/e$ point.

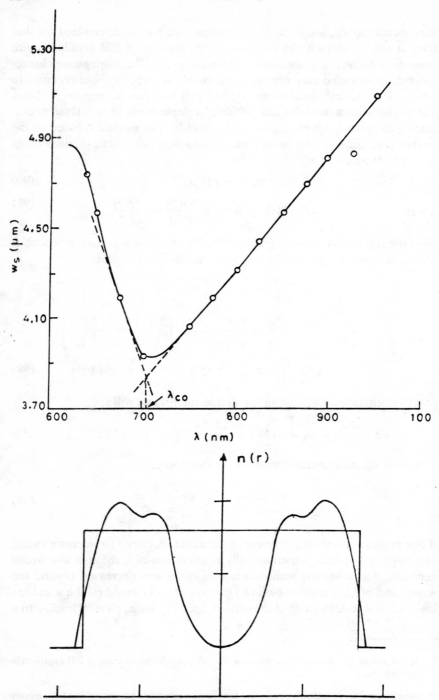

Fig. 20. (a) Typical experimental results depicting variation of spot size with wavelength of a single-mode fibre; (b) Schematic illustration of actual refractive index profile of a fibre with its ESI profile superimposed on it.

sudden increase in $1/e$ intensity width at $\lambda \lesssim \lambda_c$. An experimental plot of this behaviour of $\omega_s(\lambda)$ is shown in Fig. 20(a). Since it is known theoretically that at $\lambda = \lambda_c$, $V_s^c \equiv 2.405$ in a step index fibre, knowledge of the two quantities, namely, the spot size ω_s^c (at $\lambda = \lambda_c$) and the slope $d\omega_s/d\lambda|_{\lambda = \lambda_c}$ from the measurement of $\omega_s(\lambda)$ will yield the core radius and numerical aperture as

$$a_s = 1.0737 \left[2.1757\, \omega_s^c - 1.4507 \lambda_c \left. \frac{d\omega_s}{d\lambda} \right|_{\lambda = \lambda_c} \right], \quad (95)$$

and $\left. N.A. \right|_s = \dfrac{2.405}{2\pi} \dfrac{\lambda_c}{a_s} = 0.38277\, \dfrac{\lambda_c}{a_s} \approx 0.383\, \dfrac{\lambda_c}{a_s}.$ (96)

In fact the knowledge of the spot size (ω_s^c) at $\lambda = \lambda_c$ is sufficient to yield a_s from Eq. (90) and Eq. (91) since it is known that $V_s = V_s^c = 2.405$ at $\lambda = \lambda_c$; the so obtained a_s and the measured λ_c can be used to calculate $\left. N.A. \right|_s$ from Eq. (94). Thus, the recipe for obtaining the ESI profile corresponding to a single-mode fibre of any arbitrary profile essentially involves measurement of $\omega_s(\lambda)$ and therefrom use of Eqs. (95) and (96) will determine the equivalent step index core radius a_s and equivalent step index numerical aperture $\left. N.A. \right|_s$. An example of the actual refractive index profile along with its ESI profile (as determined through the procedure outlined above) are shown in Fig. 20(b). An alternative method of obtaining ESI core radius was suggested by Allard et al [54]. It relies on the assumption that the N.A. of the fibre is independent of λ and hence one can write

$$V_s \lambda = \text{constant}. \quad (97)$$

Thus, since $V_s = 2.405$ at $\lambda = \lambda_c$, one gets

$$V_s = 2.405\, (\lambda_c/\lambda). \quad (98)$$

After substitution of Eq. (98) in Eq. (99), one can rewrite Eq. (90) as

$$\omega_s = \frac{a_s}{\sqrt{2}} \left[0.65 + 0.434\, (\lambda/\lambda_c)^{3/2} + 0.0149\, (\lambda/\lambda_c)^6 \right]. \quad (99)$$

A direct fit of Eq. (99) to $\omega_s(\lambda)$ in the region $\lambda > \lambda_c$ will yield the parameters a_s and λ_c. Once the ESI parameters are obtained, one can use them to predict/model several of the physical characteristics of a given unknown single-mode fibre. Formulae to obtain some of these chracteristics at a particular wavelength (λ) from a knowledge of the ESI parameters and spot size at that λ are tabulated in Table 4. However, if the fibre is operated at λ's away from cut-off and also for single-mode fibres with complex refractive index profiles, the mode field profile may substantially deviate from a perfect Gaussian shape. In such cases, one normally employs two r.m.s. definitions for the LP_{01}-mode's spot size in the near field and in the far field, namely

TABLE 4

Formulae to obtain several physical quanitities of interest for single-mode fibres in terms of mode field/spot sizes and/ESI* parameters

Physical quantity	Formula	Remarks
Source to fibre coupling loss (dB)	$10 \log_{10} \left(\dfrac{\omega_0^2 + \omega_s^2}{2 \omega_0 \omega_s} \right)$	ω_0 = source spot size ω_s = fibre's mode spot size
Splice loss due to transverse offset[1] (dB)	$4.34 \, (d/\omega_s)$	d = transverse offset between the axes of two identical fibres
Splice loss due to angular offset[1] (dB)	$4.34 \, (\pi n_2 \omega_s / \lambda)^2 \theta^2$	θ = angular offset between the fibre axes; n_2 = cladding refractive index
Bend loss[2] (dB/unit length)	$4.34 \left(\dfrac{\pi}{w_s^3 R a_s} \right)^{1/2}$ $\times \dfrac{u_s}{V_s^2 K_1^2(w_s)} \exp \left[-\dfrac{4}{3} \dfrac{\Delta_s R w_s^3}{a V_s^2} \right]$	R = bend radius of curvature; u_s, w_s, V_s, a_s, Δ_s are ESI fibre parameters; K_1: modified Bessel function of order 1.
Microbend loss[3] (dB/unit length)	$2.17 A (k_0 n_1)^{2p+2}$ $\times (\omega_s)^{2+4p}$	A = coefficient of curvature power spectrum (of dimension L^{-2p-1}); p is an adjustable parameter in the range: $2 \lesssim p \lesssim 5$
Waveguide dispersion coefficient D_{wg} (ps/nm-km)	$-\left(\dfrac{n_2 \Delta_s}{c \lambda_0} \right) V_s \dfrac{d^2(V_s b)}{dV_s^2}$	$b = w_s^2 / V_s^2$

* Once the ESI parameters are obtained, namely, a_s and Δ_s, one can obtain corresponding modal propagation constant, β (and u_s, w_s) by solving Eq. [22(b)], u_s and w_s can be obtained from Eqns. [20(a)] and (17)
(1) From Ref. [55].
(2) From Ref. [35].
(3) From Ref. [52].

$$\omega_{rms} = \left[\dfrac{2 \int_0^\infty r^3 E^2(r) \, dr}{\int_0^\infty r E^2(r) \, dr} \right]^{1/2} \quad [100(a)]$$

and

$$w_{ff} = \left[\dfrac{2 \int_0^\infty p^3 E^2(p) \, dp}{\int_0^\infty p E^2(p) \, dp} \right]^{1/2}, \quad [100(b)]$$

where $E^2(r)$ represent near field intensity distribution and $E^2(p)$ the corresponding far field intensity distribution with $p = ak_0 \sin \theta$, θ being the angle in the far field at which $E^2(p)$ are defined. The far field radiation pattern of a single-mode fibre is nothing but the Fourier Transform of the fibre's near field [56]; the near field being obtainable from a knowledge of the fibre's RIP and solution of the scalar wave equation for that RIP. For example, for the step index fibres, the scalar wave equation (cf. Sec. 3) yields $J_l(ur/a)$ for $r < a$ and $K_l(wr/a)$ for $r > a$ as the near field of the fibre. Assuming the fibre exit face as a circular aperture of radius a, its far field would be given by the Fraunhofer integral from the scalar Diffraction theory [56]; in Eq. [100(b)] the far field distribution represented by $E(p)$ is precisely given by the Fraunhofer integral corresponding to a circular aperture of radius a. It can be shown that [57] ω_{ff} is nothing but inverse of the Laplace spot size in the near field defined through [58]

$$\omega_p = \left[\frac{2 \int_0^\infty r E^2(r) \, dr}{\int_0^\infty r \left(\frac{dE}{dr} \right)^2 dr} \right]^{1/2} \tag{101}$$

The subscript p-stands for Petermann who had first introduced this definition for mode spot size while formulating the theory of microbending loss in single-mode fibres and ω_p as defined above is sometimes referred to as Petermann's spot size of kind 2 in the literature. A measure of ω_{rms} enables estimation of microbending loss and joint losses due to small tilts at a single-mode fibre joint while ω_p enables estimation of joint losses due to small transverse off-sets and waveguide dispersion (from the wavelength dependence of ω_p [36]). In general $\omega_{rms} > \omega_g > \omega_p$. If, however the LP_{01}-modal field is describable by a Gaussian function, it can be shown that $\omega_{rms} = \omega_p = \omega_g$ [59]. Indeed, in system designs most of the estimations for the transmission characteristics of a single-mode fibre are usually based on Gaussian approximation for the modal near field i.e., $E(r)$. However, for a generic near field, we list below various transmission losses and also a term which is important in calculation of waveguide dispersion in single-mode fibres in terms of their characteristic spot sizes [36].

1. Splice loss due to a transverse off set of amount 'd' between the fibre axes:

$$\alpha_t \, (dB) = 4.34 \cdot (d / \omega_p)^2 . \tag{102}$$

2. Splice loss due to an angular offset (θ) between the fibre axes:

$$\alpha_\theta = 4.34 \, (\pi n_2 \omega_{rms} / \lambda)^2 \, \theta^2 . \tag{103}$$

3. Microbend loss:

$$\alpha_{mb} \, (dB / \text{length}) = \frac{4.34}{8} \, (k_0 n_1 \omega_{02})^2 \, \Phi_p (2 / k_0 n_1 \omega_{01}^2) , \tag{104}$$

where Φ_p is the curvature power spectrum and is given by

$$\Phi_p (\Omega) = \Omega^4 \sqrt{\pi} \, \sigma^2 L_c \, \exp \left\{ - \left(\frac{1}{2} L_c \Omega \right)^2 \right\} , \qquad [105(a)]$$

σ^2 and L_c being the rms deviation from straightness and the correlation length, respectively while ω_{01} and ω_{02} appearing in Eq. (104) are expressible in terms of their ratio with ω_p through the following empirical relations:

$$\omega_{01}/\omega_p = 1.0 + 2.41 (X_0 - 1) - 0.93 (X_0 - 1)^2$$

$$\omega_{02}/\omega_p = 0.99 + 0.57 (X_0 - 1) - 0.37 (X_0 - 1)^2 \quad [105(b)]$$

with $\quad X_0 = \omega_{\text{rms}}/\omega_p$.

4. The term $V(db/dV)$ required in calculating waveguide dispersion is rigorously given by [36]

$$V \frac{db}{dV} = \frac{4a^2}{V^2 \omega_p^2} \ . \tag{106}$$

7.2 Effective Cut-off Wavelength (λ_{ce})

In Eq. [25(b)], the cut-off wavelength in a single-mode fibre is defined as the wavelength such that if the fibre is operated at any $\lambda > \lambda_c$, LP_{11} and all other higher order mode would be cut-off in this fibre. This definition of cut-off wavelength is known as theoretical cut-off wavelength (λ_{ct}) in a step-index fibre. It is worthwhile to mention here that if the core RIP is different from a perfect step function, then the cut-off V-number (V_c) of the LP_{11}-mode in such fibres differs from 2.4048. Figure 21 depicts variation of V_c with core profile shape

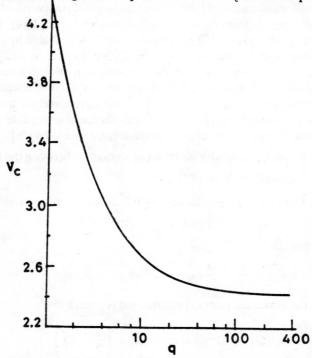

Fig. 21. Variation of cut-off V-number (V_c) with profile shape parameter 'q' (cf. Table 3).

parameter q (cf. Eq. (74)); numerical values of V_c in different q-valued fibres are tabulated in second column of Table 3. In practice, since core diameter is expressed in μm and relative core-cladding index different (Δ) in percent, one can express λ_{ct} [32] for a step index single-mode fibre for operation at 1.3 μm (at which refractive index of pure SiO_2 is about 1.447) as

$$\lambda_{ct} = \frac{\pi}{2.4048} \cdot 2a \cdot n_2 \sqrt{\Delta}$$

$$\simeq 267.2a \, (\mu m) \cdot \sqrt{\Delta \%} \,, \qquad (107)$$

while for a parabolic index core and triangular index core profiles, the coefficient 267 is replaced by 183 and 147 respectively, since corresponding V_cs are 3.5 and 4.38. Due to various perturbations like bending, microbending, twisting etc., a fibre can, in practice, be operated at a wavelength slightly below λ_{ct} because close to cut-off, the modal power in the LP_{11}-mode extends deep into the cladding (i.e. it is loosely bound to the core) and even a small perturbation can lead to a large loss. Thus, a fibre becomes effectively single moded at a $\lambda = \lambda_{ce}(<\lambda_{ct})$. In fact, the functional cut-off wavelength of the LP_{11}-mode that is measurable in practice is always an effective cut-off wavelength [10]. For operation at $\lambda \simeq 1.3 \, \mu$m, the standards committees of CCITT and EIA have recommended that $1100 < \lambda_{ce}$ (nm) < 1280. The upper limit can be extended to almost 1350 nm especially for dual operation at 1300 and 1550 nm wavelengths [36, 60].

7.3 Advanced Single-mode Fibre Designs

Single-mode fibres of today are optimised for low-loss (< 0.4 dB/km) and near zero-dispersion (< 3.5 ps/nm-km) at 1300 nm. On the other hand, as already stated earlier, fibres of today which are made of silica exhibit lowest attenuation (mainly due to low inherent scattering) at 1550 nm. If the 1300 nm-optimised fibre is operated at 1550 nm it will exhibit a nominal dispersion of ~ 18-20 ps/nm-km [36]. However, it is almost certain that in the very near future, long distance fibre optic communication systems will indeed operate at 1550 nm to exploit the lowest loss wavelength window of silica fibres-for larger repeater spacings. In order to achieve both low loss and low dispersion at 1550 nm, one can choose either of the following two routes: employ ultra-narrow in line-width (small $\Delta \lambda$) single-mode lasers, e.g., Distributed Feedback Lasers (DFB) in conjunction with 1300 nm optimised conventional single-mode fibres such that D in Eq. (61) is close to zero or use multimode laser diodes with advanced single-mode fibres which will exhibit ultra low dispersion at 1550 nm. The former route usually requires use of very expensive laser diodes. Further at high modulation rates, the laser diodes may not remain single-moded due to chirping (shift in emitted wavelength). This leaves the latter route as the best near term option.

Broadly, the advanced single-mode fibre designs are classified into two categories namely, dispersion shifted and dispersion flattened fibres. In the dispersion shifted single-mode (DS-SM) designs, the magnitude of waveguide dispersion is enhanced by tailoring the shape of the refractive index profile such that the sum of material and waveguide dispersions is made to be zero at 1550 nm while maintaining the inherent scattering loss to a minimum in the fibre. Some of the designs proposed to achieve this goal are shown in Fig. 22. One of the first designs (cf. Fig. 22a) in this direction involved a decrease in core size and corresponding increase in Δ as compared to a 1300 nm optimised matched clad* step-index fibre. With this design, however, it appeared difficult to maintain low-loss in view of excess doping of SiO_2 required to attain high Δ. This difficulty gave rise to triangular profile core design [61]. (cf. Fig. 22b). However, in this design, the cut-off wavelength (λ_{ce}) happened to be $\sim$ 1100 nm, which resulted in large sensitivity to bending for operation at 1550 nm. To increase resistance to bend induced loss, an annular ring surrounding the core was introduced in the index profile (cf. Fig. 22c) to enhance mode confinement inside the core [62]. Dual step-index core [63] is another such DS-SM fibre (Fig. 22d). As no analytical solutions are available for these DS-SM fibres, study of propagation effects in them required numerical solutions of the corresponding scalar wave equation [64].

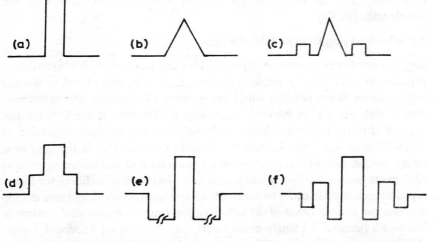

Fig. 22. Advanced single-mode fibre designs: (a) conventional step-index with relatively large Δ and small a; (b) triangular-index core; (c) segmented triangular-index core; (d) dual-shape step-index core; (e) depressed-index clad; (f) segmented core dispersion flattened.

---

* The name 'matched clad' is derived from the fact that the cladding layers in them are derived from synthetic silica layers which match the refractive index of silica support jacket.

On the other hand, dispersion flattened single-mode fibres (DF-SM) involved introduction of a low index inner cladding (called Depressed Index Clad or DIC fibres [64]) to a conventional step-index fibre or introduction of multiple claddings (called Segmented core or Segcore fibres [65]) as shown in Fig. 22 (e-f). Solution of the scalar wave equation for these index profiles reveal low dispersion (< 2 ps/nm-km) over the wavelength range 1300-1600 nm. Such low dispersion over this relatively broad low-loss wavelength window opened up the possibility of employing them as ideal transmission media for wavelength division multiplexed (i.e. simultaneous transmission of a large number of signals over several carriers in the above wavelength range) operation and thereby achieve super high transmission bandwidth for a single fibre. In Fig. 23, typical dispersion characteristics of a DS-SM fibre is depicted along with dispersion variation against wavelength of a 1300 nm optimised conventional single-mode fibre of matched clad design. Eventual mass use of these advanced design single-mode fibres will be decided by their cost effectiveness relative to conventional matched clad and DIC fibres. However, it is conceivable that as volume of commercial telecommunication markets for these fibres grow, the prices will

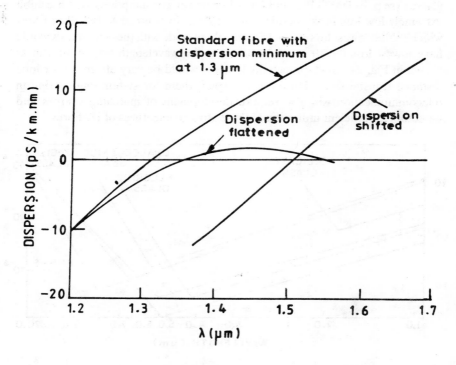

Fig. 23. Dispersion spectra of dispersion shifted and dispersion flattened fibres alongwith that of a 1300 nm optimised conventional fibre (step-index); here D_T is plotted by including the –ve sign in its definition (cf. Eq. (61)).

come down. In Table 5, some typical characteristics of commercially available multimode and single-mode fibres of today as quoted by well known fibre manufacturers in their product data sheets are displayed.

8. FUTURE DEVELOPMENTS

Although the silica based fibres are expected to dominate the telecommunication market for another ten years or so, new material systems and fabrication techniques are currently under intensive investigations in several laboratories around the world. From the new fabrication technique point of view, one which holds promise as a potentially low cost method is known as "sol-gel" process [66] in which the raw materials like $Si(OCH_3)_4$ (which are commercially available in very pure form) in the form of oxides are hydrolysed to form a wet gel which can be cast into desired shape followed by a drying step. Finally the dry gel is sintered into a solid glass rod before converting it into fibre through conventional drawing. Initial experiments with this technology yielded fibres which exhibited losses $\sim$ 5 dB/km at $\lambda \sim 0.85\,\mu m$.

New material systems which show greatest promise are the fluoride-based glasses (m.p. $\sim$ 900°C thus involving low power consumption) which exhibit extremely low loss in the mid-infrared region. In contrast to the silica fibres which exhibit lowest loss at $\lambda \sim 1.55\,\mu m$, fibres made with these materials would have lowest loss ($\sim 10^{-2} - 10^{-3}$ dB/km) at the wavelength region 4-6 μm as shown in Fig. 24. Such potentially low loss would be very attractive for long distance transmission. However, to exploit them for system engineering in telecommunication will also require development of matching sources and detectors for efficient utilisation of these attractive features of IR-fibres.

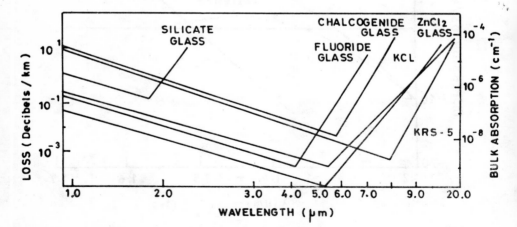

Fig. 24. Estimated attenuation spectrum of infrared fibres often referred to as V-curves due to their obvious shapes (adapted with permission from [68] © 1982 Gordon Publications. All rights reserved).

9. CONCLUSION

A semi-tutorial presentation of the fundamental principles of fibre optics relevant to telecommunication applications has been presented. Propagation characteristics of telecommunication grade fibres including source-fibre and fibre-fibre coupling characteristics have been described in sufficient details both for multimode and single-mode fibres. Advanced single-mode fibre designs which are of great interest for immediate future have been briefly discussed. Finally, the so-called infrared fibres which have the potential of enabling ultra low-loss transmission in the mid-infrared wavelength range appear to be the area of R & D interests for the immediate future.

REFERENCES

1. A.W. Snyder and Love J.D., "Optical Waveguides", Chapman and Hall, London (1984).
2. M.S. Sodha and Ghatak A.K., "Inhomogeneous Optical Waveguides", Plenum, New York (1977).
3. M.M. Butusov, Pal B.P., Galkin S.L. and Orobinsky S.P., "Fibre Optics and Instrumentation" (in Russian), Mashinostroenie Publishing House, Leningrad (1987); updated English Edition, IOPP, Bristol (in press, 1992).
4. A.K. Ghatak and Thyagarajan K., "Optical Electronics", Cambridge Univ. Press, Cambridge (1990).
5. D. Gloge, "Weakly Guiding Fibres", App. Opt. **10**, 2252-2256 (1971).
6. D. Gloge and Marcatili E.A.J., "Multimode Theory of Graded Core Fibres", Bell Syst. Tech. J. **52**, 1563-1578 (1973).
7. D. Marcuse, "Light Transmission Optics", Van Nostrand Reinhold, New York (1982).
8. F.P. Kapron and Keck D.B., "Pulse Transmission Through a Dielectric Optical Waveguide", App. Opt. **10**, 1519-1523 (1971).
9. A.C. Cherin, "An Introduction to Optical Fibres", McGraw Hill International, Japan (1983).
10. E.G. Neumann, "Single-mode Fibres: Fundamentals", Springer Verlag, Berlin (1988).
11. S.E. Miller and Chynoweth A.G. (Eds.), "Optical Fibres for Telecommunication", Academic Press, New York (1979).
12. R. Olshansky and Keck D.B., "Pulse Broadening in Graded Index Optical Fibres", App. Opt. **15**, 483-491 (1976).
13. A.K. Ghatak and Thyagarajan K., "Pulse Dispersion in Optical Fibres", App. Opt. **16**, 2538-2540 (1977).
14. G. Keiser, "Optical Fibre Communications", McGraw-Hill International, Japan (1983).
15. R. Olshansky, "Pulse Broadening caused by Deviations from Optimum Profile", App. Opt. **15**, 782-787 (1976).
16. K. Behm, "Dispersion in CVD-fabricated Fibres with a Refractive Index Dip on the Fibre Axis", A.E.U. **31**, 45-50 (1977).
17. E. Khular, Kumar A., Ghatak A.K., and Pal B.P., "Effect of the Refractive Index Dip on the Propagation Characteristics of Step Index and Graded Index Fibres", Opt. Commn. **23**, 263-267 (1977).

18. B.P. Pal, "Optical Communication Fibre Waveguide Fabrication: A Review", Fibre and Integrated Optics **2**, 195-252 (1979).
19. L.B. Jeunhomme, "Single-mode Fibre Optics: Principles and Applications", Marcel and Dekker, New York (1983).
20. D.N. Payne and W.A. Gambling, "Zero Material Dispersion in Optical Fibres", Electron. Letts. **11**, 176-178 (1975).
21. S. Kabayashi, Shibata N., Shibata S. and Izowa T., "Characteristics of Optical Fibres in Infrared Wavelength Region", Rev. of ECE (NTT) **26**, 453-467 (1978).
22. L.G. Cohen, Lin C. and French W.G., "Tailoring Zero Chromatic Dispersion into 1.5-1.6 μm Low Loss Spectral Region of Single-mode Fibres", Electron. Letts. **15**, 334-335 (1979).
23. S.K. Korotky, Eisenstein G., Gnauck A.H., Kasper B.L., Veselka J.J., Alferness R.C., Buhl L.K., Barus C.A., Huo T.C.D., Stulz L.W., Nelson K.C., Cohen L.G., Dawson R.W., and Campbell J.C., "4-Gb/s transmission experiment over 117 km of optical fibre using a Ti-LiNbO$_3$ external modulator", IEEE J. Lightwave Tech. **LT-3**, 1027-1031 (1985).
24. D. Marcuse, "Interdependence of waveguide and material dispersion", App. Opt. **18**, 2930-2932 (1979).
25. B.P. Pal, Kumar A., and Ghatak A.K., "Predicting Dispersion Minimum in a Step Index Monomode Fibre: a comparison of the Theoretical Approaches", J. Optical Common. **2**, 97-100 (1981).
26. F.P. Kapron, "Maximum information capacity of fibre optic waveguides", Electron Lett, **13**, 96-97 (1977).
27. D. Marcuse and Lin C.L., "Low Dispersion Single-mode fibre Transmission the Question of Practical Versus Theoretical Maximum Transmission Bandwidth", IEEE J. Quantum Electron, **QE-17**, 869-878 (1981).
28. K. Furuya, Miyamoto M., and Suematsu Y., "Bandwidth of Single-Mode Optical Fibres", Trans. IECE Japan, **E-62**, 305-310, (1979).
29. B.P. Pal, Kumar A., and Ghatak A.K., "Effect of Axial Refractive Index Dip on Zero Total Dispersion Wavelength in Single-Mode Fibres", Electron Lett, **16**, 505-506 (1980).
30. W.A. Gambling, Matsumura H., and Ragdale C.M., "Mode Dispersion, Material Dispersion and Waveguide Dispersion in Graded Index Single-Mode Fibres", IEE Microwaves, Optics and Acoustics, **3**, 239-246, (1979).
31. H. Murata, "Handbook of Optical Fibres and Cables", Marcel Dekker, New York (1988).
32. F.P. Kapron "Transmission Properties of Optical Fibres", in "Optoelectronic Technology and Lightwave Communications systems", Ed. by Chinlon Lin, Van Nostrand, New York (1989).
33. D. Gloge, "Bending loss in Multimode Fibres with Graded and Ungraded Core Fibres", App. Opt. **11**, 2506-2513 (1972).
34. D. Marcuse, "Curvature Loss Formula for Optical Fibres", J. Opt. Soc. Am. **66**, 216-220 (1976).
35. W.A. Gambling and Matsumura H., "Propagation in Radially Inhomogeneous Single-Mode Fibres", Opt. Quant. Electron. **10**, 31-40 (1978).
36. B.P. Pal, Das U.K., and Sharma N., "Effect of Variation in Cut-Off Wavelength on the Transmission Characteristics of Single-mode Fibres in the 1300 nm and 1550 nm Wavelength Windows", J. Opt. Commn. **10**, 108-114 (1989).
37. A. Yariv and Yeh P., "Optical Waves in Crystals" J. Wiley and Sons, New York (1984).
38. R. Olshansky, "Distortion Losses in Optical Fibres", App. Opt. **14**, 20-21 (1975).

39. M.K. Barnosky (Ed.), "Fundamentals of Optical Fibre Communications", Academic Press, N.Y. (1975).
40. M.C. Hudson, "Calculations of the Maximum Coupling Efficiency into Multimode Optical Waveguides", App. Opt. **13**, 1024-1033 (1974).
41. Y. Uematsu and Ozeki T., "Efficient Power Coupling Between a MH LED and a Multimode Fibre with a Tapered Launcher", Tech. Digest, of IOOC'77, Tokyo, 371-372 (1977).
42. T. Ozeki and Kawasaki B.S., "Efficient Power Coupling Using Taper Ended Multimode Optical Fibres", Electron. Letts. **12**, 607-609 (1976).
43. S. Horiuchi, Ikeda K., Tanaka T., and Sasuki W., "A New LED Structure with a Self-aligned Spherical Lens for Efficient Coupling to Optical Fibres", IEEE Trans, **ED-24**, 986-990 (1977).
44. R. Tewari and Pal B.P., "Mode Excitation by Lenses in Single-mode Optical Fibres Having Graded Index Profiles", J. Opt. Commun. Vol. **6**, 25-29 (1985).
45. A.K. Bhatnagar, "Studies on Fibre Fusion Splicing Machine and Characterisation of Spliced Fibres", M. Tech. (Opto-electronics) Thesis, IIT Delhi (1988).
46. Mark L. Dakass, Carlson W.J., and Benasutti J.E., "Field Installable Connectors and Splice for Glass Optical Fibre Communications Systems", 12 Annual Connector Symp, Cherry Hill, New Jersey, (1979).
47. C.M. Miller, "Loose Tube Splices for Optical Fibres", Bell Syst. Tech. J., **54**, 1215-1220 (1975).
48. C.M. Miller, Mettler S.C., and White I.A., "Optical Fibre Splices and Connectors: Theory and Methods", Marcel and Dekker, New York (1986).
49. D. Marcuse, "Gaussian Approximation of the Fundamental Modes of Graded Index Fibres", J. Opt. Soc. Am., Vol. **68**, 103-109 (1978).
50. D. Marcuse, "Loss Analysis of Single-Mode Fibre Splices", Bell Syst. Tech. J. **56**, 703-718 (1977).
51. B.P. Pal and Priye V., "Comments on Microbend Loss Calculation in Single-mode Fibres", IEEE J. Lightwave Tech. **6**, 1446-1447 (1988).
52. H. Matsumura and Suganuma T., "Normalization of Single-mode Fibres Having An Arbitrary Index Profile", App. Opt. **19**, 3151-3158 (1980).
53. C.A. Millar, "Direct Method of Determining Equivalent Step-Index Profiles for Monomode Fibres", Electron. Letts. **17**, 458-460 (1981).
54. F. Alard, Jeunhomme L., and Sansonetti P., "Fundamental Mode Spot Size Measurement in Single-mode Optical Fibres", Electron. Letts, **17**, 958-960 (1981).
55. P. Divita and Coppa G., "Single-mode fibre Measurements and Standardisation", J.I.E.T.E. (India) **32**, 227-231 (1986).
56. A.K. Ghatak and Thyagarajan K., "Contemporary Optics", Plenmum, N.Y. (1978).
57. C. Pask, "Physical Interpretation of Petermann's Strange Spot Size for Single-mode Fibres" Electron. Letts. **20**, 144-145 (1984).
58. K. Petermann and Kuhne R., "Upper and Lower Limit for the Microbending Loss in Arbitrary Single-Mode Fibres", J. Lightwave Tech. **LT-4**, 2-7 (1984).
59. A. Sharma and Ghatak A.K., "Single-Mode Fibre Characteristic", J.I.E.T.E. (India) **32**, 213-226 (1986).
60. U.K. Das, Pal B.P., and Sarkar S., "Dependence of Modal Noise on the Effective Cut off Wavelength in the 1310 nm Wavelength Window", Int. J. Optoelectron. **3**, 433-437 (1988).
61. M.A. Saifi, Tang S.J., Cohen L.G., and Stone J., "Triangular Profile Single-Mode Fibre", Opt. Letts. **7**, 43-45 (1982).

62. V.A. Bhagavatula, Ritter J.E., and Modavis R.A., "Bend-Optimised Dispersion-Shifted Single-Mode Designs", IEEE J. Lightwave Tech. **3**, 954-957 (1985).
63. M. Ohasi, Kuwaki N. and Tanaka C., "Characteristics of Bend-Optimised Convex-Index Dispersion Shifted Fibre", First Optoelectronics Conf., OEC'86, Tokyo, PDP-22 (1986).
64. R. Tewari, B.P. Pal and U.K. Das, "Dispersion-Shifted Dual-Shape Core Fibre : Optimisation based on Spot Size Definitions", IEEE J. Lightwave Technol. **10**, 1-5 (1992).
65. L.G. Cohen, Marcuse D. and Mammel W.L., "Radiating Leaky Mode Losses in Single-Mode Lightguides and Depressed Index Claddings", IEEE J. Quant. Electron. **QE-18**, 1467-1472 (1982).
66. V.A. Bhagavatula, Spotz M.S., Love W.F., and Keck, D.B., "Segmented-core Single-mode Fibres with Low Loss and Low Dispersion", Electron. Letts. **19**, 317-318 (1983).
67. K. Susana and Matsuyama I., "New Optical Fibre Fabrication Method", Electron. Letts. **18**, 499-451 (1982).
68. P. Klocek, "Infrared Fibre Optics; How good they are, how good they will be and what they are good for", Lasers and Applications, 43-46 (1982).

5

Single-Mode Fibre Designs for Telecommunications[*]

ARNAB SARKAR[**]

1. INTRODUCTION

The first commercial single-mode fibre optic telecommunication systems in the US and Japan were installed in the early eighties. Since then the acceptance of this system design for trunk applications has been very rapid. Today, it is the dominant design used in trunk applications and single-mode fibres are finding increasing use in distribution systems as well. However, single-mode fibre design and its specifications are yet to be standardized. In the US, AT & T uses a fibre design that is different from ones used by the rest of the telephone companies, who use fibres designed and manufactured by Corning Glass Works. On the other hand NTT in Japan uses specifications that are significantly different. As in multimode fibres, one expects convergence of designs and specifications world wide, but this is expected to take some amount of time.

Single-mode fibre design consists of design of the glass fibre and its coating. This design has to be done in conjunction with the system design and the manufacturing process to be used. In this paper, we assume prior knowledge of the basic principles of optical signal propagation in glass fibres, characteristics of high-silica glass materials used and familiarity with the different glass fibre manufacturing processes. We shall cover the design considerations

---

[*] Reprinted from J.I.E.T.E., Vol. 32 (Special issue on Optoelectronics and Optical Communication), July-August (1986)

[**] Light Technologies, Inc, 6737 Valjean Avenue, Van Nuys, California 91406 USA; present address : Orion Laboratories, 550 Via Alondra, CAMARILO, CA 93012 USA

of single-mode fibres, compare the different designs being used today and point out the impact of process selection on the choice of design. Finally, we shall analyse the trends in single-mode fibre designs for telecommunication applications.

2. SINGLE-MODE FIBRE DESIGN CONSIDERATIONS

Single-mode fibres for trunk applications are optimized for operation at $1.3\,\mu$m wavelength. Such fibres have low loss and low dispersions not only at this wavelength range, but also have low loss at the $1.55\,\mu$m wavelength range. The two alternate designs used for this application are called matched clad and depressed clad and are shown in Figs. 1 and 2, respectively. The matched clad design is produced by OVD [3] and VAD [4] processes. Some MCVD [5] fibres are manufactured in this design as well. The depressed clad design is manufactured only by the MCVD process. Both the designs are used for same kind of applications.

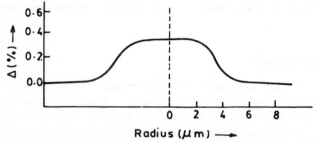

Fig. 1. Refractive index profile of an OVD matched clad single-mode fibre (after [1]).

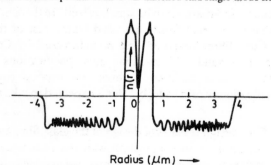

Fig. 2. Typical refractive index profile of a depressed clad single-mode fibre fabricated by MCVD process (after [2]).

In single-mode fibre design, one has to ensure that the signal is indeed propagated via a single-mode. The condition of single-mode propagation is

$$\lambda_c = 2\pi n_1 a \Delta^{1/2} / V_c, \tag{1}$$

where λ_c = cut-off wavelength (defined as the wavelength above which only the fundamental mode propagates), V_c = normalized frequency or propagation parameter V at the cut-off wavelength. The value of V_c depends on the refractive index profile of the core and is 2.404 for step index profile, n_1 is refractive index of the core, a is the core radius and Δ is the relative core-cladding refractive index difference ($= (n_1 - n_2)/n_1$).

The functional properties of single-mode fibres are attenuation, dispersion and mode-field radius. The control parameters used in fibre manufacturing, are different from the functional parameters. For ideal step-index profiles, the control parameters are relative refractive index difference (Δ), core diameter and overall fibre diameter. However, none of the fibre manufacturing processes produces ideal step-index profiles. The deviations from step-index are large enough to affect the functional properties of the fibre.

Thus, refractive index profile is an important practical control parameter for fibre manufacturing.

The theory of single-mode fibre propagation is based on ideal step-index profiles. To predict and control functional properties of fibres with arbitrary profiles, it is convenient [6] to convert the profiles in terms of equivalent step-index 'ESI' parameters (See Ch. 4). Thus for fibres with arbitrary index profiles, (1) is rewritten as

$$\lambda_c = 2\pi n \, a_{ESI} \Delta_{ESI}^{1/2} / V_c . \tag{2}$$

ESI formalism to model the propagation characteristics of single-mode fibres has been found to be very useful except probably in the prediction of dispersion properties. However, lately significant improvement in dispersion prediction has been made by modifying the ESI computation by taking into account the fourth moment of the profiles [7].

2.1 Effective cut-off wavelength

In theory, fibres used in single-mode fibre systems should have cut-off wavelength shorter than the operating wavelength all along the transmission line. If not, one would expect modal noise in the system [8]. As techniques for measurement of cut-off wavelength were being developed, it was found that the measured cut-off wavelength was dependent on sample length, as shown in Fig. 3, as well as its radius of curvature. Furthermore, this measured cut-off wavelength is significantly shorter than the theoretical cut off wavelength calculated from Eq. (2). Cut-off wavelengths measured on 2 m long relatively straight fibres are normally between 160 and 180 nm shorter than the theoretical cut-off wavelength [10]. Thus, the measured cut-off wavelength was called the effective cut-off wavelength. Initially, it was proposed that the effective cut-off wavelength of the entire fibre link had to be lower than the operating wavelength of the system to control modal noise. Others felt it is sufficient to ensure that the effective cut-off wavelength of the link is less than the operating wavelength, even if the effective cut-off wavelength of individual fibres in the

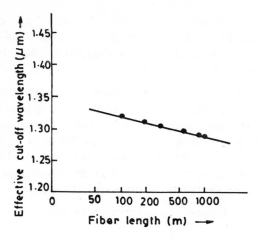

Fig. 3. Length dependence of cut off wavelength of single-mode fibres (after [9]).

link are longer than the operating wavelength. After measurement of system modal noise with fibres having high effective cut-off wavelength, Cheung et al [11] concluded that fibres with effective cut-off wavelengths as high as 1.35 μm can be used in practical systems. Use of a higher effective cut-off wavelength is particularly attractive, because it improves the microbending loss and dispersion characteristics of the fibre.

2.2 Mode field radius

Like effective cut-off wavelength, mode-field radius is a critical functional parameter for single-mode fibres. Conceptually, this parameter simply defines the near field power distribution of the propagating mode and is the controlling parameter for splice loss. Although this can be precisely calculated for a step-index fibre, Marcuse [12] used a definition for mode-field radius, w_0, as e^{-1} width of a gaussian beam which will optimally excite the single-mode fibre. He approximated w_0 by the following expression:

$$w_0/a = 0.65 + 1.619\, V^{-3/2} + 2.879\, V^{-6} . \tag{3}$$

Since normalized propagation parameter, V, is inversely proportional to wavelength, mode-field radius is also a strong function of wavelength. This functional behaviour is shown in Fig. 4.

The transmitted near-field distribution of single-mode fibres, is significantly different from that of multimode fibres. In multimode fibres, the near field is almost entirely confined to the core. In single-mode fibres, a significant amount of the signal propagates in the cladding. The fractional power propagating in the cladding is also a function of wavelength. This is shown in Fig. 5. A consequence of this is that all functional properties of single-mode fibres are wavelength dependent.

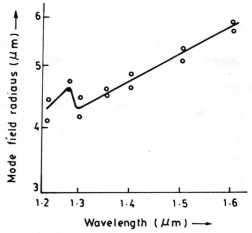

Fig. 4. Wavelength dependence of mode-field radius of single-mode fibres (after [13]).

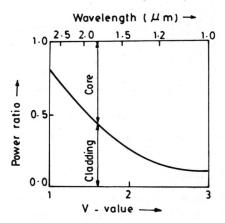

Fig. 5. Fractional power propagating in the core of a single-mode fibre as a function of its V value (after [14]).

2.3 Loss considerations

The total loss in a single-mode fibre link of length, L, can be expressed as

$$\alpha_L \,(dB) = \alpha_f L + \alpha_s (L/l - 1) + \alpha_c \tag{4}$$

where α_f = fibre loss (dB/km), α_s = unit splice loss (dB), l = distance between splices and α_c = coupling loss (dB).

Fibre loss

In today's single-mode fibres, fibre loss is primarily due to scattering loss, with small contributions from microbending and structural loss.

Scattering loss

The scattering co-efficient of single-mode fibre has been expressed as [15]:

$$\alpha(\lambda) = \left[\alpha_{core} P_{core}(\lambda) + \alpha_{clad}[1 - P_{core}(\lambda)]\right]$$

where α_{core} = scattering co-efficient of core composition, α_{clad} = scattering co-efficient of the clad composition and P_{core} = fractional power in the core.

Using this conceptual framework, the scattering co-efficient of OVD single-mode fibres with germania-silica core and silica cladding, has been accurately estimated and verified experimentally. The wavelength dependence of scattering co-efficient of OVD fibres is shown in Fig. 6.

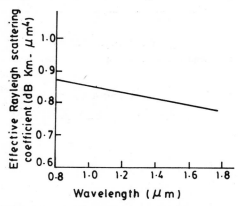

Fig. 6. Wavelength dependence of effective Rayleigh scattering co-efficient of OVD single-mode fibres (after [15]).

Bending and microbending loss

As in multimode fibres, cables are conveniently designed and manufactured without bending the fibres to a point where bending losses may become significant. Theoretical analysis of bending loss characteristics is abundant in the literature. It is suffice to say that for the existing single-mode fibre designs, bending loss is not going to be significant, provided the fibre radius of curvature is kept greater than 5 cm.

Microbending of single-mode fibres on the other hand is a very complex phenomenon, and must be understood well. Furuya and Suematsu [16] have modelled microbending of single-mode fibres and have shown good agreement with cabling losses observed in real life. Contrasting, nature of single and multi-mode fibres as to the impact of microbending loss in them are quite dramatic. In multimode fibres, one observes a wavelength independent excess loss of a small value. In single-mode fibres, a sharp increase in loss is observed at a certain wavelength. Increased microbending leads to a decrease in the low-loss operating wavelength range of the fibre. This is observed in spectral attenuation measurement at the low V value region of single-mode fibres as a function of temperature. The effect of increased microbending, at low temperature is shown in Fig. 7. A very important observation is that cabling

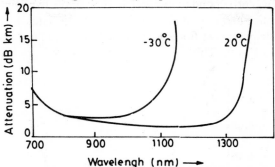

Fig. 7. Change in position of microbending loss edge as a function of the extent of microbending, caused by exposure to low temperature (after [17]).

induced microbending loss in single-mode fibres is negligible at wavelengths close to the cut-off wavelength. According to the Furuya and Suematsu model, microbending loss in the case of gaussian power spectrum can be expressed as

$$L_s = X N \overline{(1/R_1)}^2 \overline{W}^2 \Delta^{-1} \left[- Y (\overline{W} n_1 / \lambda_c)^2 \Delta^2 \right], \quad (6)$$

where N = average number of bends per metre, $\overline{(1/R_1^2)}$ = mean square of the curvature of the fibre axis, $\overline{W}$ = correlation length, λ_c = cut-off wavelength, n_1 = refractive index of core, Δ = relative refractive index difference and X and Y are wavelength dependent factors.

Coupling loss

In single-mode fibre systems, mostly laser diode sources are used. The emitted areas of these lasers are small. Therefore, one does not has to consider cross-sectional area mismatch between the laser and the fibre. Also since lasers have narrow angle of emission, the coupling losses from laser to fibre are relatively small. Source-to-fibre coupling-loss can be expressed as

$$\alpha_c = 10 \log \left[w_1 / w_0 + w_0 / w_1 \right) / 2 \right], \quad (7)$$

where w_0 = mode field radius of the fibre and w_1 = mode field radius of the source.

Splice Loss

The splice loss of single-mode fibres is governed by essentially three factors. Axial offset loss α_0, angular offset loss α_θ, and loss due to mismatch of mode field radius, α_w. Therefore splice loss, α_s, is expressed as [10]

$$\alpha_s = \alpha_0 + \alpha_\theta + \alpha_w \quad (8)$$

where $\alpha_0 = 4.3 \, (1.25/w_0)^2$,

$\alpha_\theta = 0.02 \, \theta^2$,

$$\alpha_w = 10 \log \left[(w_2/w_1 + w_1/w_2)/2 \right],$$

where w_1 and w_2 correspond to mode-field radii of the two jointed fibres. The response of splice loss due to mode field radius mismatch is shown in Fig. 8. Splice loss distributions of fibres with a mode-field radius tolerance of ± 0.5 μm is shown in Fig. 9. This is considered fully satisfactory from system considerations.

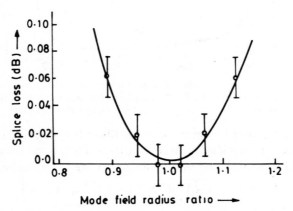

Fig. 8. Effect of mode field diameter mismatch on splice loss (after [18]).

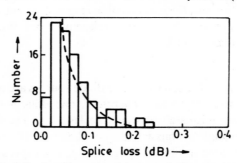

Fig. 9. Splice loss histogram of single-mode fibres (after [10]).

Loss stability

Stability of transmission loss characteristics of optical fibres was not considered an issue in this industry until recently. The first observation of fibre loss degradation in installed experimental system was in Japan. In 1983 Uchida *et al* [19] first reported this observation. Beales *et al* [20] reported similar results in experimental fibres through laboratory experiments. Both groups concluded that loss increase was an ageing phenomenon, due to migration of hydrogen into the fibre. Intensive research in the following years revealed that the sources of hydrogen in the cable structure as being silicone coating and the use of aluminium tapes. The magnitude of this effect was also found to be a function of the dopants used; co-doping of phosphorus and germania result in most adverse effects. Accelerated testing in hydrogen atmosphere and subsequent modelling of this ageing phenomenon have led to the conclusion that single-mode fibres made with proper safeguards are quite stable for long-terms field use in telecommunication [21, 22].

2.4 Dispersion considerations

The dispersive properties of single-mode fibres owe their origin to the wavelength dependence of the refractive indices of the core and the cladding glasses of the fibre structure, as well as explicit dependence of mode's propagation constant which depend on the fibre's structural characteristics and the source wavelength. Since every source exhibits some wavelength spread in its emission pattern, the above mentioned wavelength dependent effects lead to temporal broadening of the launched pulse. Accurate estimation of this broadening in a repeater section and its compatibility with the system information carrying capacity is necessary for control of bit-error rate of actual systems.

Total dispersion of a single-fibre is conventionally expressed as temporal broadening per unit length of the fibre, per unit width of the light source used. This total dispersion caused by material and structural properties of the fibre is in fact totally coupled. However, using the weakly guiding approximation Gambling et al [23] succeeded in partially separating these terms. In practice, this separation is widely used and total dispersion is expressed as

$$T = T_m + T_w, \qquad (9)$$

where T_m is material dispersion and is expressed as

$$T_m = - (\lambda/c) \left[(P_c/P)(d^2 n_1/d\lambda^2) + [1 - (P_c/P)](d^2 n_{cl}/d\lambda^2) \right] \quad (10)$$

and T_m, the waveguide dispersion as approximated by White and Nelson [24] can be expressed as

$$T_w = (4\pi n_{cl} c)^{-1} (\lambda/a^2), \qquad (11)$$

where n_1 and n_{cl} represent the core and cladding refractive indices, respectively.

The wavelength dependence of these two terms and the consequent total dispersion of a germania-silica core, silica clad nominally step index single-mode fibre is shown in Fig. 10.

2.5 Summary of design considerations

Figure 11 shows a plot of delta (Δ) versus core radius. This diagram is the conceptual framework for single-mode fibre design. But because core diameter can not be accurately measured, fibre manufacturers use delta and cut-off wavelength as the control parameters. The profile is essentially dictated by the choice of the process and one usually attempts to keep it closely controlled.

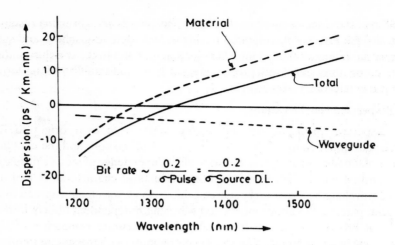

Fig. 10. Components of dispersion of single-mode fibres (after [18]).

The range of delta and cut-off wavelength that can be used to produce high-quality single-mode fibres are limited by four limits shown in Fig. 11. Attenuation limits the upper limit of delta for matched clad designs, as Rayleigh scattering co-efficient of the core goes up with increasing dopant content of the core. The dispersion limit is in the high delta and low cut-off wavelength corner of the specification box. In fact, dispersion limits the upper limit of delta to be used. This is particularly true for fibres made by OVD and VAD processes, where the core-clad interface is considerably diffused due to migration of germania during sintering. The lower limit of delta as well as the lower limit of cut-off wavelength is set by microbending considerations, where the normalized spot size (w_0/a) has the highest value. The upper limit of cut-off wavelength is of course limited by modal noise considerations already mentioned before.

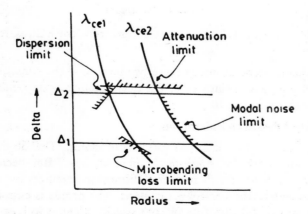

Fig. 11. Schematic of single-mode fibre design constraints.

In depressed clad designs, one has to also define two additional parameters. First, the thickness of the deposited cladding, which is primarily decided by attenuation considerations due to signal propagation in the cladding as well. Maintaining a cladding thickness to core radius ratio (t/a) of 7 ensures minimum loss. A second consideration is the magnitude of depressed index that might be permitted without having fundamental mode cut off problems causing high attenuation in the longer wavelengths. The extent of index depression permitted without affecting long wavelength attenuation is a function of the deposited cladding thickness used. For t/a of 7, index depression of more than half the fibre delta is not recommended.

3. PRESENT STATUS OF NOMINALLY STEP-INDEX SINGLE-MODE FIBRE DESIGNS

In the absence of any world-wide standardization of single-mode fibres, different users use different ranges of functional parameters, and also specify these parameters differently. The following table is the author's attempt to list the functional parameters of single-mode fibres that emerged at three different well known laboratories around the world, using a common set of terminologies.

	NTT	Corning	AT & T
Mode-field Radius (μm)	4.5—5.5	4.5—5.5	4.0—4.7
Zero Dispersion Wavelength (μm)	1.305—1.337	1.305—1.325	1.30—1.32

In order to understand the reasons for these differences, one needs to first look at the corresponding differences in the control parameters used as well as the manufacturing process used. These are listed below:

	NTT	Corning	AT & T
$\lambda_{ce}(\mu$m)	1.1—1.34	1.13—1.27	1.17—1.33
$\Delta(\%)$	0.3 ± 0.04	0.3 ± 0.04	0.37 (nominal)

Both NTT and Corning use the matched clad design. The cut-off wavelength specification of NTT reflects a recent change in upper limit from 1.28 μm to 1.34 μm based on experiments on modal noise. The total range of cut-off wavelength reflects the inherent variability cut-off wavelength of fibres made by the VAD process. The lower limit of cut-off wavelength of 1.1 μm is due to micro-bending considerations. This wide cut-off wavelength range along with a rather diffused core-clad interface of VAD single-mode fibres cause the dispersion distribution to be adversely affected. In order to use such fibres in NTT's 400 Mbts trunk line, selection of the laser was used in conjunction with span lengths to ensure efficient system performance.

The Corning specification is driven by dispersion considerations, which causes the lower limit of cut-off wavelength to be 1.13 μm. The upper limit in cut-off wavelength is yet to be changed to reflect recent improvements in understanding of modal noise considerations. This rather stringent cut-off wavelength specification is not an issue in the OVD process used by Corning, as control of refractive-index profile and dimensional tolerances are comparatively easier.

The AT & T design with its comparatively small mode field radius is based on AT & T's desire to have very high microbending protection for ease of economic cabling. However, because of use of depressed clad design in conjunction with the MCVD process, AT & T does not have to pay any attenuation penalty. This is because reduction of mode field radius is accomplished by reducing the refractive index of the cladding, which does not increase the scattering loss of the fibre. Furthermore, AT & T has demonstrated excellent splicing performance with fibres of this design. Their range of zero dispersion wavelength was selected based on the distribution of operating wavelength of their lasers. The longer cut-off wavelength range was helpful in controlling the zero dispersion wavelength within the desired range.

In summary, all the designs are functional. In the future, one expects further convergence in the specifications of the functional parameters. Acceptance of 1.33 µm as the upper limit of effective cut off wavelength is expected to be widespread, as it improves the dispersion characteristics of the fibre and at the same time reduces sensitivity to microbending induced losses.

4. TRENDS IN SINGLE-MODE FIBRE DESIGN

4.1 Silica-core fluorosilicate-clad fibre

This fibre design is potentially the lowest loss oxide glass fibre. There are two alternative refractive index profiles for this design. Matched cladding [25] and one called depressed index design, where part of the cladding is the MCVD substrate silica tube [26]. This is shown in Fig. 12. The best performance fibres of this design has been made by the MCVD process, but such fibres have also been made by the OVD and the VAD processes.

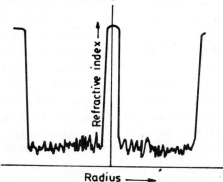

Fig. 12. Typical refractive index profile of a depressed index single-mode fibre, fabricated by the MCVD process (after [25]).

This design has the lowest scattering loss co-efficient of all oxide glass fibres, and its low-loss potential has been realized, with demonstrated loss at 1.57 μm of 0.15 dB/km [26]. The design can be made more microbending insensitive by raising its refractive index difference, without paying any penalties in attenuation. In this design, the zero dispersion is in the 1.3 μm wavelength range. If one assumes that all systems will eventually move to the 1.55 μm wavelength range to take advantage of the lowest loss window of oxide fibres, then large-scale use of this design hinges on successful development and availability of narrow linewidth laser diodes at 1.55 μm. If the operating wavelength continues to be 1.3 μm, it is safe to say that this design will dominate as soon as such fibres will meet today's fibre performance level in production.

4.2 Dispersion-shifted fibres

Design objective of such fibres is to have low dispersion and low loss in the 1.55 microns wavelength range. The alternative design approaches that permit these characteristics are many. The ones that have demonstrated the best chance of commercial use are shown in Figs. 13 and 14, respectively.

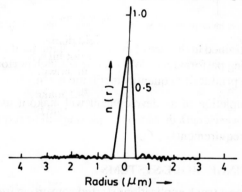

Fig. 13. Refractive index profile of triangular core dispersion-shifted single-mode fibre (after [27]).

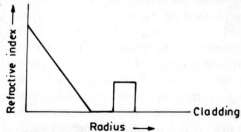

Fig. 14. Refractive index profile of SEGCOR dispersion-shifted fibre (after [28]).

The lowest loss in these designs to date have been higher than those obtained in the silica core fluorosilicate clad fibres mentioned above. These designs are also significantly more complex than today's step index fibres where need for control of refractive index profile is much less. However, fibres of this design are now commercially available and so are single-mode lasers at 1.55 μm. The degree of penetration of such systems into telecommunication trunk applications is still difficult to predict.

4.3 Dispersion-flattened fibres

The objective of this design is to have low loss and low dispersion in both the 1.3 and 1.55 μm wavelength ranges. Refractive index profiles of such fibres is shown in Fig. 15. This design is also very complex in structure. Moreover, the design tolerances for dispersion control are extremely stringent. In addition,

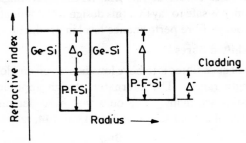

Fig. 15. Refractive index profile of SEGCOR dispersion-flattened fibre (after [29]).

the design is constrained in the size of mode-field diameter it can have, and this might affect splicing performance. Whether or not high performance fibres of this design can be produced in quantity is still unknown.

Successful completion of its development will make it useful for systems where large-scale wavelength division multiplexing will be required to meet the systems capacity requirements.

5. TRENDS IN SINGE-MODE SYSTEMS

Starting in 1982, most trunk applications started converting from multimode to step index single-mode fibres operating at 1.3 μm. Today few trunk applications are being planned for use with multimode fibres. Initial single-mode systems were 140 Mb/s, today most systems are 400-565 Mb/s. In the future upto 2 Gb/s systems are anticipated. In the near future, system operating wavelength may have to move to 1.55 μm. Also, the trend in glass composition research is in ultra low-loss fluoride glasses for operation in 2–10 μm wavelength range. A trend in system research is in coherent communications, for which single-polarization single-mode fibres (See Ch. 6) are being developed. At this point, it is difficult to predict which of these concepts and to what extent, would replace today's single-mode fibres.

6. CONCLUSION

Single-mode fibres have been instrumental in realizing essentially repeaterless optical communication systems. It has also provided predictable system performance, the lack of which had plagued multimode fibre systems. These systems are economically attractive, both from installed cost as well as from the point of view of maintenance cost.

The evolution of fibre and system design in expected to continue in the foreseable future. Increased penetration of such single-mode fibre systems in telecommunications is also expected to continue, eventually leading to widespread use of integrated communication systems.

REFERENCES

1. G.E. Berkey and A. Sarkar, *Proc OFC*, 1982, Phoenix.
2. W.T. Anderson, *et al., Proc OFC*, 1984, New Orleans.
3. D.B. Keck, P.C. Schultz and F. W. Zimar, U.S. Patent 3,737,393.
4. T. Izawa, *et al., Proc Fourth ECOC*, 1978, Genoa. Dec. 1983.
5. W.G. French, J.B. MacChesney, P.B. O'Conner and G.W. Tasker, *BSTJ*, vol 53, 1974, pp 951-954.
6. C.A. Millar, *Elec Lett*, vol 17, pp 458-460, 1981.
7. F. Martinez and C.D. Hussey, *ibid.*, vol 20, 1984.
8. S. Heckmann, *ibid.*, vol 17, pp 499-500, 1981.
9. Y. Kitayama and S. Tanaka, *Proc OFC*, 1984, New Orleans.
10. M. Tateda *et al., J. Elec Comm Soc (Japan)* vol J65, pp 324-331, 1982.
11. N.K. Cheung *et al., Elec Lett*, vol 21, pp 5-6, 1985.
12. D. Marcuse, *BSTJ*, vol 46, pp 703-718, 1977.
13. J.P. Pocholle *et al., Proc IOOC*, 1983, Tokyo.
14. M. Kawachi *et al., Elec Lett*, vol 13, pp 442-443, 1977.
15. V.A. Bhagavatula *et al., Proc OFC*, 1984, New Orleans.
16. K. Furuya and Y. Suematsu, *Appl Opt*, vol 19, 1980, pp 1493-1500.
17. A. Sarkar, to published.
18. D.B. Keck, Minitutorial *Proc OFC*, 1985, San Diego.
19. N. Uchida *et al., Proc ECOC*, 1983, Genova.
20. K.J. Beales *et al., Elec Lett*, vol 19, 1983, pp 917-919.
21. E. Miles *et al., Proc OFC*, 1984, New Orleans.
22. M. Ogani, *et al., Proc OFC*, 1984, New Orleans.
23. W.A. Gambling *et al., Microwave, Optics and Acoustics*, vol 3, 1979, pp 239-246.
24. K.I. White and B.P. Nelson, *Elect Lett*, vol 15, pp 396-397, 1979.
25. A.J. Morrow *et al.*, Ch. 2, *Optical Fibre Communications*, vol 1, Ed. T. Li, Academic Press, 1985, p 93.
26. R.C. Csencsits *et al., Proc OFC*, 1984, New Orleans.
27. M.A. Saifi *et al., Opt Lett*, vol 7, p 43, 1982.
28. V.A. Bhagaratula, *et al., Proc OFC*, 1985, San Diego.
29. V.A. Bhagavatula, *Proc OFC*, 1984, New Orleans.

6

Polarisation Maintaining Fibres and their Applications

ARUN KUMAR[*]

---

1. INTRODUCTION

The conventional single-mode fibres can in fact support two modes simultaneously which are orthogonally polarised as shown in Fig. 1. In an ideal circular-core fibre, these two modes should propagate with same phase-velocity; however, practical fibres are not perfectly circularly symmetric. As a result, the two modes propagate with slightly different phase velocities. Further, environmental factors such as bend, twist and anisotropic stress etc. also produce birefringence in the fibre, the direction and the magnitude of which keeps changing with time due to change in the ambient conditions. The above mentioned factors also couple energy from one mode to the other mode of the fibre, which can create the following problems in practice.

1. Since both the modes propagate with different group velocities, the coupling between the two modes causes the so-called polarisation mode dispersion which can limit the ultimate bandwidth of a single-mode optical communication system.

2. Since both the modes propagate with different phase velocities, the state of polarisation (SOP) of the output light keeps changing randomly with time. This is not desirable in many applications such as interferometric sensors, coherent transmission systems and for coupling to integrated optical circuits where the output SOP should be strictly maintained. One of the solutions of the above mentioned

---

[*] Department of Physics, Indian Institute of Technology Delhi, New Delhi - 110016

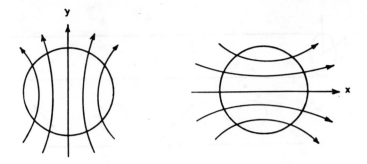

Fig. 1. States of polarisation of two orthogonally polarised modes of a conventional Single-Mode Fibre.

problems is to use a fibre which can maintain the SOP of the incident light over large distances. Such fibres are known as Polarisation Maintaining Fibres (PMF). Here we will discuss the different types of polarisation maintaining fibres and their polarisation characteristics.

2. DIFFERENT TYPES OF POLARISATION MAINTAINING FIBRES

Polarisation maintaining fibres can broadly be divided into the following two categories (i) high birefringent fibres and (ii) single polarisation single-mode (SPSM) fibres.

2.1 High Birefringent Fibres

In the case of high-birefringent fibres, the propagation constants of the two orthogonally polarised modes are made quite different from each other so that the coupling between the two modes is greatly reduced. Thus if the light is coupled to one of the polarised modes only, most of it remains in the same polarised mode and hence the SOP of the propagating light is maintained along the fibre. The polarisation holding capacity of a birefringent fibre is measured in terms of its beat length L_p which is defined as

$$L_p = \frac{2\pi}{(\beta_x - \beta_y)} = \frac{\lambda}{B}, \tag{1}$$

where β_x and β_y represent the propagation constants of the two orthogonally polarised (say along x and y directions) modes and B which is also known as the birefringence of the fibre, represents the difference between the effective indices seen by x and y polarised modes at wavelength λ. A lower value of L_p means a higher value of $(\beta_x - \beta_y)$ and hence corresponds to a fibre with high polarisation holding capacity. Physically, L_p represents the distance along the fibre during which the phase difference between the two fundamental modes becomes 2π. Thus if light is coupled to both the fundamental modes of such a

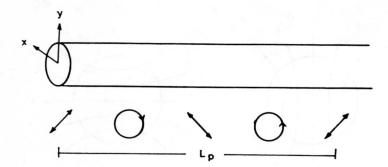

Fig. 2. Evolution of polarisation state of light guided along a birefringent fibre when both x and y-polarised modes are equally excited.

fibre, the state of polarisation will be repeated everytime after a distance L_p along the fibre (see Fig. 2).

Cross talk is an another important parameter of a PMF which indicates its practical polarisation holding capacity. It is a measure of the power coupled to the orthogonal polarisation due to random coupling along the fibre when power is launched into one of the polarised modes at the input end. For example, if P_x represents the power launched into the x-polarised mode at the input end and P_y represents the power detected in the y-polarised mode at the output end, then the cross-talk (CT) is defined as

$$CT = 10 \log \left(\frac{P_y}{P_x} \right). \tag{2}$$

The unit of CT is decibels (dB) and it has a negative value. A higher absolute value of CT represents better polarisation holding capacity of the fibre.

Coupling between the two orthogonally polarised modes is also measured in terms of what is known as the mode-coupling parameter (h) of a PMF which is defined as

$$\frac{P_y}{P_x} = \tanh(hl), \tag{3}$$

l being the length of the fibre.

The various types of most commonly used high-birefringent fibres are discussed below:

2.1.1 Elliptical Core Fibres

Such fibres have an elliptical core embedded in a circular cladding as shown in Fig. 3. The elliptical core of such fibres creates both geometrical anisotropy as

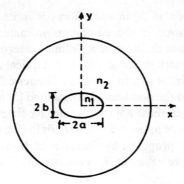

Fig. 3. Transverse cross-section of an elliptical core fibre.

well as asymmetrical stress in the core. As a result the propagation constants (β_x and β_y) of the two fundamental modes polarised along the major axis (x-direction) and minor axis (y-direction) respectively become different. The total birefringence is thus a sum of geometrical birefringence B_g (due to non-circular shape of the core) and stress induced birefringence B_s (due to asymmetrical stress produced by elliptical core).

(i) Geometrical Birefringence B_g

The geometrical birefringence of such fibres depends on the aspect ratio a/b of the core as well as on the refractive index difference Δn [$= (n_1 - n_2)$] between the core and the cladding. In the vicinity of first higher mode cut-off, the birefringence for fibres with small core ellipticities ($a/b < 1.2$) is approximately given by (as obtained by curve fitting to Fig. 1 of [1])

$$B_g \simeq 0.28 \left(\frac{a}{b} - 1\right) (\Delta n)^2. \tag{4}$$

The above equation shows that the birefringence of such fibres can be increased either by increasing core ellipticity or by increasing Δn. However it should be mentioned that neither a/b nor Δn can be increased beyond a certain limit because of the following reasons (i) the birefringence is not a sensitive function of a/b for $a/b > 2$ and for $a/b \gg 2$ it saturates to a value which is approximately given by [1]

$$B_g \simeq 0.32 \, (\Delta n)^2, \tag{5}$$

and (ii) for very large values of Δn, the core dimensions required for the single mode operation of the fibre become extremely small and the fabrication of the fibre becomes very difficult. That is why such fibres were not considered of much use initially. However, the shift to longer wavelengths for long distance communication systems has generated a great deal of interest in such fibres in recent years.

The exact calculation of B_g in such fibres is quite difficult as the wave equation has to be solved in the elliptic co-ordinates, and the eigenvalue equation so obtained is in the form of an infinite determinant [2]. Dyott et al. [3] has obtained the birefringence of different elliptical core fibres by truncating the infinite determinant to a finite order. Numerical techniques like finite element method [4] and the point matching method [5] have also been used by various authors, however, these approaches involve time consuming numerical calculations. A number of approximate methods [6-8] have also been reported amongst which the one proposed by Kumar et al. [8] is relatively simple and gives reasonably accurate values of B_g in the region of practical interest and is discussed here.

The authors [8] have shown that the birefringence of an elliptical-core fibre matches very well with that of a rectangular core waveguide having the same core area, aspect ratio and core-cladding refractive indices. Thus, the birefringence of an elliptical core fibre with major and minor axes as $2a$ and $2b$ respectively, is approximately equal to that of a rectangular core waveguide with major and minor axes given by

$$2a' = \sqrt{\pi}\, a \quad \text{and} \quad 2b' = \sqrt{\pi}\, b \tag{6}$$

and with the same core and cladding indices (n_1 and n_2) respectively. The propagation constants β_x and β_y for the two fundamental modes of this rectangular core waveguide can be obtained by using an accurate perturbation approach [9] and are given by,

$$\beta_i^2 = \left(\beta_{1i}^2 + \beta_{2i}^2\right) + k_0^2 \left(n_1^2 - n_2^2\right) P_i^2, \tag{7}$$

where i stands for x or y and β_{1i} and β_{2i} are obtained by solving the following two simple transcendental equations

$$\mu_1 \tan\mu_1 = c \left(v_1^2 - \mu_1^2\right)^{\frac{1}{2}} \tag{8}$$

and

$$\mu_2 \tan\mu_2 = \frac{1}{c} \frac{n_1^2}{n_2^2} \left(v_2^2 - \mu_2^2\right)^{\frac{1}{2}}, \tag{9}$$

where

$$c = \frac{n_1^2}{n_2^2} \text{ (or 1)}, \quad \text{for } i = x \text{ (or } y)$$

$$\mu_1 = a' \left[k_0^2 \frac{n_1^2}{2} - \beta_{1i}^2\right]^{\frac{1}{2}}, \quad \mu_2 = b' \left[k_0^2 \frac{n_1^2}{2} - \beta_{2i}^2\right]^{\frac{1}{2}}$$

$$v_1 = k_0 a' \left[n_1^2 - n_2^2\right]^{\frac{1}{2}}, \quad v_2 = \frac{b'}{a'} v_1$$

and

$$P_i^2 = \left[1 + \frac{1}{c^2}\left(\frac{v_1^2}{\mu_1^2} + 1\right)\left(\frac{2\mu_1 + \sin 2\mu_1}{1 + \cos 2\mu_1}\right)\right]^{-1}$$
$$\times \left[1 + c^2\frac{n_2^4}{n_1^4}\left(\frac{v_2^2}{\mu_2^2} + 1\right)\left(\frac{2\mu_2 + \sin 2\mu_2}{1 + \cos 2\mu_2}\right)\right]^{-1} \quad (10)$$

The birefringence $B_g[= (\beta_x - \beta_y/k_0]$ can now be obtained by obtaining β_x and β_y corresponding to E_{11}^x and E_{11}^y modes respectively, using equations (7) to (10).

Figure (4) shows the variation of the normalised birefringence $\tilde{\Delta\beta} = B_g/(\Delta n)^2$ as a function of V-parameter $[= k_0 b \, (n_1^2 - n_2^2)^{\frac{1}{2}}]$ for elliptical core fibres having $n_2 = 1.47$ and $\Delta [= (n_1^2 - n_2^2)/2n_1^2] = 1\%$ and with different ellipticities. Solid curves correspond to the numerical calculations of Dyott et al. [3] and the broken-line curves correspond to the above mentioned approach obtained by using equations (7) to (10). The vertical lines denote the first higher mode cut off taken from Reference [10]. As can be seen from the figure, the two curves match very well in the region of practical interest namely for V-values corresponding to maximum birefringence and just before the first higher mode cut off. It is also observed [8] that the values of the maximum birefringence obtained by this method for various elliptical-core fibres matches

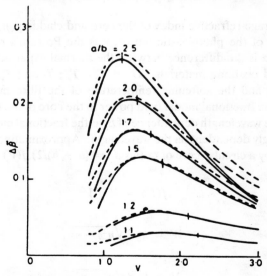

Fig. 4. Variation of normalised birefringence $\tilde{\Delta\beta}$ as a function of the V-parameter $[= k_0 b \left(n_1^2 - n_2^2\right)^{\frac{1}{2}}]$ for elliptical core fibres with different core ellipticities and with $n_2 = 1.47$ and $\Delta [= (n_1^2 - n_2^2)/2n_1^2] = 1\%$. The vertical lines denote the first higher mode cut-off. (reproduced with permission of IEE, Stevenage from ref. [8]).

very well with those obtained by Okamoto *et al.* [5] using point-matching method.

As mentioned earlier, the core dimensions of such fibres required for single mode operation with good polarisation holding capacity, become extremely small. This problem can partly be overcome by introducing a depressed inner cladding in the index profile [11] i.e. by making the profile *w*-type as the cut-off *V*-value of the first higher mode is larger in *w*-type fibres.

(ii) Stress Birefringence B_s

Since the thermal expansion coefficients of the core and cladding materials are generally different, an asymmetrical stress is frozen in elliptical core fibres during their fabrication. This stress in turn introduces a linear birefringence B_s in the fibre through the elasto-optic effect. The amount of this birefringence is obtained by calculating the difference between the stresses generated at the fibre centre along major and minor axes of the fibre and is given by [12-15]

$$B_s = f(V) \frac{c_0}{1-\nu} \Delta\alpha \, \Delta T \, \frac{(a-b)}{(a+b)}, \tag{11}$$

where

$$c_0 = \frac{n_0^3}{2} (p_{11} - p_{12})(1+\nu),$$

n_0 is the average refractive index of the core and cladding, p_{11} and p_{12} are the components of the photoclastic tensor, ν is the Poisson's ratio of the fibre material, $\Delta\alpha$ is the difference between the thermal expansion coefficients of the core and cladding materials, $\Delta T = T_r - T_s$; T_r and T_s being the room temperature and the softening temperature of the fibre material and $f(V)$ represents the fractional modal field power in the core of the fibre. The factor $f(V)$ gives the wavelength dependence of B_s as the fractional modal power in the core is strongly dependent on the wavelength. Approximating the elliptic core of the fibre by a circular core of radius $d \, (= (a+b)/2)$, $f(V)$ is approximately given by,

$$f(V) = W^2/V^2 \tag{12}$$

with

$$W = k_0 d \sqrt{n_{\text{eff}}^2 - n_2^2},$$

and

$$V = k_0 d \sqrt{n_1^2 - n_2^2},$$

where n_{eff} represents the average effective index of the fundamental modes.

2.1.2 Side-Pit and Side-Tunnel Fibres

Side-pit fibres were first proposed by Okoshi and co-workers [16]. In these fibres also, birefringence is obtained by introducing the geometrical anisotropy

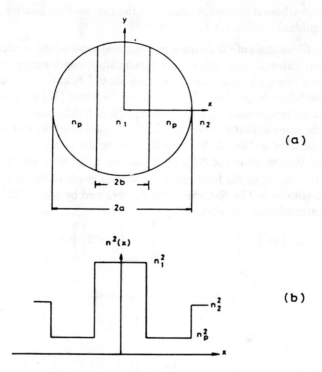

Fig. 5. (a) Transverse cross-section and (b) the refractive index distribution along x-axis of a side-pit fibre.

in the core. Figure 5 shows the refractive index distribution in a transverse cross-section of such fibres. As shown in the figure, the fibre-core consists of two-pits of refractive index n_p (< cladding index n_2) on each side of the central core. Thus the fibre has a W-type index profile along the x-axis and a step-index profile along the y-axis. If the two index pits are hollow (i.e. $n_p = 1$), the fibre is known as a Side-Tunnel fibre.

Due to the complexity of the refractive index distribution, no analytical method is available in order to analyse such fibres. As a result, Okoshi et al. [17, 18] have used the finite element method while Hayata and co-workers [19] have used a H-field finite-element method in order to estimate the polarisation characteristics of such fibres. Effective index method [20] has also been used to analyse these fibres; however, these methods are cumbersome and involve time consuming calculations. For Side-tunnel fibres, a relatively simple but approximate method was given by Varshney and Kumar [21]. In this method, the authors have approximated a side-tunnel fibre by an appropriate rectangular-core waveguide, the polarisation characteristics of which are obtained by a simple perturbation approach similar to the one used to analyse elliptical-core waveguides (see Sec. 1.1.1).

In these fibres the cut-off frequencies of x and y-polarised modes (V_{cx} and V_{cy}, say) are different and thus over a certain wavelength range they should support only one polarised mode (i.e. they should behave as truly Single-Polarisation Single Mode Fibres). Kaul and co-workers [22] have pointed out that the cut off frequencies V_{cx} and V_{cy} in side-pit and side-tunnel fibres should be nearly the same as that of a W-type planar waveguide whose refractive index profile is the same as that of the given fibre along the x-axis (as shown in Fig. 5b). Hence the cut off of the HE_{11}^x (HE_{11}^y) mode of the fibre should be determined by the cut off of the fundamental TM(TE) mode of the W-type planar waveguide (shown in Fig. 5b) which can be obtained by solving the following simple transcendental equation:

$$\tan\left(V_c \frac{b}{a}\right) + \eta \gamma \tanh\left[\gamma V_c \left(\frac{b}{a} - 1\right)\right] = 0, \tag{13}$$

where

$$\eta = \frac{n_1^2}{n_2^2} \quad \text{for TM mode,}$$
$$= 1 \quad \text{for TE mode}$$

and

$$\gamma = \left[\frac{n_2^2 - n_p^2}{n_1^2 - n_2^2}\right]^{\frac{1}{2}},$$

n_1, n_2 and n_p are the refractive indices of the central core, cladding and the pit-regions, respectively and V_c represents the value of the normalised frequency $V\left[= k_0 a (n_1^2 - n_2^2)^{\frac{1}{2}}\right]$ at cut-off. This gives a simple method for calculating the relative Single-Polarisation bandwidth S, defined as

$$S = 2\left[\frac{V_{cx} - V_{cy}}{V_{cx} + V_{cy}}\right] \tag{14}$$

for side-pit and side-tunnel fibres.

When the side-pit fibres were first proposed [16] it was predicted that the birefringence of these fibres should be quite large ($B \simeq 1-2.5 \times 10^{-4}$). Later on it was also predicted [18] that in a side-tunnel fibre with optimised channels, even larger birefringence ($B \simeq 4.5 \times 10^{-4}$) can be obtained. However, in practical side-pit and side-tunnel fibres, the birefringence is found to be $\simeq 5 \times 10^{-5}$ and $\simeq 7 \times 10^{-5}$ respectively, which is less than that of elliptical core fibres.

2.1.3 Stress Induced Fibres

A more effective method of introducing high birefringence in optical fibres is through introducing an asymmetric stress with two-fold symmetry, in the core of the fibre. The stress changes the refractive indices of the core (due to photoelastic effect) seen by the modes polarised along the principal axes of the fibre and generates birefringence. The required stress is obtained by introducing two identical stress applying parts (SAPs) centered on a diameter, one on each side of the core. The SAPs have different thermal expansion coefficient (α_3) than that of the cladding material (α_2) due to which an asymmetrical stress is applied on the fibre core after it is drawn from the preform and cooled down. The most common shapes used for the SAPs are (i) Bow-tie shape [23] and (ii) circular shape [24-26]. The corresponding fibres are known as the 'Bow-tie' and PANDA (Polarisation-maintaining and Absorption reducing) fibres respectively. The cross sections of these two types of fibres are shown in Fig. 6.

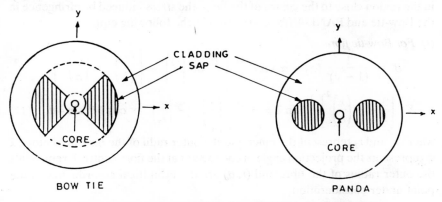

Fig. 6. Transverse cross-sections of Bow-tie and PANDA stress-induced fibres.

Modal Birefringence

As mentioned above the asymmetric stress changes the refractive indices seen by the modes polarised along the principle axes (say x and y) of the fibre. The effective refractive indices (n_x and n_y) seen by the x and y polarised modes respectively can be represented by the following equations:

$$n_x = n_{x0} + c_1 \sigma_x + c_2 \sigma_y + c_2 \sigma_z$$
$$n_y = n_{y0} + c_2 \sigma_x + c_1 \sigma_y + c_2 \sigma_z, \tag{15}$$

where $\sigma_x, \sigma_y, \sigma_z$ are the stresses generated along the x, y and z axis respectively, c_1 and c_2 are the photoelastic coefficients and n_{x0} and n_{y0} are the effective refractive indices of x and y-polarised modes in the absence of any stress. The modal birefringence B would then be given by:

$$\begin{aligned} B &= n_x - n_y \\ &= (n_{x0} - n_{y0}) + (c_1 - c_2)(\sigma_x - \sigma_y) \\ &= B_g + B_s. \end{aligned} \tag{16}$$

The first term naturally represents the shape (also known as geometrical) birefringence while the second term represents the stress induced birefringence. In the case of a circular-core fibre, the geometrical birefringence is negligibly small and hence the modal birefringence is mainly determined by the second term only, which can be obtained by calculating the distribution of the difference of stresses σ_x and σ_y in and around the core.

The stress distribution due to SAPs in these fibres is quite cumbersome to obtain. Okamota and co-workers [27] have given a finite-element method to obtain this complicated stress distribution while Varnham *et al.* [28] and Chu and Sammut [29] have given analytical methods for the calculation of thermal stress distribution and hence the birefringence for such fibres. It is shown that in the region close to the centre of the fibre, the stress induced birefringence in the Bow-tie and PANDA fibres are given by the following equations

(i) For Bow-tie fibres

$$B \simeq \frac{1}{(1-\nu)} (\alpha_2 - \alpha_3) \Delta T \frac{c_0}{\pi} \left[\left\{ 2 \ln \left(\frac{r_2}{r_1}\right) - \left[\left(\frac{r_2}{b}\right)^4 - \left(\frac{r_1}{b}\right)^4\right] \right. \right.$$
$$\left. \left. \times \left[\frac{3}{2} - 3\left(\frac{r}{b}\right)^2\right] \right\} \sin 2\varphi + \frac{3}{2} \left\{ \left(\frac{r}{r_1}\right)^2 - \left(\frac{r}{r_2}\right)^2 \right\} \cos 2\theta \, \sin 4\phi \right], \quad (17)$$

where r_1 and r_2 represent the inner and the outer radii of the bow-tie sectors, 2ϕ represents the projection angle of each sector at the fibre centre, b represents the outer radius of the fibre and (r, θ) are the cylindrical co-ordinates of the point under consideration.

(ii) For PANDA fibres:

$$B \simeq \frac{2c_0}{(1-\nu)} (\alpha_2 - \alpha_3) \Delta T \left(\frac{d_1}{d_2}\right)^2$$
$$\times \left[1 - 3\left\{1 - 2\left(\frac{r}{b}\right)^4\right\} \left(\frac{d_2}{b}\right)^4 + 3\left(\frac{r}{d_2}\right)^2 \cos 2\phi \right], \quad (18)$$

where $d_1 \,[\, = (r_2 - r_1)/2]$ represents the radius of the SAPs, $d_2 \,[\, = (r_2 + r_1)/2]$ represents the distance between the fibre centre and the centre of either of the SAPs; r_1 and r_2 being the distances of the closest and the farthest edges of SAPs from the fibre centre, 2ϕ is the projection angle of each sector at the fibre centre and (r, θ) again are the co-ordinates of the point under consideration.

A more clear picture about the dependence of the stress induced birefringence on the various fibre parameters can be obtained by calculating the birefringence at the centre ($r = 0$) of the fibre as follows:

(i) For Bow-tie fibre

$$B(r=0) \simeq \frac{1}{(1-\nu)} (\alpha_2 - \alpha_3) \Delta T \frac{c_0}{\pi} \left[2 \ln \left(\frac{r_2}{r_1}\right) - \frac{3}{2b^4} (r_2^4 - r_1^4) \right] \sin 2\phi, \quad (19)$$

which shows that B is maximised for $2\phi = \pi/2$ (the optimum sector angle). Further, B increases as r_1 decreases which means that the stress applying sectors should be as close to the core as possible. The birefringencé also increases by increasing r_2; however it reaches a maximum for $r_2 = 0.76\,b$ [28], beyond which if r_2 is increased, the birefringence B starts decreasing.

(ii) For PANDA fibre

$$B(r=0) = \frac{2c_0}{(1-\nu)}(\alpha_2 - \alpha_3)\,\Delta T \left(\frac{d_1}{d_2}\right)^2 \left[1 - 3\left(\frac{d_2}{b}\right)^4\right] \quad (20)$$

which shows that B can be increased by decreasing d_2, again indicating that SAPs should be as close to the core as possible. In this case also B can be increased by increasing r_2 (or d_1) upto an optimum value. The optimum value of r_2 is again less than the outer radius b and is given by [29],

$$\left(\frac{r_2}{b}\right) = r_0\left[1 + \frac{2R_1^{1/4}r_0^{3/4} - R_0 - r_0}{(R_1 + 5r_0)/4}\right], \quad (21)$$

where $R_1 = (r_1/b)$ and $r_0 = (1/3)^{1/4} = 0.76$ is the optimum value of (r_2/b) for Bow-tie fibres.

As discussed above, placing the SAPs closer to the core improves the birefringence of these fibres, however, they cannot be placed arbitrarily close to the core since this increases the loss of the fibre as SAPs are doped with materials other than silica. Stress-induced fibres are superior to any other type of birefringent fibre in terms of loss and polarisation holding capacity. Fibres with very low-loss and low cross-talk have been reported using optimum design parameters. For example, Sasaki *et al.* [30] have recently reported the fabrication of a PANDA fibre having 0.22 dB/km loss with birefringence $B = 3.3 \times 10^{-4}$ and cross-talk of -27 dB in a 5 km length.

2.1.4 Circularly Birefringent Fibres

All the fibre-types discussed above are linearly-birefringent fibres, because the two orthogonally polarised modes of such fibres are linearly polarised. However, it is also possible to introduce circular-birefringence in the fibre so that the two orthogonally polarised modes of the fibre are right and left-circularly polarised. The simplest way to achieve circular birefringence in a fibre is to twist it.

If a fibre is twisted, it becomes circularly birefringent [31]. It is shown theoretically that if ϕ represents the twist rate per unit length then the plane of polarisation of a linearly polarised light is rotated in the same direction with a rate of $\alpha\phi$ where $\alpha \simeq 0.07$ for silica [32]. This means that a phase difference is generated between the two circularly (right and left) polarised modes and is given by

$$\Delta\beta = 2\alpha\phi. \quad (22)$$

However, the birefringence so generated is very small as α is small. It is very difficult to obtain beat lengths less than about 10 cm using this method as the fibre will break at higher twist rates. Further, a twisted fibre is very difficult to handle.

Another way of achieving circular birefringence in a fibre is to make its path helical. In such fibres, the core follows a helical path inside the cladding as shown in Fig. 7. As a result, the propagating light is constrained to move along a helical path and experiences an optical rotation [33, 34] which is given by

$$L_r = \frac{P^3}{2\pi^2 Q^2}, \qquad (23)$$

where P is the helical pitch length, Q is the offset of the core from the fibre axis and it is assumed that $P \gg Q$.

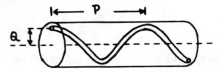

Fig. 7. A helical core fibre.

Although the birefringence is achieved only by the geometrical effects in such fibres, a large birefringence $B = 2 \times 10^{-4}$ has been reported to be achieved by Birch [35]. Such fibres are found to remain single-moded upto very large V-values ($\simeq 25$), as the higher order modes have high losses due to the small radius of curvature of the fibre. This means that the fibre core can have a large-core diameter and high launching efficiency can be achieved. Further, they are found to be less sensitive to temperature changes. These fibres should find applications in sensing electric currents through Faraday effect.

2.2 Single Polarisation Single-Mode (SPSM) Fibres

In the above, we have discussed high birefringence fibres in which the SOP of a light beam is maintained by reducing the coupling between the two fundamental modes. However, cross-talk degrades with increasing the length of such fibres due to random coupling between the two modes along the fibre length. Thus for very large distances, the fibres which support only one polarised mode (known as Single Polarisation Single Mode Fibres) seem to be more promising. A large number of schemes have been reported to achieve the SPSM fibres which can broadly be divided in the following two types:

(i) Fibres with one of the modes unguided

In this type of fibres, the two fundamental modes have different cut off wavelengths. Thus at wavelengths lying in between the two cut offs, one of the modes is not guided and the fibre behaves as a Single Polarisation Single Mode fibre. Such fibres can be achieved using geometrical anisotropy as well as stress induced anisotropy and are discussed below.

Side-pit and side-tunnel fibres [16, 17, 36] are the examples of SPSM fibres which are based mainly on the geometrical anisotropy. The predicted single polarisation bandwidth of side-pit fibres is 2.5% and for side-tunnel fibres $\simeq 7\%$. However, in practice, they are merely high-birefringent fibres as the predicted values for the single polarisation bandwidth are not yet obtained [17, 36].

Eickhoff [12] has proposed a way to make one of the modes unguided through stress effect. He has shown that for fibres with elliptical core or with elliptical inner cladding, the stress induced index profiles for the x and y polarised modes are different. Further, the direction of the fast axis of the stress induced birefringence depends on the photo-elastic constants and the thermal expansion coefficients of the fibre materials [see Eq. (11)]. Thus by selecting the fibre materials appropriately, it should be possible that the stress induced index profile cancels the dopant induced index profile (which should be the same for both x and y polarised modes) for one of the polarised modes so that the fibre does not act like a waveguide for this mode. Two such possibilities are shown schematically in Fig. 8. In the first case (see Fig. 8a), dopant induced index profile is a negative index-step while the stress induces the positive index steps such that for y-polarised mode, the two profiles cancel

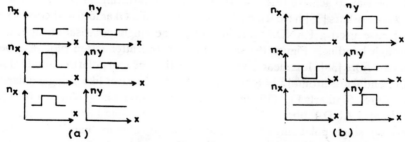

Fig. 8. Dopant induced, stress induced and the resultant index profiles in two possible schemes (a) and (b) proposed (after [12]) for achieving SPSM fibres.

each other while the stress-induced positive index step for the x-polarised mode is not exactly cancelled. As a result, the x-polarised mode is not guided. Similarly, the x-polarised mode can also be made unguided (as shown in Fig. 8b). Snyder and Ruhl [37] have proposed to use birefringent materials to obtain similar index profiles, for x and y polarised modes.

(ii) Fibres with differential loss between the two modes

Single mode single polarisation fibres can also be realised if the two polarised modes have a large difference between their losses. In the following we discuss the various methods reported in order to obtain the differential leakage loss between the two modes.

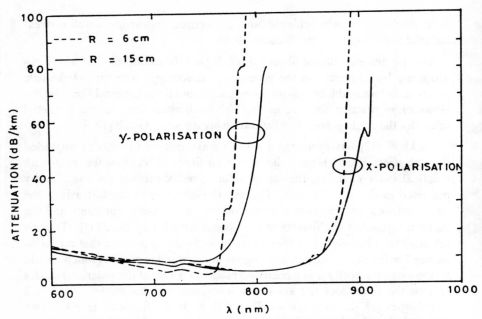

Fig. 9. Spectral attenuation of x and y-polarised modes in a Bow-tie fibre for two different bend radii (reproduced with permission of IEE, Stevenage from Ref. [38]).

One way to achieve this is by bending a high-birefringent-fibre. Varnham and co-workers [38] have used a bow-tie fibre while Okamoto and co-workers [39] have used a PANDA fibre to obtain the single-polarisation operation through bending. Figure (9) shows the spectral attenuation plot of a bent bow-tie fibre [38]. It is clear from this figure that the two polarised modes have differential loss around $\lambda = 800$ nm and the fibre behaves like a single polarisation fibre. Similarly Fig. (10) shows the spectral attenuation plot of a bent flat cladding PANDA fibre. This figure shows that around $\lambda = 1.3$ μm, fibre becomes single polarising fibre; an extinction ratio of more than 45 dB has been achieved [39].

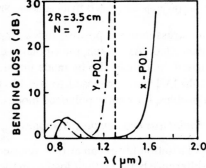

Fig. 10. Spectral attenuation of x and y-polarised modes in a bent flat cladding PANDA fibre (reproduced from ref. [39]. Copyright IEEE © 1985) Here N and R represents the number of turns and radius of the fibre coil respectively.

Simpson and co-workers [40] have used a combination of geometrical and stress effects to achieve differential loss between the two modes. They reported a SPSM fibre which uses a combination of depressed inner cladding combined with a high stress induced birefringence. The highly doped elliptical inner cladding which was surrounded by an elliptical fluorine doped outer cladding makes the fibre a W-type structure. The anisotropic stress created by the inner cladding splits the mode-effective indices of the two polarised modes such that the core index seen by one of the modes becomes less than that of the outer cladding and it suffers with a large tunnelling loss. The working principle is shown in Fig. 11, where n_1, n_2 and n_3 represent the refractive indices of the

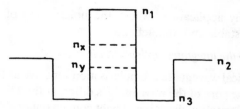

Fig. 11. Schematic refractive index profile of SPSM fibre proposed by Simpson and co-workers (after [40]). Here n_x and n_y-represent the effective indices of two polarisation modes of the fibre.

core (without taking stress into account), outer cladding and the inner cladding respectively. The refractive indices of the core seen by the x and y-polarised modes (after taking the stress into account) are marked with n_x and n_y respectively. The figure show that n_y is less then n_2 making the y-polarised mode highly leaky.

Snyder and Rühl [37] have proposed fibres composed of highly birefringent materials to obtain differential leakage loss between the two modes. Consider a fibre whose core and cladding are both made with highly birefringent material so that the refractive index profiles seen by the x and y polarised modes are like the ones shown in Fig. 12. Here n_{1x} (n_{1y}) and n_{2x} (n_{2y}) represent the core and the cladding indices for x (y) polarised modes. It is well known

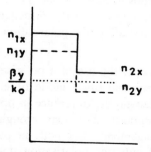

Fig. 12. Schematic refractive index profiles for x and y - polarised modes of a SPSM fibre proposed by Snyder and Rühl (after [37]).

that the effective index for the x (or y) polarised mode would then lie between n_{1x} (or n_{1y}) and n_{2x} (or n_{2y}). It can be shown easily that for V-parameters smaller than a certain value, the value of (β_y/k_0) would become smaller than n_{2x}. Since y-polarised mode has a tiny x-component (as the modes of a fibre are not exactly linearly polarised, (see Fig. 1), this mode becomes leaky as this tiny x-component behaves as if it is in a hollow waveguide. Theoretical calculations made by Snyder and Rühl [37] show that leakage losses of the y-polarised mode of the order of 1000 dB/km should be possible.

3. APPLICATIONS

There are many applications where the polarisation of the output light is required to be stable and well defined.

(i) Coupling to the integrated optical circuits

Integrated optical waveguide are polarisation sensitive as the fractional power contained in the core of the waveguide is different for TE and TM modes. As a result, the polarisation direction of light coming out of an optical fibre should be stable so that the power coupled to an integrated optical receiver is also stable. Fibre links to these devices must therefore preserve a well defined state of polarisation. The unstable polarisation state from an ordinary monomode fibre causes fading of the received signal.

(ii) Interferometric sensors

Fibre interferometric sensors also require light of a well-defined polarisation for efficient operations, since interference fringes can be produced only between beams having components polarized in the same direction. If the polarisation states of the two interfering beams are orthogonal to each other, then no interference can occur. With an ordinary monomode fibre, the output state of polarisation keeps on changing due to environmental factors such as bending and twisting etc. These fluctuations in the polarisation state are not related to the physical parameter to be sensed and produce polarisation noise and signal fading. This problem can be avoided by using a high-birefringence fibre with light launched along one of the principal axis of the fibre, since environmental factors do not affect the polarisation state of the propagating light in such fibres.

In a fibre optic gyroscope, the use of a polarisation maintaining fibre is preferred over the conventional single mode fibre. Fibre optic gyroscope measures rotation by detecting the difference in optical path lengths of two light beams travelling in opposite directions through a fibre coil. One of the requirements of this arrangement is that both the paths should be reciprocal which can be satisfied if only one polarisation state is used. If there is a difference in optical paths for the two beams, this can cause drift of the output signal – hence giving errors in the measured rotation. One solution that has been adopted is to place a polariser at the fibre ends, but in this case, the intensity may drift if the output polarisation of the fibre is not sufficiently controlled. The use of high-birefringence fibres avoids the problem of the fibre output polarisation drifting relative to the orientation of the polarizer.

(iii) Coherent communication systems

One major application of polarisation maintaining fibres is in the field of coherent communications. In coherent communication systems, a light signal transmitted through a single-mode fibre is mixed with light from a local oscillator at the receiver side to produce a beat signal. It can be shown that signal to noise ratio is much improved in this detection scheme. However, polarisation states of the transmitted signal and the local oscillator signal should constantly match in order to obtain the highest sensitivity and stability of detection. This also requires the use of polarisation maintaining fibres.

(iv) In-line fibre optic components

Optical fibres were once thought only as a transmission medium. It was proposed that any processing of the optical signal would take place outside the fibre. However, in the recent past, a large number of fibre optic in-line components have been researched and developed in which the guided light is processed inside the fibre itself. Before the development of the polarisation maintaining fibres, only a few components such as polarisation independent couplers and star couplers had been developed. However, with the development of the polarisation maintaining fibres, a wide range of fibre devices such as polarisation maintaining couplers, polarisation splitting couplers, polarizers, depolarizers, filters, multiplexers and demultiplexers etc. have been reported.

4. SUMMARY

In a large number of fibre-optic applications, one requires to use special fibres which can maintain the State of Polarisation of the guided light over long fibre lengths. In the present chapter, various schemes for realising such fibres are reviewed. In the end some of the important applications of Polarisation Maintaining Fibres are also discussed.

REFERENCES

1. D.N. Payne, Barlow A.J. and Hansen J.J., "Development of low and high birefringence optical fibres", IEEE J. Quantum Electronics, **18**, pp. 477-488, 1982.
2. C. Yeh, "Elliptical Dielectric Waveguides", J. Applied Physics, **33**, pp. 3235-3245, 1962.
3. R.B. Dyott, Cozens J.R. and Morris D.G., "Preservation of polarisation in optical-fibre waveguides with elliptical cores", Electronics Letters, **15**, pp. 380-382, 1979.
4. C. Yeh, Ha K., Dong S.B. and Brown W.P., "Single-mode optical Waveguides", Applied Optics, **18**, pp. 1490-1504, 1979.

5. K. Okamoto, Hosaka T. and Sasaki Y., "Linearly single polarisation fibres with zero polarisation mode dispersion" IEEE J. Quantum Electronics, **QE-18**, pp. 496-503, 1982.
6. J.D. Love, Sammut R.A. and Snyder A.W., "Birefringence in elliptically deformed optical fibres" Electronics Letters, **15**, pp. 615-616, 1979.
7. R.A. Sammut, "Birefringence in slightly elliptical optical fibres", Electronics Letters, **16**, pp. 728-29, 1980.
8. A. Kumar, Varshney R.K. and Thyagarajan K, "Birefringence calculations in elliptical-core optical fibres", Electronics Letters, **20**, pp. 112-113, 1984.
9. A. Kumar, Thyagarajan K. and Ghatak A.K., "Analysis of rectangular-core dielectric waveguide; an accurate perturbation approach" optics letters, **8**, pp. 63-65, 1983.
10. S.R. Rengarajan and Lewis J.E., "First higher - mode cut off in two-layer elliptical fibre waveguides", Electronics Letters, **16**, pp. 263-264, 1980.
11. R.K. Varshney and Kumar A., "Effect of depressed inner cladding on the polarisation characteristics of elliptical core fibres". Optics Letters, **9**, pp. 522-524, 1984.
12. W. Eickhoff, "Stress induced single-polarisation single-mode fibre" Optics Letters, **7**, pp. 629-631, 1982.
13. P.L. Chu, "Thermal stress induced birefringence in single-mode elliptical optical fibre" Electronics Letters, **18**, pp. 45-47, 1982.
14. J. Sakai and Kimura T., "Birefringence caused by thermal stress in elliptically deformed core optical fibres" IEEE J. Quantum Electronics, **QE-18**, pp. 1899-1909, 1982.
15. M.P. Varnham, Payne D.N., Barlow A.J. and Birch R.D., "Analytic solution for the birefringence produced by thermal Stress in Polarisation-Maintaining Optical Fibres", Journal of Lightwave Technology, LT-1, pp. 332-339, 1983.
16. T. Okoshi and Oyamada K, "Single-polarisation single-mode optical fibre with refractive-index pits on both sides of core". Electronics Letters, **16**, pp. 712-713, 1980.
17. T. Okoshi, Oyamada K, Nishimura M. and Yokota H., "Side-tunnel fibre: An approach to polarisation-maintaining optical waveguiding scheme", Electronics Letters, **18**, pp. 824-826, 1982.
18. T. Okoshi, Aihara T., and Kikuchi K., "Prediction of the ultimate performance of side-tunnel single-polarisation fibre", Electronics Letters, **19**, pp. 1980-1982, 1983.
19. K. Hayata, Eguchi M, Koshiba M. and Suzuki M, "Vectorial wave-analysis of side-tunnel type polarisation—maintaining optical fibres by variational finite elements", IEEE J. of Lightwave Technology, **LT-4**, pp. 1090-1096, 1986.
20. A. N. Kaul, Arun Kumar and Thyagarajan K., "Polarisation characteristics of side-pit and side-tunnel fibres using the effective index method", IEEE J. Lightwave Technology, **LT-5**, pp. 1610-1612, 1987.
21. R.K. Varshney and Arun Kumar, "Birefringence calculations in side-tunnel optical fibres; a rectangular-core waveguide model", Optics Letters, **11**, pp. 45-47, 1986.
22. A.N. Kaul, Arun Kumar and Thyagarajan K, "Calculation of relative single-polarisation bandwidth of side-pit fibres", Electronics Letters, **22**, pp. 786-787, 1986.
23. R.D. Birch, Payne D.N. and Varnham M.P., "Fabrication of Polarisation-Maintaining fibres using gas-phase etching", Electronics Letters, **18**, pp. 1036-38, 1982.
24. T. Hosaka, Okamoto K., Miya T., Sasaki Y. and Edahiro T., "Low loss single polarisation fibres with asymmetrical strain birefringence", Electronics Letts., **17**, pp. 530-531, 1981.
25. Y. Sasaki, Okamoto, K., Hosaka, T and Shibata, N., "Polarisation-maintaining and absorption reducing fibres", in Tech. Dig. Topical Meeting Opt. Fibre Commun. (Phoenix, AZ), 1982.

26. N. Shibata, Sasaki Y., Okamoto K. and Hosaka T., "Fabrication of polarisation-maintaining and absorption reducing fibres", J. Lightwave Technol., **LT-1**, pp. 38-43, 1983.
27. K. Okamoto, Hosaka T. and Edahiro T., "Stress analysis of optical fibres by a finite element method," IEEE J. Quantum Electron, **QE-17**, pp. 2123-2129, 1981.
28. M.P. Varnham, Payne D.N., Barlow A.J. and Birch R.D., "Analytic solution for the birefringence produced by thermal stress in polarisation-maintaining optical fibres", J. Lightwave Technology, **LT-1**, pp. 332-339, 1983.
29. P.L. Chu and Sammut R.A., "Analytical method for calculation of stresses and material birefringence in polarisation maintaining optical fibre", J. Lightwave Technology, **LT-2**, pp. 650-662, 1984.
30. Y. Sasaki, Hosaka T., Horiguchi M. and Noda J., "Design and fabrication of low-loss and low cross-talk polarisation-maintaining optical fibres", J. Lightwave Technol., **LT-4**, pp. 1097-1102, 1986.
31. A. Papp and Harms H., "Polarisation optics of index-gradient optical waveguide fibres", Applied Optics, **14**, pp. 2406-2411, 1975.
32. R. Ulrich and Simon A, "Polarisation optics of twisted single mode fibres", Applied Optics, **18**, pp. 2241-2251, 1979.
33. J.N. Ross, "The rotation of polarisation in low birefringence monomode fibres due to geometric effects", Optical and Quantum Electronics, **16**, pp. 455-461, 1984.
34. M.P. Varnham, Birch R.D. and Payne D.N., "Design of helical-core circularly birefringent fibres", Proceedings of OFC 1986, Atlanta, USA, p. 68.
35. R.D. Birch, "Fabrication and characterisation of circularly birefringent helical fibres", Electronics Letters, **23**, pp. 50-52, 1987.
36. T. Hosaka, "Single mode fibres with asymmetrical refractive index pits on both side of core" Electron. Lett., **17**, pp. 191-193, 1981.
37. A.W. Snyder and Rühl F, "New Single mode single polarisation optical fibre", Electron. Lett., **19**, pp. 185-186, 1983.
38. M.P. Varnham, Payne D.N., Birch R.D. and Tarbox E.J., "Bend behaviour of polarising optical fibres", Electron. Lett., **19**, pp. 679-680, 1983.
39. K. Okamoto, Hosaka T. and Noda J., "High-birefringence polarising fibre with flat-cladding", J. Lightwave Technology, **LT-3**, pp. 758-762, 1985.
40. J.R. Simpson, Stolen R.H., Sears F.M., Pieibel W., Macchesney J.B. and R.E. Howard, "A single polarisation fibre", J. Lightwave Technology, **LT-1**, pp. 370-373, 1983.

7

Nonlinear Processes in Optical Fibres: Phase-Matching for Second Harmonic Generation in Optical Waveguides

RAMAN KASHYAP[*]

1. INTRODUCTION

Waveguided nonlinear optics has grown rapidly since the laser was invented. Most postulated phase-matching techniques for frequency mixing have been demonstrated in bulk media and refinements peculiar to waveguides have been exploited in many different configurations. Phase-matching in second-harmonic generation refers to the condition when the fundamental wavelength and the generated second-harmonic wavelength fields propagate in a material in such directions that both experience the same refractive index. The most recent interest has been in the applications of periodic structures for phase-matching. Although phase-matching to radiation modes of a waveguide can allow the use of the largest second-order nonlinear tensor coefficient in materials, the only known practical technique for phase-matching *guided-waves* in frequency mixing experiments uses periodic structures. In this respect, periodic structures offer the most elegant phase-matching technique for use with waveguides. The confined optical fields with their correspondingly high intensities propagated over long lengths can be exploited optimally.

In this fast developing field, it is impossible to include all published experimental results in frequency mixing using periodic structures in waveguides. The purpose of this chapter is to introduce the important methods available for phase matching in waveguides. Most of the techniques have been applied, if

---

[*] BT Laboratories Martlesham Health Ipswich IP4 7RE United Kingdom.

not pioneered, in waveguides made in the well known material, lithium niobate. However, much of the subject material for phase-matching in frequency mixing experiments in lithium niobate is not covered in this review. Only theoretical aspects are considered and where appropriate, these are illustrated by examples. The review of experimental work in this article is intended to concentrate mainly on periodic structures since several published works are already available covering other aspects. Section 1.1 reviews the history of phase matching while a brief description of the nonlinear polarisation response of materials is given in Section 1.2. Phase matching is discussed in detail in Section 2, while experimental work is reviewed in Sections 3 and 4.

1.1 Historical Development

Soon after the invention of the laser in 1960, it was realised that transparent materials would no longer demonstrate a linear response in the presence of large optical fields generated by the laser. Thus nonlinear optics dates back to the famous experiments by Franken et al. [1] who demonstrated that an intense laser beam focussed into crystalline quartz could generate the second harmonic of the fundamental frequency of the laser. Thus began the activity all around the world, with the discovery of many more laser transitions and powerful sources. Frequency mixing, the generalised form of frequency doubling, was demonstrated [2] in a crystal of triglycine sulfate soon after. In frequency mixing experiments, two different frequencies mix in a nonlinear medium to generate a third (sum or difference) frequency. Early experiments quickly set the pace, verifying theoretical concepts [3] from processes such as stimulated Raman scattering in organic liquids [4], in which the intense optical beams mix with molecular vibrations to generate yet other frequencies, to newly dicovered areas such as parametric amplification and oscillation in crystalline materials [5]. Amplification requires an intense pump laser to amplify a weak signal, whereas oscillation begins from noise photons in the presence of a strong pump. These processes use the inherent electric-dipole nonlinearity in response to intense optical illumination. An electric-dipole nonlinearity only exists in crystals that do not possess inversion symmetry.

Terhune et al. [6] showed second-harmonic generation in calcite, which has inversion symmetry, both with and without the application of a dc electric field. The dc-induced second-harmonic was associated with the higher third-order nonlinear susceptibility. In this scheme, one of the four interacting fields is a static electric field which breaks inversion symmetry and induces an electric-dipole so that frequency doubling is possible. The small second-harmonic signal generated without the dc field was due to an electric-quadrupole, which is present in all materials irrespective of symmetry. Terhune et al. [6] arranged the experiment such that phase-matching was possible, along the lines of Franken et al. [1], to build up a weak effect. Giordmaine [7] showed that scattered laser radiation too can phase match in a crystal of potassium dihydrogen phosphate (KDP). A cone of radiation was observed in Giordmaine's experiment as a result of phase matching. A pulsed laser source is not monochromatic so phase-matching is affected by dispersion. Dispersion and the effect of focusing on harmonic generation were experimentally studied by Maker et al. [8] in KDP using collinear and noncollinear beams. Intrinsic anomalous dispersion [9], has also been used for phase matching. This scheme uses the fact that the refractive index close to a resonance varies rapidly (for example, near an electronic excitation), so that the phase differences between the interacting fields at various frequencies may be cancelled. Since the nonlinear susceptibility is usually described by a tensor composed of several elements, more than one element may contribute to a nonlinear process. The net contribution of the elements is calculated by the vectorial projections of the polarisation directions of the optical fields on the elements. Hobden [10] and Midwinter [11] who analysed biaxial and uniaxial crystal classes, respectively, proposed phase-matching criteria and the bulk nonlinear susceptibility vectorial projections for the calculation of effective nonlinear coefficients. Unless the fundamental wave propagates in such a way that no differential refraction occurs between it and the generated harmonic wave, 'beam walk-off' can occur limiting the useful crystal length. From an applications point of view, amongst these methods, non-critical phase-matching [12], in which, the interacting waves propagate with their wave-vectors normal to the optic axis of a birefringent crystal is recognised as probably the most important. This enables beam walk-off effects due to double refraction to be reduced considerably, and is desirable since the effective usable length of the crystal $\to \infty$.

Armstrong et al. [3] considered imperfect phase-matching, while the effect of optimum focusing and bulk crystal length were analysed by Kleinman et al. [13]. Boyd and Kleinman [14] also considered focussed Gaussian beams on second-harmonic generation and parametric processes. For a given crystal length, it has been found that a confocal parameter can be defined which produces the maximum conversion [15]. These studies highlighted a limitation of frequency mixing in bulk materials, even in the case on non-critical phase-matching. Because of focusing, the effective crystal length was shorter than its physical length.

The pursuit of new coherent sources was linked to the search for new material which, apart from having a large nonlinearity, could phase-match non-critically. Several materials were found to be candidates for different laser pump wavelengths [16], each with its own merits in terms of nonlinear coefficient, phase-matching wavelength-window and, importantly, damage threshold. For example, quartz has a small nonlinear coefficient but a very high damage threshold. Unfortunately, it is not a phase-matchable crystal. The established phase-matchable materials, such as ADP and isomorphs, also have some of the lowest second-order nonlinear coefficients. However, these crystals have a high damage threshold, allowing conversion of high power lasers. The technology has evolved to make them relatively easy to grow and polish so that they have become well established. On the other hand, $LiNbO_3$, has a large nonlinear coefficient, but is not phase-matchable at the $Nd{:}YAG$ laser wavelength of 1064 nm at room temperature and has a low damage threshold [17]. It can only be phase matched at 1064 nm wavelength by using temperature tuning. It would be desirable to use the largest nonlinear coefficient d_{33}, in $LiNbO_3$. This coefficient would provide about ×40 enhancement in conversion efficieny over the phase-matchable co-efficient d_{31}. Unfortunately it is not phase matchable in this orientation without using periodic structures (see Section 2).

Wave-guided optics changed several aspects of nonlinear optics. Firstly, there was the immediate possibility of confined optical power, which produced the very large optical-intensities required for nonlinear interactions using sources of modest power. Secondly, weak effects could be magnified since optical confinement was maintained over very long interaction lengths. Thirdly, the ensured long interaction length also meant that lower power densities could be used when optical damage was a problem. Fourthly, there was a possibility of compensating for phase mismatch by using dispersion of the guided modes.

However, the problem of phase-matching was still not an easy one to solve, since chromatic-dispersion is an intrinsic material property and can be compensated for only with some difficulty. In some applications, this was not a problem. For example, in non-parametric processes in optical fibres such as stimulated Raman scattering [18], where Stokes waves are generated from noise from an intense pump wave, dispersion effects can be small if the process is initiated close to the minimum dispersion region of a fibre at around $1.3\,\mu m$. In other stimulated processes like the soliton self frequency shift [19], the pump is arranged to be on one side of the minimum dispersion wavelength, while the Raman generated field is on the other side, such that the group velocities are matched. Using this scheme, long interaction lengths are possible. Phase-matching is required for parametric processes such as four-wave mixing where the energy difference between pump and generated photons is large, and this is achieved by using the dispersion of the guided modes supported by the fibre. Unfortunately, this is usually difficult to achieve [20, 21] and the applications are thus limited.

As well as non-parametric processes, optical fibres have been used in phase conjugation [22, 23]. Here, the process is self-phase-matching, because the two counter propagating pump beams, the probe and the generated beams, have exactly the same phase-velocities but of opposite sign. Uses of phase-conjugation in optical fibres may well be found in compensating for dispersive effects in long communication systems [24]. But apart from distortionless picture transmission in multimode fibres, not many applications have been reported.

Another scheme for phase-matching is based on the use of periodic structures. These have been known in nonlinear optics for over 20 years, when the principle of phase-matching was proposed by Bloembergen [25]. It was theoretically shown that if the sign of the second-order non-linear susceptibility was reversed at exactly the coherence length l_c, at which point the 'free-wave' and 'bound-wave' get out of phase, then the effects of dispersion can be compensated, and the material is artificially phase-matched. Somekh and Yariv (1972) [26], Tang and Bey (1973) [27], and Jaskorzynska and Arvidsson (1987) [28] too, showed that by modulating the refractive index in a periodic fashion a similar result can be achieved, albeit less efficiently. The principle of periodic phase-matching can be understood in the following way. As the fundamental wavelength field propagates through a nonlinear material, it generates a second-harmonic wavelength polarisation field which travels at the same velocity as the fundamental wavelength field and is referred to as the 'bound' wave (being bound to the fundamental wavelength field). The radiated field at the second-harmonic wavelength is referred to as a 'free' wave and travels at a phase velocity determined by the refractive index at the second-harmonic wavelength. The 'free' and 'bound' waves interfere and periodically exchange energy after an accumulated phase difference of $n\pi/2$ (where n is an integer). A $\pi/2$ phase difference is accumulated in a distance of coherence length (See Section 2). If the phase of the bound wave were changed by $\pi/2$ every coherence length, then instead of interfering destructively, the 'free' wave could be made to grow. The periodic change in the sign of the nonlinear coefficient in the fibre changes the phase of the frequency doubled 'bound' wave and therefore gives rise to 'periodic' or 'quasiphase-matching'. A method to compensate for chromatic dispersion in the case of frequency mixing has been proposed and also demonstrated using a periodic change in phase on reflection [29]. In Boyd and Patel's [29] experiment, the thickness of the crystal was such that both the fundamental and second harmonic waves were internally reflected as the fields propagated along a thin crystal. At each reflection, a 180° phase shift is introduced. The product of the three fields (fundamental, fundamental and second-harmonic) changes sign. If (as in Boyd and Patel's case) the path length before reflection is exactly equal to the coherence length, then phase-matching occurs. This scheme may be considered analogous to 'quasiphase-matching'.

Section 4 reviews the work on the application of periodic structures in second-harmonic generation in bulk materials and waveguides.

1.2 Polarisation Response of Transparent Media

The electric displacement, D, in a medium in the presence of an external applied electric field, E, (static and/or optical) can be described as follows

$$D = \varepsilon_0 E + P \qquad (1a)$$

and
$$P = \varepsilon_0 \chi E, \qquad (1b)$$

where P is the induced polarisation and χ is the material susceptibility. When the fields are intense, i.e. within a few orders of magnitude of the atomic Coulomb field ($\sim 3 \times 10^{10}$ Vm^{-1}), the response is no longer linear and must include higher order terms in the applied field to model the non-linear behaviour. For perturbations, a series expansion can be used as follows

$$P_i(\omega_0) = \varepsilon_0 \left[\chi_{ij}^{(1)}(-\omega_0;\omega_0) \cdot E_j^{\omega_1} + \chi_{ijk}^{(2)}(-\omega_0;\omega_1,\omega_2) : E_j^{\omega_1} E_k^{\omega_2} \right.$$
$$\left. + \chi_{ijkl}^{(3)}(-\omega_0;\omega_1,\omega_2,\omega_3) \vdots E_j^{\omega_1} E_k^{\omega_2} E_l^{\omega_3} + ... \right]. \qquad (2)$$

Here, the $\chi^{(n)}$ are tensors of rank $(n + 1)$, and ., :, $\vdots$ indicate tensorial dot products. The subscripts j, k, and l are the polarisation directions of the applied fields and i, that of the generated field. ω_1, ω_2 and ω_3 are the angular frequencies which mix to generate ω_0. The minus sign before the ω_0 is a convention.

$\chi_{ij}^{(1)}(-\omega_0;\omega_0)$, a second rank tensor, is the linear susceptibility and is related to the material permittivity, ε_{ij} by

$$\varepsilon_{ij} = \left[1 + \chi_{ij}^{(1)}(-\omega_0;\omega_0)\right]. \qquad (3)$$

$\chi_{ijk}^{(2)}(-\omega_0;\omega_1,\omega_2)$ is the second order susceptibility and is only present in materials lacking inversion symmetry. This tensor is responsible for sum and difference frequency mixing, the electro-optic (Pockels) effect, optical rectification and, in the degenerate frequency case ($\omega_1 = \omega_2$), second-harmonic generation. In general this tensor has 27 elements, some of which may be identical, depending on crystal symmetry and whether Kleinman symmetry [30] applies. Essentially, Kleinman symmetry proposes that, away from resonances, dispersion of the non-linear susceptibility is insignificant, and the susceptibility is unchanged by the interchange of polarisation indices, $i\,j\,k$. Thus

$$\chi_{ijk}^{(2)}(-\omega_0;\omega_1,\omega_2) = \chi_{ikj}^{(2)}(-\omega_0;\omega_1,\omega_2) = \chi_{jik}^{(2)}(-\omega_0;\omega_1,\omega_2)$$
$$= \chi_{jki}^{(2)}(-\omega_0;\omega_1,\omega_2) = \chi_{kij}^{(2)}(-\omega_0;\omega_1,\omega_2) = \chi_{kji}^{(2)}(-\omega_0;\omega_1,\omega_2). \qquad (4)$$

This reduces the number of independent tensor elements operating. Noncentrosymmetry is necessary, since the induced polarisation must follow the sign of the applied electric field (apart from a constant phase difference). But with the second order nonlinearity

$$-P_i^{(2)}(\omega_0) \neq \varepsilon_0 \chi_{ijk}^{(2)}(-\omega_0;\omega_1,\omega_2) : -E_j^{\omega_1}(-E_k^{\omega_2}). \qquad (5)$$

A change in the sign of the applied field, $E_j^{\omega_1}$ and $E_k^{\omega_2}$ will not affect the sign of the induced polarisation. Therefore, in a material with inversion symmetry

$$\chi_{ijk}^{(2)}(-\omega_0;\omega_1,\omega_2) = 0. \qquad (6)$$

It is usual to use the nonlinear 'd_{ijk}' coefficient instead of $\chi^{(2)}_{ijk}$. These are related by

$$d_{ijk} = 2\chi^{(2)}_{ijk}. \tag{7}$$

There is no such restriction for the third-order polarisability as described by Eq. (5). All materials possess a third order nonlinear optical susceptibility, and it is therefore non-zero in non-centrosymmetric media. $\chi^{(3)}_{ijkl}(-\omega_0;\omega_1,\omega_2,\omega_3)$ is described by a fourth rank tensor and has 81 independent non-zero elements in general, but in an isotropic material, these reduce to 21 of which only 3 are independent [31].

The third order polarisation, $P_i^{(3)}(\omega_0)$, from Eq. (2), is

$$P_i^{(3)}(\omega_0) = \varepsilon_0\, \chi^{(3)}_{ijkl}(-\omega_0;\omega_1,\omega_2,\omega_3) \vdots E_j^{\omega_1} E_k^{\omega_2} E_l^{\omega_3}, \tag{8}$$

and is responsible for the following important parametric (those in which the final state of the nonlinear medium remains unchanged) effects

1. Frequency tripling when $\omega_1 = \omega_2 = \omega_3$, using $\chi^{(3)}_{ijkl}(-3\omega_0;\omega_0,\omega_0,\omega_0)$

2. Optical Kerr effect with $\chi^{(3)}_{ijkl}(-\omega_0;\omega_1,-\omega_1,\omega_0)$

3. DC Kerr effect with $\chi^{(3)}_{ijkl}(-\omega_0;\omega_0,0,0)$

4. Sum and Difference frequency mixing, $\chi^{(3)}_{ijkl}(-\omega_0;\pm\omega_1,\pm\omega_2,\pm\omega_3)$

5. Self-focusing and self-phase modulation, $\chi^{(3)}_{ijkl}(-\omega_0;\omega_0,-\omega_0,\omega_0)$

6. DC induced optical rectification, $\chi^{(3)}_{ijkl}(0;\omega_1,-\omega_1,0)$

7. DC induced frequency mixing, $\chi^{(3)}_{ijkl}(-\omega_0;\pm\omega_1,\pm\omega_2,0)$

8. DC induced second-harmonic generation $\chi^{(3)}_{ijkl}(-2\omega_0;\omega_0,\omega_0,0)$

2. PHASE-MATCHING IN WAVEGUIDES

2.1 Introduction

In this Section the conditions required for phase matching and techniques which have been applied to waveguides will be reviewed. The fundamental concepts of momentum and energy conservation will be introduced and the different regimes which are possible in waveguides will be highlighted. External periodic structures will be considered for phase-matching in optical fibres. The sensitivity of phase-matching with the periodic field induced nonlinearity used experimentally in this study, on waveguide parameters such as Δn and core radius will be studied. The unusual phenomenon of internal self-written gratings in optical fibres will be compared to external grating structures.

Crystalline materials offer a large number of possibilities for phase-matching. Since crystal symmetry dictates the nonlinear coefficients of the susceptibility tensor elements which can be used, it is necessary to know the crystal point group in order to select the orientation of the crystal and the input optical polarisations. Phase-matching for frequency doubling in bulk crystals can be described as Type I or Type II. Type I phase-matching describes the condition when the long wavelength electric field polarisation propagates in the crystal to experience the extra-ordinary index while the second-harmonic electric field experiences the ordinary index. When the fundamental wavelength electric fields are orthogonally polarised as ordinary and extra-ordinary waves, the condition is known as Type II phase-matching. In three-wave mixing, where the three interacting fields are of different wavelengths, the combinations increase. Type I and Type II phase-matching can be sub-categorised according to whether the interaction takes place in a uniaxial crystal or biaxial crystal. These phase-matching considerations for the general case of three wave mixing in anisotropic crystals has recently been summarised by Morgan [32].

2.2 Phase Matching Criteria

The concept of phase-matching or conservation of momentum was first used by Giordmaine (1962) [7] and Maker et al. (1962) [8]. This is normally described by the following [33]

$$2n^\omega k^\omega = n^{2\omega} k^{2\omega} \tag{9a}$$

or,

$$\frac{4\pi n^\omega}{\lambda^\omega} = \frac{2\pi n^{2\omega}}{\lambda^{2\omega}}, \tag{9b}$$

where λ is the free space wavelength, k the free space propagation constant and n the refractive index at the superscripted frequencies. For the purposes of our discussions on periodic structures, a momentum, β^0, is introduced for the periodic structure. DC fields applied to isotropic media can allow second harmonic generation, as indicated by point 8 in Section 1.2. Spatially periodic static-electric fields can then be used for phase-matching [34]. Here, energy conservation is assumed so that electric field induced SHG process is allowed

$$2\omega - \omega - \omega + 0 = 0. \tag{10}$$

There are two basic schemes which can be used for phase-matching in optical fibres. These will be referred to as 'Modal' phase-matching [35-39], and 'Periodic' phase-matching [40, 41, 27]. Dispersion of modes can be used to allow phase-matching and is referred to as 'Modal' phase-matching. This has been extensively used in lithium niobate in-diffused waveguides. Periodic phase-matching uses a periodic structure and has not been used in second-harmonic generation advantageously until recently. However, the principle has been applied in many linear device configurations [40].

The phase-mismatch factor is

$$\Delta\beta = \left|2\beta^\omega - \beta^{2\omega}\right| - \beta^0, \tag{11}$$

with, $\beta^{i\omega}$ ($i = 0, 1, 2$), referring to the phase constant of the spatially-periodic static-field or perturbation, and the propagation constants of the fundamental and the second-harmonic wavelength modes [identical to terms in Equation (9a, b)], respectively. The propagation constants at the two wavelengths are,

$$\beta^\omega = \frac{2\pi n_{eff}^\omega}{\lambda^\omega}, \tag{12}$$

$$\beta^{2\omega} = \frac{2\pi n_{eff}^{2\omega}}{\lambda^{2\omega}}, \tag{13}$$

and n_{eff} is the effective mode index at the superscripted frequency. The coherence length, l_c, of a particular mode-interaction is

$$l_c = \frac{\pi}{\left|2\beta^\omega - \beta^{2\omega}\right|} = \frac{\lambda^\omega}{4\left|n_{eff}^\omega - n_{eff}^{2\omega}\right|} \tag{14}$$

The phase constant of the static field is given by

$$\beta^0 = 2\pi N/\Lambda, \qquad N = 1, 2, \ldots \tag{15}$$

where, Λ is the periodicity (pitch) of the static electric field (or periodic structure) and, N is the order of the spatial harmonic of the pitch.

For optical fibres, the weakly guiding approximation due to Gloge [42] can be used to calculate the effective indices of the modes. The analytical approximation to the normalised propagation constants in Gloge's paper is not accurate enough to calculate the phase-mismatch between modes, and thus the approximate characteristic equation has been solved numerically for the figures shown later. This gives better results close to cut-off of the modes, and is thought to be adequate considering the departure from step index profiles of the fibres with index dips in the middle as well as slight slopes on the core-clad boundary. Such conditions as encountered in real fibres make the accurate determination of the normalised propagation constants difficult even using techniques such as the finite-element method. It is therefore considered inappropriate to attempt a more accurate solution, given the practical uncertainties in the determination of the index difference and core radius. Since the core-clad index-difference, Δn is less than 1%, the solutions have been shown to be within 1% error of the exact solution[43], a tolerance thought to be practical for the design of devices. Despite small errors, the trends are adequately revealed using the analysis given here.

The following normalized propagation constants are used for each wavelength, λ, to calculate the effective indices seen by the modes:

$$V = \frac{2\pi a}{\lambda}\left[n_{core}^2 - n_{clad}^2\right]^{\frac{1}{2}}, \tag{16}$$

$$u = \frac{2\pi a}{\lambda}\left[n_{core}^2 - n_{eff}^2\right]^{\frac{1}{2}}, \tag{17}$$

$$w^2 = V^2 - u^2, \tag{18}$$

$$b = \frac{w^2}{V^2}, \tag{19}$$

the core-cladding index-difference,

$$\Delta n = n_{core} - n_{clad}, \tag{20}$$

and a is the core radius of the fibre. The refractive indices of the core and cladding are n_{core} and n_{clad}. The weakly guiding approximation gives rise to nearly transversely polarised modes since the electric field component (E_z) of the mode in the propagation direction, z, is small compared to the transverse component, E_x, or E_y. The ratio of the electric fields, $E_z/E_{transverse}$, is of the order of $2\Delta n$. Therefore these modes can be considered linearly-polarised (LP). For each value of u, n_{eff} can be calculated to give dispersion relations for the fundamental and SH wavelength modes from the characteristic eigenvalue equation [42]. In the following, Sellmeier data from Fleming [44] is used for index calculations as function of wavelength in a germania-doped silica fibre.

Equation (19) is used to calculate the modal dispersions as a function of the core radius of both fundamental and second-harmonic wavelengths shown in Fig. 1. The computation is for a fundamental wavelength of 1.064 μm and an index-difference, Δn of 4.5×10^{-3}. For a fibre with a core radius of 4μm, only two fundamental wavelength modes, LP_{01}^{ω} and $LP_{11}^{2\omega}$ are supported whereas five second-harmonic modes can propagate. Figure 1 also shows the large phase-mismatch between the fundamental and harmonic modes, so that phase-match-

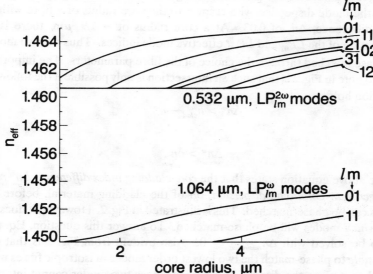

Fig. 1. Dispersion of the fundamental and second harmonic wavelength modes as a function of the core radius for a fibre core cladding index difference of 4.5×10^{-3}.

ing is not normally possible using this index-difference for any of the modes. This condition can be altered by birefringence, but in optical fibres, it is not possible to induce large enough birefringence to allow phase-matching and this will be discussed in Section 2.2.1 C. However, as has been shown by Terhune and Weinberger [71], it is possible to phase-match modes in germania doped silica fibre by using 'modal' phase matching.

2.2.1 Modal Phase-Matching

In terms of measurable quantities, the index mismatch is equal to the chromatic dispersion plus the difference between the normalised propagation-constants of the interacting modes, the condition being given by

$$\Delta n_{mismatch} = n_{clad}^{2\omega} - n_{clad}^{\omega} + b^{2\omega}\Delta n^{2\omega} - b^{\omega}\Delta n^{\omega}, \qquad (21)$$

where Δn is the absolute core-cladding index difference at the superscripted frequency, and b the normalised-propagation constant as described in Eq. (12). In 'modal' phase-matching the index mismatch vanishes by appropriate choice of waveguide parameters. In the following sections, three important cases of 'modal' phase-matching are considered.

A. Modal Phase Matching: Isotropic Circular Cored Fibres

Terhune and Weinberger [71] have computed from n_{eff} data, the index (or phase) matching curves for germania doped silica fibres. They have also calculated the phase-matching condition of particular mode combinations for a given set of fibre parameters for frequency mixing. Unfortunately, fibre parameters governing the phase-matching conditions are not given. Figure 2 shows the mode dispersion with respect to the core radius of a fibre with an index difference, Δn, of .0.03. At a core radius of ~ 1.9 μm, there is an intersection of the LP_{01}^{ω} and $LP_{31}^{2\omega}$ effective mode indices. Thus the two modes are phase matched through the choice of the fibre parameters. The important point to note in Fig. 2 is that such an intersection is only possible if the following condition holds:

$$\Delta n^{\omega} > n_{clad}^{2\omega} - n_{clad}^{\omega}$$

or,

$$\Delta n^{\omega} > \delta n. \qquad (22)$$

The above equation states that the *core-cladding index difference, Δn^{ω} must be larger than chromatic dispersion, δn of the cladding material*, before any modes can be phase-matched. This is illustrated in Fig. 2. However, it does not state *which* modes will be phase-matched. To answer this question, Eq. (21) has to be solved with $\Delta n_{mismatch} = 0$. Also evident from Fig. 2 is that it is *impossible* to phase-match the two lowest order modes in isotropic fibres using this scheme, since the dispersion curves for those two modes cannot intersect for *any* value of a.

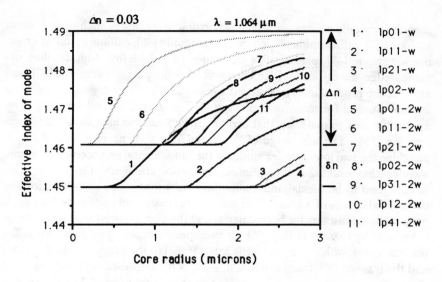

Fig. 2. Mode dispersion diagram for a fibre in which mode-matching is possible because Δn is greater than the chromatic dispersion, δn.

The advantages of using a fibre with $\Delta n^\omega > \delta n$ have also been dealt with in detail by Terhune and Weinberger [71]. They have shown that for a given core radius e.g. of 3.0 μm, the phase-matched frequency doubling interaction of selected modes is insensitive to the actual index difference at *a particular wavelength*. They have also shown that for specific value of the core radius, 3.8 μm, and a mole fraction of germania of 0.13 in the core, there is a possibility of broadband phase matched interaction of LP_{01}^ω and $LP_{21}^{2\omega}$ modes at a wavelength of ~ 2μm (~ 5000 cm^{-1}). Unfortunately, with this core-dopant concentration, the phase matched interaction is sensitive to the exact value of *a*. The practical demonstration of such a scheme is faught with difficulties, since it would be almost impossible to design a fibre which is phase-matched over a *very long length* (> 100 mm) for a *particular mode interaction and at a given wavelength*. The two requirements together imply the control of the fibre parameters to such a high degree of accuracy as to render it almost unusable except over short lengths [45]. However, it is not difficult to demonstrate this principle by using a tunable laser source, which gives the experiment the much needed degree of freedom and this technique has been demonstrated recently [46]. Temperature tuning is usually difficult for this scheme, since the core and cladding materials generally have refractive index temperature coefficients which are similar in magnitude and sign, unless the core and cladding materials are dissimilar.

B. Modal Phase Matching: Anisotropic Fibres

This is by far the commonest and probably the most useful technique of modal phase-matching. There have been several demonstrations of phase-matched second harmonic generation using anisotropic waveguides such as in lithium niobate [47], gallium phosphide ribbon waveguide [48], gallium arsenide channel waveguide [49], planar silica waveguide [50], 2-Methyl 4-nitroaniline organic wedge-waveguide [51], ZnS on polycarbonate planar waveguide [52], and also recent reports in crystal cored fibres [53-54]. The crystal cored fibre system is useful for many organic crystalline materials but till now has been applicable only to materials that can be grown from the melt without decomposition [55]. The advantage of this technique is in the potential of phase matching the two lowest order modes at the fundamental and second harmonic wavelengths, resulting in the optimum conversion efficiency. Organic crystals usually tend to be biaxial and computation of propagation constants for the orthogonal modes of a generalised waveguide is extremely difficult. However, propagation constants for both uniaxial and the biaxial cored fibres have been calculated by Cozens (1976) [56] for two specific cases. The first case is for uniaxial cores, with the crystal optic axis parallel to the propagation direction, and the transverse refractive index, $n_x = n_y \approx n_z$. In this case, the solutions of the propagation constants are similar to the isotropic fibre. For the biaxial core, an analytic solution is not possible and the propagation constants have to be evaluated numerically, although a coupled mode approach can provide βs quite easily [56].

It has been shown recently by Nayar, Kashyap and White (1986) [38] that phase matching of fundamental modes in anistropic optical fibre waveguides with crystalline cores is possible provided the following criteria can be satisfied for the modes $\left(HE_{11}^{\omega}\right)_x$ and $\left(HE_{11}^{2\omega}\right)_y$.

$$n_{clad}^{2\omega} - n_{clad}^{\omega} = \frac{1}{8n_{clad}^{\omega}}\left\{\left[\left(n_x^{\omega}\right)^2 - \left(n_y^{\omega}\right)^2\right] + 3\left[\left(n_x^{\omega}\right)^2 - \left(n_{clad}^{\omega}\right)^2\right] + 8n_x^{\omega}\left(n_x^{2\omega} - n_x^{\omega}\right)\right\}, \qquad (23)$$

where the x and y subscripts refer to the laboratory cartesian frame. The left hand side of Eq. (23) is the dispersion of the cladding glass, and allows the selection of a cladding glass with the appropriate dispersion. In principle, the refractive indices of the core material at the appropriate wavelengths can be measured and then Eq. (23) used for the selection of the cladding glass in conjunction with published data on commercial glasses. Equation (23) is independent of core radius and relies solely on the bulk refractive indices of the waveguide materials. It should be stressed that phase-matching holds only for the condition described in Eq. (23) at a particular temperature, since each element of the equation is usually temperature sensitive. Thus, despite the simplicity of the relationship described in the equation, the problems are as severe for this scheme as for the isotropic waveguide. In anisotropic waveguides, however, temperature tuning is always possible, since here the temperature coefficients can vary widely in magnitude, and also in sign. A theoretical paper describing the dispersion relations is given elsewhere by White, Nayar and Kashyap (1988) [39]. Core materials must fall into one of 9 crystal classes [38] listed in Table 1 to be useful for second harmonic generation in crystal cored fibres. Along with this restriction, the growth of the crystal must also have the correct orientation for optimum interactions. These two requirements are difficult to satisfy and thus severely limit the number of useful materials for growth in capillaries. The phase-matching curve for a fibre with an anisotropic core is shown in Fig. 3. The intersection of the phase index curves for the fundamental modes at ω and 2ω can be seen for arbitrary values of cladding glass and a birefringent core.

TABLE 1

Crystal classes useful for second harmonic generation and parametric amplification as cores of crystal cored fibres using phase-matched interactions between non-degenerate modes of the fibre using The Institute of Radio Engineers (IRE) convention

Crystal System	Crystal Class	Phase Matching Requirements
BIAXIAL		
Triclinic	1	no preferred orientation
Monoclinic	m	either x- or z-axis transverse to guide and n_x or n_z < other transverse index
	2	y-axis transverse to the guide and n_y < other transverse index
Orthorhombic	mm2	z-axis transverse to guide and n_z < other transverse index
	222	not useful
UNIAXIAL		
Tetragonal	4	z-axis transverse to guide and –ve uniaxial
	$\bar{4}$	z-axis transverse to guide and –ve uniaxial
	422	not useful
	4mm	not useful
	$\bar{4}2m$	not useful
Trigonal	3	z-axis transverse to guide and –ve uniaxial
	3m	z-axis transverse to guide and –ve uniaxial
	32	not useful
Hexagonal	6	z-axis transverse to guide and –ve uniaxial
	6mm	z-axis transverse to guide and –ve uniaxial
	$\bar{6}$	not useful
	$\bar{6}m2$	not useful
	622	not useful

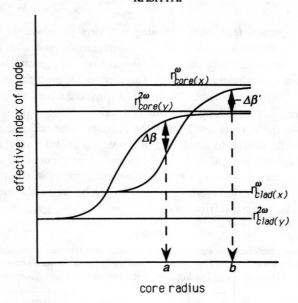

Fig. 3 Positive and negative phase mismatch at two different core radii, 'a' and 'b' in a birefringent waveguide where the 2ω mode is polarised along the axis of lower refractive index. The intersection point shows the core radius at which modal phase matching occurs

C. Effect of Ellipticity of the Core

Ellipticity of the core in a fibre causes the fibre to become birefringent, and the orthogonal modes of the waveguide are then no longer degenerate. The induced normalised birefringence has been calculated by Dyott, Cozens and Morris (1979) [57] and the following simple relationship for a fibre with a V-value of 1 and an infinite major to minor axis ratio can be derived from their data for the fundamental modes

$$\frac{n_x^\omega - n_y^\omega}{\Delta n^\omega} \approx 0.4 \times \Delta n^\omega, \tag{24}$$

where n_x^ω and n_y^ω are the effective indices of the orthogonal modes, and Δn^ω is the core cladding index difference. Equation (24) gives the *maximum* normalised birefringence induced in an optical fibre and this happens if the fibre core approximates to a slab shape. Using Eq. (24) for a fibre with a $\Delta n^\omega = 0.04$, the maximum change in the effective indices of the orthogonal modes is 6.4×10^{-4}, which is 1.6% of Δn^ω. Increasing Δn^ω to 0.1, still induces only 4% change. Therefore, using ellipticity for phase matching is not very useful, and can only be used for 'fine-tuning' provided the phase mismatch is already small.

Propagation in these waveguides has been considered by several workers [57-59], but approximations used in the model for the numerical calculations cause large inaccuracies in mode indices.

D. Modal Phase Matching: Radiation Field

Radiation field [60] or Cerenkov radiation phase matching is the simplest and least critical of all techniques. Coupling to the radiation field occurs if the phase velocity of a guided mode exceeds the velocity of light in the material. The design of the fibre requires that the dispersion of the cladding material be greater than that of the core at 2ω, or that the cladding index be larger than the long wavelength mode index. Thus the following relationship must be satisfied:

$$n_{clad}^{2\omega} \geq n_{core}^{2\omega} \tag{25}$$

or,

$$n_{clad}^{2\omega} \geq n_{eff}^{\omega} \tag{26}$$

while for waveguiding,

$$n_{core}^{\omega} > n_{clad}^{\omega}. \tag{27}$$

Equation (25) states that the fibre is no longer a waveguide at 2ω and some of the modes at the second harmonic wavelength are cut off, however the fibre is able to guide at the fundamental wavelength as shown by Eq. (27). The phase match condition is described as

$$n_{eff}^{\omega} = n_{clad}^{2\omega} \cos\theta, \tag{28}$$

where, θ is the angle made by the radiated second harmonic ray with the direction of propagation. For Eqs. (25)-(26) to hold, the core and cladding

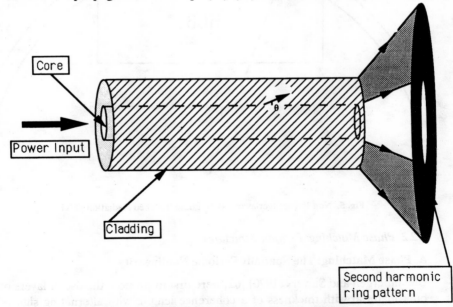

Fig. 4. Phase-matched radiation mode second harmonic generation

must be made from different materials which makes Δn^ω, and therefore n^ω_{eff} a function of wavelength and temperature. Equation (28) is the momentum matching condition which also shows that the only effect of tuning the wavelength or temperature is an alteration of the emergence angle, θ. The advantage of this method is that it is potentially broadband since there is a continuum of radiation modes to which the second harmonic can couple.

One disadvantage of this method is that the growth of the second harmonic power is dependent only on the length of the waveguide, L [61], rather than on L^2. A schematic of radiation field phase matching in an optical fibre is shown in Fig. 4. It is possible to use an overlay material for radiation field phase matching with optical fibres, provided the chromatic dispersion in the overlay nonlinear material satisfies Eq. (27). A schematic of the device is shown in Fig. 5.

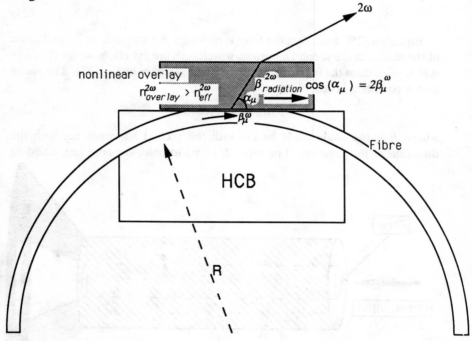

Fig. 5. Non-linear Overlay on HCB: Phase-matched Radiation SHG

2.2.2 Phase Matching: Periodic Structures

A. Phase Matching: The Spatially Periodic Nonlinearity

Bloembergen and Sievers (1970) [62] were first to propose the use of layers of bulk material with thickness of a coherence length, with alternating sign of non-linearity to allow phase matching. Somekh and Yariv (1972) [63] have examined phase matching by the periodic modulation of the second-order non-linear susceptibility for waveguide applications. They concluded that the effective second-order non-linear susceptibility, d_{eff}, is reduced by $\pi/2$ in the case of periodic modulation between $-d$ and $+d$.

In electric field induced second harmonic generation (EFISH), the second harmonic power generated in a waveguide of length L, with an input power of P^ω in the fundamental wavelength mode is given by [34]

$$P_i = \left[\kappa P^\omega E_i^0 L \operatorname{sinc}(\Delta \beta L / 2) \right]^2, \tag{29}$$

where the coupling coefficient

$$\kappa = -\frac{3}{4} \varepsilon_0 \omega \chi^{(3)}_{ijkl} I. \tag{30}$$

I is the overlap integral of the interacting fields, while E_i^0 is the periodic static field. From Eq. (29) it can be seen that the second-harmonic power is a maximum when $\Delta \beta L/2 = 0$. In EFISH the effective second order nonlinearity, $\chi^{(2)}_{eff}$, can be described as

$$\chi^{(2)}_{eff} = \chi^{(3)}_{ijkl} E_x^0 (\beta^0 z), \tag{31}$$

where the static field, $E_x^0 (\beta^0 z)$, is spatially periodic, and so consequently is $\chi^{(2)}_{eff}$. Therefore, the description in Eq. (11) is quite general and applicable to any such interaction where d_{eff} has a spatial periodicity. For a second-order non-linear waveguide, $3\chi^{(3)}_{ijkl} E_i^0$ in Eq. (29) can be replaced by d_{eff}. From Eq. (11) the following condition for phase-matched operation holds:

$$\beta^0 = \left| 2\beta^\omega \pm \beta^{2\omega} \right|, \tag{32}$$

and therefore at phase-match

$$\frac{2\pi N}{\Lambda} = \left| 2\beta^\omega \pm \beta^{2\omega} \right|. \tag{33}$$

The power in the second harmonic wave grows as the square of the fibre-length. Equation (33) describes two important features. Firstly, the $-$ and $+$ signs signify *forward* and *backward* propagating second harmonic wave coupling, respectively. For the forward wave, the period of the spatially modulated nonlinearity is coarse, whereas for the backward propagating wave, the variation in phase is additive and thus the period becomes very small. For a typical phase mismatch in an optical fibre of $n_{eff}^{2\omega} - n_{eff}^\omega \approx 0.025$, the forward wave phase matching period, $\Lambda \sim 40\,\mu\text{m}$, while for backward second harmonic wave coupling, $\Lambda \sim 0.17\,\mu\text{m}$.

Secondly, Eq. (33) allows for the phase mismatch to be positive or negative. In devices of $\Delta n >$ *chromatic dispersion*, this has significance because modal dispersion can allow either condition to hold depending on the effective indices of the interacting modes. Figure 6 shows the two conditions in the mode

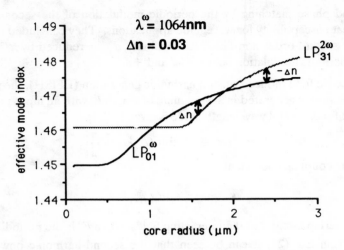

Fig. 6. Mode dispersion diagram showing change in the sign of the phase mismatch on either side of the mode match point.

dispersion diagrams. When $a < 1.9\,\mu$m, then $\beta^{2\omega} < 2\beta^\omega$ and for $a > 1.9\,\mu$m, $\beta^{2\omega} > 2\beta^\omega$. Standard telecommunications type fibres are in the normally dispersive regime where the difference, $2\beta^\omega - \beta^{2\omega}$ is negative.

B. Periodic Modulation of Waveguide Dimensions or Refractive Index

The first published reference on the use of a periodic modulation of the linear guiding properties of a waveguide for phase matching is by Somekh and Yariv (1972) [26]. They considered the application to a thin film waveguide with a corrugated boundary and concluded that effective non-linearity, d_{eff}, for a material with a second order non-linear coefficient, d, can be described as

$$d_{\text{eff}} \approx d\,J_1(M_0^\omega), \tag{34}$$

where J_1 is Bessel-function of order one, and the argument

$$M_0^\omega = \frac{(k^\omega)^2\,|\xi^\omega|^2\,\left[(n_{\text{guide}}^\omega)^2 - 1\right]\,a\,\Lambda}{4\pi\,\beta^\omega}, \tag{35}$$

where $2a$ is the peak-to-peak modulation of the corrugation. Equation (34) is for the best interaction involving the fundamental modes of the waveguides and assumes that the overlap of modes is unity. A consequence of using the corrugated boundary is a large reduction of the effective non-linearity. Despite this penalty, the authors observe that the technique is worth pursuing since the resulting non-linearity for some materials is still large enough for useful conversion efficiencies to be achieved. The effect of etching a grating in the linear part of a waveguide is similar to modulating the refractive index. Direct periodic modulation of the refractive index alone is difficult but this will be considered in conjunction with the modulation of the non-linear susceptibility in the next Section.

C. Simultaneous Periodic Modulation of Nonlinearity and Refractive Index

Tang and Bey (1973) [27] studied the combined effect of periodic modulation of both the refractive index and nonlinearity in detail. However, Jaskorzynska and Arvidsson (1986) [28] have examined the practical case of lithium niobate waveguides in which titanium indiffusion was used to cause a spatial modulation of the linear properties. Their method of analysis highlights the differences between the modulation of refractive index, nonlinearity, or a combination of both. To summarise their results, there is a reduction in the conversion efficiency, η, if a periodic scheme is used and this reduction is given by

$$r_{n,d} = \frac{\eta_{periodic}}{\eta_{nonperiodic}} = \left\{ \frac{\delta A}{2} \frac{O_{\mu\nu}(h)}{O_{\mu\nu}(1)} \left(J_0(B) + J_2(B) \right) \right.$$

$$\left. - J_1(B) \left[1 - \delta \frac{O_{\mu\nu}(h)}{O_{\mu\nu}(1)} \right] \right\}^2, \quad (36)$$

where,

$$B = \alpha \Lambda R_{\mu\nu} / (2\pi). \quad (37)$$

$O_{\mu\nu}(h)$ is the transverse field overlap integral for modes μ and ν; the periodic modulation in the non-linear coefficient with a modulation depth of, δ, and transverse distribution, h; $O_{\mu\nu}(1)$, the transverse field overlap integral in the case of the unperturbed waveguide with uniform longitudinal nonlinearity and refractive index. α is the modulation depth of the permittivity, ε_r, and Λ is the period of the perturbation. A is effectively the Fourier component of the order of perturbation period used. $R_{\mu\nu}$ is the normalised phase difference between the two frequencies as a result of the perturbation, giving rise to a phase offset. This arises from the non-uniform transverse distribution in the permitivitty. In the case of the modulation of refractive index alone when $\delta = 0$, the reduction in the efficiency compared to the non-periodic case is

$$r_{n,0} = J_1^2(B). \quad (38)$$

When only the nonlinearity is modulated, $\alpha = 0$ and

$$r_{0,d} = \left[\frac{\delta A}{2} \frac{O_{\mu\nu}[h]}{O_{\mu\nu}[1]} \right]. \quad (39)$$

For a uniform transverse modulation, with $\delta = 1$ and $h = 1$, (h is a function of y) the highest conversion results, so that the reduction from the case of uniform nonlinearity is then

$$r_{0,d} = \frac{A^2}{4}. \quad (40)$$

Finally, for modulation of both the nonlinearity and refractive index, it can be seen from Eq. (36) that depending on the magnitudes of the terms in the overlap integrals, there could be complete cancellation. It can also be seen from Eq. (39) that since the reduction in the case of the modulation of the nonlinearity is dependent on the ratio of the overlap integrals; it is generally more effective.

D. Sensitivity of Period of Phase-matching Structure on Waveguide Parameters

An important consideration when designing a phase-matched device using periodic phase-matching is the sensitivity of period of such parameters as index-difference and core-cladding index difference. This has been studied [34], and it is interesting to note that a type of 'non-critical' phase-matching is possible in waveguides using periodic structures. For specific core radii and core-cladding index differences in optical fibres, it has been shown that random variations in these two parameters cause little change in the period of the phase-matching structure[34]. This is of great importance since small fluctuations in the core-cladding index difference and the core radius are impossible to eliminate. Since the period of the phase-matching structure is inversely proportional to the difference in the propagation constants of the modes at the fundamental and second-harmonic wavelengths [See Eq. (33)], the condition for non-critical phase-matching requires that the differential of the propagation constants at the two wavelengths with respect to the waveguide parameters be identical. The propagation constants at the two wavelengths disperse in such a way that the condition is almost always met for certain waveguide parameters. The same type of non-critical phase-matching with respect to waveguide dimensions should also be possible for other waveguide structures (channel, ridge etc.) since the underlying mechanisms are similar.

2.3 Phase-matching via Self-written Gratings in Optical Fibres

Recently, an interesting phenomenon was discovered by Hill *et al.* (1978)[64] who noted that optical fibres could frequency double light when pumped with high power-densities despite being centro-symmetric. This effect was then investigated by Österberg (1986) [65] who showed that not only was the fibre capable of frequency doubling, but that the frequency doubled light *grew* with time to levels of a *few percent* of the pump power. This observation could only occur if the process was phase-matched, and was later postulated to be through a periodically written second-order susceptibility, $\chi^{(2)}$, grating[66] which had a wave-vector equivalent to the momentum mismatch between the pump and the frequency doubled light.

Thus there is now another mechanism for phase matching in fibres, based on an extremely powerful scheme which is *self generating and self correcting*, since any longitudinal variation in waveguide parameters causes the grating period to alter. Second harmonic generation (SHG) in optical fibres has been observed by many researchers [64, 65, 67-69]. It continues to be a curiosity since the inversion symmetry of the fibre material does not allow second-order dipole assisted interaction. As well as being interesting scientifically, the effect could be used to frequency double semiconductor diode lasers to produce sources of blue radiation. A theory put forward suggests electric-quadrupole assisted second-harmonic generation [66] enhanced by a self-written grating of color-centres through a magnetic-dipole contribution. Recently, fibres were 'seeded' with the second-harmonic of 1064 nm radiation [70]. This process allowed the phase-matching grating to be written into the fibre within a matter of a few minutes. The cause of the nonlinearity was not explained, although it was speculated that colour-centres may be a contributing factor. It has also been proposed that the core-cladding interface breaks the inherent symmetry of the glass and allows an electric-quadrupole contribution [71] to generate the second-harmonic as a result of the gradient of the longitudinal electric field of the mode. This work by Terhune and Weinberger [71] is intended to quantify the maximum conversion efficiency possible from an inherent nonlinearity, and does not propose any explanation of the growth in the second-harmonic power as a function of time when pumping with the fundamental wavelength field. Conversion efficiencies for specific phase-matched mode-interactions are predicted but are much lower than those observed. The theory of self-generated second-harmonic as proposed by Payne [72] assumes contributions from both the electric-quadrupole and magnetic-dipole moments and shows good agreement with the observations. The underlying reasons behind saturation of the generated second-harmonic power in fibres has also been examined [73]. What is clear is that for self-written gratings the particular mode-interaction is difficult to select, this being determined by the field overlaps of the fundamental and second-harmonic modes excited during the writing process. Many questions remain unanswered; for example, the selection rules governing a particular mode-interaction are dependent on symmetry properties and on the type of nonlinearity, and although appreciable conversion efficiencies have been achieved, there is concern over the observation of mode-combinations forbidden by symmetry [71].

Hill et al. (1978) [64] observed weak three wave mixing in an optical fibre and soon after, Fujii et al. [67] noted sum frequency components as a combination of the pump frequency and Raman generated frequencies as well as second harmonic frequency of the pump, all of which appeared as cladding modes in the fibre. They noted that these observations were most unusual and were unable to explain them. Phase-matched guided-mode sum frequency generation was recognised by Sasaki and Ohmori (1981) [68] in optical fibres, however the frequency doubled component of the pump wavelength was small, and no explanation was given. It was Österberg and Margulis (1986) [65] who then reported the *growth* of the second harmonic frequency after many hours of pumping an optical fibre. Phase-matching via an optically written grating of colour centres was then proposed to be the probable cause for this unusual process by Farries *et al.* (1987) [66] and by Stolen and Tom (1987) [74] [70] for external seeding by second harmonic concurrently with the pump wavelength. In their model, Stolen and Tom showed that the pump and its weak frequency doubled seeding wavelength mix via the third order susceptibility to generate a spatially periodic static polarisation in the fibre which they then deduced somehow polarises the material over a period of time, allowing the nonlinearity to increase by this self organising process. The dc polarisation was shown to be

$$P^0 = \frac{3\varepsilon_0}{4} \operatorname{Re} \left[\chi^{(3)} (0: -\omega, -\omega, 2\omega) (E^\omega)^* (E^\omega)^* E^{2\omega} e^{(i\Delta\beta z)} \right], \quad (41)$$

where E^ω is the pump field, $E^{2\omega}$ is the second harmonic seed field and the phase mismatch $\Delta\beta$

$$\Delta\beta = 2\beta^\omega - \beta^{2\omega}. \quad (42)$$

The physical origins giving rise to the second-order susceptibility are not understood, but thought to be through the redistribution of carriers and trapping at defect sites. However symmetry is broken, and the second order susceptibility is assumed to be proportional to

$$\chi^{(2)} = \alpha_{SH} P^0 = \tfrac{3}{4} \alpha_{SH} \varepsilon_0 \chi^{(3)} (0; -\omega, -\omega, 2\omega) |E^\omega|^2 |E^{2\omega}| \cos(\Delta\beta z + \phi_p), \quad (43)$$

where α_{SH} is some proportionality constant which is dependent on the microscopic process effecting the formation of the susceptibility, $\chi^{(2)}$, and ϕ_p depend on the relative phase between the pump and its frequency doubled seed. Once this non-linear grating is formed, the fibre is able to frequency double when probed with the pump field alone. A coupled mode analysis shows that there is a parametric exchange of energy from the pump to the second harmonic frequency and the second harmonic power in the undepleted pump regime is [66]

$$P^{2\omega}(L) = \left| \gamma P_0^\omega L \right|^2 \operatorname{sinc}^2 (\kappa L / 2) \quad (44)$$

where,

$$\gamma = \frac{3\omega}{4n^\omega c} \varepsilon_0^2 \alpha_{SH} f_{\omega,\omega,2\omega} \chi^{(3)} (0; -\omega, -\omega, 2\omega) |E^\omega|^2 |E^{2\omega}|. \quad (45)$$

Here $f_{\omega,\omega,2\omega}$, is the overlap integral of the interacting modes, and P_0^ω is the fundamental wavelength field incident at the input of the fibre. Notice that the term γP_0^ω effectively describes a *fifth* order process since five fields are involved, although it is cascaded and temporally separated [75]. It has also been shown that long coherence length gratings can be written by narrowing the spectral width of the writing wavelengths [76]. Because of chromatic dispersion, a source with a finite spectral width generates gratings with slightly varying periods of $2\pi/\Delta\beta$, for each frequency component. For a given spectral distribution, the static polarisation has to be calculated by integrating over the spectrum of the pump so that for a Gaussian spectrum [76]

$$P^0_{eff} = P^0 e^{-(z/l_c)^2}, \qquad (46)$$

where

$$l_c = \frac{2}{(a-b)\,\delta(\omega)}, \qquad (47)$$

and

$$a = \frac{d\beta}{d\omega}\bigg|_{\omega},$$

$$b = \frac{d\beta}{d\omega}\bigg|_{2\omega}. \qquad (48)$$

It is assumed that dispersion at the two frequencies are linear functions at both wavelengths. Self phase modulation of the pump as it propagates along the fibre has a detrimental effect since it broadens the spectrum and therefore reduces the effective coherence length, l_c.

Many questions remain unanswered. The initial frequency doubled light is assumed to be due to a quadrupole interaction, but has not been proved to be so conclusively[70]. Further, recent observation of uniform-dipole, mode-symmetry forbidden, second harmonic generation from $LP^{\omega}_{01} \to LP^{2\omega}_{31}$ by Kashyap (1989)[45] seems to suggest that there is spontaneous organisation by simple irradiation of the fibre with strong visible radiation, linking it to be earlier photosensitive phenomenon responsible for reflection grating formation first reported by Hill *et al.* (1978) [77].

The self seeded or externally seeded scheme in optical fibres is wavelength sensitive and the sensitivity to frequency double rapidly tails off at wavelengths longer than 1.064 μm. However, the best second harmonic conversion efficiencies reported are 13% for 1 KW peak pump at 1.064 μm wavelength and summarised by Farries *et al.* (1989) [78]. This active research area continues to progress rapidly.

2.4 Electric Field Induced SHG (EFISH) in Fibres

The study of phase-matched second-harmonic generation by the deliberate application of electric fields across fibres, may be a route to understanding of the mechanisms behind this unusual process. The data on phase-matching and also field overlap thus acquired provides valuable information on waveguide and symmetry properties of nonlinearity and the interacting modes.

Electric-field induced second-harmonic generation (EFISH) [79] is a technique used to induce an asymmetry in an otherwise symmetric material as a probe of the molecular second-order nonlinearity, commonly known as the molecular hyperpolarisability. This technique can also be applied to optical fibres for second-harmonic by inducing an electric-dipole. Two theoretical studies [80] [71] of EFISH in fibre have been published. In one of the studies [71], the interaction between the fundamental and the second-harmonic is phase-matched by designing a step index fibre such that the fundamental long wavelength mode has the same effective index as the second-harmonic mode. By making the index difference between the core and cladding larger than the chromatic dispersion, the phase velocities of the two interacting modes can be made equal. The problems with this method are the tight tolerances required for parameters such as core-radius and refractive index difference. Phase-matching is very sensitive to small changes in these parameters, and this makes the technique almost impractical. Not only is phase-matching crucial for efficient interaction, the overlap of the interacting fields determines how strong the interaction will be (See Section 2.3). It is difficult to optimise both these parameters, since the most efficient interaction between the fundamental modes of long and short wavelengths is not possible without birefringence in the fibre. In contrast, the other technique [80] allows independent optimisation of both parameters, by the use of a periodic electrode structure which separates the phase-matching requirement from the design of the fibre. The momentum (phase) mismatch between the fundamental and the generated waves can be compensated by inducing a spatial periodicity in the effective second-order nonlinearity by the application of a periodic electric field. There are excellent reviews and several papers on the use of periodic structures [40, 41, 27, 81] for phase-matching in bulk materials and wave-guided integrated-optics. They survey the techniques used for linear and non-linear coupling using passive and active (electro-optic and modulation of second-order non-linear *d* coefficient) periodic-structures and show their effectiveness in wave-guided wavelength filters, couplers and mode-converters. However, apart from one application [82] of the modulation of the third-order non-linearity, $\chi^{(3)}$, in wave-guides, no other studies have taken place other than bulk measurement of molecular hyper-polarisability in gases [83].

3. PHASE-MATCHED HARMONIC GENERATION IN WAVEGUIDES: NON-PERIODIC STRUCTURES

Phase-matching to the radiation field was first demonstrated in waveguide by Tien, Ulrich and Martin (1970) [84] in a waveguide made of a linear film of ZnS on a non-linear substrate of ZnO. Phase-matched SHG from the TE_0 mode to the radiation field over an interaction length of 0.1 cm was achieved. The first demonstration of radiation mode phase matching in a crystal cored fibre was reported by Nayar (1982) [98]. Here a uniaxial organic crystal, benzil, was grown in a small bore capillary made of SKN18 Schott glass. Typical bore diameters of 15 to 60 μm were used and radiation mode phase matched harmonic generation was demonstrated from a Q-switch *Nd:NAG* pump source operating at a wavelength of 1.064 μm. The interaction was not optimised and the conversion efficiencies were reported to be low. However, the principle of phase-matching in a crystal cored cylindrical waveguide was demonstrated.

Radiation mode SHG has also been demonstrated in optical fibre [85] which was single mode at the frequency doubled wavelength. Cladding modes at the second harmonic wavelength were generated from the fundamental wavelength at 1.064 μm by the intrinsic fibre nonlinearity. Many of the non-periodic phase matching schemes introduced in the last Section have been demonstrated in waveguides and are summarised in Table 2.

4. REVIEW OF SHG USING PERIODIC STRUCTURES

The physical basis of periodic coupling between the fundamental and second-harmonic waves was proposed theoretically by Bloembergen and Sievers (1970) [25] using a laminated structure in GaAs, and also to an earlier one claimed by Ashkin and Yariv (1961) [115] who considered a periodic modulation of the refractive index. Both these proposals were for bulk crystalline media lacking a centre of symmetry, and the first published paper (again theoretical) using waveguides was by Somekh and Yariv (1972) [26]. They studied the nonlinear coupling in waveguides with a corrugated boundary effecting the mode evanescent field. This is similar to modulating the refractive index of the guide. They further extended their work [63] to include the modulation of the non-linear coefficient, showing that there is a reduction in the effective non-linear coefficient. At about the same time, Tang and Bey (1973) [27] in a theoretical paper studied the coupling of forward and backward harmonic waves with the forward propagating fundamental wave in a waveguide. Some of the ideas considered in that paper were later demonstrated experimentally [50] using a tunable laser source in a periodic etched quartz waveguide over-coated with Al_2O_3. The experimental details are shown in Fig. 7, while the tuning curve obtained for their experiment can be seen in Fig. 8. The modulation of both the nonlinearity and, unavoidably in their experiment, the refractive-index, caused a broadening of the tuning curve. However, scattering into radiation and other guided modes as a result of index modulation is also a very serious problem, causing complete coupling of the guided second-harmonic power to the radiation field within the first few mm of the device. The sub-harmonics of the perturbation frequency required to couple the two desired interacting waves generally falls within the band of frequencies required to couple between guided and radiation modes. It is difficult to eliminate low frequency jitter in the fabrication of the periodic structure. Thus it is likely that such a technique will be of limited practical

TABLE 2

Summary of results: non-periodic phase-matched second harmonic generation in waveguide

Waveguide	$\lambda^\omega (\mu m)$	Modes	l_{eff}	Non-linear medium	$\eta(P_\lambda^\omega)$	Reference
ZnS -ZnO	1.064	$TE_0 \to$ radiation	10mm	substrate	$\sim 10^{-5}(2W)$	Tien [84]
GaAs(Slabs)	10.6	$TE_1 \to TE_2$	1mm	film	$\sim 10^{-2}(1W)$	Anderson [86]
ZnO - glass	1.06	$TE_0 \to TM_1$	20μm	film	$\sim 10^{-9}(8kW)$	Zemon [87]
7059 taper-α-quartz	1.06	$TE_0 + TM_0 \to TM_2$	12mm	substrate	$\sim 10^{-4}(300W)$	Suematsu [104] [89]
Liquid - TiO_2 -α-quartz	1.06	$TE_0 \to TM_0$	3.5mm	substrate	$10^{-4}(100W)$	Burns [90]
Al_2O_3 -α-quartz	0.6	$TE_0 \to TE_2$	~100μm	substrate	$10^{-5}(300W)$	Chen [91]
ZnS -BK7	1.05	$TE_1 \to TM_4$	~35μm	film		Ito [93]
AlGaAs-GaAs-AlGaAs	2.0	$TE_0 \to TM_2$	focused	film		van der Ziel [110]
7059-α-quartz	0.58			substrate	$10^{-9}(2W)$	Chen [95]
Air-TiO_2-ZnS glass	0.93	$TE_0 \to TM_1$	1mm	film	$10^{-3}(100mW)$	Ito [92]
GaP ribbon	1.06	$TE \to TE$	50μm	film		Stone [48]
ZnO-SiO_2	1.06	$TE \to TE$	2.5mm	film	$5 \times 10^{-6}(150mW)$	Shiosaki [96]
Si-α-quartz	1.06	$TE_0 \to$ radiation		substrate	$10^{-9}(20mW)$	Peuzin [97]
Crystal cored fibre	1.064	$LP_{01}^\omega \to$ radiation	30mm	Benzil core	2×10^{-5}	Nayar [98]
Parachlorophenylurea-	0.84	$TM_0 \to TM_2$		film	$0.3 P_\lambda^\omega$	Hewing [99]
Corning 7059	0.92	$TM_0 \to TE_2$				
MNA-Corning 7059-fused silica	1.06	$TE_m \to TE_n$ $m \neq n$	0.5mm	cladding		Saski [51]
GeO_2:silica	1.064	$LP_{01}^\omega \to$ radiation	~1m	optical fibre	$10^{-7}(1kW)$	Lucek [85]

TABLE 2

$LiNbO_3$ waveguides				Comment		
$ZnS\ film\ -LiNbO_3$	1.06	$TE \to TE$	8mm	platelets	$10^{-3}(10W)$	Reutov [100]
	1.11	$TE_0 \to TE_2$	~40 μm	substrate	$10^{-7}(70W)$	Ito [101]
channel	1.06	$TE_{00} \to TE_{00}$	10mm		$1.5 \times 10^{-4}(2\ mW)$	Uesugi [105]
channel	1.09	$TE_{00} \to TM_{00}$	20mm		$0.7 \times 10^{-2}(60\ mW)$	Uesugi [103]
planar	1.08	$TM_0 \to TE_1$	17mm		$0.25(45\ W)$	Sohler [104]
channel	1.09	$TE_{00} \to TM_{00}$	17.3mm	E-O tuning		Uesugi [105]
	1.06		4mm	preparation effects		Noda [106]
p-e/Ti-indiffused	1.08	$TM_0 \to TE_1$		effect on		Micheli [107] [108]
	1.24	$TM_7 \to TE_8$		phase-matching		
resonant channel	1.06		24mm		$10^{-3}(1\ mW)$	Sohler [109]
xz-planar	1.06	$TE_1 \to TM_2$	12 mm		$10^{-4}(15\ mW)$	Zolotov [110]
channel	1.06	$TE_{11} \to TM_{21}$		E-O/temp. tuning	$3 \times 10^{-5}(2\ mW)$	Zolotov [111]
		$TE_{11} \to TM_{13}$				
		$TE_{11} \to TM_{15}$				
coupled 2 channels	1.06	$TM_{11} \to TE_{13}$	5mm	E-O tuning		Bozhevol'nyi [112]
$LiNbO_3$-Sapphire	0.8695	$TM_0 \to TM_2$	150μm	$LiNbO_3$ thin film	$10^{-3}(100\ W)$	Hewig [64]
$Ti:LiNbO_3$; planar	1.06	$TE_0 \to TM_0$	small	non-collinear	3%	Gridin [113]
		$TE_0 \to TM_1$				
$Ti:MgO;LiNbO_3$	1.06	$TM_{00} \to TE_{00}$	16 mm		$0.032(1.35\ W)$	Laurell [114]
		$TM_{00} \to TE_{10}$			$0.03(1\ W)$	
		$TM_{00} \to TE_{20}$			$0.021(1\ W)$	

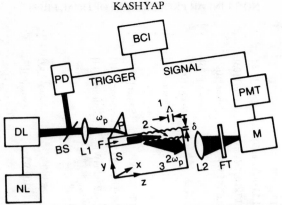

Fig. 7. Non-linear quartz substrate with aluminium oxide thin film waveguide deposited on corrugated surface [after Chen *et al.* [50]]. Glass prism coupling was used for launching fundamental wavelength light from a dye laser into the waveguide, while the guided mode SH light was coupled into the radiation field as a result of the perturbation.

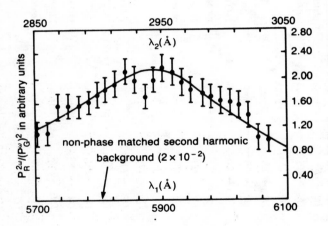

Fig. 8. Tuning curve for the device in Fig. 7. [after Chen *et al.* [50]].

value. Modulation of the refractive index has this great disadvantage. On the other hand, modulation of the non-linear coefficient in crystalline materials can be of significance. However, an inevitable side-effect of implementing such a technique is also the modulation of the refractive index as shown by Jaskorzynska and Arvidsson (1987) [28] who used a lithium niobate waveguide with periodic indiffusion of titanium. This produced a spatial modulation of the refractive index in a channel waveguide. The scheme is shown in Fig. 9. They too, reported a reduction in the effective second-order nonlinearity as a result of the modulation of the refractive index.

Periodic alteration in the orientation of crystalline media to produce artificial phase matching has been demonstrated by several researchers, with Dewey

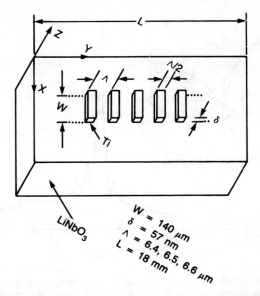

Fig. 9. Schematic of the titanium pattern used for producing the indiffused periodically varying titanium concentration [after Jaskorzynska and Arvidsson [28]].

and Hocker (1975) [116] using rotationally twinned ZnSe crystals to generate continuously tunable radiation between 4 and 21 μm. Piltch et al. (1976) [117] used upto 5 Brewster's angle plates of CdTe, each approximately 5 coherence lengths thick (together giving a total interaction length of 5 coherence lengths), to generate 5.3 μm radiation from a CO_2 laser. Thompson et al. (1976) [118] also frequency doubled a CO_2 laser emission using essentially the same technique but with GaAs plates. Okabda et al. (1976) [119] doubled 1.064 μm radiation from a *Nd: YAG* laser using a stack of 30 quartz and also 6 lithium niobate plates. Frequency doubling experiments in rotationally twinned zinc blende compounds is summarised by Dewey (1977) [120].

SHG in a periodic waveguide was first demonstrated by Ziel et al. [49] using gallium arsenide. They generated the second-harmonic of 2.12 μm wavelength from the emission of a parametric oscillator. A fine corrugated structure of $Al_{0.3}Ga_{0.7}As$ with a 6 μm period was grown as a boundary to a gallium arsenide waveguide. Phase-matched interaction was observed with a peak at the second-harmonic emission wavelength of 1.06 μm. Figure 10 shows the periodic structure and the tuning curve. No attempt was made to measure either the conversion efficiency or the phase-matching bandwidth. As predicted by Tang et al. [27], a cancellation in the effective non-linearity occured owing to the modulation of both the refractive index and the non-linear coefficient.

Laminates of non-linear materials with alternating signs of the nonlinearity of the required coherence length offer a good solution to the phase-matching

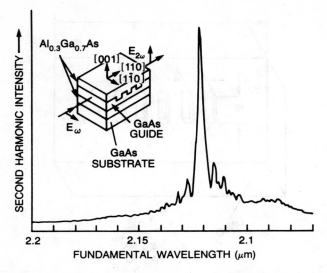

Fig. 10. Phase-matched second harmonic generation using a grating period of 6 μm in a AlGaAs waveguide [after Ziel *et al.* [49]]. Polarisation directions of the fundamental wavelength and the SH wave are shown in the inset.

problem. This method allows access to the largest normally non-phase-matchable non-linear coefficient in materials such as lithium niobate. However, scattering at boundaries may remain a limiting problem with an inevitable reduction in conversion efficiency. Recent results in generating periodic alternation in domains in lithium niobate [121-123] show the potential of this technique. Koidl (1985) [122] has demonstrated second harmonic generation in a chromium doped lithium niobate periodic structure using the largest second-order nonlinearity, d_{33}. Figure 11 shows the growth of the signal as a function of the number of domains; the departure from a straight line in the figure highlights the difficulty in growing exactly the correct domain size equal to the coherence-length in the material. This is a demanding requirement and great control in crystal growth and domain formation is needed. However, the power of this technique is evident from the various experiments carried out.

More recently, a number of studies have been reported using periodic domain reversal in lithium niobate waveguides. Frequency doubling into the green and blue wavelengths from a dye laser has also been demonstrated in titanium in-diffused, domain reversed, proton ion exchanged planar waveguides [124, 125] and lithium out-diffused, proton-exchanged channel waveguides [126]. Different techniques have been used in each case. Lim *et al.* (1989) [124] used a periodic pattern of titanium and in-diffused it into the +z face of a wafer of lithium niobate. Subsequent heat treatment domain-reversed the patterned regions, creating periodically poled sections. The domains are shown in Fig. 12. A pattern with sections of varying pitch was used and a planar waveguide was subsequently formed using proton exchange. The sections of

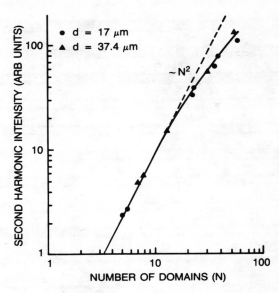

Fig. 11. Second harmonic generation as a function of number of domains in lithium niobate [after Feisst & Koidl [122]]. $d/l_c = 5(\bullet)$ and $11(\triangle)$. The departure from the ideal case can be seen for large number of domains. Quasi phase-matching allowed the use of the d_{33} non-linear coefficient.

varying pitch ensured phase-matching in one of them. Periodic phase-matched frequency doubling was achieved at 1.064 μm, using a domain period of three coherence lengths. The conversion efficiency was measured to be 5% per W-cm². Frequency doubling was also achieved into the blue wavelength spectrum using a cw Styryl-9 dye laser as the pump, with a reported efficiency of 2.4% per W-cm². In both cases, the launched pump power was 25 mW. Arvidsson et al. (1989) [126] achieved a similar result by periodic out-diffusion of lithium at 1100°C to periodically pole the domains. A periodic silica mask was used to prevent out-diffusion of lithium in the masked regions. The result was a spatially periodic domain reversed wafer. Subsequently, proton ion-exchange was used to form a channel waveguide. Frequency doubling was demonstrated using 1.064 μm and 0.83 μm pump wavelengths. A device length of 1 cm was used and a conversion efficiency of approximately ×0.5 of the calculated theoretical maximum was achieved (~ 0.1% at 532 nm for 0.1 W pump). Phase-matched SHG from the red into the blue has also been demonstrated and is shown in Fig. 13 (Arvidsson, *unpublished*). Khanarian et al. [127-128] recently demonstrated periodic-electrically-poled polymer waveguides which allowed phase-matched frequency doubling from a wavelength of 1.064 μm for the two lowest order modes.

Levine et al. (1975) [82] used a periodic static electric field (EFISH) on a nitrobenzene cored planar-waveguide as shown in Fig. 14, to induce a non-

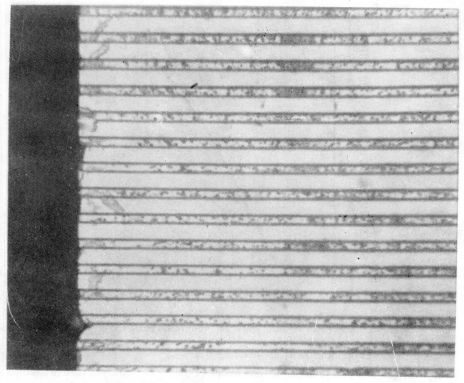

Fig. 12. Surface of lithium niobate wafer etched on the c-face with hydrofluoric acid reveals periodic ferroelectric domains as a result of titanium indiffusion. The period of the grating is 10 microns. [After Lim *et al.* [125]].

linearity and phase-match guided modes. Electrodes, one periodic (pitch of 13.8 μm) and another plane, were deposited on the glass surface and on the hypotenuse surface of a prism. These surfaces were overcoated with magnesium fluoride which formed the cladding of the core material, nitrobenzene. The 1.064 μm emission from a *Nd:YAG* laser was coupled into the waveguide via prism-coupling. On applying an electric field between the electrodes and rotating the periodic grating, phase-matching was observed. Three mode-interactions were reported and can be found in Table 3. The guide used for these experiments was intentionally designed with a large index-difference in order to probe the microscopic hyperpolarisability of a series of molecules of different liquids with widely varying indices of refraction. For this reason, the mismatch between the fundamental TE_0^ω and $TE_0^{2\omega}$ was too large for a reasonable pitch length to be used for phase-matching. The only interactions possible were for a higher order fundamental wavelength mode to a lower-order harmonic mode. The smallest phase mismatch was between the TM_2^ω and $TM_4^{2\omega}$ modes which was the only interaction requiring the lowest order of the

TABLE 3

Summary of results on phase-matched second-harmonic generation in waveguides using periodic structures

Waveguide	$\lambda^\omega(\mu m)$	Modes	l_{eff}	Comment	$\eta(P_\lambda^\omega)$	Reference
Nitrobenzene on glass	1.064	$TM_2^\omega \to TM_4^{2\omega}$	20 mm	periodic dc-shg		Levine [80]
	1.064	$TM_3^\omega \to TM_1^{2\omega}$				
	1.064	$TM_4^\omega \to TM_5^{2\omega}$				
Al_2O_3 on quartz substrate	$0.57 \to 0.61$	$TE_0^\omega \to TE_1^{2\omega}$ →radiation	72 μm	grating etched in quartz	2.5×10^{-7}(40 kW)	Chen [43]
GaAs planar	2.121	$TE^\omega \to TM^{2\omega}$	1 mm	GaAs/AlGaAs grating		Ziel J P van der [47]
Ti: $LiNbO_3$ planar	1.064	$TE_0^\omega \to TM_0^{2\omega}$	18 mm	*periodic Ti: indiffusion*		Jaskorzynska [29]
Silica fiber	1.064	Several modes	9.8 mm	periodic dc-shg	4×10^{-6}(1.4 kW)	Kashyap [130] [61]
Polymer	1.34	$TM_0^\omega \to TM_2^{2\omega}$	~294 μm	periodically poled		Khanarian [128]
Planar proton-ion exchanged $LiNbO_3$	1.06	$TM_{00} \to TM_{00}$	1 mm	Periodic Ti: indiffusion	5×10^{-7}(1 mW)	Lim [125]
Proton ion-exchanged channel $LiNbO_3$	0.820	Fundamental	1 mm	Periodic Ti: indiffusion	6.4×10^{-5}(14.7 mW)	Lim [142] [124]
Ti: $LiNbO_3$ channel	0.833	$TM_{00} \to TM_{00}$	14 mm	LiO_2 out-diffusion	2×19^{-5}(3 mW)	Webjörn [143]
Rb: KTP channel	1.064	$TM_{00} + TE_{00} \to TE_{00}$	5 mm	Periodic Rb guide	1.5% (0.1 W)	Bierlein [139]
Rb: KTP channel	0.852	$TM_{00} \to TM_{00}$	5 mm	Periodic Rb 1st order	0.44% (0.45W)	Poel [140]
	0.801	$TE_{00} \to TM_{00}$	5 mm	2nd order grating	0.125%(0.1W)	

Fig. 13. Blue light generated by frequency doubling laser diode light in a lithium niobate waveguide using quasi-phase-matching. [Photograph by kind permission of G Arvidsson and F Laurell, unpublished].

pitch. The other two interactions were second-order interactions. The tuning curve for one of the interactions is shown in Fig. 15. Conversion efficiency measurements were not reported.

From 1976 on, for a period of ten years, no work was reported using second-order non-linear waveguiding devices with periodic structures. In 1986, the use of a periodic electrode structure with a capillary fibre filled with a highly non-linear liquid, such as nitrobenzene, in a similar scheme used by Levine *et al.* [82], to modulate the third-order susceptibility was proposed [80]. The conversion efficiency for such a device was calculated to be a few percent for $1W$ pump power, which is within the reach of commercially available semiconductor diode sources [129]. It should be remembered that the best second-order material, lithium niobate, suffered from optical damage until recently [130], and that it was *not possible* to launch 1W cw light into a waveguide device without photo-refractive [131] damage. Despite the large second-order nonlinearity, the best results reported for a lithium niobate guided wave frequency doubler are in the region of 25% [104] for *pulsed* operation.

Lateral confinement of optical power, such as in a ridge, channel [132], or fibre waveguide, has the great advantage of increasing the power-density in the waveguide. For example 1 mW of power propagating in a guide of 10 μm diameter produces a power density of 10^7 Wcm^{-2}. It is difficult to control the

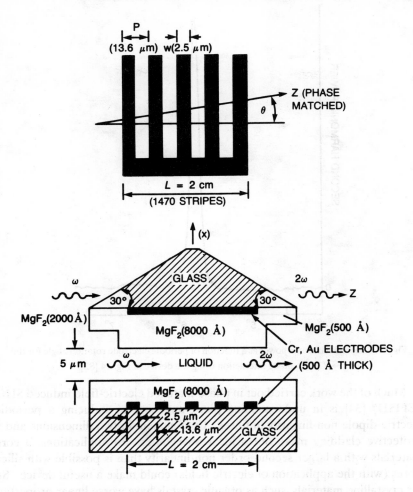

Fig. 14. Waveguide geometry used for periodic electric field induced SHG in a nitrobenzene waveguide [after Levine et al. [82]].

dimensions of ridge and channel waveguides to the accuracy needed for phase-matching over the length of the guide. In addition, the etched side surfaces of a ridge waveguide are usually rough and scatter the power out of the guided mode. In contrast, glass fibre waveguides have the advantages of long interaction length, ease of handling, case of coupling light in and out, and protection of fragile core materials, as in the case of crystalline cores [133, 134]. Organic crystal cored fibres offer the potential advantage of large non-linearities, low power operation and high efficiencies [135]. However, before devices of acceptable quality can be fabricated, problems such as optical throughput, uniformity of core, defects, kinetics of confined growth and handling need to be researched. Long term stability and optical damage of these compounds are other questions which have not yet been addressed.

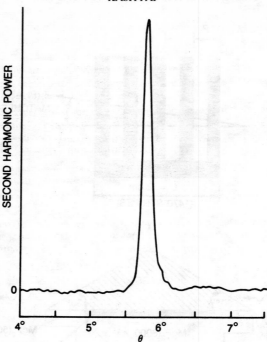

Fig. 15. Second harmonic power as a function of periodic electrode rotation angle for the $TM_2^{\omega} \rightarrow TM_4^{2\omega}$ mode interaction using the device in Fig. 14 [82].

Much of the work carried out in phase-matched electric-field induced SHG (EFISH) [34] is in understanding phase-matching by inducing a periodic electric-dipole non-linearity in fibres. Confinement in two dimensions and a protective cladding make fibres ideal waveguides. In applications, a core materials with a larger second-order non-linearity than is possible with silica fibres (with the application of electric fields) could make a useful device. So far crystalline materials, such as organic crystals have worse linear properties than optical fibres. This makes the task of making good waveguides very difficult. Given that the control of fibre dimensions is probably the best available for any waveguides, the question remains to be answered: is the control good enough for second-order parametric interactions using the fibre structure? Glass optical fibre are generally isotropic and possess a centre of symmetry. Thus in order to study second-order processes, the symmetry has to be broken and this is done by an external electric field. Recently, electric field induced SHG has been produced by the orientation of defects in fibres [45]. Periodic EFISH in optical fibres has also been reported [34]. This work may also help to gain an understanding of the mechanism behind self-generated second-harmonic. By applying a periodic electric field across fibres with periodic gratings already written-in, it should be possible to probe the symmetry of the written-in non-linearity.

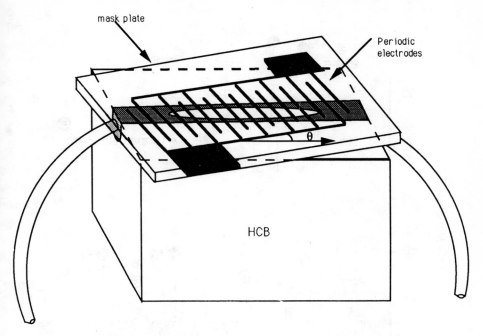

Fig. 16. HCB with electrode overlay mask plate for electric-field induced second-harmonic generation (EFISH)

The scheme used for EFISH in optical fibres in shown in Fig. 16. A fibre is imbedded in a silica block in such a way that it curves away from the surface of the block. By polishing the surface, the core region was nearly exposed over a section of the fibre. An interdigitated periodic electrode structure on a glass mask is then lowered onto the surface. By rotating the mask, the fibre core experiences an electric field with a spatial periodicity determined by the rotation angle and the pitch of the periodic electrode. Using this device, phase-matched SHG was demonstrated to virtually all propagating modes, as shown in Fig. 17, from a fundamental wavelength of $1.064\,\mu$m. Phase-matched interactions over approximately 10 mm were reported [34].

This technique has also been used to permanetly periodically pole optical fibres [136, 137] using periodic static fields of the correct period for phase-matching. Poling was achieved by simultaneous application of the periodic static field and propagation of $1.064\,\mu$m QS radiation. The second-harmonic field generated through EFISH appears to aid the poling process. After the poling field was switched off, the fibre could frequency double $1.064\,\mu$m radiation. Frequency doubled radiation can then act as seeding radiation for the subsequent length of fibre beyond the poled region.

Periodic electric field poling should be possible for a variety of ferroelectric crystalline materials such as lithium niobate, potassium niobate, KTP analogues etc, although this has so far not been reported.

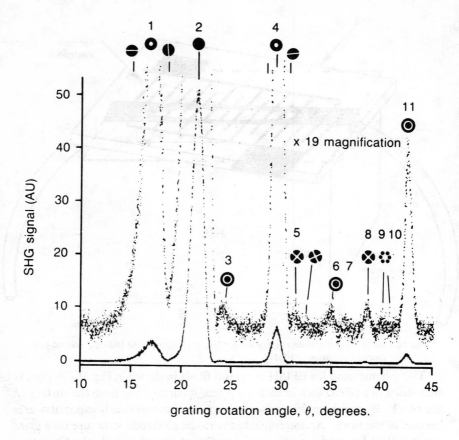

Fig. 17. Electric-field induced SHG signal for a fibre device as a function of rotation angle, θ, of priodic-electrode. The orientation of the SHG modes is shown, along with the expanded tuning curve to highlight the phase-matched peaks (From Kashyap [34]).

Another scheme for poling crystalline material has been reported by Key et al. [138]. They have demonstrated domain reversal in the negative c-face of lithium niobate for the first time by selective ion-beam bombardment of the negative c-face of the crystal while a poling field of 10 V-cm^{-1} was applied at 580°C, well below the Curie temperature. A periodic gold mask was used to allow selective bombardment to generate a first-order grating suitable for optimised frequency doubling in a proton-exchanged waveguide. Although frequency-doubling was not demonstrated, domain reversal was realised. This technique appears to be promising since titanium in-diffusion is not necessary, nor poling at elevated temperatures is required. It should be possible to apply this technique to other ferroeletric crystals which suffer from degradation when heated to the Curie temperature for poling.

Finally, in an unusual approach, Bierlein et al. [139] have reported a variation on quasi phase-matching. This scheme is different to standard periodic phase-matching previously discussed since it uses pairs of sections, one of which compensates for phase-mismatch in the other. The device has periodic indiffusion of rubidium to produce domain-reversed Rb-exchanged KTP waveguides. The regions between the indiffused regions are not waveguides. The length of the waveguiding region is designed to be shorter that the coherence length. The non-waveguiding region is kept shorter than the Rayleigh length so that the loss is insignificant. The length is tailored to bring the free and bound waves exactly into phase at the beginning of the next waveguiding section. The overall effect is equivalent to phase-matching by domain reversal as in conventional quasi phase-matching, combined with linear dispersion correction. This allows a long length to be phase-matched. A schematic of the device is shown in Fig. 18 while a photograph of a section of the device is shown in Fig. 19. Harmonic generation from a fundamental

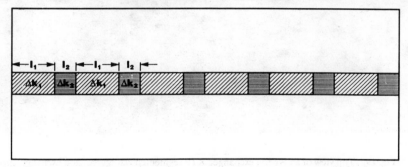

Fig. 18. Segmented waveguide structure showing alternate regions. Each pair, l_1 and l_2, forms a fully phase-matched section [After Bierlein et al. [139]].

wavelength of 1.064 μm was demonstrated in a 5 mm long sample with a normalised conversion efficiency of 15% W^{-1} cm^{-2}.

In a more recent publication, Poel et al. [140] have demonstrated a first-order grating Type I phase-matching in a similar device with a conversion efficiency of 76% W cm^{-2} (approximately 1.5% for a guided fundamental wavelength power of 100 mW), the highest reported so far in the literature for a periodic waveguide. The fundamental wavelength was 834 nm from a tunable Ti: Sapphire laser. They also reported a second-order grating interaction for $TE \to TM$ Type I interaction at a fundamental wavelength of 852 nm (450 mW) with a conversion efficiency of approximately 0.44%. As is typical in waveguides which use periodic structures for phase-matching, the tolerance on waveguide parameters, such as index-difference and waveguide dimensions, is considerably relaxed [34], making phase-matching non-critical.

Results on SHG in waveguides using periodic structures for phase-matching are summarised in Table 3.

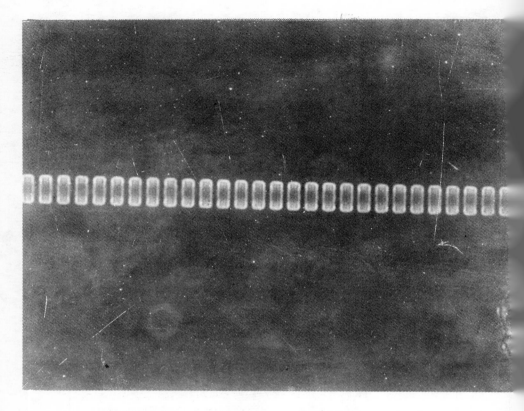

Fig. 19. Top view of an Rb ion exchanged KTP segmented waveguide. Ion exchange into KTP is highly anisotropic and results in very sharp edges, seen clearly in the figure. The period is 4 microns [After Bierlein *et al*. [139]].

5. CONCLUDING REMARKS

This chapter has attempted to cover the rapidly advancing field of periodic structures in frequency doubling waveguide devices. Although it cannot claim to be an exhaustive review of phase-matching techniques, it should provide a flavour of a multitude of techniques which have been used experimentally and especially highlight the powerful concept of phase-compensation by periodic structures. As the field progresses, we should expect many exciting results and possibly for the first time, frequency conversion efficiencies approaching 100% in waveguides using periodic phase-matching and applications in parametric amplification.

I would like to thank C.A. Millar, B.K. Nayar, S.T. Davey and G.D. Maxwell for constructive comments on the manuscript.

REFERENCES

1. P.A. Franken, Hill A.E., Peters C.W. and Weinreich G, "Generation of optical harmonics", *Phys Rev lett* 7(4), 118-119, 15 August 1961.
2. M. Bass, Franken P.A., Hill A.E, Peters C.W. and Weinreich G, "Optical mixing", *Phys Rev letts*, 8(1), 1 January 1962.
3. J.A. Armstrong, Bloembergen N, Ducuing J. and Pershan P.S., "Interactions between light waves in a non-linear dielectric", *Phys Rev*, 127(6), 1918-1939, September 1962.
4. G. Eckhardt, Hellwarth R.W., McClung F.J., Schwarz, Weiner D. and Woodbur E.J., "Stimulated Raman scattering from organic liquids", *Phys Rev Letts*, 9(11), 455-457, 1 December 1962.
5. J.A. Giodmaine and Miller R.C., "Tunable coherent parametric oscillation in $LiNbO_3$ at optical frequencies" *Phys Rev Letts* 14(24), 973-976, 1965.
6. R.W. Terhune, Maker P. D. and Savage C.M, "Optical harmonic generation in calcite", *Phys Rev Letts* 8(10), pp 404-406, May 15, 1962.
7. J.A. Giordmaine, "Mixing of light beams in crystals", *Phys Rev Letts*, 8(1), 19-20, 1 January 1962.
8. P.D. Maker, Terhune R.W., Nisennoff and Savage C.M., "Effect of dispersion and focusing on the production of optical harmonics", *Phys Rev Letts* 8(1), 21-22, 1 January 1962.
9. D.C. Hanna, Yuratitch M.A. and Cotter D, in *Non-linear optics of free atoms and molecules*, Springer series in Optical Sciences 17, Ed MacAdam D L, pp 258-261, 1979.
10. M.V. Hobden, "Phase-matched second-harmonic generation in biaxial crystals", *J Appl Phys*, 38(11), 4365-4372, October 1967.
11. J.E. Midwinter and Warner J, "The effect of Phase-matching method and of uniaxial crystal symmetry on the polar distribution of second-order non-linear optical polarisation", *Brit J of Appl Phys*, 16, 1135-1142, 1965.
12. R.C. Miller, Boyd G.D. and Savage A, "Non-linear optical interactions in $LiNbO_3$ without double refraction", *Appl Phys Lett*, 6(4), 77-79, 15 February 1965.
13. D.A. Kleinman, Ashkin A. and Boyd G.D. "Second harmonic generation of light by focussed laser beams", *Phys Rev*, 145 (1), 338-379, 6 May 1966.
14. G.D. Boyd and Kleinman D.A., "Parametric interactions of focused Gaussian beams", *J Appl Phys*, 39 (8), 3597-3639, July 1968.
15. R.L. Byer "Optical parametric oscillators, in *Treatise in Quantum Electronic, Vol. I*, part B, H. Rabin and Tang C.L. Eds., New York: Academic, 1975.
16. J.T. Lin and Chen C, "Choosing a non-linear crystal", *Lasers and Optronics*, November 1987.
17. See for example, *Proceedings of the Fourth European Conference on Integrated Optics*, (ECIO 87), Eds: C.D.W. Wilkinson, Lamb J, Glasgow, May 11-13, 1987.
18. R.H. Stolen "Fibre Raman Lasers", in *Fibre and Integrated Optics, Vol 3*(1), Crane Russak & Co, 1980.
19. K.J. Blow and Doran N.J., "Non-linear effects in optical fibres and fibre devices", *IEE Proc 134* part J(3), 138-144, June 1987.
20. R.H. Stolen Bösch and Chinlon Lin, "Phase-matching in birefringent fibres" *Optics Lett* 6 (5), 213-215, May 1981.
21. Chinlon Lin, Reed W.A., Pearson A.D. and Shang H.T., "Phase-matching in the minimum-chromatic-dispersion region of single-mode fibres for stimulated four-photon mixing", *Optics Letts* 10 (6), 493-495, October 1981.
22. R.W. Hellwarth, "Optical beam phase conjugation by four-wave mixing in a waveguide", *Optical Engineering* 21 (2), 263-265, March/April 1982.

23. D.M. Pepper, "Non-linear optical phase conjugation", *Optical Engineering* 21 (2), March/April 1982.
24. A. Yariv, Au Yeung J., Fekete D. and Pepper D.M., "Image phase compensation and real-time holography by four-wave mixing in optical fibres", *Appl Phys Letts* 32, 635, 1978.
25. N. Bloembergen and Sievers A. J., "Non-linear optical properties of periodic laminar structures", *Appl Phys Letts* 17, (11), 483-485, 1 December 1970.
26. S. Somekh and Yariv A., "Phase-matchable non-linear optical interactions in periodic thin films", *Appl Phys Lett 21* (4), 140-141, 15 August 1972.
27. C.L. Tang and Bey P.P., "Phase-matching in second-harmonic generation using artificial periodic structures", *IEEE J. Quant Electron, QE*-9 (1), 9-17, January 1973.
28. B. Jaskorzynska and Arvidsson G., "Periodic structures for phase-matching in second-harmonic generation in titanium lithium niobate waveguides", in Proceedings of *The Third International Symposium on Optical and Opto-electronic Applied Science and Engineering: Integrated Optical Circuit Engineering III, SPIE Vol.* 651, Innsbruck, Austria, 1987.
29. G.D. Boyd and Patel C.K.N. "Enhancement of optical second-harmonic generation (SHG) by reflection phase-matching in ZnS and GaAs", *Appl Phys Lett* 8, 313, 1966.
30. D.A. Kleinman, "Non-linear dielectric polarisation in optical media", *Phys Rev 126* (6), 1977-1979, 15 June 1962.
31. For a complete description of symmetry and properties of the non-linear optical tensors, see P.N. Butcher and Cotter D., "Elements of Non-linear Optics", Cambridge University press, 1990.
32. R.A. Morgan "Phase-matching considerations for generalised three-wave mixing in non-linear anisotropic crystals", *Appl Opts* 29 (9), 1259-1264, 20 March 1990.
33. N. Bloembergen, "Conservation laws in non-linear optics", *J. Opt Soc Am* 70 (12), 1429-1436, 12 December 1980.
34. R. Kashyap, "Phase-matched periodic electric-field-induced second-harmonic generation in optical fibres", *J. Opt Soc Am B* 6 (3), 313-328, March 1989.
35. A. Azéma, Botineau J., Gires F. & Saissy A, "High efficiency generation of second-harmonic in plane waveguides", *Proc SPIE 213*, 26-29, 1979.
36. H. Suche, Ricken R. and Sohler W., "Integrated optical parameteric oscillator of low threshold and high power conversion efficiency", *in Proceedings of the 4th European Conference on Integrated Optics* (ECIO 87), Ed. Wilkinson C.D.W. and Lamb J., pp 202, Glasgow, 11-13, May 1987.
37. G. Arvidsson and Laurell F., "Second harmonic generation in channel waveguides fabricated by titanium indiffusion in magnesium doped lithium niobate", *in Proceedings of the 4th European Conference on Integrated Optics* (ECIO 87), Ed. Wilkinson C.D.W. & Lamb J., pp 198, Glasgow, 11-13, May 1987.
38. B.K. Nayar, Kashyap R. and White K.I., "Design of efficient organic crystal cored fibres for parameteric interactions: phase-matching requirements", SPIE 651, pp 235-237, Innsbruck, Austria, 1986.
39. K.I. White, Nayar B.K. and Kashyap R., "Amplification and second harmonic generation in non-linear waveguides", *Opt and Quant Electron* 20, 339-342, 1988.
40. C. Elachi, "Waves in active and passive periodic structures: a review", *Proc IEEE 64* (12) 1666-1698, Dec. 1976.
41. A. Yariv and Nakamura M, "Periodic structures for integrated optics", *IEEE J. Quant Electron QE-13* (4), 233-153, April 1977.
42. D. Gloge, "Weakly guiding fibres", *Appl Optics 10*, 2252-2258, 1971.
43. A. Snyder and Love J.D., *Optical Waveguide Theory*, Chapman and Hall, 1983.

44. J.W. Fleming, "Material dispersion in lightguide glasses", *Electron Lett* **14** (11), 326-328, 25 May 1978.
45. R. Kashyap, "Photo induced enhancement of second harmonic generation in optical fibres", *Topical Meeting on Non-linear Guided Wave Phenomenon: Physics and Applications, 1989, Technical Digest Series*, Vol 2, held on February 2-4, 1989, Houston, USA, (Optical Society of America, Washington DC 1989), 255-258.
46. M.E. Fermann, Li L., Farries M.C. and Payne D., "Phase-matched second harmonic generation in permanently poled optical fibres", *in Proc of the XIV European Conference on Optical Communications, IEE Conf Proc 292* (1), 135-138, Brighton, UK, 11-15 September 1988.
47. M. C. Micheli, Zyss J. & Azéma A., "Possibilities of optically non-linear thin films", *Proc SPIE 401*, 216-225, 1983.
48. J. Stone, "Phase-matched second harmonic generation in vapour grown GaP ribbon waveguides", *J. Appl Phys* **50** (12), 7906-7913, Dec. 1979.
49. Ziel J.P. van der, Ilegems M, Foy P. W. and Mikulyak R. M., "Phase-matched second harmonic generation in a periodic GaAs waveguide", *Appl Phys Lett* **29** (12), 775-777, 1976.
50. B.U. Chen, Ghizoni C.C. and Tang C. L., "Phase-matched second-harmonic generation in solid thin films using modulation of non-linear susceptibilities", *Appl Phys Letts* **28** (11), 1 June 1976.
51. K. Sasaki, Kinoshita T. & Karasawa N., "Second harmonic generation of 2-methyl-4-nitroaniline by a neodymium: yttrium aluminium garnet laser with a tapered slab-type optical waveguide", *Appl Phys Lett* **45** (4), 333-334, 15 August 1984.
52. G.H. Hewig and Jain K., "Phase-matched frequency doubling in a $LiNbO_3$-sapphire waveguide", *J. Non-Cryst Solids* **47** (2), 271-273, 1982.
    G.H. Hewig and Jain K., "Frequency doubling in a $LiNbO_3$ thin film deposited on sapphire", *J Appl Phys* **54** (1), 57-61, January 1983.
53. S. Umegaki, Takahashi Y., Manabe A. & Tanaka S., "Optical second harmonic generation in an organic crystal core fibre", *Extended Abstracts of Symposium O an Non-linear Optical Materials*, Materials Research Society (Pittsburg), Ed. Miller D.A.B., 1985 Fall meeting Boston MA, pp 97-99.
54. R. Heckingbottom, Hill J.R., Holdcroft G.E., Dunn P.L., Pantelis P. and Rush J.D., "Second order optical non-linearities in organic polymer films and crystal cored fibres", in Proc of the Interational Symposium on Non-linear Optics of Organics and Semiconductors, Ed. Kobayashi T., Tokyo, Japan, pp 284-291, July 25-26, 1988.
55. B.K. Nayar, "Optical fibres with organic crystalline cores" in Springer Proc in Physics 7, *Non-linear Optics: Materials and Devices*, Ed. Flytzanis C. & Oudar J.L., 142-153, 1986.
56. J.R. Cozens, "Propagation in cylindrical fibres with anisotropic crystal cores", *Electronics Letters* **12** (16), 413-415, 5 Aug 1976.
57. R.B. Dyott, Cozens J.R. and Morris D.G., "Preservation of polarisation in optical-fibre waveguides with elliptical cores" *Electronics Letters* **15**, 13, 380-382, 23 May 1979.
58. C. Yeh and Manshadi F., "On weakly guiding single-mode optical waveguides", *IEEE J Lightwave Technol. LT-3* (1), 199-205, February 1985.
59. A. Kumar, Shenoy M. R. and Thyagarajan K., "Polarisation characteristics of elliptical-core fibres with stress induced anisotropy", *J. Lightwave Tech. LT-5* (2), 193-198, February 1987.
60. P.K. Tien, Ulrich R. and martin R.J., "Optical second harmonic generation in form of coherent Cerenkov radiation from a thin film waveguide", *Appl Phys Lett* **17**, 477, 1970.

61. B.K. Nayar, "Optical second harmonic generation in crystal-cored fibres", in OSA Digest of the 6th Topical Mtg. on Integrated and Guided Wave Optics, ThA2, 1982.
62. N. Bloembergen and Sievers A. J., "Non-linear optical properties of periodic laminar structures", *Apply Phys Lett 17*, (11), 483-485, 1 December 1970.
63. S. Somekh and Yariv A., "Phase-matching by periodic modulation of the non-linear optical properties", *Opt Comm 6* (3), 301-304, November 1972.
64. K.O. Hill, Kawasaki B.S., Johnson D.C. and MacDonald R.I., "CW three wave mixing in single mode optical fibres" *J. Appl Phys. 49* (10), 5098-5106, October 1978.
65. U. Ōsterberg and Margulis W., (a) *Digest of XIV International Quantum Electronics Conference (OSA)*, Paper WBB2, pp 102, Washington DC, 1986.
   (b) "Dye laser pumbed by $Nd:YAG$ laser pulses frequency doubled in glass optical fibre". *Optics Letts*, *11*, 516-518, 1986.
66. M. Farries, St. J. Russell P., Fermann M.E., and Payne D.N, "Second harmonic generation in an optical fibre by a self-written $\chi^{(2)}$ grating", *Electron Lett 23* (7), 322-324, 26 March 1987.
67. Y. Fujii, Kawasaki B.S., Hill K.O. and Johnson D.C., "Sum frequency generation in optical fibres", *Opts Lett 5* (2), 48-50, February 1980.
68. Y. Sasaki and Ohmori Y., "Phase-matched sum-frequency light generation in optical fibres", *Appl Phys Lett 39* (6), 466-468, 15 September 1982.
69. B. Valk, Kim E.M. and Salour M.M., "Second harmonic generation in Ge-doped fibres with a mode-locked Kr^+ laser", *Appl Phys Lett, 51* (10), 722-724, 7 September 1987.
70. R. Stolen and Tom H.W.K., "Self-organised phase-matched harmonic generation in optical fibres", *Optics Lett 12* (8), 585-587, August 1987.
71. R.W. Terhune and Weinberger D.A., "Second harmonic generation in fibres", *J. Opt Soc Am B 4* (5), 661-674, May 1987.
72. F.P., Payne, "Second harmonic generation in single-mode optical fibres", *Electron Lett 23* (23), 1215-1216, 5 Nov. 1987.
73. A. Krotkus and Margulis W., "Investigations of the preparation process for efficient second-harmonic generation in optical fibres", *Appl Phys Lett 52* (23), 1942-1944, 6 June 1988.
74. R.H. Stolen and Tom H.W.K., "Self organisation in optical fibres for efficient phase-matching of harmonic generation", in *Technical digest of Conference on Lasers and Electro-Optics, Vol. 17*, paper THL2, 274-276, 1987.
75. P. Chmela, "On the phenomenology of self-organised quasi-phase-matched second harmonic generation in optical fibres", *J Mod Optics* 37 (3), 327-338, March 1990.
76. H.W.K. Tom, R.H. Stolen, Aumiller and Pleibel W., "Preparation of long coherence length second harmonic generating optical fibres by using mode-locked pulses", *Optics Letts. 13* (6), 512-514, June 1988.
77. K.O. Hill, Fujii Y, Johnson D.C. & Kawasaki B.S. "Photosensitivity in optical fibre waveguides: Application to reflection filter fabrication", *Appl Phys Lett.* 32 (10), 647-649, 15 May 1978.
78. M.C. Farries, Fermann M.E. & Russell P., St J, "Second harmonic generation in optical fibres", *Topical Meeting on Non-linear Guided Wave Phenomenon: Physics and Applications, 1989, Technical Digest Series, Vol. 2*, held on February 2-4, 1989, Houston, USA, (Optical Society of America, Washington DC 1989), 246-249.
79. S. Kielich, "Optical second-harmonic generation by electrically polarised isotropic media", IEEE J *Quant Electron QE-5* (12), 562-568, 1969.
80. R. Kashyap, "Non-linear optical interactions in waveguides with cylindrical geometry", in the Proceedings of *Molecular and Polymeric Optoelectronic Materials: Fundamentals*

and Applications, Society of Photo-Optical and Instrumentation Engineers *SPIE Vol. 682*, 170-178, San Diego, California, USA, 1986.

81. R.C., Alferness, "Waveguide electrooptic modulators", *IEEE Trans on Microwave Theory and Techniques, MTT-30* (8), 1121-1137, 8 August 1982.

82. B.F. Levine, Bethea C.G. and Logan R.A., "Phase-matched second harmonic generation in a liquid filled waveguide", *Appl Phys Lett 26* (7), 375-377, 1 April 1975.

83. D.P. Shelton and Buckingham A.D., "Optical second harmonic generation in gases with low power", *Phys Rev A 26* (5), 2787-2798, November 1982.

84. P.K. Tien, Ulrich R. and Martin R.J., "Optical second harmonic generation in form of coherent cerenkov radiation from a thin-film waveguide", *Appl Phys Lett 17*, 447-450, 1970.

85. J. Lucek, Kashyap R., Davey S.T. and Williams D.L., "Second harmonic generation in optical fibres", *J. Mod Opts 37* (4), 533-543, April 1990.

86. D.B. Anderson and Boyd R.T., "Wide band CO_2 laser second harmonic generation phase-matched in *GaAS* thin-film waveguides", *Appl Phys Lett 19*(8), 266-268, October 1971.

87. S. Zemon, Alfano R.R., Shapiro S.L. and Conwell E., "High-power effects in non-linear optical waveguides", *Appl Phys Lett 21* (7), 327-329, 1 October 1972.

88. Y. Suematsu, Sasaki Y. and Shibata K., "Second harmonic generation due to a guided wave structure consisting of a quartz coated with a glass film", *Appl Phys Lett 23* (3), 137-138, 1 August 1973.

89. Y. Suematsu, Sasaki Y., Furuya K., Shibata K. and Ibukuro S., "Optical second-harmonic generation due to guided-wave structure consisting quartz and glass film", *IEEE J of Quant Electron EQ-10* (2), 222-229, February 1974.

90. W.K. Burns and Lee A.B., "Observation of non-critically phase-matched second-harmonic generation in an optical waveguide", *Appl Phys Letts 24* (5), 222-224, 1 March 1974.

91. B.U. Chen, Tang C.L. and Telle J.M., "CW harmonic generation in the uv using a thin-film waveguide on a non-linear substrate", *Appl Phys Letts 25* (9), 495-498, 1 November 1974.

92. H. Ito and Inaba H., "Efficient phase-matched second-harmonic generation method in four-layered optical-waveguide structure", *Optics Lett 2* (6), 139-141, June 1978.

93. H. Ito, Uesugi N. and Inaba H., "Phase-matched guided optical second harmonic wave generation in oriented ZnS polycrystalline thin-film waveguide", *Appl Phys Letts 25* (7), 385-387, 1 October 1974.

94. J.P. Ziel van der, Miller R.C., Logan R.A., Norland Jr W.A. and Mikulyak R.M., "Phase-matched second-harmonic generation in *GaAS* optical waveguides by focussed laser beams", *Appl Phys Letts 25* (4), 238-240, 15 August 1974.

95. B.U. Chen and Tang C.L., "Generation of second-harmonic uv radiation using a non-linear substrate coated with a uv absorbing thin-film", *IEEE J Quant Electron QE-11* (5), 177-178, April 1975.

96. T. Shiosaki, Fukuda S., Sakai K., Kuroda H. and Kawabata A., "Second harmonic generation in As-sputtered ZnO optical waveguide", *Jap J. Appl Phys 19* (12), 2391-2394, December 1980.

97. J.C. Peuzin, Olivier M., Cuchet R. and Chenevas-Paulle A, "How to use bulk materials in integrated optics", *Proc SPIE 400*, 139-140, 1983.

98. B.K. Nayar, "Non-linear optical interactions in organic crystal cored fibres", in *Non-linear optical properties of organic and polymeric materials, ACS Symposium series 233*, 153-166, Ed Williams D.J., 184th meeting of the American Chemical Society, Kansas City, Missouri, September 12-17 1983.

99. G.H. Hewig and Jain K., "Frequency doubling in an organic waveguide", *Opt Comm 47* (5), 347-350, 1 October 1983.
100. A.T. Reutov and Tarashchenko P.P., "Frequency multiplication of coherent radiation in an all optical micro-waveguide with non-linear layer of lithium niobate", *Opt Sepctrosc 37* (4), 447-448, October 1974.
101. H. Ito and Inaba H., "Phase-matched guided, optical second-harmonic generation in non-linear ZnS thin-film waveguide deposited on non-linear $LiNbO_3$ substrate", *Opt Comm 15* (1), 104-107, September 1975.
102. N. Uesugi and Kimura T., "Efficient second-harmonic generations in three-dimensional $LiNbO_3$ optical waveguide", *Appl Phys Letts 29* (9), 572-574, 1 November 1976.
103. N. Uesugi, Diakoku K. and Fukuma M., "Tuning characteristics of parametric interaction in a three-diamensional *$LiNbO_3$* optical waveguide", *J Appl Phys 49* (9), 4945-4946, September 1978.
104. W. Sohler and Suche H., "Second-harmonic generation in Ti-diffused *$LiNbO_3$* optical waveguides with 25% conversion efficiency", *Appl Phys Letts 33* (6), 518-520, 15 September 1978.
105. N. Uesugi, Diakoku K. and Kubota K., "Electric field tuning of second-harmonic generation in a three-dimensional $LiNbO_3$ optical waveguide", *Apply Phys Letts 34* (1), 60-62, 1 January 1979.
106. J. Noda, Fukuma M. and Ito Y., "Phase-matching temperature variation of second-harmonic generation in Li out-diffused *$LiNbO_3$* layers", *J Appl Phys 51* (3), 1379-1384, March 1980.
107. Micheli M. De, Botineau, Neveu S., Sibillot P., Ostrowsky D.B. and Papuchon M., "Extension of second-harmonic phase-matching range in lithium niobate guides", *Opts Lett 8* (2), 116-118, February 1983.
108. Micheli M. De, "Non-linear effects in TIPE-$LiNbO_3$ waveguides for optical communicaions", *J Opt Comm 4* (1), 25-31, January 1983.
109. W. Sohler and Suche H., " Frequency conversion in Ti:$LiNbO_3$ optical waveguides", *Integrated Optics III, SPIE 408* 163-171, 1983.
110. E.M. Zolotov, Mikhalevich V.G., Pelekhatyi V.M., Prokhorov A.M., Chernykh V.A. and Scherbakov E.A., "Second-harmonic generation in a diffused $LiNbO_3$ waveguide", *Sov Tech Phys Letts 4* (2), 89-90, February 1978.
111. E.M. Zolotov, Prokhorov A.M. and Chernykh v.A., "Electrooptic and temperature tuning of phase-matching in second-harmonic generation in Ti:$LiNbO_3$ channel waveguides", *Sov Tech Phys Letts 7* (3), 129-130, March 1981.
112. S.I. Bozhevol'nyi, Buritskii K.S., Zolotov E.M. and Chernykh V.A., "Second-harmonic generation in coupled Ti:$LiNbO_3$ channel waveguides", *Sov Tech Phys Letts 7* (6), 278-279, June 1981.
113. V.A. Gridin, Mavritski O.D. and Petrovskii A.N., "Second harmonic generation by picosecond laser pulse in a diffused waveguide with vector phase-matching", *Sov Tech Phys Letts 9* (10), 509-510, October 1983.
114. F. Laurell and Arvidsson G., "Frequency doubling in Ti:MgO:$LiNbO_3$ channel waveguides", *J Opt Soc Am B 5* (2), 292-299, February 1988.
115. A. Ashkin and Yariv A., Bell Telephone Laboratories Technical Memorandum, 13 November 1961 *see comment in next reference.*
116. C.F. Dewey Jr and Hocker L.O., "Enhanced non-linear optical effects in rotationally twinned crystatls", *Appl Phys Letts 26*, (8) 442-444, 15 April 1975.
117. M.S. Piltch, Cantrell C.D. and Sze R.C., "Infrared second harmonic generation in non-birefringent cadmium telluride", *J Appl Phys 47* (8), 3514-3517, Aug 1976.
118. D.E. Thompson, McMullen J.D. and Anderson D.B., "Second harmonic generation in GaAs "stack of plates" using high power CO_2 laser radiation", *Appl Phys Letts 29* (2), 113-115, 15 July 1976.

119. M. Okada, Takizawa K. and Ieiri S, "Second harmonic generation by periodic laminar structure of non-linear optical crystal", *Opt Comm 18* (3), 331-334, Aug 1976.
120. C.F. Dewey Jr, "Non-linear optical effects in rotationally twinned crystals: an evaluation of CdTe, ZnTe and ZnSe", *Révue de Physique Appliquée 12*, 405-409, February 1977.
121. Duan Feng, Nai-Ben Ming, Jing-Fen Hong, Yong-Shun Yang, Jin-Song Zhu, Zhen Yang and Ye-Ning Wang, "Enhancement of second harmonic generation in $LiNbO_3$ crystals with periodic laminar ferro-electric domains", *Appl Phys Lett 37* (7), 607-609, 1 October 1980.
122. A. Feisst and Koidl P., "Current induced periodic ferroelectric domain structures in $LiNbO_3$ applied for efficient non-linear optical frequency mixing", *Appl Phys Lett 47*, (11), 1125-1127, 1 Dec. 1985.
123. R.L. Byer, "Progress and new concepts in inorganic non-linear materials", Extended Abstracts of Symposium O on *Non-linear optical materials*, Ed. Miller D.A.B., pp 103, Fall meeting of the Materials Research Society, 1985.
124. E.J. Lim, Fejer M. and Byer R.L., "Second-harmonic generation of blue and green light in periodically-poled planar lithium niobate waveguides", Postdeadline papers of Symposium on *Non-linear guided-wave phenomenon: physics and applications*, Optical Society of America (Washington DC), Huston, Texas, USA, paper PD3-1, 2-4 February 1989.
125. E.J. Lim, Fejer M.M. and Byer R.L., "Second harmonic generation of green light in periodically poled planar lithium niobate waveguide", *Electron Lett 25* (3), 174-175, 2 February 1989.
126. G. Arvidsson and Jaskorzynska B., "Periodically domain-inverted waveguides in lithium niobate for second harmonic generation: influence of the shape of the domain boundary on the conversions efficiency", in Abstracts of the International conference on *Materials for non-linear and electro-optics*, Inst of Phys Conf Series *103* (1), 47-52, 1989, Ed Lyons M.H., Held 4-7 July 1989, Girton College, Cambridge, United Kingdom.
127. G. Khanarian, Hass D., Keosian R., Karim D. and Landi P, "Phase-matched second-harmonic generation in a polymeric waveguide", in Proceedings of *Conference on Lasers and Electro-Optics (CLEO 89) Baltimore, USA*, paper THB1, pp 254-255.
128. G. Khanarian and Norwood R.A., "Quasi-phase-matched frequency doubling over several millimeters in poled polymer waveguides", in Technical Digest of Integrated Photonics Research 1990, (Optical Society of American, Washington DC, 1990), Vol. 5, Post Deadline paper PD11.
129. L. Goldberg and Weller J.F., "Injection locking and single-mode fibre coupling of a 40 element laser diode array", *Appl Phys lett 50*, 1713-1715, 1987.
130. W.J. Kovlovsky, Gustafson E.K., Eckhardt R.C. and Byer R.L, 'Efficient monolithic MgO $LiNbO_3$ singly resonant optical parametric oscillator", *Optics Lett 13* (12), 1102-1104, December 1988.
131. C. Walther and Gunter P., "Photoinduced TE-TM mode conversion in $Ti:LiNbO_3$ waveguides", *Proc Int Soc Mats and Tech*, pp 381-384, Springer Verlag, Erice, Italy, 1986.
132. R.G. Hunsperger, in *Integrated Optics: Theory and Technology*, 2nd Ed, Springer Verlag Series in Optical Sciences, 1984.
133. J.L. Stevenson and Dyott R.B., "Optical fibre waveguide with single crystal core", *Electron Letts 10*, (22), 449-450, 31 October 1974.
134. B.K. Nayar, "Optical fibres with organic crystalline cores", in Springer Proc in Physics 7, *Non-linear Optics: Materials and Devices*, Ed Flytzanis C. and Oudar J.L., 142-153, 1986.

135. K.I. White, Nayar B.K. and Kashyap R. "Amplification and second-harmonic generation in non-linear waveguides", *Opt and Quant Electron 20*, 339-342, 1988.
136. R. Kashyap, "Phase-matched second-harmonic generation in periodically poled optical fibres", *App Phy Lett 58* (12), 25 March 1991.
137. R. Kashyap, Davey S.T. and Williams D.L., " Phase-matched SHG in periodically poled optical fibres" in the proceedings of *The Annual Meeting of the Optical Society of America Boston 2-4 November 1990*.
138. R.W. Keys, Loni a, de la Rue R.M., Ironside C.N., Marsh J.H., Luff B.J. and Townsend P. D., *"Fabrication of domain reversed gratings for SHG in LiNbO$_3$ by electron beam bombardment"*, *Electron Letts 26* (3), 188-189, 1 February 1990.
139. J.D. Bierlein, Laubacher D.B., Brown J.B. and Poel C.J. van der, "Balanced phase-matching in segmented KTiOPO$_4$ waveguides", *Appl Phys Lett* 56, 1725, 30 April 1990.
140. C.J. Poel van der, Colak S., Bierlein J.D. and Brown J.B., " Efficient Type I blue second harmonic generation in periodically segmented KTiOPO$_4$ (KTP) waveguides", *in Post Deadline Digest of Conference on Lasers and Electro-Optics*, Paper CPDP33, Anaheim, California, USA, 21-25, May 1990.
141. R. Kashyap, "Phase-matched electric field induced second harmonic generation in optical fibres", in Technical Digest of the *Sixteenth International Conference on Quantum Electronics*, paper MP25, 110-111, Tokyo, Japan, 18-21 July, 1988.
142. E.J. Lim, Fejer M., Byer R.L. and Kovlovsky W.J., "Blue light generation by frequency doubling periodically poled lithium niobate channel waveguide", *Electron Lett 25* (11), 731-732, 25 May 1989.
143. J. Webjörn, Laurell F. and Arvidsson G., "Laser diode light frequency doubled to blue using a lithium niobate channel waveguide", *CLEO'89 paper PD10*, Baltimore, Maryland, USA, April 24-28, 1989.

6. EPILOGUE

Since the submission of this article, several development have taken place in the field of quasiphase matching for nonlinear optical frequency mixing. There are now several research laboratories actively exploiting periodic phase-matching techniques in waveguides; as this Epilogue is being written, further advances in the field continue to take place, and consequently a truely up to date review is impossible. In the following we describe some of the latest noteworthy developments.

Lim et al [1] have experimentally demonstrated the existence of broad-band or non-critical phase-matching in lithium niobate waveguides. This method of phase-matching had been predicted in [34] and has been discussed in § 2.2.2D. Lim et al fabricated proton-exchanged $LiNbO_3$ planar waveguides in the range of 0.33-0.57 μm in thickness. Measurement of the propagation constants were made by the prism coupling technique at fundamental and second-harmonic wavelenths of 916 nm and 458 nm, respectively. They showed that for a waveguide thickness of approximately 0.4 μm, the variation in the phase-mismatch, $\Lambda\beta$ as a function of waveguide thickness is indeed zero in the first order. Although they have shown the principle indirectly, frequency doubling which is intolerant to the fluctuations in the waveguide dimensions has yet to be demonstrated experimentally.

Also demonstrated recently, is the equivalent of periodic electric field poling reported earlier in optical fibre [136]. Khanarian et al [128] had recently demonstrated *half-period* poling, such that a poled section was followed by an unpoled region in polymeric waveguides. Matsumoto et al [2] have now demonstrated periodically poled lithium tantalate waveguides using electric fields, such that *alternating* domains were created (as in optical fibres [136]), but phase-matching was possible over 1 mm but using a third-order interaction ($\Lambda/2 = 3l_c$). Although this is a step in the right direction, further work needs to be done in order to optimise the fabrication process to eliminate problems of loss of nonlinearity.

An extension of electric-field poling of $LiNbO_3$ using electron beams is reported by Itoh et al [3]. Keys et al [138] had originally demonstrated poling at elevated temperatures under the simultaneous influence of an electric field and an e-beam. Now Itoh et al [3] have shown that poling is possible at room temperature, *without* a DC bias, and demonstrated first order phase-matching with a domain reversal period of 7.9 μm. This *non-waveguided* interaction was over a length of 1.5 mm and using a 7.5 mm focal length lens gave a conversion efficiency of $1.75 \times 10^{-3}\%$ (40% $W^{-1}cm^{-2}$) for 20mW of pump power at 861.1 nm. Also demonstrated was quasiphase matching in several samples with domain period between 7 and 79 μm, giving phase matching between wavelengths of 806 and 861.1 nm.

Yamamoto et al [4] have shown frequency doubling in $LiTaO_3$ waveguides, quasi-phase matched by domain inversion achieved by heat treatment below the Curie temperature: lithium out-diffusion of selected regions by proton exchange reduces the Curie temperature, so that the periodic ion-exchanged regions alone are domain inverted. Subsequently, proton ion-exchange defined a single-mode waveguide for 850 nm operation. Third-order periodic phase-matching was achieved at wavelengths between 835 and 870 nm with periods of 10.5-12 μm. A conversion efficiency of 2.42% was reported for a pump power of 99 mW at 848 nm in a device 9 mm long. They also reported a temperature tuning bandwidth of 3.2°C, demonstrating good temperature stability.

Since the process of domain inversion in $LiNbO_3$ by titanium in-diffusion also causes a periodic refractive index change [See § 2.2.2C], if the period for quasi-phase matching happens to be the same or an odd multiple of the period required to meet the Bragg condition for reflection at the pump wavelength, then oscillation of a laser should be possible with the waveguide acting as distributed Bragg reflector. This clever technique of 'double phase matching' is precisely what was proposed and also demonstrated by Shinozaki et al [5]. They showed that if the domain period is around 13 μm, both these condition can be met for phase matched frequency doubling and Bragg reflection at 1.327 μm in $LiNbO_3$, for first order SHG and 43rd-order DBR operation! Pumped by an anti-reflection coated laser diode, the laser oscillated at the DBR wavelength with an output of 60 μW, and a frequency doubled power of 6.52 pW was measured from a self-quasiphase matched device. The normalised conversion efficiency reported was a respectable 4.1% $W^{-1}cm^{-2}$.

Until recently the only report of phase-matching in poled waveguides using the fundamental-modes *and* a first order phase-matching grating was in optical fibres (1990)[136]. Armani (1992) *et al* [6] have now demonstrated for the first time first-order quasi-phase matching by titanium in-diffusion to create periodic domain reversal with periods between 3.5 and 4 μm in $LiNbO_3$. Single-mode channel waveguides for 800-900 nm operation were then fabricated using proton ion-exchange. A conversion efficiency of 0.84% (11% $W^{-1}cm^{-2}$) was reported for a pump power of 67 mW at 870 nm. The authors report an interaction length of 10 mm, although the conversion efficiency from the 10 mm sample was lower than that achieved by the 4 mm cutback device. This discrepancy is attributed to inhomogeneities in the waveguide.

Finally, as an indication of goals in this field, we consider a recent result by Jundt et al [7]. A 1:24 mm long bulk sample of MgO-doped $LiNbO_3$ with a domain period of 6.94 μm was placed in a bow-tie cavity, frequency locked to a lamp-pumped Nd:YAG laser. Although they did not use waveguides in their experiments, they reported an *external* conversion efficiency of 40.5% when pumped with 4.2 W of 1.064 μm radiation. This translates to an *internal* conversion efficiency of 67%, allowing for mode mismatch of the laser and the frequency doubling cavities is an indication of the immense power of the technique of periodic poling. The 0.8 mm diameter samples were grown by the laser-heated pedestal growth technique [8] to form 178 domain-reversed periods, demonstrating the great degree of control possible. This is a tremendous technological achievement and should open the avenues for further innovative research for device application.

Acknowledgements

Thanks are due to the many researchers for their valuable contributions in this area. I would also like to apologise to others for the inevitable oversight on my part in writing a long review, for not including their work.

REFERENCES

1. E.J. Lim, Matsumoto S and Fejer M.M., "Noncritical phase matching for guided-wave frequency conversion", *Appl Phys Lett 57* (22), 2294, 1990.
2. S. Matsumoto, Lim E.J., Hertz H.M. and Fejer M.M., "Quasiphase-matched second harmonic generation of blue light in electrically-poled lithium tantalate waveguides", *Electron Lett 27* (22), 2040, 1991.
3. H. Ito, Takyu C and Inaba H., "Fabrication of periodic domain grating in $LiNbO_3$ by electron beam writing for application of nonlinear optical processes", *Electron Lett 27* (14), 1221, 1991.
4. K. Yamamoto, Mizuuchi K and Taniuchi T., "Milliwatt-order blue-light generation in a periodically domain-iverted $LiTaO_3$ waveguide", *Opts Lett 16* (15), 1156, 1991.
5. K. Shinozaki, Fukunaga T., Watanabe K and Kamijoh, "Self-quasi-phase-matched second-harmonic generation in the proton exchanged $LiNbO_3$ optical waveguide with periodically domain-inverted regions", *Appl Phys Lett 59* (5), 510, 1991.
6. F. Armani, Delacourt D., Lallier F., Papuchon M., He Q, De Micheli M and Ostrowsky D.B., "First order quasiphase matching in $LiNbO_3$", *Electron Lett 28* (2), 139, 1992.
7. D.H. Jundt, Magel G.A., Fejer M.M. and Byer R.L., "Periodically poled $LiNbO_3$ for high-efficiency second-harmonic generation", *Appl Phys Lett 59* (21), 2657, 1991.
8. M.M. Fejer, Nightingale J.L., Magel G.A. and Byer R.L., *Rev Sci Instrum 55*, 1791, 1984.

8

Non-linear Integrated Optics

M.P. DE MICHELI[*]

1. INTRODUCTION

Experimental non-linear optics was born just after the discovery of the laser, with the first experiment of second harmonic generation performed by Franken and co-workers in 1961 [1]. After this pioneering experiment, the field developed in two directions:

- study of new phenomena such as stimulated Raman and Brillouin scattering, two photon absorption, etc.
- improvement of the conversion efficiency for the phenomena where new frequencies are created, such as harmonic generation, up conversion, parametric oscillation.

Indeed, the very interesting practical applications of these phenomena, creating coherent sources at various wavelengths where no lasers are available, were demonstrated very rapidly, and a lot of effort was made to optimize these interactions.

In this chapter, we shall concentrate on the phenomena employing second order non-linearities, usually called parametric interactions. Depending on the phenomena, the efficiency is given by a different formula but is always a growing function of:

- the non-linear coefficient of the material
- the power confinement (the power density multiplied by the length of interaction)

---

[*] Laboratoire de Physiqué de la Matière Condensée, U.R.A. CNRS N°. 190, Université de NICE, Parc Valrose, 06034 NICE Cedex FRANCE.

- the phase-matching term
- the overlap of the interacting waves.

All these parameters have to be optmised to get a high conversion efficiency.

In the early stages, the experiments were run with high power lasers and a lot of materials were tested in order to select the most efficient ones. Some effort was also expended to optimize the power confinement, which led to the confocal configuration, which is currently used in the commercial frequency doublers. But the attractive features of optical waveguides, as far as power confinement is concerned, were recognized at the very beginning of integrated optics, giving rise to a new field: non-linear integrated or fibre optics. It is in this configuration that it has become possible to use low power pump sources such as diode laser, opening the way to compact, reliable devices. This is the reason for the rapidly growing interest in non-linear integrated optics.

2. SECOND HARMONIC GENERATION IN INTEGRATED OPTICS

2.1 Bulk-waveguide comparison

The guided wave configuration allows concentrating the optical fields in long structures with cross-sections of the order of a wavelength. Very effective interactions are then possible, and the efforts to take advantage of this configuration split in two directions:

- investigation of effects in material with weak non-linear coefficients but able to give very long waveguides (typically silica fibres) (cf. Chapter 7)
- reduction of the power necessary to obtain a reasonable conversion efficiency in integrated optical structures,which have a reasonable non-linear coefficient.

The first direction of investigation, which produces very interesting phenomena such as solitons in fibres, Raman shifting, etc. will not be developed here. We shall focus on the second direction. Indeed the reduction of the pump power necessary to obtain a reasonable effect allows us to conceive practical devices where the pump can be a laser diode giving full significance to the concept of hybrid non-linear integrated optics as both the source and the non-linear component are realised in a compact and practical format.

In this chapter, we shall take second harmonic generation (SHG) as a paradigm, to describe the possibilities of non-linear integrated optics (NLIO). To begin our discussion we shall compare the case of SHG in both bulk and guided wave configurations.

The efficiency of such a process, is defined as $\eta = \dfrac{P_{2\omega}}{P_\omega}$, the ratio between the power generated at the harmonic frequency, $P_{2\omega}$, and the pump power P_ω.

For a plane wave interaction at efficiencies less than 20% we have

$$\eta \propto d^2 \frac{L^2 P_\omega}{S} \sin^2\left(\frac{\Delta k \cdot L}{2}\right) I_R, \qquad (1)$$

where, I_R is the overlap between the fundamental and the harmonic waves, d is the non-linear coefficient, P_ω represents fundamental optical power, L the interaction length, S the beam cross-section and $\Delta k = k_{2\omega} - 2k_\omega$. This formula will be derived later, but let us consider the result in order to point out some basic features of NLIO. The term $\frac{L^2 P_\omega}{S}$ in Eq. (1) describes the influence of power density confinement over the interaction length and the $\sin^2$ term describes the effect of optical phase-matching between the fundamental and the harmonic waves.

Integrated optics provides an important means of optimizing both of these terms. As far as the phase-matching term is concerned, guided wave optics allows to obtain the phase-match ($\Delta k = 0$) via modal dispersion, in cases where the natural crystal birefringence does not. Since phase-matching is essential for an efficient interaction, this is a major advantage permitting the use, in integrated optics, of materials that cannot be used in the bulk configuration[2]. However, to continue our comparison, we shall consider cases where phase-matching is possible in both the bulk and guided wave configurations. In this case, the essential advantage of the guided wave arises from the waveguides' ability to confine the optical power density $\frac{P_\omega}{S}$ over long interaction lengths. In the bulk case, diffraction leads to a compromise between a small value of S, and hence a high power density, and a large value of L. For Gaussian beams, the optimum configuration is close to that called confocal focusing where $S = \frac{\lambda L}{2n}$. If we compare this confocal focusing case to that of guided waves, the ratio of the guided wave term (GW) to that of the confocal term (CF), is given by

$$\frac{\text{GW}}{\text{CF}} = \frac{1}{2n}\frac{\lambda L}{S},$$ where n is the refractive index of the medium.

For $n = 2, \lambda = 1\,\mu\text{m}, L = 2$ cm and $S = 25\,\mu\text{m}^2$ this leads to an impressive factor of improvement of about 200 for the guided wave case. However, the efficiency of the guided wave interaction may be reduced by the overlap integral of the interacting waves, which is proportional to $\int_{NL} E_{2\omega}^2 \cdot E_\omega \, ds$. In the case of bulk interaction, both fundamental and harmonic beams are Gaussian and thus the overlap is rather good. In the case of guided wave interactions, depending on the interacting modes, the overlap factor can take on very different values and in some cases practically cancel the power confinement advantage. A careful study of the geometry of the waveguide is then necessary if one wants to preserve the advantage of IO.

2.2 Choice of a medium

In the case of NLIO the material and the configuration optimization are no longer independent. To choose the right medium, the value of the non-linear coefficient is no longer the only relevant parameter. We have to also consider the techniques available to realize the waveguides and the resulting optical quality. How does one choose a material for NLIO ?

This material must allow the realisation of very good quality waveguides and present a high non-linear coefficient. Up to now, the best compromise was realized by $LiNbO_3$ whose non-linear coefficients have reasonable values and where we know at least two techniques for realizing good waveguides: Titanium Indiffusion (TI)[3], and Proton Exchange (PE)[4]. Those techniques may be superimposed to adjust the index profile [5] or to improve the quality of the guide [6] during the fabrication of Integrated Optical Waveguides, and we will now focus on the description of SHG in $LiNbO_3$ waveguides.

3. SHG IN INTEGRATED OPTICS

3.1 Theory

3.1.1 Introduction

The calculation we shall present in this section is adapted to the kind of waveguides we actually use in our own laboratory experiments. These are $LiNbO_3$ waveguides. In order to simplify the calculation, we shall consider a graded index planar waveguide, uniformly excited on a width W, rather than a real strip waveguide. Moreover this guide is realised in a birefringent medium. To conform with most of the experimental studies we shall take the fundamental wave as a TM mode.

3.1.2 Calculation of the harmonic fields

Following the conventions shown in Fig. 1 the fundamental field is given by

$$\vec{H}_\omega = \begin{pmatrix} 0 \\ 0 \\ H_z = H_0(y)\, e^{j(\omega t - \beta_\omega x)} \end{pmatrix}. \tag{2}$$

The electric field is then given by the Maxwell's equations and yield two non zero components:

$$\vec{E}_\omega = \begin{pmatrix} E_x = \dfrac{-j}{\omega \varepsilon_\omega} \dfrac{d H_0(y)}{dy} e^{j(\omega t - \beta_\omega x)} \\ E_y = \dfrac{\beta_\omega}{\omega \varepsilon_\omega} H_0(y)\, e^{j(\omega t - \beta_\omega x)} \\ 0 \end{pmatrix}. \tag{3}$$

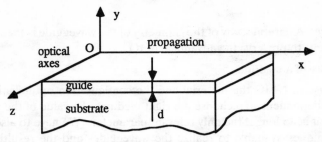

Fig. 1. Orientation convention

The only dielectric constant used at the frequency ω is $\varepsilon_1(\omega)$. We have, therefore, dropped the subscript 1 in Eq. (3)

In the case of non-linear media, the electric displacement is given by $\vec{D} = \varepsilon \vec{E} + \vec{P}_{NL}$ where $\vec{P}_{NL}$ is the non-linear response of the medium. The wave equation of the harmonic wave, deduced from Maxwell's equations, is then

$$\frac{\partial^2 E_z}{\partial y^2} + \left(4\omega^2 \mu_0 \varepsilon_z - \beta_{2\omega}^2\right) E_z = -4\omega^2 \mu_0 P_{NLz}. \tag{4}$$

The non-linear polarization is given by

$$\vec{P}_{NL} = \begin{pmatrix} P_{NLx} \\ P_{NLy} \\ P_{NLz} \end{pmatrix} \text{ with } \begin{cases} P_{NLx} = 2d_{31} E_x E_z - 2d_{22} E_x E_y \\ P_{NLy} = -d_{22} E_x^2 + d_{22} E_y^2 + 2d_{31} E_y E_z \\ P_{NLz} = d_{31} E_x^2 + d_{31} E_y^2 + d_{33} E_z^2 \end{cases} \tag{5}$$

Due to phase-matching possibilities, the only interesting component of $\vec{P}_{NL}$ is P_z, which can be calculated using Eqs. (3) and (5) as

$$P_z = d_{31} (E_x^2 + E_y^2). \tag{6}$$

Equation (6) can be written as

$$P_z = U(y) e^{2j(\omega t - \beta_\omega x)}. \tag{7}$$

with
$$U(y) = \frac{d_{31}}{\omega^2 \varepsilon_\omega^2} \left[\beta_\omega^2 H_0^2(y) - \left(\frac{dH_0(y)}{dy}\right)^2\right] \tag{8}$$

$$= \frac{d_{31}\mu_0}{\varepsilon_\omega} \left[H_0^2(y) - \frac{1}{\omega^2 \varepsilon_{eff,\omega} \mu_0} \left(\frac{dH_0(y)}{dy}\right)^2\right], \tag{9}$$

where $\varepsilon_{eff,\omega}$ is the dielectric constant associated with the effective index, n_{eff} at ω.

We are looking for a solution polarized along the Oz-axis of the form

$$\vec{E}_{2\omega} = \sum_{m=0}^{N-1} \vec{E}_{2\omega,m}(x,y), \tag{10}$$

where m is the index and N the number of guided modes at 2ω in the structure. In the case of weak interactions we are able to suppose that the transverse shape of the mode is not modified by the non-linear polarization and that the amplitude is now a growing function of x. That is to say

$$\vec{E}_{2\omega, m}(x, y) = \begin{pmatrix} 0 \\ 0 \\ E_z = A_m(x) E_m(y) e^{j(2\omega t - \beta_{2\omega, m} x)} \end{pmatrix}. \quad (11)$$

Then the only dielectric constant useful at the frequency 2ω is $\varepsilon_3(2\omega)$. In the following we shall drop the subscript 3.

To calculate $\vec{E}_{2\omega}$ we need to substitute Eqs. (10) and (11) into the wave equation, which gives, after using the mode orthogonality

$$-2j\beta_{2\omega,n} \frac{dA_n}{dx} \int_{-\infty}^{+\infty} E_n^2(y) dy + 4\omega^2 \mu_0 e^{j\Delta\beta_n x} \int_{-\infty}^{\infty} U(y) E_n(y) dy = 0 \quad (12)$$

with

$$\Delta \beta_n = \beta_{2\omega,n} - 2\beta_{\omega,i} \quad (13)$$

and $\beta_{\omega,i}$ being the propagation constant of the mode i at the frequency ω.

Equation (12) can be integrated to obtain

$$A_n(L) = -j\omega \sqrt{\frac{\mu_0}{\varepsilon_{eff\,n,\,2\omega}}} L \frac{\sin\left(\Delta\beta_n \frac{L}{2}\right)}{\Delta\beta_n \frac{L}{2}} e^{j\Delta\beta_n \frac{L}{2}} \frac{\int_{-\infty}^{+\infty} U(y) E_n(y) dy}{\int_{-\infty}^{+\infty} E_n^2(y) dy}. \quad (14)$$

The harmonic field created at the abscissa x is then

$$E_n(x,y) = -j\omega d_{31} \frac{\mu_0^{3/2} \varepsilon_{eff,\,\omega}}{\sqrt{\varepsilon_{eff\,n,\,2\omega}}} x \frac{\sin\left(\Delta\beta_n \frac{x}{2}\right)}{\Delta\beta_n \frac{x}{2}} J e^{j\Delta\beta_n \frac{x}{2}} E_n(y) e^{j(2\omega t - \beta_{2\omega,n} x)} \quad (15)$$

with

$$J = \frac{\int_{-\infty}^{Y_0} \varepsilon_\omega^{1/2} \left(H_0^2(y) - \frac{1}{\omega^2 \varepsilon_{eff,\omega} \mu_0} \left(\frac{dH_0}{dy}\right)^2 \right) E_n(y) dy}{\int_{-\infty}^{+\infty} E_n^2(y) dy}, \quad (16)$$

where Y_0 is the limit of extent of the non-linear medium.

The magnetic field at 2ω is given by

$$\vec{H}_{2\omega, n}(x, y) = \begin{pmatrix} H_x = \dfrac{j}{2\omega \mu_0} \dfrac{\partial E_n(x,y)}{\partial y} \\ H_y = \dfrac{\beta_{2\omega,n}}{2\omega \mu_0} E_n(x,y) \\ 0 \end{pmatrix}. \quad (17)$$

3.1.3 Calculation of the harmonic power

The power propagating at the frequency ν, through a section S, along Ox is defined as

$$P_\nu = \iint_S \mathrm{Re}\left(\frac{\vec{E}_\nu \times \vec{H}_\nu^*}{2}\right)_x dy\, dz. \tag{18}$$

The power created at 2ω in the mode n of a waveguide of length L is then

$$P_{2\omega,n}(L) = \frac{\beta_{2\omega,n}}{4\omega\mu_0} \iint E_n(L,y) E_n^*(L,y)\, dy\, dz. \tag{19}$$

Replacing $E_n(x,y)$ by Eq. (15) we obtain:

$$P_{2\omega,n}(L) = \frac{\omega^2 \mu_0^{5/2} \varepsilon_{\mathit{eff},\omega}^2}{2\sqrt{\varepsilon_{\mathit{eff},2\omega}}} d_{31}^2\, J^2 L^2 \left(\frac{\sin\Delta\beta_n \frac{L}{2}}{\Delta\beta_n \frac{L}{2}}\right)^2 W \int_{-\infty}^{+\infty} E_n^2(y)\, dy. \tag{20}$$

3.1.4 Calculation of the conversion efficiency

We can now calculate the conversion efficiency defined by:

$$\eta = \frac{P_{2\omega,n}(L)}{P_\omega}. \tag{21}$$

Using the definition in Eq. (18) we have

$$P_\omega = \frac{\beta_\omega}{2\omega} \iint \frac{H_0^2(y)}{\varepsilon_\omega(y)}\, dy\, dz \tag{22}$$

which, in the case of a planar waveguide uniformly excited over a width W, can be written as

$$P_\omega = \frac{\sqrt{\varepsilon_{\mathit{eff},\omega}\,\mu_0}}{2} W \int \frac{H_0^2(y)}{\varepsilon_\omega(y)}\, dy. \tag{23}$$

Because of the small index increase in the case of diffused waveguides (weak guiding), we can consider ε_ω to be constant.

Then Eq. (23) becomes:

$$P_\omega = \frac{1}{2}\sqrt{\mu_0/\varepsilon_\omega}\, W \int H_0^2(y)\, dy. \tag{24}$$

According to Eqs. (20), (21) and (24), the conversion efficiency is:

$$\eta = \omega^2\, \frac{\mu_0^2\, \varepsilon_{\mathit{eff},\omega}^{3/2}}{\varepsilon_{\mathit{eff},2\omega}^{1/2}}\, d_{31}^2\, L^2 \left(\frac{\sin\left(\Delta\beta_n \frac{L}{2}\right)}{\Delta\beta_n \frac{L}{2}}\right)^2 \Psi, \tag{25}$$

with $\Psi = \dfrac{\left(\int_{-\infty}^{x_0} \dfrac{1}{\varepsilon_\omega^2(y)} \left(H_0^2(y) - \dfrac{1}{\omega^2 \varepsilon_{\text{eff},\omega} \mu_0} \left(\dfrac{dH_0}{dy}\right)^2\right) E_n(y)\, dy\right)^2}{\int_{-\infty}^{\infty} E_n^2(y)\, dy \int_{-\infty}^{\infty} \dfrac{H_0^2(y)}{\varepsilon_\omega}\, dy}$ (26)

In the case of weakly guiding waveguides

$$\dfrac{1}{\omega^2 \varepsilon_{\text{eff},\omega} \mu_0} \left(\dfrac{dH_0}{dy}\right)^2 \ll H_0^2(y) \quad (27)$$

and in such a case

$$\eta = 2\omega^2 \dfrac{\mu_0^{3/2}}{\varepsilon_\omega \varepsilon_{2\omega}^{1/2}} d_{31}^2 L^2 \dfrac{P(\omega)}{W} \left(\dfrac{\sin\left(\Delta\beta_n \dfrac{L}{2}\right)}{\Delta\beta_n \dfrac{L}{2}}\right)^2 I_R \quad (28)$$

with

$$I_R = \dfrac{\left(\int_{-\infty}^{x_0} H_0^2(y) E_n(y)\, dy\right)^2}{\int_{-\infty}^{\infty} E_n^2(y)\, dy \left(\int_{-\infty}^{\infty} H_0^2(y)\, dy\right)^2} \quad (29)$$

which can be written as:

NL coef.	Power confinement	Phase matching	Overlap integral	
$\eta = C$	d_{31}^2	$\dfrac{L^2 P(\omega)}{W}$	$\left(\dfrac{\sin\left(\Delta\beta_n \dfrac{L}{2}\right)}{\Delta\beta_n \dfrac{L}{2}}\right)^2$	I_R	(30)

In this formula we can recognize the non-linear coefficient, the power confinement, and the phase-matching terms, which were briefly discussed in the second paragraph in order to point out the advantage of NLIO and of LiNbO$_3$. We now have to discuss in more detail the overlap problem and the phase-matching possibilities of LiNbO$_3$ waveguides.

3.2 SHG in LiNbO$_3$ waveguides

3.2.1 Optimization of the overlap integral

Looking at the expression of I_R, one would say that the best overlap is obtained when interaction takes place between two zero order modes. In fact this is not always true, and each particular case has to be examined very carefully.

The optimization of the overlap is one of the key problems of NLIO when one wishes to achieve efficient interactions. The distribution of the interacting fields is determined by the waveguide structure, that is to say its index profile, and the order of the interacting modes. Therefore, the optimization of overlap depends on parameters such as waveguide thickness and index, substrate and superstrate index. One important point to notice is that the overlap is proportional to the integral over the product of field amplitudes, which may become positive and negative along the waveguide cross-section. In the case of step index waveguides, the interaction between odd and even order modes may result to a vanishing overlap such as in Fig. 2. In the case of gradex index waveguides, the situation may be quite different. The example presented here

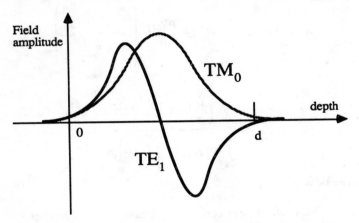

Fig. 2. Mode overlap in the case of a step index waveguide.

in detail, is currently the most used type of waveguide in active IO : Titanium Indiffused (TI) lithium niobate waveguides. Because of the dispersion characteristic of the medium, and the amplitude of the index increase due to the Ti indiffusion, the only way to fulfill the phase-matching condition for a fundamental wavelength in the near infrared, is to use the birefringence to compensate for the dispersion, with the fundamental propagating as an ordinary mode and the harmonic as an extraordinary mode. The fine tuning is then realised using the temperature dependence of the refractive indices.

For a typical TI waveguide whose Ti concentration profile is given by:

$$C_{Ti} = [Ti]_0 \exp\left(-\frac{y^2}{b^2}\right) \text{ with } [Ti]_0 = 1.5 \cdot 10^{21} \text{ cm}^{-3} \text{ and } b = 4.5\,\mu\text{m}. \quad (31)$$

The field distributions of the modes susceptible to interact in a SHG experiment are represented in Fig. 3. One can see that the best overlap is then obtained for the interaction of a TM_0 mode for the pump and a TE_0 mode for the harmonic. This situation is not optimal and can be improved in two ways.

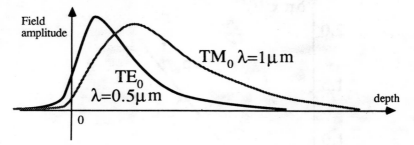

Fig. 3. Mode overlap in the case of a Titanium indiffused $LiNbO_3$ waveguide with high Ti concentration

First, one can replace the upper part of the waveguide by a linear layer presenting nearly the same refractive index as the corresponding Ti doped region. As the integration in Eq. (24) is to be taken only over the non-linear part of the waveguide, it no longer suffers the reduction due to the change of sign of the harmonic field amplitude. In the case of a TI waveguide, the convenient linear material is Nb_2O_5, and an improvement by a factor 4 was demonstrated using this technique [7].

Another way to optimize the overlap is to study the dependance of the modes field distribution on the shape of the index profiles, that is to say on the fabrication parameters. In that case, Ti concentration is the relevant parameter. The dependence of the indices of refraction on the Ti concentration is shown in Fig. 4. In this figure, we can see that the extraordinary index increase δn_e is a linear function of the Ti concentration while the ordinary index increase δn_o is a non-linear function of that parameter. In addition to that we can see that in the low concentration region δn_e is smaller than δn_o, a situation which is reversed in the high concentration region. As the shape of the pump mode is fixed by the ordinary index profile when the harmonic field distribution is given by the extraordinary one, this different behaviour as a function of the Ti concentration allows us to adjust the Ti concentration in order to optimize the overlap of the zero order modes. The best overlap is obtained in the low concentration regime. This is illustrated in Fig. 5. This technique was used by W. Sohler and his co-workers to realise an integrated optical frequency doubler [8]. They also implemented this device by depositing mirrors at the end of the waveguide, creating a resonator at the pump frequency [9]. This allows one to increase the pump power in the waveguide by about a factor 10, which according to Eq. (30), increases the conversion efficiency by the same factor.

3.2.2. Phase-matching possibilities

To complete our discussion of NLIO we still have to examine the use of the modal dispersion as a means of compensating the material dispersion.

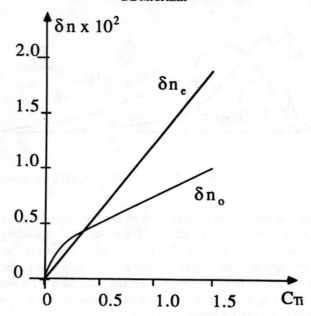

Fig. 4. Index increase in a Titanium in-diffused LiNbO$_3$ waveguide as a function of the Ti concentration.

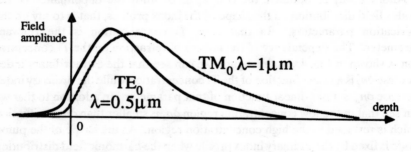

Fig. 5. Mode overlap in the case of a Titanium indiffused LiNbO$_3$ waveguide with low Ti concentration

The effective index of refraction of an optical waveguide takes a value between the substrate index and the maximum index of the guiding region, which usually occurs at the surface. We can, therefore determine the total phase-matching range of the pump wavelength at a given temperature by constructing a diagram of the type shown in Fig. 6, for a typical TI guide realized on a X- or a Y-cut LiNbO$_3$ plate [10]. Along the abscissa we represent the fundamental pump wavelength λ, and along the ordinate, the index of refraction. One pair of curves represents the evolution of n_o substrate and n_o surface (TM waves) as a function of λ. The other pair represents the evolution of n_e substrate and n_e surface (TE waves) as a function of $\lambda/2$. Phase-matching of the fundamental (λ) and second harmonic ($\lambda/2$) waves can occur only when the effective indices can be made equal, that is, for certain discrete values of λ

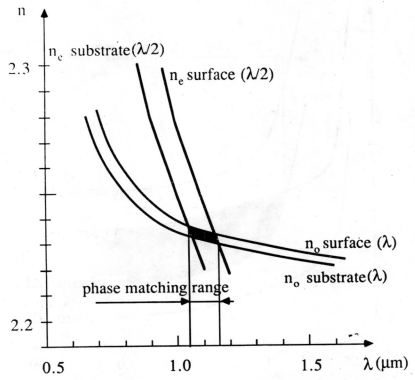

Fig. 6. Phase-matching diagram of a Titanium indiffused LiNbO3 waveguide.

(which depend on the respective mode spectra at the two wavelengths) in the region where the two pairs of curves overlap. At room temperature, the phase-match can be attained for a range of λ from 1.07 to 1.10 μm.

In the case of TI waveguide, the values of δn_e and δn_o are quite small, and the modal dispersion is not able to modify very much the phase-matching possibilities of the waveguide as compared to those in bulk. But, to realise waveguides in LiNbO$_3$, there is another technique, Proton Exchange (PE), which produces much higher index modifications. For example at $\lambda =$ 632.8 nm, the extraordinary index increase may be as high as $\delta n_e = 0.14$. This figure is to be compared to the value of the material dispersion of LiNbO$_3$, between the near infrared and the visible which is about 0.1. This comparison shows that, in this case, it is even possible to compensate the material dispersion, without using the birefringence of LiNbO$_3$. Indeed the phase-matching diagram of a typical PE waveguide (Fig. 7) shows that the phase-matching range of such a waveguide is very extended. Moreover this configuration allows us to use the non-linear coefficient d_{33}, which is six times larger than the d_{31} coefficient which was used in the other configuration. Nevertheless, the phase matching diagram shows also that to be phase-matched, the interaction

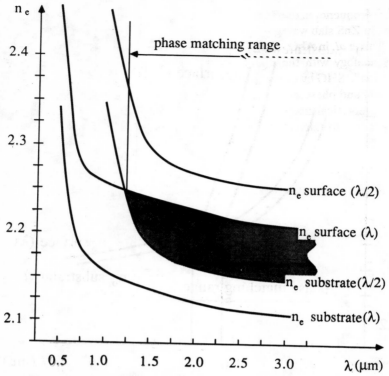

Fig. 7. Phase-matching diagram of a Proton exchanged LiNbO3 waveguide.

has to take place between a low order mode for the fundamental (n_{eff} nearly equal to n_{max}) and a high order mode for the harmonic (n_{eff} nearly equal to n_{min}). According to our discussion in the previous paragraph, these interactions are not very efficient because of the poor overlap in this situation. This problem completely cancels the advantage of using a better non-linear coefficient.

These waveguides are nevertheless very interesting, because they permit to demonstrate that there is a possibility to conceive very efficient interactions between guided and radiating modes. This configuration occurs when the non-linear polarization created at 2ω, has a phase velocity higher than to the phase velocity of the free light (plane wave) in this medium. We shall now discuss the possibilities of this configuration.

4. "CERENKOV" CONFIGURATION SHG

4.1 Introduction

Interactions between guided and radiating modes occur when the non-linear polarization in a waveguide, travels at a velocity faster than that of a free wave at the same frequency in the substrate. This configuration was first observed by Tien *et al.* in ZnS slab waveguide on ZnO substrate with YAG laser [11], and then by Chen *et al.* in Al_2O_3 slab waveguide on a quartz substrate with Ar laser [12]. By analogy with the physics of the particles, they call it "Cerenkov configuration". SHG by Cerenkov radiation is very attractive because it can be very efficient and phase-matching is satisfied automatically. In this section, we propose a theoretical analysis of Cerenkov SHG in step index waveguides, and apply this theory to calculate the possibilities of PE slab waveguides [13].

4.2 Theory

4.2.1 Propagation equation

Figure 8 shows the waveguide structure, and the crystal orientation that we are now considering. In this part we shall use the subscripts f for the fundamental fields and h for the harmonic fields. The subscripts 1, 2 and 3 denote the waveguide, the substrate and the cladding. The fundamental field is a guided

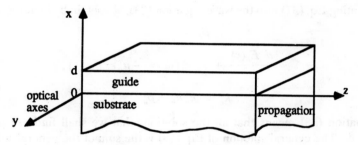

Fig. 8. Orientation convention.

wave, but the harmonic field is free to radiate into the substrate with a Cerenkov angle θ with respect to z direction. The mathematical expression of this condition added to the Cerenkov relation between the phase velocities at ω and 2ω is

$$k_h n_{h3} < 2\beta_f < k_h n_{h2} \quad \text{with} \quad k_h = \frac{2\pi}{\lambda_h}. \tag{32}$$

In the coordinate system of Fig. 8, a TE polarized harmonic field E_h obeys the wave equation

$$\nabla^2 E_h + k_h^2 n_h^2 E_h = -\mu_0 \omega_h^2 P_{NL} \tag{33}$$

with the conditions at the interfaces

$$x = 0, \quad E_{h1} = E_{h2}, \quad \frac{\partial E_{h1}}{\partial x} = \frac{\partial E_{h2}}{\partial x},$$
$$x = d, \quad E_{h1} = E_{h3}, \quad \frac{\partial E_{h1}}{\partial x} = \frac{\partial E_{h3}}{\partial x}. \tag{34}$$

4.2.2 Solution of the wave equation

As in the case of the interaction between guided waves, we suppose that the amplitude of the fundamental field is constant. This hypothesis is justified by the fact that the conversion efficiency is small. Under this condition, the non-linear polarization can be written as

$$P_{NL}(x,z) = \varepsilon_0 d_{NL} E_f^2(x,z) = \varepsilon_0 d_{NL} A_f^2 E_f^2(x) e^{-j2\beta_f \cdot z}, \tag{35}$$

where d_{NL} is the non-linear coefficient and A_f is the amplitude of the fundamental field.

Equation (35) can be written as

$$P_{NL}(x,z) = P_{NL}(x) e^{-j2\beta_f \cdot z} \quad \text{with} \quad P_{NL}(x) = \varepsilon_0 d_{NL} A_f^2 E_f^2(x). \tag{36}$$

We are looking for a solution for the harmonic field of the form

$$E_h(x,z) = E_h(x) e^{-j\beta_h \cdot z}. \tag{37}$$

Substituting Eq. (37) into the wave equation (33), we obtain by identification

$$\beta_h = 2\beta_f \tag{38}$$

and

$$\frac{d^2 E_h(x)}{dx^2} + \kappa_h^2 E_h(x) = -\mu_0 \varepsilon_h^2 P_{NL}(x), \tag{39}$$

with

$$\kappa_h^2 = k_h^2 n_h^2 - \beta_h^2. \tag{40}$$

Equation (38) means that all the solutions that we shall find are phase-matched. The general solution of Eq. (39) is the sum of the general solution without the RHS term and a particular solution [14]. Thus,

$$E_h(x) = G(x) + F(x), \tag{41}$$

where $F(x)$ is a particular solution of Eq. (39) and $G(x)$ is the general solution of the equation without RHS term:

$$\frac{d^2 G(x)}{dx^2} + \kappa_h^2 G(x) = 0. \tag{42}$$

The solution of this equation is well known:

$$G(x) = \begin{cases} D e^{-\kappa_{h3} x} + D' e^{\kappa_{h3} x} & d < x \\ A e^{-j\kappa_{h1} x} + B e^{j\kappa_{h1} x} & 0 < x < d \\ C e^{j\kappa_{h2} x} + C' e^{-j\kappa_{h2} x} & x < 0 \end{cases} \tag{43}$$

with

$$\kappa_{h1}^2 = k_h^2 n_{h1}^2 - \beta_h^2 \,;\, \kappa_{h2}^2 = k_h^2 n_{h2}^2 - \beta_h^2 \,;\, \kappa_{h3}^2 = \beta_h^2 - k_h^2 n_{h3}^2 \tag{44}$$

and

$$F(x) = \begin{cases} F_3(x) & d < x \\ F_1(x) & 0 < x < d \\ F_2(x) & x < 0 \end{cases} \tag{45}$$

Then we can write the harmonic field as

$$E_h(x,z) = \begin{cases} \left[D\, e^{-\kappa_{h3}(x-d)} + D'\, e^{\kappa_{h3}(x-d)} + F_3(x)\right] e^{-j\beta_h z} & d < x \\ \left[A\, e^{-j\kappa_{h1}x} + B\, e^{j\kappa_{h1}x} + F_1(x)\right] e^{-j\beta_h z} & 0 < x < d \\ \left[C\, e^{j\kappa_{h2}x} + C'\, e^{-j\kappa_{h2}x} + F_2(x)\right] e^{-j\beta_h z} & x < 0 \end{cases} \quad (46)$$

The constants in Eq. (46) can be determined by using the limiting conditions.

- When x tends to infinity, E_h has to remain finite, thus $D' = 0$.
- The term $C'\, e^{-j\kappa_{h2}x}$ represents a plane wave coming from $-\infty$. In our case, there are no sources at $-\infty$, so $C' = 0$.
- To determine the four remaining constants, we apply the continuity conditions at the interfaces.

Substituting Eq. (46) into Eq. (34), we get

$$\begin{cases} A + & B - & C & = F_{12} \\ -j\kappa_{h1}A + & j\kappa_{h1}B + & -j\kappa_{h2}C + & = F'_{12} \\ e^{-j\kappa_{h1}d}A + & e^{j\kappa_{h1}d}B - & & D = F_{13} \\ -j\kappa_{h1}e^{-j\kappa_{h1}d}A + & j\kappa_{h1}e^{j\kappa_{h1}d}B + & & \kappa_{h3}D = F'_{13} \end{cases} \quad [4(7a)]$$

with

$$F_{12} = F_2(0) - F_1(0), \quad F'_{12} = \left.\frac{dF_2}{dx}\right|_{x=0} - \left.\frac{dF_1}{dx}\right|_{x=0},$$

$$F_{13} = F_3(d) - F_1(d), \quad F'_{13} = \left.\frac{dF_3}{dx}\right|_{x=d} - \left.\frac{dF_1}{dx}\right|_{x=d}. \quad [47(b)]$$

The principal determinant of Eq. (47) is

$$\Delta = \begin{vmatrix} 1 & 1 & -1 & 0 \\ -j\kappa_{h1} & j\kappa_{h1} & j\kappa_{h2} & 0 \\ e^{-j\kappa_{h1}d} & e^{j\kappa_{h1}d} & 0 & -1 \\ -j\kappa_{h1}e^{-j\kappa_{h1}d} & j\kappa_{h1}e^{j\kappa_{h1}d} & 0 & \kappa_{h3} \end{vmatrix} \quad (48)$$

In the case of Cerenkov radiation, the condition $\beta_h < k_{h2}$, implies that Δ is non-zero in the real domain. Equation [47(a)] has then the following solution:

$$A = \frac{\Delta_A}{\Delta}, \quad B = \frac{\Delta_B}{\Delta}, \quad C = \frac{\Delta_C}{\Delta}, \quad D = \frac{\Delta_D}{\Delta}, \quad (49)$$

with

$$\Delta_A = \begin{vmatrix} F_{12} & 1 & -1 & 0 \\ F'_{12} & j\kappa_{h1} & j\kappa_{h2} & 0 \\ F_{13} & e^{j\kappa_{h1}d} & 0 & -1 \\ F'_{13} & j\kappa_{h1}e^{j\kappa_{h1}d} & 0 & \kappa_{h3} \end{vmatrix}, \quad (50)$$

$$\Delta_B = \begin{vmatrix} 1 & F_{12} & -1 & 0 \\ -j\kappa_{h1} & F'_{12} & j\kappa_{h2} & 0 \\ e^{-j\kappa_{h1}d} & F_{13} & 0 & -1 \\ -j\kappa_{h1}e^{-j\kappa_{h1}d} & F'_{13} & 0 & \kappa_{h3} \end{vmatrix}, \quad (51)$$

$$\Delta_C = \begin{vmatrix} 1 & 1 & F_{12} & 0 \\ -j\kappa_{h1} & j\kappa_{h1} & F'_{12} & 0 \\ e^{-j\kappa_{h1}d} & e^{j\kappa_{h1}d} & F_{13} & -1 \\ -j\kappa_{h1}e^{-j\kappa_{h1}d} & j\kappa_{h1}e^{j\kappa_{h1}d} & F'_{13} & \kappa_{h3} \end{vmatrix}, \quad (52)$$

$$\Delta_D = \begin{vmatrix} 1 & 1 & -1 & F_{12} \\ -j\kappa_{h1} & j\kappa_{h1} & j\kappa_{h2} & F'_{12} \\ e^{-j\kappa_{h1}d} & e^{j\kappa_{h1}d} & 0 & F_{13} \\ -j\kappa_{h1}e^{-j\kappa_{h1}d} & j\kappa_{h1}e^{j\kappa_{h1}d} & 0 & F'_{13} \end{vmatrix}. \quad (53)$$

4.2.3 Choice of the particular solution $F(x)$

The particular solution $F(x)$ depends directly on the non-linear polarization $P_{NL}(x)$, and $P_{NL}(x)$ is determined by the fundamental field $E_f(x)$.

The fundamental field is a guided TE mode which has the following form:

$$E_f(x) = \begin{cases} A_f \cos(\kappa_{f1}d - \phi)e^{-\kappa_{f3}(x-d)} & d < x \\ A_f \cos(\kappa_{f1}x - \phi) & 0 < x < d \\ A_f \cos\phi \, e^{\kappa_{f2}x} & x < 0 \end{cases} \quad (54)$$

with
$$\text{tg}\,\phi = \frac{\kappa_{f2}}{\kappa_{f1}} \; ; \; \text{tg}(\kappa_{f1}d - \phi) = \frac{\kappa_{f3}}{\kappa_{f1}} \quad (55)$$

$$\kappa_{f1} = \sqrt{k_f^2 n_{f1}^2 - \beta_f^2} \; ; \; \kappa_{f2} = \sqrt{\beta_f^2 - k_f^2 n_{f2}^2} \; ; \; \kappa_{f3} = \sqrt{\beta_f^2 - k_f^2 n_{f3}^2}\,.$$

By assuming that the medium 3 is air and that the non-linear coefficient is the same for both the film and the substrate, we can write the non-linear polarisation as

$$P_{NL}(x) = \varepsilon_0 \mathbf{d}_{NL} E_f^2 = \begin{cases} 0 & d < x \\ \varepsilon_0 d_{NL} A_f^2 \cos^2(\kappa_{f1}x - \phi). & 0 < x < d \\ \varepsilon_0 d_{NL} A_f^2 \cos^2\phi \, e^{2\kappa_{f2}x} & x < 0 \end{cases} \quad (56)$$

In Eqn. (39), we can now replace $E_h(x,z)$ and $P_{NL}(x)$ by their expressions (46) and (56). This gives three equations from which we can deduce $F_1(x)$, $F_2(x)$ and $F_3(x)$. The particular solution $F(x)$ is then:

$$F(x) = \mu_0 \varepsilon_0 \omega_h^2 d_{NL}^2 A_f^2 f(x) \qquad 57(a)$$

with

$$f(x) = \begin{cases} 0 & d < x \\ \dfrac{1}{\kappa_{h1}^2 - 4\kappa_{f1}^2} \left[2 \dfrac{\kappa_{f1}^2}{\kappa_{h1}^2} - \cos^2(\kappa_{f1} x + \phi) \right] & 0 < x < d \\ \dfrac{-\cos^2(\phi)}{\kappa_{h2}^2 + 4\kappa_{f2}^2} e^{2\kappa_{f2} x} & x < 0 \end{cases} \qquad 57(b)$$

4.2.4 Calculation of the harmonic power and the conversion efficiency

In the case of a planar waveguide uniformly excited over a width W [cf. Eq. 18], the harmonic power is

$$P_h = \frac{W}{2} \operatorname{Re} \int_{-\infty}^{+\infty} (\vec{E}_h \times \vec{H}_h^*) \cdot \vec{z}_0 \, dx, \qquad (58)$$

where $\vec{z}_0$ is the unit vector in the Oz direction.

The limits of the integral are $+\infty$ and $-\infty$ for an infinitely long waveguide. For a waveguide of length L, we take into account only the contribution of a portion of an infinitely long waveguide. Equation (58) then becomes

$$P_h = \frac{W}{2} \operatorname{Re} \left[\int_{-L \operatorname{tg}\theta}^{0} (\vec{E}_{h2} \times \vec{H}_{h2}^*) \cdot \vec{z}_0 \, dx + \int_0^d (\vec{E}_{h1} \times \vec{H}_{h1}^*) \cdot \vec{z}_0 \, dx \right. \\ \left. + \int_d^{+\infty} (\vec{E}_{h3} \times \vec{H}_{h3}^*) \cdot \vec{z}_0 \, dx \right], \qquad (59)$$

where θ is the angle with respect to the z direction. The second and third integrals are independent of L, and have a limited value ($\vec{E}_{h3}$ and $\vec{H}_{h3}$ are evanescent). The first integral increases with L, and when L is long enough ($L \gg d$, depth of the waveguide), it will be much greater than the other two integrals. Physically, for the harmonic generated in Cerenkov configuration, the main part of the power is in the substrate, and the power in the film and in the air can be neglected, then

$$P_h \simeq \frac{W}{2} \operatorname{Re} \int_{-L \operatorname{tg}\theta}^{0} (\vec{E}_{h2} \times \vec{H}_{h2}^*) \cdot \vec{z}_0 \, dx. \qquad (60)$$

In the case of a TE polarized field with

$$\vec{E}_{h2} = E_{h2} \vec{y}_0 \qquad (61)$$

$$\vec{H}_{h2} = \frac{1}{j\omega_h \mu_0} \vec{\nabla} \times \vec{E}_{h2}, \qquad (62)$$

Equation (60) becomes

$$P_h \approx \frac{W\beta_h}{2\omega_h \mu_0} \operatorname{Re} \int_{-L\,tg\theta}^{0} E_{h2} E_{h2}^* \, dx. \qquad (63)$$

Introducing the expression of E_{h2} (cf. Eqn. (46)) into Eqn. (63) we obtain

$$P_h = \frac{W\beta_h}{2\omega_h \mu_0} \operatorname{Re} \int_{-L\,tg\theta}^{0} \Big(|C|^2 + |F_2(x)|^2 + CF_2^*(x) e^{jk_{h2}x} $$
$$+ C^* F_2(x) e^{-jk_{h2}x} \Big) \, dx. \qquad (64)$$

According to Eqn. (57) $F_2(x)$ is an evanescent field, then the three last terms of (64) may be neglected. It remains

$$P_h = \frac{W\beta_h}{2\omega_h \mu_0} \operatorname{Re} \int_{-L\,tg\theta}^{0} |C|^2 \, dx, \qquad [65(a)]$$

which after integration yields

$$P_h = \frac{W\beta_h \, |C|^2 \, L\,tg\theta}{2\omega_h \mu_0}. \qquad [65(b)]$$

From Eqs. (48), (49), (52) and (57), we have

$$|C|^2 = \mu_0^2 \varepsilon_0^2 \omega_h^4 A_f^4 |c|^2 \qquad [66(a)]$$

with

$$|c|^2 = \frac{\left[\begin{array}{c} \kappa_{h1}\left(\kappa_{h3}f_{13} + f'_{13}\right) + \kappa_{h1}\left(\kappa_{h1} \sin \kappa_{h1} d - \kappa_{h3} \cos \kappa_{h1} d\right)f_{12} \\ -\left(\kappa_{h1} \cos \kappa_{h1} d + \kappa_{h3} \sin \kappa_{h1} d\right) f'_{12} \end{array}\right]^2}{\kappa_{h2}^2\left(\kappa_{h1} \cos \kappa_{h1} d + \kappa_{h3} \sin \kappa_{h1} d\right)^2 + \kappa_{h1}^2\left(\kappa_{h3} \cos \kappa_{h1} d - \kappa_{h1} \sin \kappa_{h1} d\right)^2} \qquad [66(b)]$$

In the above equation, f_{12}, f_{12}', f_{13} and f_{13}' are defined in the same fashion as in Eq. [47(b)].

Following the definition, the fundamental power is

$$P_f = \frac{W\beta_f A_f^2}{4\omega_f \mu_0} \left(\frac{1}{\kappa_{f2}} + d + \frac{1}{\kappa_{f3}}\right). \qquad (67)$$

Then, using Eqs. (65) and (67), the conversion efficiency is

$$\eta = \frac{P_h}{P_f} = \frac{8\mu_0^3 \varepsilon_0^2 \omega_h^3}{\beta_h} \frac{LP_f}{W} \frac{tg\theta \, |c|^2}{\left(\dfrac{1}{\kappa_{f2}} + d + \dfrac{1}{\kappa_{f3}}\right)^2}. \qquad (68)$$

4.2.5 Discussion

Equation (68) shows that the conversion efficiency in the case of Cerenkov configuration is very different to what we obtain in the case of interaction between guided modes. Nevertheless we can recognize the power confinement term, $\frac{LP_f}{W}$ which is similar to that of the guided case but the dependence on L is no longer quadratic but linear.

From Eqn. [66(b)], we know that $|c|^2$ is a function of f_{12}, f_{12}', f_{13} and f_{13}' which are growing functions of the discontinuities at the interface guide-substrate of the particular solution $F(x)$, i.e. the forced field. Those discontinuities increase monotonically with the amplitude of the refractive index step. With a graded index waveguide, or a low step index waveguide SHG in Cerenkov does exist but with a very small efficiency. In the next paragraph we shall apply the theory we have just developed to step index waveguides, realized on x-cut lithium niobate by proton exchange.

4.3 SHG in Cerenkov configuration in PE waveguides

4.3.1 Model

In order to present the possibilities of PE waveguides we shall study the influence of the depth and of the amplitude of the index increase on the efficiency of conversion. This study is done for a pump power of 1 W at a wavelength of 1.06 μm or 0.84 μm, in a waveguide of width 10 μm, and length 1 cm.

In Fig. 9 we have drawn, for two pump wavelengths, the influence of the refractive index step on the conversion efficiency for a PE waveguide. We can see on this graph that, for each wavelength, the efficiency is nearly proportional

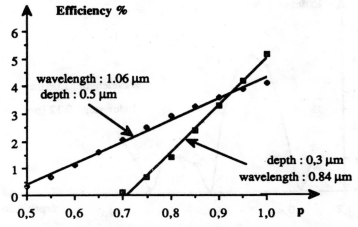

Fig. 9. Cerenkov second harmonic generation efficiency as a function of the index increase in a Proton Exchanged LiNbO$_3$ waveguide.

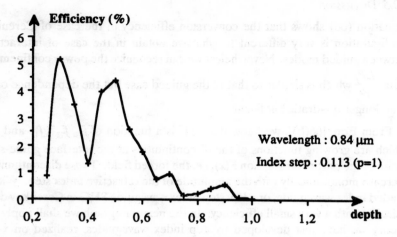

Fig. 10. Cerenkov second harmonic generation efficiency as a function of the depth of a Proton Exchanged LiNbO$_3$ waveguide.

to the index increase but that the slope of the curve depends very much on the wavelength. The index step is represented by the parameter p, defined by $\delta n_e(\lambda) = p\, \delta n_{e0}(\lambda)$, where $\delta n_{e0}(\lambda)$ is the maximum index increase possible by PE at the wavelength λ. In Figs. 10 and 11, we show the influence of the depth of the waveguide on the conversion efficiency. We can see that the behaviour is rather complicated, but that for some well defined depth the efficiency may be higher than 5%. If we try to look at the influence of both parameters together we obtain the three dimensional graph plotted in Fig. 12 which illustrates the necessity to control very accurately the index profile in order to get a high efficiency.

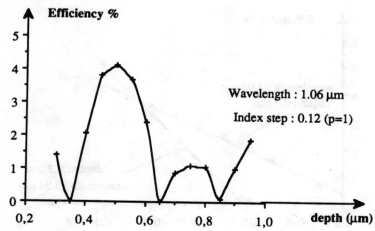

Fig. 11. Cerenkov second harmonic generation efficiency as a function of the depth of a Proton Exchanged LiNbO$_3$ waveguide.

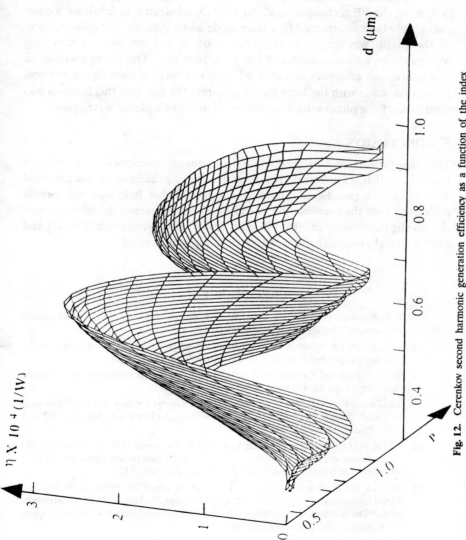

Fig. 12. Cerenkov second harmonic generation efficiency as a function of the index increase and the depth of a Proton Exchanged LiNbO$_3$ waveguide.

4.3.2 Experiment

The validity of this model has been verified on multimode planar waveguides [13], but these structures were not realized to optimize the efficiency of conversion. Optimized structures were realized in the stripe waveguide technology [15], using the PE technique on Z-cut LiNbO$_3$ substrates, to fabricate a compact visible laser constituted by a laser diode and a doubler. Coupling 65 mW of the pump wavelength in the TM$_0$ mode of a 2 μm wide, 6 mm long waveguide, an harmonic output of 1 mW was obtained. This figure correspond to a conversion efficiency of 1.06% which is not very far from the theoretical prediction made with the formula (68), despite the fact that this formula was established for a pump wave coupled to TE mode of a planar waveguide.

5. CONCLUSION

This chapter does not pretend to be an extensive description of all what is possible in NLIO. I just wanted to present the basic principles used in that field with SHG as a paradigm. The device aspect of that field was only briefly addressed with the presentation of the Cerenkov frequency doubler. Nevertheless the potentiality of NLIO to yield devices is very important [16, 17] and each material progress brings new possible applications [18].

REFERENCES

1. P.A. Franken, Hill A.E., Peters C.W. and Weinreich G., "Generation of optical harmonique", Phys. Rev. Letters. **7**, 118 (1961).
2. R.H. Stolen and Bjorkholm J. E., "Parametric Amplification and Frequency Conversion in Optical Fibers", IEEE J. Quant. Elec. **QE-18**, 1062 (1982).
3. R.V. Schmidt and Kamminow I.P., "Metal-diffused optical waveguides in LiNbO$_3$", Appl. Phys. Lett. **25**, 458-460 (1974).
4. J.L. Jackel, Rice C.E. and Veselka J., "Proton exchange for high-index waveguides in LiNbO$_3$", Topical Meeting on Guides Wave Optics, Asilomar, CA, January 1982; and Appl. Phys. Letters **41**, 607 (1982).
5. M. De Micheli, Botineau J., Neveu S., Sibillot P., Ostrowsky D.B., and Papuchon M., "Fabrication and characterisation of titanium indiffused proton exchanged (TIPE) waveguides in lithium niobate", Opt. Comun. **42**, 101-103 (1982).
6. M.J. Li, De Micheli M., Ostrowsky D.B. and Papuchon M., "High index low loss LiNbO$_3$ waveguides", Opt. Comm. **62**, 17 (1987) or M. De Micheli, Li M.J., Ostrowsky D.B., "The double proton exchange in LiNbO$_3$", Proceedings of the Fourth European Conference on Integrated Optics (1987).
7. M. De Micheli, "Generation de deuxieme harmonique en optique intégrée", Thèse de 3ème cycle, Université, de Nice, 1982
8. W. Sohler, "Non-linear integrated optics", in "New direction in guided wave and coherent optics", Edited by D.B. Ostrowsky and E. Spitz, Martinus Nijhoff Publisher, The Hague, 1984.
9. W. Sohler and Suche H.,"Frequency conversion in Ti: LiNbO$_3$ waveguides", SPIE Proceedings 408, p. 163 (1983).

10. M. De Micheli, "Non-linear interaction in LiNbO$_3$ guides", in, "New direction in guided wave and coherent optics", Edited by D.B. Ostrowsky and E. Spitz, Martinus Nijhoff Publisher, The Hague, 1984 or M. De Micheli, Botineau J., Neveu S., Sibillot P., D.B. Ostrowsky, and M. Papuchon, "Extension of second harmonic phase-matching range in lithium niobate guides", Opt. Lett., **8**, 116 (1983).

11. P.K. Tien, Ulrich R., and Martin R.J., "Optical second harmonic generation in form of coherent Cerenkov radiation form a thin-film waveguide", Appl. Phys. Lett., **17**, 447-450 (1970).

12. Bor-Uei Chen, Tang C.L. and John M. Telle, "CW harmonique generation in the uv using a thin-film waveguide on a non-linear substrate", Appl. Phys. Lett. **25**, 495-498 (1974).

13. Ming Jun Li, "Génération de deuxième harmonique en configuration Cerenkov dans les guides d'ondes réalisés par échange protonique dans le niobate de lithium". Thése de 3ème cycle, Université de Nice, 1988.

14. J.E. Sipe and Stegeman G.I., "Comparision of normal mode and total field anaysis techiques in planar integrated optics" J. Opt. Soc. Am. **69**, 1676-1683 (1979).

15. T. Taniuchi and Yamamoto K., "Second harmonique generation in proton-exchanged optical LiNbO$_3$ waveguides", "Advance Optoelectronic Technology", Cannes, 1987 SPIE Procs., 864-09, 1987.

16. Gorge I. Stegeman and Colin T. Seaton, "Non-linear integrated optics", J. Appl. Phys. **58(12)**, 57-78 (1985).

17. W. Sohler, Hampel B., Regener R., Rivken R., Suche H. and Volk R., "Integrated optical parameteric devices", IOOC-ECOC'85, 29-37.

18. M.J. Li, M. De Micheli, Ostrowsky D.B., Lallier E., Breteau J.M., Papuchon M. and Pocholle J.P., "Optical waveguide fabrication in neodymium doped lithium niobate" Electron. Lett. **24**, 914-915 (1988).

6. EPILOGUE* : QUASI PHASE MATCHING FOR SHG

Phase matching between the fundamental and the second harmonic wave is very critical for efficient second harmonic generation. In the preceding discussion, we saw how automatic phase matching can be obtained by using the Cerenkov configuration wherein the fundamental is a guided mode and the second harmonic appears as a packet of radiation modes radiating into the substrate. Since radiation fields are involved, the efficiency of conversion to second harmonic scales as length rather than as square of length.

Quasi phase matching (QPM) [1] is a technique by which the dispersion in the phase velocities at fundamental and second harmonic can be compensated by periodically modulating either the refractive index or the nonlinearity with period Λ so that

$$\Delta\beta = \beta(2\omega) - 2\beta(\omega) = \pm \frac{2\pi}{\Lambda},$$

---

* This section is authored by my colleague K. Thyagarajan at my request to bridge the delay between the date of receipt of the original manuscript and its final publication. — Editor

where $\beta(\omega)$ and $\beta(2\omega)$ are the propagation constants of the fundamental and second harmonic guided mode, since $\pi/\Delta\beta$ corresponds to the phase coherence length (L_c), the periodicity is twice the coherence length. This is referred to as first order QPM. When the nonlinearity is not modulated, the second harmonic grows till L_c at which point the waves at ω and 2ω are in phase opposition and the newly created second harmonic interferes destructively with the existing wave thus reducing the second harmonic wave. Thus if the sign of nonlinear coefficient is reversed at this point, then the second harmonic will continue to grow till $2L_c$ when again the sign of nonlinearity is reversed so that further constructive addition takes place.

Since typical coherence lengths are 1-2 μm, third order QPM wherein the period is thrice the coherence length is usually used although at the expense of reduced efficiency.

Since phase mismatch is compensated by a periodic modulation of non-linear coefficient, QPM can be used to obtain SHG in even isotropic crystals at any temperature using any component of the nonlinear tensor over the whole transparency range of the material. Thus using QPM, guided-guided mode SHG in LiNbO$_3$ using the largest d_{33} coefficient has been demonstrated; normalised efficiencies of 170% / W-cm^2 have been reported for generation of blue light. This would imply that if 100 mW of fundamental power is input, an interaction length of 1 cm leads to 17 mW of blue light. Some of the best figures using QPM include 230%/W-cm^2 in KTP waveguides. In the future one can expect efficiencies greater than 1000%/W-cm^2 using organic nonlinear materials. QPM techniques can also be used for generation of difference frequencies which lie in the near infrared.

Various techniques are being developed for periodic reversal of domains in LiNbO$_3$, LiTaO$_3$ or KTP waveguides. These include Ti indiffusing in +c face of LiNbO$_3$ crystal [2], Li out-diffusion [3], electron beam writing [4, 5, 6] etc.

REFERENCES

1. M.M. Fejer, Magel, G.A., and Lin, E.J., "Quasi phase matched interactions in lithium niobate", Proc. SPIE vol. 11&8, pp. 213-224 (1989).
2. E.J. Lim, Fejer M.M., and Byer R.L., "SHG of green light in periodically polar planar LiNbO3 waveguides", Electron Letts 25, pp. 174-175 (1988).
3. J. Webjorn, Lawell F. and Arvidsson G., "Blue Light generated by frequency doubling in LiNbO3 channel waveguide", IEEE Photonics Tech. Letts. 1, pp. 316-318 (1989).
4. M. Yamada and Kishima K., "Fabrication of periodically reversed domain structure for SHG in LiNbO3 by direct electron beam lithography at room temperature", Electron. Letts. 27, pp. 828-829 (1991).
5. H. Ito, Takyu C. and Inaba H., "Fabrication of periodic domain grating in LiNbO3 by electron beam writing for application of nonlinear optical processes", Electron. Letts. 27, pp. 1221-1222 (1991).
6. R.W. Keys,, Loni A., De La Rue, R.M. Ironside, C.N., Marsh J.H., Luff B.J. and Townsend P.D., "Fabrication of domain reversed gratings for SHG in LiNbO3 by electron beam bombardment", Electron. Letts. 26, pp. 188-190 (1989).

9

Fabrication Techniques of Optical Fibres[*]

H. KARSTENSEN[**]

1. INTRODUCTION

With the enormous progress in optical fibre technology in the last few years, large capacity optical fibre transmission systems were installed all around the world with many thousands of fibre route kilometers. The basic technological problems are already solved and a great increase in the annual fibre production rate is expected in the near future. For single-mode fibres average losses of 0.6 dB/km at 1.3 μm wavelength and 0.3 dB/km at 1.55 μm have already been achieved under routine large scale production environment. These losses are very near to the theoretical limits, so the reduction of loss and the bandwidth of the fibres are no longer the main aim of the present fibre optic research. Instead the increase in the rate of fibre production and subsequent reduction in the cost of fibre manufacture are the major tasks facing current R and D in fibre technology.

As already discussed in previous chapters, an optical fibre, in general, consists of a core, a cladding and a coating or jacket. Core and cladding are made of silica glass in high quality fibres. Their refractive index mismatch leads to the light guiding properties of the fibre. The coating, which usually consists of a special kind of polymer, serves to protect the bare fibre against chemical and mechanical attacks. Especially the humidity of the normal atmosphere causes microcracks on the fibre surface to grow — which eventually degrades the inherent high tensile strength of the fibre. Therefore the coating is applied

---

[*] Updated reprint from J.I.E.T.E, vol. 32 (Special issue on Optoelectronics and Optical Communication), July-August (1986).

[**] Siemens Research Laboratories, Otto-Hahn-Ring 6, D-8000 Muenchen 83, Germany.

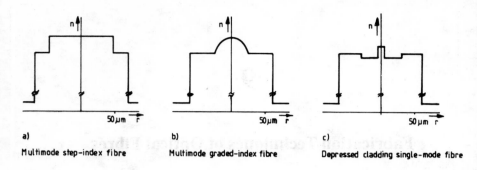

a) Multimode step-index fibre   b) Multimode graded-index fibre   c) Depressed cladding single-mode fibre

Fig. 1. Refractive index profiles of commonly used optical fibres

in line with the fibre drawing process to prevent any kind of such attack. A necessary condition for waveguidance is that the refractive index of the cladding is lower than that of the core. Within the core, the refractive index can be either uniform or it can vary with radius. Figure 1 shows some examples of the refractive index profiles of some important types of optical fibres. In a graded-index fibre, for example, the refractive index is maximum on the axis and decreases towards the cladding region. Of special importance for telecommunication applications are the graded-index fibres with a nearly parabolic index distribution (Fig. 1b), and monomode fibres with depressed cladding (Fig. 1c) which show the lowest propagation losses.

To fabricate fibres, one can either use different glasses with different refractive indices for core and cladding or use fused silica (quartz glass) whose refractive index is modified by doping. Figure 2 shows the refractive index change as a function of the doping level for some widely used doping materials. Uptill now fibres made of fused silica show the lowest attenuation coefficients with values down to below 0..2 dB/km at 1.55 μm wavelength.

To date, several techniques for the production of high quality optical fibres have been developed. The purpose of this paper is to give an overview of fibre fabrication and numerous process technologies. For more thorough discussions on this subject, the reader is referred to some excellent review papers and books [1-12].

2. FIBRE DRAWING

In most cases, the fibre production process is split into two fabrication steps: first the fabrication of a thick (typically 10 to 25 mm in diameter) and short (maximum length: few meters) glass rod called the fibre preform, and second, the drawing of fibre from the preform. The techniques used for preform fabrication will be described below. In the drawing process the preforms must be converted to fibres with diameters in the range of 100 to 400 μm; in most cases a value of 125 μm is used as standard overall fibre diameter. Moreover,

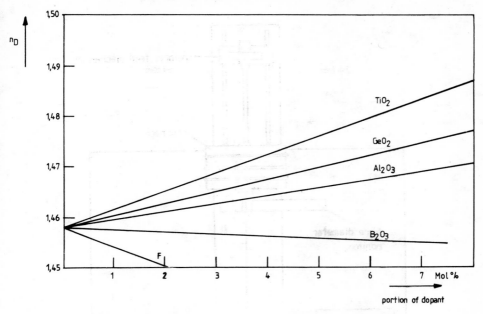

Fig. 2. Refractive index of doped fused silica as a function of the mole fraction of some usual dopants

the fibres must be embedded in one or more polymer coatings to protect them against humidity and external mechanical forces. It is worth mentioning that during fibre drawing, the refractive index profile of the preform is retained. A typical apparatus is shown schematically in Fig. 3. In practice, such drawing towers could be almost upto 18 m in height. For the drawing of high-silica glass fibres, furnaces with maximum temperatures upto as high as 2200°C are needed. It is important that the temperature is controlled very precisely because temperature fluctuations directly result in diameter fluctuations of the drawn fibre. Most widely used are the graphite resistance furnaces and the zirconium induction furnaces, respectively [12]. The graphite furnace has the advantage that it can be very easily controlled. However, a protective noble gas like argon is necessary in order to prevent any degradation or even a burning of the graphite heater. Turbulences in the protective gas flow can cause fibre distortions. The zirconium induction furnace does not need a protective gas but the temperature control and its handling are relatively much more involved.

Directly below the furnace, a diameter monitor is mounted which forms an integral part of the control unit for fibre drawing. As illustrated in Fig. 4 where the neck-down region of a preform is shown, the law of conservation of mass leads to the inter relationship that exists between the fibre diameter d_f, the preform diameter d_p, the fibre pulling velocity v_f and the preform feed velocity v_p of the preform. This relation is given by

$$d_f = d_p(v_p/v_f)^{1/2} \qquad (1)$$

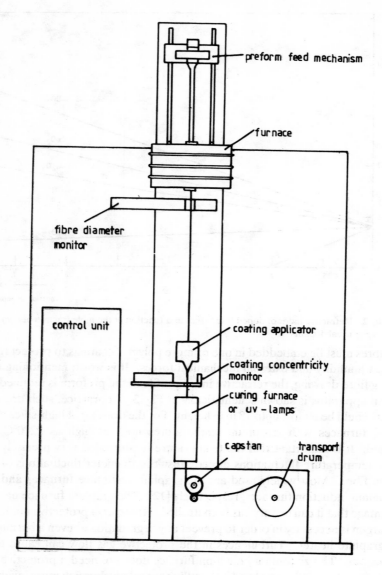

Fig. 3. Schematic of a fibre drawing apparatus

(However, a small portion of the glass rod will be evaporated within the hot zone). Usually, the diameter is controlled by varying the pulling velocity v_f of the fibre, i.e., by varying the speed of rotation of the capstan (cf. Fig. 3).

In the intermediate region between the diameter monitor unit and the capstan drive, the fibre is coated. The coating materials are in most cases polymers which are usually applied by passing the fibre through a coating bath

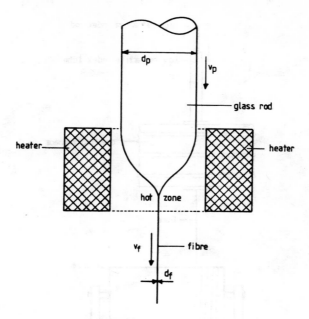

Fig. 4. Neck-down region of a fibre preform

with a nozzle at the bottom. If the fibre pulling velocity exceeds 5 m/s, the coating bath must be pressurized in order to minimize shear rates, thus avoiding slipping of the fibre. The applied polymer is then cured either by heating or by UV radiation. It should be noted that it is the minimum tolerable curing time which eventually determines the required height of a fibre drawing tower. The higher the pulling speed, the higher the tower must be. The coating concentricity is monitored and, if necessary, corrected by varying the position of the coating nozzle [13]. Finally, the coated fibre is wound directly from the capstan onto a transport drum.

3. MULTICOMPONENT TECHNOLOGY

3.1 Rod-in-tube method

Until sixties two methods for fabricating fibres were widely used, the rod-in-tube method and the double crucible method [14]. As shown in the schematic diagram of the rod-in-tube method (Fig. 5), a glass rod having a higher refractive index is placed in a glass tube of lower refractive index of compatible material. This combination is fed at a constant speed into the furnace in which the glass softens, the tube shrinks due to its surface tension and fuses with the rod. Finally, under gravitational pull, the melted glass tapers into a thin fibre which is then fixed to a rotating and slowly traversing take-up winding drum for spooling the fibre. Depending on the choice of the rod and the tube glasses, fibres with relatively high numerical apertures can be fabricated. In order to

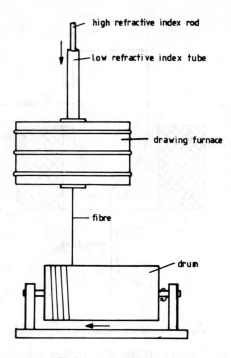

Fig. 5. Schematic diagram of the rod-in-tube method

achieve low losses, the adjacent surfaces of rod and tube have to be carefully cleaned and polished so that the interface between the core and the cladding of the drawn fibre is free of any particles or voids. With high quality fused silica as core rod and a fluorine doped silica tube as cladding, fibre losses ~ 3 dB/km have been reported [15]. However, fibres made of multicomponent glasses usually exhibit higher losses and hence this fabrication method is not used for manufacturing communication grade fibres of today. It is mentioned here for the sake of completeness because this method shows the basic principles of fibre drawing from a solid glass rod.

3.2 Double crucible method

In the double crucible method, as shown in Fig. 6, at first two very high purity glass rods with different refractive indices are prepared with conventional techniques. These rods consist of, for example, sodium borosilicate (Na_2O – B_2O_3 – SiO_2) with different compositions for core and cladding [3, 14-16]. To draw cladded-core fibres from these glass rods, two cylindrical crucibles are arranged concentrically as shown in Fig. 4. The rods are fed slowly downwards into the heated crucibles where they are melted filling up the crucibles with glass melt. The height of the melt must be constant. The crucible temperature of 850 to 1100°C depends on the choice of the glass composition. At the base, the crucibles taper down to concentric nozzles through which the melted glasses come out and are pulled into a fibre. The pulling speed is generated by the rotating drum onto which the fibre is wound.

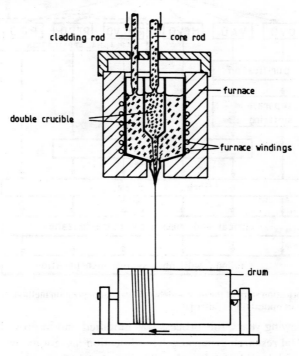

Fig. 6. Schematic diagram of a double crucible apparatus

With this method, a fibre can be continuously fabricated from the molten glass in only one fabrication step. In principle, fibres of unlimited length can be manufactured. Due to the low cost of this process, the double-crucible method was improved mainly by the British Post Office and Nippon Sheet Glass Company in Japan which resulted in fibres with total attenuation below 5 dB/km [16]. This method is even capable of fabricating graded-index fibres [15-17]. To form the desired graded index profile, an ion exchange between the core and the cladding glass is exploited during fibre pulling. Bandwidth of 400 to 900 MHz-km can be obtained in this way [5].

4. VAPOUR-DEPOSITION METHODS

A major milestone in the progress of optical fibre technology was achieved in 1970 when researchers at Corning Glass Works in the USA had developed a fibre with loss below 20 dB/km (measured at 633 nm wavelength) [18]. This fibre was fabricated by a new process: the inside-vapour-deposition method, also called inside-vapour-phase-oxidation method.

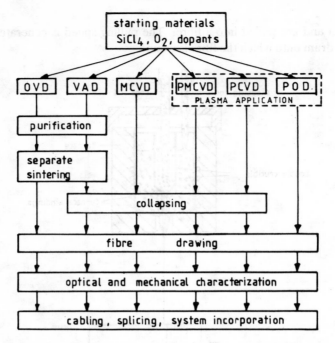

Fig. 7. Fabrication steps of the most widely used vapour deposition methods for fabrication of optical fibres (after [10])

In the following years, this process was modified and improved again and again by several research groups-each modification has got its own name and at last fibres with total losses less than 0.2 dB/km at 1.55 μm wavelength could be fabricated, thereby reaching nearly the theoretical minimum loss of silica fibres [19, 20].

Figure 7 shows an overview of the essential vapour deposition methods as used today together with the individual process steps [10]. These methods will be described in more detail below.

Most of these methods need a collapsing step for fabricating a solid glass rod from the substrate tube which is done by heating the substrate tube with a very slowly traversing burner or ring oven to a temperature above 2000°C so that the glass softens and the surface tension of the glass can contract the tube. The collapsing rate depends mostly on the pressure difference between the outside and the inside of the tube, the tube dimensions, and on the glass viscosity which itself is strongly dependent on the temperature and the composition of the glass. The rate of collapse K is roughly given by [11, 21].

$$K = (\Delta P + \sigma/r_i + \sigma/r_o)/\eta_{eff} . \qquad (2)$$

Here ΔP is the pressure difference between the outside and the inside of the tube, σ is the surface tension, r_i and r_o are the inside and the outside radii of the tube, respectively, and η_{eff} is the effective viscosity of the glass.

Care must be taken to avoid any elliptical deformation of the resulting preform and this requires several passes of the burner or oven. the collapsing process therefore needs a considerable fraction of the total preform processing time.

Due to the high temperature during this step, nearly all the dopants in the glass at the internal surface of the substrate tube vaporise. The result is a decreased refractive index in the preform and fibre axis, commonly known as the refractive index dip. As this dip deteriorates the bandwidth of a graded-index multimode fibre substantially, many attempts have been made to avoid it (see Sec. 4.4).

4.1 Inside-vapour-deposition (IVD) method

In the original process of the Corning Glass Works [1, 3], high purity $SiCl_4$ vapour mixed with a few percent of $TiCl_4$ vapour as dopant were made to react with oxygen in an open flame. Generally, this process is called flame hydrolysis. The raw materials were oxidized forming tiny particles of doped silica glass of about 0.01 to 0.1 μm in diameter. These particles were then pumped through a substrate tube of fused silica, where a small fraction of the glass particles get deposited on the inner wall. In a second process step, the deposited soot was sintered with the help of a moving burner to a thin transparent glass film of slightly higher refractive index (due to doped silica) than the fused silica of the tube. In a third step, the tube is collapsed and finally drawn to a fibre.

Glasses which are fabricated by this process are generally called high-silica glasses because they usually consist of more than 80% of SiO_2. The remaining are the dopants, e.g. GeO_2, P_2O_5, B_2O_3 and fluorine which are most frequently used. As described, the IVD process consists of four steps: (i) forming of glass particles (soot) from the raw materials in a gas phase reaction, (ii) subsequent sintering of the soot to a transparent glass, (iii) collapsing the tube to a fibre preform, and (iv) fibre drawing from the preform. Most of these process steps are also used in the improved versions of fibre fabrication which were developed since the IVD process was developed.

4.2 Outside-vapour-deposition (OVD) method

One great disadvantage of the IVD process was that only fibres with very small core diameters could be fabricated. With the development of the OVD method in 1973, researchers at Corning Glass Works could overcome this problem. This new method was originally called outside-vapour-phase-oxidation (OVPO) method [3, 4, 22, 23]. A schematic diagram of the process is shown in Fig. 8. The process starts with the fabrication of a low density porous glass body, the soot preform. As shown in Fig. 8, the glass forming vapours $SiCl_4$, $GeCl_4$, $POCl_3$, etc. are fed through a oxygen-hydrogen burner. The burner consists of at least three concentric orifices: the center orifice is used for the halide vapours, while the outer one for feeding the fuel gas and oxygen

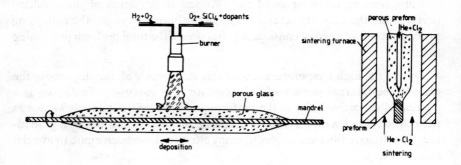

Fig. 8. Schematic of the OVD method

mixture. The orifice in between serves for shielding gas to prevent a premature reaction of the halides which could form a glass deposit on the burner face itself. In the flame, the halides react with oxygen, forming tiny particles of doped silica glass. The burner is directed against a slowly rotating target mandrel made of Al_2O_3 or graphite with a typical diameter of 0.5 cm where a fraction of the glass particles is collected. During deposition, the burner traverses back and forth parallel to the mandrel axis so that one layer of doped silica glass is deposited per pass. Soot preforms with upto 1000 layers have been fabricated. These are about 80 cm in length, 11 cm in diameter and have a weight of 1800 g. Average deposition rates are 1 to 4 g/min [23].

In the next process step, the target mandrel is removed from the porous soot preform. This is easily accomplished by cooling the soot preform, as the expansion coefficients of mandrel and glass soot differ. Then the soot preform is sintered or consolidated to a dense glass rod by passing it vertically through the hot zone of a special annular furnace. During this step, the temperature is raised to approximately 1500°C. By using an atmosphere of helium with a few percent of chlorine within this furnace, OH-ions could be removed very effectively resulting in very low OH-content fibre preforms. The preforms are then as usual drawn to fibres in the next process step.

The OVD method is a very efficient technique for fabricating high quality optical fibres and is extensively used by the Corning Glass Works for large scale production. The capability of this method is clearly demonstrated by the fabrication of single-mode fibre with 0.16 dB/km attenuation at 1.55 μm wavelength [23], the best reported attenuation value to date. For the fabrication of multiple clad dispersion flattened single-mode fibres, the normal OVD process has the drawback that with fluorine doping only relatively low relative refractive index depressions Δ^- of about 0.27% can be achieved. However, recently relative refractive index depressions Δ^- of more than 0.60% have been obtained with a slightly modified OVD process [24].

4.3 Chemical vapour deposition (CVD) method

In 1973, researchers of Bell Laboratories (USA) had reported a new fibre fabrication technique: the CVD method. In this process, silane (SiH_4) and germane (GeH_4) or diborane (B_2H_6) as dopants were made to react with oxygen in a substrate tube of fused silica at temperatures of about 1000 to 1300°C [3, 25, 26,]. In the first experiments, the substrate tube was placed inside an electrical furnace, but later the substrate tube was moved back and forth through a stationary ring shaped furnace to yield a thin layer of transparent glass in every pass. The refractive index of the layers could be controlled by the composition of the atmosphere inside the substrate tube. After the deposition of a sufficient number of layers, the tube was collapsed to a solid glass rod, i.e., the preform. This was done by moving the tube very slowly through the hot zone of the furnace. The temperature of the hot zone was simultaneously raised so that the substrate tube softened and collapsed by its own surface tension. This preform was then drawn to a fibre.

The basic difference of this method as compared to the IVD and OVD processes is that the deposition and sintering of the glass particles are not two different fabrication steps but occur nearly simultaneously in each pass of the burner, i.e., in the fabrication of every single layer. Because of their highly explosive nature, the reagents are diluted to a fraction of less than 1% of the total gas flow rate. So the deposition rate is very low resulting in a large fabrication time. e.g., the fabrication of a preform for 1 km of fibre length would take 24 to 36 hours.

4.4 Modified chemical vapour deposition (MCVD) method

The Bell Laboratories achieved a tremendous progress in 1974 with an improvement of the CVD process which was called modified CVD method [3, 21, 27-30]. In this process, $SiCl_4$, $GeCl_4$, $POCl_3$, BCl_3 or BBr_3 are used as raw materials. Compared to the old CVD process, the concentrations of reagents as well as the reaction temperatures were considerable increased. As shown in Fig. 9, controlled amounts of $SiCl_4$ vapour as dopant alongwith oxygen are fed into a rotating fused silica substrate tube. A traversing oxy-hydrogen burner heats a short zone of the tube to a temperature of about 1600°C. In this hot zone, the chemicals react forming glass particles which are subsequently deposited downstream on the inner wall of the tube. The heat from the slowly traversing burner which followed, sinters this soot to a transparent glass layer. Once the burner has arrived at the downstream end of the tube, it is traversed back very fast to the upstream end. This is done because the deposition is only carried out while the burner traverses in the direction of the gas flow. In this way, the desired amount of material is built up, layer by layer, by repeated traversal of the burner. The composition and with it the refractive index of the layers can be varied by changing the gas mixture. After deposition of typically

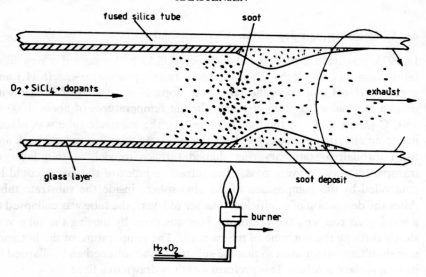

Fig. 9. Schematic of the MCVD method

40 to 100 layers, the tube is collapsed to a glass rod, the preform, which can be later drawn to a fibre.

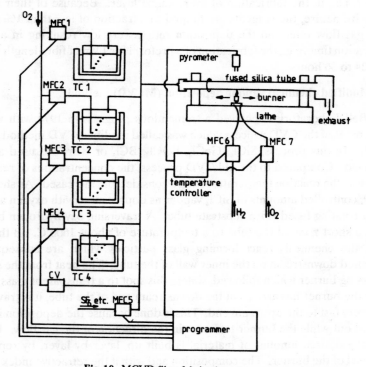

Fig. 10. MCVD fibre fabrication apparatus

Because of its simplicity and high performance, a lot of companies are manufacturing fibres with the MCVD method. It is therefore the most widely used fibre fabrication method throughout the world. For years, the best fibres were fabricated with this method and it was a MCVD-fibre which first showed an attenuation of 0.2 dB/km at 1.55 μm wavelength [19].

Figure 10 shows a schematic diagram of a complete conventional MCVD apparatus as is currently used. The main parts are the delivery system for the reagents through control valves, mass flow controllers, temperature controlled bubblers, and the carrier gas source; the glass working lathe with burner and pyrometer, the exhaust with the scrubber for neutralization of chlorine, and the central control system which controls the process data. The fabrication starts with cleaning a high quality fused silica tube with conventional cleaning procedures including a degreasing step to remove organic impurities and an acid step to remove metals. The tube must have a good dimensional uniformity and should be free of bubbles, inclusions or other defects because the strength, optical attenuation and the fibre geometry are strongly affected by these defects. Most frequently used are Heralux waveguide grade tubes having a outer diameter of 25 mm and an inner diameter of 19 mm. This starting tube is mounted in the lathe with perfectly aligned chucks and connected to the chemical delivery system by means of a swivel connector. The tube is then aligned so that it rotates exactly around its axis. This is a very critical step because even small deviations from perfect alignment will lead to deviations of circular symmetry in the final preform in the subsequent high temperature steps. Next, the substrate tube is fused to an exit tube with larger diameter which is designed to collect the excess glass particles without clogging. A fire polishing at very high temperatures often combined with a gas phase etching of the tube with fluorine usually precedes the deposition steps in order to remove surface irregularities from the inner tube wall.

For deposition, precisely controlled amounts of vapours of $SiCl_4$ and required dopants together with oxygen are passed through the substrate tube. In most cases, oxygen serves simultaneously as a reagent and a carrier gas for the halides. To control the gas composition, the flow rates of the carrier gas through the bubblers (Fig. 10) are controlled by mass flow controllers (MFC). The carrier gas bubbles through the liquid halides thereby getting saturated with the vapour of the reagent. This mixture is then fed through valves (frequently made of teflon) to the substrate tube. The chemicals which are normally used are liquid halides, having high vapour pressures at the room temperature. Therefore, enough halide vapour is carried with the gas even at low gas flow rates. The flow rate Q_v of the vapourised halide is directly proportional to the flow rate Q_g of the carrier gas according to

$$Q_v = k_s \cdot p_v/(p - p_v) \cdot Q_g . \tag{3}$$

Here, k_s is the degree of saturation, p_v is the vapour pressure of the halide and p is the total pressure [31]. Because p_v depends exponentially on temperature, the temperature of the liquids must be precisely controlled. Vapour pressure for SiCl$_4$ at 20°C is $p_{vs} = 228$ HPa, for GeCl$_4$ is $p_{vg} = 89$ HPa, and for POCl$_3$ $p_{vg} = 17$ HPa [12]. Assuming complte saturation ($k_s = 1$), the halide flow rates are given by

$$Q_{vs} = 0.30 \ Q_{gs} \text{ for SiCl}_4,$$

$$Q_{vg} = 0.10 \ Q_{gs} \text{ for GeCl}_4, \text{ and}$$

$$Q_{vp} = 0.02 \ Q_{gp} \text{ for POCl}_3.$$

In practice, the assumption of complete saturation may not be valid always. However, if the degree of saturation is kept constant, this is not absolutely necessary. For the purpose of a constant gas saturation, special kinds of bubblers are constructed and sometimes two bubblers are used in tandem to yield a nearly complete saturation. Gaseous dopants like SiF$_6$, SF$_6$, CF$_4$, CCl$_2$F$_2$ etc. are directly controlled by MFCs.

In the short hot zone above the traversing burner, these halide vapours react with oxygen forming glass particles according to the following reaction equilibria

$$\text{SiCl}_4 + \text{O}_2 \rightleftharpoons \text{SiO}_2 + 2\,\text{Cl}_2$$

$$\text{GeCl}_4 + \text{O}_2 \rightleftharpoons \text{GeO}_2 + 2\,\text{Cl}_2$$

$$2\,\text{POCl}_3 + \tfrac{3}{2}\text{O}_2 \rightleftharpoons \text{P}_2\text{O}_5 + 3\,\text{Cl}_2$$

$$2\,\text{BBr}_3 + \tfrac{3}{2}\text{O}_2 \rightleftharpoons \text{B}_2\text{O}_3 + 3\,\text{Br}_2.$$

At usual reaction temperatures, the equilibrium is far to the right for SiCl$_4$, so nearly all SiCl$_4$ is converted to SiO$_2$. For GeCl$_4$, however, the equilibrium lies somewhere in between and depends strongly on temperature and partial pressures of oxygen and chlorine [21, 32].

In the case of fluorine doping, the principal reaction is given by the following equilibrium [11, 33]:

$$3\,\text{SiO}_2 + \text{SiF}_4 \rightleftharpoons 4\,\text{SiO}_{1.5}\text{F}.$$

The incorporation of fluorine and hence the decrease in the refractive index depends on the partial pressure p_f of SiF$_4$ according to $p^{0.25}$ [11, 33]. Unfortunately, SiF$_4$ is a very aggressive gas and therefore seldom directly used. Commonly, it is formed inside the substrate tube as a reaction product of other fluorine containing gases. Frequently used are CCl$_2$F$_2$ and SF$_6$ whose reactions are described by the equilibria [34]

$$2\,\text{CCl}_2\text{F}_2 + \text{SiCl}_4 + \text{O}_2 \rightleftharpoons \text{SiF}_4 + 2\,\text{CO}_2 + 4\,\text{Cl}_2,$$

$$2\,\text{SF}_6 + 3\,\text{SiCl}_4 + \text{O}_2 \rightleftharpoons 3\,\text{SiF}_4 + 2\,\text{SO}_2 + 6\,\text{Cl}_2.$$

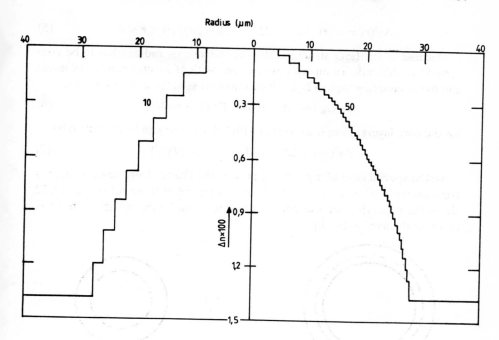

Fig. 11. Approximation of a parabolic index profile with 10 and 50 staircase steps

In the usual MCVD process, the incorporation of fluorine is limited by these thermal equilibria to mole fractions below 1.5%, so the maximum relative refractive index depression Δ^- is less than 0.4% [34].

The glass particles deposit downstream the reaction zone as a glass soot of well defined composition. The soot is sintered, by the slowly traversing burner that follows, to a thin glass layer of a definite refractive index. The MCVD process is well suited for the realisation of nearly any refractive index profile but due to the layer structure, the desired profile can be well approximated by a staircase function. In Fig. 11, the approximation of a parabolic profile by an ideal staircase function with 10 and 50 stairs is shown, in order to demonstrate the distortion of the original function. Calculations have shown that more than 40 layers are needed to prevent a degradation of the fibre bandwidth [15]. The desired index profile of the fibre is given by

$$n^2(r) = n_c^2 [1 - 2\Delta f(r)], \ 0 < r < a \quad (4)$$
$$= n_c^2(1 - 2\Delta) = n_{cl}^2, \ r > a$$

with n_c and n_{cl} as refractive indices on axis and in the cladding, respectively, Δ is the relative refractive index difference and a the core radius. In order to calculate the process parameters, the index profile is expressed as (squared) refractive index difference relative to the cladding glass

$$\delta n^2(r) = n^2(r) - n_{cl}^2 = 2\Delta \cdot n_c^2 \left(1 - f(r)\right), 0 < r < a \ . \tag{5}$$

Because of the layer structure, the dependence on radius r has to be converted to a dependence on layer number m. With M as total number of layers and the assumption that all layers have equal cross-section, it follows that

$$\delta n^2(m) = 2\Delta \cdot n_c^2 \left(1 - f^{1/2}\left(1 - m/M\right)\right) \tag{6}$$

for the core layers. Simple q-profiles with $f(r) = (r/a)^q$ can be described by

$$\delta n^2(m) = 2\Delta \cdot n_c^2 \left[1 - (1 - m/M)^{q/2}\right] . \tag{7}$$

In the special case of a parabolic profile, the change from layer to layer is constant with $\delta n^2 = 2 \cdot \Delta \cdot n_c^2/M$. As an illustration of these relations, Fig. 12 shows the refractive index distribution before and after the collapsing process (without the axial index dip).

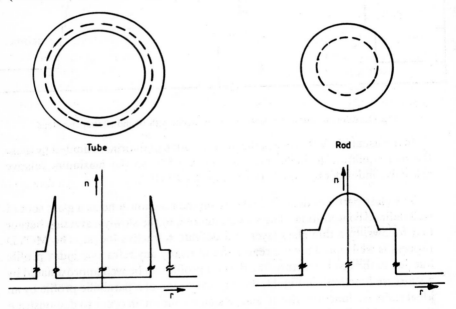

Fig. 12. Refractive index profiles of the substrate tube after deposition and after the collapse step, i.e. of the preform (without dip)

The essential deposition mechanism of the MCVD process was shown to be thermophoresis [21]. With a mathematical model for thermophoretic deposition [35], the deposition efficiency ε was approximately estimated to be

$$\varepsilon = 0.8 \left(1 - T_e/T_r\right) \tag{8}$$

with T_e as equilibrium temperature of the gas and the tube wall downstream of the hot zone and T_r as the temperature at which the chemical reaction to form the glass particles occurs; temperatures are in Kelvin. T_r is nearly a constant but T_e depends strongly on deposition length, burner traverse velocity, ambient temperature and wall thickness. A high deposition efficiency is reached if T_e is much lower than T_r. This can be achieved by cooling the substrate tube with water or air downstream the hot zone and by adding a considerable amount of helium to the gas flow in order to increase the thermal diffusivity of the gas mixture [21].

The collapsing of the tube is performed as described in Sec. 3.4. The high temperatures involved during this step lead to a decreased refractive index along axis of the preform and the fibre, the so called refractive index dip. Many attempts have been reported in the literature to avoid it. Most promising are methods which remove the dopant depleted internal surface by a gas phase etching process with fluorine just before or during the final collapsing stage [36, 37].

4.5 Plasma enhanced MCVD (PMCVD) method

The PMCVD method is an extension of the MCVD method which combines the advantages of the conventional MCVD method with a large increase in the deposition rate. The basic difference is that an isothermal rf-plasma ($f \simeq 3$ MHz) at atmospheric pressure is used in this method. During the initial studies since 1977, this plasma itself had to initiate the chemical reaction in order to sinter the deposited material. However, plasma instabilities limited the performance of the process. In the improved versions, the plasma only initiates the chemical reaction and deposits the glass particles while for sintering of the soot, an extra oxygen hydrogen burner is installed. As a result, plasma and the burner are traversing independently which allows independent optimisation of the processes.

A PMCVD apparatus differs only nominally from a conventional MCVD one, and usually the same reagents are used. Only a rf-generator and a rf-coil are needed as additional components. The diameters of the substrate tubes are

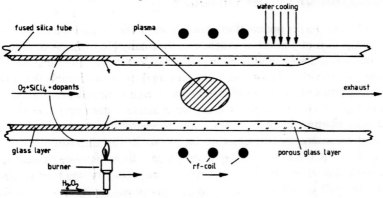

Fig. 13. Schematic of the PMCVD method

drastically increased with typical diameter as 46 mm for inside and 50 mm for the outside diameter [38, 11]. These diameters are necessary because the penetration depth of the rf-field into the plasma should be smaller than the tube radius. As the penetration depth is inversely proportional to the square root of the frequency, low frequencies require large tube diameters [39].

Figure 13 shows a schematic of the PMCVD process. With a rf-power of about 12 kW an oxygen rf plasma is built up at normal pressures. The plasma is centered in the substrate tube with a gap of more than 1 cm between the visible edge of the plasma and the tube. The plasma temperature is approximately 10000°C. This high temperature leads to a high temperature gradient to the tube wall, and the glass particles suffer high thermophoretic forces. The tube surface at the plasma zone is continuously cooled with water in order to prevent hot spots which can lead to tube distortion. This cooling also enhances the plasma centering. Furthermore, as the equilibrium temperature T is decreased, the deposition efficiency gets increased. Deposition efficiencies of 90% for pure SiO_2 and more than 50% for GeO_2 and P_2O_5 have already been obtained [38, 11]. The deposited glass particles are sintered to a 45 to 80 μm thick transparent glass layer by a conventional burner. Because the sintering process is separated from the chemical reaction and deposition, the sintering temperature can be decreased to avoid vapourisation of the deposited GeO_2. With this technique even pure GeO_2 can be deposited and sintered, and fibres with numerical apertures as large as 0.5 were already fabricated [40]. The collapsing process is similar to that of the conventional method, but due to the large tube diameters, this process step must be done very carefully.

To date, deposition rates of 6 g/min have been obtained but the fibre losses are typically 0.25 dB/km at 1.55 μm wavelength which is a bit higher than the losses of the fibres which are fabricated by the conventional MCVD-method [41].

4.6 Plasma activated CVD (PCVD) method

The use of a plasma for optical fibre fabrication was first established in 1975 by a research group of Philips at Aachen in Germany [10, 42]. Although the PCVD method is also a modification of the MCVD method, the mechanisms for chemical reactions, deposition and sintering differ considerably due to the special properties of the non-isothermal low pressure microwave plasma. As shown in Fig. 14, the substrate tube is placed inside a stationary furnace at a temperature of approximately 1200°C. This temperature is necessary in order to prevent microcracks and chlorine inclusions in the deposited layer. A rotation of the tube is not necessary in principle. But in practice, it is done to avoid profile distortions due to small temperature gradients over the tube cross section. Instead of a burner, an annular microwave cavity ($f \simeq 2.5$ GHz) traverses the furnace repeatedly. At a pressure of 10 to 50 HPa, a non-isothermal or cold plasma is generated in the microwave resonator inside the tube. This plasma initiates a heterogenous reaction on the inner wall of the tube, which forms a transparent dense glass layer directly on the inner wall. Any sintering is redundant. The deposition efficiency depends on the layer composition. It is about 80% for GeO_2 and reaches 100% for SiO_2 [43], which is much higher than in other fibre fabrication methods. Unlike the MCVD process, the reaction zone can be moved very fast back and forth the tube with depositions taking place in both directions. A great number of layers can be deposited which results in a very good approximation to the desired index profile. In practice, upto 1000 layers are deposited. With graded-index multimode fibres, bandwidth of 8 GHz-km have already been measured [37]. Another advantage of this method is the capability of a high fluorine incorporation in the glass layer because the reaction and deposition are not limited by the thermal equilibria inherent in the MCVD process [44, 45]. So depressed index clad fibres of index profiles with deep index depressions can be realised, and near optimum dispersion flattened single-mode fibres can be fabricated [46, 10].

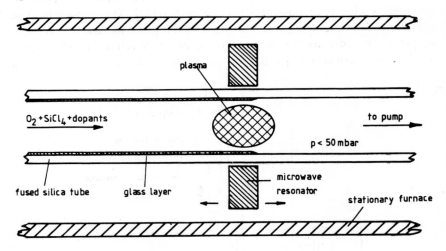

Fig. 14. Schematic of the PCVD method

4.7 Plasma outside deposition (POD) method

The POD method of Heraeus Quarzschmelze in Germany which was introduced in 1975 [47] uses a plasma torch [12] for chemical reaction and deposition. With this method, big transparent preforms of more than 1 kg weight can be fabricated without the need of a sintering or collapsing process [48]. The raw materials $SiCl_4$ vapour, CCl_2F_2 and oxygen are fed into a plasma torch where they react at temperatures of 1800 to 2300°C. The reaction product namely, particles of fluorine doped silica are then deposited as dense transparent glass on the surface of a rotating Suprasil W-rod which traverses the plasma torch. With every traversal of the burner in each pass, a new glass layer of desired refractive index is deposited. This procedure is repeated until the desired dimensions of the mother preform are obtained. So far, preforms with 60 mm diameter and 200 mm length have been fabricated. These rather short and thick mother preforms are subsequently lengthened to rods of 10 mm diameter and 7 m length. These rod preforms can be drawn to fibres in a separate step. Due to the large fluorine doping possible, this method is very attractive for the fabrication of dispersion flattened multiple clad single-mode fibres [49]. Unfortunately, to date, fibres made with this process contain a rather large amount of OH-ions which are responsible for the relatively high attenuation exhibited by theses fibres at 1.3 and 1.55 μm wavelengths.

4.8 Vapour phase axial deposition (VAD) method

In 1977, researchers at Nippon Telegraph and Telephones in Japan introduced a new fibre fabrication technique which at first was named "vapour phase vernuil method" [50], later the expression "vapour phase axial deposition" or VAD method has been introduced for this method [51]. This method has two important advantages compared with other vapour phase methods: no collapsing process is necessary, thus it avoids the undesirable refractive index dip on the fibre axis, and secondly it has the capability of fabricating preforms in a

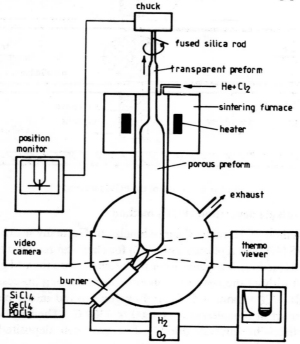

Fig. 15. Schematic of a VAD fibre fabrication apparatus

quasi-continuous process. The highest bandwidth of a graded index fibre to date was yielded by the VAD process, which exhibited a bandwidth of 9.7 GHz. km [52]. Preforms, from which as long as 580 km fibre could be drawn have been also reported [53].

In principle, the VAD process is quite similar to the OVD process with the difference that the deposition is made axially instead of laterally. A schematic of a VAD apparatus is shown in Fig. 15. Essentially, the apparatus consists of: (i) a water cooled stainless steel box as the deposition chamber with the burner and the exhaust as the lower part and the sintering furnace with the annular graphite heater as the upper part, (ii) the chemical delivery and exhaust system which is nearly the same as for other fabrication methods, and (iii) the control system with controllers for position and rotation of the preform and for the temperature profiling of the preform endface. Very often, the deposition chamber and the sintering furnace are separate units which enable better optimisation of both the processes independently.

The raw materials $SiCl_4$, $GeCl_4$, $POCl_3$, BBr_3 are vapourised in the usual way (see Fig. 10) and fed to the burner. The burner consists of at least three concentric orifices: the center orifice for the oxygen, $SiCl_4$ and dopant vapour, the intermediate for pure $SiCl_4$ vapour plus oxygen, and the outer one for the fuel gases namely, hydrogen and oxygen. The halide vapours oxidize and form tiny glass particles which are blown to the end surface of the rotating fused silica rod. Beginning at this surface, the porous preform grows continually in axial direction. During the deposition, the distance between the burner and the endface must be kept constant in order to achieve a uniform deposition of the glass particles and a definite temperature profile across the preform endface. The position of the endface can be kept to an accuracy of about 50 μm by pulling the preform upwards at exactly the same rate as the preform growth rate. For monitoring the position, either a video camera [51, 54] or a laser scanning unit is used [13]. The temperature distribution is measured with a scanning pyrometer [55]. It is controlled by varying the flow rates of the burner gases.

The incorporation of dopants and the refractive index profile that results are essentially determined by the temperature profile of the preform endface and the dopant distribution in the flame. In the beginning, it was assumed that only the appropriate mixing of dopants and $SiCl_4$ in the flame forms the desired refractive index profile. However, recent experiments have definately shown that the temperature during the deposition process is the most important parameter for fixing the GeO_2 concentration in the deposited glass [56, 57, 11]. This is the reason why such simple coaxial burners as described above can be used. The profile parameter q of a graded-index profile can be varied from 1 to 10 by adjusting the temperature distribution on the preform endface. The temperature distribution itself is mainly given by the angle between the axes of the burner and the preform. To obtain a parabolic index profile, this angle should be about 45° [6, 11]. A fine adjustment is then done by varying the flow rate ratio of the burner gases namely, hydrogen and oxygen [6, 58, 11] and the mixing of the dopants. At the same time when the porous preform is growing at its lower end, the upper end is consolidated in the sintering furnace and pulled out of the box as a transparent glass rod (Fig. 15). However, due to the combustion of hydrogen and oxygen, the porous preform contains a high OH-contamination of about 200 ppm. Fibres made of such preforms would still contain 5 to 30 ppm of OH-ions without special drying process and would exhibit an attenuation of more than 300 dB/km at a wavelength near 1.4 μm. In order to reduce this contamination, the porous preform is exposed to an atmosphere of about 95% helium and 5% chlorine at raised temperatures prior to sintering [59, 60]. The reduction of the OH-content depends strongly on the drying time. With simultaneous deposition and sintering, as shown in Fig. 15, this time is rather short, and to date, a minimum content of 18 ppb of OH-ions was achieved, which corresponds to an absorption loss of 1 dB/km at 1.4 μm wavelength [54]. Much better results are obtained when deposition and sintering are separate process steps, because, in that event, much longer drying times are possible. In this way, nearly water free fibres with less than 1 ppb OH-content, corresponding to an absorption loss of less than 0.05 dB/km, have already been fabricated [59, 60].

If fluorine is used as dopant, the VAD process has the same disadvantages as the conventional OVD process: the maximum relative index depression Δ^- is limited to about 0.27%. Thus, for fabricating dispersion flattened multiple clad single-mode fibres, the VAD method is not so well suited as the plasma methods or even the MCVD method.

5. COMPARISON OF PERFORMANCE

In order to economise on the size of the table, in which the performance data are shown, only those methods are compared which are capable of fabricating fibres with very low loss in the wavelength range 1.3 to 1.6 μm: OVD, MCVD, PMCVD, PCVD and VAD methods. In Table 1, the published results are shown starting with the data concerning process economy: deposition rate and efficiency. The attenuation at 1.3 μm and in the region between 1.5 and 1.6 μm and the bandwidth of multimode graded-index fibres with a numerical aperture of about 0.2 follow. Finally, the attenuation values of single-mode fibres at 1.3 μm and 1.5 to 1.6 μm wavelength are presented. The data in this table represent the best reported values, mostly achieved in laboratories. Nevertheless, they are considered as best performance criteria, as the values achieved in the field of large scale production are more and more approaching these values.

TABLE 1
Performance data of fibre fabrication techniques

Process	OVD	MCVD	PMCVD	PCVD	VAD
Deposition rate: Core	4.2	>1.3	6	2	4.5
(g/min) Cladding	6	2.3	6	—	6
Efficiency (%)	50	50	50-90	80-100	70
Multimode graded-index: (NA 0.2)					
Attenuation (dB/km)					
1.3 μm	0.52	0.40	0.7	0.5	0.41
1.5-1.6 μm	0.30	0.29	0.5	<0.5	0.25
3-dB Bandwidth (GHz. km)	6.0	4.7	—	8	9.7
Single-Mode:					
Attenuation (dB/km)					
1.3 μm	0.27	0.30	0.55	0.38	0.31
1.5 - 1.6 μm	0.16	0.16	0.25	0.20	0.17

A comparison of the data for the different manufacturing techniques shows that no method can be regarded as the best. The future will show which method is capable of combining best quality and efficiency in large scale production environments. Due to the fact that the companies, in general, favour their own methods, all the different fabrication processes are expected to be used in parallel.

ACKNOWLEDGEMENT

The author wishes to thank Dr. E. Klement and Dr. H. Schneider from Siemens Research Laboratories for valuable discussions and J. Urhahne for preparing the figures.

REFERENCES

1. R.D. Maurer, "Glass Fibres for Optical Communications", Proc. IEEE, Vol. 61, pp. 452-462 (April 1973).
2. T.G. Giallorenzi, Optical Communications Research and Technology: Fibre Optics, Proc, IEEE, Vol. 66, pp. 744-780 (July 1978).
3. B.P. Pal, "Optical Communication Fibre Waveguide Fabrication: A Review", Fibre and Integrated Optics, Vol. 2, pp. 195-252 (1979).
4. I.W. Versluis and Peelen J.G.J., "Optical Communication Fibres. Manufacture and Properties", *Philips Telecomm. Rev.*, Vol. 37, pp. 215-230 (1979).
5. P.C. Schultz, "Progress in Optical Waveguide Process and Materials", Appl. Opt., Vol. 18, pp. 3684-3693 (1979).
6. G.J. Koel, "Technical and Economic Aspects of the Different Fibre Fabrication Processes", Proc. 8th ECOC, pp. 1-8, Cannes (1982).
7. H. Karstensen, "Fabrication Techniques of Optical Waveguides (in German)", *Laser u. Optoelektronik*, Vol. 4, pp. 13-32 (1982).

8. H. Schneider and Zeidler G., "Manufacturing Process and Designs of Optical Waveguides", Telcom Report, Vol. 6, Special Issue, "Optical Communications", pp. 27-33 (1983).
9. Th. Huenlich, Bauch H., Kersten R. Th., Paquet V. and Weidmann G.F., "Fibre-Preform Fabrication Using Plasma Technology: A Review", *J. Opt. Comm.*, Vol. 8, pp. 122-129 (1987).
10. P. Bachmann, "Review of Plasma Deposition Applications: Preparation of Optical Waveguides", *Pure & Appl. Chem.*, Vol. 57, pp. 1299-1310 (1985).
11. T. Li (Ed.)," Optical fibre Communications", Volume 1: Fibre Fabrication, Academic Press, Orlando (1985).
12. S.E. Miller and Chynoweth A.G. (Eds.), "Optical Fibre Telecommunications", Academic Press, New York (1979).
13. L.S. Watkins, "Control of Fibre Manufacturing Process", Proc. IEEE, Vol. 70, pp. 626-634 (1982).
14. N.S. Kapany, "Fibre Optics, Principles and Applications", Academic Press, New York, (1967).
15. H.G. Unger, "Planar Optical Waveguides and Fibres", Clarendon press, Oxford, (1977).
16. K.J. Beales, Day C.R., Dunn A.G. and Partington S., "Multicomponent Glass Fibres for Optical Communications", Proc. IEEE, Vol. 68, pp. 1191-1194 (1980).
17. K. Koizumi, Ikeda Y., Kitano I., Furukawa M. and Sumimoto T., "New Light-Focusing Fibres, Made by a Continuous Process", *Appl. Opt.*, Vol. 13, pp. 255-260 (Feb. 1974).
18. F.P. Kapron, Keck D.B. and Maurer R.D., "Radiation Losses in Glass Optical Waveguides", *Appl. Phys. Lett.*, Vol. 17, pp. 423-425, (1970).
19. T. Miya, Terunume Y., Hosaka T. and Miyashita T., "Ultimate Low Loss Single-mode Fibre at 1.55 μm", *Electr., Lett.*, Vol. 15, pp. 106-108 (1979).
20. S. Tomaru, Yasu M., Kawachi M. and Edahiro T., "VAD Single-mode Fibre with 0.2 dB/km Loss", *Electr, Lett.*, Vol. 17, pp. 92-93 (1981).
21. S.R. Nagel, MacChesney J.B. and Walker K.L., "An Overview of the Modified Chemical Vapour Deposition (MCVD) Process and Performance", IEEE *Journ. Quant. Electr.*, Vol. QE-18, pp. 459-476 (1982).
22. P.C. Schultz, "Fabrication of Optical Waveguides by the Outside Vapour Deposition Process", Proc. IEEE, Vol. 68, pp. 1187-1190 (1980).
23. M.G. Blankenship and Deneka, C.W., "The Outside Vapour Deposition Method of Fabricating Optical Waveguide Fibres," *IEEE Journ. Quant. Electr.*, Vol. QE-18, pp. 1418-1423 (Oct. 1982).
24. G.E. Berkey, "Fluorine-doped Fibres by the Outside Vapour Deposition Process", *Techn. Dig.* OFC'84, pp. 20-21, New Orleans (1984).
25. W. French, Pearson A.D., Tasker G.W. and MacChesney J.B., "Low Loss Fused Silica Optical Waveguide with Borosilicate Cladding", *Appl. Phys. Lett.*, Vol. 23, pp. 338-339 (1973).
26. J.B. MacChesney, Jaeger R.E., Pinnow D.A., Ostermayer F.W., Rich T.C. and Van Uitert L.G., "Low-loss Silica Core-Borosilicate Clad Fibre Optical Waveguide", *Appl. Phys. Lett.*, Vol. 23, pp. 340-341 (1973).
27. W.G. French, MacChesney J.B., O'Connor P.B. and Presby H.M., "Optical Waveguides with Very Low Losses", Bell Syst. Tech. J., Vol. 53, pp. 951-954 (1974).
28. J.B. MacChesney, O'Connor P.B. and Presby H.M., "A New Technique for the Preparation of Low-Loss and Graded-Index Optical Fibres", Proc. IEEE, Vol. 62, pp. 1280-1281 (Sept. 1974).

29. G.W. Tasker and French W.G., "Low-Loss Optical Waveguides with Pure Fused SiO$_2$ Cores", Proc. IEEE, Vol. 62, pp. 1281-1282 (Sept. 1974).
30. J.B. MacChesney, O'Connor P.B., Di Marcello F.V., Simpson J.R. and Lazay P.D., "Preparation of Low-Loss Optical Fibres using Simultaneous Vapour Phase Deposition and Fusion", Proc. 10th, Intern. Congr. on Glass, pp. 6-40-45, Kyoto, (1974).
31. W.A. Gambling, Payne D.N., Hammond C.R. and Norman S.R., "Optical Fibres Based on Phosphosilicate Glass", Proc. IEE, Vol. 123, pp. 570-576 (1976).
32. K.B. McAffee Jr. et al., "Equilibria Concentrations in the Oxidation of SiCl$_4$ and GeCl$_4$ for Optical Fibres", IEEE Journ. Lightwave Techn., Vol. LT-1, pp. 555-561 (Dec. 1983).
33. K.L. Walker, Csencsits R. and Wood D.L., "Chemistry of Fluorine Incorporation in the Fabrication of Optical Fibres", Techn. Dig. OFC'83, pp. 36-37, New Orleans (1983).
34. A. Kawana et al., "Fluorine Sources for Single-Mode Fibres, Trans. IECE Japan, Vol. E65, pp. 529-533 (Sept. 1982).
35. K.L. Walker, Geyling F.T. and Nagel S.R., "Thermophoretic Deposition of Small Particles in the Modified Chemical Vapour Deposition (MCVD) Process, J. American Ceram. Soc., Vol. 63, pp. 552-558 (1980).
36. H. Schneider, Deserno U., Lebetzki E. and Meyer A., "A New Method to Reduce the Central Dip and the OH-Content in MCVD-Preforms", Proc. 8th ECOC, pp. 36-40, Cannes (1982).
37. J. Peelen, Pluyms R., and Koel G., "Dipless Multimode and Monomode Fibres Manufactured by the PCVD-Process", Techn. Dig. OFC'84, pp. 68-69, New Orleans (1984).
38. J.W. Fleming and Raju V.R., "Low-loss Single-mode Fibres Prepared by Plasma-enhanced MCVD", Electr. Lett. Vol. 17, pp. 867-868 (1981).
39. R.E. Jaeger, MacChesney J.B. and Miller T.C., "The Preparation of Optical Waveguide Preforms by Plasma Deposition", Bell Syst. Tech. J., Vol. 57, pp. 205-210 (1978).
40. P.B. O'Connor, Fleming J.W., Atkins R.M. and Raju V.R., "Plasma-enhanced Modified Chemical Vapour Deposition: a Versatile High-rate Process", Tech. Dig. OFC'85, pp. 100-101, San Diego (1985).
41. S.R. Nagel and Fleming J.W., "Latest Developments in Fibre Manufacture by MCVD and PMCVD", Techn. Dig., OFC'84, pp. 52-53, New Orleans (1984).
42. P. Geittner, Kuppers D. and Lydtin H., "Low Loss Optical Fibres Prepared by Plasma-activated Chemical Vapour Deposition (PCVD)", Appl. Phys. Lett., Vol. 28, pp. 645-646, (Nov. 1976).
43. P. Bachmann, Geittner P. and Wilson H., "The Deposition Efficiency for the GeO$_2$-doping in Optical Fibre Preparation by Means of Low pressure PCVD", Proc. 8th ECOC, pp. 614-617, Cannes (1982).
44. P. Bachmann, Huebner H., Lennartz M., Steinbeck E. and Ungelenk J., "Fluorine Doped Single-mode and Step Index Fibres Prepared by the Low Pressure PCVD Process", Proc. 8th ECOC, pp. 66-69, Cannes (1982).
45. R. Setaka, Takahashi H., Sato T. and Yoshida S., "Fluorine-doping Levels in the Low-pressure PCVD process", Trans. IECE Japan, Vol. E-67, pp. 333-334, (June 1984).
46. P.K. Bachmann, Leers D., Wehr H., Weirich F., Wiechert D., Van Steenwijk J.A., Tjaden D.L.A. and Wehrhahn E., "PCVD DFSM-Fibres: Performance, Limitations, Design Optimization", Proc. 11th ECOC, pp. 197-200, Venice (1985).

47. A. Muehlich, Rau K., Simmat F. and Treber N., "A New Doped Fused Silica as Bulk Material for Low-loss Optical Fibres", Proc. 1st ECOC, London (1975).
48. A. Muehlich, Rau K., and Treber N., "Preparation of Fluorine Doped Silica Preforms by Plasma Chemical Technique", Proc. 3rd ECOC, pp. 10-11, Munich(1977).
49. W. Lieber, Loch M., Etzkorn H., Heinlein W.E., Klein K.F., Bonewitz H.U. and Muhlich A., "Three-Step-Index Strictly Single-mode Only F-doped Silica Fibre for Broadband Low Dispersion", Proc. 11th ECOC, pp. 201-204, Venice (1985).
50. T. Izawa, Kobayashi S., Sudo S. and Hanawa F., "Continuous Fabrication of High Silica Fibre Preform", *Techn. Dig.* IOOC', 77, p. 375, Tokyo (1977).
51. T. Izawa, and Inagaki N., "Materials and Processes for Fibre Preform Fabrication-Vapour Phase Axial Deposition", Proc. IEEE, Vol. 68, pp. 1184-1187, (1980).
52. M. Horiguchi, Nakahara M., Inagaki N., Kokura K., Yoshida K., "Transmission Characteristics of Ultra-wide Bandwidth VAD Fibres", Proc. 8th ECOC, pp. 75-80, Cannes (1982).
53. H. Suda, Shibata S., Nakahara M. and Miya T., "Double-flame VAD Process for High-rate Deposition", Proc. 10th ECOC, pp. 296-297, Stuttgart (1984).
54. S. Suda, Kawachi M., Suda H., Nakahara M. and Edahiro T., "Refractive-Index Profile Control Techniques in the Vapour-Phase Axial Deposition Method", Trans. IECE Japan, Vol. E64, pp. 536-543 (1981).
55. S. Sudo, Kawachi M., Edahiro T. and Chida K., " Transmission Characteristics of Long-Length VAD Fibres", Trans. IECE Japan, Vol. E64, pp. 175-180 (1981).
56. M. Kawachi, Sudo S., Shibata N. and Edahiro T., "Deposition Properties of SiO_2—GeO_2 Particles in the Flame Hydrolysis Reaction for Optical Fibre Fabrication", *Jpn. J. Appl. Phys.*, Vol. 19, pp. L69-L71 (1980).
57. T. Edahiro, Kawachi M., Sudo S. and Tomaru S., "Deposition Properties of High Silica Particles in the Flame Hydrolysis Reaction for Optical Fibre Fabrication", *Jpn. J. Appl. Phys.*, Vol. 19, pp. 2047-2054 (1980).
58. N. Yoshioka, Tanaka G., Kuwahara T., Sato H., Kyoto M., Hoshikawa M., Okamoto K. and Edahiro T., "Graded Index Profile Formation and Transmission Characteristics of VAD Fibre", Proc, 6th ECOC, pp. 10-33, York (1980).
59. T. Moriyama, Fukuda O., Samada K., Inada K., Edahiro T. and Chida K., "Ultimate Low OH-content VAD Optical Fibres", *Electr. Lett.*, Vol. 16, pp. 698-699 (1980).
60. F. Hanawa, Sudo S., Kawachi M., Nakahara M., "Fabrication of Completely OH-free VAD Fibre", *Electr. Lett.*, Vol. 16, pp. 699-700 (1980).

10

Characterisation of Optical Fibres for Telecommunication and Sensors — Part I : Multimode Fibres[†]

B.P. PAL[*], K. THYAGARAJAN[*] and A. KUMAR[*]

---

1. INTRODUCTION

In the two decades since the first low-loss optical fibre of attenuation ~20 db/km was reported [1], the field of fibre optics and optical communication has seen vigorous research and developmental efforts worldwide. The rapid progress made in recent years in the technology of optical fibres [2] has depended in part upon the availability of high-precision measurement techniques to characterise fibres of near ultimate quality and performance. During the last decade and a half, numerous fibre characterisation techniques have been proposed in the literature. This has generated a growing interest in establishing standard measurement techniques to meet the needs of industry and system engineers alike.

From the point of view of system engineering, the three major components of any fibre optical communication system are (as shown in Fig. 1) the electrooptic transmitter (LED/laser diode), the optical transmission medium (fibre) and the optoelectronic receiver (PIN/APD-photodetector). Of these,

---

[†] This topic distributed over two chapters as parts I and II is based on two papers under the same title and authorship which was originally published in INTERNATIONAL JOURNAL OF OPTOELECTRONICS VOL. 3, NO. 1 pp. 45-71 (1988) and VOL. 3, NO. 2, pp. 153-176 (1988). The article in its present form differs little from the original papers except for some minor changes. The permission granted by publishers TAYLOR and FRANCIS (UK) to reproduce these two papers here is gratefully acknowledged.

[*] Department of Physics, Indian Institute of Technology Delhi, New Delhi 110016, India.

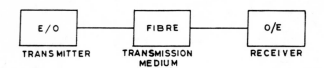

Fig. 1. Basic components of an optical fibre telecommunication system.

prior knowledge of fibre characteristics is most important for the system engineer because, once installed, the fibre cannot readily be replaced. Having realistic fibre parameters is of great interest to manufacturers too, because they need to assess the effects of material composition and fabrication conditions on fibre performance. As is true with any transmission medium, the two most important characteristics of a fibre are attenuation and transmission bandwidth, though other fibre parameters should be considered as well; numerical aperture, core/cladding diameter, refractive index profile being prime examples. While for single-mode fibres, in particular, quantities such as the mode field diameter or spot size, the ESI parameters of conventional single-mode fibres and the cut-off wavelength of the first higher-order (LP_{11}) mode are of principal importance. Furthermore, for birefringent fibres, which are expected to be employed in coherent communication [3] and in fibre optic sensors [4], beat length or modal birefringence are important. Table I gives a summary of the inter-relationship between measured fibre parameters and the related system characteristics that eventually determine overall performance.

In this chapter, we review the most widely used characterisation techniques for optical fibres which are now evolving as standards for applications in telecommunication and sensor systems. The subject of fibre characterisation is now so vast that it is difficult to do justice to this area of research (see[5] and [6], for example) within the space of a few pages. As a compromise, while providing an outline of most standard measurement techniques, wherever possible, details will be presented, in particular, on those techniques which have been studied in some depth in our Fibre Optics Laboratory at IIT-Delhi. Though no attempt is made to review the field exhaustively, our aim is primarily to present a unified account of current standard fibre characterisation procedures.

In the following sections, we first describe some key experimental factors that are common to all fibre measurements and this is followed by a discussion of the characterisation of multimode fibres. In chapter 11 (Part 2) we will present details of the characterisation of single-mode fibres.

2. KEY EXPERIMENTAL FACTORS COMMON TO FIBRE CHARACTERISATION TECHNIQUES

2.1 Fibre-end preparation

The first step in evaluating the parameters of an optical fibre is to obtain optically flat endfaces perpendicular to the fibre axis. Several commercially

TABLE 1
Inter-relationship between measured fibre parameters and system characteristics or quality in fibre technology

Fibre type	Measured parameter	Prediction of system characteristics or quality of fabrication technology
Multimode	Numerical aperture	Source-fibre coupling efficiency; splice loss
	Geometry (core and cladding diameters, core eccentricity)	Splice loss; connector designs
	Refractive index profile	Bandwidth prediction; source-fibre coupling efficiency; number of modes; splice loss
	Attenuation	Repeater distance
	Dispersion/bandwidth	Bit rate; repeater distance
	Differential mode loss	Presence of index dip and localized faults in core profile; interface irregularities; lossy jackets
	Differential Mode delay	Overall and localized departure from optimum profile shape
Single-mode	Cut-off wavelength	Single-mode operating regime
	Attenuation	Repeater distance
	Total dispersion	Bit rate; repeater distance
	Mode field diameter or spot size	Source-fibre coupling efficiency; splice loss; microbend loss; waveguide dispersion
	ESI parameters	Bend loss; splice loss; total dispersion, cut-off wavelength
	Beat length	Polarisation dispersion; birefringence, polarisation-maintaining characteristics

available fibre-cutting tools provide smooth mirror-like endfaces, free from 'heckles' and 'mists', by means of controlled fracture under tension; all are essentially based on the theory developed by Gloge *et al.* [7]. The principle employed in any such fibre-cutting tool is to initiate a crack at the fibre surface by scribing the bare fibre (after removal of the plastic jacket from the portion to be cut) under tension. The scriber is typically a diamond pen or tungsten-carbide-edged blade, and tension is induced in the fibre by placing it over a curved surface, whose radius of curvature is ~57 mm for a 125 μm silica fibre (See Fig. 2). Once the fibre is scored, a gentle pull will break the fibre. In the absence of a commercial tool, and with a little experience — one can normally obtain a good-quality fibre endface simply by placing the bare fibre on a finger tip (the curved surface), initiating the fracture by scoring the fibre with the scriber, and pulling gently either side. Alternatively, one can initiate microcracks at the fibre surface by exposing a bare portion of the fibre to a well-localized flame (such as that from a cigarette lighter) and giving a gentle

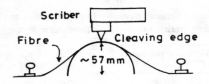

Fig. 2. Schematic diagram of the principle behind the 'bend and score' technique of fibre end-face preparation.

pull [8]. One UK based company named York Technology employs a ultrasonically driven blade to cleave a fibre with smooth end-face.

A quick and effective qualitative assessment of the quality of cleaved fibre face can be made by launching light from a He-Ne laser into the fibre and observing the output light in the far field on a screen. Poor cleavage is revealed by scattered light from the sides of the far field spot in the form of a 'comet tail' from an otherwise circular pattern on the screen [8]. In addition, if the fibre is poorly cleaved, the fibre end is usually seen to glow when viewed transversely, as a result of scattering. Alternatively, endface quality can be judged by illuminating the far end of the fibre with a white-light source and observing the transmitted near field of the fibre under a microscope. White-light illumination has the advantage that it allows an easy and convenient visual check for relatively large fibre endfaces. The presence of any irregularity of the fibre face is automatically revealed. Sometimes dust and small glass fragments, in particular, adhere to the fibre endface as a result of residual electrostatic charges; they appear as dark spots on the fibre endface image. Gently touching the fibre endface with Scotch Tape is an easy and convenient means of clearing dust and other contaminants [9]. Although somewhat crude, in practice, this works very well most of the time without destroying the fibre end.

2.2 Launching light into fibres

Almost all fibre characterisation measurements require efficient optics between the light source and the fibre to launch the beam. For certain measurements, a laser (such as a He−Ne gas laser, tunable fibre Raman laser or a semiconductor laser diode) is used as a source; while measurements of attenuation, cut-off wavelength, ESI parameters and so on, often require a white-light source like a tungsten halogen or xenon arc lamp. A typical arrangement for launching light into a fibre is shown in Fig. 3(a). Light from the source is collimated by means of a microscope objective[†] and then refocused by another microscope objective onto the input face of the fibre. Into this scheme one has the option of including such elements as a variable-aperture diaphragm or shutter to control the launch NA and the size of the focused spot, a plate or

---

† In the case of illumination by a He—Ne laser, the laser beam is emitted almost as a collimated beam and hence this objective is redundant.

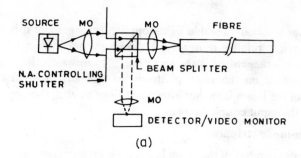

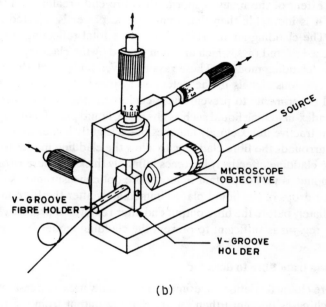

Fig. 3. (a) Schematic diagram of an optical beam launch arrangement from a laser diode/LED/apertured tungsten halogen lamp with a variable NA and spot size at the fibre input end; the beam splitter enables monitoring of the input spot size and position on fibre core; MO, microscope objective. (b) A typical stable source to fibre coupling arrangement.

cube beam-splitter to direct the collimated input beam along the two mutually orthogonal paths (one of which provides a reference enabling one to monitor the stability of the light source), or a set of narrow bandpass interference filters (often mounted on a filter wheel), which can be inserted in between the two microscope objectives to derive isolated wavelengths from the white-light source. If a monochromator is used in place of the interference filters at the fibre input or output to obtain single wavelengths, the signal levels usually are very low and a lock-in amplifier is normally used to improve the signal-to-noise ratio through a.c. detection. Since use of a lock-in amplifier requires use of a mechanical chopper, the chopper can also be inserted between the two objectives. In order to obtain a stable light-coupling arrangement it is essential to avoid any possible relative movement between the second objective and the fibre. Such an arrangement is shown in Fig. 3(b) in which the objective and fibre are mounted on the same platform, so that any external mechanical disturbance of the beam launcher affects the fibre as well, thereby minimizing variations in the input coupled power.

2.3 Cladding-mode stripping

Usually the effective NA of the beam launcher is greater than the fibre NA and the spatial extent of the focused spot at the fibre end greater than the core diameter. It is inevitable then that some optical power is injected into the cladding. The cladding-air interface can form a total reflecting surface for some of these injected rays, which are then guided by the cladding-air boundary, to form 'cladding modes'. These rays may eventually reach the detector, leading to spurious signals in the detection system [10]. It is necessary in virtually all experiments to prevent these cladding modes from reaching the detector. Index-matching liquid such as glycerine, liquid paraffin or eucalyptus oil, whose refractive index closely matches but is slightly higher than the fibre cladding, surrounds the fibre cladding, to allow the cladding modes to radiate away. (The cladding effectively becomes infinitely thick for these rays.) Such 'mode stripping' is very important in most fibre optic measurements. Usually two or three drops of the liquid placed immediately after the fibre input end and immediately before the fibre output end after removal of protective plastic from these regions is sufficient to prevent the cladding modes from reaching the detector.

2.4 Coupling from fibre to detector

In most fibre characterisation experiments one usually uses a large-area detector. In such cases one must then ensure that the output from the fibre fills roughly 60-70% of the detector's active area, so that the detector is not saturated locally and so that any variations in detector surface response are averaged out.

When the coupling is via a lens/microscope objective and for measurements that involve very small changes of power, it is essential that the beam is also incident normally on the detector area. However, there still exist problems regarding calibration of detector-power meter combinations and the National Institute of Standards and Technology in U.S.A. provides guidelines in defining proper calibration procedures to tackle this issue.

3. MEASUREMENT TECHNIQUES

3.1 Refractive index profile (RIP)

Even in the early days of telecommunication fibre development it was known that pulse dispersion, which ultimately determines the bit rate, is critically dependent upon the core refractive index profile. For example, pulse dispersion in the so-called optimum refractive-index-profiled multimode fibre is almost three orders of magnitude less than in an equivalent step-index fibre having the same relative core-cladding index difference Δ and core radius a. Once this fact had been appreciated, manufacturers made a widespread effort to fabricate high-bandwidth (BW) graded-core fibres. This called for index-profile measurement techniques of high spatial resolution many of which were proposed, and have been reported in the literature (see, for example, [5] and [6]). However, of these many different techniques, those which are more or less standard are the following:

(i) transmitted near field (TNF):

(ii) refracted near field (RNF): and

(iii) transverse interferometry (TI)

and we shall consider each in turn.

(i) Transmitted near-field (TNF) method

The TNF method [11] essentially involves a scanning measurement of intensity at the output end of a short length (typically one or two metres) of multimode fibre in which all possible guided modes are excited equally by means of an incoherent source, such as a tungsten halogen lamp (or an LED). As shown in Fig. 4(a), in practice a magnified image of the fibre output end is projected onto the plane of an apertured (pin-hole) photo-detector driven by a stepper motor so as to scan the image along its diameter. The detector output, connected to an *XY*-recorder, directly yields the near-field intensity distribution as the 'near-field profile' or NFP. It is known from theory [12] that the NFP bears a close resemblance to the refractive-index profile of a multimode fibre through

$$\frac{P(r)}{P(0)} = \frac{n^2(r) - n_2^2}{n_1^2 - n_2^2} = \frac{n^2(r) - n_2^2}{(\text{axial NA})^2} \quad , \tag{1}$$

where n_2 is the cladding index and n_1 the axial refractive index. Here $P(r)$ is the power at a distance r from the axis ($r = 0$) of the fibre. The result of a typical measurement on a standard 50 μm core fibre is shown in Fig. 4 (b). As is apparent, this provides only a relative measure and no information on the absolute values of the core refractive index. An independent measurement of the fibre axial NA [see Eq. (1)] in the far field can be used to calibrate the NFP measurement.

The most attractive feature of the method is that it provides a fast and relatively simple means to get a quick estimate of the core index profile and, for this reason alone, is employed by practically all fibre manufacturers. Nevertheless there is still some controversy over the accuracy. It is argued that since a

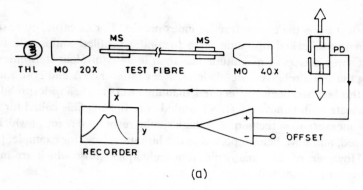

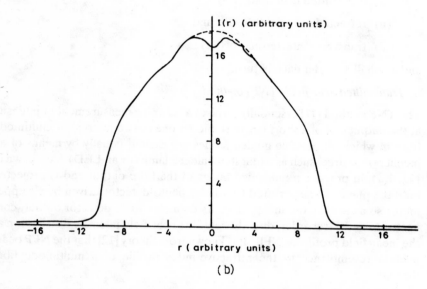

Fig. 4. (a) Schematic diagram of the set-up for the transmitted near-field (TNF) method for index profiling: THL tungsten halogen lamp; MS, mode strippers; PD, photodetector. (b) Measured NFP of a graded-core fibre [13].

relatively short fibre sample is typically excited in overfilled (both spatial and angular) launched conditions, the measured near field may include a large contribution from leaky rays. One may then be required to take account of this by incorporating a correction factor [14] on the right-hand side of (1) in order to convert the measured near-field intensity distribution to the fibre refractive index profile. The presence of leaky rays leads to fibre length dependence in the measured near-field intensity distribution. Adams et al. [14] have provided curves for the leaky ray correction factor as a function of fibre radius over a range of fibre index profiles. One apparent drawback of their method is that one has to know before hand the index exponent q that governs the so-called power-law class of graded-index profiles [12] — which is the same thing as knowing the refractive index profile itself! This problem is not as serious as at first may seem because the leaky ray correction factors for a variety of profiles and fibre lengths are found to be almost identical up to a normalized radius (r/a, where a is the fibre radius) of 0.8, and differed by roughly ±6% up to r/a = 1. Adams *et al.* [14] therefore, suggest that correction factors corresponding to a median value of $q = 3$ may be applied to the range of profiles of major interest, namely those having q~2. In a typical NFP measurement on a graded fibre, this procedure corrects a nominal q-value of 2.05 to 1.96 [9]. Whether or not the leaky ray correction factor is absolutely necessary is still open to debate. And the correction factors were based on assumptions, namely (i) the absence of mode coupling over the measured fibre length, (ii) perfectly circular core (in fact a deviation as small as one part in 10^7 in core circularity of a parabolic fibre may lead to the almost total absence of leaky rays in the near-field profile [15]), (iii) the excitation of all possible leaky rays and (iv) truly multimode fibres. If any of these conditions is not met, applicability of these correction factors become uncertain [9]. In fact, for a quick general estimate of the index profile, most workers in practice ignore leaky rays in the NFP by taking fibre samples of a few metres length (within which, supposedly, most of the leaky rays, if excited, would have been lost through radiation).

Since, the derivation of (1) is based on the assumption that all modes carry equal amounts of power, care should be taken to avoid (i) any unnecessary stress at the fibre mounts and (ii) sharp bends along the length of the fibre sample [9]. Care should be also taken to centre the detector scan path on the fibre axis. Many laboratories use a vidicon with a monitor to scan the near field. However, the near-field image of CVD fibres is usually characterised by an axial index dip [16], which helps in defining the fibre centre. This dark spot also helps the experimenter to check for optimal imaging of the fibre endface at the detector plane by means of small adjustments in the (Z-direction) of the position of the fibre end in front of the imaging objectives so as to attain maximum contrast by monitoring the NFP on the XY-plotter [9].

The inherent resolution afforded by the method is [9]

$$\Delta r \simeq \frac{\lambda}{\pi \text{ (fibre NA)}}. \qquad (2)$$

For typical multimode fibres NA ~ 0.2 at λ~1.3 μm, giving a resolution of about 2 μm. On the other hand, detector spatial resolution is set by the size of the pinhole in front of it. If the fibre image size is l (mm) at the detector plane and the pinhole diameter is d (mm), then the number of independent measurement points on the image plane is roughly l/d; thus the resolution limit due to the pinhole is $Y = 2a/(l/d)(\mu m)$, a being the core radius in μm. For high measurement precision, Y should be less than or of the order of the theoretical resolution given by (2); this can be attained by having a relatively large image produced by a combination of microscope objective and an eyepiece [9].

However, since in a graded-core fibre the local fibre NA decreases from the core centre towards the core-clad interface, there is loss of resolution in the edges of the core profile [17], consequently the core edges of the NFP are rounded (see Fig. 4 (b)). Another point which may be noted while implementing this method is that the emission spectrum of the optical source must be sufficiently broad (hence lasers cannot be used) that the finite number of guided modes (typically 150 at $\lambda \sim 1.3 \mu m$) in practical fibres do not interfere to produce speckle [18]. This is seen as ripples in the NFP whose period is inversely proportional to the normalised waveguide parameter-defined by

$$V = \frac{2\pi a}{\lambda_0} \text{NA}, \tag{3}$$

where a is the core radius, λ_0 the free-space wavelength and NA the numerical aperture of the fibre. Maximum possible overall error in the measurement of $n(r)$ has been estimated to be only a few parts in 10^5[6]. It may be worth while to point out that for fibres having a power-law profile given by

$$n^2(r) = \begin{cases} n_1^2[1 - 2\Delta(r/a)^q], & r < a \\ n_1^2[1 - 2\Delta] & r \geq a, \end{cases} \tag{4}$$

Eq. (1) reduces to

$$1 - \frac{P(r)}{P(0)} = \left(\frac{r}{a}\right)^q. \tag{5}$$

Thus a plot of log $[1-P(r)/P(0)]$ against log (r/a) should be a straight line, the slope of which would yield the profile exponent q. However, any deviation in the NFP (the axial dip, for example) from the ideal power-law distribution complicates this recipe to obtain a unique q for the entire core cross-section [9]. An estimate of the profile shape parameter can always be obtained by a least-squares curve fitting (with q and a as adjustable parameters) of the power-law equation (4) to the measured NFP between the 10 and 80 per cent intensity points.

(ii) Refracted near-field (RNF) method

This method is complementary to the above, in the sense that it relies on the measurement of rays lost sideways through refraction at the core-cladding interface rather than using the guided rays of the TNF method. Two major advantages are gained in using the RNF method: (i) the complications introduced by the possible presence of leaky rays in the TNF are used to advantage in the RNF; and (ii) the RNF can also yield the index profile of single-mode fibres (the TNF yields profiles of multimode fibres only). It is implemented by scanning the core of the test fibre, immersed in an index-matching liquid (index n_2), with a small spot He−Ne laser beam focused through a microscope objective. The NA of the objective is chosen to far exceed the fibre axial NA, so that at any point r across the fibre core, the focused spot will inject power into guided as well as refracted rays [as shown in Fig 5 (a)]. Since the fibre is immersed in an index-matching liquid, the refracted rays effectively propagate in an unbounded medium of index n_2 and thus escape completely from the fibre.

A ray incident on any point r at an angle θ' (greater than the acceptance angle) will exit at an angle θ'' from the cladding. The angles θ' and θ'' are related through

$$n_2 \sin\theta' = \left[n^2(r) - n_2^2 + n_2^2 \sin^2\theta'' \right]^{\frac{1}{2}}, \qquad (6)$$

where $n(r)$ is the local refractive index of the fibre at the point of incidence. In the actual experiment, the refracted rays from a short piece of the test fibre are focused into the detector by means of a large diameter lens.

Under such an overfilled launch condition, leaky rays are also excited and escape the fibre, though at relatively smaller angles than the refracted rays. However, they can easily be stopped from reaching the detector by an opaque screen of appropriate size [see Fig. 5(b)]. If θ'_{max} denotes the angle corresponding to the launch NA (of the microscope objective) and θ_s'' is the smallest angle defined by the stop used to block off the leaky rays [see Fig. 5 (b)] then it can be shown that the refracted power $P(r)$ reaching the detector will be related to $n(r)$ through [5]

$$n^2(r) - n_2^2 = n_2^2 \left(\sin^2\theta'_{max} - \sin^2\theta_s' \right) \frac{P(a) - P(r)}{P(a)}. \qquad (7)$$

Here $P(a)$ represents the power detected when the laser is focused on the cladding and serves as a reference value, θ'_{max} is related to θ''_{max} through (6) which varies with the position r of the illuminating spot.

The experimental set-up is shown schematically in Fig. 5(c)[19]. Light from an unpolarized[†] He–Ne laser is spatially filtered by means of a ×20 microscope objective and a 25 μm pinhole. Another ×40 objective (NA = 0.55) is used to focus the filtered beam onto the entrance face of the test fibre which is mounted in a glass cell which in turn is mounted on an *XYZ* translation stage. The front end of the cell contains a coverslip of ~100 μm thickness. The beam-splitter and the translucent screen shown in Fig. 5 (c) enable one to see the position of the focused laser spot on the fibre cross-section by monitoring the reflected laser spot from the fibre end, which is back-illuminated with a white-light source such as a tungsten lamp. The focused laser spot is scanned across the fibre endface by moving the cell (along with the fibre) laterally by

---

[†] If a linearly polarized light is used a $\lambda/4$ plate is required to convert the beam to circular polarization, otherwise the reflectivity of the beam at the air-glass interface will depend on angle and polarization.

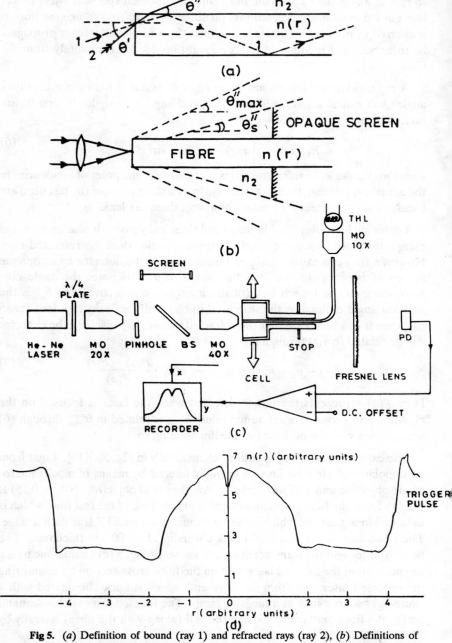

Fig 5. (a) Definition of bound (ray 1) and refracted rays (ray 2), (b) Definitions of angles used in equation (7), (c) Schematic diagram of set-up for the refracted near-field (RNF) method of index profiling [19]: BS, beam-splitter. (d) Index profile as measured by the RNF method of the fibre used in Fig. 4 (b)[19].

means of a motor. The divergent cone of refracted rays is focused onto a photo-detector with the help of a Fresnel lens. The size of the stop is chosen to be such that it subtends an angle at the fibre entrance face which is slightly greater than the NA of the fibre. The experiment is carried out by scanning the input end of the fibre and measuring the refracted power. The output of the motor and the photo-detector are connected to an XY recorder. As the cell containing the fibre is moved across the focal plane of the second microscope objective the refractive index profile is displayed directly on the XY recorder.

Figure 5(d) shows the result of index profiling with the RNF method described above using the same fibre as that in Fig. 4(b). It is clear that the RNF method yields more detailed information than the TNF. For example, the layered structure of the CVD fibre can be seen in the fine detail of the core profile, as can the borosilicate barrier layers at the core edges. Unlike the TNF, it is the optical set-up, not the fibre, that determines spatial resolution. When properly implemented with well-designed optics, submicron ($\sim 0.35\,\mu$m) spatial resolution can be achieved and changes in refractive index of only 4 parts in 10^5 have been demonstrated [19]. In fact the resolution and accuracy offered by the RNF method is higher than any other fibre index-profiling technique proposed to date in the literature. However, to attain such high measurement resolution certain care has to be exercised in performing the experiment [19]: the input face of the fibre should be flat within the depth of focus (typically $\sim 2-4\,\mu$m corresponding to an endface angle of $\sim 2-4°$) of the microscope objective and the axis of the input cone of light should be parallel to the fibre axis to an angle better than $\sim 0.3°$. Appropriate focusing of the laser spot on the input face also requires extreme care (for details, see [19]).

Since the refracted rays form a continuum of radiation modes (in contrast to finite number of guided modes on which the TNF method is based) the RNF method is independent of the modal nature of propagation. Accordingly, the measured intensity distribution does not suffer from interference ripples even though a laser is used as the illuminating source, and it can be used to profile low-moded or even monomode fibres.

(iii) Transverse interferometry

This, transverse, method does not require one to cleave an endface and is thus used for non-destructive index profiling of optical fibres [19]. An interference microscope objective and an on-line computer are required for evaluating the measured data. A short section of a bare fibre (without the protective plastic jacket) is immersed in an index-matching liquid and viewed transversely under an ordinary microscope fitted with a $\times 10$ interference objective. The index-matching liquid helps in preventing very large fringe shift by the fibre cladding. As shown in Fig. 6, the fibre section under test is placed on a metallized optical flat serving as a mirror. (This can be heated to obtain a perfect index match between the liquid and the fibre cladding.) The method essentially relies on

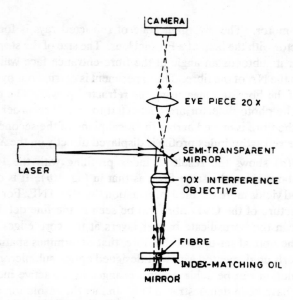

Fig. 6. Schematic diagrm of set-up for the transverse interferometric technique [after 6,20]

the assumption that the fibre core cross-section consists of a large number of concentric cylinders of stepwise uniform refractive index (staircase approximation). The refractive index distribution is inferred by determining the index values of each successive layer in terms of a fringe displacement relative to the stationary fringes in the cladding as viewed through the camera. In order to reconstruct the refractive index profile from the interferogram, one has to solve an integral equation with the help of a computer, which makes the whole system rather expensive [5]. However, the method has the advantage that relatively large but localized refractive-index fluctuations like index dips, bumps, wiggles, etc., can be well resolved by this technique.

3.2 Geometrical measurements

The geometrical characterisation of a multimode optical fibre essentially involves measurement of the core and cladding diameters, together with the concentricity and eccentricity of the fibre cross-section. The numerical aperture (NA) is also a quantity of great interest since (along with core diameter) it largely determines optical power launched into a multimode fibre.

Core diameter is usually defined in terms of reference points on the core RIP [21]. According to the standards recommended by CCITT and EIA the core diameter of a multimode fibre is defined on the basis of following relation:

$$n_3 = n_2 + K(n_1 - n_2). \tag{8}$$

Here n_2 is the cladding index and n_3 and n_1 are as shown in Fig. 7 for a typical graded core profile. The value of the constant K, which is still under study by

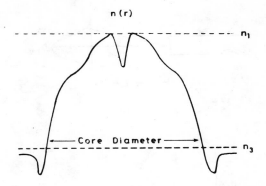

Fig. 7. A typical refractive index profile showing how the core diameter is defined.

the standards committees, depends on the choice of the RIP measurement though it normally lies within the range $0 < K < 0.05$. The results of one extensive study devoted to this question lead to the recommendation that the value of $K = 0.02$ be used in defining the core diameter [21]. The EIA is more specific. If recommends two alternatives; in particular, if RIP is used to determine the core diameter: one sets $K = 0.025$; the other involves a curve fit of the power-law RIP expression (4) to the measured TNF between its 10 and 80 per cent intensity points. These procedures usually yield agreement to within 1 μm between TNF and other refractive-index profiling techniques.

Core and cladding diameters can also be measured by viewing the fibre endface under a transmitted light microscope. In industry this procedure can be automated by digitizing video scans of the endface and processing of the data by a microcomputer, which can yield all the spatial characteristics of the fibre cross-section. Core and/cladding diameters can then be estimated by determining the diameter of the numerically best fitted circles to the measured cross-section.

3.3 Numerical aperture

The NA of a fibre is usually obtained by measuring the angular dependence of the far field of the fibre projected onto a screen (as shown in Fig. 8 (*a*)).

In contrast to the near field, which represents the field distribution at the plane of the fibre endface, the far field of a fibre is defined to be at any distance $Z \gg (2a)^2/\lambda$ analogous to definitions used in the diffraction theory of antenna, here *a* represents the core radius of the fibre. In actual measurements one often encounters problems in defining a base line, owing to smoothing of the intensity curve at the core-cladding edges; to circumvent such problems one common practice is to define the NA as the sine of the half-angle at which the far-field intensity falls to 5% of its maximum value. Alternatively one can obtain the NA in terms of the core-cladding index difference directly from the measured RIP. It should be mentioned that over relatively short fibre lengths

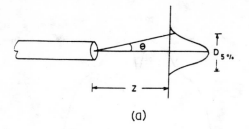

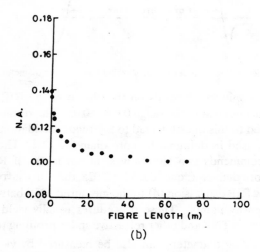

Fig. 8. (a) Measurement of NA from the far field. (b) Measured NA as a function of length [after 22].

the NA can exhibit a fibre length dependence, depending somewhat on the fibre RIP. This can be attributed to the excitation of leaky and higher-order modes, which lose power at a much faster rate than the relatively low-order guided modes, with propagation resulting in a transient nature of the measured NA, as shown in Fig. 8(b). Only after some distance of propagation may the modal power distribution within the fibre attain a steady-state condition in which distribution remains unchanged with length. NA measured under this condition will yield a steady-state value. In the next section on attenuation measurements, we shall revert to a more detailed discussion on simulation of steady-state conditions within a relatively short fibre length.

3.4 Total attenuation

In common with any transmission system, the two most important characteristics of a fibre are its attenuation and dispersion. In fact it was the reduction in attenuation offered by optical fibres that lead to their emergence as a viable alternative to conventional wire communication. Formally, the attenuation coefficient is expressed as

$$\alpha = \frac{10}{L} \log_{10} \frac{P_i}{P_o} \text{ (dB km}^{-1}\text{)} , \qquad (9)$$

where P_i and P_o denote input power and output power, respectively and L denotes the length of the fibre, which is usually expressed in kilometers for telecommunication purposes. In laboratory measurements of attenuation, one usually takes a test sample of about one or two kilometers in length to obtain a representative value to use in determining the total attenuation over a link length, which typically runs into 10-30 km, between repeaters. When applied to multimode fibres, one problem frequently encountered in this procedure is that each of several hundred propagating modes has a different loss characteristic and loss measured over short sections may not necessarily scale linearly with length to yield loss for longer lengths. This is evident from Fig. 9(a), which shows that only after the so-called coupling distance (z_c), does the transmitted power decay uniformly with length. The distance z_c is defined to be the fibre length after which a steady-state (SMD) or equilibrium mode distribution (EMD) is established, and in practice this varies from fibre to fibre. In high-quality fibres (of which telecommunication-grade fibre is a good example) that are free from imperfections such as diameter fluctuations and microbending, the SMD/EMD may be established after only a few kilometers.

The most widely used technique for attenuation spectrum measurements is known as the cut-back method [5, 6]. In this method, power is measured as a function of wavelength once with a long length of the test fibre (P_o) and then a second time on a much shorter piece (called the 'reference' length) of the same fibre (P_i) without changing the launch conditions. Since this is clearly not a non-destructive technique, one would naturally wish to sacrifice as little length as possible in this second measurement. However, for realistic measurements, SMD/EMD must be established within this reference length so that attenuation calculated on the basis of (9) can be extrapolated to arbitrary fibre lengths. In practice, this is done by simulating the SMD/EMD within a short length of the test fibre through either of the following two approaches:

(a) *The limited phase space* (LPS) *approach*: by controlling the launch beam optics so that the launch spot and launch NA are limited to 70 ± 5% of the fibre core diameter and fibre NA respectively; and

(b) *The mandrel wrap/mode filter approach*: using overfilled launch conditions followed by subsequent filtering of the higher order modes by winding a short section of the fibre around a cylinder (typically, five turns on a 1-1.5 cm diameter rod [see Fig. 9 (b)]). In order to qualify as an adequate mode filter, EIA recommends that the following condition be satisfied: the far-field angle θ_s (at 5% intensity points) on a 2 m length of the test fibre measured with the mode filter should

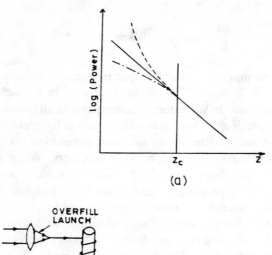

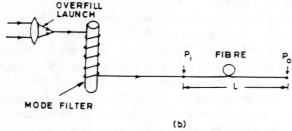

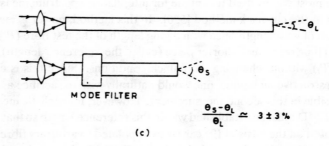

Fig. 9. (a) Decrease in guided power with length of fibre, illustrating the onset of the steady-state power distribution with propagation distance, $z > z_c$ corresponds to uniform attenuation rate; —— steady state excitation, ———— overfilled launch, and —·—·— underfilled launch. (b) Schematic diagram illustrating mode-filter approach to simulate a steady-state mode distribution within a short distance from the fibre input end; P_o corresponds to power measured on the test fibre and P_i is the power measured after the fibre is cut to a short piece. (c) Definition of the criterion that a good mode filter must satisfy.

be within $3 \pm 3\%$ of the corresponding far-field angle θ_L measured on a los(10)ng length of the fibre [see Fig. 9 (c)].

Either of these approaches is expected to replicate SMD/EMD launch conditions within a short length of the fibre to simulate the modal power distribution that would prevail at the end of a long length of fibre.

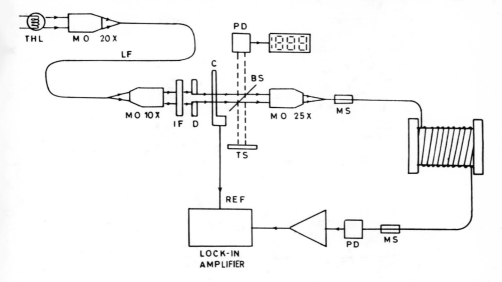

Fig. 10. Schematic diagram of the set-up for attenuation measurement by the cut-back method through LPS approach: THL, halogen lamp; LF, large-core fibre; IF, interference filter; D, diaphragm; C, chopper; BS, beam-splitter; TS, translucent screen; MS, mode stripper; PD, photo-diode.

A typical experimental set-up used to carry out attenuation measurements by the cut-back method is shown in Fig. 10. Most often one uses tungsten-halogen lamp as the light source followed by a monochromator or interchangeable interference filters and conventional launch optics to focus the input beam at the fibre endface. However, in such an arrangement it is difficult to get a uniformly illuminated circular spot at the fibre input face because the focused spot is always affected by the filament structure of the tungsten lamp. To overcome this problem we have used a short piece of large core (200 μm) step-index fibre as the effective source (as shown in Fig. 10), which yields a uniformly illuminated circular spot at the fibre input end. The size and NA of the focused spot was controlled by means of a diaphragm and suitable choice of microscope objectives [23]. Pinholes of various sizes along with various settings of the diaphragm can be used in combination to vary the spatial and angular extent of the launched spot [24]. To ensure that the focused spot meets the LPS criterion mentioned above, the fibre under test is back-illuminated by means of an He–Ne laser and the focused circular spot (image of the illuminating fibre) is viewed against the laser-illuminated fibre end on a translucent screen by exploiting the 4% Fresnel reflection of the input beam at the fibre input face (Fig. 10). The detector on the other side of the beam-splitter is used to monitor the stability of the optical source during the experiment.

A typical result is shown in Fig. 11 (a) for a step-index fibre of NA 0.17. The figure clearly shows the effect of varying the launch NA: an overfilled launch

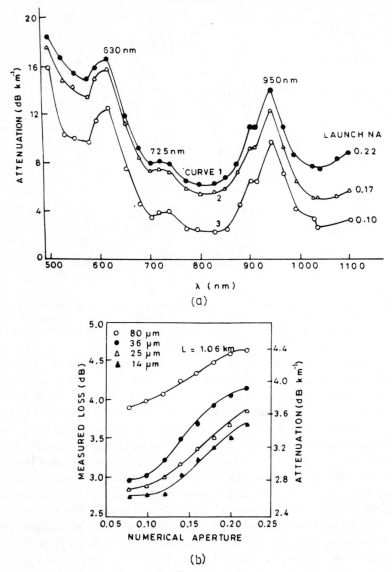

Fig. 11. (a) Measured attenuation for a step-index fibre (core size 200 μm, fibre NA = 0.17; launch spot size = 130 μm) as a function of launch NA [23]. (b) Measured attenuation on a graded fibre (Corning E; L = 1.06 km) as a function of NA for various launch spot sizes [25]: O 80 μm; ● 36 μm; Δ 25 μm; ▲ 14 μm

yielding higher attenuation at all wavelengths owing to the excitation of higher order modes. This behaviour is again found in Fig. 11 (b) taken from [25]. In Fig. 12, we have plotted the results of measurements made on a graded-core fibre using mode-filter approach. It is now well known that attenuation in current telecommunication fibres can be approximately modelled [23] through

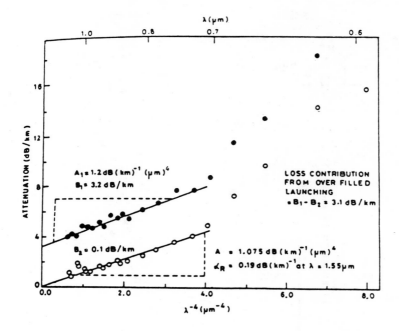

Fig.12. Measured loss as a function of λ^{-4} on a graded-core fibre (Corning II) with (o) and without (●) mode filter [23].

$$\alpha \simeq B + A\lambda^{-4} \qquad (10)$$

by assuming the absence of absorptive impurities. Thus the slope and y-intercept of the plots in Fig. 12 will yield respectively the Rayleigh scatter loss coefficient and the wavelength-independent contribution to loss. Values of A and B so determined for the cases with and without the mode filter are shown in the inset of the figure. These figures compare favourably with theoretical estimates of loss. Well known laboratories in the U.S.A. using standard measurement procedures outlined above have achieved a precision ~0.1-0.15 dB km^{-1}, while round robin results give a precision between laboratories of ~0.2-0.3 dB km^{-1} over a test length of 2 km. These may be taken as representative figures of merit in attenuation measurements.

3.5 Scattering loss and differential mode loss

In the early days of development of telecommunication-grade fibres, knowledge of the scattering loss was important because it yielded information on the extent of waveguide imperfections such as localized inhomogeneities and non-uniformities in the core profile. Scattering loss can be measured by launching say He—Ne laser light into the fibre under test and having the fibres pass through the axis of an integrating sphere whose inner surface is coated with a highly reflecting diffusive material. The scattered power is measured by

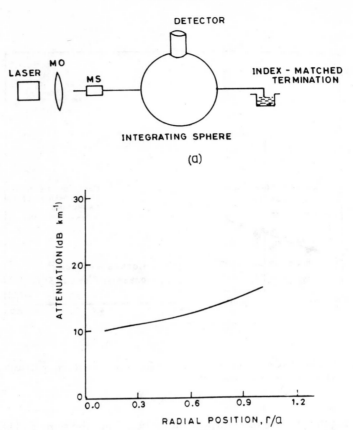

Fig. 13. (*a*) Schematic diagram of set-up for scatter-loss measurements. (*b*) Typical measured results illustrating variation of differential mode attenuation with radial position [26].

a detector fixed into the wall of the detector [see Fig. 13 (*a*)] from which the scatter loss coefficient can be evaluated by dividing detected power by the diameter of the integrating sphere. In the modern high-quality fibres, since impurities are virtually absent, Rayleigh scatter loss essentially determines the total loss and independent scatter loss measurements are practically no longer necessary.

On the other hand, knowledge of the differential mode loss may be important for high-performance fibres where one is concerned with the fibre material and waveguide properties. In such applications, one is interested to find out which mode groups suffer the greatest loss. This can be determined by butt-coupling a single-mode fibre to a given multimode fibre at different radial positions across its core cross-section, each time measuring the total power exiting the multimode fibre, then repeating the procedure on a short piece of the same fibre, in analogy with the cut-back method. Since in power-law profiled fibres one can describe propagation in terms of a compound mode number (rather than with an azimuthal mode number l for a fixed radial mode number m), for a group of modes associated to a particular radial position [26], the above mentioned excitation condition will exhibit the loss behaviour of different mode groups [see Fig. 13 (b)]. Accordingly, through such measurements one can determine the differential scatter loss contributed by various doping concentrations in different regions of the core profile. Information about differential mode loss can also be determined from the TNF measurements [28]. Differential mode loss measurement techniques are useful in the study of differential mode delays, as well.

3.6 Non-destructive loss measurements (OTDR)

Optical time-domain reflectometry or OTDR, named after its electrical analogue, TDR, is used to obtain a measure of fibre attenuation non-destructively [29]. This technique essentially involves launching an optical pulse of half-power width ~5-10 ns (the width will determine length resolution of fibre length measurement, e.g. 10 ns ≅ 1 metre) at one end of the fibre under test. As the pulse propagates, it undergoes continuous scattering along the length of the fibre owing to both Rayleigh scattering and waveguide imperfections such as diameter fluctuations, bubbles, etc., thereby, producing a low-level reflected signal. The former is isotropic, i.e. emitted at all angles while imperfections tend to scatter light preferentially in the forward and backward directions. Some of the reflected and back-scattered light is guided by the fibre, back to the launching end where it is detected, normally by an APD and displayed directly on an oscilloscope [see Fig. 14 (a)]. If there are poor quality splices or joints, connectors or breaks, etc., along the length of the fibre, they appear as distinct peaks in the measured back-scattered power since they produce reflected signals much larger than the continuous back-scatter signals as shown in Fig. 14 (b), thereby revealing the location of such faults or local variations in fibre attenuation. Accordingly, the OTDR results are often said to contain 'fault signatures' of an optical communication link [30]. The technique has the advantage that it can easily be deployed in field measurements as it requires access to only one end of the fibre. The value of attenuation is given by the slope of the curve in Fig. 14 (b). It may be noted that the value of attenuation yielded by this method is not as accurate as the cut-back method since the back scattered power not only depends on the power at that point but also on the backscatter coefficient. Normally the OTDR technique is used only to measure loss at a fixed wavelength and not for spectral loss measurement. Since the backscatter signal is usually weak, expensive signal-averaging equipment, such as the box-car integrator, is usually required to process the reflected 'signature signal'. Measurement accuracy is continuously being improved [30], and several companies now market OTDR measurement instruments as a compact unit.

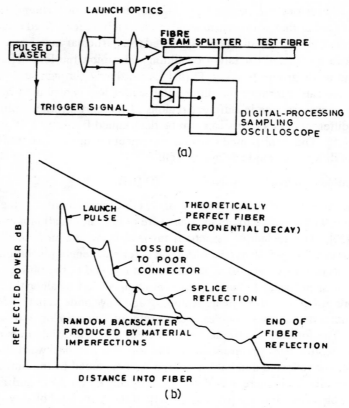

Fig. 14. (a) Schematic diagram of the set-up for an OTDR measurement. (b) Schematic diagram of an OTDR signal (adapted from a PHOTODYNE Datasheet).

3.7 Transmission bandwidth and dispersion

The repeater distance in an optical fibre link is determined both by attenuation and dispersion. In fact, in ultra-low-loss fibres, dispersion is the dominant factor in determining maximum repeater distance. The temporal dispersion in a multimode fibre, as is well known, has two components, intermodal and intramode dispersions, of which the former is the dominant factor, except in the so-called optimum profiled fibres at $\lambda \sim 1.3\,\mu$m [32, 33]. In typical measurements, one essentially measures the total dispersion, i.e. sum of intermodal and intramodal dispersion. Though it is possible to make an estimate of the value of the dispersion in a multimode fibre on the basis of index-profile measurements alone, for a system engineer, a direct measurement of the fibre's transmission bandwidth BW $\sim$ (Dispersion)$^{-1}$ over a relatively long length of the fibre is more important. This is understandable, because ultimately, the installed system is based on a relatively long length of fibre and there may be several other length-dependent propagation effects such as mode coupling, differential mode loss and so on, which could significantly affect the fibre BW [34].

Typically temporal pulse dispersion in a fibre is experimentally determined by launching short optical pulses (FWHM ≃ 100-300 ps) from a laser diode into the test fibre and detecting the same by a fast APD at the output end followed by display on a storage oscilloscope working in sampling mode. As with the cut-back method used in attenuation measurements, the measurement is repeated under the same launch conditions on a short length of fibre, which yields an input reference pulse. These two output pulse widths are compared. If the input and output pulses are assumed to be Gaussian with r.m.s. widths of σ_i and σ_o, then a measure of pulse-broadening dispersion can be obtained from

$$\sigma = (\sigma_o^2 - \sigma_i^2)^{\frac{1}{2}}, \qquad (11)$$

where σ represents the r.m.s. width of the impulse response. A delayed trigger pulse (electrical) from the laser drive-circuits triggers the oscilloscope (see Fig. 15). According to EIA recommendations, the 3-dB optical bandwidth should be taken as the transmission BW of the fibre. The 3-dB optical BW is calcu-

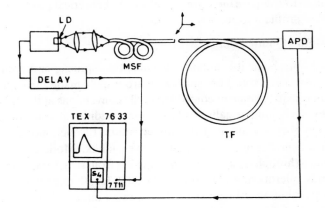

Fig. 15. Schematic diagram of the set-up for pulse dispersion measurements; LD, laser-diode; MSF, mode-scrambled fibre; TF, test fibre; TEX, Tektronix oscilloscope with a sampling head.

lated from the following theoretical relation obtainable from a Fourier transform analysis of the temporal pulse:

$$(\text{3-dB optical BW}) \times (\text{RMS impulse width}) \simeq 187 \text{ MHz ns}, \qquad (12)$$

where it is assumed that the input and output pulses can be approximated by Gaussian functions. In practical measurements, one gets a value between about 169 and 174 depending on the type of fibre, the cabling and the mode mixing present [35]. On the other hand, if FWHM of the pulses is used in place of r.m.s. widths to obtain dispersion, then for a Gaussian pulse, a relation equivalent to (12) is given by:

$$(3\text{-dB optical BW}) \times (\text{FWHM in pulse width}) \simeq 440 \text{ MHz ns}. \tag{13}$$

As in the case of attenuation measurements, one must be sure that EMD/SMD has been established within the reference length of the fibre, so that the injected light pulse is independent of the spatial distribution of the source. Thus, to obtain an EMD excitation condition at the input end of the fibre, one usually employs an overfilled launch condition followed by use of what is commonly known as a *mode scrambler*. Though several types of mode scrambler have been proposed in the literature, a sequence of three short pieces of fibres — step, graded and step — cascaded through two intermediate splices is widely accepted as the most useful in BW measurements [36]. In typical measurements, the optical pulse from the laser diode is injected into this mode scrambler, the output of which effectively acts as the source which is then butt-jointed to the test fibre for final measurements. Unless, such steps are taken during dispersion measurements one may get unrealistic results and it would be futile to compare measured results with theoretical predictions on the basis of index-profile information alone [37].

3.8 Bandwidth of jointed fibres

In long-haul fibre links, it is almost inevitable that splicing and jointing have been used. Any such splice or joint which are usually made through electric-arc fusion may lead to some uncontrolled mode conversions at the splice point. Though any such joint usually leads to some additional transmission loss, redistribution of power amongst the guided modes through mode conversion may lead to a decrease in dispersion of the overall link. Mode conversion may also take place through coupling of power arising from structural perturbations such as bends, microbends and so on. In fact, the length dependence of modal delay in long multimode fibre links (with or without intermediate joints) can be described by the following relation:

$$\tau = \tau_o L^\gamma, \tag{14}$$

where γ typically lies in the range $0.5 < \gamma < 0.7$ and τ_o is the delay measured on a short piece of the fibre relatively free from mode-coupling effects. Though controversy prevails over the precise modelling of propagation effects in jointed fibre links [38], the exact value of γ seems to vary from fibre to fibre as well as on the quality of the splices. In the laboratory, it is difficult to simulate precisely field conditions of concatenated fibre links owing to the unpredictable nature of the mode conversion that may be introduced at a splice point and the extent of mode coupling, if any, that may be introduced along the fibre length.

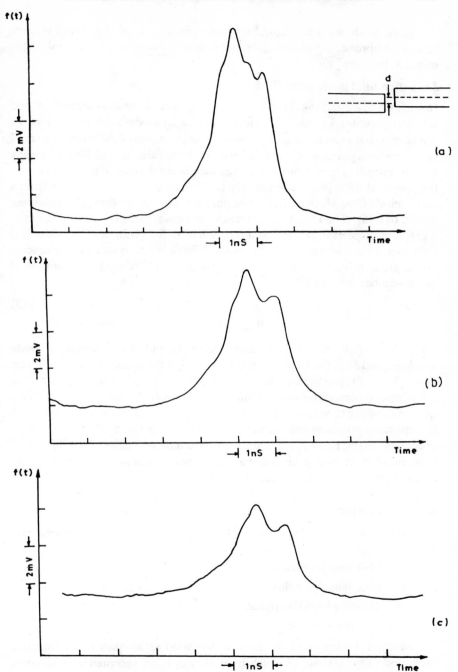

Fig. 16. Typical output pulse shapes from a concatenated link between two graded-core ($a \simeq 25\,\mu$m) fibres shown as a function of various transverse off-sets at the joint [39]: (a) $d = 0.0\,\mu$m; (b) $d = 2.0\,\mu$m; (c) $d = 16.5\,\mu$m.

Figure 16 shows some typical measurement results of dispersion in concatenated fibre segments having deliberately introduced measurable misalignments at the joint [39].

3.9 Differential mode delay (DMD)

In §3.5, we discussed the measurement of differential mode losses, where one selectively excites a group of modes. Differential mode-delay measurements are carried out in an analogous manner; focusing a short light pulse as a small spot at various points across the diameter of the multimode test fibre by means of well controlled launch optics and then measuring the output pulse and hence the temporal delay [40]. Alternatively, the launched beam can be taken from a single-mode fibre which scans the test fibre across its diameter while remaining butted to it [5]. The DMD characteristics are usually displayed in the form of a graph of dispersion (DMD) as a function of radial position of the launched light spot. Since, in near parabolic profiled fibres, in which such measurements are of greatest interest, radial position can be simply related to a compound mode number through [5]

$$\frac{M}{M_{max}} = \left(\frac{r}{a}\right)^2, \tag{15}$$

where $M = (2m + l + 1)$, m and l being the radial and azimuthal mode numbers, and M_{max} the largest possible value of M for guided modes in the test fibre. One can relate the measured DMD versus radial position to the DMD as a function of compound mode number. In optimally profiled fibres, DMD is quite insensitive to variation in M and any departure of the index profile from the optimum profile would therefore be immediately revealed in DMD measurements. In this respect, it is a more sensitive indicator of local profile defects than direct index profile measurements and hence is extremely useful in fibre fabrication to monitor fibre performance.

4. CONCLUSION

After discussing the key experimental factors that affect precise fibre measurements, namely

    (i) fibre-end preparation,
    (ii) fibre launch conditions,
    (iii) cladding mode stripping, and
    (iv) detector-fibre coupling

we surveyed a number of the most widely used techniques to characterize multimode fibres, some of which have already been accepted as standards. Provided care is taken in preparing the experimental conditions, these techniques can yield precise measurements of

    (i) refractive index profile,

(ii) fibre geometry,
(iii) numerical aperture,
(iv) attenuation and fibre losses, and
(v) transmission bandwidth and dispersion.

In chapter 11 as part II we continue with a survey of techniques that apply to single-mode fibres.

REFERENCES

1. F.P., Kapron, Keck, D.B., and Maurer, R.D., "Radiation losses in glass optical waveguides", Appl. Phys, Lett., 17, pp. 423-425 (1970).
2. V.A. Bhagavatula, Berkey, G.D., and Sarkar, A., "Scattering loss in single-mode fibres by outside process", Proc. OFC '83, Los Angeles (Washington: Optical Society of America), p. 22 (1983).
3. T.F. Hodgkinson, Smith, D.W., Wyatt, R., and Malyon, D.J., "Coherent optical fibre transmission systems", Br. Telecom Technol. J., 3, No. 3 (1985).
4. G.D. Pitt, Extance, P., Neat, R.C., Batchelder, D.N., Jones, R.E., Barnett, J.A., and Pratt, R.H., "Optical fibre sensors", Proc. IEE, Pt. J., 132, pp. 214-248 (1985).
5. D. Marcuse, "Principle of Optical Fibre Measurements", New York: Academic Press (1979).
6. G. Cancellieri and Ravaioli, U., "Measurements of Optical Fibres and Devices: Theory and Experiments", Massachusetts: Artech House Inc. (1984).
7. D. Gloge, Smith, P.W., Bisbee, D.L., and Chinnock, E.L., "Optical Fibre end Preparation for Low Loss Splices", Bell. Syste. Tech. J. 52, pp. 1579-1588 (1973).
8. B.P. Pal and Kersten, R. Th., "Teaching optical waveguides: A contemporary course with demonstration experiments", IEEE Trans. Education, 28, pp. 46-52 (1985).
9. B.P. Pal, "Refractive Index Profile Measurements in Optical Fibres", Electronics Research Laboratory Report STF44A 77208, Trondheim, Norway (1977).
10. D. Marcuse, "Attenuation of Cladding Modes", Bell Syst. Tech. J., 50, pp. 2565-2583 (1971).
11. F.M.E. Sladen, Payne, D.N., and Adams, M.J., "Determination of Optical Fibre Refractive Index Profiles by a Near Field Scanning Technique", Appl. Phys. Lett., 28, pp. 255-258 (1976).
12. D. Gloge and Marcatili, E.A.J., "Multimode Theory of Graded Core Fibres", Bell Syst. Tech. J. 52, pp. 1563-1578 (1973).
13. K. Dilip, "Experimental Studies on Pulse Dispersion in Optical Fibres", M. Tech. (Applied Optics) Thesis, IIT Delhi (1981).
14. M.J. Adams, Payne, D.N., and Salden, F.M.E., "Correction Factor for the Determination of Optical Fibre Refractive Index Profiles by the Near-Field Scanning Technique", Eelctron. Lett., 12, pp. 281-283 (1976).
15. K. Petermann, "Leaky Mode Behaviour of Optical Fibres with Non Circularly Symmetric Refractive Index Profile", AEÜ, 31, pp. 201-204 (1977).
16. B.P. Pal, "Optical Communication Fibre Waveguide Fabrication: a Review", Fibre Int. Opt., 2, pp. 195-252, (1979).
17. J.P. Hazan, "Intensity Profile Distortion Due to Resolution Limitation in Fibre Index Profile Determination by Near Field", Electron. Lett., 14, pp. 158-160 (1978).

18. M.J. Adams, Payne, D.N., Salden, F.M.E., and Hartog, A., "Resolution Limit of The Near Field Scanning Technique", Proc. Third European Conf. on Optical Communication (ECOC'3), Munich (Berlin: VDE-Verlag), pp. 25-27 (1977).
19. T.V.B. Subrahmanyam, "Refracted Near Field Scanning Technique for Refractive Index Profiling of Optical Fibres", M. Tech. (Applied Optics) Thesis, IIT, Delhi (1983).
20. M. Boggs, Presby, H.M., and Marcuse, D., "Rapid Automatic Index Profiling of Whole Fibre Samples: Part I", Bell Syst. Tech. J., **58**, pp. 867-882 (1979).
21. A.H. Cherin, Rich, P.J., and Mettler, S.C., "Measurement of the Core Diameter of Multimode Graded Index Fibres: a Comparison of the Transverse Near Field and Index Profiling Techniques", J. Lightwave Technol., 1, pp. 302-311 (1983).
22. M. Maeda and Yamada, S., "Leaky Moes in W-fibres: Mode Structure and Attenuation", Appl. Optics, **16**, pp. 2196-2203 (1977).
23. A.K. Gupta, "Attenuation Measurement in Optical Fibres", M.Tech (Optoelectronics and Optical Communication) Thesis, IIT, Delhi (1984).
24. G.T. Holmes and Hauk, R.M., "Limited Phase Space Attenuation Measurements of Low Loss Optical Waveguides", Optics Lett., **6**, pp. 55-57 (1981).
25. M. Eriksrud, et al., 1982, "Experimental Investigation of Attenuation as a Function of Spot Size and Numerical Aperture", Internal Study Report STF 44 A 82188, Electronics Research Laboratory, Trondheim, Norway (1982).
26. N. Rao, "Measurement of Differential Mode Attenuation in a Multimode Optical Fibre", M. Tech. (Applied Optics) Thesis, IIT, Delhi (1982).
27. R. Olshansky and Oaks, S.M., "Differential Mode Attenuation Measurements in Graded Index Fibres", Appl. Optics, 17, pp. 1830-1835 (1978).
28. Y. Daido, Miyauchi, E., Iwama, T., and Otsuka, T., "Determination of Modal Power Distribution in Graded Index Optical Waveguides from Near Field Patterns and Its Application to Differential Mode Attenuation Measurement". Appl. Optics **18**, pp. 2207-2213 (1979).
29. M.K. Barnosky and Jensen, S.M., "Fibre Waveguides: A Novel Technique for Investigating Attenuation Characteristics", Appl. Optics, **15**, pp. 2112-2115 (1976).
30. M. Eriksrud, Lauritzen, S., and Ryen, N., 1980, "Fibre waveguides: a Novel Technique for Investigating Attenuation Characteristics", Electron. Letts. **16**, pp. 877.
31. M.P. Gold and Hartog, A.H., "Long-range Single Mode OTDR: Ulimate Performance and Potential Uses", Proc. Tenth European Conf. on Optical Communication (ECOC '10), Stuttgart (Berlin: VDE-Verlag) (1984).
32. M.M. Butusov, Pal, B.P., Galkin, S.I., and Orobinsky, S.P., "Fibre Optics and Instrumentation", (Leningrad: Mashinostroenie Publishing House) (in Russian), 1987.
33. B.P. Pal, "Fundamental Characteristics of Telecommunication Optical Fibres: a Review", I.E.T.E. (India) Tech. Rev., 3, pp. 347-371 (1986).
34. D. Marcuse, "Light Transmission Optics" (New York: Van Nostrand Reinhold) (1982).
35. R.L. Gallawa and Franzen, D.L., "Progress in Fibre Test Standards", Photonic Spectra, April, pp. 55-68 (1983).
36. W. Love, "Novel Mode Scrambler for Use in Optical Fiber Bandwidth Measurements", Technical Digest: Topical Meeting on Optical Fibre Communication, Washington (1979).
37. A.D. Sharma, Subrahamanyam, T.V.B., Pal, B.P., Thyagarajan K., Kumar, A. and Goyal, I.C., "Bandwidth Prediction in Optical Fibres: Correlation of Index Profile and

Dispersion Measurement" Proc. All India Symp. on Communication, Bangalore, pp. 117-122 (1984).
38. S. Cibboto and Someda, C.G., "Bandwidth Concatenation in Multimode Fibres: a Review", J. Inst. Electron. Telecom. Engrs. (India), **32**, pp. 253-258 (1986).
39. B.P. Pal, "Experimental Study on Effects of Fibre Joints in Pulse Dispersion", Internal Study, Electronics Research laboratory, Trondheim, Norway (1983).
40. R. Olshansky, and Oaks, S.M., "Differential Mode Delay Measurements", Proc. Fourth European Conf. on Optical Communication (ECOC'4), Genoa, Italy, pp. 128-130 (1978).

11

Characterisation of Optical Fibres — Part II: Single-Mode Fibres

K. Thyagarajan*, B.P. Pal* and A. Kumar*

1. INTRODUCTION

In recent years the focus of attention in fibre optics telecommunication and sensor development has shifted to single-mode fibres. This can be attributed to a number of factors [1]: (i) the attainment of ultra-low loss (~0.15 (dB-km^{-1}) at 1.55 μm, which is close to the theoretical limit set by Rayleigh scattering [2]) and very high bandwidth, (ii) the development of high-quality machinery to splice fibres with loss of less than 0.1 dB/splice, (iii) the unpredictable nature of the bandwidth of concatenated multimode fibre links, and (iv) the ultrahigh sensitivity of interferometric fibre phase sensors. Such developments have lately generated an intense interest in the characterization of single-mode fibres. The main parameters characterizing single-mode fibres are listed in the table given in Part I (Ch. 10). In the following sections, we discuss how these parameters can be measured for single-mode fibres.

2. MEASUREMENT TECHNIQUES

2.1 Attenuation

Unlike multimode fibres, in single-mode fibres differential mode attenuation and the problems of attaining steady-state conditions do not play a part. Consequently the measurement of fibre attenuation before installation is a fairly straightforward matter. The most common technique used in determining the spectral attenuation is the cut-back technique using the set-up shown in Fig. 10 of Part I. The signal level typical of single-mode fibre is 10 dB below

---

* Department of Physics, Indian Institute of Technology Delhi, New Delhi-110016, India

that of a multimode fibre, so the signal is normally detected by means of a phase-sensitive detection scheme in conjunction with a PIN photodetector; at longer wavelengths, a cooled detector (e.g. Ge-detector) may be necessary. Loss spectrum of a typical single-mode fibre is shown in Fig. 1. The OH-absorption peak at $\lambda \sim 1.39$ μm is clearly evident. This fibre had a cut-off wavelength ~ 1.2 μm; hence, the broad loss peak at $\lambda \sim 1.15$ μm due to high attenuation suffered by the LP_{11}- mode, being so near to its cut-off wavelength. The repeatability of this technique is mainly determined by the repeatability of the coupling between fibre and detector, and here standard deviations of ≤ 0.05 dB are typical. For good measurement repeatability, the input light spot should both spatially and angularly overfill the fibre core cross-section. Round robin inter-laboratory measurement precision of ~ 0.03 dB/km over 2 km test fibre lengths have been reported in the literature by the cut-back method.

For installed fibres, link attenuation at a specific wavelength can be determined using optical time-domain reflectometry (OTDR).

2.2 Refractive index profile (RIP)

If the RIP of a single-mode fibre is precisely known, one can simulate almost all aspects of propagation within the fibre by solving the scalar wave equation numerically. However, owing to the very small core dimensions (diameter $\sim 4-10$ μm) and low relative core cladding index difference ($\Delta \sim 0.1-0.3\%$) of single-mode fibres, accurate determination of the RIP of these fibres demands very high precision apparatus and thorough measurement procedure. Although a variety of techniques for this purpose have been proposed in the

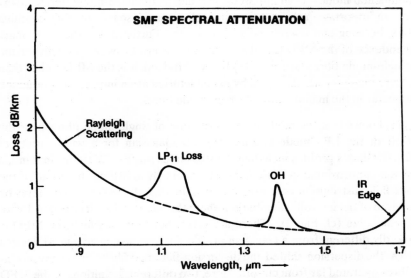

Fig. 1. Attenuation Spectrum of a typical single-mode fibre (adapted from [3]).

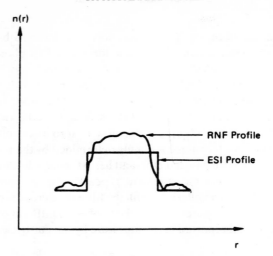

Fig. 2. Measured refractive-index profile on a quasi-step single-mode fibre [5] and its equivalent step-index (ESI) profile [17].

literature [4], the one which has come to be most widely accepted is the RNF technique described in Part I in connection with multimode fibres. For single-mode fibres, measurement precision as high as $0.2\,\mu$m has been reported with this technique. A typical result is shown in Fig. 2[5].

2.3 Mode field diameter

The mode field diameter (MFD) — a measure of the spatial extent of the fundamental mode — is an important parameter for single-mode fibres. MFD plays an important role in estimating splice losses, source-to-fibre coupling losses, bending and microbending losses, etc. Furthermore, the wavelength dependence of the MFD can also be used to determine waveguide dispersion in single-mode fibres (see Eq. (18) below). Indeed, it is the MFD rather than the core diameter which is used by manufacturers as an important production parameter in the manufacture of single-mode fibres.

It is known that the mode field distribution of single-mode fibres near the cut-off of the LP_{11} mode is approximately Gaussian for a wide variety of refractive-index profiles including the step-index profile. With this in mind, a Gaussian approximation is used almost universally in MFD estimation. If the mode field distribution were exactly Gaussian, then the MFD would simply be the width of the intensity distribution where the intensity has dropped to $1/e^2$ of its peak value (cf. Fig. 3). There are cases, however, in which the MFD is markedly different from a Gaussian profile; for instance, in advanced designs such as the dispersion-shifted and dispersion-flattened fibres, or in conventional fibres operated far from cut-off. Here, two different definitions of the MFD are required [6], namely the near-field r.m.s. spot size and the Petermann 2 spot

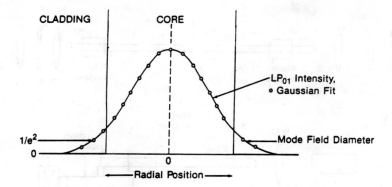

Fig. 3. Gaussian nature of the LP_{01}-mode of a typical step-index fibre (adapted from [3]).

size (which is simply the inverse of the far-field intensity width). For a Gaussian distribution both definitions are identical with the $1/e^2$ intensity diameter.

A variety of techniques are available for the measurement of MFD. These include near-field and far-field scanning, the variable-aperture far field and transverse offset techniques. None of these methods has yet been adopted universally as a standard so it is essential that their results are consistent with one another [7].

Figure 4(a) shows the experimental arrangement used in the transverse-offset technique. Light from a tungsten halogen lamp is coupled into a short length (~1 m) of the test fibre. A second piece of the same fibre is placed in close proximity with the output end of the first fibre. The relative transmitted power $P_t(d)$ across the splice as a function of the transverse displacement d between the fibre ends is measured. The measured $P_t(d)$ is fitted to a Gaussian variation:

$$P_t(d) = P_0 e^{-d^2/w_0^2} \qquad (1)$$

and the MFD read from the $1/e$-points as $2w_0$. To measure wavelength dependence of the MFD, a monochromator or interference filter can be placed between the output fibre and the detector.

Although this method appears quite straightforward, the following points must be observed:

(a) The fibre end faces must be of good quality, having an end angle of less than 1°;
(b) Cladding-mode strippers must be used;
(c) The separation between the fibre end faces must be less than ~5 μm;
(d) The power coupled into the test fibre must be relatively insensitive to the position of the input end (which can be achieved by having an

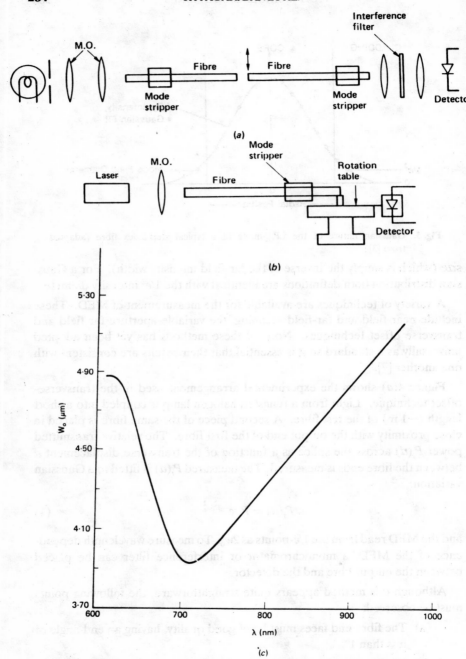

Fig. 4. Schematic diagram of (*a*) MFD measurement set-up for the transverse offset technique, (*b*) MFD measurement set-up for the far-field method, (*c*) Measured variation of spot size with wavelength used in determining the ESI profile through the transverse offset technique [17].

overfilled launch condition as in the case of attenuation measurement); and

(e) For measurement very close to the cut-off wavelength, higher-order modes must be filtered out (by bending the fibre sufficiently on either side of the coupled ends).

The MFD of a single-mode fibre can be measured in the near field by imaging the fibre output end by means of a microscope objective on a scanning photodetector as in the case of transmitted near field employed for measuring the near field of a multimode fibre. The measured near-field intensity profile $I(r)$ can be fitted to a Gaussian function $\sim I(O) \exp(-2r^2/w_1^2)$; hence MFD = $2w_1$. The fitting criterion employed involves maximisation of the overlap integral between an arbitrary Gaussian function and the measured distribution [3]. For example, if

$$E_g(r) = E(0)e^{-r^2/w_1^2} \qquad (2)$$

represents the Gaussian modal field, then w_1 and hence MFD ($= 2w_1$) will be yielded by maximisation of the following overlap integral:

$$I = \frac{\left[\int_0^\infty E_g(r) E_m(r)\, rdr\right]^2}{\int_0^\infty E_g^2\, rdr \int_0^\infty E_m^2\, rdr}, \qquad (3)$$

where $E_m^2(r)$ is the measured power distribution.

The far-field technique [shown in Fig. 4 (b)] by contrast, measures the angular distribution of the far-field intensity. Assuming this distribution is Gaussian, one can obtain the MFD, W_{ff}, by first fitting the angular distribution of power, $P(\theta)$, with [3]

$$P(\theta) = P_o \exp[-2(\theta/\theta_{ff})^2], \qquad (4)$$

and then using

$$2W_{ff} \simeq \frac{2\lambda}{\pi \tan\theta_{ff}}. \qquad (5)$$

For non-Gaussian modal fields, one can use a similar procedure with more exact calculated far fields [8].

There exist two other alternative techniques to measure the MFD from the far-field of a single-mode fibre; these are known as Variable Aperture Far-Field (VAFF) method [9] and the Knife Edge method [10]. In the former a disc-containing various apertures sizes are placed in the far-field and the power transmitted through these apertures are measured as a function of aperture size [11, 12]. The power collected in a cone of angle θ is fitted to the following expression [3]:

$$P(\theta) = P_{max}(1 - e^{-m\tan^2\theta}). \tag{6}$$

In the above P_{max} and m are varied to obtain best least square fit to the experimental data. The MFD is yielded by calculating [3]: $\lambda\sqrt{2m}/\pi$. In the Knife Edge method [10], a knife edge is motor-driven laterally through the far-field and the power transmitted past the knife edge is recorded. The MFD is calculated by fitting a complementary error function to the measured data or by making a Gaussian fit to derivatives of the measured data [10].

Recently an alternative and more straightforward technique has been proposed which avoids the need to perform a fitting procedure in processing the far-field data [13, 15]. The technique consists in identifying the –3 dB and –40 dB far-field angles from the measured far-field angular intensity distribution, and using these values in an empirical formula described in [13, 14]. Unlike a Gaussian approximation for the modal near-field all the way from the core into the cladding, this method relies on a much more accurate representation of the LP_{01}-mode near-field in terms of a combination of Gaussian and modified Bessel function. One major advantage of this method from the point of view of measurement convenience is that it involves measurement of intensity levels much above the tails of the far-field main lobe, which are almost 60-70 dB down from the intensity level at the center. Furthermore, the method yields an accurate estimate of the LP_{01}-modal near-field without recourse to prior knowledge of the fibre's refractive index profile; this is important in predicting the performance of fibre devices like directional couplers constructed out of such fibres. Measurement details and results can be found in [15]. One of the main advantages of the far-field technique is that the quality of end face of the fibre is not critical. It is, in addition, a direct method: one not involving imaging optics. On the other hand, since one is measuring the far-field, the signal being detected is comparatively weak. It is therefore, difficult to measure the wavelength variation of the MFD using incoherent sources.

2.4 Equivalent step-index (ESI) profile

Propagation characteristics of an ideal step-index single-mode fibre can be described in terms of just two parameters, namely the core radius and core-cladding index difference, Δ. Knowledge of these two parameters, along with the wavelength of operation, can be used to compute the V-number ($V = 2\pi a NA/\lambda_o$) of the fibre and also its MFD, since the modal field of such ideal fibres can be shown analytically to be a zeroth-order Bessel function in the core and a modified Bessel function in the cladding. In practice, however, it is difficult to fabricate the ideal step-index fibre (some rounding at the edges is unavoidable) and the modal fields cannot be determined analytically. However, the shape of the LP_{01} mode field in such fibres can, fortuitously, be well approximated to a Gaussian of appropriate width. When this observation is linked to the fact that, within the normal operating range of single-mode fibres ($2.1 < V < 3.0$), the modal field distribution of a step-index single-mode fibre closely matches a Gaussian function, one can conclude that it should be possible to ascribe an equivalent step-index profile (ESI) to practical single-mode fibres. It would be 'equivalent' in the sense that the step-index fibre thus modelled would yield an almost identical LP_{01}-mode intensity profile and propagation constant to that of the test fibre [1]. A variety of techniques have been proposed and studied in the literature [16, 17]. One which seems to be very popular is the transverse offset technique as used for MFD measurement (§2.3). The same measurement set-up can be employed with a tungsten halogen lamp as the source and variation of spot size with wavelength can accordingly be determined. A typical result of such a measurement is shown in Fig. 4(c). It is found that the $1/e$ width, W_o, of the curve of power transmission through the splice as a function of transverse offset decreases with wavelength in the single-mode regime. This is due to greater mode confinement. As one approaches the cut-off of the LP_{11} mode, however, W_o increases abruptly, owing to the greater spatial extent of this mode field. Above the cut-off, one can use the formula of Marcuse [18] for the spot size variation with λ:

$$W_o = (a_e/\sqrt{2})\,[0.65 + 0.434\,(\lambda/\lambda_c)^{3/2} + 0.0149(\lambda/\lambda_c)^6]. \tag{7}$$

Here a_e is the core radius of the equivalent step-index fibre. By fitting the experimental plot of W_o against λ to Eq.(7) (for $\lambda > \lambda_c$) one can find the values of a_e and λ_c. The numerical aperture of the ESI fibre can then be obtained from

$$(NA)_{ESI} = \frac{2.405\lambda_c}{2\pi a_e} \simeq 0.383\,\frac{\lambda_c}{a_e}. \tag{8}$$

The $(NA)_{ESI}$ and a_e thus obtained can be used to reconstruct the form of the modal field of the equivalent step-index fibre and, subsequently, to model those propagation characteristics of the fibre that depend explicitly on the propagation constant (the modal field, for instance). However, care should be exercised in attempting to extend the ESI results to temporal modal effects, such as dispersion phenomena, which involve derivatives of β with respect to wavelength. Equation (8), for example, assumes that the numerical aperture is independent of wavelength. Figure 2 shows the ESI profile obtained by applying this method to a single-mode fibre of near step-index profile.

2.5 Mode cut-off wavelength and the single-mode operating regime

It is well known that as the V-number (see Eq. (3) of Part I) of an optical fibre drops below a critical value, the fibre can support only a single mode. Single-mode operation offers high bandwidth transmission, due to the absence of intermodal dispersion [18]. To use the large information-carrying capacity effectively, one must employ a source of very narrow spectral width so that chromatic dispersion is low. And here there is a further motive for operating with a single mode: coherent light propagating in a number of different modes will interfere at the output end of the fibre which may be affected by external factors such as temperature changes, vibrations and splice quality, introducing what is known as 'modal noise'.

Since the V-number is inversely proportional to wavelength, a fibre will be single mode above a certain critical wavelength, λ_c, the cut-off wavelength. Owing to various perturbations (bending, microbending, etc.) the fibre becomes effectively single mode somewhat below the theoretical value λ_c. This happens because close to the cut-off point the modal power distribution extends deep into the cladding and there even a small perturbation can lead to a large loss. The advantage of operating a fibre at the largest possible V value for *effective* single-mode propagation is that one can reduce sensitivity to microbending loss.

Several schemes for the measurement of the effective cut-off wavelength λ_{ce} have been proposed. They are all based on determining the wavelength dependence of a particular parameter, such as the modal spot size [19] the near or far field [20], the transmitted power [21], the refracted power [22] or bending loss [23]. Here we shall discuss in detail those methods to measure λ_{ce} based on spectral variation of the transmitted power through a bent fibre together with those exploiting spectral variation of the modal spot size.

Bending technique

This technique relies on the fact that a greater fraction of the modal power is contained in the cladding as the mode cut-off is approached. Above λ_c only the fundamental mode is present and, being far from its cut-off, is well confined. As λ is decreased, a second mode is excited whose field distribution extends

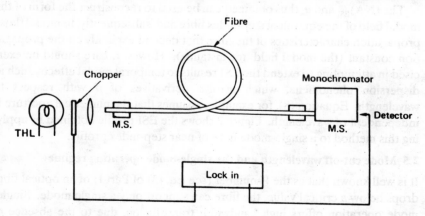

Fig. 5. Schematic of the experimental set-up for measurement of the effective cut-off wavelength by the bending technique.

well into the cladding. Here bending will result in high loss. As the wavelength is further decreased, this second mode becomes increasingly confined, and the loss progressively decreases. Thus there should be a loss peak corresponding to the cut-off of the mode. The cut-off wavelength will be close to that when the loss starts to increase as λ is decreased.

Figure 5 shows the experimental arrangement used in the measurement of λ_{ce} (the set up is the same as that used for the measurement of attenuation; Part I) [24, 25]. Light from a tungsten halogen lamp is passed through a chopper and focused by means of a microscope objective onto the input end of the fibre. The output end of the fibre is held at the entrance aperture of a monochromator whose output is fed to a photomultiplier or a cooled Ge detector, and thence to a low-noise amplifier. The signal from the detector is fed to the lock-in amplifier for phase-sensitive detection (the chopper providing the reference signal).

To measure loss due to bending, the spectral dependence of the transmission, $P_1(\lambda)$, is first measured on a ~2 m sample of the test fibre bent into a circle of radius 140 mm. Care is exercised to ensure that the fibre is straight elsewhere. Then without changing the input launch conditions, the fibre is looped again with a radius of curvature sufficiently small that the LP_{11} mode is filtered out. Typically this radius is 30 mm. The spectral transmission ($P_2(\lambda)$) is then measured again. A measure of the loss induced by bending is then calculated as follows:

$$R(\lambda) = 10 \log [P_1(\lambda)/P_2(\lambda)]. \tag{9}$$

Figures 6 (a) and 6 (b) show plots of $R(\lambda)$ against λ for a typical fibre (York Technology, U.K.) for two radii of curvatures [25]. As can be seen, the bend-induced loss is negligible at large wavelengths, but starts to increase as the wavelength approaches the cut-off value. According to EIA standards, the cut-off wavelength is defined as the wavelength at which the long-wavelength edge of the bend-induced attenuation rises by 0.1 dB from the long-wavelength baseline. As can be seen from Figs. 6(a) and (b) the cut-off wavelength as determined by this technique is almost independent of the bend radius. For too small a radius of curvature (Fig. 6(b)) the bend also induces loss in the LP_{01} mode at large wavelengths.

Another equivalent method relies on the fact that if one measures the transmission through a single-mode fibre from a spectrally flat source the transmitted power increases abruptly at the cut-off wavelength of the LP_{11} mode due to the extra power propagating through the second-order mode. Since practical sources are not spectrally flat over a broad range, in order to measure the cut-off wavelength using this technique, one has to compare the transmission spectrum $P_s(\lambda)$ of a single-mode fibre of about 2 m length bent

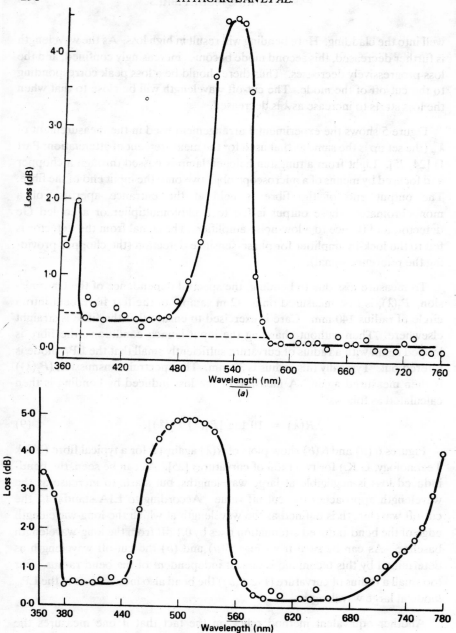

Fig. 6. Variation of the ratio $R(\lambda)$ given by Eq. (9) with λ for (a) bend radius of 30 mm and (b) bend radius of 11 mm for a single-mode fibre (York Technology, U.K.). Notice that for sufficiently small bend radius, corresponding to (b), the ratio $R(\lambda)$ again starts to rise due to the bend-induced loss of the LP_{01} mode [25]. From (a) we obtain $\lambda_{ce}(LP_{11}) = 576$ nm, $\lambda_{ce}(LP_{02}) = 376$ nm.

into a single arc of radius 140 mm with the transmission spectrum $P_m(\lambda)$ of a similarly prepared multimode fibre. Figure 7 is a plot of

$$R(\lambda) = 10 \log [P_s(\lambda)/P_m(\lambda)] \qquad (10)$$

for the same fibre as was used in plotting Fig. 4. In order to calculate λ_{ce} using this technique, we draw a line parallel to the long-wavelength branch of $R(\lambda)$ with a separation of 0.1 dB as shown. The value of λ at the point of intersection of this line with the $R(\lambda)$ curve gives the effective cut-off wavelength. Table 1 provides a comparison of the values of the cut-off wavelengths obtained from three different techniques applied to a single-mode fibre from York Technology (U.K.).

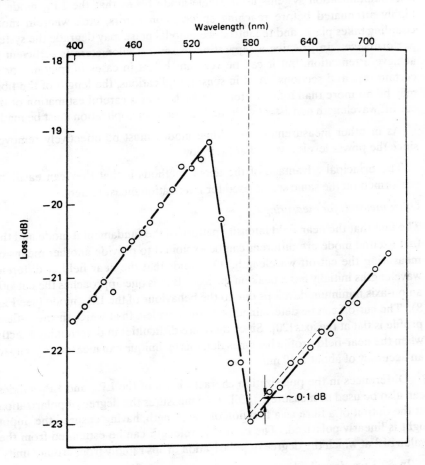

Fig. 7. Variation of $R(\lambda)$ given by Eq. (10) with λ using a multimode fibre reference [25]: bend radius 140 mm; $\lambda_{ce} = 578$ nm.

TABLE 1
Comparison of effective cut-off wavelengths measured using different techniques on a single-mode fibre (York Technology, U.K.) [25]

Method	λ_{ce} (nm)
Bending technique	
Single-mode reference	576 ± 4
Multimode reference	578 ± 4
Polarization technique	610

The effective cut-off wavelength determined by the above methods equates 'cut-off' with ~20 dB attenuation of the LP_{11} mode. It is very important in telecommunication systems using single-mode fibres that the LP_{11} mode be highly attenuated before reaching splices, connectors, etc., wherein mode coupling takes place, and the introduced modal noise may degrade the system performance. In practice a separation of several kilometers is sufficient to achieve attenuation, but it can be very much less in cases of 'pigtails' or in certain repaired sections. And in sensor applications, the length of the fibre may be no more than a few meters. In such cases a careful estimation of the cut-off wavelength consistent with the conditions of application must be made.

As in other measurements, cladding modes must be effectively removed since the power levels involved are very low.

The principal advantage of the above methods is that they can easily be performed on the same set-up used for attenuation measurement.

Other methods for measuring λ_{ce}

The fact that the near-field intensity pattern of the fundamental mode and the first excited mode are different can be exploited to provide another means of measuring the cut-off wavelength. One finds that the near field at different wavelengths initially has a peak on the axis. But as one approaches the cut-off, an on-axis minimum develops due to the behaviour of the LP_{11} mode (see Fig. 8). The cut-off can be determined from the wavelength at which the near-field profile is flat at the axis [20]. Since there are difficulties in determining exactly when the near-field profile becomes flat, this technique can measure λ_{ce} only to an accuracy of about ±10 nm.

Differences in the polarization characteristics of the LP_{01} and LP_{11} modes can also be used to measure λ_{ce} [26]. One measures the degree of polarization at the output of a fibre as a function of wavelength, having ensured the input light is linearly polarized. The cut-off wavelength can be estimated from the value of λ at which the degree of polarization drops rapidly from around unity.

In §2.3 we described a technique for the measurement of the mode field diameter (MFD) of a single-mode fibre from the measurement of the transmitted power variation across a joint. Measurement of the wavelength dependence of the MFD can in fact be used to characterize a fibre in terms of an

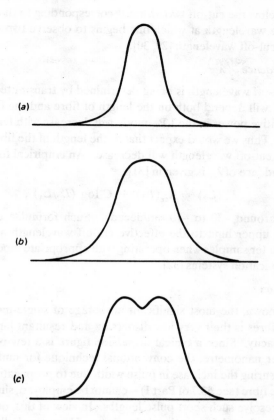

Fig. 8. Schematic representation of near-field pattern for (a) $\lambda > \lambda_{ce}$ (b) $\lambda \sim \lambda_{ce}$ and (c) $\lambda < \lambda_{ce}$, where λ_{ce} is the effective cut-off wavelength.

equivalent step-index fibre [17, 19]. It is well known that the transverse spatial extent of the LP_{11} mode is greater than the LP_{01} mode; thus the cut-off wavelength can also be estimated using the spectral variation of the MFD. In the single-mode regime, as the wavelength is reduced, the MFD decreases since the mode becomes more and more confined. As one approaches the cut-off of the LP_{11} mode, the MFD follows an upward trend as shown in Fig. 4(c). The point of intersection of the tangents to the decreasing and increasing portions of $W(\lambda)$ can be taken as the cut-off wavelength of the fibre. This method can also be used in estimating the cut-off wavelength of elliptic-core fibres [27, 28].

In §2.7 we describe a prism-coupling technique for the measurement of beat length in birefringent single-mode fibres. This can also be used to measure cut-off wavelength. Thus if one couples white light into the fibre and observes the output coupled light, then above the cut-off one would observe a single

m-line and below the cut-off two m-lines (corresponding to the LP_{01} and LP_{11} modes). The wavelength at which one begins to observe two m-lines can be taken as the cut-off wavelength [29, 30].

Length dependence of λ_{ce}

Since the cut-off wavelength is being determined by transmitted power measurements, λ_{ce} will depend both on the length of fibre and the fibre curvature, since the relative power in the LP_{11} mode will decrease with length, especially near cut-off. Thus we would expect that as the length of the fibre is increased, the effective cut-off wavelength will decrease. An empirical formula for such length dependence of λ_{ce} is given in [31]:

$$\lambda_{ce}(L_2) = \lambda_{ce}(L_1) + C \log (L_2/L_1), \qquad (11)$$

where C is around -50 to -70 nm/decade. Such formulae can be used to estimate the upper limit to the effective cut-off wavelength as measured on standard 2 meter samples when operating with appropriate modal noise penalty in communication systems [32].

2.6 Dispersion

As is well known, the most significant advantage of single-mode fibres over multimode fibres is their very low dispersion and resultant large information carrying capacity. Since a typical dispersion figure is a few picoseconds per kilometre per nanometre, the conventional technique for multimode fibre — directly measuring the increase in pulse width due to propagation in kilometre lengths of the fibre (see §3.7 of Part I) — cannot be employed, since photodetectors cannot resolve such short pulse lengths. In view of this, other techniques have been developed for single-mode fibres. Here we shall briefly discuss some of the more important of these.

Unlike multimode fibres, single-mode fibre does not suffer intermodal dispersion: the only mechanism at work is intramodal dispersion [33]. As discussed in detail in [33] intramodal dispersion consists essentially of two parts: the material dispersion and waveguide dispersion.

Material dispersion in silica-based fibres passes through a zero at about 1300 nm and so the waveguide dispersion can be tailored to counterbalance the effects of material dispersion to achieve zero dispersion anywhere within the range 1.3-1.6 μm. In principle there is another type of dispersion, profile dispersion, which depends on $d\Delta/d\lambda$, where Δ is the relative core-cladding index difference [34]. However, its contribution to total dispersion in a single-mode fibre is relatively insignificant. In a birefringent single-mode fibre the two orthogonally polarized LP_{01} modes have different group velocities, leading to what is known as polarization mode dispersion. This dispersion effect, however, is of little significance in conventional telecommunication-grade **single-mode** fibres.

Since single-mode fibres are operated at, or very close to, the zero-dispersion wavelength, its value (λ_o) together with the slope (S_o) of a plot of dispersion against wavelength can be used to characterise a fibre's dispersion.

The group velocity of a mode is given by $v_g = (d\beta/d\omega)^{-1}$ and hence the group delay time for propagation through a fibre of length L is simply

$$\tau(\lambda) = L \, d\beta/d\omega. \qquad (12)$$

Since dispersion is dependent on the variation of $\tau(\lambda)$ with λ, one can obtain the dispersion by directly measuring the propagation time through the fibre of short pulses at different wavelengths. The measurement technique consists essentially in propagating narrow pulses from a fibre Raman laser source which, in combination with a monochromator, can generate short pulses over a wide range of wavelengths (1.1 to 1.7 μm). The time delays through the test fibre are compared with those through a short reference fibre. Short pulses at discrete wavelengths can also be generated by using semiconductor lasers having different peak emission wavelengths. Although the latter cannot generate the whole delay spectrum, it is atleast portable (unlike the fibre Raman laser) and thus can be used to characterise installed fibre cables.

The measured $\tau(\lambda)$ against λ curve is fitted to an analytical function such as the three-term (S3) or five-term (S5) Sellmeier fit or polynomial fits (P3 or P5):

S3: $\quad \tau(\lambda) = A\lambda^{-2} + B + C\lambda^2,$ \qquad (13)

S5: $\quad \tau(\lambda) = A\lambda^{-4} + B\lambda^{-2} + C + D\lambda^2 + E\lambda^4,$ \qquad (14)

P3: $\quad \tau(\lambda) = A + B(\lambda - D)^2 + C(\lambda - D)^3,$ \qquad (15)

P5: $\quad \tau(\lambda) = A + B(\lambda - D) + C(\lambda - D)^2 + E(\lambda - D)^3.$ \qquad (16)

The coefficients $A, ..., E$ are determined by a least-squares fit between the measured data and the expansion formula. Having obtained the coefficients, we can easily calculate λ_o and S_o. It has recently been found that although the three-term Sellmeier fit can be used to fit the group time delay curve of matched-clad single-mode fibre designs that operate at around 1.3 μm, for dispersion-shifted and dispersion-flattened fibre designs, the five-term Sellmeier fit is much more accurate [35, 36].

The variation of $\tau(\lambda)$ with λ can also be obtained by measuring the phase shift suffered in propagation through the fibre of light from a sinusoidally modulated source as a function of wavelength [37, 39]. Thus, if f is the source modulation frequency the phase shift suffered by the wave in propagating through the fibre is $\phi(\lambda) = 2\pi f \tau(\lambda)$. Hence a measurement of $\phi(\lambda)$ essentially gives us $\tau(\lambda)$. In the experiment, an LED is modulated at a frequency of about 30 MHz and the output detected by an avalanche photodetector which is connected to a vector voltmeter. The vector voltmeter measures the relative phase of the output with respect to the electrically modulated input. The wavelength can be varied over a suitable range by means of a monochromator. Measurement of $\phi(\lambda)$ gives $\tau(\lambda)$ from which λ_o and S_o can be obtained, as discussed earlier.

Dispersion in a single-mode fibre of only a meter or so can also be measured by an interferometric technique [40-42]. In this technique, light from the source is divided into two parts: one propagates through the fibre; the other through a variable delay line. The output of the fibre is made to interfere with the output from the variable delay line. By a suitable data processing, one can measure dispersion in the fibre.

The total chromatic dispersion is usually written as a sum of material and waveguide dispersion, although the two mechanisms are in fact intricately related to one other. For example, the material dispersion is itself a composite, depending on both core and cladding indices. It also changes with spot size, which changes the fraction of the modal power propagating in the core. However, after a careful study [43] it has been shown that a good estimate of the total chromatic dispersion can usually be obtained by separately estimating material and waveguide dispersion and then simply summing them. Since waveguide dispersion is governed by variation of the propagation constant with V value, one can indirectly measure waveguide dispersion by measuring the wavelength dependence of the modal spot size. Indeed if W_p represents the Petermann spot size, defined by

$$W_p^2 = 2\int \psi^2 r\, dr \Big/ \int (d\psi/dr)^2 r\, dr, \qquad (17)$$

where $\psi(r)$ is the fundamental-mode field distribution, then the waveguide dispersion is given by [44-46]

$$\frac{d\tau}{d\lambda} = \frac{\lambda}{2\pi^2 c n_1} \frac{d}{d\lambda}\left(\frac{\lambda}{W_p^2}\right). \qquad (18)$$

Thus measurement of the wavelength variation of W_p^2 can yield the modal dispersion. If one can suitably approximate material-dispersion effects, then the total dispersion can also be estimated.

2.7 Birefringence measurement

A perfectly circular-core single-mode fibre actually supports two modes having orthogonal polarization but the same propagation constant. In practice, single-mode fibre cores are not perfectly circular, and, with the presence of stress, twists, bends etc., such fibres are slightly birefringent. This small difference in propagation constants leads to a coupling of power between the two modes even under the smallest external perturbation. So for nominally circular core fibres it is found that a linearly polarized light quickly becomes elliptically polarized after propagating over a short length of fibre. In addition, this state of polarization at the fibre output is sensitive to temperature, wavelength or external perturbation. For applications of single-mode fibres in coherent communications, fibre optic phase sensors or in applications wherein the fibre is coupled to polarization-sensitive devices, such as integrated optic devices, this effect causes problems. In order to cater for such applications, new fibre designs have been developed to achieve large linear birefringence so that small perturbations will not couple power between the two polarization states and incident light polarized along one of the birefringent axes of the fibre remains in that state of polarization as it propagates though the fibre (see Ch. 6 in this book).

The birefringence in the fibre is a measure of the difference in the effective indices n_e of the two orthogonal modes and is defined by

$$B = \delta n_e = \delta(\beta/k_o) = (\lambda_o/2\pi)\delta\beta, \qquad (19)$$

where $\delta\beta$ is the difference in the two propagation constants and λ_o is the free-space wavelength. If linearly polarized light is launched into the fibre with its polarization direction at 45° to the birefringent axes of the fibre, then the difference in propagation constant of the two modes yields a difference in phase of the two components with propagation. Whenever the phase difference is an integral multiple of π, the state of polarization is linear and whenever it is an odd integral multiple of $\pi/2$, it is circular. Thus the light beam, as it propagates through the fibre alternately becomes circularly then linearly polarized (see Fig. 9). The minimum length of the fibre over which the state of polarization comes back to the original state is called the 'beat length' and is given by

$$L_b = 2\pi/\delta\beta = \lambda_o/B. \qquad (20)$$

At distances $L_b/2$, $3 L_b/2$, etc., the state of polarization is linear but perpendicular to the initial state.

Thus the beat length is a good measure of the birefringence of the fibre. In order to maintain the state of polarization, the fibre should have a large modal birefringence B. So such fibres require small beat length L_b (~2-5 mm).

Hence the polarization-maintaining capability of a birefringent fibre can be described in terms of its modal birefringence B or the beat length L_b. We will now describe three different techniques that have been used in our Fibre Optics Laboratory for the measurement of L_b, and one technique that has very recently been developed for direct measurement of $\delta\beta$.

Rayleigh scattering technique

Rayleigh scattering is a fundamental mechanism which scatters light out of the fibre. It is well known that for a linearly polarized light undergoing scattering, there is no scattered wave along the axis of the dipole, i.e. along the direction of polarization, and the scattering is maximal in a direction perpendicular to the dipole axis. This effect too can be used in obtaining the beat length of fibres.

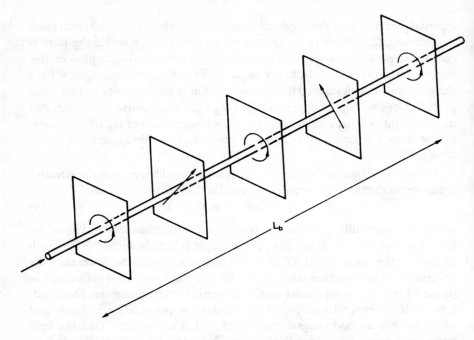

Fig. 9. If circularly polarized light or linearly polarized light with its polarization direction at 45° to the birefringent axes is incident on a birefringent fibre, then as it propagates, it changes its state of polarization periodically. The distance between two points where the state of polarization repeats itself defines the beat length L_b.

Let us consider a circularly polarized light beam launched into a birefringent fibre. Since the eigenpolarizations are linear along two orthogonal directions (say, x and y directions) the incident light beam excites both the eigenpolarizations with equal amplitudes. Since the propagation constants of the two modes are different, the x- and y-polarizations develop a phase difference and the state of polarization changes as shown in Fig. 10(a) (bottom). Now, if one observes the scattered radiation escaping from the fibre in a direction orthogonal to the fibre axis and along a direction making an angle of 45° with the eigen-axes, then alternate bright and dark bands will be seen along the length of the fibre [Fig. 10(a) (top)] [47, 48]. This arises because whenever the state of polarization is linear and along the direction of observation there is no scattering, and whenever it is linear and perpendicular to the observation direction, the scattered light intensity is a maximum. A measurement of the separation of adjacent bright and dark regions gives us the beat length at that wavelength directly.

The experimental set-up is very simple and is shown in Fig. 10 (b). The $\lambda/4$ plate is used to select circularly polarized input. One could equally well have chosen an incident linearly polarized beam polarized at 45° to the axes. The latter would of course require alignment of the fibre axes with respect to the

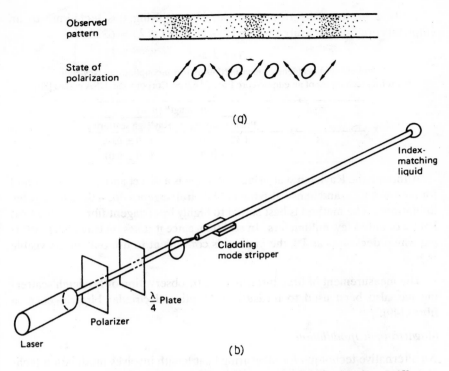

Fig. 10. (a) Schematic diagram of the observed beat pattern along the direction at 45° to the eigen-axes. (b) Experimental set-up for the measurement of beat length using the Rayleigh scattering technique.

incident polarization direction. As the scattering is usually very weak, it is essential that maximum possible light intensity is coupled into the fibre, and the cladding modes must be removed in order to obtain good contrast. In addition, since a fraction of the light may be coupled back into the fibre by reflection at the fibre end, it too will produce a beat pattern (although of very much lower intensity). This pattern, in general, will not be in phase with the other beat pattern, and the output end of the fibre may be dipped into an index-matching liquid to reduce the end reflection itself.

The following two observations will confirm that the pattern that one observes is due to the change in the state of polarization along the fibre:

(i) If one looks along either of the eigen-polarization directions, no beat pattern will be observed, since the component of the propagating beam involved in the scattering is always of the same amplitude.

(ii) If one couples light linearly polarized along either of the eigen-polarization directions, then there will again be no beat pattern, irrespective of the direction of observation.

Table 2 presents the results obtained in applying this technique to an elliptical core fibre measured with a He−Ne laser at $\lambda_o = 6328$ Å.

TABLE 2

Comparison of beat lengths measured using prism-coupling and Rayleigh scattering techniques of an elliptic-core fibre (Andrew Corporation, USA make) [30]

Trial	Beat length (mm)	
	Prism coupling	Rayleigh scattering
1	1.52 ± 0.08	1.50 ± 0.01
2	1.50 ± 0.08	1.50 ± 0.01

Although the Rayleigh scattering technique is a direct and accurate method for measuring L_b, and hence also the modal birefringence $\delta\beta$, it does have some limitations. The method is best suited to highly birefringent fibres with a beat length of only a few millimeters. In addition, since it relies on Rayleigh scattering which decreases as λ^{-4}, the method is convenient to use only in the visible region.

The measurement of fibre birefringence by observation of Rayleigh scattering has also been used to measure twist-induced circular birefringence in fibres [49].

Magneto-optic modulation

An alternative technique for measuring beat length involves modulation techniques whereby the birefringence of the fibre is modulated locally at a point along its length and the resultant change in the state of polarization of the output light is monitored. One of the simplest techniques to change locally the birefringence is through the introduction of circular birefringence via the Faraday effect [50-51]. Thus if the fibre is passed through a small electromagnet which provides a magnetic field along the axis of the fibre, then at the point of application of the magnetic field, the Faraday effect would lead to an externally induced circular birefringence. The effect of the induced circular birefringence on the polarization depends on the state of polarization of the propagating beam in the fibre. If the polarization is circular, then the induced circular birefringence has no effect. But if it is linear it produces maximum effect. Thus if the electromagnet is moved along the length of the fibre and the modulation in the state of polarization is measured at the output end of the fibre, one observes a periodic variation in the depth of modulation. The separation of two adjacent positions of the electromagnet at which one observes maximum and minimum modulation corresponds to half a beat length. The experimental arrangement used to measure the beat length by means of magneto-optic modulation is shown in Fig. 11(*a*). Light from a polarized He−Ne laser is coupled into the single-mode fibre; the polarization of the laser is chosen to be at 45° to the eigen-axes of the fibre (or equivalently one may choose the input state of polarization to be circular). A microscope objective collimates the output light and the beam is passed through a $\lambda/4$ plate followed by a Soleil-Babinet compensator (SBC) and a linear analyser (see Fig. 11(a)). The $\lambda/4$ plate is so adjusted that in the absence of a magnetic field, the polarization of the output light from the fibre is made linear. The SBC is then mounted with its slow axis aligned with the linear state of polarization emerging from the $\lambda/4$ plate and the linear analyzer is aligned at 45° to the slow axis of the SBC. Such an arrangement results in an analyzer which is most sensitive to changes in the state of polarization. The phase retardation of the SBC is adjusted for every position of the solenoid so as to maximise the modulation signal, which is then detected using a lock-in amplifier.

Figure 11(b) shows the variation of the amplitude of the modulation signal as a function of the position of the solenoid for a linearly and circularly polarized input states. The sinusoidal modulation is very clear, as too is the phase shift of half a period between linear and circular inputs. The beat length of the fibre is twice the period, which in the present case is about 41.5 cm.

For good resolution, one must have an electromagnet capable of generating a strong magnetic field over a short distance. For beat lengths less than 1 mm, it may be difficult to design a magnet to satisfy the above requirement. The magneto-optical method cannot be used to measure the circular birefringence of a fibre. Here one may employ electro-optic effects through application of high electric fields or by applying transverse stress which will induce linear birefringence.

Prism-coupling technique

The methods discussed above are based on the measurement of the beat length in estimating modal birefringence. Recently a novel technique has been demonstrated wherein one can measure the modal birefringence directly [30]. The method proposed is the same as the conventional prism-film coupling technique used in integrated optics in which a prism whose refractive index is greater than the maximum refractive index of the guiding structure is positioned close to the waveguiding structure. Since such a structure is a leaky waveguide, light propagating in the film enters the prism. Owing to the phase-matching condition, the angle at which the beam emerges from the prism is representative of the value of the propagation constant of the propagating mode. In a normal cladded fibre, the modal field is not accessible from the outside (due to the thick cladding). However, if most of the cladding is removed by polishing, it is then possible to couple light out of the fibre using the prism-coupling arrangement shown in Fig. 12. Since in a birefringent fibre, the two fundamental orthogonally polarized modes have different propagation constants, they couple out along different directions, and one can then calculate the modal birefringence through a measurement of the corresponding propagation constants.

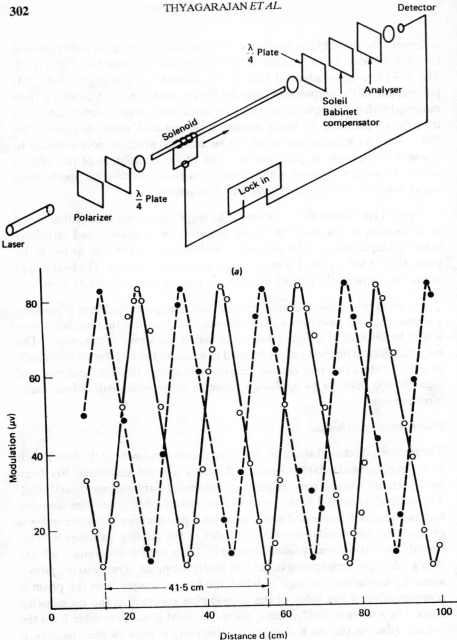

Fig. 11. (a) Experimental set-up for measurement of beat length using magneto-optical modulation. (b) Variation of the amplitude of the modulation signal with position of the solenoid along the length of the fibre. Maximum modulation signal corresponds to a linear state of polarization at the position of the solenoid [51]. ○ Input linearly polarized. ● input circularly polarized.

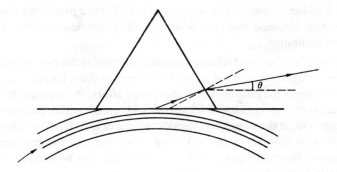

Fig. 12. Experimental arrangement for measurement of propagation constant and modal birefringence in a single-mode fibre [30].

Figure 13 shows the pattern of the output from a prism placed on a polished elliptic-core fibre (850417 A; Andrew Corporation, USA) with its major axis perpendicular to the polished surface. The two lines are orthogonally

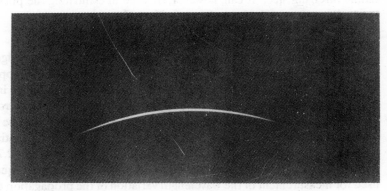

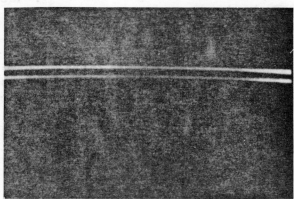

Fig. 13. Output 'm-lines' at a distance of (a) ~15 cm and (b) ~2 m from the polished semiplate. The two 'm-lines' in (b) are orthogonally polarized [30].

polarized and are separated by an angle of ~4'. Table 2 provides a comparison of the results obtained from the prism-coupling technique and the Rayleigh-scattering technique.

Since the above method relies on an ability to resolve the two output coupled lines, it is used only in cases of a high birefringence (beat length ≤ a few mm). In addition, since the polishing operation itself affects the waveguide structure somewhat, sufficient cladding thickness should remain so that this perturbation is not large and, at the same time, the decoupling of light from the fibre is good. This technique has also been used in our laboratory for the measurement of birefringence in stress birefringent fibres and can also be used in estimating polishing-induced changes in birefringence of fibres.

Other techniques are available, like the cut-back method or the wavelength-sweeping technique [52, 53] which have been suggested for measurement of fibre birefringence characteristics. The cut-back technique relies on the measurement of the state of polarization at the end of the fibre which is repeatedly cut to observe the evolution of the state of polarization along its length. The wavelength-sweeping technique requires knowledge of the dispersion of the modal birefringence. It can, however, be used in estimating the polarization-mode dispersion in highly birefringent fibres [54].

2.8 Measurement of the propagation constant of the fibre mode

There are a large number of applications that require knowledge of the propagation constant of the propagating mode. For example, in applications requiring phase-matched coupling using periodic structures [55] or in obtaining the characteristics of non-identical fibre directional-coupler filters [56, 57]. The prism-coupling technique described in the above section can be used for the direct measurement of the propagation constant of the propagating mode [29, 30]. A detailed description of the technique was given in §2.7 for the measurement of birefringence in a single-mode fibre. The technique can be used for measurement of the wavelength variation of the propagation constant, the cut-off wavelength and so on.

3. CONCLUSION

In chapters 10 and 11 we have presented a review of the various important and standard characterisation techniques for both multimode and single-mode fibres. Here in chapter 11 we focused on the precise measurement of

    (i)   attenuation,
    (ii) refractive index profile,
    (iii) mode field diameter,
    (iv) cut-off wavelength,
    (v) dispersion, and
    (vi) birefringence,

all of which are important for the characterisation of single-mode fibres. From the panoply of existing characterisation techniques, details of the more standard ones have been presented, especially those techniques suited to applications of optical fibres in telecommunications and sensors. It is hoped that this review will be useful in building up of R&D facilities in fibre characterisation and also for developing a university-level curriculum in the area of fibre optics.

ACKNOWLEDGMENTS

We are grateful to Professor M.S. Sodha and Professor A.K. Ghatak for their constant encouragement and interest. In fact, much of this review resulted from the work carried out as a team during the development of the Fibre Optics Laboratory at IIT, Delhi since 1980 under the overall leadership of Professor A.K. Ghatak who was instrumental in the generation of team spirit. We are indeed fortunate to be associated with this wonderful team of which special mention must be made of our colleagues: Professor I.C. Goyal, Dr. A. Sharma, Dr. A.D. Sharma, Dr. M.R. Shenoy, Mr. M.R. Ramadas and Mr. Verghese Paulose, who have made important contributions to this activity.

Many of the facilities providing the infrastructure of our laboratory have received active support from funding agencies like Department of Science and Technology, Department of Electronics, Council of Scientific and Industrial Research, Telecommunications Research Centre, and National Institute of Standards and Technology (formerly NBS), USA, through sponsored projects besides support complemented through the R&D programmes of our Institute namely the program on Optical Communication Engineering and the Laser Applications Program.

REFERENCES

1. B.P. Pal, "Fundamental Characteristics of Telecommunication Optional Fibers: A Review", IETE (India) Tech. Rev. Vol. 3, p-347, 1986.
2. V.A. Bhagavatula, Berkey, G.D., and Sarkar, A., Proc. OFC 1983, New Orleans, p. 22, 1983.
3. D.L. Franzen, "A Short Course on Single-mode Fibre Measurements", presented at OFC '90, San Francisco, Calif. (1990).
4. D. Marcuse, "Principles of Optical Fibre Measurement", (New York: Academic Press), 1979.
5. T.V.B. Subrahmanyam, "Refracted Near Field Scanning Technique for Refractive Index Profiling of optical fibres", M. Tech. (Applied Optics) Thesis, IIT Delhi, 1983.
6. P. Di Vita, and G. Coppa, "Single Mode Fibre Measurements and Standardization", J. Inst. Electron. Telecommun. Engrs. (India) Vol. 32, p-227, 1986.
7. D.L. Franzen, and R. Srivastava, "Determining the Mode-Field Diameter of A Single-mode Optical Fibre: An Inter laboratory Comparision", J. Lightwave Technol., Vol 3, p-1073, 1985.

8. R. Tewari, "Propagation Effects in Single mode Optical Fibres and Couplers", Ph.D. Thesis, IIT, Delhi, 1985.
9. J.M. Dick, R.A. Modavis, J.G. Rachi, and W.A. Westuig, "Automated Mode Radius Measurement Using the Variable Aperture Method in the Far Field," Tech. Digest OFC '84, New Orleans (1984).
10. W.G. Otter, F.P. Kapron. and T.C. Olson, "Mode-Field Diameter of Single-mode Fibre by Knife Edge Scanning in the Far-Field", IEEE J. Lightwave Tech. Vol. **LT-4** pp. 1576-1579 (1986).
11. R. Srivastava and D.L. Franzen, "Signle-mode Optical Fibre Characterisation", NBS (Boulder), Colorado, Internal Report (July, 1985).
12. Y. Gogia, "Mode Field Radius Measurement in the Far-Field by Variable Aperture in the Far-Field", M. Tech. (Opto-electronics and Optical Communication) Thesis, IIT Delhi, 1986.
13. A.K. Ghatak, T.T. Tjung and E.K. Sharma, "Accurate emperical formulae on the estimation of the modal field from far-field intensity distributions in single-mode fibres", J. Opt. Comm. 6, p-147, 1985.
14. B.P. Pal, V. Priye, R. Tewari, and K. Thyagarajan "Estimation of Spot Sizes from Far-Field Measurement of a Single-Mode Fibre", J. Opt. Comm. Vol. **8**, p-70, 1989.
15. V. Priye, Y. Gogia, and B.P. Pal, "Simple Technique to Estimate Spot Sizes From Far-Fields of Single-mode Fibres", Electron. Letts. Vol. **24**, pp. 1400-1401 1988.
16. M. Fox, "Calculation of Equivalent Step-Index parameters For Single-mode Fibres", Opt. Quant. Electron. Vol.**15**, p. 451, 1983.
17. T.V. Mahadevan, "Determinaton of Equivalent Step Index parameters of Single-Mode Fibers", M.Sc. (Physics) Thesis, IIT, Delhi, 1985.
18. D. Marcuse, "Light Transmission Optics", (New York: Van Nostrand Reinhold), 1982.
19. C.A. Miller, "Direct Method of Determining Equivalent Step Index Profiles for Monomode Fibre", Electron. Lett. Vol. **17**, pp. 458-460, 1981.
20. Y. Murakami, A.K. Kawana, and H. Tauchiya, "Cut-off Wavelength Measurements For Single-mode Optical Fibres", Appl. Optics Vol. **18**, p. 1101, 1979
21. R. Worthington, J. Phys. E, Vol. **4**, p. 1052, 1971.
22. V.A. Bhagavatula, W.F. Love, D.B. Kech, and R.A. West, "Refracted Power Technique for Cut-off Wavelength Measurement in Single-mode Waveguides", Electron. Lett. Vol. **16**, p. 695, 1980.
23. Y. Katsuyama, M. Tokuda, N. Uchida, and M. Nakahara, "New Method For Measuring V-value of A Single Mode Optical Fibre", Electron. Lett., Vol. **12**, P. 699, 1976.
24. C.G. Ramadevi, "Cut-off Wavelength Measurement of Single-mode fibres: Comparison of Various Standard Techniques", M. Tech. (Opto-Electronics and Optical Communication) Thesis, IIT, Delhi,, 1986.
26. G. Coppa and P. Di Vita, "Polarisation Measurement of Cut-off Wavelength in Monomode Fibres", Proc. ECOC'9, Geneva, pp. 193-196, 1983.
27. A.V. Belov, E.M. Dianov, A.S. Kurkov, V.B. Neu Struev, and A.V. Chicolini, "Equivalent Step Index (ESI) Profile of Elliptical Core Single-mode Fibres", Op. Commun. Vol. **56**, p. 93, 1985.
28. K. Thyagarajan, S.N. Sarkar, and B.P. Pal, "Equivalent Step Indnex (ESI) Model for Elliptical Core Fibres", J. Lightwave Technol, Vol. **5**, p. 1041, 1987.
29. W.V. Sorin, B.Y. Kim, and H.J. Shaw, "Phase Velocity Measurement Using Prism-output coupling for Single and Few Mode Optical Fibres", Optics, Lett. Vol. **11**, p.106, 1986.

30. K. Thyagarajan, M.R. Shenoy, and M.R. Ramdas, "Prism Coupling Technique: A Method for Measurement of Propagation Constants and Beat Length in Single-Mode Fibres," Electron. Lett. Vol. 22, p. 832, 1986.
31. W.T. Anderrson and T.A.Lenahan, "Length Dependence of the Effective Cut-off Wavelength in Single Mode Fibres", J. Lightwave Technol., Vol. 2, p. 238, 1984.
32. L. Wei, C. Saravanos, and R.S. Lowe, "Practical Upper Limits to Cut-off Wavelength for Different Single Mode Fibre Designs", Technical digest, NBS Symposium on Optical Fibre Measurements, Boulder, Colorado, p. 121, 1986.
33. I.C. Goyal, "Dispersion in Telecommunication Optical Fibres: A Tutorial Review", J. Inst. electron Telecommun. Engrs. (India), Vol. 32, p. 196, 1986.
34. W.A. Gambling, H. Matsumura, and C.M. Ragdale, "Mode Dispersion, Material dispersion and Waveguide Dispersion in Graded Index and Single Mode fibres", IEE Microwaves, Opt Acoust, Vol. 3, p. 239, 1979.
35. W.T. Anderson, "Status of Single-mode fibre Measurement", Technical digest, NBS Symposium on Optical Fibre Measurements, Boulder, Colorado, p. 1, 1986.
36. W.A. Read and D.L. Philen, "Study of Algorithms Used to Fit Group Delay Data for Single Mode Optical Fibres", Technical Digest, NBS Symposium on Optical Fibre Measurements, Boulder, Colorado, p. 7, 1986.
37. P.J. Vella, P.M. Garel-Jones, and R.S. Lowe, "Measurement of Chromatic Dispersion of Long Spans of Single Mode Fibre: A Factory and Field Test Method", Electron. Lett. Vol. 20, p. 167, 1984.
38. B. Costa, M. Puleo, and E. Vezzoni, "Phase Shift Technique For the Measurement of Chromatic Disperison in Single Mode Optical Fibres Using LED'S", Electron. Lett., Vol. 19, p. 1074, 1983.
39. M. Fujise, M. Kuwazaru, N. Nunokawa, and Y. Iwamoto, "Chromatic Dispersion Measurement Over a 100 km Dispersion Shifted Single Mode Fibre by A New Phase Shift Technique", Electron. Lett. Vol. 22, p. 570, 1986.
40. M. Tateda, N. Shubata, and S. Sakari, "Interferometric Method for Chromatic Dispersion Measurement in a Single Mode Optical Fibre", IEEE J. Quant. Electron., Vol. 17, p. 404, 1981.
41. H. Shang, "Chromatic Dispersion Measurement by White Light Interferometry on Meter-Length Singlemode Optical Fibres", Electron. Lett. Vol. 17, p. 603, 1981.
42. J. Stone and L.G. Cohen, "Minimum Dispersion Spectra of Single Mode Fibres Measured With Subpicosecond Resolution by White Light Cross Corelation", Electron. Lett., Vol. 18, p. 716, 1982.
43. D. Marcuse, "Interdependence of Waveguide and Material Dispersion", Appl. Optics Vol. 18, p. 2930, 1979.
44. K. Petermann, "Constraints For Fundamental Mode Spot Size For Broadband Dispersion — Compensated Single Mode Fibres", Electron. Lett. Vol. 19, p. 712, 1983.
45. C. Pask, "Physical Interpretation of Petermann's Strange Spot Size For Single Mode Fibres", Electron. Lett. Vol. 20, p. 144, 1984.
46. C.D. Hussey and F. Martinez, "Approximate Analytic Forms For The Propagation Characteristics of Single Mode Optical Fibres", Electron. Lett., Vol. 21, p. 1103, 1985.
47. A. Papp and H. Harms, "Polarization Optics of Index Gradient Optical Waveguide Fibres", Appl. Optics Vol. 14, p. 2406, 1975.
48. L.B. Jeunhomme, "Single Mode Fiber Optics" (New York: Marcel Dekker), 1983.
49. P. Graindorge, K. Thyagarajan, H. Arditty, and M. Papuchon, "Scattered Light Measurement of The Circular Birefringence In A Twisted Single Mode Fibre", Opt. Commun. Vol. 41, p. 164, 1982.

50. A. Simon and R. Ultrich, "Evolution of Polarization Along A Single Mode Fibre", Appl. Phys. Lett. Vol. **31**, p. 517, 1977.
51. A.K. Bhat, "Measurement of Beat Length of Single Mode Fibres Using Magneto Optic Modulation", M. Tech. (Applied Optics) Thesis, IIT Delhi, 1984.
52. P.V. Leela, "Beat Length Measurement of Highly Birefringent single Mode Fibres", M. Sc. (Physics) Thesis, IIT Delhi, 1986.
53. S.C. Rashleigh, "Measurement of Fibre Birefringence by Wavelength Scanning Effect of Dispersion", Optics Lett. Vol. **8**, p. 336, 1983.
54. G.W. Day, "Birefringence Measurements in Single Mode Optical Fibre", Proc. SPIE, Vol. **425** (Bellingham: SPIE), 1983.
55. J.N. Blake, B.Y. Kim, and H.J. Shaw, "Fibre Optic Modal Coupler Using Periodic Micro-bending", Optics Lett., Vol. **11**, p. 177, 1986.
56. D. Marcuse, "Directional Coupler Filter Using Dissimilar Optical Fibres", Electron. Lett., Vol. **21**, p. 726, 1985.
57. R. Tewari, K. Thyagarajan, and A.K. Ghatak, "A Novel Method For Characterisation of single Mode fibres and Prediction of Cross Over Wavelength and Bandpass in Non-identical Fibre Directional Couplers", Electron. Lett. Vol. **22**, p. 792, 1986.

12

Optical Fibre Cables

A. HORDVIK*

---

1. INTRODUCTION

The optical fibre is vulnerable to mechanical and environmental stresses. The purpose of the cable structure is to provide protection for the fibre while at the same time preserve its optical properties. In addition, a cable should meet several practical requirements. It should be easy to install, and the design should facilitate fibre splicing and the attachment of connectors.

As fibre optics has found wider and wider use, a great variety of cable types have been developed, each suited for certain applications. In this article we will, however, mainly limit the discussion to topics which are of general interest to the design of fibre optic cables. Emphasis will be placed on discussing the optical core of the cable and the various core designs which have been developed.

The optical property which can be most affected by the cable is attenuation, and the various mechanisms which can induce losses will be discussed in Section 2. Properties like bandwidth and numerical aperture are related to the fibre design, and the effect of mechanical and environmental conditions on these properties can usually be regarded as negligible, and will thus not be considered any further.

The fibre is subjected to mechanical stresses during manufacturing, installation and operation of the cable. It is of course essential that the fibre should withstand these forces. The mechanical properties of the fibre will be discussed in Section 3. This is a large topic in itself, and the discussion will of necessity be somewhat brief and simplified.

---

* Optolan as, Brøsetveien 168, Moholtan, N-7002 Trondheim, Norway.

Various cable core designs will be discussed in Section 4, and examples of various cable types will be briefly presented in Section 5.

For a more detailed and thorough description of fibre optic cables including cables for various applications and a discussion of the manufacturing process, the interested reader is referred to recent books on the subject [1, 2].

2. LOSS MECHANISMS IN FIBRE OPTIC CABLES

In a cable, a fibre can suffer bends due to the cable design itself as well as external factors. One usually distinguishes between macrobends and microbends as discussed in a companion chapter (chap. 4) and both types of bends can cause losses.

Although quartz is a very stable material, there are two ageing phenomena which may increase the fibre attenuation. Hydrogen will, if present, diffuse into the fibre core resulting in a wavelength dependent loss, and ionizing radiation can create absorbing colour centers in the fibre.

In this chapter, we will discuss the bend loss mechanisms and the hydrogen induced losses. The effect of ionizing radiation on the fibre attenuation can usually be neglected provided the fibre is only subjected to the natural background radiation. Further, since this effect is really not much related to the cable design, it will not be discussed any further. Readers with special interest in the topic are referred to ref. [3] and the cross-references therein.

2.1 Macro-and microbending losses

By microbends are meant microscopic perturbations of the fibre lay with spatial wavelengths of the order of a millimeter. An example would be the bends suffered by a fibre pressed against a rough surface.

The term macrobends is used when the fibre follows a curved path. The radius of curvature will normally change slowly along the fibre. A somewhat special example would be a length of the fibre wound into a coil.

It can sometimes be difficult to clearly designate a specific bending pattern to one or the other category, and a fibre might well suffer perturbations which are a mixture of the two. Still, it has been found practical to define this distinction.

Macro-and microbending losses affect single-mode fibres and multimode fibres somewhat differently. It is therefore, convenient to treat the two fibre types separately.

2.1.1 Macrobending losses in single-mode fibres

A fibre's sensitivity to macrobending losses is well described by considering its losses when bent into a constant radius of curvature. In a curved fibre, the phase velocity will increase as one moves away from the center of curvature. At a sufficiently large distance, the velocity would have to exceed the speed of light, and the power outside this position will radiate away.

Marcuse [4] has derived an expression for the loss increase experienced by a single-mode fibre with a constant bend radius. The fibre has a step-index profile and matched cladding. The refractive index in the core is n_1 and in the cladding n_2, and it is assumed that $(n_1-n_2) \ll n_1$.

The bend induced loss coefficient α is then given by

$$\alpha = \frac{\sqrt{\pi}\,(U)^2 \exp\left[-\dfrac{4(W)^3 \cdot \Delta \cdot R}{3\,aV^2}\right]}{2(W)^{3/2} V^2 \sqrt{(Ra)}\,[K_1(W)]^2}, \tag{1}$$

where

$U^2 = a^2(n_1^2 k^2 - \beta^2)$,

$W^2 = a^2(\beta^2 - n_2^2 k^2)$,

$k = \dfrac{2\pi}{\lambda}$, where λ is the wavelength,

β is the propagation constant of the guided LP_{01}-mode in the straight fibre,

$V^2 = U^2 + W^2 = k^2 a^2 (n_1^2 - n_2^2)$,

$\Delta = (n_1 - n_2)/n_1$

a = core radius,

R = radius of curvature of the bent fibre, and

$K_1(x)$ is the modified Bessel function.

In Fig. 1 is shown the calculated loss increase versus bend radius assuming the operating wavelengths to be 1300 nm and 1550 nm for a typical matched clad single-mode fibre. As can be seen, the variation in loss with radius is approximately exponential as expected from the form of Eq. (1), and the loss also varies strongly with the wavelength.

From curves similar to those in Fig. 1 it is possible to determine the minimum permissible bend radius to avoid a certain excess loss at a specific wavelength of operation.

It should be noted that for fibres with depressed cladding, the variation in bend loss with radius of curvature is somewhat different [5].

2.1.2 Macrobending losses in multimode fibres

In a multimode fibre, the bend loss at a certain radius is different for the various modes. This means that the energy distribution between the various modes will vary as the signal propagates along the curved fibre, and the loss coefficient will be length dependent.

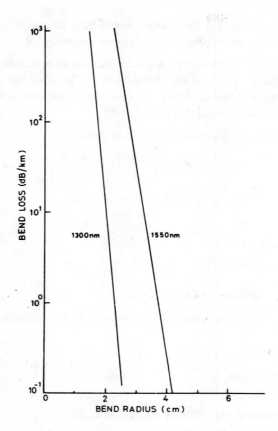

Fig. 1. Bend loss as a function of bend radius for a matched clad single-mode fibre. Core radius a is 4.3 μm, and $\Delta = (n_1-n_2)/n_1 = 3.0 \times 10^{-3}$.

The results of a bend loss experiment made on a standard 50 μm core, NA = 0.2 graded index multimode fibre is shown in Fig. 2 [6]. Depicted is the increase in bend loss per turn (in dB) as a function of bend diameter and with the number of turns N as a parameter. The figure demonstrates clearly that the loss increase per turn decreases as the number of turn increases.

It is in principle possible to calculate the bend loss in a multimode fibre, but it is not very practical. Instead, a different approach has been taken by Marcuse et al. [7]. They show that the effective number of modes N_{eff} which is supported by a curved parabolically graded index multimode fibre is given by

$$N_{\text{eff}} = N_\infty \left[1 - \frac{1}{\Delta} \left(\frac{2a}{R} + \left(\frac{3}{2n_1 kR} \right)^{2/3} \right) \right] \quad (2)$$

Here N_∞ is the total number of modes in a straight fibre (i.e. $R = \infty$) and the other variables are defined in 2.1.1.

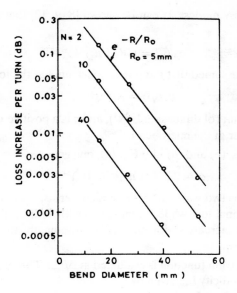

Fig. 2. Increase in bend loss per turn (dB) as a function of bend diameter. N is number of turns.

From this expression, one can calculate the minimum bend radius which allows a certain percentage of the modes to effectively propagate. If as an example, we assume that we want $N_{eff} > 0.9 N_\infty$, then for a $2a = 50 \mu m$, $NA = 0.2$ graded index fibre, the radius of curvature R must be larger than 55 mm.

2.1.3 Microbending losses in single-mode fibres

When a fibre is subjected to microbends, the field distribution in the fibre will constantly have to change to fit the changing fibre path. This results in a loss of power. Microbending losses in single-mode fibres have been discussed quite extensively in the literature. We will here limit the discussion to the treatment given by Petermann [8, 9] for fibres with a step-index profile. Fibres with more complicated profiles have been discussed by Petermann and Kühne [10].

Let us assume that the distribution of the electric field $E(r)$ is Gaussian

$$E(r) = E(0) \exp(-r^2/2 W_0^2), \quad (3)$$

where r is the radial position and W_0 is the radius at which the power per unit area has fallen to $1/e$ of the maximum value. For a Gaussian field the mode field diameter is $2\sqrt{2} W_0$. Then the loss coefficient α is given by [8, 9].

$$\alpha = 0.5 (k n_1 W_0)^2 \Phi(\Delta\beta) \quad (4)$$

Here $\Phi(\Delta\beta)$ is the power spectrum of the curvature function. The power spectrum is evaluated at the spatial frequency $\Delta\beta$ which is the difference between the propagation constant of the straight fibre and the propagation constant of the lossy modes. It can be shown [8, 9, 10] that for single-mode fibres

$$\Delta \beta_{sm} = (W_0^2 k n_1)^{-1} . \tag{5}$$

Very often it is assumed that the power spectrum has the form [11]

$$\Phi(\Delta\beta) = K/(\Delta\beta)^{2p} , \tag{6}$$

where K is a constant (of dimension L^{-2p-1}), and p is a positive number depending on the character of the microbend.

Substituting Eqs. (5) and (6) into Eq. (4), one gets

$$\alpha = 0.5 K (k n_1)^{2+2p} (W_0)^{2+4p} . \tag{7}$$

Equation (7) has two wavelength dependent terms. The wave number k is inversely proportional to λ while W_0 increases with λ. The net result is, however, that the loss increases with increasing wavelength; the larger p is, the faster is the increase.

The power spectrum function is evaluated at $\Delta\beta_{sm}$. This means that perturbations with a periodicity L_{sm} given by

$$L_{sm} = 2\pi/\Delta\beta_{sm} \tag{8}$$

will be particularly undesirable. For a typical single-mode fibre $L_{sm} = 0.5-0.6$ mm.

2.1.4 Microbending losses in multimode fibres

Marcuse has shown that for graded index multimode fibres with a parabolic profile the loss coefficient α is given by [12]

$$\alpha = \frac{5.8\Delta}{a^4} \frac{1}{(\Delta\beta)^4} \Phi(\Delta\beta) , \tag{9}$$

where Δ is the relative refractive-index difference between the core axis and the cladding and a the core radius. $\Phi(\Delta\beta)$ is as before the power spectrum of the fibre curvature function, and here $\Delta\beta$ is the difference in propagation constant between the neighbouring modes. For a parabolic index fibre, this difference is independent of mode number and is given by

$$\Delta\beta = \frac{\sqrt{2\Delta}}{a} . \tag{10}$$

It can thus be seen that the microbend loss for a multimode fibre is independent of wavelength. The dependence on a and Δ will depend on the form of the power spectrum function.

The critical periodicity L_{cr} is again given by

$$L_{cr} = 2\pi/\Delta\beta . \tag{11}$$

For a graded index fibre with a 50 μm core and NA = 0.20, one gets L_{cr} = 1.1 mm. This is about twice the length found for a typical single-mode fibre.

2.2 The Effect of Hydrogen on Fibre Losses

Hydrogen, if present, will diffuse into the fibre core and cause an increase in the fibre attenuation. For a standard 125 μm outer diameter fibre, the time for hydrogen to diffuse into the center of the core is of the order of one week at 20°C.

The first observation of hydrogen induced losses in an installed fibre optic cable was reported in 1983 [13]. This discovery led to a substantial research effort with the aim both to understand the causes for the loss increase and to find ways to eliminate the problem. The discussion to be given here will be brief. For a more complete treatment, the interested reader is referred to a review article by Stone [14] and the references therein.

There are several possible origins for hydrogen in a cable. It might be generated by one or more of the cable materials. It can be generated by electrolysis in case the cable contains dissimilar metals and water gets into the cable. Finally, the cable might be used in an environment where hydrogen is present, and the gas can diffuse through the cable wall into the fibre.

There are several mechanisms which will cause a loss increase. There will be absorption from molecular hydrogen which diffuses into interstitial sites in the silica network. The diffusion process is reversible. Thus, if the fibre after exposure to hydrogen is placed in a hydrogen free environment, the interstitial hydrogen will diffuse out of the fibre, and the corresponding absorption will disappear.

Hydrogen may also react with various glass constituents to form absorbing centers, e.g., OH-groups, which will remain in the glass even if the fibre surroundings become hydrogen free later. The corresponding absorption must, therefore, be regarded as permanent.

In Fig. 3 is shown an example of the loss increase caused by interstitial hydrogen [15]. This fibre has been exposed to 1 atm hydrogen for 23 days at

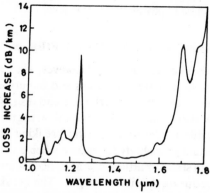

Fig. 3. Loss increase due to interstitial hydrogen. The fibre has been exposed to 1 atm. hydrogen for 23 days at 15°C ([15], © 1985 IEEE).

15°C. This means that the hydrogen concentration is near the saturation level. The marked absorption line at 1.25 μm is caused by the first overtone of the hydrogen molecule, and the other lines are due to various combination frequencies with the silica lattice. The fibre used here was a multimode fibre, but both the magnitude and spectral character of the loss can for practical purposes be regarded as being independent of the type of fibre.

The excess loss caused by interstitial hydrogen is given by the following expression:

$$\alpha(\lambda) = A(\lambda)\, p\, \exp(E_a/RT). \quad (12)$$

Here $A(\lambda)$ is a wavelength dependent scaling factor; at 1300 nm A (1300) = 0.018 dB/km-atm. while at 1550 nm, A (1550) = 0.038 dB/km-atm. [16]. Further, p is the hydrogen pressure in atm., E_a the activation energy (E_a = 1550 cal/mol), R the universal gas constant and T is the absolute temperature.

There are two prevalent losses caused by the chemical reactions between hydrogen and the silica glass. Firstly, the formation of Si−OH and Ge−OH bonds which have absorption peaks at 1.39 μm and 1.41 μm, respectively. Secondly, there is an increase in the UV-absorption with a tail into the IR-region. The magnitude of the losses caused by these mechanisms depends on the dopants used and on the manufacturing process. In particular, the use of phosphorous is detrimental.

In recent years, a substantial effort has ben made to modify the fibre designs and manufacturing processes so as to reduce the hydrogen induced losses. As a result, it can, for to-day's commercial fibres, usually be said that at normal operating temperatures it is the interstitial hydrogen which dominates the hydrogen induced increase in loss at 1300 nm and 1550 nm wavelengths.

Under these assumptions, one finds from Eq. (12) that if we require that the hydrogen induced loss be less than 0.01 dB/km at 1550 nm and 20°C, then p must be less than 0.018 atm. By proper selection of the cable materials, the hydrogen pressure can be kept well below this level.

3. MECHANICAL STRENGTH OF OPTICAL FIBRES

Quartz has a very high inherent strength. However, when manufacturing long lengths of fibre, flaws such as surface cracks and impurities imbedded in the surface will appear along the fibre. It is the size and distribution of these flaws which will eventually determine the mechanical strength of the fibre. As a result, it is not possible to define a specific breaking strength for a fibre. Rather, there is a certain probability for failure when a certain length of fibre is subjected to a certain strain. Further, when the fibre is under strain, the flaws will increase in size reducing the fibre strength. The breaking strength is thus also a function of the duration of the load.

For more detailed discussions of the mechanical properties of optical fibres reference can be made to an article by Kalish et al. [17].

For the discussion to be presented here we will take as a starting point that it is now common practice to proof test each fibre before it is shipped from the fibre manufacturer. This means that the whole fibre has been subjected to a specified load or strain for a specified time, e.g. 1 sec., and fibres which do not pass the test, are discarded. Typical proof test levels as of today are 0.5% to 1%.

The fibre might be exposed to strain during the cabling process, installation and when in operation. To determine which fibre strain can be allowed, one must know the relationship between the short term proof strain and the long term strain. This relationship can be derived from crack growth theory.

It can be shown [17] that the time to failure t for a fibre subjected to a strain ε is given by:

$$\varepsilon^n t = \text{const.} \qquad (13)$$

Here n is the stress corrosion susceptibility factor, and the value of the constant on the R.H.S. of Eq. (13) depends on the character of the failure mechanism. Thus, if a fibre has survived a proof test strain ε_p for a time t_p, then the minimum survival time t_s for the fibre under strain ε_s is given by

$$t_s > t_p (\varepsilon_p/\varepsilon_s)^n \qquad (14)$$

(under the approximation that the fibre is unaffected by the proof test [17]).

The stress corrosion constant n is strongly dependent on the environment surrounding the fibre. In particular, high humidity and also high temperature will result in a lowering of n. Typically, n-values of 15-20 are used for cables which are to be operated at normal temperatures. If we assume $n = 15$ and t_p 1 sec and $\varepsilon_p = 1\%$, then from Eq. (14) we find that a fibre should tolerate a strain of about 25% of the proof test level for about 30 years.

Recently fibres having hermetic coatings have become available. For such fibres n-values as high as about 200 have been reported [18, 19].

4. CABLE CORE STRUCTURES

When the fibre is being pulled from the preform, it is given a protective coating. The coating may have one or two layers, and for a 125 μm fibre it has typically a thickness of about 60 μm. In the cable, the fibre is given additional protection, and there are usually two types of structures being used, a tight jacket or a loose tube, as shown in Fig. 4. There are some cable designs like the use of ribbon structures (see Sec. 5) that do not quite fit into this picture, however, we will restrict our discussion here to these two types.

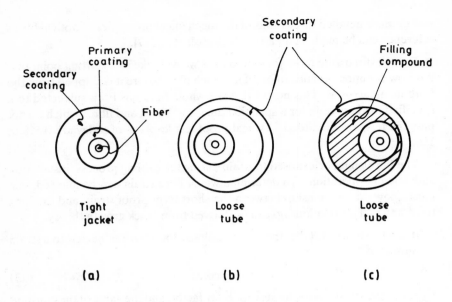

Fig. 4. Secondary coatings (buffers) for primary coated fibres: (a) tight jacket (b) loose tube, air filled (c) loose tube with filling compound.

4.1 Loose tube structures

In a loose tube, the fibre is well shielded from being affected by the surrounding cable elements. As we shall see, by stranding the tube, it is also possible to design in strain relief from both temperature changes and mechanical forces. For these reasons, several variations of the loose tube structure have found widespread use. The loose tube cables are, however, more difficult to terminate and connectorize as compared to the tight jacketed ones.

The tube can be air filled, but very often a filling compound is used, see Fig. 4 (c). The purpose of the filling compound is to prevent water ingress, and it also restricts fibre movements since the compound is much more viscous than air. If water is present, it can freeze and cause microbends. Water has also a detrimental effect on the mechanical strength of the fibre.

The tube is normally made of a plastic and has an inner diameter of typically 4-5 times that of the primary coated fibre. Sometimes two or more fibres are placed in the same tube.

Silica fibres have a very low coefficient of thermal expansion ($\approx 5 \times 10^{-7}$ °C^{-1}). The various other cable materials typically have an expansion coefficient of $10^{-5} - 10^{-4}$ °C^{-1}. This means that when the temperature changes, the fibre and the cable structure expand (contract) by different amounts.

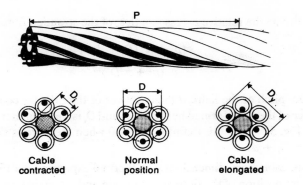

Fig. 5. Stranded optical cable. The fibre can move between the inner and outer curve resulting in a strain-free window.

By stranding the tubes around a central strength member, (see Fig. 5) the fibres will be able to move freely a small distance in the radial direction to compensate for changes in cable length either due to temperature variations or mechanical forces. Normally, the fibres are centered in the tube at zero strain and average operating temperature. However, this position can be selected as required by making proper adjustments in the manufacturing process. If the cable is elongated due to increasing temperature or a tensile load, the fibres move inward. If the cable contracts due to a decrease in temperature or a compressive load, the fibres move outward.

Another configuration which provides a similar protection for the fibre and achieves the same compensation for changes in cable length, is the slotted core structure as shown in Fig. 6. Here the slots or grooves can follow a helical path [20] or an oscillating path [2, 21]. Although the manufacturing process for the loose tube structure and slotted core structure is somewhat different, their basic properties are much the same, and the discussion to follow will, in general, be applicable to both.

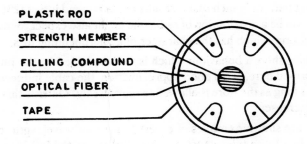

Fig. 6. Example of a slotted core structure.

We will define the relative excess length ε_x of the stranded fibre by

$$\varepsilon_x = 1 - \frac{\left(p^2 + \pi^2 (D_i + d)^2\right)^{1/2}}{(p^2 + \pi^2 D^2)^{1/2}}. \tag{15}$$

Here p is the pitch of the helix, d the diameter of the primary coated fibre, D the diameter of the helix formed by the fibre, and D_i is the diameter of the inner curve (see Fig. 5). With this definition $\varepsilon_x = 0$ when the fibre is closest to the center ($D = D_i + d$).

When the cable core dimensions, the thermal expansion coefficient of the cable, and the position of the fibre at zero strain and a certain temperature are known, Eq. (15) can be used to calculate the temperature window for strain free operation. Conversely, if a certain strain free window is desired, Eq. (15) can be used to determine the appropriate dimensions and fibre position at zero cable strain. Strain free operation for elongations of several tenths of one percent is easily achievable.

Due to the stranding, the fibre will have a constant curvature. One must take care when determining the cable core geometry, that the curvature does not cause unacceptable fibre losses (see Sec. 2.1).

If the cable is elongated outside the strain window, the fibre will be pressed against the wall of the tube or groove. This will result in microbends due to the roughness of the wall and non-uniformities in the fibre coating. The character of the microbends will depend on the properties of the wall and fibre coating. In Sec. 2.1, the power spectrum of the curvature distribution function was expressed by a simple function [see Eq. (6)]. There have been several reports on the experimental determination of the exponent p. Typical reported values are around 3-4 [22, 23].

When the cable contracts sufficiently, the fibre will have to buckle to fit into the tube or slot. In a somewhat idealized model, this will happen when the actual excess length of the fibre becomes longer than that corresponding to $D = D_y - d$ in Eq. (15). Here D_y is the diameter of the outer curve (see Fig. 5). The buckling pattern is such that it results in macrobend losses [23]. However, since there will be a distribution of radii, one might also expect to find a smaller loss contribution, which has the character of a microbending loss.

Real cables show a behaviour which is well explained by the rather simple model described here. To a good approximation, the cable induced losses are negligible as long as the fibre is allowed to move freely. Outside this region, the losses will increase.

In Fig. 7 are shown examples of excess loss versus wavelength for a single-mode fibre in a loose tube structure. Fig. 7(a) shows the loss increase at low temperatures, and the labelling parameter is the relative fibre length with respect to the outer curve. The fact that losses are observable at negative

OPTICAL FIBRE CABLES

relative lengths, must mean that the fibre starts to buckle before it reaches the outer curve. The rapid change in loss with increasing wavelength is quite typical for macrobend losses in single-mode fibres.

In Fig. 7(b) are shown curves for excess losses at high temperatures. Here the fibre is pressed against the inner wall. The parameter is the actual fibre

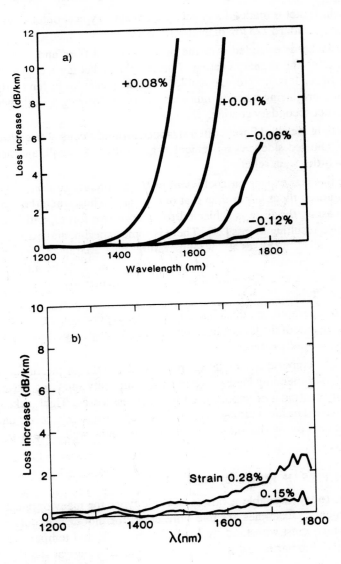

Fig. 7. Examples of excess losses for a single mode fibre in a loose tube structure. See text for detailed explanation. (a) Macrobending loss (b) Microbending loss

strain. It can be seen that the losses increase rather slowly with increasing wavelength. This is typical for microbend losses in single-mode fibres.

Comparing the two curves we see that the loss increases much more rapidly on the low side of the temperature window than on the high side.

4.2 Tight jacket secondary coating

The tight jacket is made of a plastic, e.g., nylon. Typical outer diameter is 0.9 mm for a standard 125 μm diameter fibre.

The jacket is extruded over the primary coated fibre and usually adheres quite well to the primary coating. Further, the jacket gives a good protection for the fibre. It is therefore, possible to use the jacket for fastening the fibre by clamping or glueing when terminating or connectorizing a cable which has a tight jacket secondary coating.

The tight jacketed fibre is often stranded around a central strength member, but the allowed sideways movement is generally very small, and accordingly there is little strain relief.

The thermal expansion coefficient of the secondary coating is typically at least hundred times larger than that of the fibre. Thus, when the temperature is decreasing, the fibre is subjected to compressive forces, and buckling will take place resulting in loss [24]. There is some experimental evidence that for the tight jacket the losses at low temperatures have character of macrobending losses [25]. However, one would expect that there could be a distribution of radius of curvature, which will also result in a smaller loss contribution of the microbend type.

The temperature at which buckling occurs, depends on the thermal expansion coefficient of the jacket, its thickness and Young's modulus as well as how uniform and well centered it is.

At high temperature the jacket puts tensile strain on the fibre. This can result in microbending losses. Again good uniformity and concentricity of the secondary coating is of importance for low excess losses. Tight jacketed fibres are often used inside buildings, e.g., to link the transmitter/receiver units to the termination box of the outside cable which might well have a loose tube structure.

5. VARIETIES OF FIBRE OPTIC CABLES

There is, today, available a variety of cable types for many applications and environments. Examples are one fibre cords for indoor use, outdoor aerial cables which must withstand ice loads, heavy winds and temperatures below −40°C, and submarine cables for transoceanic crossings at depths down to 7000-8000 m.

A fibre optic cable has often a central core element containing the fibres. The core is surrounded by a plastic jacket of e.g., polyethylene (PE) or

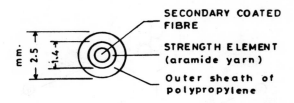

Fig. 8. Cross-section of a simple one fibre cable (courtesy STK Alcatel).

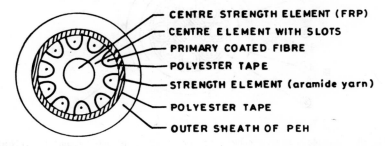

Fig. 9. Cross-section of a slotted core cable for duct installation (courtesy STK Alcatel).

polyvinyl chloride (PVC). If the cable is specified to tolerate only small tensile forces, it can be sufficient to use just a central strength member of steel or fibre reinforced plastic (FRP), or use a layer of e.g. aramide yarn between the cable core and the outer jacket. If the cable is specified to tolerate larger tensile forces, one or may be even two layers of steel armour can be required outside the inner jacket. A protective outer jacket is often applied over the steel armour.

In Fig. 8 is shown an example of a simple one fibre cable. It consists of a secondary coated fibre (tight jacket) surrounded by aramide yarn, which acts as a strength member, and an outer protective sheath.

In Fig. 9 is shown a cable for duct installation. It uses a slotted core design which makes it quite resistant to lateral pressure. The tensile strength is provided partly by the central strength member of FRP and partly by the aramide yarn between the core element and the outer jacket. The cable shown has ten fibres, one in each slot. But each slot could have 2 or 3 fibres without affecting the cable performance to any appreciable extent.

An example of an aerial cable is shown in Fig. 10. The cable uses a loose tube construction with two fibres in each tube. The tubes are stranded around the center strength element. The main tensile strength is provided by the steel wires located in the upper part of the figure eight structure.

A number of submarine cables have been developed. Examples of two somewhat different designs are shown in Fig. 11 [26, 27]. They are used in the American (a) and the UK section (b) of the TAT-8 transatlantic cable.

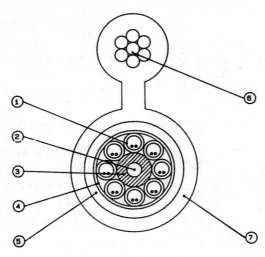

Fig. 10. Example of an aerial cable using a figure eight construction (courtesy EB/Norsk Kabel); (1): Loose, jellyfilled tube with 2 fibres, (2): Central strength element (FRP), (3): Sheath over strength element (PE), (4): tape, (5): Inner jacket, (6): Strength member (Steel), (7): Outer jacket.

The cable in Fig. 11(a) is built around core in which the fibres are embedded in an elastomer that provides a cushion against microbend losses. The fibres are stranded around the center steel core. Surrounding the strength members is a copper cylinder which serves as a power conductor for the repeaters along the cable and also acts as a water and hydrogen barrier. By using such a tube, the fibres are also protected from the hydrostatic pressure.

In the other cable [Fig. 11(b)], the fibres have a tight jacket, but here only the core unit is contained within the hermetically sealed copper tube. The tube is filled with a water-blocking material to prevent penetration of water into the cable in case of breakage. The cable shown here is a so called lightweight cable meant for deep-water regions. For more shallow water where there is a risk that the cable might be caught by an anchor or fishing gear, additional armour is used.

Finally, two examples of cables with many fibres are shown. In Fig. 12 is shown a cable using a stack of ribbons [28]. In this particular cable, each ribbon has 12 fibres, and the stack contains 12 ribbons giving a total of 144 fibres in the cable. The main advantages of this design are that it gives a high packing density and facilitates mass splicing of the fibres.

A combination of the slotted core structure and ribbon design is shown in Fig. 13 [29]. Here each ribbon contains 5 fibres, 4 ribbons are placed in each slot and there are 5 slots in each rod. The 100-fibre units are then stranded around a central strength member. In the figure the cable has six such units

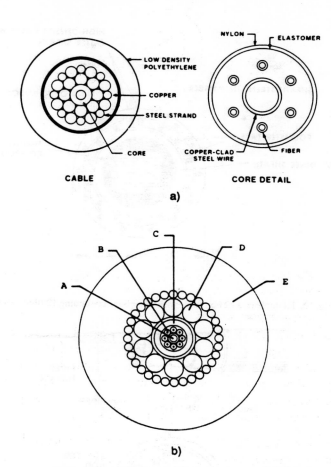

Fig. 11. Examples of submarine cable (a) Cross-section of cable used in the American section of TAT-8, (b) Cross-section of cable used in the British section of TAT-8; A: Kingwire, B: eight fibre, C: copper tube, D: steel wires and E: polyethylene ([26-27], © 1986 IEEE).

resulting in a 600-fibre cable. Cables with even larger number of fibres have been considered [2]. Cables with many fibres are mainly intended for applications in optical subscriber networks.

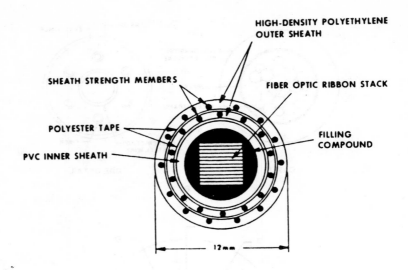

Fig. 12. Example of a ribbon cable with 12 ribbons containing 12 fibres each [28].

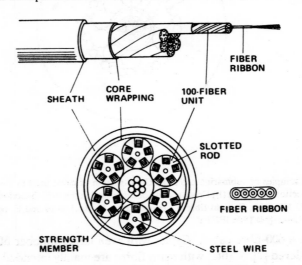

Fig. 13. Cross-section of a 600-fibre cable [29].

REFERENCES

1. H. Murata, "Handbook of optical fibres and cables", Marcel Dekker, Inc. (1988).
2. G. Mahlke and Gössing P., "Fibre Optic Cables-Fundamentals, Cable Technology, Installation Practice", John Wiley & Sons Ltd. (1987).
3. E.J. Friebele, Long K.J., Askins, C.G., Gingerich M.E., Marrone M.J. and Griscom D.L., "Overview of radiation effects in fibre optics", Radiation Effects in Optical Materials, SPIE, Vol. 541, pp. 70-88 (1985).

4. D. Marcuse, "Curvature loss formula for optical fibres", J. Opt. Soc. Amer., Vol. 66, pp. 216-220 (1976).
5. S.B. Andreasen, "New bending loss formula explaining bends on loss curve", Electron. Lett., Vol. 23, pp. 1138-1139 (1987).
6. M. Eriksrud, Private communication.
7. D. Marcuse, Gloge D. and Marcatili E.A.J., "Guiding properties of fibres", in Optical Fibre Telecommunications, Ed. by S.E. Miller and A.G. Chynoweth, Academic Press, pp. 37-100 (1979).
8. K. Petermann, "Theory of microbending loss in monomode fibres with arbitrary refractive index profile", Arch. Elektr. Übertr., Vol. 30, pp. 337-342 (1976).
9. K. Petermann, "Fundamental mode microbending loss in graded-index and W fibres", Opt. and Quant. Electron., Vol. 9, pp. 167-175 (1977).
10. K. Petermann and Kühne R., "Upper and lower limits for the microbending loss in arbitrary single-mode fibres", J. Lightwave Technol., Vol. LT-4 pp. 2-7 (1986).
11. R. Olshansky, "Mode coupling effects in graded-index optical fibres", Appl. Opt., Vol. 14, pp. 935-945 (1975).
12. D. Marcuse, "Losses and impulse response of a parabolic index fibre with random bends", Bell Syst. Tech. J., Vol. 52, pp. 1423-1437 (1973).
13. K. Mochizuki, Namihira Y. and Yamamoto H., "Transmission loss increase in optical fibres due to hydrogen permeation", Electron. Lett., Vol. 19, pp. 743-745 (1983).
14. J. Stone, "Interactions of hydrogen and deuterium with silica optical fibres: A review", J. Lightwave Technol., Vol. LT-5, pp. 712-733 (1987).
15. K. Noguchi, Shibata N., Uesugi N. and Negishi Y., "Loss increase for optical fibres exposed to hydrogen atmosphere", J. Lightwave Technol., Vol. LT-3, pp. 236-243 (1985).
16. S.R. Barnes, Pitt N.J. and Hornung S., "A model for predicting hydrogen degradation in optical cables", Proc. 11th European Conf. on Optical Comm. (ECOC), Venezia, 1-4 Oct. pp. 897-900 (1985).
17. D. Kalish, Key P.L., Kurkjian C.R., Tariyal B.T. and Wang T.T., "Fibre Characterization-Mechanical", in Optical Fibre Telecommunications, Ed. by S.E. Miller and A.G. Chynoweth, Academic Press, pp. 401-433 (1979).
18. K.E. Lu, Glaesemann G.S., Vandewoestine R.V. and Kar G., "Recent developments in hermetically coated optical fibre", J. Lightwave Technol., Vol. LT-6, pp. 240-244 (1988).
19. R.G. Huff, DiMarcello F.V. and Hart A.C., Jr., "Amorphous carbon hermetically coated optical fibres", OFC'88, New Orleans, La, 25-28 Jan., Technical Digest, Paper Tu G2 (1988).
20. G. Le Noane, De Vecchis M. and Hulin J.P., "Experimental results of cylindrical V grooved structure optical cables laid in ducts and spliced", Fourth Europ. Conf. on Optical Comm. (ECOC), Genova, Italy, 12-15 Sept., pp. 218-223 (1978).
21. T.S. Swiecicki, King F.D. and Kapron F.P., "Unit core cable structures for optical communication systems", Proc. 27th, Int. Wire and Cable Symp. pp. 404-410 (1978).
22. S. Hornung, Doran N.J. and Allen R., "Monomode fibre microbending loss measurements and their interpretation", Opt. Quantum Electron. Vol. 14, pp. 359-362 (1982).
23. A. Hordvik and Eriksrud M., "Loss mechanisms in single-mode fibres jacketed with a loose jelly-filled tube", J. Lightwave Technol., Vol. LT-4, pp. 1178-1182 (1986).
24. T. Yabuta, Yoshizawa N. and Ishihara K., "Excess loss of single-mode jacketed optical fibre at low temperature", Appl. Opt., Vol. 22, pp. 2356-23362 (1983).

25. Y. Katsuyama, Mitsunaga Y., Ishida Y. and Ishihara K., "Transmission loss of coated single-mode fibre at low temperatures", Appl. Opt., Vol. 19, pp. 4200-4205 (1980).
26. P.K. Runge and Trischitta P.R., "The SL undersea lightwave system", Undersea Lightwave Communication, Edited by P.K. Runge and P.R. Trischitta, IEEE Press, New York, pp. 51-67 (1986).
27. P. Worthington, "Cable design for optical submarine systems", ibid pp. 251-262.
28. B.R. Eichenbaum, Santana M.R., Tate L.D. and Sabia R., "Design and performance of a filled, high-fibre-count, multimode optical cable", Proc. 31st, Int. Wire and Cable Symp. pp. 396-400 (1982).
29. S. Hatano, Katsuyama Y., Kokubun T. and Hogari K., "Multi-hundred-fibre cable composed of optical fibre ribbons inserted tightly into slots", Proc. 35th Int. Wire and Cable Symp., pp. 17-23 (1986).

Part II
Optical Sources, Detectors and Communication Systems

Part II

13

Optical Sources, Detectors and Engineered Systems — Introductory Remarks

B.P. PAL[*]

The 'PHOTOPHONE' experiment performed by Graham Bell as early as in 1880 contained all the essential features of a modern optical communication system. Sun rays as the optical carrier wave along with the reflecting diaphragm of a microphone constituted the transmitter, air was the transmission medium and a Selenium photocell acted as the receiver. A modern optical transmitter is formed out of a semi-conductor laser, or in certain situations, a light emitting diode(LED), which is intensity modulated by the signal (input as electrical signals) fed directly to the laser/LED drive circuit. Optical output from the modulated laser/LED is launched into an optical fibre, which is the transmission medium in this case. Finally, a PIN/APD photodetector, once again, a semi-conductor device (usually followed by an amplifier) forms the receiver in a modern optical communication system. The first four chapters (Chs. 14, 15, 16 and 17) in this part of the book are concerned with generation and detection of optical signals for fibre optical communication system. The following chapter (Ch. 18) on fundamentals of digital communication is purported to be an introduction to principles of digital communication especially for those not possessing a background in communication theory. Loss and bandwidth budgeting are topics of prime concern in any fibre optical system design. These aspects are discussed in the Ch. 19. Coherent optical communication which had to wait for sometime due to non-availability of single frequency semi-conductor lasers until recently forms the subject matter of the Ch. 20 by Hodgkinson et al. Coherent optical communication systems are

---

[*] Department of Physics, Indian Institute Technology Delhi, New Delhi-110016, India

expected to provide an additional system margin of 10-20 dB in signal reception as compared to direct detection systems thereby allowing longer repeater spacings. In some situations, it is also possible that the laser is operated in its cw (continuous wave) mode and the signal (in electrical form) instead of being fed directly to the laser drive circuit is fed to an integrated optical waveguide modulator, which is kept on the path of the propagating laser beam. The laser beam exits from the modulator as a modulated beam which eventually is launched into the fibre for demodulation at the other end by a PIN/APD optical receiver. This scheme is usually employed when transmission bandwidth requirement is very high. Integrated optic devices form the subject matter of Ch. 21 of this part of the book.

14

Semiconductor Lasers for Optical Fibre Communications[†]

Govind P. Agrawal[*]

---

1. INTRODUCTION

During the last ten years or so, the field of optical fibre communications has experienced an unprecedented growth[1-4]. Whereas the early lightwave systems installed around 1978 operated at a bit rate of 50-100 MHz with a repeater spacing of 10-15 km, a data transmission rate of 4 Gb/s over 117 km of repeaterless fibre length has already been demonstrated[5]. Advances in both the fibre-fabrication technology[6] and the development of long-wavelength semiconductor lasers[7] have contributed to this remarkable improvement in the performance of optical communication systems. This chapter briefly describes the working principles and the operating characteristics of semiconductor lasers with particular attention paid to their applications in optical fibre communications. The presentation is pedagogical in nature in the hope of making it accessible to a relatively wide readership.

2. THE p-n JUNCTION

At the heart of a semiconductor laser is the p-n junction[8], the same element used in transistors and responsible for the microelectronics revolution. As one may recall, the electrical properties of an intrinsic semiconductor material can be modified by adding impurities. Depending on the valency of the impurity atom, the resulting material is termed an n-type or p-type semiconductor. In

---

[†] Reprinted from J.I.E.T.E., Vol. 32 (Special issue on Optoelectronics and Optical Communication), July-August (1986)

[*] The Institute of Optics, University of Rochester, Rochester, New York 14627, USA.

an n-type semiconductor, impurity atoms, called donors, contribute one of their electrons to the conduction band and become positively charged. In a p-type semiconductor, impurity atoms, called acceptors, acquire an electron from the valence band and become negatively charged. The absence of an electron or the vacancy leaves the valence band with a "hole" that acts as if it were a positive charge. In a p-n junction, the p-type and n-type semiconductors are brought in contact with each other. When they first come in contact in the vicinity of the junction, electrons in the n-region diffuse toward the p-region while the reverse occurs for holes. Their recombination in the junction region leaves the p side with an excess of negative charge and the n side with an excess

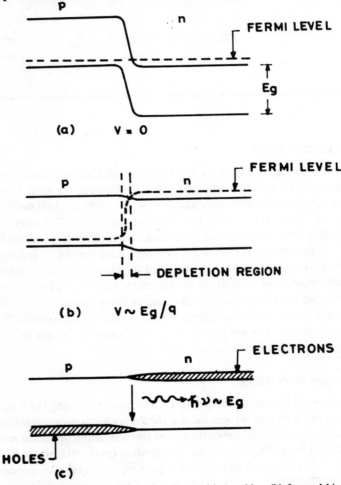

Fig. 1. Energy-band diagram of a p-n junction at (a) zero bias, (b) forward bias and (c) schematic representation of the electron and hole populations under forward bias. Radiative recombination of electrons and holes in the narrow depletion region generates light.

of positive charge. In the steady state, the resulting built-in electrical field opposes further diffusion of electrons and holes (see Fig. 1(a)).

When the p-n junction is used as a semiconductor laser, a forward-bias voltage is applied across the junction. This lowers the built-in electric field (Fig. 1(b)) and the steady-state balance between electrons and holes is disturbed. Once again, electrons move from the n-side to the p-side while the reverse occurs for holes. Their movement manifests as the electric current flowing through the p-n junction. In a narrow region, known as the depletion region, electrons and holes are simultaneously present, and their recombination can lead to emission of photons (Fig. 1(c)). For a further understanding of the lasing process, it is necessary to consider the mechanisms through which electrons and holes recombine in the junction region.

The recombination mechanisms can be classified as being radiative or non-radiative. During each recombination event, an energy approximately equal to the band-gap energy E_g is released. Several processes contribute to the rate of non-radiative recombination (defect recombination, surface recombination, Auger recombination)[7] and the released energy is eventually lost to the lattice vibrations. By contrast, during each radiative recombination event, the released energy appears in the form of a photon whose frequency ν or the wavelength λ is given by

$$h\nu = hc/\lambda \simeq E_g, \tag{1}$$

where h is the Planck constant and c is the velocity of light in vacuum. A p-n junction is, therefore, capable of generating electromagnetic radiation whose wavelength is governed by the band gap associated with the semiconductor material used to make the p-n junction.

An essential characteristic of a laser is that it emits coherent radiation in a narrow spectral range. To understand the origin of coherence, it is important to realize that a radiative recombination event can occur through two fundamental optical processes known as spontaneous emission and stimulated emission, shown schematically in Fig. 2. In the case of spontaneous emission, photons are emitted in random directions with no apparent phase relationship. Stimulated emission, by contrast, is initiated by a photon whose energy nearly equals the band-gap energy. The remarkable thing, however, is that the emitted photon matches the original photon not only in its wavelength and direction of propagation, but also in phase. It is the fixed phase relationship between the incident and emitted photons that renders the light emitted by a laser coherent.

The preceding discussion shows that stimulated emission can amplify the optical field propagating through the p-n junction. However, it has to compete against an absorption process that attenuates the field. In the absorption process, an electron in the valence band absorbs the incident photon. Because

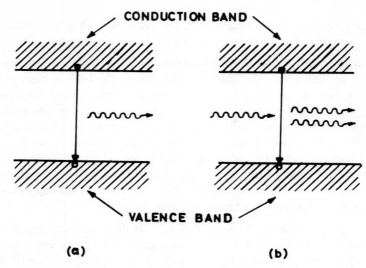

Fig. 2. Schematic illustration of the radiative recombination mechanisms. (a) spontaneous emission (b) stimulated emission.

of the acquired energy, it gets promoted to the conduction band leaving behind a hole in the valence band. In the absence of an external current flowing through the p-n junction, the electron population in the valence band far exceeds that of the conduction band, and the absorption process dominates. Under the forward bias, the carrier injection increases the conduction-band population, thereby enhancing the probability of stimulated emission. At some value of the current, the stimulated-emission and absorption processes balance each other. A further increase in current leads to a condition known as "population inversion" wherein the population of charge carriers in the conduction band exceeds that of the valence band. The p-n junction is now capable of amplifying the electromagnetic radiation whose wavelength satisfies Eq. (1). It is said to exhibit optical gain whose magnitude can be controlled through the current. Since population inversion is achieved using current injection, semiconductor lasers are sometimes called injection lasers. The name laser diode is also used for a semiconductor laser to emphasize the use of a p-n junction.

3. FABRY-PEROT CAVITY

It takes more than optical gain to make a laser. The other necessary ingredient is an optical feedback mechanism. In most lasers it is accomplished by placing the gain medium between two mirrors which form a Fabry-Perot cavity, named after its inventors. Semiconductor lasers do not require external mirrors The cleaved facets perpendicular to the p-n junction provide significant natural reflectivity because of a relatively large refractive index of the semiconductor material ($n \simeq 3.5$). In effect, the facets form a Fabry-Perot cavity which provides feedback for photons traveling parallel to the p-n junction.

The role of a Fabry-Perot cavity in lasers is two-fold. As mentioned before, spontaneously emitted photons have no direction selectivity and travel in all directions. The cavity confines the photons travelling along its axis, allowing others to escape. As these confined photons travel back and forth across the gain region, they induce an avalanche of photons generated by stimulated emission. In other words, the Fabry-Perot cavity provides a direction of propagation for the laser radiation. The laser output appears in the form of a narrow beam coming out from both of the facets.

The other role of the Fabry-Perot cavity is to provide wavelength selectivity, the original purpose of its invention in 1896. A consideration of multiple reflections occurring at the facets shows that the optical feedback is strongest only for wavelengths which fit evenly inside the cavity, i.e., the cavity length L and the wavelength λ must satisfy the relation

$$nL = m\lambda/2, \qquad (2)$$

where m is an integer and n is the refractive index of the semiconductor material. The corresponding wavelengths are called resonant wavelengths and constitute the longitudinal modes of the Fabry-Perot cavity. Which of the longitudinal modes are selected by the laser depends on the wavelength range over which the p-n junction is able to amplify the electromagnetic radiation. Generally speaking, the wavelength range is sufficiently narrow that only few longitudinal modes can be excited. The optical spectrum therefore consists of few uniformly spaced discrete lines whose wavelengths satisfy Eq. (2). The center wavelength is approximately governed by the bandgap energy [see Eq. (1)]. Semiconductor lasers used for optical communications, depending on the material used for the p-n junction, typically radiate at $\lambda \sim 1\,\mu$m. The cavity length L is typically $250\,\mu$m. Using these values and $n \simeq 3.5$ in Eq. (2), one can see that the integer $m \sim 1000$.

4. HETEROSTRUCTURE SEMICONDUCTOR LASERS

The feasibility of stimulated emission in semiconductor lasers was demonstrated as early as 1962 using a GaAs p-n junction[9-11]. Practical utility of these earlier devices was, however, limited since liquid-nitrogen cooling was required together with very high values of the current. There were two basic reasons for the poor performance exhibited by the p-n junctions. First, there was no mechanism for the confinement of charge carriers near the junction interface and the concentration of diffusing electrons and holes decreased exponentially from its highest value occurring at the interface. The resulting optical gain was non-uniform and had a significant value only over a relatively narrow region ($\sim$ a few nanometers). Second, the confinement of the optical field generated by stimulated recombination was also very poor. The penetration of the optical field beyond the gain region introduced additional losses which had to be compensated by increasing the device current.

It turned out that both of the above problems can be remedied by a single solution. The basic idea is to sandwich a thin layer of a different semiconductor with a relatively narrow band-gap in between the p-type and n-type semiconductors. However, the two semiconductors should match in their lattice spacing in order to realize good quality interfaces. A junction of two dissimilar semiconductors is called a heterojunction and devices employing such junctions are referred to as the heterostructure semiconductor lasers. Even though the initial suggestions for the use of heterojunctions were made already in 1963, the realization of heterostructure semiconductor lasers, operating continuously at room temperature, had to wait for the technological advances in the device design and in the epitaxial growth techniques[2] and was not achieved until 1970[12]. Figure 3 shows schematically a heterostructure semiconductor laser. Typical dimensions of the device are also indicated to emphasize that the physical size of a semiconductor laser is comparable to that of a grain of salt. Stimulated recombination occurs only inside the thin active layer (thickness ~ 0.1 μm). Its bandgap is slightly lower compared with that of the p-type and n-type cladding layers.

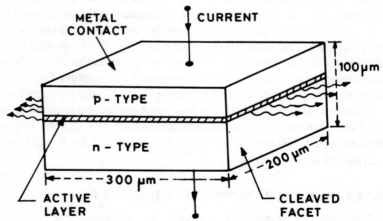

Fig. 3. Heterostructure semiconductor laser shown schematically. The active layer and the p- and n-type cladding layers are made of two different semiconductor materials. Typical dimensions of a semiconductor laser are also shown (not to scale).

The physics behind the simultaneous confinement of charge carriers and photons by the heterostructure scheme can be understood by referring to Fig. 4. Since the narrow central region has a lower band-gap, electrons from the p-region and holes from the n-region can move freely to the active layer

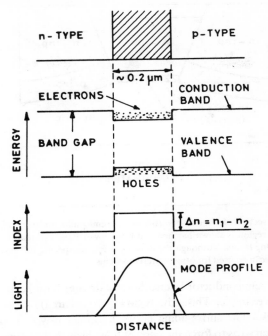

Fig. 4. Schematic illustration of the simultaneous confinement of the charge carriers and photons offered by the heterostructure scheme. In the thin active region (hatched area), the smaller band-gap confines electrons and holes. At the same time the higher refractive index confines the emitted radiation.

However, once there, their cross-over to the other side is restricted by the potential barrier arising from the band-gap difference of two semiconductors. This allows for the build-up of electron and hole populations in the conduction and valence bands, respectively. The light confinement occurs because of a fortunate coincidence: A semiconductor with a smaller band-gap also has a higher refractive index, thereby allowing for the total internal reflection to occur at the heterostructure interfaces. The active layer, in effect, acts as a dielectric waveguide and the refractive-index difference Δn helps to confine the optical mode within the gain region.

The use of a heterostructure for the semiconductor laser of Fig. 3 solves the confinement problems only in the direction perpendicular to the junction plane. Such a laser is often referred to as a broad-area laser since both the current and the optical field spread over a wide region ($\sim 100 \mu m$) along the junction plane. Relatively high (~ 1 A) values of the current are required to operate a broad-area laser. The heterostructure concept can, of course, be generalized to confine the charge carriers and photons in the lateral direction as well, albeit at the expense of considerable fabrication complexity. The active region then is in the form of a rectangular slab, and is buried on all sides in a higher band-gap semiconductor material. Such lasers are called buried

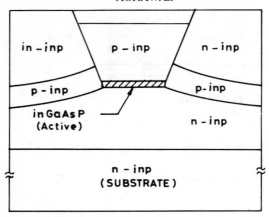

Fig. 5. Cross-section of a buried-heterostructure semiconductor laser (schematic). The active region is shown hatched and has a lower band-gap than that of the surrounding layers. Although the drawing is for the specific InP system, the same design can be used for other material systems.

heterostructure semiconductor lasers. Several designs have been proposed, and Fig. 5 shows one example. The active region (hatched area) is typically $2\,\mu$m wide and $0.2\,\mu$m thick. The surrounding layers are either n-type or p-type and are judiciously chosen so as to force the current flow through the active region.

5. SEMICONDUCTOR MATERIALS

The choice of semiconductors used for fabrication of semiconductor lasers depends on several factors. To start with, the semiconductor must have a direct band-gap, i.e., the conduction-band minimum and the valence-band maximum should correspond to the same value of the electron and hole momenta. In indirect band-gap semiconductors, the radiative recombination of an electron-hole pair requires the assistance of phonons (quanta of lattice vibrations) to conserve the momentum during the process. This makes radiative recombinations much less probable compared with non-radiative recombinations and excludes the use of semiconductors such as Si and Ge. A large number of direct band-gap semiconductors have been found suitable for the laser purpose, and taken together, these materials cover the optical spectrum from near ultraviolet to far infrared. Two semiconductors which have attracted most attention are GaAs and InP and emit radiation in the near infrared region from 0.8 to $1.6\,\mu$m.

Heterostructure semiconductor lasers require two semiconductors with different band-gaps, but with almost identical lattice spacings. This is achieved by the use of ternary or quaternary alloys. In the case of GaAs semiconductor, Al is used to replace a fraction of Ga atoms. The resulting AlGaAs alloy semiconductor has a higher band-gap compared with that of GaAs. In the case of InP semiconductor, Ga and As are used to replace fixed fractions of In and P atoms, respectively. The resulting InGaAsP alloy semiconductor has a lower band-gap compared with that of InP and is used for the active layer (see Fig. 5). Such InGaAsP lasers can be made to emit light in the wavelength range of 1.1-1.6 μm by suitably adjusting the relative concentrations of the four elements. This is the wavelength range of interest for optical fibre communications.

A great deal of effort has recently been spent for the development of practical InGaAsP lasers to be used as a light source in optical communication systems. The technology has matured enough that these lasers are now commercially available and are currently being employed in lightwave transmission systems. In the remainder of this article the emission characteristics of semiconductor lasers are discussed from the standpoint of their application in optical fibre communications.

6. EMISSION CHARACTERISTICS

The performance of a semiconductor laser is characterised through its static, dynamic, and spectral characteristics. Under continuous operation, an important role is played by the light-current (L-I) characteristics which describe the variation of the light output with the current applied to the semiconductor laser. Figure 6 shows, by way of an example, the L-I curves at different temperatures for a buried-heterostructure-type semiconductor laser emitting at the wavelength of 1.3 μm. At a given temperature, almost no light is emitted until the current reaches a critical value, referred to as the threshold current.

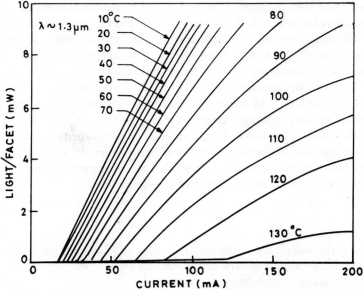

Fig. 6 Light-current characteristics of a buried-heterostructure semiconductor laser emitting light at 1.3 μm. Different curves are obtained at different temperatures and show that the performance degrades with an increase in temperature.

With a further increase in the current, the output power increases linearly with the current and eventually starts to saturate. The amount of power delivered by a semiconductor laser is strongly temperature dependent and the device of Fig. 6 does not even reach threshold when the temperature exceeds 130°C. Nonetheless, at room temperature, properly designed semiconductor lasers routinely emit ~ 10 mW of output power at a current ~ 100 mA. Since the voltage across the p-n junction is about 1 volt, a semiconductor laser converts electrical energy into optical energy with an efficiency ~ 10%. This is a very respectable figure compared to most of the other kinds of lasers and is even more remarkable in view of the device compactness.

To understand the origin of the laser threshold, it is helpful to reconsider the processes of spontaneous and stimulated emissions discussed earlier (see Fig. 2). As the current is progressively increased, the density of electrons and holes injected into the active layer builds up, and eventually population inversion occurs. Because of the feedback provided by the cleaved facts, the number of photons travelling perpendicular to the facets increases. However, some photons are lost through the partially transmitting facets and some get scattered or absorbed inside the cavity. If the loss exceeds the gain, stimulated emission can not sustain a steady supply of photons. This is precisely what happens below threshold and the output consists of mainly spontaneously emitted photons with power levels of a few microwatts. At threshold, gain equals loss and stimulated emission begins to dominate.

In a narrow current range in the vicinity of the threshold current, the output power jumps by several orders of magnitude and the spectral width of the emitted radiation narrows considerably because of the coherent nature of stimulated emission. At that point, the laser has reached threshold and the injected carrier density is pinned to its threshold value. With a further increase in the current, the internal quantum efficiency becomes nearly 100%, and almost all electrons and holes injected into the active region recombine through stimulated emission. However, not all generated photons constitute the output power, since some of them are internally absorbed or scattered. An useful measure of efficiency is the differential quantum efficiency which is the ratio of the rate of emitted photons to the rate of generated photons. It is related to the slope of the L-I curve by the simple relation

$$\eta_d = \frac{2q}{h\nu} \frac{dL}{dI}, \qquad (3)$$

where q is the magnitude of the electron's charge. The differential quantum efficiency is typically 50%. The corresponding value of the slope dL/dI at 1.3 μm is about 0.24 mW/mA/facet.

It is evident from Fig. 6 that the performance of a semiconductor laser degrades with an increase in the device temperature. The threshold current that is fairly low ($\simeq 15$ mA) around room temperature increases almost exponentially with the temperature rise. This behavior can be qualitatively understood by recalling that electrons and holes in a semiconductor can recombine either radiatively or non-radiatively. For long-wavelength semiconductor lasers ($\lambda > 1 \mu$m), the non-radiative recombination is dominated by the Auger process. The rate of Auger recombination increases exponentially with temperature [7] leading to the observed increase in the threshold current. Figure 6 also shows that the differential quantum efficiency decreases with the temperature rise and the L-I curves roll-off at much lower output powers. Several factors may contribute to this behaviour. One possible factor is the increase in the "leakage current" when the total device current is increased. To understand its origin, note in Fig. 5 that not all current passes through the active region. The part of the current that flows around the active region is the leakage. As attempt is made to minimize the leakage current through the use of current-blocking junctions. However, these junctions become less effective at high currents and lead to the roll-off observed in the L-I curves shown in Fig. 6.

A valuable feature of semiconductor lasers is that their output can be modulated directly by modulating the applied current. This feature is important for optical communication systems since the information coded in the form of electrical pulses can be converted directly into the optical form. The information-transmission capacity is measured in terms of bits per second and depends on how quickly the laser responds to electrical changes. The present-day technology requires transmission rate in the range of Gigabits/sec. To meet this demand, the semiconductor laser should have a response time of a fraction of a nanosecond.

When the current applied to a semiconductor laser is changed suddenly, the laser does not reach its new steady state immediately. Rather, the electron and photon populations undergo oscillations before attaining their final steady-state values. These oscillations are referred to as relaxation oscillations and are manifestation of an intrinsic resonance during which electrons and photon exchange energy periodically. The frequency of relaxation oscillations plays an important role and, as a general rule, the semiconductor laser cannot be modulated at frequencies exceeding significantly this internal resonance frequency. Under typical operating conditions, the relaxation-oscillation frequency is about 3-5 GHz, and semiconductor lasers can be used in optical communication systems operating at several Gigabits/sec.

The spectral width of the emitted light is another important parameter used to characterise the performance of a semiconductor laser. Below threshold, the output is dominated by spontaneous emission with a spectral width ~ 20-30 nm (see Fig. 7). In the above-threshold regime, the optical spectrum consists of few equispaced discrete lines corresponding to the longitudinal modes of the Fabry-Perot cavity discussed earlier. For a 250 μm-long laser cavity, the mode

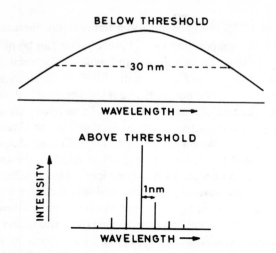

Fig. 7. Spectrum of the emitted light shown schematically. It is relatively wide below threshold. In the above threshold regime, the light is emitted at discrete wavelengths corresponding to the longitudinal modes of the Fabry-Perot cavity.

spacing is about 150 GHz which corresponds to a wavelength spacing of about 1 nm at the 1.3 μm wavelength. The spectral width of each longitudinal mode is, however, much smaller and is typically ~100 MHz when the laser is operating continuously.

7. SINGLE-FREQUENCY SEMICONDUCTOR LASERS

We have seen that when the semiconductor laser operates in several longitudinal modes simultaneously, its spectral width is a few nanometers. Although remarkably small, such a spectral width is still sometimes unacceptable for high-speed optical communication systems. Clearly it can be substantially reduced if a semiconductor laser can be forced to emit light in a single longitudinal mode. Such lasers are referred to as single-frequency semiconductor lasers and have attracted considerable attention recently [13]. To understand their usefulness, it is best to begin by describing how the spectral width affects the performance of an optical communication system.

When a light pulse is transmitted through an optical fibre, two thing happen: Pulse energy is reduced because of absorption and pulse width increases because of chromatic dispersion. The extent of pulse broadening is related to the spectral width. Physically, different wavelengths have different speeds inside the fibre with the consequence that different wavelength components of the pulse disperse with respect to each other as they travel. The broadening for a pulse of spectral width $\Delta\lambda$ is given by

$$\Delta\tau = DL\Delta\lambda , \qquad (4)$$

where L is the fibre length and D is the dispersion coefficient expressed in units of ps/km/nm. In order not to interfere with the neighbouring pulse, $\Delta\tau$ should not exceed the time slot allocated to a single pulse (or bit). If B is the bit rate, the system parameters must satisfy the condition $B\Delta\tau < 1$, or using Eq. (4),

$$BLD\Delta\lambda < 1 . \qquad (5)$$

To reduce the system cost, an attempt is made to maximize the fibre length L before it becomes necessary to amplify the transmitted signal. This is achieved by operating at the wavelength for which silica fibres have the lowest loss. However, at the minimum-loss wavelength of 1.55 μm, these fibres have considerable dispersion with $D = 15\text{-}20$ ps/nm/km. If we take the fibre length $L = 100$ km and consider transmission at the bit rate $B = 1$ Gb/s, the condition (5) implies that $\Delta\lambda < 1$nm. This is a fairly stringent requirement and is not satisfied by semiconductor lasers emitting in several longitudinal modes. Such multi-longitudinal-mode lasers are, therefore, used at a wavelength of about 1.3 μm, where silica fibres happen to have negligible dispersion.

Two alternatives are being pursued to design optical communication systems operating at the minimum loss wavelength of 1.55 μm. In one scheme, the dispersion coefficient D is reduced by modifying the fibre design[14,15]. Such dispersion-shifted fibres can then be used with multi-longitudinal-mode semiconductor lasers. The other possibility is to use conventional fibres but force the semiconductor laser to emit in a single longitudinal mode so that the spectral width $\Delta\lambda \ll 1$ nm. This is the reason for the current interest in single-frequency semiconductor lasers.

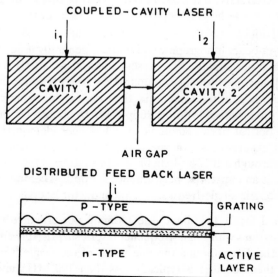

Fig. 8. Schematic illustration of the two schemes employed for single-frequency semiconductor lasers.

Two mechanisms have been used for controlling the longitudinal modes of a semiconductor laser. These are referred to as the coupled-cavity and distributed-feedback (DFB) mechanisms. In the coupled-cavity case, considerable attention has been paid to the cleaved-coupled-cavity or C^3 laser [16], made by cleaving the conventional semiconductor laser approximately in the middle. The resulting C^3 laser has two sections which are optically coupled but electrically isolated from each other (see Fig. 8). The current in each section can be individually adjusted. The physical mechanism behind the mode selectively is optical interference: only the longitudinal mode that is simultaneously resonant with respect to both cavities has the lowest overall loss and is selected by the device. In contrast to the optical spectrum of a single-cavity laser shown in Fig. 7, the side modes are suppressed in power by two or three orders of magnitude and the optical spectrum of a C^3 laser consists of a single line for all practical purposes.

In the case of DFB lasers, a grating is etched along the cavity length on the surface of a cladding layer (see Fig. 8). The grating leads to an effective spatial modulation of the refractive index. The phenomenon of Bragg diffraction occurring in such a medium couples the forward and backward propagating waves and is responsible for the distributed feedback. The wavelength selectivity arises from the fact that Bragg diffraction occurs only at a particular wavelength λ determined by the grating period Λ though the relation

$$\Lambda = m\lambda/2n , \qquad (6)$$

where n is the refractive index and m is the order of Bragg diffraction. For $\lambda = 1.55\,\mu$m and $n \simeq 3.4$, the grating period $\Lambda \simeq 0.23\,\mu$m for the first-order grating ($m = 1$) and doubles for the second-order grating ($m = 2$). The holographic or the electron-beam lithographic technique is generally used to etch a grating with submicron periodicity. The present state-of-the-art DFB semiconductor lasers[17,18] show performance comparable to the Fabry-Perot-type semiconductor lasers while, at the same time, their optical spectrum exhibits a single longitudinal mode with side modes suppressed by about a factor of 1000. Their potential in optical fibre communication has been demonstrated in several transmission experiments. In one experiment[19] the signal at 4 Gb/s was transmitted through a 103-km-long fibre with a bit-error rate of less than one in billion. This is an impressive performance for optical communication systems employing 1.55 μm single-frequency semiconductor lasers.

The critical factor that limits the ultimate performance of optical communication systems with single-frequency lasers, surprisingly enough, is still the spectral width. Even though the line width of the single longitudinal mode selected by these devices is typically less than 100 MHz under continuous operation, it broadens by more than a factor of 100 when semiconductor lasers are directly modulated in the Gigahertz range. The exact amount of broadening depends on several parameters such as the bit rate and the amplitude of the modulation current. Under typical operating conditions and at the bit rate $B \sim 1$ Gb/s, the broadened line width is ~ 10 GHz or ~ 0.1 nm on the wavelength scale.

The physical mechanism behind the line broadening is called frequency chirping. Its origin lies in the fact that in semiconductor lasers, the power or amplitude modulation (AM) is always accompanied by the frequency modulation (FM). More specifically, when the current is modulated, the population of the injected charge carriers inside the active region varies periodically. The resulting gain change not only affects the amplitude but also the phase of the emitted light through refractive-index variations. Since dynamic variation of the phase is equivalent to shifting the instantaneous optical frequency, both AM and FM occur simultaneously in a semiconductor laser. Since frequency chirping is a fundamental phenomenon, it cannot be altogether avoided in directly modulated semiconductor lasers. One possibility is to operate the semiconductor laser continuously and then modulate the emitted beam externally. The use of external modulators is being pursued as a viable option for 1.55 μm optical communication systems. The record-breaking transmission experiment at 4 Gb/s over 117 km made use of a C^3 laser with external modulator [5].

8. SEMICONDUCTOR LASERS FOR COHERENT SYSTEMS

At present, there is a growing interest[20] in adopting the coherent detection techniques for lightwave systems since their use leads to a considerable improvement in the receiver sensitivity (by 10-20 dB). In a coherent fibre-optic communication system, the received signal is first mixed with the output of a local oscillator (a high-power wavelength-stabilized semiconductor laser) and then detected using homodyne or heterodyne technique. The coherent lightwave systems are particularly attractive when the information is transmitted by modulating the frequency (frequency-shift keying) or the phase (phase-shift keying) of the carrier wave rather than the amplitude. The use of frequency-division multiplexing then has the potential of providing a large number of independent, narrow communication channels distributed over the entire fibre bandwidth in a manner analogous to the FM radio transmission.

Semiconductor lasers should meet several requirements for their use in coherent lightwave systems. Both the transmitting laser and the local oscillator at the receiving end should provide their output in a single longitudinal mode with well-stabilized wavelengths. Their wavelengths should either exactly coincide (homodyne detection) or differ by a small amount (heterodyne detection), typically a fraction of a nanometer. Ideally, one would like to have a high power, narrow linewidth semiconductor laser. The linewidth requirements are particularly stringent when the phase-shift keying is used for the carrier modulation. It is estimated[21] that the linewidth should be lower by a factor of 300 or more in comparison to the bit rate. For a bit rate of ≤ 1 Gb/s, the required linewidth is below 1 MHz and may be difficult to realize with single-cavity single-frequency lasers such as DFB semiconductor lasers.

A laser that meets the above-mentioned requirements of stable single-longitudinal-mode operation, a wide tuning range, and a narrow linewidth is the external-cavity semiconductor laser [22]. As shown schematically in Fig. 9, the basic idea is simple and requires the coupling of a multimode semiconductor laser to an external grating. The wavelength-selective nature of the feedback provided by the grating selects a single longitudinal mode whose wavelength can be tuned over a wide range ($\gtrsim 50$ nm) by simply rotating the grating. A narrow linewidth is obtained since the external cavity increases the Q-factor of the coupled cavity. It can be shown that the linewidth $\Delta\nu$ of the external-cavity laser approximately obeys the relation

$$\frac{\Delta\nu}{\Delta\nu_0} = \left(\frac{nL}{L_e}\right)^2 \tag{7}$$

where $\Delta\nu_0$ is the linewidth of a solitary semiconductor laser of length L, $L_e \gg L$ is the external-cavity length and n is the refractive index of the semiconductor material. For a 250 μm-long semiconductor laser, $nL \simeq 1$ mm. If we choose a typical value of $L_e = 10$ cm, Eq. (7) predicts that the linewidth of the external-cavity semiconductor laser would be reduced by a factor of 10^4 compared with that of the semiconductor laser alone. Since $\Delta\nu_0$ is typically ~ 100 MHz, $\Delta\nu \sim 10$ kHz can be readily achieved [22].

As is evident from the above description, external-cavity semiconductor lasers have potential for their use in coherent communication systems and have been used worldwide in laboratory demonstrations. In a recent experiment[23] repeaterless transmission over a distance of 148 km at a bit rate of 1 Gb/s was demonstrated. It should however be stressed that the coherent communication technology is far from being mature. Further improvements in the perfor-

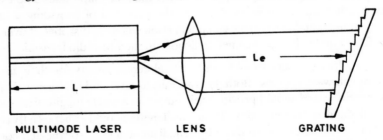

Fig. 9. Schematic illustration of an external-cavity semiconductor laser. The laser facet facing the grating has an antireflection coating to maximize the coupling between the laser cavity and the external cavity.

mance of semiconductor lasers would be required before its deployment in actual systems.

9. CONCLUDING REMARKS

In this chapter I have discussed the physics and the performance of semiconductor lasers from the standpoint of their application in optical fibre communications. The semiconductor laser has proved its usefulness in this field and soon a large fraction of telephone calls and data transmission will involve the use of a semiconductor laser. This is however not the only application of semiconductor lasers. Because of their compact size, these lasers are being used in optical video and audio disks, and in laser printers. Using lead-salt compounds as the semiconductor material for the gain medium, semiconductor lasers can be made to emit in the far infrared region of the optical spectrum and have potential application in ultrahigh-resolution molecular spectroscopy. Another application of semiconductor lasers is their use for optical pumping of solid-state lasers such as a YAG laser. Phase-locked GaAs arrays can be made to emit a few watts of power and are ideally suited for this purpose. Other applications of semiconductor lasers will be investigated simply because these lasers are extremely compact and highly efficient; the properties not shared by other kinds of lasers.

REFERENCES

1. T. Li, "Advances in optical fibre communications: An historical perspective," IEEE J. Selected Areas Commun., Vol. 1, pp. 356-372 (April 1983).
2. Y. Suematsu, "Long-wavelength optical fibre communication", Proc. IEEE, Vol. 71, pp. 692-721 (June 1983).
3. H. Kogelnik, "High-speed lightwave transmission in optical fibres", Science, Vol. 228, pp. 1043-1048 (May 1985).
4. J.E. Midwinter, "Current status of optical communications technology", J. Lightwave Tech., Vol. LT-3, pp. 927-930 (Oct. 1985).
5. S.K. Korotky, Eisenstein G., Gnauck A.H., Kasper B.L., Veselka J.J., Alferness R.C., Buhl L.L., Burrus C.A., Huo T.C.D., Stulz L.W., Nelson K.C., Cohen L.G., Dawson R.W., and Campbell J.C., "4-Gb/s transmission experiment over 117 km of optical fibre using a Ti-LiNbO$_3$ external modulator", J. Lightwave Tech., Vol. LT-3, pp. 1027-1031 (Oct. 1985).
6. T. Li, ed., "*Optical Fibre Communications: Fibre Fabrication,*" Vol. 1, Academic Press, New York (1985).
7. G.P. Agrawal and Dutta N.K., "*Long-wavelength Semiconductor Lasers*", Van Nostrand Reinhold, New York (1986).
8. S.M. Sze, "Physics of Semiconductor Devices", John Wiley, New York, Chapter 2 (1981).
9. R.N. Hall, Fenner G.E., Kingsley J.D., Soltys T.J., and Carlson R.O., "Coherent light emission from GaAs junctions", Phys. Rev. Lett., Vol. 9, pp. 366-368 (Nov. 1982).

10. M.I. Nathan, Dumke W.P., Burns G., Dill, F.H., Jr., and Lasher G., "Stimulated emission of radiation from GaAs p-n junctions", Appl. Phys. Lett., Vol. 1, pp. 62-64 (Nov. 1962).
11. T.M. Quist, Rediker R.H., Keyes R.J., Krag W.E., Lax B., McWhorter A.L., and Zeiger H.J., "Semiconductor Maser of GaAs", Appl. Phys. Lett., Vol. 1, pp. 91-92 (Dec. 1962).
12. I. Hayashi, Panish M.B., Foy P.W. and Sumski S., "Junction lasers which operate continuously at room temperature", Appl. Phys. Lett., Vol. 17, pp. 109-111 (Aug. 1970).
13. T.E. Bell, "Single-frequency semiconductor lasers", IEEE Spectrum, Vol. 20, pp. 38-45, (Dec. 1983).
14. U.C. Paek, Peterson G.E., and Carnevale A., "Dispersionless single-mode lightguides with α index profiles", Bell Syst. Tech. J., Vol. 60, pp. 583-598 (1981).
15. T.D. Craft, Ritter J.E. and Bhagavatula V.A., "Low-loss dispersion-shifted single-mode fibre manufactured by the OVD process", J. Lightwave Tech., Vol. LT-3, pp. 931-934 (Oct. 1985).
16. W.T. Tsang, "The cleaved-coupled-cavity (C^3) laser, in *Semiconductors and Semimetals*", Vol. 22, Part B, Academic Press, New York, Chapter 5 (1985).
17. M. Kitamura, Yamaguchi M., Murata S., Mito I. and Kobayashi K., "High-performance single-lingitudinal-mode operation of InGaAsP/InP DFB-DC-PBH LD's", J. Lightwave Tech., Vol. LT-2, pp. 363-369 (Aug. 1984).
18. G.P. Agrawal and Dutta N.K., "Distributed feedback InGaAsP lasers", Chapter 15.
19. A.H. Gnauck, Kasper B.L., Linke R.A., Dawson R.W., Koch T.L., Bridges T.J., Burkhardt E.G., Yen R.T., Wilt D.P., Campbell J.C., Nelson K.C. and Cohen L.G., "4-Gb/s transmission over 103 km of optical fibre using a novel electronic multiplexer/demultiplexer", J. Lightwave Tech., Vol. LT-3, pp. 1032-1035 (Oct. 1985).
20. D.W. Smith, "Coherent fibreoptic communciations", Laser Focus, Vol. 11, pp. 92-106 (Nov. 1985).
21. G. Nicholson, "Probability of error for optical heterodyne DPSK system with quantum phase noise", Electron. Lett., Vol. 20, pp. 1005-1007 (Nov. 1984).
22. M.R. Matthews, Cameron K.H., Wyatt R. and Devlin W.J., "Packaged frequency-stable tunable 20 kHz linewidth 1.5 μm InGaAsP external cavity laser", Electron. Lett., Vol. 21, pp. 113-115 (Jan. 1985).
23. B.L. Kasper, Olsson N.A., and Linke R.A., presented at the European Conference on Optical Communication (ECOC/IOOC'85), Venice, Italy (Oct. 1985).

15

Distributed Feedback InGaAsP Lasers[†]

GOVIND P. AGRAWAL[*] AND NILOY K. DUTTA[‡]

---

1. INTRODUCTION

The development of InGaAsP semiconductor lasers [1-3] has led to phenomenal progress in the field of optical fibre communications. Although conventional semiconductor lasers are generally adequate for lightwave systems operating near the 1.3 μm wavelength where fibre dispersion is negligible, their use at the fibre-loss-minimum wavelength of 1.55 μm is limited to systems employing dispersion-shifted fibres. The reason behind it is that a conventional semiconductor laser does not emit light in a single longitudinal mode. In general, the mode closest to gain peak is most intense and a few per cent of the output power is carried by other longitudinal modes lying close to the gain peak [4]. In the presence of fibre-chromatic dispersion, the unwanted side modes limit the information transmission rate by reducing the fibre bandwidth. It is therefore, desirable to devise means so that semiconductor laser emits light predominantly in a single longitudinal mode even under high-speed modulation. Such lasers are referred to as single-frequency lasers [5] and have been of much recent interest in view of their potential application in optical communication systems operating at the 1.5 μm wavelength where the loss of silica fibres is minimum.

In conventional Fabry-Perot (FP) type semiconductor lasers, the feedback is provided by facet reflections whose magnitude remains the same for all

---

[†] Reprinted from J.I.E.T.E., Vol. 32 (Special issue on Optoelectronics and Optical Communication), July-August (1986).

[*] The Institute of Optics, University of Rochester, Rochester, New York 14627, USA.

[‡] AT&T Bell Laboratories, Murray Hill, New Jersey 07974, USA.

longitudinal modes. The only longitudinal-mode discrimination in such a laser is provided by the gain spectrum itself. Since the gain spectrum is usually much wider than the longitudinal-mode spacing, the resulting mode discrimination is poor. One way of improving the mode selectivity is to make the feedback frequency dependent so that the cavity loss is different for different longitudinal modes. Two mechanisms have been found useful in this respect and are known as the distributed feedback and the coupled-cavity mechanisms [5]. Distributed feedback (DFB) lasers are described in this chapter.

In a DFB semiconductor laser, the feedback necessary for the lasing action is not localized at the cavity facets but is distributed throughout the cavity length. This is achieved through the use of a grating etched on a cladding layer along the cavity length (see Fig. 1). The resulting periodic perturbation of the refractive index provides feedback via Bragg scattering which couples the forward and backward running waves. Mode selectivity of the DFB mechanism results from the Bragg condition which states that coherent coupling between counter-propagating waves occurs for wavelengths such that the grating period $\Lambda = m\lambda_m/2$, where λ_m is the wavelength inside the laser medium and the integer m represents the order of Bragg diffraction induced by the grating. By choosing Λ appropriately, such a device can be made to provide distribution feedback only at selective wavelengths.

Kogelnik and Shank [6, 7] were the first to observe and analyse the lasing action in a periodic structure that utilized the DFB mechanism. Since then, DFB semiconductor lasers have attracted considerable attention both ex-

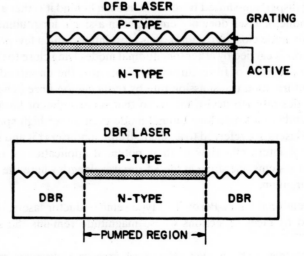

Fig. 1. Schematic illustration of a distributed feedback (DFB) and a distributed Bragg reflector (DBR) semiconductor laser. Different refractive indices on two sides of the grating result in a periodic index perturbation which is responsible for the distributed feedback. Shaded area shows the active region of the device.

perimentally and theoretically. Although most of the early work was related to GaAs lasers, the need of a single-frequency semiconductor laser operating at the fibre-loss-minimum wavelength of 1.55 μm has resulted in the development of DFB-type InGaAsP lasers [8-11]. From the viewpoint of device operation, semiconductor lasers employing the DFB mechanism can be classified into two broad categories (see Fig. 1) referred to as DFB lasers and distributed-Bragg-reflector (DBR) lasers. In DBR lasers, [11] the unpumped corrugated end-regions act as effective mirrors whose reflectivity is of DFB origin and is, therefore, wavelength dependent. We consider only DFB lasers in this chapter.

The chapter is organized as follows. Next section deals with the theory of DFB lasers. Starting from the coupled-wave equations, we obtain the wavelengths and the threshold gain of the DFB-longitudinal modes for the general case wherein both the etched grating and the cleaved facets participate in providing the feedback necessary for the lasing action. We then discuss the expected lasing characteristics such as the threshold current density, the differential quantum efficiency, the mode-suppression ratio, and the linewidth. Various kinds of DFB laser structures and their fabrication techniques are described in the next section. The experimental results for 1.55 μm InGaAsP DFB lasers are then presented. In particular, we discuss the light-current characteristics, the wavelength tuning, the modulation response, and the linewidth characteristics.

2. THEORY

2.1 Coupled-wave Equations

To understand the operating characteristics of a DFB laser, it is first necessary to consider wave propagation in periodic structures. The grating-induced dielectric perturbation leads to a coupling between the forward and backward waves associated with a particular laser mode. In the coupled-wave approach [7], the coupled set of two equations corresponding to the counter propagating waves is solved subject to the appropriate boundary conditions.

The starting point of the analysis is the time-independent wave equation

$$\nabla^2 E + \varepsilon(x,y,z) k_0^2 E = 0, \tag{1}$$

where $k_0 = \omega/c$ and ω is the mode frequency. It is useful to write

$$\varepsilon(x,y,z) = \bar{\varepsilon}(x,y) + \Delta\varepsilon(x,y,z), \tag{2}$$

where $\bar{\varepsilon}(x,y)$ is the average value of ε and the dielectric perturbation $\Delta\varepsilon$ is non-zero only over the grating region whose thickness is equal to the corrugation depth. Since $\Delta\varepsilon$ is periodic in z, it can be expanded in a Fourier series

$$\Delta\varepsilon(x,y,z) = \sum_{m \neq 0} \Delta\varepsilon_m(x,y) \exp(2im\pi z/\Lambda). \tag{3}$$

We assume that the field E can be approximately written as

$$E(x, y, z) = \hat{X} U(x,y) \left[A \exp(i\beta_0 z) + B \exp(-i\beta_0 z) \right], \quad (4)$$

where $\hat{X}$ is the polarization unit vector, $U(x,y)$ is the waveguide-mode distribution, and

$$\beta_0 = m\pi / \Lambda \quad (5)$$

is the Bragg wavenumber corresponding to the mth order grating, i.e., the mth order Fourier component in Eq. (3) provides strongest coupling between the forward and backward waves. Using Eqs. (1-4), the coupled-wave equations can be written as

$$\frac{dA}{dz} = i\Delta\beta A + iKB, \quad [6(a)]$$

$$-\frac{dB}{dz} = i\Delta\beta B + iKA, \quad [6(b)]$$

where

$$\Delta\beta = (\bar{\mu}k_0 - \beta_0) - i\bar{\alpha}/2 = \delta - i\bar{\alpha}/2, \quad (7)$$

$\bar{\mu}$ is the effective mode index, and $\bar{\alpha}$ is the mode gain. The parameter δ provides the detuning of the waveguide mode from the Bragg wavelength. The coupling co-efficient K is given by

$$K = \frac{k_0^2}{2\beta} \int \int_{-\infty}^{\infty} \Delta\varepsilon_m(x,y) U^2(x,y) \, dx \, dy \bigg/ \left(\int \int_{-\infty}^{\infty} U^2(x,y) \, dx \, dy \right) \quad (8)$$

and, in general, it depends on the lateral and transverse profiles of the waveguide mode, together with the grating characteristics such as the corrugation shape and depth.

A general solution of the coupled-wave equations is of the form

$$A(z) = A_1 \exp(iqz) + A_2 \exp(-iqz), \quad (9)$$

$$B(z) = B_1 \exp(iqz) + B_2 \exp(-iqz), \quad (10)$$

where q is the complex wave number to be determined from the boundary conditions. The constants A_1, A_2, B_1 and B_2 are interdependent. If we substitute the general solution in Eq. (6) and equate the co-efficients of $\exp(\pm iqz)$, we obtain

$$(q - \Delta\beta)A_1 = K B_1, \; (q + \Delta\beta) B_1 = -K A_1, \quad (11)$$

$$(q - \Delta\beta) B_2 = K A_2, \; (q + \Delta\beta) A_2 = -K B_2. \quad (12)$$

These relations are satisfied with non-zero values of A_1, A_2, B_1 and B_2 if the possible values of q obey the dispersion relation (obtained by setting the determinant of the coefficient matrix to be zero)

$$q = \pm [(\Delta\beta)^2 - K^2]^{1/2} . \tag{13}$$

The plus and minus signs correspond to the forward and backward propagating waves, respectively. Further, we can define the DFB reflection co-efficient

$$r(q) = (q - \Delta\beta)/K = -K/(q + \Delta\beta) , \tag{14}$$

and eliminate A_2 and B_1 in Eqs. (9) and (10) in favour of $r(q)$. The general solution of the coupled-mode equation then becomes

$$A(z) = A_1 \exp(iqz) + r(q) B_2 \exp(-iqz) , \tag{15}$$

$$B(z) = B_2 \exp(-iqz) + r(q) A_1 \exp(iqz) . \tag{16}$$

Since $r = 0$ if $K = 0$, it is evident that $r(q)$ represents the fraction of the forward-wave amplitude that is reflected back toward the backward wave and vice versa. The sign ambiguity in the expression (14) for $r(q)$ can be resolved by choosing the sign of q such that $r(q) \leq 1$. Mode selectivity of DFB lasers stems from the q dependence of r. The eigenvalue q and the ratio B_2/A_1 are determined by applying the appropriate boundary conditions at the laser facets.

Before considering a finite-length DFB laser, consider wave propagation in an infinite-length periodic structure. The dispersion relation (13) governs the possible q values in such a medium. In general, q is complex since $\Delta\beta$ is complex in the presence of gain or loss associated with the medium. If we neglect $\bar{\alpha}$ in Eq. (17), $\Delta\beta = \delta$ is real. According to Eq. (13), q is purely imaginary for $|\Delta\beta| \leq K$ and the medium cannot support a propagating wave. Consequently, an infinite periodic structure has a stop band of width $2K$ centered at β_0 i.e., only waves whose propagation constant β satisfies $|\beta - \beta_0| > K$ can propagate inside the medium. It will be seen later that the stop band exists even for a finite-length DFB laser, although under specific conditions, the device may support a "gap-mode" which lies inside the stop band.

2.2 Longitudinal Modes of a DFB Laser

For a finite length DFB laser, the boundary conditions at the facets are satisfied only for discrete values of q. The real and imaginary parts of q give respectively the DFB-longitudinal-mode frequency and the corresponding threshold gain. Consider a laser of length L. Since L is not necessarily an exact multiple of the grating period Λ, the last period of the grating close to the facet is generally not complete. Even though no significant distributed feedback occurs over these incomplete grating periods, the phase shift in this region plays an important role in determining DFB-laser characteristics and should be properly accounted for. A simple way to do this is to assume that the effective facet reflection co-efficient is complex,

$$r_j = \sqrt{R_j} \exp(i\phi_j) , \quad j = 1, 2 \tag{17}$$

where R_j is the facet reflectivity and ϕ_j is the round trip phase shift encountered by the field in travelling the extra distance (a fraction of Λ) corresponding to the last incomplete grating period. In applying the boundary conditions, the mirrors can then be assumed to be located at the last complete grating period. The constant phase shift of about 180° occurring during facet reflections can also be incorporated into ϕ_j. A characteristic feature of the DFB semiconductor laser is that the phase shifts ϕ_1 and ϕ_2 are expected to vary from device to device since the relative distance between the cleaved facet and the last complete grating period (a small fraction of 1 μm) is uncontrollable at present except through the use of controlled etching or coating of individual facets.

The boundary conditions at the two facets are

$$B(0) = r_1 A(0) , \quad A(L) = r_2 B(L) . \tag{18}$$

If we use them in the general solution given by Eqs. (15) and (16), we obtain two homogeneous equations for the unknown constants A_1 and B_2,

$$(r_1 - r)B_2 - (1 - r r_1)A_1 = 0 , \tag{19}$$

$$(r_2 - r) \exp(2iqL)A_1 - (1 - r r_2)B_2 = 0 . \tag{20}$$

These equations have a consistent nontrivial solution only for values of q satisfying the eivenvalue equation [12, 13]

$$[(r_1 - r)(r_2 - r)/\{(1 - r r_1)(1 - r r_2)\}] \exp(2igL) = 1 . \tag{21}$$

This is the threshold condition for DFB lasers. It is similar to the threshold condition obtained for an FP laser and reduces to it if the DFB contribution is neglected by setting $r = 0$. On the other hand, if facet reflectivities are neglected by setting $r_1 = r_2 = 0$, we obtain

$$r^2 \exp(2igL) = 1 , \tag{22}$$

implying that r is the effective reflectivity for a DFB laser. The allowed eigenvalues q are obtained after noting that r given by Eq. (14) itself depends on q. Each eigenvlaue q is complex and can be used to calculate

$$\Delta\beta = \delta - i\bar{\alpha}/2 = \pm(q^2 + K^2)^{1/2} . \tag{23}$$

The values of δ corresponding to different eigenvalues indicate how far that longitudinal mode is displaced from the Bragg wavelength λ_0. The DFB longitudinal modes are situated symmetrically with respect to λ_0. In contrast to FP lasers, however, different modes have different threshold gains $\bar{\alpha}$ as determined by the eigenvalues equation (21). The mode gain $\bar{\alpha}$ can be thought of as the gain required to overcome the distributed losses and plays the same role as the mirror loss

$$\alpha_m = (0.5 / L) \ln (R_1 R_2)^{-1} \tag{24}$$

for an FP laser. The thresholds for the DFB and FP lasers can thus be compared through $\bar{\alpha}$ and α_m provided the internal loss α_{int} is assumed to be identical for both kinds of lasers. Using cleaved-facet reflectivities $R_1 = R_2 \simeq 0.32$, we obtain $\alpha_m L \simeq 1.1$, and a DFB laser would have a lower threshold compared to an FP laser if $\bar{\alpha} L < 1.1$ for the lowest-threshold mode.

Numerical solution of the eigenvalue Eq. (21) is generally required to obtain the longitudinal-mode frequencies and their respective threshold gains of a DFB laser. The ideal case wherein the facet reflectivities $r_1 = r_2 = 0$ has been discussed by Kogelnik and Shank [7]. In practice, at least one facet is generally left as cleaved. It is thus important to study how the DFB-mode spectra is modified by the facet reflectivities R_j and the grating phases $\phi_{\alpha j}$ [see (17)].

We first consider the case $R_1 = R_2 = 0.32$ to include reflections from both cleaved facets. This is the case commonly encountered. Figure 2 shows the mode gain $\bar{\alpha} L$ and the detuning δL for six DFB modes as ϕ_1 is varied over its 2π range for a fixed value of ϕ_2. Two curves for different values of KL are shown. The horizontal dotted line shows the FP laser threshold in the absence of distributed feedback ($K = 0$). The open circles denote the location of DFB modes for $\phi_1 = 0$. As ϕ_1 increases, the modes shift to the left on the curve.

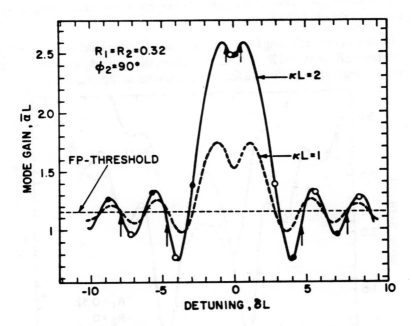

Fig. 2. Normalized threshold gain αL and detuning δL of the six longitudinal modes of a DFB laser with two cleaved facets. Continuous curves are obtained by changing ϕ_1 over its entire range from 0 to 2π. DFB modes for $\phi_1 = 0$ (open circles), $\phi_1 = \pi$ (solid dots), and $\phi_1 = \pi/2$ (arrows) are shown explicitly on the solid curve. The horizontal dashed line shows the threshold gain of an FP laser with cleaved facets (no distributed feedback).

Their position for the specific case of $\phi_1 = \pi$ is marked by closed circles. Arrows denote the modes for the intermediate case of $\phi_1 = \pi/2$.

Several conclusions can be drawn from Fig. 2. It is evident that facet reflections break the gain degeneracy that occurs in the absence of reflections for the two modes separated by the stop band [7]. Which of the two modes has a lower threshold gain depends on the relative values of the phases ϕ_1 and ϕ_2. For the case of $\phi_2 = \pi/2$ shown in Fig. 2, the mode on the left side of the Bragg wavelength has a lower gain for $-\pi/2 \leq \Phi_1 \leq \pi/2$ while the reverse occurs for $\pi/2 < \phi_1 \leq 3\pi/2$. The mode spectrum is degenerate for $\phi_1 = \pi/2$ and $\phi_1 = 3\pi/2$. Arrows in Fig. 2 mark the location of degenerate modes for $\phi_1 = \pi/2$. Another feature arising from facet reflections is the existence of a high-gain mode inside the stop band.

It is of interest to compare the threshold of the lowest-threshold-gain DFB mode with that of an FP laser. Figure 2 shows that the DFB-threshold is always lower than the FP-threshold and the gain margin increases with the coupling co-efficient. It should be stressed, however, that the gain peak is assumed to occur in the vicinity of the Bragg wavelength. When the two are significantly far apart, an FP-mode close to the gain peak may have lower threshold gain and would reach threshold first.

The case of one reflecting facet and one non-reflecting facet is shown in Fig. 3 after taking $R_2 = 0$, and $R_1 = 0.32$. As before, the phase ϕ_1 is varied over its entire 2π range. The qualitative features are similar to the case of two cleaved

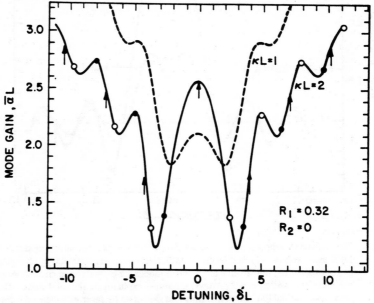

Fig. 3. Same as for Fig. 2 except that the case of DFB laser with one cleaved facet and one non-reflecting facet is considered.

facets. The most significant difference occurs for the higher order DFB modes which have much higher thresholds compared to the case shown in Fig. 2. For a given KL, the threshold of the lowest-gain mode is also higher when compared with the case of two cleaved facets. The main advantage of a non-reflecting facet is that the FP modes are effectively suppressed, resulting in stable DFB-mode operation.

Recently, phase-shifted DFB lasers wherein the grating phase is shifted by a constant amount over a small region along the cavity length has drawn considerable attention [14,17]. In such devices, the lowest-threshold mode lies inside the stop band. The optimum phase-shift is $\pi/2$ for which the gap mode occurs exactly at the Bragg wavelength ($\delta = 0$). Several different approaches have been followed to introduce the required phase shift in the middle of the laser cavity. In one approach, a small uncorrugated groove region (typical length 10 μm) is introduced in the centre such that the optical phase changes by $\pi/2$ during wave propagation over this region [15]. In another approach, the phase shift is introduced indirectly by using a non-uniform active width [16]. A different active width over a central region changes the effective mode index which in turn changes the optical phase. Eda et al [17] have discussed the longitudinal modes of a phase-shifted DFB laser.

2.3 Operating Characteristics

The performance of a DFB laser depends on the magnitude of the coupling co-efficient given by Eq. (8). The numerical value of K depends on the grating parameters such as the shape, the depth, and the period of corrugation. Further, since the coupling occurs as a result of the interaction between the grating and the evanescent part of the transverse mode, K is also affected by the thickness and composition of the active and cladding layers. Equation (8) can be used to study the variation of K with the design parameters. Experimentally, K is estimated from a measurement of the stop band and typically $K = 40$-80 cm^{-1}.

For a given value of KL, the DFB-mode threshold gain $\bar{\alpha}$ is obtained using the eigenvalue (21) and (23). The threshold material gain is then given by

$$g_{th} = (\bar{\alpha} + \alpha_{int})/\Gamma, \qquad (25)$$

where Γ is the confinement factor and α_{int} takes into account the internal loss arising from the processes such as free-carrier absorption and scattering at various interfaces. If we assume that the material gain varies linearly with the carrier density

$$g(n) = a\,(n - n_0), \qquad (26)$$

the threshold value of the carrier density is given by

$$n_{th} = n_0 + (\bar{\alpha} + \alpha_{int})/(a\Gamma). \qquad (27)$$

Here a is the gain co-efficient and n_0 is the transparency value of the carrier density corresponding to the onset of population inversion. If carrier diffusion is ignored, the threshold current density is given by

$$J_{th} = qdn_{th}(A_{nr} + Bn_{th} + Cn_{th}^2) , \qquad (28)$$

where d is the active-layer thickness, B is the radiative co-efficient and C is the Auger co-efficient. The parameter A_{nr} accounts for the non-radiative recombination due to other processes occurring at various surfaces and impurity sites. The strong temperature dependence of the Auger co-efficient C makes the light-current characteristics of an InGaAsP laser highly temperature dependent, as discussed later.

After the threshold has been reached, the output power varies almost linearly with the current. The differential quantum efficiency is approximately given by

$$\eta_d = \eta_i \bar{\alpha} / (\bar{\alpha} + \alpha_{int}) , \qquad (29)$$

where η_i is the internal quantum efficiency, and is nearly 100% when stimulated emission dominates in the above threshold regime. The internal loss α_{int} incorporates losses from all possible mechanisms. For a DFB laser, two additional mechanism may contribute to increase α_{int} and decrease η_d compared to its value for an FP laser. These are (i) scattering losses due to grating imperfections and (ii) radiation losses from periodic structures. The latter mechanism does not occur for a first-order grating, but may contribute a loss ~ 5 cm^{-1} in the case of a second-order grating since first-order Bragg diffraction then radiates perpendicular to the junction plane. It should also be noted that for a DFB laser with cleaved facets both J_{th} and η_d are device dependent because of uncontrolled grating phases at the facets. As a result, the light-current characteristics of such DFB lasers obtained from a single wafer are expected to exhibit more device-to-device variations than similar FP lasers.

The optical spectrum of a DFB laser consists of a single longitudinal mode with the lowest threshold gain $\bar{\alpha}$. The gain margin $\Delta\alpha$ between this mode and the other neighbouring modes (see Fig. 2) dictates how much the side modes are suppressed. In general, $\Delta\alpha \gtrsim 10$ cm^{-1} is large enough to suppress the side-mode power by 30 dB or more. Further, this side-mode-suppression ratio is maintained even under high-speed modulation. It is this property of DFB semiconductor lasers that makes them attractive for optical communication systems.

The linewidth of a single-frequency semiconductor laser is given by the modified Schawlow-Townes formula [18]

$$\Delta\nu = n_{sp}(h/2\pi)\omega(1 + \beta^2)\eta_{1d}\gamma^2/(8\pi P) , \qquad (30)$$

where $\eta_{sp} \simeq 2$ is the spontaneous-emission factor, $\beta \simeq 5$ is the linewidth enhancement factor, P is the output power, and

$$\gamma = v_g(\alpha_m + \alpha_{int}) \tag{31}$$

is the photon-decay rate inside the cavity (v_g is the group velocity). For typical laser parameters $\Delta v \simeq 100$ MHz at a power level of 1 mW and decreases inversely with an increase in the output power.

3. DEVICE STRUCTURE AND FABRICATION

Semiconductor laser structures can be classified into two groups, gain guided and index guided. In the gain-guided structure, the width of the optical mode along the junction plane is principally determined by the width of the optical gain region which is turn is determined by the width of the current-pumped region (typically in the range 5-10 μm). In index-guided lasers, a narrow central region of relatively higher index confines the lasing mode to that region. The index-guided lasers can, in general, be divided into two subgroups, weakly index-guided and strongly index-guided. In weakly index-guided lasers, the active region is continuous and the effective index discontinuity is provided by a cladding layer of varying thickness. The strongly index-guided lasers, by contrast, employ a buried heterostructure. In these lasers, the active region is bounded by epitaxially grown layers of lower index both along and normal to the junction plane. The lateral index difference is $\sim$0.01-0.03 for weakly index-guided lasers while it is $\sim$0.2 for strongly index-guided buried heterostructure lasers.

Although both gain-guided and index-guided laser structures can be used to achieve single frequency operation through a distributed feedback mechanism, the index guided structure is preferable since its fundamental mode is more stable with increasing operating current than that of gain guided lasers. Furthermore, strongly index-guided lasers are generally superior to weakly index-guided lasers. The schematic cross-sections of three commonly used DFB laser structures are shown in Fig. 4. They are (a) the ridge waveguide (RWG) DFB laser, [19-21] (b) the double-channel planar buried heterostructure (DCPBH) DFB laser [9,10] and (c) the etched mesa buried heterostructure (EMBH) DFB laser [8]. The ridge waveguide laser is a weakly index-guided laser and the DCPBH and EMBH are strongly index-guided lasers. In this chapter, we concentrate principally on the fabrication and performance characteristic of DCPBH type DFB lasers.

The frequency-selective feedback in a DFB laser is provided by a grating etched in one of the cladding layers. The direct etching of the active layer is generally not preferred since it can increase the rate of non-radiative recombination by introducing defects in the active region. This would affect the device performance through a higher threshold current. Since only the evanes-

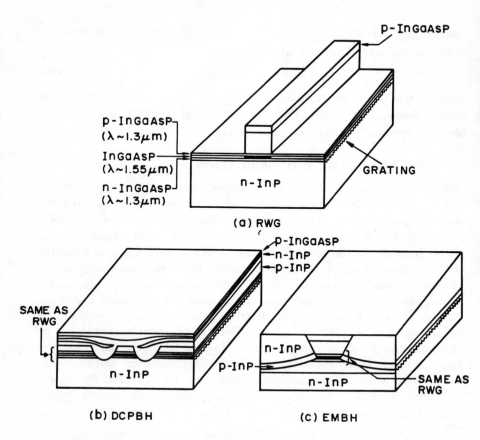

Fig. 4. Schematics of three laser structures employed in DFB laser fabrication.

cent field associated with the fundamental transverse mode interacts with the grating, the exact location of the grating with respect to the active layer and the corrugation depth are critical in determining the effectiveness of the grating. The grating period is determined by the device wavelength in the medium and the order of Bragg diffraction used for the distributed feedback. The Bragg condition for the mth order coupling between the forward and backward propagating waves is

$$\Lambda = m\lambda/2\bar{\mu}, \qquad (32)$$

where $\bar{\mu}$ is the effective-mode index and $\lambda/\bar{\mu}$ is the wavelength inside the medium. For a 1.55 μm InGaAsP laser $\Lambda \simeq 0.23$ μm, if we use $m = 1$ (first-order grating) and a typical value $\bar{\mu} \simeq 3.4$. For a second-order grating $\Lambda \simeq 0.46$ μm.

The fabrication of a 1.55 μm InGaAsP DFB-DCPBH laser involves the following steps. First, a grating with a periodicity of 0.46 μm is formed on a

(100)-oriented n-InP substrate using a holographic technique. In this technique, the substrate is covered with photoresist and placed in the path of two interfering beams obtained from a He-Cd laser emitting at 4420Å. The interference pattern produces a grating on the developed photoresist which is then transferred to the susbstrate using wet chemical etching. Electron-beam writing followed by ion melting or chemical etching has also been used to fabricate gratings on the substrate. Example of first-and second-order gratings holographically farbicated on n-InP with periodicities of 0.23 μm and 0.46 μm are shown in Fig. 5. Although the depth of the grating grooves for the second-order grating is higher than that for the first order, second-order grating is generally less effective in providing frequency-selective feedback compared with a first-order grating because of a lower value of the coupling co-efficient [see Eq. (8)].

The grating fabrication is followed by a planar epitaxial growth on the substrate using either the liquid-phase-epitaxy (LPE) or the vapour-phase-epitsaxy (VPE) growth technique. The five layers grown are n-InGaAsP ($\lambda \sim 1.3$ μm) waveguide layer, undoped InGaAsP ($\lambda \sim 1.55$ μm) active layer, p-InGaAsP ($\lambda \sim 1.3$ μm) antimelt-back layer, p-InP cladding layer, and a p-InGaAsP layer. The last layer protects the cladding layer from thermal decomposition. The growth temperature is $\sim 630°C$. The protection layer is removed by etching before further processing. Two channels parallel to the (110) direction are then etched on the wafer using wet chemical etching through a SiO_2 mask deposited by standard photolithographic techniques. A scanning electron micreoscope cross-section of a double heterostructure wafer grown over a second-order grating is shown in Fig. 6.

The wafer goes through a second LPE growth step which involves the growth of four epitaxial layers. The successive layers grown are p-InP, n-InP and p-InGaAsP ($\lambda \sim 1.3$ μm) respectively. The last layer serves as a contact layer and the other layers confine the current to the active region. The wafer then goes through several processing steps including deposition of p and n contact metals, thinning and cleaving. The final laser chips are 250 μm long and have cleaved facets.

4. DEVICE PERFORMANCE

The light-current characteristic of a typical 1.55 μm DFB DCPBH laser is shown in Fig. 7 under continuous-wave (CW) operation at room temperature. The inset shows the emission spectrum. The mode suppression ration, i.e. the ratio of the intensity of the dominant mode to the next most intense mode, is greater than 30 dB for this device.

The measured threshold current as a function of temperature is shown in Fig. 8. The threshold current I_{th} as a function of temperature T may be written

DFB GRATING FABRICATION

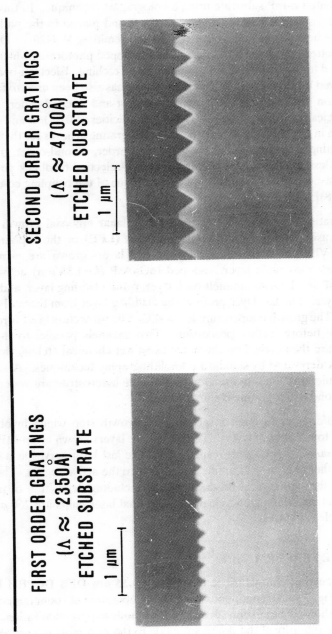

Fig. 5. SEM photomicrographs of first-order and second-order gratings etched on the InP substrates for the fabrication of 1.55 μm DFB lasers.

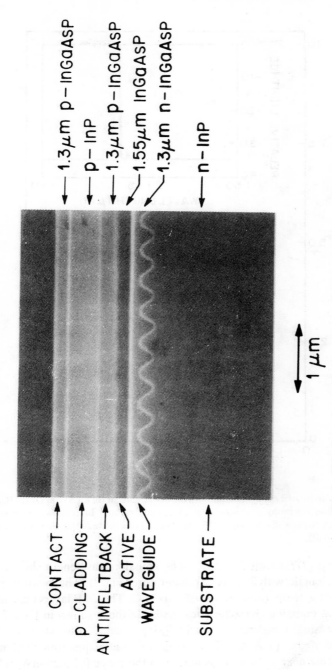

Fig. 6. SEM photomicrograph of a 1.55 μm DFB InGaAsP laser showing thickness and composition of various layers grown by liquid-phase epitaxy on a grating-etched substrate.

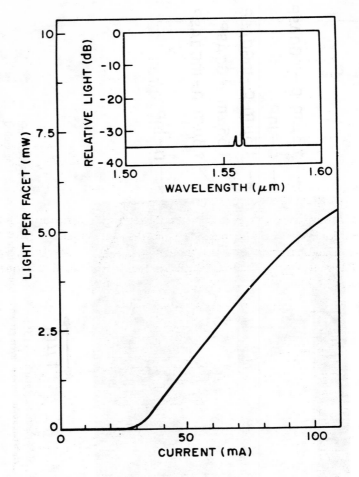

Fig. 7. Variation of the light output with the device current for a 1.55 μm DFB-DCPBH laser operating continuously at room temperature. Inset shows the emission spectrum with a dominant DFB mode and a side mode suppressed by more than 30 dB.

as $I_{th} \sim I_0 \exp(T/T_0)$ with $T_0 \sim 55K$. The emission wavelength shifts towards longer wavelengths with increasing temperature (Fig. 8). The measured shift is 1.08 Å/°C in the temperature range 20°C to 70°C. The DFB lasers emitting near 1.3 μm show emission characteristics similar in those shown in Figs. 7 and 8. With increasing temperature, their lasing modes also shift to longer wavelengths at about 1 Å/°C. The external differential quantum efficiencies of lasers emitting near 1.3 μm are generally in the range 0.2-0.25 mW/mA/facet compared to values in the range 0.1-0.15 mW/mA/facet for lasers emitting near 1.55 μm.

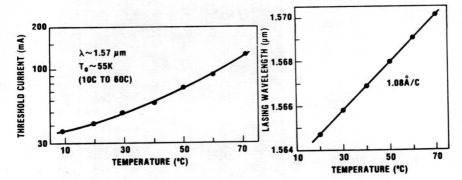

Fig. 8. Variation of threshold current and lasing wavelength with temperature for a 1.57 μm DFB-DCPBH laser.

For lightwave system applications, the modulation characteristics are of considerable interest. The measured small-singal amplitude-modulation response using 2 mA sinusoidal modulation at various bias levels in shown in Fig. 9. Figure 9 suggests that the DFB laser can be modulated at frequencies in excess of 2 Gb/s.

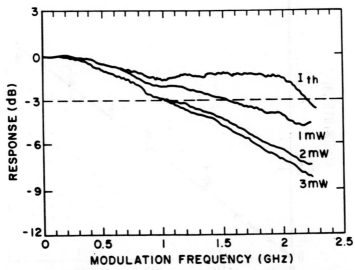

Fig. 9. Small-signal modulation characteristics of a DFB-DCPBH laser. Modulation power (normalized) is plotted as a function of the modulation frequency for several bias levels. The intersection of the dashed line with the solid curves determines the 3-dB modulation bandwidth.

The spectral width of the lasing mode under modulation was also measured. Figure 10 shows the linewidths of the lasing mode both under CW operation and under a 20 mA square-wave current modulation of 0.5 GHz superimposed on a CW bias of 40 mA. The CW linewidth in the tope trace is limited by the

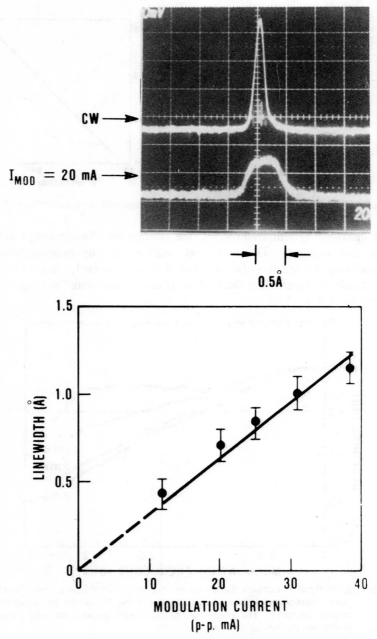

Fig. 10. Frequency-chirp characteristics of a DFB-DCPBH laser under square-wave modulation at 500 MHz. The mode linewidth in shown on the left under CW operation (top curve) and under 20 mA modulation at 500 MHz (bottom curve). Variation of the chirp linewidth with the modulation current is shown on the right.

resolution of the spectrometer (0.2Å). The broadening observed under modulation is due to the modulation of the refractive index by the injected current which modulates the spectral position of the longitudinal modes of the cavity. The spectral linewidth under modulation is also known as the frequency chirp. The chirp linewidth as a function of the modulation current is shown in Fig. 10. The chirp increases linearly with increasing modulation current.

The large laser linewidth under modulation limits the maximum transmission distance (repeater spacing) in lightwave systems. The limitation is due to pulse spreading caused by fibre dispersion. The maximum repeater spacing is approximately given by

$$L \sim 1/(4D\,\sigma\,B)\,,\tag{33}$$

where D is the fibre dispersion, σ is the linewidth, and B is the bit rate. For conventional silica fibres, $D \sim 16$ ps/nm km near $\lambda \sim 1.55\,\mu$m. Using $\sigma \sim 1$Å, Eq. (33) suggests that the maximum allowable repeater spacing is ~ 80 km for a system operating at 2 Gb/s.

The linewidth of an injection laser under CW operation is an important parameter when the laser is used as a source in a coherent lightwave transmis-

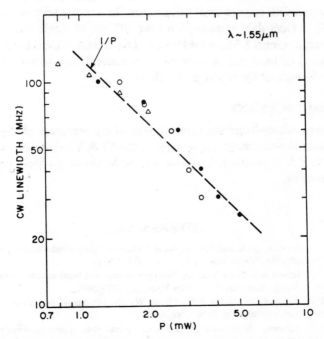

Fig. 11. Variation of the measured CW linewidth with the output power for a DFB-DCPBH laser. Different symbols correspond to three different lasers from the same wafer.

sion system. The linewidth arises from spontaneous-emission-induced phase fluctuations occurring inside the laser cavity. The CW linewidth is measured using a Fabry-Perot interferometer. The linewidth as a function of the output power is shown in Fig. 11 for three 1.55 μm DFB lasers with the first order gratings. The linewidth at an output power of 1 mW is ~100 MHz and is found to vary inversely with the output power in agreement with (30).

5. CONCLUDING REMARKS

In this chapter, we have reviewed the theory and the operating characteristics of DFB InGaAsP lasers. These lasers are at the development stage and their performance is continually being improved. Output powers of ~20 mW at room temperature under continuous operation have been achieved for 1.55 μm DFB lasers with a threshold current of about 20 mA [9,22]. In general, the performance of DFB semiconductor lasers is comparable to similar Fabry-Perot lasers as far as the operating characteristics such as the temperature dependence of the threshold current, the modulation response, and the frequency-chirp behaviour are concerned. At the same time, however, the DFB semiconductor lasers have the additional advantage of maintaining stable single-longitudinal-mode operation under high-speed modulation. This property makes them an ideal candidate for 1.55 μm optical communication systems employing conventional silica fibres. In a recent transmission experiment [23], 4 Gb/s data transmission over 103 km of single-mode fibre was demonstrated using a 1.54 μm DFB laser. The InGaAsP DFB lasers also have the potential of their use in coherent communication systems. Work is in progress throughout the world in this direction.

ACKNOWLEDGEMENT

The authors acknowledge the contributions of the members of the Semiconductor Laser Development Department of the AT & T Laboratories. They are thankful to P.J. Anthony, J.E. Geusic and R.W. Dixon for their support and encouragement.

REFERENCES

1. Y. Suematsu, Iga K. and Kishino K., in "GaInAsP Alloy Semiconductors", ed by T.P. Pearsall, John Wiley, New York, pp. 341-378 (1982).
2. R.J. Nelson and Dutta N.K., in "Semiconductors and Semi-metals", Vol. 22c, ed by W.T. Tsang, Academic Press, New York, pp. 1-59 (1985).
3. G.P. Agrawal and Dutta N.K., "Long-Wavelength Seminconductor Lasers", Von Nostrand Reinhold Co., New York (1986).
4. G.P. Agrawal, "Semiconductor lasers for optical fibre communication", chapter 14 (this book)..
5. T.E. Bell, "Single-frequency semiconductor lasers", IEEE Spectrum, Vol. 20, pp. 38-45 (Dec. 1983).

6. H. Kogelnik and Shank C.V., "Stimulated emission in a periodic structure", Appl. Phys. Lett., Vol. 18, pp. 152-154 (Feb. 1971).
7. H. Kogelnik and Shank C.V., "Coupled-wave theory of distributed feedback lasers", J. Appl. Phys., Vol. 43, pp. 2327-2335 (May 1972).
8. K. Sakai, Utaka K., Akiba S. and Matsushima Y., "1.5 μm range InGaAsP/InP distributed feedback lasers", IEEE J. Quantum. Electron., Vol. QE-18, pp. 1272-1278 (Aug. 1982).
9. M. Kitamura, Yamaguchi M., Murata S., Mito I. and Kobayashi K., "High-performance single-longitudinal-mode operation of InGaAsP/InP DFB-DC-PBH LD's", J. Lightwave Tech., Vol. LT-2, pp. 363-369 (Aug. 1984).
10. D.P. Wilt, Yen R., Wessel T., Besomi P., Brown R.L., Cella T., Dutta N.K. and Anthony P.J., "Proc. Int. Electron Dev. Meeting", pp. 154, San Franscisco (Dec. 1984).
11. Y. Suematsu, Kishino K., Arai S. and Koyama F. in, "Semiconductors and Semimetals", Vol. 22B, ed by W.T. Tsang, Academic Press, New York, pp. 205-255 (1985).
12. W. Streifer, Burnham R.D. and Scifres D.R., "Effect of external reflectros on longitudinal modes of distributed feedback lasers", IEEE J. Quantum Electron., Vol. QE-11, pp. 154-161 (1975).
13. R.F. Kazarinov and Henry C.H., "Second-order distributed feedback lasers with mode selection provided by first-order radiation losses", Ibid, Vol. QE-21, pp. 144-150 (Feb. 1985).
14. K. Sekartedjo, Eda N., Furuya K., Suematsu Y., Koyama F. and Tan bun-Ek T., "1.5 μm phase shifted DFB lasers for single-mode operation", Electron. Lett., Vol. 20, pp. 80-81 (Jan. 1984).
15. S. Koentjoro, Broberg B., Koyama F., Furuya K. and Suematsu Y., "Active distributed reflector lasers phase adjusted by groove region", Jpn. J. Appl. Phys. Vol. 23, pp. L791-L794 (Oct. 1984).
16. H. Soda, Wakao K., Sudo H., Tanahashi T. and Imai H., "GaInAsP/InP phase-adjusted distributed feedback lasers with a step-like non-uniform stripe width structure", Electron. Lett., Vol. 20, pp. 1016-1018 (Nov. 1984).
17. N. Eda, Furuya K., Koyama F. and Suematsu Y., "Axial mode selectivity in active distributed reflector for dynamic-single-mode lasers" J. Lightwave Tech., Vol. LT-3, pp. 400-407 (Apr. 1985).
18. C.H. Henry, "Theory of the linewidth of semiconductor lasers", IEEE J. Quantum Electron., Vol. QE-18, pp. 259-264 (Feb. 1982).
19. L.D. Westbrook, Nelson A.W., Fiddyment P.J. and Evans J.S., "Continuous-wave operation of 1.5 μm distributed feedback ridge-waveguide lasers", Electron. Lett., Vol. 20, pp. 225-226 (Mar. 1984).
20. H. Temkin, Dolan G.J., Olsson N.A., Henry C.H., Logan R.A., Kazarinov R.F. and Johnson L.F., "1.55 μm InGaAsP ridge waveguide distributed feedback laser", Appl. Phys. Lett., Vol. 45, pp. 1178-1180 (Dec. 1984).
21. W.T. Tsang, Logan R.A., Olsson N.A., Johnson L.F. and Henry C.H., "Heteroepitaxial ridge overgrown distributed feedback laser at 1.5 μm", Appl. Phys. Lett., Vol. 45, pp. 1272-1274 (Dec. 1984).
22. H. Nagai, Matsuoka T., Noguchi Y., Suzuki Y. and Joshinkuni Y., "InGaAsP/InP DFB-BH lasers with both facets cleaved structure", IEEE J. Quantum Electron. (to be published).
23. A.H. Gnauck, Kasper B.L., Lenke R.A., Dawson R.R., Koch T.L., Bridges T.J., Burkhardt E.G., Yen R.T., Wilt D.P., Campbell J.C., Nelson K.C. and Cohen L.G., "4-Gbit/s transmission over 103 km of optical fibre using a novel electronic multiplexer/demultiplexer", J. Lightwave Tech., Vol. LT-3, pp. 1032-1035 (Oct. 1985).

16

Some Fundamentals of Photodetectors for Fibre Optics

B.P. Pal[*] AND G. Umesh[*]

---

1. INTRODUCTION

Photodetectors are an integral part of an optical receiver which is an optoelectronic device used at the receiving end of a fibre optic communication system. The function of an optical receiver is to detect the optical signal exiting an optical fibre, convert it into an electrical signal (voltage or current) and amplify it, if necessary, for demodulation and further processing. In this article we shall confine our attention only on photodetectors, which basically convert the optical signal into an electrical signal.

Broadly speaking, photodetectors function on either external or internal photoelectric effect. Photomultiplier tubes are good examples of external photoemissive devices: these do not find applications in fibre optics. All the detectors used for fibre optic applications rely on internal photoelectric effect: there is no photoemission as such. These detectors are invariably semiconductor p-n junction devices based on silicon, germanium, gallium arsenide, indium phosphide etc. The devices may be operated either in the *photovoltaic mode* in which the junction is treated as an open circuit to measure the voltage across the junction (as in solar cells) or in the *photoconductive mode* as a reverse-biased photodiode in which the current is made to flow externally to complete the circuit between the two junctions. The latter configuration is preferred for better sensitivity and faster response.

---

[*] Department of Physics, Indian Institute of Technology Delhi, New Delhi-110016, India.

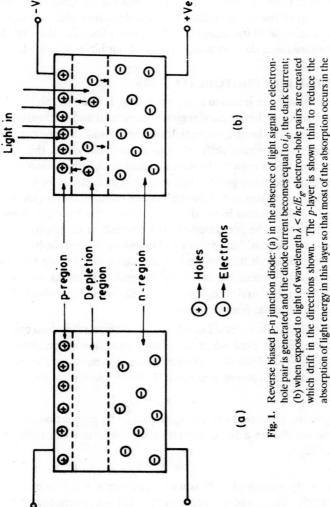

Fig. 1. Reverse biased p-n junction diode: (a) in the absence of light signal no electron-hole pair is generated and the diode current becomes equal to i_d, the dark current; (b) when exposed to light of wavelength $\lambda < hc/E_g$, electron-hole pairs are created which drift in the directions shown. The p-layer is shown thin to reduce the absorption of light energy in this layer so that most of the absorption occurs in the depletion region (adapted with permission from [1]).

Another class of detectors, called photoconductive detectors, are doped semiconductors (not junction devices) wherein the generation of electron-hole pairs increases the conductivity of the device. In these devices, the excitation of an electron (hole) occurs from the donor (acceptor) level into the conduction (valence) band. Consequently, the excitation energies are much less than the band-gap of the semiconductor. Such detectors are, therefore, used for detecting radiation of wavelength greater than a few μm. We shall not discuss them anymore since the wavelength range of our interest is 0.8-1.6 μm.

2. REVERSE BIAS PHOTODETECTORS

A reverse biased p-n junction is shown in Fig. 1. Due to the bias voltage, the potential barrier in the junction region is increased and the barrier thickness is also increased. The large electric fields drive away all the free carriers that may get created there or may drift into this region; therefore, this region is also called the depletion region. In such a situation the absorption of a photon of energy $h\nu \geq E_g$, the band-gap energy, results in an electron being excited to the conduction band leaving a hole in the valence band. Thus an electron-hole pair is created. If this occurs in the depletion region, both the electrons as well as holes contribute to the generated electric current. If the photon is absorbed in the p-region (n-region) but within a diffusion length from the barrier edge, then the electron (hole) drifts into the barrier region and is then swept across the junction thereby generating the signal current. Electrons and holes created well away from the barrier region are very likely to recombine before they get anywhere close to the barrier region.

Since the device is reverse biased, a small reverse saturation current i_d, flows through the device even when no optical radiation is incident on it. This current is also called the dark current. Thus, the net detector current i, is the sum of i_d and i_p, the current generated by the incident photons, i.e.,

$$i = i_d + i_p . \qquad (1)$$

The photon generated current, i_p, is seen to be proportional to the incident optical power P over a large intensity range of several decades. It is easy to show that [2],

$$i_p = \eta a P/h\nu , \qquad (2)$$

where $h\nu$ is the energy of a photon of frequency ν, h is Planck's constant, a is the electronic charge and η is the quantum efficiency (see Sec. 2.2).

The circuit diagram shown in Fig. 2 represents a photodiode to a very good accuracy for the applications being discussed here. The device behaves as a low pass filter with a freequency cut-off i.e. bandwidth $(BW) \approx (1/2\pi R_L C_d)$, R_L being the load resistance and C_d the internal capacitance of the device ($\approx$1pF). Thus, in order to increase the bandwidth, R_L and C_d may be minimised. R_L behaves as a current limiting resistor and cannot be chosen arbitrarily small. Further, if i_p is very small, R_L has to be large enough to produce a reasonable

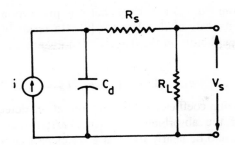

Fig. 2. Equivalent circuit representing a reverse bias photodetector. Here, R_S is the diode internal resistance and R_L is the load resistance; usually $R_L \gg R_S$.

voltage signal for measurement. The junction capacitance C_d may be reduced by decreasing the junction area and increasing the barrier thickness. But increasing the barrier thickness increases the transit time of the charge carriers crossing the barrier, which, in turn, reduces the response time (typically $\approx$ 1 ns) of the device. Junction capacitance can also be reduced by increasing the reverse bias. Another factor which limits the response speed of a detector is the time taken by carriers created outside the depletion region to diffuse to that region. All these factors are important design features which determine the eventual speed of a photodiode.

2.1 Dark Current

As mentioned above the dark current i_d is the detector current when the incident optical power is zero and is equal to the reverse saturation current of the p-n diode; the saturation is reached even at small reverse voltages. The magnitude of this current is strongly dependant on the device temperature. In silicon diodes the dark current increases 20-fold if the device temperature increases by $\approx$ 50°C above room temperature [4]; the temperature (T) variation is essentially like $\exp[-E_g/k_B T]$, where E_g is the band-gap energy and k_B is the Boltzmann's constant.

The dark current generation is random in character and, therefore, sets a noise floor for the detectable signal. This noise level may be minimised by careful device design and fabrication [6]. Dark current in optical telecommunication grade silicon PIN photodiodes is $\approx$ 100 pA while in silicon APDs it is $\approx$ 10 pA. On the other hand, in germanium APDs, the dark current is $\approx$ 100 nA [7]. In InGaAs based PIN photodiodes and APDs, the dark current is typically 2-5 nA [7]. Dark currents could pose serious problems in germanium detectors (specially APDs) unless the device is cooled to appropriate temperatures.

2.2 Quantum Efficiency

Quantum efficiency η is formally defined as the ratio of the number of photogenerated electrons to the number of incident photons and is expressed in percentage. From Eq. (2) it is clear that η is proportional to (i_p/P), the Responsivity or Sensitivity of the detector [8]. In terms of the absorption coefficient, α, of the photons in the detector, the quantum efficiency is given by [4]

$$\eta = (1 - r)[1 - e^{-\alpha W}], \qquad (3)$$

where r is the reflection coefficient of the photons at the detector surface and W is the width of the absorbing layer (the barrier region). In silicon photodetectors, η could be nearly 100% at a wavelength of $\approx 0.82\ \mu$m, while germanium detectors have a maximum η of ≈ 55% at 1.55 μm; sensitivity of germanium detectors is quite flat over the wavelength range 1.1 – 1.6 μm. Detectors based on InGaAsP or InGaAs have an $\eta \approx 60$-80% in the wavelength range 1.0 – 1.6 μm [1]. In general, the quantum efficiency is large for any detector only in a certain wavelength range and falls-off rapidly outside this range. This happens because the absorption of photons in semiconductors is characterised by a certain cut-off wavelength λ_c give by

$$\lambda_c = hc/E_g, \qquad (4)$$

where E_g is the band-gap energy. If the wavelength of the incident radiation is longer than λ_c, the photons cannot be absorbed and the material is almost transparent to this radiation. But for wavelengths shorter than λ_c, the absorption is so strong that most of the photons are absorbed at/near the surface itself. The generated electrons and holes are created too far away from the barrier region and they recombine before being able to reach the barrier region. Thus, every photodetector is characterised by an operating wavelength range, e.g. for silicon it is 400-1100 nm, for germanium it is 700-1600 nm and for InGaAsP/InP (or InGaAs/InP) it is 1000-1600 nm.

The responsivity of a detector, defined as the ratio of the detector current or the corresponding voltage across the load resistor R_L to the input optical power, may be expressed in units of amperes/watt or volts/watt. In fibre optics it is usually expressed as A/W or μA/μW. From Eq. (2) it follows that the responsivity R is given by

$$R = \eta a/h\nu = \eta a\lambda/hc \text{ A/W}, \qquad (5)$$

where c is the speed of light in vacuum and λ is the wavelength of the incident radiation. With good quality silicon diodes having $\eta > 90$%, $R \approx 0.5$-0.6 A/W in the 0.8-0.9 μm range for λ.

2.3 Signal to Noise Ratio (SNR)

The signal to noise ratio is a very important figure of merit of an optical receiver system. It is defined for analog signals as

$$\text{SNR} = \frac{\text{Signal Power}}{\text{Noise Power}}, \qquad (6)$$

and is expressed in decibel units (dB) which is simply the logarithm of SNR as defined above. The SNR is a measure of the sensitivity of the detection system. The one having larger SNR is a better receiver system.

On the other hand, in a digital communication system, the signal detection is basically a binary process, i.e. the signal is either high (value = 1) or low (value = 0). The accuracy of measurement is now characterised by a parameter called the "Bit Error Rate" (BER). BER is essentially a measure of the probability of erroneous detection of the Bits of digital information transmitted, i.e. the ones and zeros. For a given data rate, smaller the received power, greater is the BER value. In optical telecommunication a BER of 10^{-9} is accepted as a tolerable value; this number implies that a single error could occur after the detection of every 10^9 bits of information. The BER value is decided by factors such as noise level, detector sensitivity etc. (see Ch 18 by J. Das in this book).

In optical receiver systems, noise originates from several sources. Shot-noise can be traced to either the fluctuations in the number of photons arriving at the detector surface or to the randomness in the generation of carriers due to the absorption of photons. The origin of these fluctuations is quantum mechanical in nature [3, 4]. These fluctuations are imposed on the detector output also. The shot-noise in the frequency interval ν and $\nu + \Delta\nu$ is characterised by a mean square current amplitude [9]

$$\overline{i}^2_{N1}(\nu) = 2q\overline{I}\Delta\nu, \qquad [7(a)]$$

where q is the electronic charge and $\overline{I}$ is the average detector current; the bar over the variable i_{N1}^2 indicates averaging over time. Avalanche photodiodes (APDs) have an internal current gain factor M (see Sec. 3) and therefore, for such detectors the shot-noise level is given by [9]

$$\overline{i}^2_{N1}(\nu) = 2qM^2\overline{I}\Delta\nu F(M). \qquad [7(b)]$$

The factor $F(M)$ accounts for the non-ideal behavior of the current gain process and is called the excess noise factor [5]. In ideal situations, $F(M)$ is unity [4].

The dark current also adds to the shot-noise generated by the input optical signal. In this case too the noise level is given by Eq. 7(a) or 7(b) except that $\overline{I}$ is replaced by the dark current i_d.

The other major sources of noise are the resistive elements in the circuit and the amplifier noise. Due to the finite temperature of the resistor, thermal fluctuations in the charge carrier velocities are reflected as fluctuations in the voltage drop across the resistor. This constitutes the so called Johnson noise and the mean square noise amplitude is given by [9]

$$\overline{i}^2_{N2}(\nu) = (4k_BT_e/R)\Delta\nu, \qquad (8)$$

where T_e is chosen such that the RHS of Eq. (8) accounts for the noise generated by the resistor R as well as that generated by the amplifier following the photodiode [9].

Thus, the SNR at the output of the detection system may be written as

$$\text{SNR} = \frac{\overline{i}_p^2}{\overline{i}_{N1}^2 + \overline{i}_{N2}^2}, \qquad (9)$$

where $\overline{i}_p$ is the current generated by the incident photons. It may be noted that the optical signal itself may have some noise level and this should be added to the denominator of Eq. (9).

2.4 Noise Equivalent Power (NEP)

This quantity is also called the minimum detectable power and is the value of the incident optical power which corresponds to an SNR = 1. In other words, it is the optical power, incident on the detector, required to produce a detector current signal equal to $\sqrt{(\overline{i}_{N1}^2 + \overline{i}_{N2}^2)}$. It is expressed in units of $W/\sqrt{H_z}$ and is dependant on the signal modulation rate as well as the noise bandwidth.

2.5 Speed of Response and Bandwidth

Response speed essentially represents the time constant of the detector and is also called the rise time. The rise time of a detector is normally defined as the time required by the output signal to rise from 10% to 90% of its final value for a step function optical input. Likewise one can ascribe a fall time to a detector's response: the response time of a detector is generally taken as the larger of these two times [1]. For silicon photodetectors used in fibre optic communication links, the response time is typically 0.5-1.0 ns while in specially prepared fast photodectors it could be a few ps. As already stated, the speed of response is decided by the carrier transit time across the depletion region in the diode: smaller the transit time faster is the response. On the other hand, high quantum efficiency requires wider depletion region implying, thereby, that a design trade-off exists between the quantum efficiency and the device speed. The APDs are typically half as fast as the PINs [4]. Inverse of the response time gives the frequency response, i.e. the bandwidth of the detector. In the case of digital fibre optic transmission, the maximum Bit transmission rate is decided by the smaller of two quantities, namely, the detector bandwidth and the fibre bandwidth (which is dictated by the pulse dispersion characteristics of the fibre).

2.6 Linearity and Dynamic Range

As is true with many electronic and optoelectronic devices, photodetectors also exhibit non-linear response beyond a range of input optical powers. Detector current nominally varies linearly with input power over approximately 6 decades in well designed photodetectors [4,8]. The range of optical power within which the detector current shows linear variation with the input optical power determines the dynamic range of the photodetector. If the optical signal power level exceeds the dynamic range of the photodetector, the detector output would no longer be a linearly proportional measure of the input optical signal. On the other hand in digital transmission, this will lead to an increase in the BER. In power budgeting during fibre optic link designs, this is an aspect which requires attention. If the transmission loss is lower than the estimated value, the received power may overshoot the dynamic range of the detector. This problem can be overcome by inserting optical attenuators in front of the detector. Dynamic range of a particular detector is determined by the device parameters like its resistivity, contact resistance etc. [8].

3. TYPES OF DETECTORS

Semiconductor p-n junction detectors are of basically three types, i.e. p-n junction photodiodes, the PIN photodiodes and the avalanche photodiodes (APDs). The PIN photodiode is an improvement on the simple p-n junction in that there is an extra layer of very lightly doped semiconductor in between the p- and the n-layers as shown in Fig. 3. This new layer is practically an intrinsic material and, therefore, has a high resistivity; this explains the "I" in the

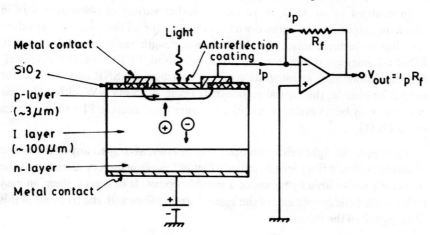

Fig. 3. Schematic of a PIN photodiode with a pre-amplifier (after [2]).

notation "PIN". Since there would be very few charge carriers in the intrinsic region, the I-layer essentially serves to physically extend the potential barrier region. To maximise the sensitivity of these detectors, one would like most of the incident photons to be absorbed in the barrier region so that all the generated electrons and holes contribute to the signal current [4]. Thus, the PIN diode, which has a wider barrier region, will be more sensitive compared to a p-n junction diode. It may be noted that the entire reverse bias voltage appears across the highly resistive I-layer in the PIN detector. The PIN detectors can have response times < 1 ns and a dynamic range $\approx$ 50 db [1].

An APD has the additional feature of an internal gain over and above the other two varieties. This gain results when the reverse bias voltage is raised to a value just below the diode breakdown voltage at which the electric field is typically $\approx 3 \times 10^5$ V/cm [4]. In this situation, every absorbed photon produces a primary electron-hole pair which gains enough energy from the electric field in the barrier region to cause further ionisation by electron impact. Thus, secondary electron-hole pairs are created. In short, under such large reverse bias voltages, an avalanche carrier multiplication, by a factor M, occurs. Therefore, the photogenerated current is amplified from i_p to $<M> i_p$, where $<M>$ is the mean value of the gain factor M [4]. The gain factor M depends on the operating bias voltage, the device breakdown voltage and also on the junction temperature. Bias voltage may range from tens of volts to over 100 volts and the bias circuitry often requires automatic gain control to compensate for temperature variations [1]. At high gains in APDs, noise increases at a rate slightly faster than the detector signal power, which varies as $<M^2>$ [9]. Accordingly, for optimum operation, M typically ranges from 30 to 100.

In contrast to an APD, there exists another variety of photodetectors in which an amplifier is integrated with a PIN detector on the same semiconductor chip to reduce noise. In the long wavelength range, i.e. 1.3–1.6 μm, PIN-FET integrated receivers are extensively used; FET standing for field-effect transistor. Such an integration greatly enhances the SNR and the detector output amplitude, the typical responsivity being $\approx$ 15 mV/μW, although their rise time may be an order of magnitude larger than discrete PIN photodiodes or APDs [1].

To couple the light exiting an optical fibre conveniently to any of the above photodetectors, a fibre length, called "Pigtail", is often factory attached to the detector's active area by means of a suitable epoxy. It becomes, then, an easy task to splice the output end of the signal carrying fibre with the free end of this fibre pigtail of the detector.

4. CONCLUSIONS

In this chapter, we have discussed the physical principles of operation of junction type semiconductor photodetectors operated under reverse bias. These detectors are extensively used in fibre optic communication systems. Important performance characteristics of photodetectors have also been discussed. This chapter is expected to be a useful prerequisite for understanding the contents of the next chapter which gives details of the electronic circuitry involved in the design of fibre optic receivers.

REFERENCES

1. J. Hecht, the chapter on receivers in "Understanding Fibre Optics", Howard W. Sams, Indiana (1987).
2. H. Melchior, "Detectors for Lightwave Communication", Phys. Today, **30**, pp. 32-39 (Nov., 1977).
3. S.D. Personick, "Receiver Design for Digital Fibre Optic Communication Systems, Part I", *Bell Syst. Tech. J.*, **52**, pp. 843-874 (1973).
4. R.G. Smith, "Photodetectors for Fibre Transmission Systems", *Proc. IEEE*, **68**, pp. 1247-1253 (1980).
5. R.J. McIntyre, "Multiplication Noise in Uniform Avalanche Diodes", *IEEE Trans. Electron Devices*, **ED-13**, pp. 164-168 (1966).
6. H. Melchior, Hartman A.R., Schinke D.P. and Seidel T.E., "Planar Epitaxial Silicon Avalanche Photodiode", *Bell Syst. Tech. J.*, **57**, pp. 1791-1807 (1978).
7. T. Li, "Advances in Optical Fibre Communications: A Historical Perspective", *IEEE J. Selected Areas in Commn.*, **SAC-1**, pp. 356-372 (1983).
8. D.P. Schinke, Smith R.G. and Hartman A.R., "Photodetectors" in Semiconductor Devices for Optical Communication, Ed. H. Kressel, Springer-Verlag, Berlin (1982).
9. A. Yariv, "Optical Electronics", 3rd Edition, Holt-Saunders International Editions, Tokyo (1985).

17

Receivers for Digital Fibre Optic Communication Systems[†]

SUJATHA CHANDRAMOHAN[*] AND J.P. RAINA[††]

1. INTRODUCTION

Following the realization of the laser in 1960 and subsequent work in the area of optical communications, the first serious proposal to employ the fibre as a telecommunications transmission medium appeared in 1966 [1]. The development of low-loss fibres of 20 dB/km and lower [2], and of reliable semiconductor injection lasers, LEDS and sensitive photodetectors paved the way for useful communication application.

The earliest receivers resembled those that were used for nuclear particle counters [3] and television cameras [4]. Edwards [5] recognized the high impedance design for a fibre optic receiver, but failed to recognize the use of equalization. Many optical communication theorists [6, 7, 8, 9] have in the past often used the criterion '$RC \leq T$' loading down the front end amplifier so as to have adequate bandwidth without equalization therein incurring an unnecessary noise penalty. Personick [10, 11] and Goell [12] were among the first to put forth ideas regarding the theoretical and practical aspects of digital fibre optic receiver design.

[†] Updated reprint from J.I.E.T.E., Vol. 32 (Special issue on Optoelectronics and Optical Communication), July-August (1986).

[*] Department of Applied Mechanics, Indian Institute of Technology, Madras-600 036.

[††] Department of Electrical Engineering, Indian Institute of Technology, Madras-600 036.

2. BASIC COMPONENTS OF DIGITAL FIBRE OPTIC COMMUNICATION RECEIVER

The receiver is basically an optical to electrical converter. The chief aim in the design of a receiver is to maximize the sensitivity of the optical receiver for a given BER. The block diagram of a receiver [10, 13] is as shown in Fig. 1. It can be made out from the figure that the signal voltage at the output of the linear channel is

$$V_S(\omega) = Z_T(\omega) \cdot I_S(\omega), \quad (1)$$

where $Z_T(\omega)$ is the total system transfer function including equalizer. $I_S(\omega)$ is the Fourier transform of $I_S(t)$ where $I_S(t)$ is the current produced due to electrons generated within the photodiode.

3. DETECTORS FOR DIGITAL FIBRE OPTIC RECEIVERS

These detectors must have high sensitivity at the wavelength of operation, sufficient speed of response to accommodate the information rate, minimum noise, low susceptibility of performance characteristics to changes in ambient conditions and proper coupling arrangement to the fibre and ensuing electronics. The choice of a photodiode [14, 15, 16, 17] depends mainly on the wavelength of operation. Silicon diodes are used in the uv, visible and near infrared part of the spectrum upto 1μm. With Ge diodes, the response can be extended to over 1.5μm.

For the 0.8-0.9 μm region, silicon photodiodes, both PIN and APD have sufficiently high sensitivity and fast response. These diodes have a typical quantum efficiency of 85% in this wavelength range. They have low dark currents and bandwidth from dc to 1 Gb/s.

For the 1.1-1.7 μm range, Ge diodes as well as those fabricated from the III-V alloy semiconductors are suitable. Germanium is sensitive to wavelengths below 1.8 μm and has a broad-band quantum efficiency of about 40%, but exhibits a high multiplication noise level. The bandgap of the III-V materials can be tailored to the desired wavelength resulting in low dark currents. These materials can also be fabricated in heterojunction structures so as to enhance high speed operation.

For the 1.0-1.4 μm range, InGaAs or GaAlSb photodiodes have been used and for the 1.3-1.6 μm range, Ge and several alloys including $In_xGa_{1-x}As_yP_{1-y}$ and $Ga_xAl_{1-x}Sb_y As_{1-y}$ have been used. Major effort has gone into the development of the InGaAsP system which can be grown lattice matched to the binary InP with an energy gap continuously variable between 1.35 eV and 0.75 eV corresponding to wavelengths between 0.92 and 1.65 μm.

3.1 The PIN Diode

The most widely used photodetector is the PIN diode because of simplicity in design, simplicity of operation and low cost. But this device suffers from low

Fig. 1. Block diagram of a digital fibre optic receiver.

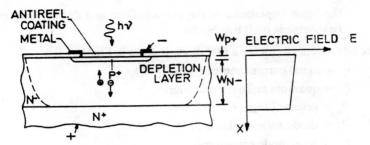

Fig. 2. Typical structure of a PIN diode.

responsivity in terms of its overall efficiency and response to transients. In order to improve its speed the depletion layer is optimized to give higher speed and efficiency. The typical structure of a PIN diode is shown in Fig. 2. The optical energy impinging on the window is partly absorbed and partly reflected by the surface (inspite of anti-reflection coating). The photon energy hv is transferred to the electron in the valence band. If $hv > E_g$ (where E_g is the band-gap energy), these electrons pass to the conduction band, while holes are formed in the valence band. The carriers drift towards the electrodes under the influence of the electric field E. If P is the optical power impinging on the photo-diode per unit time,

$$P = \frac{n_f h v}{\Delta t}, \quad (2)$$

where n_f is the number of photons, hv their energy and Δt is the time interval. The signal current is given by

$$I_s = SP, \quad (3)$$

where S is the responsivity defined by

$$S = \frac{q\eta}{hv} = \frac{\lambda q \eta}{hc}, \quad (4)$$

where q = electronic charge ($q = 1.602 \times 10^{-19}$ C),
  λ = wavelength of incident light,
  h = Planck's constant (6.626×10^{-34} Js),
  c = velocity of light (3.00×10^8 m/s),
  η = quantum efficiency = $\dfrac{\text{generated electron–hole pairs}}{\text{No. of incident photons}}$.

The responsivity of the detector depends on two major factors:

1. Reflectivity at the outer surface and the interface between the outer surface and the semiconductor.
2. Physical and geometrical characteristics of the device. The frequency response of the device is another very important parameter for the device and is governed by two basic parameters, namely, transit time and device capacitance. The above capacitance comes in shunt to

the input impedance of the pre-amplifier and defines the receiver (uncompensated) bandwidth.

The equivalent circuit for a PIN diode is given in Fig. 3.

Here i_s = signal current source ($<i_s>$ = I_s),

i_n = quantum noise current source,

R_p = reverse junction resistance,

R_d = diode series resistance,

C_T = total diode capacitance,

Z_L = load impedance due to the amplifier.

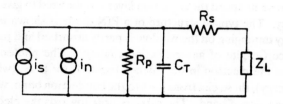

Fig. 3. Equivalent circuit for a PIN diode.

3.2 The Avalanche Photo Diode (APD)

It is known that if a p-n junction is operated under high reverse electric field, the impinging photon-generated carriers can be amplified because of the internal current multiplication mechanism. The amplification can be of the order of 100 or more in well designed APDs. Unfortunately, the internal current gain is not a linear function of the applied reverse voltage and is dependent on temperature. The avalanche process introduces noise in excess of the shot noise due to the current flowing in the device. The excess noise is dependent on the following:—

1. Material characteristics,
2. Device structures,
3. Gain of the device,
4. Illuminating conditions and,
5. Ambient temperature of the device.

3.2.1 Response of APD

As mentioned above, the current gain is a non-linear function of the electric field E as shown in Fig. 4. The responsivity is expressed by

$$S = \left[\frac{M(V_R) q}{h\nu}\right] \eta, \qquad (5)$$

where η is the device quantum efficiency at unity gain when $M(V_R = 20V) = 1$; $M(V_R)$ at any value other than $V_R = V$ is given by Miller equation

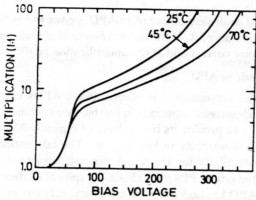

Fig. 4. Photo-current multiplication vs reverse bias voltage.

$$M(V_R) = \frac{1}{\left(1 - \dfrac{V_R}{V_{br}}\right)^n}, \quad (6)$$

where V_{br} is the reverse break-down voltage and n is an empirically determined exponent. It may be noted that as V approaches the reverse break-down value V_{br}, the avalanche gain increases and the device speed increases with high efficiency.

The speed of response of an APD depends on (i) the time for carriers to complete the process of recombination and multiplication and (ii) the transit time of the carriers across the depletion width. In order to achieve the maximum speed, the carriers must move with saturation limited velocities and this is achieved at an electric field of $E > 10^4$ V/cm. APDs exhibit an unsymmetrical pulse shape with fast rise time and slow fall time due to electron-hole movement. The typical performance figures for silicon and germanium are

5×10^{-12} s for silicon and,

5×10^{-13} s for germanium.

The upper speed limit of an APD is 3 to 4 orders of magnitude higher than that of a PIN diode. Unfortunately, the APD response is limited by the RC time constants of the diode and the external circuit.

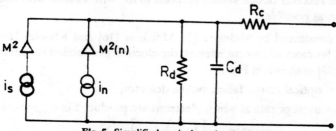

Fig. 5. Simplified equivalent circuit of APD.

The simplified equivalent circuit of an APD is given in Fig. 5. Here R_C is the series contact resistance, R_d is the diode reverse bias resistance, i_s the signal current, i_n the noise current and M the multiplication factor.

3.3 The PIN Diode vs APD

Both these diodes are operated in reverse bias. APDs have economic and operational disadvantages, especially in low bit-rate systems. The APD needs a dc-dc converter to provide its bias voltage of typically 200-400 V and additional circuitry to control its avalanche gain. The fabrication of the APD is complicated, particularly for the 'reach-through' structure and so is more expensive than the simple PIN photodiode. In spite of its disadvantages, the Si reach-through APD has high quantum efficiency, very low excess noise resulting from the avalanche process and a very low bulk leakage current. The Si reach-through APD gives 15-20 dB higher receiver sensitivity than a Si PIN photodiode. Front illuminated Si-PIN diodes have very fast response and high quantum efficiency. Recently a PIN diode with a 30 ps rise time has been reported [18]. Table 1 shows some of the general characteristics of commonly used photodiodes.

TABLE 1
General characteristics of commonly used photodiodes [14, 15, 16, 17]

Type of diode	Wavelength region	Quantum efficiency	Response time	Dark current
Si PIN	0.4-1.1 μm	90%	0.1-5 ns	1-50 nA
Si APD	0.4-1.1 μm	85%	0.05-0.3 ns	20-100 nA
Ge APD	0.6-1.6 μm	85%	0.1 ns	20-100 nA
Ge PIN	1-1.6 μm	60%	2.5 ns	100 nA
III-V	0.92-1.65 μm	50-70%	10s of ps	10^{-5} A/cm^2 to 10^{-1} A/cm^2

3.4 Noise and Signal Modelling of a Photodiode

A reasonable small-signal model of an APD [10] with a biasing circuit is shown in Fig. 6, and its equivalent circuit in Fig. 7. Here C_d represents the junction capacitance of the diode. The production of charges by optical and thermal generation and collision ionization in the diode high-field region is represented by $i(t)$. The receiver circuit should respond to $i(t)$ with as little distortion and added noise as possible.

Studies conducted by McIntyre [7], Melchior [16] and Klauder [19] have shown that for cases of current interest, the electron production process can be modelled [10] as shown in Fig. 8.

Here $p(t)$ = optical power falling on the detector,

$\lambda(t)$ = average rate at which electrons are produced in response to $p(t)$.

Then
$$\lambda(t) = \left[\left(\frac{\eta}{h\Omega}\right) p(t)\right] + \lambda_0. \tag{7}$$

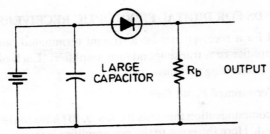

Fig. 6. A small signal model of an APD with biasing circuit.

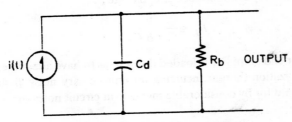

Fig. 7. Equivalent circuit of biased detector.

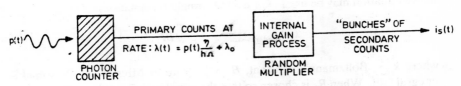

Fig. 8. Model of electron production process.

Here η = quantum efficiency of photon counter,

λ_0 = Dark current 'counts'/second.

$h\Omega$ = photon energy.

For this Poisson electron production process in interval T, the probability that exactly N electrons will be produced at the appropriate times $t_1 \pm 1/2\,\Delta$, $t_2 \pm 1/2\,\Delta$, $\cdots\cdots\cdots$, $t_N \pm 1/2\,\Delta$ (where Δ is very small) is given by

$$P\left[N,\{t_k\}\right] = \left\{e^{-\Lambda} \prod_{1}^{N}\left[\lambda(t_k)\Delta\right]/N!\right\} + 0(\Delta), \tag{8}$$

where $\Lambda = \int_{t_0}^{t_0+T} \lambda(t)\,dt$ and $0(\Delta)/\Delta = 0$ in the limit $\Delta \to 0$.

Each of the primary electrons produced by the photon counter is multiplied by a value g. For a PIN diode, the above noise model hoqslds good with $g = 1$.

Goell [20] has modelled the detector and biasing circuit slightly differently. He has taken into account the effect of the diode spreading resistance.

4. FRONT ENDS FOR DIGITAL FIBRE OPTIC RECEIVERS

The front end for a receiver may be a straight terminated one, a high-input impedance amplifier or a transimpedance amplifier. Each of these types is dealt with in the following sections.

4.1 Straight Terminated Front End

Many optical communication theorists [6, 7, 8, 9, 21] have in the past used this type of front end. Here the input to the pre-amplifier is terminated with a load resistor R_L such that in conjunction with the input capacitance C_T, the bandwidth of the input admittance is $\geq$ bit rate B.

i.e.
$$R_L \leq \frac{1}{2\pi B C_T}. \qquad (9)$$

Here the front end amplifier is loaded down so as to have adequate bandwidth without equalization-therein incurring an unnecessary noise penalty. Thus simplicity is paid for by considerable increase in circuit noise over that potentially available.

The circuit may be designed by using that value of R_L which satisfies the equality in the above equation. Alternately a 50 Ω termination satisfying the above equation may be used. For a 50 Ω straight termination,

$$\langle i^2 \rangle_{\text{load resistor}} = \frac{4kT}{R_L} B I_2, \qquad (10)$$

where k = Boltzmann's constant, B = bit rate in Mb/s, I_2 = Personick's integral [10]. When R_L is chosen so that the equality in Eq. (9) is satisfied,

$$\langle i^2 \rangle_{\text{load resistor}} = 4kT (2\pi C_T) B^2 I_2. \qquad (11)$$

Comparing Eq. (11) with the minimum circuit noise achievable with a Field Effect Transistor (FET), we get

$$\frac{\langle i^2 \rangle_{\text{Load resistor}}}{\langle i^2 \rangle_{\text{minFET}}} = \frac{g_m \cdot I_2}{\Gamma (2\pi C_T) I_3 B}, \qquad (12)$$

where I_2 and I_3 are Personick's integrals
Γ = 0.7 for Si FET's [23]
= 1.1 for GaAs FET's [22].

The equivalent relation for a bipolar with small r_{bb}', at optimum collector current is given by

$$\frac{\langle i^2 \rangle_{\text{Load resistor}}}{\langle i^2 \rangle_{\text{min, bipolar}}} = \beta_0^{1/2} (I_2/I_3)^{1/2}. \qquad (13)$$

Thus we see that the minimum noise achievable with a straight forward termination exceeds that achievable with an FET or an optimized bipolar transistor. Thus simplicity is paid for by a considerable increase in circuit noise over that potentially achievable.

4.2 High input impedance type front end amplifier

In this approach to the design of a front end, all sources of noise are reduced to an absolute minimum. Whether using a bipolar or an FET input device, the input capacitance is decreased by the selection of low capacitance, high frequency devices, by selecting a detector with low dark current and by decreasing the thermal noise contributed by the biasing resistors [10, 11, 12, 13, 20, 24, 25, 26, 27, 28, 29, 30]. The latter is accomplished using a large R_L. Hence the name high impedance front end. With large input resistance, the input admittance is dominated by capacitance C_T and the signal is integrated by the capacitor. Hence this amplifier is also called the integrating type.

The equalizer restores the input pulse shape which is distorted due to the limited bandwidth of the input admittance. The equalizer in the simplest form is usually a simple differentiator. The high impedance equalized amplifier is thus capable of decreasing circuit noise to an absolute minimum while retaining a transfer function that preserves the information contained in the signal.

4.2.1 Noise and signal modelling of a high input impedance amplifier for digital signal reception

Here the amplifier (Fig. 9) is modelled as an ideal high-gain infinite impedance amplifier with an equivalent shunt capacitance C_A and resistance R_A at the input and with two noise sources $i_a(t)$ and $e_a(t)$ referred to the input. The assumptions made are: (i) the noise sources are white, Gaussian and uncorrelated, (ii) the amplifier gain is sufficiently high so that noise contribution of the following stages is negligible, (iii) the power falling upon the detector [31] assumes the form of a digital pulse stream given by

$$p(t) = \sum_{-\infty}^{\infty} b_k h_p (t - kT). \tag{14}$$

The average voltage at the equalizer output has been worked out by Personick [10] as

$$\langle V_{out}(t) \rangle = \frac{A\eta \langle g \rangle q p(t)}{h\Omega} * h_I(t) * h_{eq}(t), \tag{15}$$

where A is an arbitrary constant,

$$h_I(t) = F\left\{ \frac{1}{\frac{1}{R_T} + j\omega(C_d + C_A)} \right\}, \tag{16}$$

$$\frac{1}{R_T} = \frac{1}{R_b} + \frac{1}{R_A}, \tag{17}$$

$$C_T = C_d + C_A, \tag{18}$$

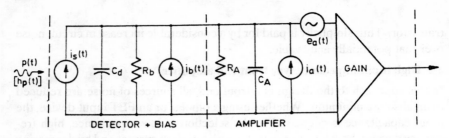

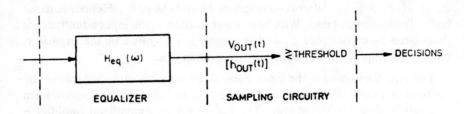

Fig. 9. Receiver for a digital fibre optic communication system.

$h_{eq}(t)$ = equalizer impulse response.

$$\langle V_{out}(t) \rangle = \sum_{-\infty}^{\infty} b_k h_{out}(t - kT), \tag{19}$$

$$V_{out}(t) = \sum_{-\infty}^{\infty} b_k h_{out}(t - kT) + n(t), \tag{20}$$

where $n(t)$ = output noise = $V_{out}(t) - \langle V_{out}(t) \rangle$.

Therefore, output noise = $n_s(t) + n_R(t) + n_I(t) + n_E(t)$. (21)

The worst case noise, $NW(b_0)$ for each of the two possible values of b_0 is defined as follows:

$$NW(B_0) = \max_{\{b_k\}, k \neq 0} \left[\langle V_{out}^2(0) \rangle - \langle V_{out}(0) \rangle^2 \right]. \tag{22}$$

Converting everything to the frequency domain and making $h_{out}(t)$ equal to unity at $t = 0$,

$$NW(B_0) = \max_{\{b_k\}, k \neq 0} \left[\left(\frac{1}{2\pi}\right)^2 \cdot \int_{-\infty}^{\infty} \frac{\langle g^2 \rangle h \Omega}{\langle g \rangle^2 \eta} H_p(\omega) \left(\sum_{-\infty}^{\infty} b_k e^{j\omega kT} \right) \right]$$
$$\times \left(\frac{H_{out}(\omega)}{H_p(\omega)}\right) * \left(\frac{H_{out}(\omega)}{H_p(\omega)}\right) d\omega$$

$$+ \frac{\left(\frac{h\Omega}{\eta}\right)^2}{2\pi \langle g \rangle^2 q^2} \left(\frac{2k\theta}{R_b} + S_I + q^2 \langle g^2 \rangle \lambda_0\right) \int_{-\infty}^{\infty} \left|\frac{H_{out}(\omega)}{H_p(\omega)}\right|^2 d\omega$$

$$+ \frac{\left(\frac{h\Omega}{\eta}\right)^2}{2\pi \langle g \rangle^2 q^2} S_E \int_{-\infty}^{\infty} \left|\frac{H_{out}(\omega)\left(\frac{1}{R_T} + j\omega C_T\right)}{H_p(\omega)}\right|^2 d\omega, \qquad (23)$$

where S_I = two-sided spectral density of amplifier current noise source $i_a(t)$,

S_E = two-sided spectral height of amplifier voltage noise source $e_a(t)$,

$H_p(\omega)$ = input power pulse transform,

$H_{out}(\omega)$ = output power pulse transform and,

$$\frac{1}{2\pi} \int_{-\infty}^{\infty} H_{out}(\omega) \, d\omega = 1. \qquad (24)$$

From Eq. (23) it is inferred that to reduce the noise, (i) regardless of the choice of $H_{out}(\omega)$, R_b should be made as large as possible, (ii) the amplifier input resistance should be as large as possible and the amplifier shunt capacitance should be as small as possible and (iii) all b_k (except b_0) should assume the smaller of the two possible values.

Goell [20] has modelled the amplifier differently taking into account the effect of diode spreading resistance and load noise. The application of this method to cases in which the noise of more than one stage is significant is possible. It is found that [20, 32] the noise contribution of subsequent stages is minimized by employing a transistor configuration with both a high output impedance and a high short circuit current gain. Hence a common emitter (CE) or common source (CS) configuration should be employed in the first stage of low noise pre-amplifier circuits.

4.2.2 FET high input impedance amplifier

The effective noise for FET amplifiers is described for receivers operating in the 1 Mb/s to 1 Gb/s range in [20]. From the expression for noise current [20] it is clear that from the standpoint of thermal noise, R_b should be as large as possible. Assumption R_b being large and the detector leakage current to be negligible gives the simple relation [13]

$$\langle i^2 \rangle_{FET} \simeq \left(\frac{4kT}{R_L}\right) I_2 B + 4kT\Gamma \frac{(2\pi C_T)^2}{g_m} I_3 B^3, \qquad (25)$$

$$\langle i^2 \rangle_{FET, min} = 4kT\Gamma \frac{(2\pi C_T)^2}{g_m} I_3 B^3, \qquad (26)$$

where $\frac{\sqrt{g_m}}{C_T}$ is a figure of merit for the FET. Hence effective noise current varies as B^3 and also as C_T^2/g_m.

The noise as a function of bit-rate for various values of R_L and I_{gate} are shown as dashed lines in Fig. 10 below. Assume that a system is to be operated at data rate of 1 Mb/s for which minimum FET noise is approximately $10^{-22} A^2$. Here

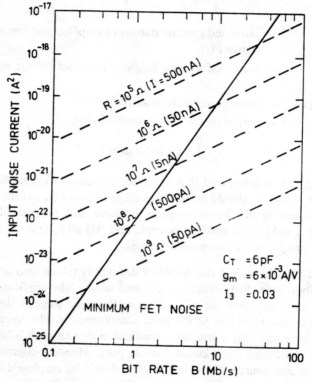

Fig. 10. Effective input noise current vs. bit rate for an FET front end.

noise is equal to that of a resistor of $10^8 \Omega$ or a leakage current of 500 pA. Thus to reach the minimum achievable noise level, limited by the channel noise of the FET, it is necessary that $R_L \gg 10^8 \Omega$ and $I_{gate} \ll 500$ pA. Non-multiplied dark current of the detector contributes noise with the same function and bit rate dependence as the gate leakage current and thus same constraints apply. At higher bit rates, the restrictions on R_L and leakage currents become less severe due to the differing bit rate dependences of the noise sources.

4.2.3 Bipolar high input impedance amplifier

Bipolar transistor amplifiers for receivers operating in the 1 Mb/s to 1 Gb/s range are discussed in [20]. The chief difference between a FET and a bipolar transistor is that for a Bipolar Junction Transistor (BJT), the optimum bias depends on frequency. Assuming that most of the noise orginates from the first stage ($R_L \to \infty$), the optimum bias current is approximately

$$I_{c \text{ optimum}} = \frac{kT}{q} 2\pi C_T \beta_0^{1/2} B \left(\frac{I_3}{I_2}\right)^{\frac{1}{2}} \left[1 + \frac{I_2 I_3}{(2\pi B C_T R_L)^2}\right]^{\frac{1}{2}}, \quad (27)$$

where $\beta_0 = h_{FE}$.

The minimum circuit noise [13] for bipolar transistor is

$$\langle i^2 \rangle_{\text{bipolar min}} = 8\pi kT \left(\frac{C_T}{\beta_0^{1/2}}\right) (I_2 I_3)^{\frac{1}{2}} B^2$$
$$+ 4kT r_{b'b} (2\pi)^2 \left(C_d + C_s'\right)^2 I_3 B^3 \quad (28)$$

A figure of merit for a bipolar transistor is $\sqrt{\beta_0}/C_T$. The optimum collector current and minimum circuit noise for a CE amplifier in the limit $R_L \to \infty$ are shown in Figs. 11 and 12 using the following parameters:

$C_T = 6 \text{ pF}, I_2 = 0.5, C_d + C_s = 4 \text{ pF}, I_3 = 0.03, \beta_0 = 100, r_{b'b} = 100$. The effect of the base resistance noise is seen to become significant for $B \simeq 500$ Mb/s.

4.2.4 FET vs BIPOLAR front ends

The effective noise for bipolar transistors and FETs without leakage have a different dependence on frequency. The noise in the case of an FET is

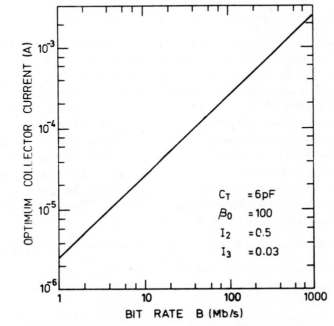

Fig. 11. Optimum collector current for bipolar front end.

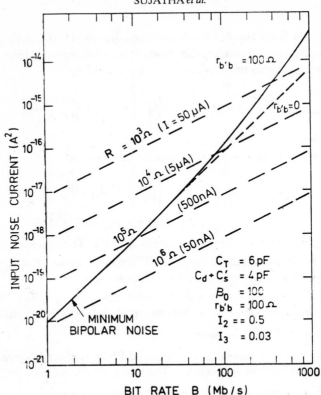

Fig. 12. Effective input noise current vs. bit rate for an optimized bipolar front end.

proportional to B^3 whereas for a bipolar transistor, it is proportional to B^2, where $r_{bb'}$ is small and is proportional to B^3 when base resistance noise dominates. Hence at low bit rates, the FET is superior whereas at higher bit rates, the BJT is superior. Fig. 13 illustrates [20] the dependence of the noise of Si FETs, GaAs FETs and BJTs as a function of frequency. The leakage current noise, shown separately must be added to the device noise to get the total circuit noise, especially for GaAs FETs which have large leakage currents. For Si FETs, high frequency operation is limited by the finite gain of the first stage (fixed g_m) which increases the contribution of subsequent stages to the total noise. Bipolar transistors operated well above the reverse saturation current have better temperature stability than FETs. This could be important in applications employing DC coupling.

4.2.5 Advantages and disadvantages of the high input impedance type front end

This type of front end provides means for achieving the ultimate in low noise front end design. Maximum sensitivity is obtained with this type of amplifier. But it has lower dynamic range and increased complexity of equalizer design. Although the equalizer subsequently restores the pulse shape, build up of low frequency components within the circuit prior to equalization leads to prema-

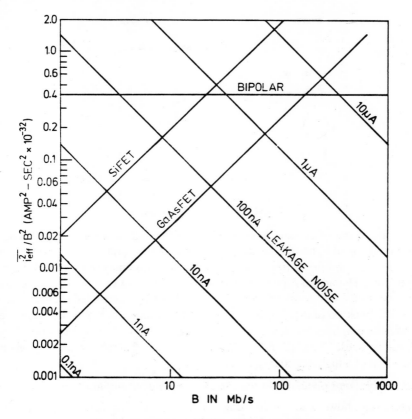

Fig. 13. Comparison of FET and Bipolar amplifiers.

ture saturation of the amplifier at high input signal levels. The zero of the equalizer has to be positioned to compensate for the zero of the input admittance. The location of the zero depends upon C_T and R_T which in turn depend on parasitic capacitances and the β of the transistor. Hence individual amplifiers may need to be individually equalized. This equalization may need temperature compensation. Thus the complexity and cost increase.

4.3 Transimpedance Front-end

This is the most commonly used front end in Fibre Optic Applications [13, 33, 34, 35, 36, 37, 38, 39, 40]. It is basically a current-to-voltage converter (Fig. 14). In the limit of large loop gain, the relation between output voltage and input current is given by a simple relation

$$V_{\text{out}} = -Z_f I_{\text{in}}. \qquad (29)$$

Assuming that the loop gain is large, frequency response depends on that of Z_f, i.e. on R_f and C_f. Since C_f can be made small, bandwidth is greater than for a similar amplifier using non-feedback techniques. The noise performance of

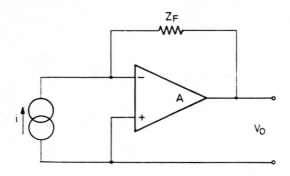

Fig. 14. Transimpedance amplifier.

this type of amplifier is not as good as that of the high impedance type because of the effects of R_f on the frequency response of the amplifier. In practice, the gain of the amplifier is finite and at high frequencies may not be much higher than 100. The actual transfer function is composed of two or more poles and may contain a few zeros. For a fixed open loop gain and fixed C_f, increase in the feedback resistance tends to make pole locations complex and may even make the amplifier oscillate. Therefore there is a practical limit to how large R_f can be made. Thus the noise of a transimpedance amplifier is always higher than that of a high impedance type.

4.3.1 Noise and signal modelling of a transimpedance amplifier

Figure 15 shows a schematic diagram [37] of the optical transimpedance amplifier. The current to voltage transfer function for this amplifier is

$$H_f(\omega) = \frac{R_f}{1 + j\omega \{R_f C_T / A\}}, \qquad (30)$$

where C_T = total input capacitance as defined earlier.

The 3 dB bandwidth $\qquad \omega_h = A/R_f C_T. \qquad (31)$

For minimum noise, the first stage must employ the active device in the common source/emitter mode. Currently GaAs MESFETs offer the best noise

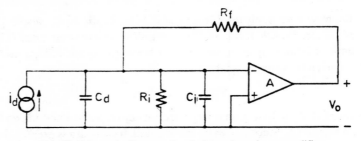

Fig. 15. Schematic diagram of optical transimpedance amplifier.

performance. The noise performance of GaAs MESFETs is discussed in [41]. The input equivalent noise current spectral density for an amplifier incorporating a MESFET common-source first stage is [42,43]

$$S_{eq} \simeq 2kT \left| \frac{1}{R_f} + \frac{\Gamma \omega^2 C^2}{g_m} \right|, \qquad (32)$$

where $C = C_d + C_{gs} + C_{gd}$. Equation (32) accounts for the noise contribution of the feedback resistance and the thermal channel noise of the FET. The gate shot noise, the frequency dependent gate coupling noise and the noise contribution of subsequent stages are negligible in comparison. In Figs. 16 and 17 are shown the amplifier current to voltage transfer function and the theoretical amplifier output noise as a function of ω. To minimize the amplifier noise, R_f and g_m/C^2 should be maximized. Maximization of R_f while maintaining a given bandwidth implies maximization of A and minimization of C. Circuit design considerations and the stability of the closed loop determine the maximum achievable value of A. Minimization of C is effected by the choice of a low capacitance photodiode and FET, while minimization of C also requires a circuit configuration chosen to minimize Miller capacitance. Furthermore,

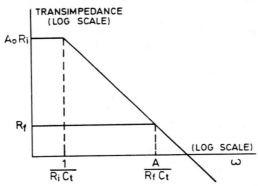

Fig. 16. Amplifier current-to-voltage transfer function.

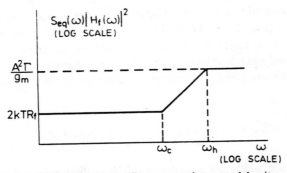

Fig. 17. Theoretical amplifier output noise spectral density.

since hybrid IC technology with GaAsMESFETs yields very low C values, the noise associated with R_f tends to dominate the amplifier noise performance. Hence the choice of R_f and second stage design is critical.

Hullett [32, 42] has outlined a simple and systematic method of calculating the input equivalent noise current spectral density of a feedback amplifier. The approach is complementary to Goell's [20]. Several design examples including the feedback cascode and the voltage feedback pair are considered.

4.3.2 Feedback cascode amplifier

Many of the early transimpedance amplifiers were of the classical feedback type-Fig. 18 [36, 43, 44]. The feedback cascode comprises a classical cascode amplifier with current feedback from output to input. The CE stage is required at the input for optimal noise performance. The common base stage is chosen because it has a low input impedance and helps to minimize Miller capacitance. However, this stage restricts the maximum value of A achievable and this in turn limits R_f. For the common base second stage, the i/p impedance is

$$Z_{iCB} \simeq \frac{1}{g_{m2}}. \qquad (33)$$

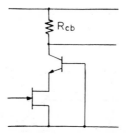

Fig. 18. Classical cascode configuration.

4.3.3 Three stage amplifier with shunt feedback in the second stage

This type of amplifier has gained popularity owing to some inherent disadvantages of the cascode amplifier. Figure 19 shows the first two stages of this amplifier. The first stage is similar to that of the cascode amplifier. The shunt feedback second stage has the same low input impedance as the common base stage thus minimizing Miller capacitance. Thus for the shunt feedback stage, we have

$$Z_{iSf} \simeq \frac{R_{Sf}}{g_{m2}(R_{Sf}//R_{C2})} \simeq \frac{1}{g_{m2}} \text{ for } R_{C2} \gg R_{Sf}. \qquad (34)$$

The shunt feedback stage contributes significantly to gain A thus permitting a larger value of R_f to be employed for a given bandwidth. A circuit advantage

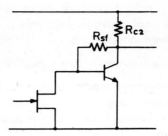

Fig. 19. Common source FET and shunt feedback stage.

of the shunt feedback configuration follows from the use of localized feedback which provides for greater open loop stability in the overall amplifier.

The choice of a shunt feedback second stage necessitates the use of a third stage within the overall feedback loop. This stage can serve to increase the gain. A suitable third stage is the series feedback circuit and the pre-amplifier configuration is then as shown in Fig. 20. The loop gain for this configuration is

$$A_{SF} \simeq - g_m R_{Sf} \frac{R_{C3}}{R_{E3}}. \tag{35}$$

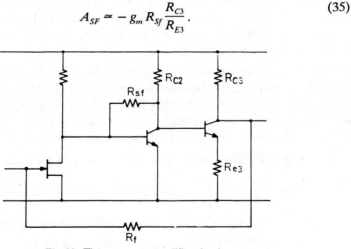

Fig. 20. Three stage preamplifier circuit.

4.3.4 Cascode feedback amplifier vs. three stage feedback amplifier

The three stage feedback amplifier has the advantage over the cascode of larger gain and hence larger R and the use of localized feedback which provides for greater open loop stability in the overall amplifier. Besides it has the same low input impedance as the common base stage.

4.3.5 BJT. vs. MESFET front ends in transimpedance preamplifiers

Though the GaAs MESFET undoubtedly offers the best noise performance, very little noise penalty is incurred [38] if a BJT is used in place of the

MESFET. The actual penalty depends on the quality of the transistors used. In applications where the noise margin is not critical, such an amplifier would be less expensive and more robust.

4.3.6 Advantages and disadvantages of transimpedance amplifiers

This type of amplifier has larger bandwidth than the high impedance type. It provides a greater dynamic range and has a noise performance which can approach that of high impedance front end. Equalizer design is also relatively simpler for this front end.

5. EQUALIZERS FOR OPTICAL COMMUNICATION RECEIVERS

In a typical fibre optic link, the received pulse shape $h_p(t)$ is dependent upon the transmitted pulse shape $h_{tr}(t)$ and the fibre impulse response $h_f(t)$

$$h_p(t) = h_{tr}(t) * h_f(t) . \qquad (36)$$

If $H_p(\omega)$ and other parameters [10] regarding the photodiode and the preamplifier like $\langle g \rangle$, $\langle g^2 \rangle$, S_I, S_E, R_b, R_A, C_d and C_A are known, it is possible to determine the transfer function $H_{eq}(\omega)$ of the equalizer which produces at the output of the receiver the equalized pulse shape $H_{out}(\omega)$ for each value of b_0 that minimizes the worst case noise.

In practice, other considerations in addition to the noise are also of interest. It is important that the intersymbol interference be low at the nominal decision times $\{kT\}$ and that it be sufficiently small at times offset from $\{kT\}$ to allow for timing errors in the sampling process. From Eq. (23), it is clear that the noise depends on the input and equalized output pulse shapes.

Referring to Eq. (23), the following definitions [10] are made: —

$R_T = \left| \dfrac{1}{R_b} + \dfrac{1}{R_A} \right|^{-1}$ = total detector parallel load resistance,

$C_T = C_d + C_A$ = total detector parallel load capacitance, $\qquad (37)$

b_{max} = larger value of b_k, b_{min} = smaller value of b_k.

$$H'_p(\omega) = H_p\left(\frac{2\pi\omega}{T}\right) .$$

$$H'_{out}(\omega) = \frac{1}{T} H_{out}\left(\frac{2\pi\omega}{T}\right) , \qquad (38)$$

$H_p'(\omega)$ and $H_{out}'(\omega)$ depend only upon the shapes of $H_p(\omega)$ and $H_{out}(\omega)$, not upon the time slot width T.

Also $\qquad \displaystyle\int_{-\infty}^{\infty} h_p(t)\, dt = 1 ,$

$$\int_{-\infty}^{\infty} H_{out}(t) \, df = 1. \tag{39}$$

As a result of the above mentioned normalizations Eq. (23) becomes

$$NW(b_0) = \left(\frac{h\Omega}{\eta}\right)^2 \left\{ \frac{\langle g^2 \rangle}{\langle g \rangle^2} \frac{\eta}{h\Omega} \left[b_0 I_1 + b_{max} [\Sigma_1 - I_1] \right] + \frac{T}{(\langle g \rangle e)^2} \right.$$

$$\left. \left[S_I + \frac{2k\theta}{R_b} + \langle g^2 \rangle e^2 \lambda d + \frac{S_E}{R_T^2} \right] I_2 + \frac{(2\pi C_T)^2 S_E I_3}{T(\langle g \rangle e)^2} \right\}, \tag{40}$$

where $I_1 = \int_{-\infty}^{\infty} H'_p(f) \left[\frac{H'_{out}(f)}{H'_p(f)} * \frac{H'_{out}(f)}{H'_p(f)} \right] df$,

$$\Sigma_1 = \sum_{k=-\infty}^{\infty} H'_p(k) \left[\frac{H'_{out}(k)}{H'_p(k)} * \frac{H'_{out}(k)}{H'_p(k)} \right],$$

$$I_2 = \int_{-\infty}^{\infty} \left| \frac{H'_{out}(f)}{H'_p(f)} \right|^2 df, \tag{41}$$

$$I_3 = \int_{-\infty}^{\infty} \left| \frac{H'_{out}(f)}{H'_p(f)} \right|^2 f^2 \, df.$$

In order to obtain the noise for various input and equalized output pulse shapes, I_1, I_2, I_3 and Σ_1 are to be calculated.

Three families of input shapes are depicted in Table 2.

TABLE 2

Input pulse	$h_P(t)$	$H_P'(t)$
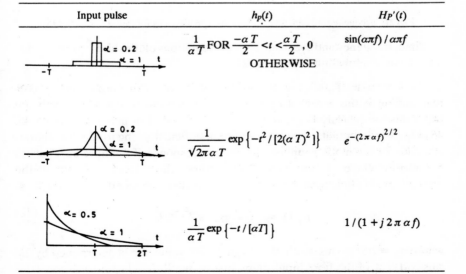	$\frac{1}{\alpha T}$ FOR $\frac{-\alpha T}{2} < t < \frac{\alpha T}{2}$, 0 OTHERWISE	$\sin(\alpha \pi f)/\alpha \pi f$
	$\frac{1}{\sqrt{2\pi}\,\alpha T} \exp\{-t^2/[2(\alpha T)^2]\}$	$e^{-(2\pi \alpha f)^2/2}$
	$\frac{1}{\alpha T} \exp\{-t/[\alpha T]\}$	$1/(1 + j2\pi \alpha f)$

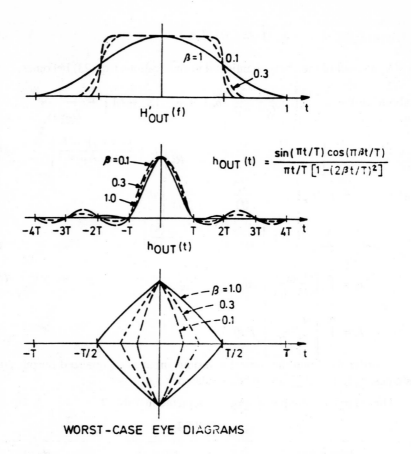

Fig. 21. Frequency domain, time domain and eye diagrams for raised cosine family.

Time, frequency and eye diagram representations of the raised cosine family are shown as a function of β in Fig. 21.

To determine $H_{eq}(\omega)$, the pulse shape of the detector output current corresponding to the received optical signal is chosen in agreement with the experimental measurements and the results reported in previous papers [45, 46, 47]. It follows that, in multimode fibres sufficiently long to reach a stationary state for the mode coupling, the impulse response of the fibre approaches a Gaussian shape. From these considerations, the current $i_{in}(t)$, due to the optical signal falling upon the detector [25], has been assumed to be of the form

$$i_{in}(t) = \frac{1}{\sqrt{2\pi}\alpha T} e^{-t^2/2(\alpha T)^2}, \qquad (42)$$

where $\sigma_{in} = \alpha T$ = rms width of the signal. The value of α is determined by the rms width σ_e of the fibre input pulse and by the fibre delay distortion σ_f (this distortion results from material dispersion and from the spread in the group delays of different fibre modes). Thus

$$\sigma_{in}^2 = \sigma_e^2 + \sigma_f^2. \tag{43}$$

The receiver output signal pulse shape, as mentioned earlier, must bring about minimum intersymbol interference. Raised cosine pulses satisfy this condition ideally, but Gaussian pulses may also be said to satisfy the condition since the intersymbol interference produced by them may be fixed a-priori at a small value. With this pulse shape, intersymbol interference is only slightly deteriorated when sampling time is moved from the central point.

Therefore, the receiver output voltage can be expressed as

$$V_{out}(t) = e^{-t^2/2(\beta T)^2}, \tag{44}$$

where $\sigma_{out} = \beta T$ = rms width of the signal. The intersymbol interference introduced by the signal of the form described in Eq. (44) is controlled by β. Therefore, this parameter must properly be chosen to assume a trade-off of intersymbol interference and output noise which minimizes the error rate in the received signal recognition.

$$V_{out}(\omega) = I_{in}(\omega) H_a(\omega) H_{eq}(\omega), \tag{45}$$

where $H_a(\omega)$ = transfer function of the circuit before the equalizer.

Therefore, $$H_{eq}(\omega) = \frac{V_{out}(\omega)}{I_{in}(\omega) \cdot H_a(\omega)}, \tag{46}$$

where $$V_{out}(\omega) = \sqrt{2\pi} \, \beta \, T \, e^{-(\beta T \omega)^2/2}, \tag{47}$$

$$H_a(\omega) = \left(\frac{1}{R_T} + j\omega \, C_T\right)^{-1}, \tag{48}$$

$$I_{in}(\omega) = e^{-(\alpha T \omega)^2/2} \tag{49}$$

$$H_{eq}(\omega) = \sqrt{2\pi} \, \beta T \left(\frac{1}{R_T} + j\omega \, C_T\right) e^{-(\beta^2 - \alpha^2)(T\omega)^2/2}. \tag{50}$$

From this expression, we can note that, for filtering the noise introduced by the circuit preceding the equalizer,

(i) β should be $> \alpha$ to have the transfer function $H_{eq}(\omega)$ a low-pass function.

(ii) It is desirable that the difference $(\beta^2 - \alpha^2)$ is as large as possible, taking into account the acceptable intersymbol interference.

More information about equalization can be obtained from references [31, 48 and 49].

6. PIN-FET RECEIVERS FOR LONGER WAVELENGTH COMMUNICATION SYSTEMS

Smith [50] analysed the effects of photodiode bulk leakage current and amplifier noise on receiver sensitivity and found that the sensitivity of a receiver using a PIN photodiode can be greatly improved by employing a high performance microwave FET in the input stage to the point where its remaining technical disadvantages in comparison with a silicon APD receiver at 800-900 nm may be offset by economic and operational attractions. The sensitivity of an optical receiver with a simple PIN photodiode can be significantly improved by reducing the pre-amplifier noise. Smith et al [51] have demonstrated that hybrid PIN-FET receivers for 2 Mb/s and 140 Mb/s have a sensitivity only 5 dB less than a conventional Si APD.

Silicon 'reach-through' APDs normally give a 15-20 dB higher receiver sensitivity than a silicon PIN photodiode. An APD, however, is relatively expensive and has several operational disadvantages, particularly for low bit rate systems. A dc-dc converter is required to supply a bias voltage of 100-400 V and additional circuitry is needed to control the avalanche gain. For systems in the 1.1-1.6 μm range, the optical receiver could be designed with either a Ge or III-V APD. Ge APDs have a high bulk leakage current and a high excess avalanche noise. Similarly, III-V APD structures show a much higher excess noise than Si APDs and to obtain micro plasma free operation raises significant difficulties. Receiver sensitivity for a PIN diode can be improved by reducing pre-amplifier noise with a hybrid integrated PIN photodiode and FET pre-amplifier having very low input capacitance and leakage currents. The FET should also have a high transconductance. Comparable performance to an APD receiver is achieved by choosing the FET preamplifier input resistor or feedback resistor for a transimpedance first stage sufficiently large so that thermal noise is small. The series resistance of the photodiode should also be small ($< 10\ \Omega$) so that its noise may be negligible. Similarly photodiode and FET gate leakage currents should be low such that the shot noise is small compared to the dominant noise, i.e., the FET channel noise [44] which is proportional to C_T^2/g_m. Various aspects of hybrid PINFET receivers are discussed in references [50-55] and are summarized in Table 3.

The equivalent input noise current and voltage spectral densities of the receiver for an FET first stage are given approximately [50] by

$$S_E = \frac{2k\theta\Gamma}{g_m}, \qquad (51)$$

and
$$S_I = qI_L, \qquad (52)$$

assuming that the induced gate noise is negligible.

TABLE 3

Characteristics of PIN-FET hybrids

Device	Capacitance	Transconductance	Leakage current	Quantum efficiency	Comment
Si FET Si MESFET	2 pF 0.15 pF	10-15 mmho 9 mmho	0.1 nA		Reduced receiver noise particularly for low bit rate applications
Ga As MESFET	0.1-0.3 pF	10-20 mmho	10 nA		Ideally suited for higher bit rate systems; very small 1/f noise contribution above 30 MHz
Si PIN diode	0.2 pF	—	10^{-8} A/cm^2	70%	Suited for systems in 800-900 nm range
III-V PIN	0.2 pF	—	10^{-6} to 10^{-8} A/cm^2	50-70%	Suited for systems in 1.2-1.4 μm range

$\Gamma = 0.7$ for Si and GaAs FETs at the minimum noise bias conditions, I_L = gate leakage current. I_L could also include photodiode surface leakage current. However, surface leakage can normally be successfully eliminated by careful surface passivation or a blocking ring.

Both silicon and GaAs FETs have been made with very low capacitance, high transconductance and low leakage currents. Dual-gate silicon FETs with input capacitance of $\simeq$ 2pF and 10-15 mmho transconductance are commercially available. However, a much better figure of merit is achieved with a narrow gate silicon MESFET, for which a gate source capacitance of 0.15 pF and 9 mmho transconductance have been reported for a gate length of 0.5 μm and width of 200 μm. Silicon FETs have very low 1/f noise and gate-leakage currents <0.1 nA are readily attainable. Consequently these devices could significantly reduce the receiver noise, particularly for low bit-rate applications.

GaAs MESFETs are ideally suited for higher bit-rate systems. These FETs have gate-source capacitances of 0.1-0.3 pF and transconductances of 10-20 mmho. They have gate-leakage currents of less than 10 nA and only a small 1/f noise contribution above 30 MHz.

As far as the PIN photodiode requirements are concerned, for systems operating in the 800-900 nm wavelength range, small area silicon PIN photodiodes can be fabricated relatively easily to have a high quantum efficiency, very low leakage current and low capacitance. Diodes with quantum ef-

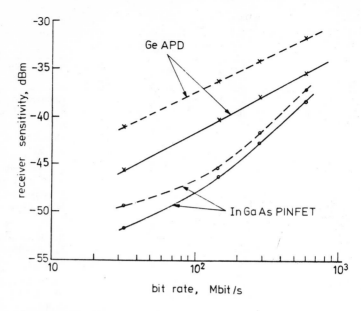

Fig. 22. Comparison of performance of PINFET hybrid receiver with that of receiver using GE APD.

ficiency > 70% and with capacitance < 0.2 pF can be fabricated easily. Such a diode would have a bulk leakage current < 10^{-8} A/cm^2.

For systems operating in the longer wavelengths region, $1.2-1.4 \mu$m, either a Germanium or III-V PIN photodiode such as GaInAs, GaAsSb or GaInAsP are suitable. Germanium, however, has a high bulk leakage current, $\simeq 25 \times 10^{-4}$ A/cm^2. A low capacitance of 0.2 pF and leakage current of 10^{-6} to 10^{-8} A/cm^2 are achievable successfully from III-V diodes.

Smith et al. [51] have demonstrated the feasibility and advantages of PIN-FET hybrid receivers for both low (2-8 Mbit/s and high (140-280 Mbit/s) bit rate applications. Smith [55] has compared experimentally the sensitivity of a PINFET hybrid optical receiver with that of a receiver using a Ge APD. Fig. 22 shows the comparision. The InGaAs PINFET receiver makes use of an integrating front end design while the Ge APD receiver uses a transimpedance amplifier.

7. TYPICAL PRACTICAL RECEIVERS

Table 4 lists out the salient features of some of the receivers reported. Generally for low speed applications, one can use a PIN diode or an APD with SI-FET high input impedance amplifier and for high speed applications, one can use a PIN diode with a BJT or GaAs FET transimpedance amplifier. For long-haul applications, APDs have an edge over the PIN diode owing to increased sensitivity.

TABLE 4

Sl. No	Wave length in nm	Photodiode type	Type of input amplifier	Amplifier input device	Bit rate Mb/S	Signal Format Input	Signal Format Output	P in dBm for 10^{-9} BER	Reference No.
1.	850	Si PIN	high input impedance	FET	25	NRZ	NRZ	-44.89	[11]
		Si APD (opt. G)	high input impedance	FET	25	NRZ	NRZ	-57.85	[11]
2.	900	PIN with $R_L = 1M\Omega$	high input impedance	JFET	6.3	RZ	RZ	-52.5	[28]
		PIN with $R_L = 4k\Omega$	high input impedance	JFET	6.3	RZ	RZ	-44.5	[28]
		APD (Opt. G)	high input impedance	JFET	6.3	RZ	RZ	-63.5	[28]
3.	850	Si PIN	high input impedance	FET	34.368	NRZ	NRZ	-39.75	[27]
4.	850	Si PIN	high input impedance	FET	50	RZ	NRZ	-41.5	[30]
5.		Si APD	Transimpedance	Si JFET	1.5	NRZ	NRZ	-56.6	[35]
		Si PIN	Transimpedance	Si JFET	1.5	NRZ	NRZ	-57.2	[35]
6.	820	Si PIN	Transimpedance (Cascode)	GaAsFET	274	NRZ	NRZ	-35	[36]
	Optimized at 1320 and measured at 820	Ge PIN	Transimpedance (Cascode)	GaAsFET	274	NRZ	NRZ	-32	[36]
7.	1100-1600	III-V PIN	Transimpedance (Cascode)	Si JFET	2	NRZ	NRZ	-65	[62]
	1100-1600	III-V PIN	Transimpedance (Cascode)	GaAsMESFET	2	NRZ	NRZ	-45	[62]

Sl. No	Wave length in nm	Photodiode type	Type of input amplifier	Amplifier input device	Bit rate Mb/S	Signal Format Input	Signal Format Output	P in dBm for 10^{-9} BER	Reference No.
8.	1300	III-V PIN	PINFET hybrid	GaAsFET	280	NRZ	NRZ	-38.9	[53]
	1500	III-V PIN	PINFET hybrid	GaAsFET	280	NRZ	NRZ	-39.4	[53]
9.	850	Si PIN	PINFET hybrid	GaAsFET	140	NRZ	NRZ	-45.8	[63]
	850	Si PIN	PINFET hybrid	GaAsFET	280	NRZ	NRZ	-42.4	[63]
10.	1300-1500	InGaAsPIN	Transimpedance shunt series feedback	BJT	140	NRZ	NRZ	-30.95	[57]
11.	1300-1550	InGaAsPIN	PIN-FET hybrid high impedance	GaAsMESFET	34	NRZ	NRZ	-52	[55]
	1300-1550	InGaAsPIN	PIN-FET hybrid high impedance	GaAsMESFET	140	NRZ	NRZ	-47	[55]
	1300-1550	InGaAsPIN	PIN-FET hybrid high impedance	GaAsMESFET	280	NRZ	NRZ	-42.5	[55]
	1300-1550	InGaAsPIN	PIN-FET hybrid high impedance	GaAsMESFET	565	NRZ	NRZ	-38	[55]
	1300-1550	GeAPD	PINFET Transimpedeance shunt feedback	BJT	34	NRZ	NRZ	-46	[55]
	1300-1550	GeAPD	PINFET Transimpedance shunt feedback	BJT	140	NRZ	NRZ	-41	[55]
	1300-1550	GeAPD	PINFET Transimpedance shunt feedback	BJT	280	NRZ	NRZ	-37.5	[55]
	1300-1500	GeAPD	PINFET Transimpedance shunt	BJT	565	NRZ	NRZ	-35	[55]

8. CONCLUSION

The theoretical and practical aspects of digital fibre optic receiver design have been analysed. Different types of photodiodes and front-ends have been dealt with, explaining the limitations in the design and performance capabilities of each. As an overall review of the state-of-the-art in front end optical receiver design, the attempt was to present a comprehensive overview of the possible techniques with pointers towards possible shortcomings in each case, for further research.

REFERENCES

1. K.C. Kao and Hockham G.A., "Dielectric fibre surface waveguides for optical frequencies", Proc. IEE., Vol. 113, pp. 1151-1158, July 1966.
2. F.P. Kapron, Keck D.B. and Maurer R.D., "Radiation losses in glass optical waveguides", Appl. Phys. Lett. Vol. 17, pp. 423-425, Nov. 15, 1970.
3. A.B. Gillespie, "Signals noise and resolution in Nuclear counter amplifiers", New York: Mcgraw Hill, 1953.
4. O.H. Schade, Sr., "A solid state low noise pre-amplifier and picture tube drive amplifier for a 60 MHz Video System", RCA Review 29, No. 1 p. 3, March 1968.
5. B.N. Edwards., "Optimization of pre-amplifiers for detection of short light pulses with photodiodes", Appl. Opt. 5, No. 9, pp. 1423-1425, September 1966.
6. L.K. Anderson and McCurtry B.J., "High speed photodetectors", Proc. IEEE. No. 54, pp. 1335-1349 (Oct. 1966).
7. R.J. Mc Intyre, "Multiplication noise in Uniform avalanche diodes", IEEE Trans. Electron Devices, ED-13, No. 1, pp. 164-168, Jan. 1966.
8. M. Chown and Kao. K.C., "Some broadband fibre-system design considerations", ICC 1972 Conf. Proc., Philadelphia, June 19-21, pp. 12-1 to 12-5, 1972.
9. M. Ross., "Laser Receivers", New York, John Wiley and Sons, p. 328, 1967.
10. S.D. Personick, "Receiver design for digital fibre optic communications systems, I", Bell System Technical Journal, Vol. 52 (6) pp. 843-874, July August 1973.
11. S.D. Personick, "Receiver design for digital fibre optic communications systems", II, Bell System Technical Journal, Vol. 52 (6) pp. 875-886, July August 1973.
12. J.E. Goell, "A repeater with high impedance input for optic fibre transmission systems", presented at the Conference on Laser Engineering and applications, Washington D.C., May 30th to June 1st, 1973, abstract in IEEE J.O.E., Vol. 39, pp. 641-642, June.
13. R.G. Smith and Personick S.D., "Receiver design for optical fibre communication systems", Topics in Applied Physics: Semiconductor Devices for optical communication: edited by H. Kressel, Vol. 39, pp. 89-160, 1980.
14. S.E. Miller, Tingye Li and Marcatili E.A.J., "Research toward optical fibre transmission systems part II: Devices and systems consideration", Proc. IEEE, Vol. 61, No. 12, pp. 1726-1751, Dec. 1973.
15. H. Melchior, "Semiconductor detectors for optical communication", Presented at the 1973 Conference on Laser Engg. and applications, Washington D.C. (abstract in IEEE Journal of Quantum Electronics), Vol. QE9, p. 659, June 1973.
16. H. Melchior, Fisher M.B. and Arams F.R., "Photodetectors for optical communication systems", Proc. IEEE, Vol. 58, No. 10, pp. 1466-1486, October 1970.

17. R.G. Smith, "Photodetectors and receivers-an update", Topics in Applied Physics semiconductor devices for optical communications, edited by R. Kressel, Vol. 39, pp. 293-299, 1980.
18. T.P. Lee, Burrus C.A., Ogawa K., Dentai A.G., "Electron Lett", 17, p. 431, 1981.
19. J.R. Klauder, and Sudarshan E.C.G., "Fundamentals of quantum optics", New York: W.A. Benjamin, Inc. pp. 169-178, 1968.
20. J.E. Goell, "Input amplifiers for optical PCM receivers", B.S.T.J., Vol. 53, No. 9 pp. 1771-1793, Nov. 1974.
21. W.M. Hubbard, "Comparative performance of twin channel and single channel optical frequency receivers", IEEE Trans. Communication, COM. 20, No. 6, pp. 1079-1086, 1972.
22. W. Baechtold, "Noise Behaviour of GaAsFETs with short gate lengths", IEEE Trans. on Electron Devices, Ed-19, pp. 674-680, 1972.
23. Mukunda. B. Das, "FET Noise sources and their effects on amplifier performance at low frequencies", IEEE Transactions on Electron Devices, Vol. ED-19, No. 3, pp. 338-348, March 1972.
24. S.D. Personick, "Receiver design for optical fibre systems", Proc. IEEE. Vol. 65, No. 12, pp. 1670-1678, Dec. 1977.
25. F. Lombardi and Orlandi G., "Receiver design and equalization for digital fibre optic communication systems", Alta Frequenza, Vol. XLV, N. 4, pp. 235-240, April 1976.
26. D.R. Smith and Garrett I., "A simplified approach to digital optical receiver design", Optical and Quantum Electronics 10, pp. 211-221, 1978.
27. F. Lombardi and Orlandi G., "Design considerations for a receiver for digital fibre optic communication systems", Electronics Letters, Vol. 11, No. 18, pp. 439-440, Sept. 1975.
28. J.E. Goell, "An Optical repeater with high impedance input amplifier", B.S.T.J., Vol. 53, pp. 629-643, April 1974.
29. Witcowicz, "IEEE J. SSC" -13, pp. 195 and 722, 1978.
30. P.K. Runge, "An experimental 50 Mb/s fibre optic PCM repeater", IEEE Transactions on communications: Vol. COM-24, No. 4, pp. 423-418, April 1976.
31. S.D. Personick, "Baseband linearity and equalization in fibre optic communication systems", BSTJ, Vol. 52, pp. 1175-1194, Sept. 1973.
32. S. Moustakas and Hullett J.L., "Noise modelling for broadband amplifier design", IEE Proc., Vol. 128, pt. G., No. 2, pp. 67-76, April 1981.
33. Y. Ueno and Ohgushi. Y., "A 40 Mb/s and a 400 Mb/s repeater for Fibre Optic Communications", IEEE Journal of quantum Electronics, QE-11, pp. 78D and 79D, Sept. 1975.
34. J.L. Hullett and Muoi T.V., "A feedback received amplifier for optical transmission systems", IEEE Transaction on communications, pp. 1176-1185, Oct. 1976.
35. W.M. Muska, "An experimental optical fibre link for low bit rate applications", B.S.T.J., pp. 65-75, Jan. 1977.
36. K. Ogawa and Chinnock. E.L., "GaAsFET transimpedance front end design for a wideband optical receiver", Electronics Letters, Vol. 15, No. 20, pp. 650-652, Sept. 1979.
37. J.L. Hullett and Moustakas. S., "Optimum transimpedance broadband optical pre amplifier design", Optical and Quantum Electronics' 13, pp. 65-69, 1981.
38. S. Moustakas and Hullett. J.L., "Comparison of BJT and MESFET front ends in broadband optical transimpedance pre-amplifiers", Optical and Quantum Electronics 14, pp. 57-60, 1982.

39. R. Mazzucco, Pierobon G.L., Pupolin S.G., "Computer aided design of receiver amplifiers for fibre optic digital transmission systems", Fourth European Conference on Optical Communication, pp. 483-489, 1978.
40. Papers which were presented at any of the European Conferences on Optical Communications.
41. K. Ogawa., "Noise caused by GaAs MESFET in optical receivers", BSTJ, Vol. 60, No. 6, pp. 923-928, July-August 1981.
42. J.L. Hullett and Muoi. T.V., "Referred impedance noise analysis of feedback amplifiers", Electronic Letters, Vol. 13, No. 13, pp. 387-389, June 1977.
43. D.R. Smith, Hooper R.C. and Webb R.P., "International Symposium on Circuits and Systems", Tokyo (1979).
44. D.R. Smith, Hooper R.C., Webb R.P. and Saunders M.P., "5th European Conference on Optical communication", Amsterdam (1979).
45. D. Gloge, "Impulse response of clad optical multimode fibres", BSTJ, Vol. 52, No. 6, pp. 801-816, July-August 1973.
46. S.D. Personick, Hubbard W.M. and Holden W.S., "Measurements of the baseband frequency response of 1 Km fibre", Applied Optics, Vol. 13, No. 2, February 1974.
47. S.D. Personick, "Optimal trade-off of mode mixing optical filtering and index difference in digital fibre optic communication systems", BSTJ., Vol. 53, No. 5, pp. 785, May-June 1974.
48. R.A.M. Rugemalira, "Optimum Linear equalization of a digital fibre optic communication system", Optical and Quantum Electronics, 13, pp. 153-163, 1981.
49. D.H. Henderson, "Dispersion and equalization in fibre optic communication systems", BSTJ, Vol. 52, No. 10, pp. 1867-1876, Dec. 1973.
50. D.R. Smith, Hooper R.C., Garrett I., "Receivers for Optical Communications: A comparision of avalanche photodiodes with PIN-FET hybrids", Optical and Quantum Electronics 10 pp. 293-300, 1978.
51. D.R. Smith, Hooper R.C., Webb R.P. and Saunders M.P., "PIN photodiode hybrid optical receiver", 5th European Conference on Optical communication, Amsterdam, pp, 13.4.1-13.4.3, Sept. 1979.
52. D.R. Smith, Hooper R.C., Ahmad K., Jenkins D., Mabbit A.W. and Nicklin R., "PIN-FET hybrid optical receiver for longer wavelength optical communication systems, Electronics Letters, 17th January, Vol. 16, No. 2, pp. 69-71, 1980.
53. D.R Smith, Chatterjee A.K., Rejman M.A.Z., Wake D. and White B.R., "PIN-FET hybrid optical receiver for 1.1-1.6 μm optical communication systems", Electronics Letters, Vol. 16, No. 19, pp. 750-751, Sept. 1980.
54. R.F. Letheny, Nahory R.E., Pollack M.A., Ballman A.A., Beebe E.D., Dewinter J.C., Martin R.J., "Integrated InGaAs PIN-FET photo receiver", Electronics Letters, Vol. 16, No. 11, pp. 353-355, May 1980.
55. D.R. Smith, Hooper R.C., Smyth P.P. and Wake D., "Expermintal comparision of a Ge APD and InGaAs PINFET Receiver for longer wavelength optical communication systems", Electronics Letters, Vol. 40, No. 11, pp. 453-454, 1982.
56. Tingye Li, "Optical fibre communication-The state of the art", IEEE Transactions on communications, Vol. COM 26, No. 7, pp. 946-955, July 1978.
57. T.L. Maione, Sell D.D. and Wolaver D.H., "Practical 45 Mb/s regenerator for light wave transmission", B.S.T.J., Vol. 57, pp. 1837-1856, July-Aug. 1978.
58. R.W. Blackmore and Fell P.H., "8.448 Mb/s optical fibre system", 1st European Conference on Optical Fibre Communication, pp. 182-184, 1975.
59. Y. Yamagata, Senmoto S., Inamura Y., Kaneko H. and Takahashi T., "A 32 Mb/s regenerative repeater for fibre cable transmission", IEE Conference Publication

No. 132, 1st European Conference on Optical Fibre Communications, pp. 144-146, Sept. 1975.
60. J.L. Hullett and Muoi T.V. and Moustakas S., "High speed optical pre-amplifiers", Electronics Letters. Vol. 13, No. 23, pp. 688-690, Nov. 1977.
61. R.T. Unwin, "High speed optical receiver", Optical and Quantum Electronics 14, pp. 61-66, 1982.
62. Taufii and Yamamoto T. and Takahara M., "Optimum Receiver for digital signal with multiplication noise, additive noise and intersymbol interference", Electronics and Communications in Japan, Vol. 59A, No. 9, 1976.
63. M.J.O. Mahony, "Duobinary transmission with PIN-FET Optical Receivers", Electronics Letters, Vol. 16, No. 19, pp. 752-753, Sept. 1980.

18

Fundamentals of Digital Communication

J. DAS[*]

---

1. INTRODUCTION

Classical information signals, e.g., speech and video, are lowpass in nature and may be directly transmitted through various cable systems only. For transmission through radio channels and for frequency division multiplexing (FDM), it is necessary to shift the information in the frequency domain by superimposing the low frequency signals on a high frequency carrier, thus generating a bandpass signal. Analog modulation schemes, e.g., Amplitude Modulation (AM), Frequency Modulation (FM) and Phase Modulation (PM), have been used for these purposes. With the formulation of the Sampling Theorem, due to Nyquist and Shannon, analog pulse modulation systems, e.g. Pulse-Amplitude (PAM), Pulse-length (PLM) and Pulse Position modulation (PPM) were developed. The information now could be transmitted in a time-discrete form and the interlacing of the samples from multiple signal sources resulted in what is known as the Time-Division Multiplexed (TDM) Systems.

In the above analog modulation schemes, the channel noise is cumulative over repeatered long distance circuits and this requires careful control of the noise and distortion per repeater section. If on the other hand, the signal samples are quantized and coded, say, in binary form (i.e. the samples are represented by a group of 1/0 pulses), then it is possible to regenerate the coded signals at every repeater and eliminate the disturbing noise (if the noise level is below a certain threshold) along a long distance transmission link. This

---

[*] Retired Professor of Electrical and Communication Engineering, IIT-Kharagpur and IIT-Kanpur. Present address: Flat A/2, 164/78, Lake Gardens, Calcutta.

technique of digital representation of analog signals is known as Pulse Code Modulation (PCM).

PCM scheme has many advantages over the analog modulation schemes viz. AM, FM, PAM, PPM etc. Since the signal is now digital in form, all digital signal processing techniques, e.g., digital filtering, band compression, error control, digital equalization, TASI, Echo cancellation etc., may be applied for making the communication system most efficient. It is well known that active digital devices, such as, digital MSI and LSI chips, are most efficient and these may be employed for processing and regeneration of digital signals. In particular in fibre-optic communication systems, the optical sources, namely LED's and Laser diodes, show non-linear characteristics and are generally unsuitable for linear modulation [1]. Similarly, the optical receivers which use photo detectors of the type PIN and APD, have non-linear characteristics. In addition, currently there is a growing trend for digitization of various telecommunication services around the world. Naturally, deployment of fibre optics in digital transmission systems became on obvious choice due to their mutual compatibility and this fact is responsible for the ever increasing popularity of fibre optical links in both short and long distance telephone networks. This chapter discusses the fundamentals of digital communication and is written with the aim to help in the understanding of companion chapters in the area of fibre optic communication systems in the book.

2. ANALOG MODULATION [2]

2.1 Amplitude Modulation (AM)

Assume that the low frequency information signal $x(t)$ is contained within a bandwidth of W Hz. To shift the signal to a higher frequency band, we vary one of the parameters of a high frequency sinusoidal carrier of frequency ω_0, given by

$$e(t) = A(t) \cos [\omega_0 t + \phi_0(t)].$$

When the amplitude $A(t)$ is varied according as $x(t)$, then the process is known as Amplitude modulation. The resultant bandpass signal $y(t)$ would have either the carrier suppressed (DSB-SC) or the carrier retained (DSB-AM); — thereby yielding

$$y(t) = x(t).A \cos (\omega_0 t + \phi_0)..... \text{(DSB-SC)},$$
$$y(t) = A [1 + mx(t)] \cos (\omega_0 t + \phi_0)..... \text{(DSB-AM)}, \quad (1)$$

where m is the modulation index. It is possible to generate single side band (SSB) signals by filtering the DSB-SC signals, DSB-SC signals are generated by using product modulators and DSB-AM signals by using non-linear devices, e.g., a class-C amplifier. In the receiver, $x(t)$ is recovered through appropriate demodulators and lowpass filters. The spectrum of the modulated signal is now contained within $(f_0 \pm W)$Hz and suitable bandpass filters are used both at the transmitter and receiver to eliminate undesired interference and noise. Figure 1 shows the AM-process and the relevant spectra.

AM systems are simple and are widely used in radio and television broadcasts, long distance FDM carrier systems and HF communication circuits. In the FDM system, multiplexing is carried out by generating SSB signals using different carriers $\omega_1, \omega_2, ..., \omega_n$, having $\Delta f = \omega_2 - \omega_1 = \omega_3 - \omega_2 = ... = \omega_n - \omega_{n-1} \geq W$ as shown in Fig. 2. The hierarchy of longhaul FDM transmission systems for speech and data is shown in Fig. 3, where it is seen that upto 10,800 speech channels may be accommodated in a coaxial cable with a bandwidth of 60 MHz.

Signal-to-noise ratio (SNR) is one of the most important characteristics of a communication system. For the same total transmitted power, the input-output SNR relations for the above three schemes are given by:

(a) SSB : $S_0/N_0 = S_i/N_i$,

(b) DSB-SC : $S_0/N_0 = 2 S_i/N_i$, (2)

(c) DSB-AM : $S_0/N_0 = \frac{2}{3} S_i/N_i$,

where, S_0/N_0 = SNR at the receiver output and S_i/N_i = SNR at the receiver input. AM, in general, does not provide any significant SNR improvement as is obtainable in FM systems.

2.2 Frequency Modulation (FM)

Frequency modulated signal $y(t)$ is generated by changing the instantaneous carrier frequency ω_0 in accordance with the variation of $x(t)$ giving

$$y(t)_{FM} = A \cos [\omega_0 t + \phi_0 + m_f \int x(t)\, dt]\,, \quad (3)$$

where the instantaneous frequency ω_i is

$$\omega_i = \omega_0 + m_f x(t)\,.$$

For sinusoidal modulation by a signal $x(t) = A_m \cos \omega_m t$,

$$y(t) = A \cos \left[\omega_0 t + \phi_0 + \frac{m_f A_m}{\omega_m} \sin \omega_m t\right], \quad (4)$$

where $m_f A_m\ (=2\pi f_D)$ represents the maximum frequency deviation of the carrier. The deviation ratio D is now defined as

$$D = \frac{f_D}{f_m} = \frac{m_f A_m}{\omega_m}\,. \quad (5)$$

Equation (4) may now be expanded using Bessel functions as (neglecting ϕ_0),

$$\begin{aligned}y(t) &= A [\cos \omega_0 t \cos (D \sin \omega_m t) - \sin \omega_0 t \sin (D \sin \omega_m t)] \\ &= A [J_0(D) \cos \omega_0 t + J_1(D) \cos (\omega_0 + \omega_m)t + J_2(D) \cos (\omega_0 + 2\omega_m)t \\ &\quad - J_1(D) \cos (\omega_0 - \omega_m)t + J_2(D) \cos (\omega_0 - 2\omega_m)t - ...]\,, \quad (6)\end{aligned}$$

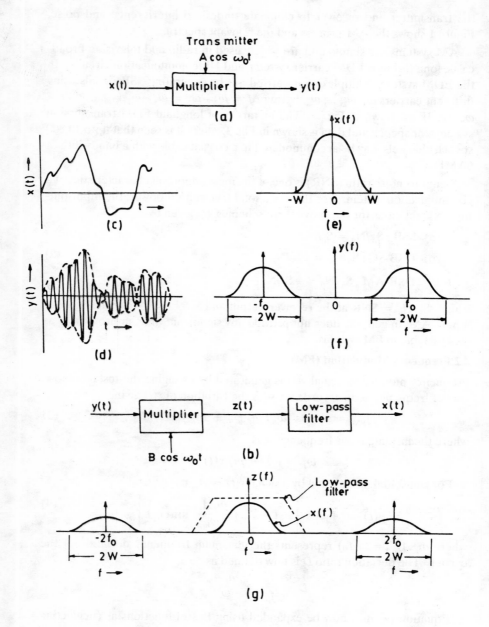

Fig. 1. AM-process and relevant spectra. (a) product modulator. (b) demodulator, (c) modulating signal $x(t)$, (d) modulated signal $y(t)$, (e), (f), (g): Spectra of $x(t)$, $y(t)$ and $z(t)$, respectively.

FUNDAMENTALS OF DIGITAL COMMUNICATION 419

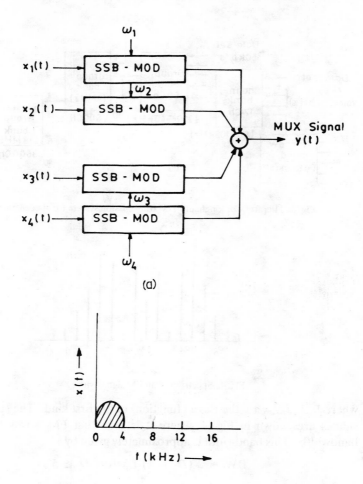

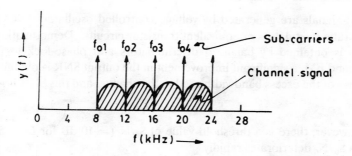

Fig. 2. (a) Generation of AM-MUX signal, (b) spectra of a 4 channel SSB-MUX where $x(f)$ has a bandwidth <4 kHz and the subcarriers are 8, 12, 16 and 20 kHz, respectively. The overall BW of $y(t)$ is 16 kHz.

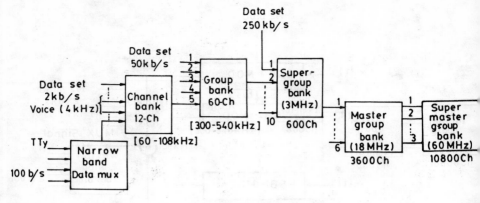

Fig. 3. Hierarchy of longhaul FDM transmission system for speech and data.

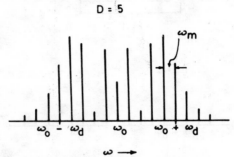

Fig. 4. Spectrum of an FM signal with D = 5.

where $J_0, J_1, J_2 \ldots$ are the Bessel functions of the first kind. The spectra of FM signals are shown in Fig. 4, where it is seen that FM waves have a large bandwidth. This bandwidth is approximately given by

$$\text{BW} \simeq 2(D+1)f_m \quad \text{for } D \geq 5. \tag{7}$$

For narrowband FM signals with $D < 0.6$, the BW $\simeq 2 f_m$.

FM signals are generated by voltage controlled oscillators (VCO), reactance tube modulators or equivalent transistor circuits. Demodulation of FM waves is obtained by balanced discriminators and phase-lock loops. For wideband FM, a significant improvement in the output SNR is obtained at the expense of the excess bandwidth used in the channel and this SNR is given by

$$S_0/N_0 = 3D^3(S_i/N_i). \tag{8}$$

However, there is a threshold value of S_i/N_i ($\simeq 10$ dB for $D = 5$) below which S_0/N_0 deteriorates rapidly.

2.3 Phase Modulation (PM)

Phase-modulated signals are obtained by changing the instantaneous phase of the carrier directly as $x(t)$. Accordingly Eq. (3) is now modified as

$$y(t)_{PM} = A\cos[\omega_0 t + \phi_0 + m_f x(t)], \quad (9)$$

where the instantaneous frequency ω_i is given by

$$\omega_i = \omega_0 + m_f \frac{d[x(t)]}{dt}.$$

It is thus seen that PM waves are similar to FM waves and have large spectra similar to that given by Eq. (6).

3. DISCRETE REPRESENTATION OF ANALOG SIGNALS [3, 4]

3.1 Pulse modulation schemes

In the above, we have discussed briefly the analog methods of information transmission by using AM, FM and PM. As a prelude to the development of digital transmission of information, pulse modulation systems, e.g., pulse-amplitude (PAM), pulse-length (PLM) and pulse-position (PPM) modulation were developed. The theoretical basis for these modulation schemes lies in the sampling theorem, due to Nyquist and Shannon, which states that:

> "A real-valued band limited signal, having no spectral components above a frequency of W Hz, is determined uniquely by its value at uniform intervals spaced no greater than 1/(2W) seconds apart".

This statement is a sufficient condition such that an analog signal can be completely reconstructed from a set of uniformly spaced discrete samples in time. The signal samples $x_s(t)$ are usually obtained by multiplying $x(t)$ by a train of narrow pulses $p_T(t)$, with a time period $T = 1/f_s \le 1/2W$, thus giving,

$$\begin{aligned} x_s(t) &\approx x(t) \cdot p_T(t) \\ &\approx x(t) \cdot \sum_{n=-\infty}^{+\infty} \delta(t - nT) \\ &= \sum_{-\infty}^{+\infty} x(nT)\,\delta(t - nT), \end{aligned} \quad (10)$$

where it is assumed that the sampling pulses are ideal impulses. It may now be shown, using Fourier series expansion of $x_s(t)$, that in the frequency domain, the spectrum $|X_s(f)|$ of $x_s(t)$, contains the same signal spectrum $x(f)$. In addition $|X_s(f)|$ contains repetitions of $X(f)$ around the harmonics nf_s and these sidebands are weighted by the amplitudes of the Fourier coefficients of $P_T(t)$. Now the original spectrum $X(f)$ can be retrieved from $X_s(f)$ by an ideal lowpass filter or by a bandpass filter around nf_s. The sampling process and the relevant spectra are shown in Fig. 5. The reconstruction of $x(t)$ is now carried out by an ideal lowpass filter (LPF) or by a sample-and-hold circuit along with an equalizer-filter. Assuming the impulse response of the LPF to be

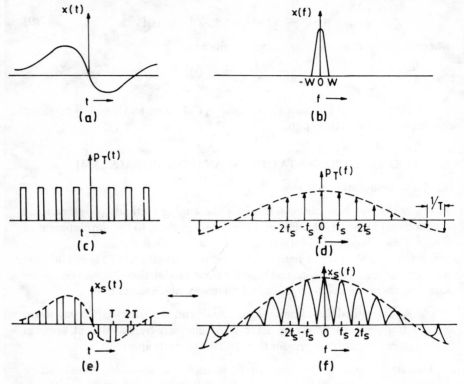

Fig. 5. The sampling process and the relevant spectra (a) information signal $x(t)$, (b) spectrum of $x(t)$, (c) sampling pulses $p_T(t)$, (d) Fourier series of $p_T(t)$, (e) sample signal $x_s(t)$, (f) spectrum of $x_s(t)$.

$$h(t) = 2W \cdot \frac{\sin(2\pi Wt)}{2\pi Wt}$$

the reconstructed signal is given by

$$\tilde{x}(t) = 2W \sum_{n=-\infty}^{+\infty} x(nT) \cdot \frac{\sin[2\pi W(t-nT)]}{2\pi W(t-nT)} \tag{11}$$

$$= x(t)/T, \quad T = 1/2W$$

This reconstruction is shown in Fig. 6.

The sampling theorem now leads to three possibilities: (a) any parameter of a pulse train, such as, amplitude, duration or position of pulses, may be varied according as the sampled values of $x(t)$ leading to PAM, PLM and PPM respectively; (b) the samples may now be quantized and coded in binary or m-ary format and transmitted as a digital signal, leading to Pulse Code Modulation (PCM); (c) it is possible to interlace in time the signal samples from multiple sources and transmit them as a single composite signal leading to Time-division multiplexing (TDM), as shown in Fig. 7.

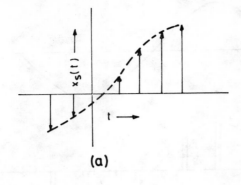

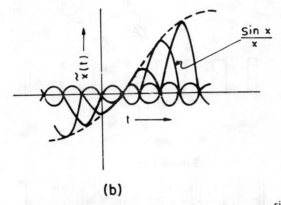

Fig. 6 (a) sampled signal $x_s(t)$ (b) reconstructed signal $x(t)$ using $\frac{\sin x}{x}$ filter.

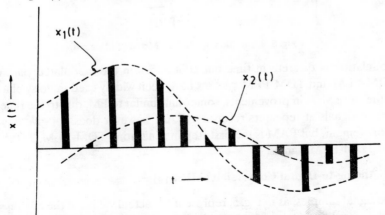

Fig. 7. Time multiplexing of two PAM signals producing PAM-TDM.

PAM, PLM and PPM signals in the time domain are shown in Fig. 8, and it should be noted that the (usually linearly) modulated parameter may have any of a continuum of values within the range of allowed values. Thus pulse

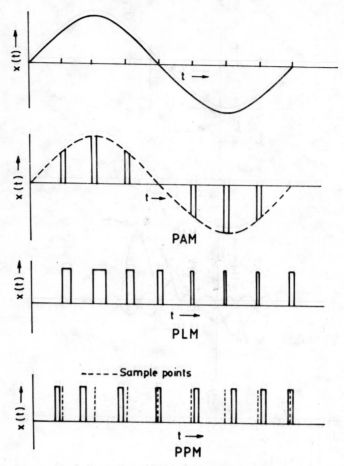

Fig. 8. Illustration of PAM, PLM and PPM signals.

modulation is discrete in time but continuous in the modulated parameter. TDM-PLM and TDM-PPM systems have been widely used in noisy channels, as they give SNR improvements, somewhat similar to FM, due to the regeneration of signals at repeaters/receivers. PAM however does not give any SNR improvement, but PAM is the first step in generating TDM-PLM, TDM-PPM and TDM-PCM signals.

3.2 Analog-to-Digital Conversion (ADC) [4]

The signal samples, as in Fig. 5, represent the actual values of the analog signal at the sampling instants. In a practical communication system or in a realistic measurement setup, the received/measured values can never be absolutely correct, because of the channel noise present or the inaccuracies of the measuring devices. It is therefore, sufficient for practical purposes to transmit/record only the quantized values of the signal samples. The quantized values, say, the nearest digit (i.e. 7 for 6.8 or 7.3) may then be represented in a binary form or in any coded form using only 'one or zero' pulses. As for example, the digit '7' may be written in a binary form as (0111). This technique of digital representation of analog signal is known as Pulse Code Modulation (PCM). PCM has many advantages over AM or FM schemes. The principles of PCM is further developed by considering the following example. Assume that the sample values of a signal is given as correct to two decimal places, as shown in Table 1; however, we may decide to transmit the values correct to only one decimal place. Then the total range of amplitude variation is quantized to 16 levels (0-15) only (since in Binary code, the number of levels L has to be a power of 2, i.e. $L = 2^B$ for a B-bit binary code). The approximated sample values and their equivalent binary codes are given in Table 1. It is seen that for a quantization step size $\Delta = 0.1$, the maximum error in a 4-bit PCM representation will be $.04/1.6 = 2.5\%$.

TABLE 1
PCM codes for signal samples

Sample values	Quantized values $\Delta = 0.1$	Equivalent Binary code (4-bit)	Sample values ×40 (in expanded scale)	Finer Quantized values $\Delta = 1$	6-bit Binary code
0.11	0.1	0001	4.4	4	000100
0.92	0.9	1001	36.8	37	100101
1.52	1.5	1111	60.8	61	111101
1.08	1.1	1011	43.2	43	101011
0.56	0.6	0110	22.4	22	010110
0.28	0.3	0011	11.2	11	001011
0.27	0.3	0011	10.8	11	001011

To minimise the maximum error in the PCM code, we may use finer quantization, say, $L = 2^6 = 64$, $B = 6$. Then the same samples will have the quantized values as shown in column 5 of Table 1 and the equivalent 6-bit code is shown in column 6. The maximum error is now $(0.4/64) = 0.6\%$ only.

3.3 Pulse Code Modulation (PCM)

A simple technique of generating PCM signals is to use successive approximations of $x(n)$, as illustrated in Fig. 9. Here a sample value of 85 units is successively compared with reference values of 64, 32, 16, 8, 4, 2, 1, and whenever the reminder is greater than the reference, digit 1 is recorded (otherwise, zero is recorded), thus giving the binary equivalent of 85 as: (1010101). The technique is mechanised by using a feedback type coder which uses a D/A converter (DAC) in the feedback circuit, as shown in Fig. 10(a). The DAC shown in Fig. 10(b), consists of an R-2R ladder network which produces binary weighted currents in response to the switch positions controlled by the steering pulses, generated at the comparator output. The signal sample is now compared repeatedly with the summed output of the DAC, and the comparator

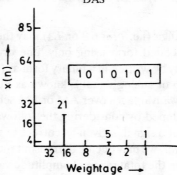

Fig. 9. Binary encoding using successive approximations.

output gives the PCM bits which are the binary equivalent of the signal sample. In the PCM receiver, a complementory DAC with pulse-steering circuits is used to generate the quantized approximation $\tilde{x}(t)$. The DAC and ADC are now available as LSI chips. PCM signals may also be generated by other techniques, such as, (i) counter type ADC (ii) parallel coder and (iii) cascade coder. However, the most commonly used type is the feedback coder of Fig. 10.

The linear PCM coder essentially consists of three operations-sampling, quantizing and coding. The message $x(t)$ is sampled as $x(n)$, at a rate f_s, where

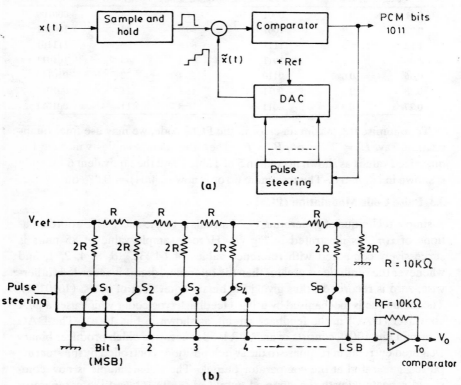

Fig. 10. (a) Feedback type ADC, (b) Schematic of DAC.

f_s is slightly greater then 2W, and each sample is quantized to one of L levels, covering the total range of the signal amplitude. If $x(t)$ is converted to B-bit PCM and Δ = quantum step, then we have number of bits

$$B = \log_2 L; \qquad L = 2^B; \qquad (12)$$

and $L = (X_{max} - X_{min})$, where $X_{min} \leq x(t) \leq X_{max}$.

Quantization is obtained through a 'staircase transducer' (equivalent to a fixed-value quantizer), having the input-output relations, as shown in Fig. 11(a). If a linear waveform (say, a saw-tooth voltage as shown) is superimposed on the transducer, the error $e(t)$ due to quantization will be as shown in Fig. 11(b). The error power in $e(t)$ is known as the quantization noise N_q and is calculated to be

$$\overline{e^2(t)} = N_q = \Delta^2/12. \qquad (13)$$

The signal power S_0 of the liner waveform in the range of $\pm X_0$ is given by

$$S_0 = X_0^2/3 .$$

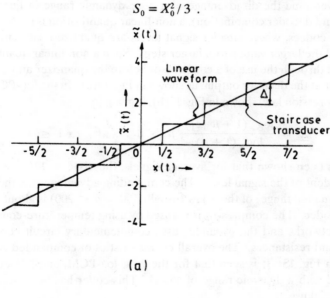

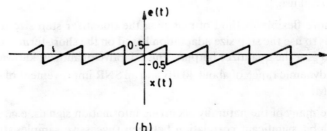

Fig. 11. (a) Staircase transducer, (b) Error voltage with a linear waveform as input

The SNR at the decoder output is

$$S_0/N_q = 4X_0^2/\Delta^2 = L^2\Delta^2/\Delta^2 = L^2 = 2^{2B} \qquad (14)$$
$$\equiv 6B \text{ dB}.$$

Since S_0 is dependent on the signal waveform, while N_q is almost constant for all arbitrary waveforms; the SNR for different signals is calculated as:

(a) For sinusoidal signals, $S_0/N_q = (6B + 2)$ dB, \qquad (15)

(b) For random signals, $S_0/N_q = (6B-7)$ dB.

The above values of SNR, given by Eqs. (14) and (15), are for optimum input to the coder. However, for signals with a dynamic range of R dB, e.g. in speech, the SNR value degrades by R dB when the input is at its minimum level. Thus it is necessary to modify the quantizer characteristics to accommodate large dynamic ranges of $x(t)$. The S_0/N_q values may be equalized for varying input levels by using compandors or by adaptive techniques.

To overcome the disadvantage of a poor dynamic range in linear codecs (coder and decoder combination), a non-linear quantization is preferably used in PCM codecs, where smaller signal levels are quantized with smaller step sizes and the larger values with larger steps. Such a non-linear quantization is obtained through the use of a compressor before the quantizer and a matching expander at the decoder output, as shown in Fig. 12(a). In the log-PCM coder, the compression law, shown in Fig. 12(b), is given by

$$y = \frac{\log_e(1 + \mu|x|)}{\log_e(1 + \mu)}, \mu > 0 \quad \text{and} \quad |x| \le 1. \qquad (16)$$

It has been shown that for average signals, the SNR = $12/\Delta^2$, and is now independent of the signal level. The companding advantage (i.e. the increase in the dynamic range) of the μ-law (usually $100 < \mu < 200$) is about 26 dB for a 7-bit codec. The compressor is realised by using temperature-compensated diode networks and the expander uses complementary circuit consisting of diodes and resistances. The overall characteristics of companded codecs are shown in Fig. 13. It is seen that for the 7-bit log-PCM, an SNR of 33 dB is obtained with a dynamic range of 45 dB. This coder has been accepted for commercial use.

A more flexible method of matching the quantizer step size to the signal power is to use the step size adaptation based on the short-term average signal power $\overline{\sigma_n^2}$ at the quantizer output. This adaptive quantizer, known as APCM, gives a dynamic range of about 40 dB and an SNR improvement of 4-5 dB over log-PCM.

Since many of the naturally occurring information signals, e.g., speech and TV, exhibit significant correlation between successive samples, the variance (average power) of the first difference $<d_n^2(1)>$ is smaller than the variance

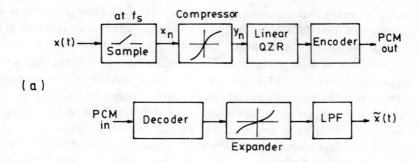

(a)

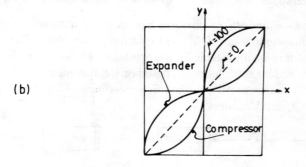

(b)

Fig. 12. (a) Block diagram of a companded PCM codec, (b) Compressor-expander characteristic.

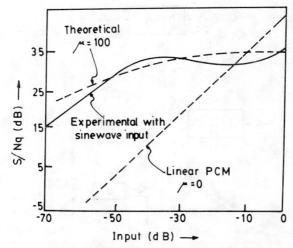

Fig. 13. SNR values for 7-bit log-PCM codec.

of σ_x^2 of the signal itself ($<>$ indicates ensemble average of the variables). Such a differential input to a quantizer will behave more as a random signal and hence an optimized quantizer may be used. The technique is known as Differential PCM (DPCM) and the block schematic of the codec is shown in Fig. 14. In the coder, the predicted sample $\tilde{x}_n$ is subtracted from x_n and the difference d_n is used as the input to the quantizer. In the decoder, a complementary operation is performed. It is also possible to use a multi-tap predictor in the feedback circuit and this gives a better result. DPCM with a single-tap predictor gives SNR gains of 3-6 dB over PCM and for multi-tap predictors, SNR gains are 6-10 dB. DPCM has been extensively used for coding of TV and picture-phone signals. It is also possible to further refine the coder by providing adaptive quantization along with the predictive feedback, and the system is now called ADPCM. ADPCM is the most efficient of the coders developed so far, and gives SNR gains of 10 dB or more over log-PCM.

Over the years, various other techniques have been developed for waveform coding, e.g., Delta-modulation (DM) and its adaptive version ADM. While PCM and DM coders are applicable for any arbitrary waveform, special coders have been developed for speech signals, based on the parametric properties of speech. These are known as Adaptive predictive coding (APC), Linear predictive coding (LPC), and Vocoders.

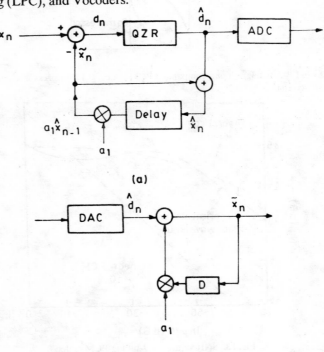

Fig. 14. Block schematic of a companded DPCM codec.

4. TDM-MUX [5]

Multiplexing of signals on FDM basis has been shown in Figs. 2 and 3, and Time-division multiplexing (TDM) of PAM signals has been illustrated in Fig. 7. TDM techniques are used for multiplexing and multiaccessing pulse and digital signals. Two primary digital MUX systems are 24-ch and 30-ch PCM-TDM for speech signals and their relevant parameters are given in Table 2. In the 24-ch PCM-TDM, twenty four 8-bit voice channels are time multiplexed to give 192 bits per frame of 125 μsec, and one extra bit is inserted to give frame synchronization, yielding 193 bits per frame. Using a sampling rate of 8 KHz, the overall clock rate is 1.544×10^6 bits/sec. Signalling information is usually transmitted over the 8th bit of the code word. The block diagram of a 24-ch. PCM coder is shown in Fig. 15(a) and the composite waveform at the coder

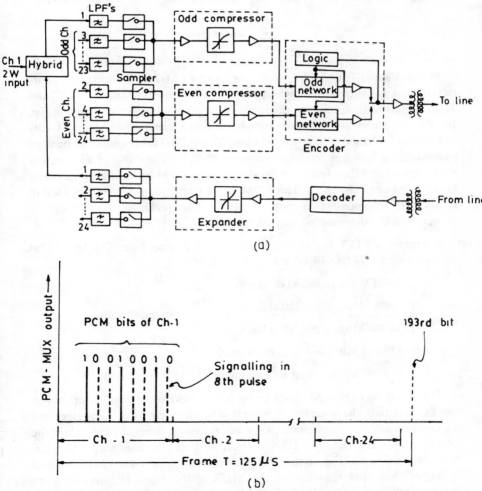

Fig. 15. (a) 24-ch. log-PCM system, (b) Composite waveform for PCM-TDM.

output is shown in Fig. 15(b). The alternative 30-ch PCM-MUX operates at a clock rate of 2.048 Mb/s and its parameters are also given in Table 2.

TABLE 2
Parameters of PCM-TDM Systems

System parameter	24-ch PCM	30-ch PCM
Sampling rate	8 KHz	8 KHz
Frame time	125 μs	125 μs
No. of bits per sample	7+1	8
Signalling	8th bit	31st channel
Synchronisation (word)	193rd bit in the frame	32nd channel
Companding	μ-low, $\mu = 100$	A-low, A = 87.6
Companding gain	26 dB	24 dB
Clock rate	1.544 Mb/s	2.048 Mb/s
Dynamic range with $S/Nq = 32$ dB	50 dB	48 dB

In the AT and T digital hierarchy, the 24-ch PCM-TDM, known as T_1-carrier, is used as the basic system and higher order channel banks, T_2, T_3, T_4 and T_5 are obtained by combining the lower order channel banks as shown in Fig. 16. Picture-phone and television signals are also accommodated in the T-carrier system for long distance transmission. The PCM-TDM signals can be used to modulate a high frequency carrier using ASK, FSK or PSK (as discussed in next section) and then further multiplexed on an FDM-basis to accommodate larger number of channels. This FDM/TDM technique is sometimes used in fibre-optic communication systems where the technique is known as Wavelength division multiplexing (WDM) [1].

Currently, CCITT has recommended the following hierarchy for digital transmission levels (at 64 Kb/s per speech channel):

(1) 2.048 Mb/s, giving 30 channels,

(2) 8.448 Mb/s, giving 120 channels,

(3) 34.368 Mb/s, giving 480 channels,

(4) 139.264 Mb/s, giving 1920 channels,

(5) 564.992 Mb/s, giving 7680 channels.

TDM may be organised either by bit interleaving or by word interleaving. For PCM signals, the word interleaving is generally used, as in T_1-carriers; but for Delta-modulated bit streams, the bit interleaving is generally used, since the latter requires less high speed digital electronics and the hardware is simpler. It is observed that in Fig. 16 that the bit rates at the higher-level MUX output is higher than that required by the MUX if the input bit streams were synchronous (e.g., in T_2-MUX, 6.312 Mb/s are greater than 4×1.544 Mb/s).

FUNDAMENTALS OF DIGITAL COMMUNICATION

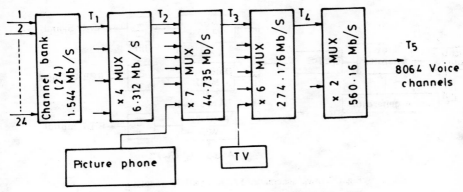

Fig. 16. The T-carrier TDM-PCM hierarchy of AT and T.

Thus for multiplexing asynchronous data, it is necessary to use higher clock rates at the MUX-outputs and, at the same time, use the 'pulse-stuffing' technique (also known as 'justification') to prevent loss of data from the primary bit streams. As an example, the T_2-master frame has a length of 1176 bits, which is unrelated to the frame bits of the T_1-carrier. Of these, 1148 are information bits (287 per T_1-input), 12 framing bits, 12 stuffing control bits and 4 timing bits. Similarly, the T_3-master frame has 4760 bits and the T_4 master frame 4704 bits. Pulse stuffing is provided only in the designated positions in the frame, so that these can be removed at the DEMUX output.

5. DIGITAL SIGNALLING [3]

The digital data, either from an ADC coder or from a computer terminal, has a lowpass spectrum, but the transmission medium e.g., voice-grade circuits, radio links, is generally bandpass. As such, it is necessary to match the data with the medium through a modulation process. These modulation techniques are mainly: (i) Amplitude Shift Keying (ASK/on-off); (ii) Frequency Shift Keying (FSK) and (iii) Phase Shift Keying (PSK).

For binary inputs, the speed of transmission is generally 1 bit/Hz. But for higher speeds of transmission, the binary data may be converted to multilevel bauds and each baud then modulates the carrier leading to multilevel ASK, multi-frequency FSK, multiphase PSK and multilevel QAM (APK). There are other spectrally efficient modulation schemes, e.g., CPFSK, OFFSET-QSK, MSK and PRS, which are being used in newer systems.

The signalling waveforms for the above modulation schemes are:

(i) ASK: $\left.\begin{array}{l} S_1(t) = A \cos \omega_0 t \\ S_0(t) = 0 \end{array}\right\}$, $\quad 0 \leq t \leq T$ \hfill (17)

(ii) FSK: $\left.\begin{array}{l} S_1(t) = A \cos \omega_1 t \\ S_0(t) = A \cos \omega_0 t \end{array}\right\}$, $\quad 0 \leq t \leq T$

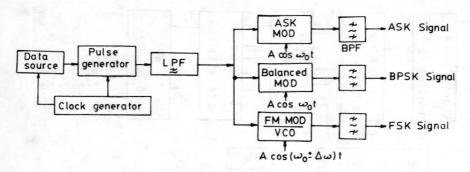

Fig. 17. Block schematic of ASK/FSK/PSK modulators.

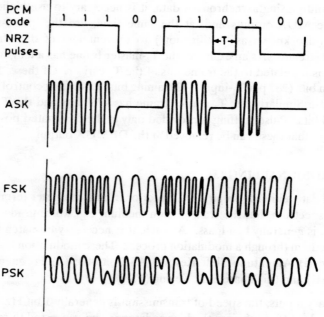

Fig. 18. Waveforms for ASK, FSK and PSK signals.

where $(\omega_1 - \omega_0)$ and $(\omega_1 + \omega_0)$ are multiples of π/T (T = bit duration) and the two signals are orthogonal.

$$\text{(iii)} \quad \text{BPSK:} \quad \left. \begin{array}{l} S_0(t) = A \cos \omega_0 t \\ S_1(t) = A \cos (\omega_0 t \pm \pi) \end{array} \right\}, \quad 0 \le t \le T \qquad (18)$$

The block schematics of these modulators are given in Fig. 17 and the corresponding waveforms are shown in Fig. 18. These bandpass signals have approximate bandwidths of:

For ASK and PSK, BW (ch) $\simeq 1.5 f_r$, $f_r = 1/T$

and roll off factor $\alpha = 0.5$

For FSK, BW (ch) $\simeq 3f_r$; $\alpha = 0.5$. (19)

The receivers for the modulated signals may use either coherent or incoherent detection, but the coherent detection gives better performance. The detection schemes are shown in Fig. 19(a-d). It is seen that the coherent detectors are basically correlators with the phase-synchronous carrier and clock available at the receiver. Equivalently, the detector may be realized

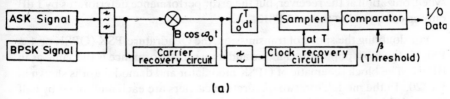

(a)

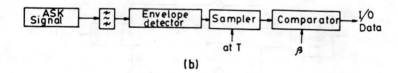

(b)

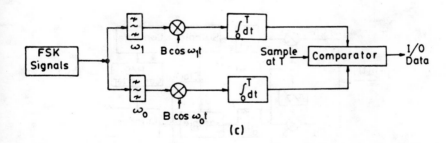

(c)

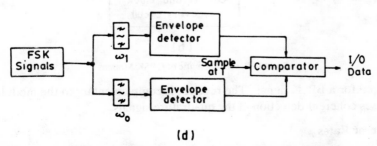

(d)

Fig. 19. Optimum detectors for digital signals. (a) coherent detector for ASK and PSK, (b) incoherent detector for ASK, (c) coherent detector for FSK, (d) incoherent detector for FSK.

through a matched-filter (MF) correlator. It should be noted that in Fig. 19(a), the BPSK signal will give bipolar (±1) output, whereas for ASK, the output will be unipolar (1/0) only.

The incoherent detection of BPSK signals is realized by using the phase of $(k-1)^{th}$ pulse as the reference for the kth pulse in the coherent detector. To obtain the correct data bits at the receiver, the transmitter data are precoded differentially, generating DPSK signals. This avoids the difficulties of carrier synchronization at the receiver, but the error performance deteriorates by 1 dB only as compared to that of CPSK.

For doubling the speed of transmission, the Quadrature-PSK (QPSK or 4-ϕ PSK) is the most popular technique and its error performance is the same as in BPSK. The block schematic of QPSK modulator and demodulator is shown in Fig. 20. In the modulator, two quadrature carriers are each multiplied by half the data bits, thus making $T = 2T_b$, and the channel bandwidth half that

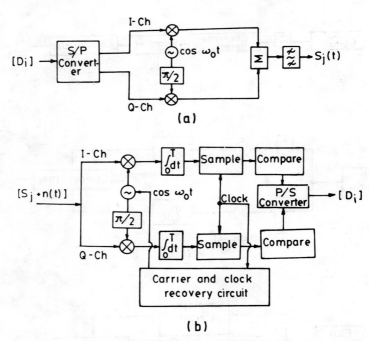

Fig. 20. Block schematic of QPSK Modem.

required for a BPSK signal. The receiver is complementary to the modulator and uses coherent detection of the quadrature signals.

5.1 Error Rates

Using detection theory, it has been proved that the MF (or cross-correlator) is the optimum receiver for coherent detection. But for incoherent detection, envelope detector along with a comparator is used as shown in Fig. 19(d). The error probability at the receiver output now depends only on the average signal energy E_{av}, the noise spectral density n_0 ($n_0 = N_i/W_{ch}$, W_{ch} is the channel bandwidth and noise is Gaussian) and the correlation ρ between the two signalling waveforms S_1 and S_0 representing 1 and 0. The error probability P_e is independent of the signalling waveform and is given by

$$P_e = erfc\sqrt{(1-\rho)E_{av}/n_0}, \quad (20)$$

where $\quad E_{av} = (E_1 + E_2)/2$

and $\quad erfc(x) = [1-erf(x)]$

$$erf(x) = \int_{-\infty}^{x} \frac{1}{\sqrt{2\pi}} \exp\left[-\frac{z^2}{2}\right] dz.$$

For practical binary systems of Figs. 18 and 19, P_e reduces as follows:

(a) ASK:

   (i) Coherent detection:

$$P_e = erfc(\sqrt{E/2n_0})$$
$$\simeq \sqrt{\frac{n_0}{\pi E}} \exp(-E/4n_0), \quad \frac{E}{n_0} \gg 1 \quad (21)$$

  (ii) Incoherent detection:

$$P_e \simeq \frac{1}{2} \exp(-E/4n_0), \quad \frac{E}{n_0} \gg 1 \quad (22)$$

(b) FSK:

   (i) Coherent detection:

$$P_e = erfc(\sqrt{E/n_0})$$
$$\simeq \sqrt{\frac{n_0}{2\pi E}} \exp(-E/2n_0), \quad \frac{E}{n_0} \gg 1 \quad (23)$$

  (ii) Incoherent detection:

$$P_e = \frac{1}{2} \exp(-E/2n_0), \quad (24)$$

(c) (i) CPSK and QPSK:

$$P_e = erfc(\sqrt{2E/n_0})$$
$$\simeq \frac{1}{2}\sqrt{\frac{n_0}{\pi E}} \exp(-E/n_0), \quad \frac{E}{n_0} \gg 1 \quad (25)$$

  (ii) DPSK:

$$P_e = \frac{1}{2} \exp(-E/n_0). \quad (26)$$

The above results are shown graphically in Fig. 21, where it is seen that CPSK and QPSK have the minimum error rate for a given E/n_0. DPSK is seen to have 3 dB lesser E/n_0 requirements than the incoherent FSK and has similar P_0 for large E/n_0. A reference $P_0 = 10^{-5}$ is obtained for $E/n_0 = 9.6$ dB in PSK, for $E/n_0 = 12.6$ dB in FSK and for $E/n_0 = 15.6$ dB in ASK, using coherent detection. For incoherent detection, the E/n_0 requirement is only 1 dB more in respective systems.

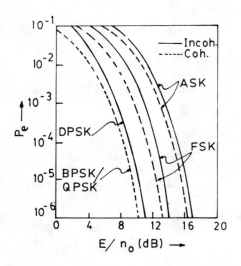

Fig. 21. Error rates in binary communication systems.

5.2 Carrier and Clock Synchronization [5]

In coherent detection of Fig. 19, it has been assumed that a synchronized carrier and a synchronized sampling pulse (for bit recovery) are available at the receiver. But in ASK and PSK, suppressed carrier modulation is used to minimize the transmitted power; and as such, a ready carrier reference is not available in the received signal. For ASK and BPSK, the carrier, however, may be re-generated through a simple squaring circuit and a local VCO may then be locked to this frequency through a phase-locked loop (PLL). Usually a second order PLL is used and this achieves both frequency and phase lock for a constant frequency input, but having random phase, under the condition that the phase error is much less than $\pi/2$. The PLL is generally used for tracking carrier frequencies, since its steady state phase error tends to zero at $t \to \infty$, if the initial frequency offset is within a limit. The initial acquisition of the unknown inputs frequency may be obtained by a swept VCO. In a practical carrier recovery circuit, shown in Fig. 22, a bandpass limiter is used to remove the input level variation and this gives an SNR degradation of 1 dB only for small input CNRs. Further to simplify the multiplier circuit of the PLL, the VCO output is hard-limited and a switched multiplier is used without any degradation of the loop performances. In case the initial frequency offset is not too large, PLL may be operated in both acquisition and tracking mode, by using a large PLL bandwidth for acquisition and a small bandwidth for tracking. Alternatively, the well known Costas loop may be used for carrier recovery.

The problem of clock (bit or symbol) synchronisation for digital modulation systems is somewhat similar to that of the carrier synchronisation. Although sophisticated techniques, such as a MAP estimator (based on maximum a posteriori probability), may be used for special systems, simple squaring or similar loops operate quite efficiently with reasonably acceptable CNRs at the receiver input. Two such sub-optimal circuits are shown in Fig. 23. The squaring loop of Fig. 23(a) is similar to that of Fig. 21 and has an equivalent performance. The circuit of Fig. 23(b) produces a line spectrum at $f_r = 1/T_b$ and this may be filtered to give the required clock frequency. The variance of the timing error in above clock recovery circuits is worse by only a few percent as compared to the optimum MAP synchronizers, but the circuits have the advantage of much simpler implementation.

For PCM-TDM multiplex systems, it is also necessary to have word (or channel) synchronisation. This is achieved by using a long unique identification code in 193rd bit in 24-ch PCM frame or in the 32nd channel of 30 ch. PCM-MUX.

6. DIGITAL TRANSMISSION [5]

A modern digital communication system consists of many sub-blocks, and Fig. 24 shows some of the important ones. The input to such a system is usually a digital signal, i.e. PCM/DM bits from A/D converters or computer data from a terminal. The basic blocks, as shown, are: A/D converter, error control circuit and digital modulator in the transmitter; amplifier-equalizer, detector, decoder and D/A converter in the receiver. For long distance circuits, regenerators are used as line repeaters.

The digital data, either from an ADC coder or from a computer peripheral, has a lowpass spectrum, and in an ideally lowpass channel, unipolar (1/0) or bipolar (±1) data may be directly transmitted and received. However, a random data sequence occupies a large bandwidth and it is necessary to use a lowpass filter after the data generator to restrict the required bandwidth. Moreover the low frequency response in a long line is restricted due to the use of coupling transformers and other devices in the repeater amplifiers. Thus in a voice grade data circuit which is a bandpass medium, one of the modulation schemes has to be used. Even in coaxial cable systems, which are typically wideband lowpass media, the low frequency response is poor and such techni-

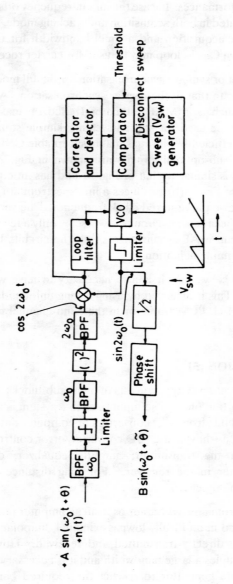

Fig. 22. Carrier acquisition and tracking circuit.

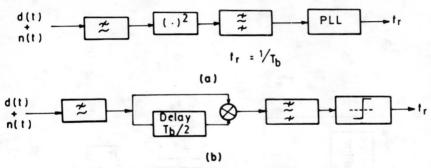

Fig. 23. Suboptimal clock recovery circuits.

ques as AMI (Alternate Mark Inversion), pseudo-ternary and PRS (Partial Response Signalling) are used to overcome this problem.

The power spectral density (PSD) of a random binary signal is given by

$$G(f) = \frac{\sigma^2}{T} |S(f)|^2 + \frac{a_0^2}{T^2} \sum_{k=-\infty}^{\infty} \left| S\left(\frac{k}{T}\right) \right|^2 \delta(f - kf_r), \qquad (27)$$

where $S(f)$ is the Fourier transform of the pulse waveform $S(t)$, σ^2 = variance of the data symbols, a_0 = mean amplitude of the symbols and $f_r = \frac{1}{T}$ = repetition rate of the pulses. For bipolar symmetrical RZ (return to zero) random pulses with $a_0 = 0$, and $p(1) = p(0) = 0.5$, $G(f)$ is modified as

$$G(f) \text{ [BSR]} = \frac{A^2}{T} \left(\frac{\sin \pi f \tau}{\pi f} \right)^2, \qquad \tau = \text{pulse width} \leq T. \qquad (28)$$

From Eq. (28), it is seen that the main lobe of $G(f)$ extends upto $f = 1/\tau$, and thus, for maximum energy transfer through the channel, the channel bandwidth $BW_{(ch)} \geq 1/\tau$. Since $\tau \leq T$, the minimum $BW_{(ch)}$ is $1/T$ Hz for NRZ (not return to zero) pulses. On the other hand, the sampling theorem tells us that random narrow pulses, $\tau \ll T$, can be transmitted through an ideal lowpass channel if the channel cut off frequency $f_c = \frac{1}{2T}$ Hz. The ideal lowpass filter produces $(\sin x / x)$ type of impulse response with zeros at $t_k = k/2f_c = kT$, and the successive pulses can be distinguished without any error. However, ideal filters are not physically realisable and the zeros of the filter response will no longer be at $t_k = kT$. This gives rise to the wellknown intersymbol interference (ISI) problem in practical systems. As a partial solution to this, the signalling pulses are shaped with raised cosine or PRS filters, such that the osillatory tail of the response is reduced, thereby reducing the ISI. The raised-cosine filter has an amplitude response

$$A(\omega) = \frac{1}{2} \left[1 + \cos \frac{\pi \omega}{2 \omega_c} \right] = \cos^2 \frac{\pi \omega}{4 \omega_c}, \qquad \omega \leq 2\omega_c$$

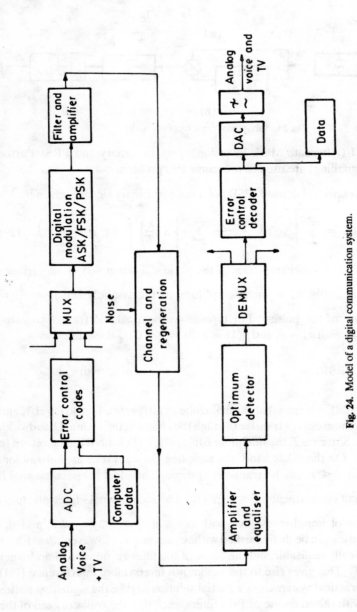

Fig. 24. Model of a digital communication system.

and its impulse response is

$$h(t) = \left(\frac{\tau\omega_c}{\pi}\right) \frac{\sin 2\omega_c t}{2\omega_c t \left[1 - (2\omega_c t/\pi)^2\right]}, \quad \tau \le \frac{T}{2}. \quad (29)$$

The above $A(\omega)$ and $h(t)$ are shown in Fig. 25, where the rolloff factor $\alpha \le 1$. In practice, $0.25 \le \alpha \le 1$ is used. For finite pulse width τ comparable to T, $A(\omega)$ is modified by the factor $(\omega\tau/2) \sin (\omega\tau/2)$ to obtain the same responses as in Fig. 25. In practical systems, a random pulse train with $\tau = T/2$ is transmitted through the modified raised cosine filter and the overall ISI due to the filter plus medium is corrected by proper equalization at the receiver. The BW requirement of the system is now given by Eq. (19).

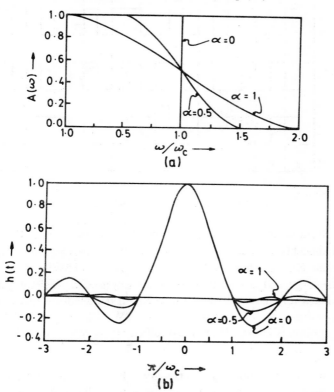

Fig. 25. Raised cosine spectral shaping : (a) Filter characteristics, (b) Impulse responses for $\alpha = 0, 0.5, 1$.

6.1 Line Codes [6]

Apart from the NRZ and RZ pulses discussed above, PCM and data signals are transmitted in several different formats to suit the requirements of channel constraints. Some of the useful formats are AMI, Manchester code and Miller code (delay modulation) as shown in Fig. 26. Power spectral densities of the

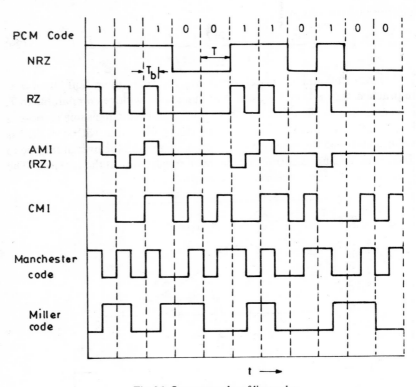

Fig. 26. Some examples of line codes.

waveforms are shown in Fig. 27. It is observed that the spectrum of RZ pulse has a spread almost twice that of NRZ pulses, and both have large low frequency components. On the other hand, AMI, Manchester and Miller codes have negligible l.f. energy, and as such, they are extensively used in practical systems. Miller code has a very narrow bandwidth and AMI bandwidth is restricted to $1/T_b$ Hz approximately. Further the potential BW efficiency of NRZ (and Miller code) pulses is 2 bits/s/Hz but BW efficiency in other codes is only 1 bit/s/Hz.

For efficient synchronization, it is necessary to have as many 1/0 transitions as possible. Moreover, repetitive data patterns generate line spectra which are undesirable from the interference point of view. The data is, therefore, randomized by using scramblers at the input to the data modems and multiplexers. Usually a PN-Sequence is used to multiply the data at the transmitter and the data is recovered at the receiver by dividing the received pattern by the same PNS. Data scramblers, however, donot prevent long string of zeros in the data stream. They simply ensure that the relatively short repetitive patterns are randomized and avoid the line spectra in the cable/radio medium.

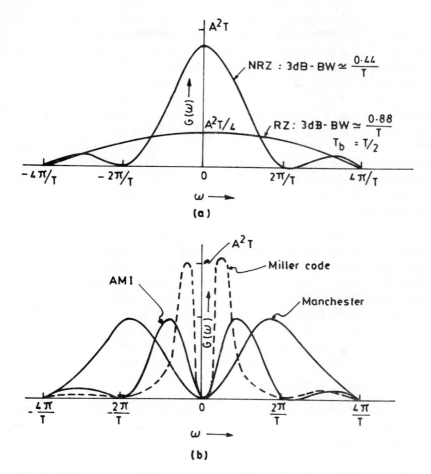

Fig. 27. PSD of various coded waveforms. (a) NRZ and RZ, (b) AMI, MILLER and Manchester.

Special line coding techniques are used to avoid long strings of zeros and to ensure the removal of d.c. in the transmitted signal. In these line codes, a '1' is normally inserted whenever three or more zeros occur in succession. Further binary to ternary encoding is used to obtain a balanced signal (as in AMI) and to compress the signal bandwidth. The useful line codes are: AMI, CMI, BNZ5, HDB3, 4B-3T, and mBnB.

The AMI encoding does not solve the problem of the undesirable long strings of zeros, but in the Coded Mark Inversion (CMI), the problem is solved at the expense of larger BW requirement. In CMI, as shown in Fig. 26, the zeros are encoded as a halfcycle squarewave of one particular phase, whereas the successive '1's are coded as NRZ pulses with opposite phase. Both CMI and Manchester codes have strong timing transitions and there is no d.c.

wander. CMI has been used as an interface code for 140 Mb/s coaxial systems. Alternatively, zero-substitution codes, known as BNZS (binary N-zero substitution) are used along with AMI. The BNZS generally used are B3ZS and B6ZS, where three or six consecutive zeros are replaced by a ternary code. As for example, in B3ZS code, the string '000' is replaced by one of the codes (00–), (+0+), (00+), (–0–), depending upon the polarity of the preceding pulse and the number of bipolar pulses since the last substitution. Consequently, the $\{D_i\} = \{10100011100001\}$, is converted by the B3ZS code (alongwith AMI) as, $\{T_i\} = \{+0$–00–$+$–$+00+0$–$\}$. Another popular BNZS is the high density bipolar coding, known as HDB3, where strings of four zeros are replaced by one of the codes (000–), (+00+), (000+), (–00–). Use of BNZS and HDB3 codes, however, produces bipolar violations (as required in AMI) in the resultant signal bits.

In the above AMI and BNZS coding, the higher transmission capacity of ternary codes is not utilized, but only efficient timing information is used. Higher transmission rate is obtained by using 4B-3T and 6B-4T codes in coaxial cables. In 4B-3T code, 4 binary bits are coded into 3 ternary bauds, thus saving 25% of the BW. Similarly in 6B-4T code, 33% of BW is saved. In Fibre optic cables, the above ternary codes cannot be used, and as such, a redundant binary line code, say, 7B-8B including the Mark parity control, is used. This facilitates the error detection in the receiver by digital sum violations or mark parity violations. However, the transmission rate is now increased by 14% only.

6.2 ISI and Equalizers [5]

In practical channels, there are considerable amplitude and delay distortions introduced and the received pulses have pre-echoes and post-echoes, as shown in Fig. 28 (for voice-grade circuits). As a result, there is considerable inter-

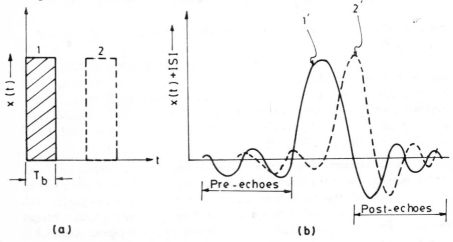

Fig. 28. (a) Transmitted pulses, (b) Received pulses-showing ISI.

ference from one pulse to the other, called Inter Symbol Interference (ISI), and the S_1/N_1 required for a given P_e is now more. The ISI may be measured quantitatively from the impulse response of the channel, but the 'Eye-patterns' on an oscilloscope also give good quantitative indication of the ISI, as shown in

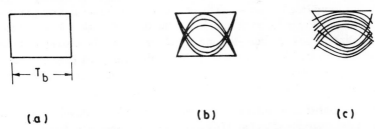

(a)   (b)   (c)

Fig. 29. Eye patterns for PCM: (a) Wideband, no noise; (b) ISI due to BW limitation, no noise; (c) ISI and noise present.

Fig. 29. It will be seen that the eye closes more as more ISI and noise are added to the signal. The best sampling instant is when the eye opening is maximum, and the relative amount of closure at the best sampling time gives an indication of the degradation caused by ISI. For example, the eye closing by 20% represents an equivalent S/N degradation of 2 dB. Thus the system would require an additional 2 dB of $(S/N)_i$ above that required for a specified error rate and zero ISI. Thus it is necessary to minimize the ISI through the use of equalizers in the receiver front end.

Among the many types of equalizers that are used in practice, the transversal equalizer using a tapped delay line (TDL) as the basic element, is the most versatile and easily implemented. Figure 30 shows the block diagram of a transversal equalizer, where the delay line is tapped at intervals of T sec, T = symbol duration. Assuming that the number of taps is $2N+1$ and the corresponding weight, $a_{-N},..., a_0,..., a_N$ the summed and sampled output y_k is given by

$$y_k = \sum_{n=-N}^{N} a_n x_{k-n}. \tag{30}$$

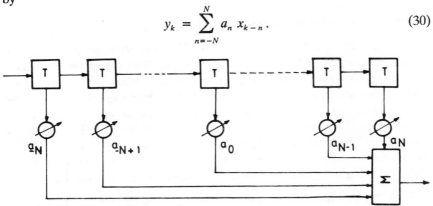

Fig. 30. Block diagram of a Tansversal filter.

Equation (30) yields $(2N+1)$ independent equations in terms of the a_n. Solution of these for the condition that (for zero ISI)

$$y_k = \begin{cases} 1 & \text{for } k = 0 \\ 0 & \text{for } k = \pm 1, \pm 2, ..., \pm N \end{cases} \quad (31)$$

yields the weighting factors a_n. Thus the equalizer forces the output y_k to be zero at N sample points on either side of the desired peak at $k = 0$, and thus, the equalizer is called a zero-forcing equalizer. It may be also shown that the equalizer is optimal in the sense that it minimizes the peak ISI. The results of a simple 3-tap equalizer is shown in Fig. 31, where it is seen that the unsymmetric input pulse have been modified to have zeros at $k \neq 0$.

More sophisticated iterative techniques have been developed to implement (i) automatic (Preset) and (ii) adaptive equalizers, which are used extensively in high speed modems for data transmission.

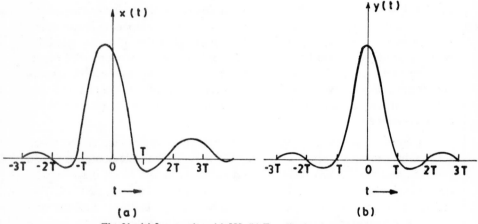

Fig. 31. (a) Input pulse with ISI, (b) Equalized pulse waveform.

7. DATA TRANSMISSION SYSTEMS [5]

Historically, PCM transmission at 1.544 Mb/s was introduced for interexchange trunk traffic over the older audio cables. With the growth of higher-order digital systems, it was necessary to introduce coaxial cables for digital transmission and quite often the cables, used for high capacity analog FDM systems, were used for high capacity digital systems as well. The currently manufactured high capacity systems in the range of 34-565 Mb/s, have been listed in section 4 and the repeater spacings in these are 2 km for smaller cables and 4.5 km for larger cables. Similar digital systems are also being used in LOS MW (line of sight microwaves) and satellite communication systems.

With the development of optical fibres, the digital systems have now a unique medium where thousands of voices and tens of television signals in digital formats may be transmitted simultaneously. Many such systems at 8/34/140/565 Mb/s are already installed and many more are planned in many countries. One such typical system is shown in Fig. 32. Its overall specifications are:

System capacity: 140 Mb/s (1920 telephone channels or 2 digital TV channels)

Source: Semiconductor Laser $\lambda = 0.85\,\mu$m, with power coupled = -3 to 0 dBm

Medium: a pair of GI multimode fibres with 3.5 dB km loss and 1 ns/km dispersion at $\lambda = 0.85\,\mu$m

Detector: APD with a gain of 100 and sensitivity of -48 dBm,

Maximum allowable path loss: 32 dB,

Overall receiver gain: 50 dB,

Repeater Spacing: 9 km,

Power-feeding span: 40 km,

Maximum path length: 2300 km,

Overall error rate: better than 10^{-9},

Line code: input CMI; output 7B-8B,

Transmission rate: 160 M baud,

System supervision provided for 256 terminals/repeaters,

Alarms provided for signal failure, excessive error rate, laser drive failure, excessive laser power and laser operated at currents exceeding permissible limits.

The processing of signals in the transmit path is somewhat similar to that in coaxial cable systems, except that here, a redundant binary line code, 7B-8B, with mark parity control is used. The transmission rate is now increased to 160 Mbauds and the laser is modulated to give a sequence of half width light pulses for transmission via the fibre. The terminal receiver consists of an APD receiver, regenerator, line code decoder, descrambler and the CMI coder. The APD receiver has the APD closely coupled to a low noise transimpedance preamplifier followed by an AGC (automatic gain control) amplifier, with a total control of 25 dB. A PLL circuit converts the 160 Mbaud line signal to the data signal at 140 Mb/s. This circuit also acts as a jitter reducer with a BW of 50 Hz and a maximum jitter gain of 0.5 dB. Certain fibre cables contain metallic wire pairs which are used for transmitting and receiving supervisory signals and also for power feed to the unattended repeaters.

Further developments in fibre-optic technology will give transmission capability of 400-1000 Mb/s with repeater spacings of 30-100 km. Transoceanic fibre cables are being installed for Trans-Atlantic Transmission (TAT) and these use single-mode fibres at 1.3 μm, with a capacity of 274 Mb/s and a

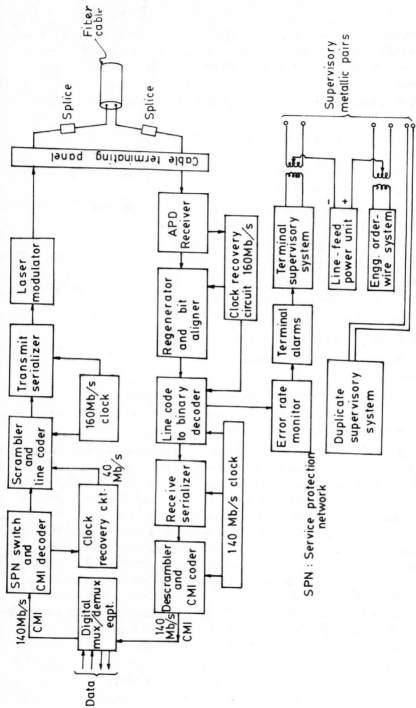

Fig. 32. Block diagram of a 140 Mb/s optical transmission system.

repeater spacing of 35 Km. Systems at 400 Mb/s and 565 Mb/s are also being installed and the repeater spacing is also expected to increase to 100 Km at 1.55 μm. To increase the fibre capacity further, various WDM (similar to FDM) techniques are being investigated and as many as ten optical channels have been already experimented with at $\lambda \sim 1.55\ \mu$m through a single fibre cable (68.5 km long) achieving a phenomenal bit rate of 1.37 Tera-bit-km/s. Use of coherent detection techniques hold key to the widespread use of WDM systems as discussed in Ch. 20 [7].

REFERENCES

1. G.P. Agrawal, "Semiconductor lasers for optical fibre communication" — Ch. 14 in this book.
2. A.B. Carlson, "Communication Systems", McGrawHill, N.Y. (1975).
3. J. Das. Mullick S.K., Chatterjee P.K., "Principles of Digital Communication", Wiley Eastern, New Delhi (1986).
4. N.S. Jayant, "Digital Coding of speech waveforms: PCM, DPCM and ADM Quantizers", Pro. IEEE, Vol. 62, p. 611 (1974).
5. J. Das, "Review of Digital Communication", Wiley Eastern, New Delhi (1988).
6. J. Bellamy, "Digital Telephony", John Wiley & Sons, N.Y. (1982).
7. T.G. Hodgkinson et al., "Coherent Optical Fibre Transmission Systems", Ch. 20 in this book.

19

Purpose Designing of Optical Fibre Systems[†]

M. Chown[*]

---

1. INTRODUCTION

Some of the first practical applications of optical fibre for telecommunications have been as direct replacement for coaxial cable within a previously designed system. However, in order to realise the full technological and economic benefits of optical fibre, it is necessary to go beyond the concept of a simple replacement for a wire cable, and instead match the whole system design to the optical technology. This means understanding the production and user environment from the earliest design stages.

This chapter is concerned with design steps involved in progressing from the laboratory feasibility stage of optical links to robust and practical systems; it is not intended as a review of the current state of the art. The chapter is based on practical design experience covering a variety of applications, and is centred on power budget and dispersion criteria. For successful design it is necessary to match the technology to the full system requirement, and to incorporate an understanding of production and user environment.

2. SYSTEM CONFIGURATION AND POWER BUDGET

The essentials of an optical transmission sub-system (Fig. 1) are: an optical transmitter (Tx) modulated with the information to be sent, an optical receiver (Rx), from which the information is recovered, and a fibre optical waveguide to connect them. A two-way link would consist of two such sub-systems (East-

---

[†] Updated reprint from J.I.E.T.E., Vol. 32 (Special issue on Optoelectronics and Optical Communication), July-August (1986).

[*] Standard Telecommunication Laboratories, London Road, Harlow, Essex, England.

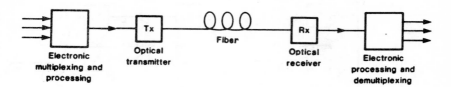

Fig. 1. Optical transmission system-Basic.

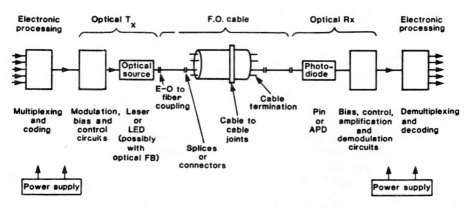

Fig. 2. Essentials of a point-to-point fibre optic link.

West and West-East), the fibres being protected within a common cable; similarly with "spatial multiplexing", several links in each direction can be accommodated. The electronic processing may feed the optical source with digital pulse streams (e.g. pulse position modulation) or other analogue signals including continuously variable signals for optical intensity modulation.

The basic one-way link is shown again in Fig. 2, with a little more detail, and indicating some of the parameters that must be specified.

A repeatered system generally consists of a number of such links in tandem, as represented in Fig. 3, although the electronic processing would not be the same as at a terminal. (Until all-optical regeneration becomes available), the single point-to-point link of Figs. 1 and 2 contains the main elements for the purposes of this chapter.

The heart of the system design is the power budget. Let us define SL as Source Level, expressed in dBm (i.e. dBs relative to 1 mW), which is the mean optical power launched into propagating modes or mode of the fibre, PL as the total Path Loss in dBs, and RL as Received Level, defined as the mean optical power available at the receiver. The received power is then given by

$$RL = SL - PL. \qquad (1)$$

(As a typical example, with SL = −3 dBm (0.5 mW), PL = 30 dB (1000 : 1), then RL = −33 dBm).

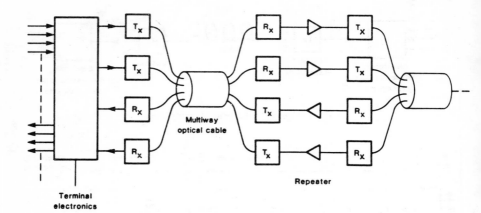

Fig. 3. Repeatered optical system.

2.1 Path loss

The path loss is the total of optical losses, including fibre, splices, connectors and any other passive optical devices. For example, for 5 km of fibre at 1.2 dB/km together with 6 connectors at 1 dB each and 10 splices at 0.2 dB each, the total path loss PL is $5 \times 1.2 + 6 \times 1 + 10 \times 0.2 = 14$ dB. More generally,

$$PL = A_f L + A_c N_c + A_s N_s + A_o, \qquad (2)$$

where A_f (dB/km) is fibre loss (as cabled), L(km) is path length, A_c, A_s and A_o (dB) are losses of each connector, splice or other component and N_c, N_s, N_o are the number of these components.

2.2 Working margins

We require the received light level to be within the acceptable range for the receiver, i.e. greater than the sensitivity level and less than the overload level. Furthermore, to allow for tolerances and variability, we would include margins rather than designing up to the theoretical limit.

To put these criteria more precisely, consider first the sensitivity limit. The condition for system operation is

$$RL > RS \qquad (3)$$

at all times, where RS, Receiver Sensitivity, is the mean power level below which the output electrical signal quality can fall below the standard specified for the system. As both RL and RS can vary (with time, or from unit to unit), we require that

$$(RL)_{min} - (RS)_{max} > M_p, \qquad (4)$$

where M_p is the power margin.

Similarly, to avoid signal degradation by overload; we require

$$RL < RO \qquad (5)$$

at all times, where RO, Receiver Overload Level, is the power level above which signal quality will degrade below specified quality. To accommodate variations as before we require

$$(RO)_{min} - (RL)_{max} < M_o, \qquad (6)$$

where M_o is the overload margin.

In principle, the system design works for $M_p = 0$, $M_o = 0$, but in practice, margins will be needed for foreseeable situations (such as repair allowances or degradation) and perhaps also some to cover the unforeseen.

In a full system design, one may wish to break down these margins, e.g. to allocate a definite proportion to extra repair splices.

2.3 Power budget diagram

The foregoing is summarized in Fig. 4, which shows how the signal level varies through the system. The idea is extended in Fig. 5 to include the effects of variations (from production spreads, changes in service, or variations between installations of a given system). This shows how the margins M_p and M_o can be greatly eroded, to the extent that the effect of variations can be much more limiting than the straight power budget based on typical parameters.

Table 1 indicates some sources of variability in each of the essential components. Some general points can be added to this.

1. Ageing is potentially common to all components, until it is proved otherwise. For a new technology, the challenge is to produce sufficient long-term data for a particular component, in order to support reliability claims. This data must not simply be blindly accumulated

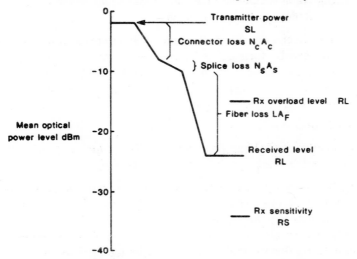

Fig. 4. Power budget diagram (notional example).

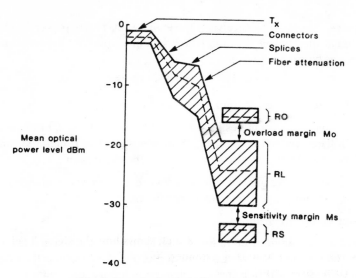

Fig. 5. Power budget including variations.

device-hours, but must include an understanding or potential failure mechanisms. Only then can a basis be established for TRUSTED accelerated life testing, giving the opportunity to prove adequate levels of reliability in realistic time scales.

2. A source of variation common to all parameters is the measurement accuracy of component acceptance testing; quite demanding accuracy may be necessary, or the sum of the inaccuracies of all the relevant tests may further erode the working margins. For this reason, development of test methods should be followed with interest as the technology progresses.

3. The sheer scale of test effort and facilities are easily underestimated, and can be formidable for a large system even though the tests themselves may be simple in principle. This is particularly true whenever fibre dispersion (bandwidth limitation) is considered, as discussed below.

2.4 Power budget examples

Four examples of power budget are shown in Fig. 6, which are used to illustrate how the concepts are applied to a variety of system types.

A long-haul high-capacity system, such as a single-mode undersea system [1, 2, 3] is shown in Fig. 6(a). Here, the maximum possible allowance goes into the fibre loss, which, together with the lowest available fibre attenuation will achieve repeater spacings of over 50 km, and more in 1500 nm systems. A few points to note are that the launched power is limited by single-mode fibre and the need for lasers to operate in high reliability conditions. The latter point

TABLE 1

Some sources of variation in power budget

Parameter	Symbol	Sources of Variation
Fibre Attenuation	A_f	—Production spread of fibre loss. —Additional cabling loss (if any) may be variable. —Variability of operating wavelength. —Modal distribution (not for single-mode fibre). —Additional loss due to ionizing radiation (relevant to some military systems).
Connector Attenuation	A_c	—Production spread. —Imperfect repeatability on reconnection, especially in difficult environments (e.g. military tactical). —Sensitivity to environment (e.g. temperature or vibration).
Splice Attenuation	A_s	—Production spread, which depends partly on fibre geometry tolerance. —Variability would be greater where field repairs are needed in difficult environments.
Losses of any other Passive Optical device	A_0	—Production spread, wavelength, temperature, modes, etc.
Path Length	L	—May be governed by required location of terminals or repeaters, e.g. in a military tactical system, the same equipment may be used over full length one day and very short the next. In others, once set, it is fixed.
Number of Connectors or Splices in Path	N_c, N_s	—Similar considerations as for L. —Also variable number of bulkheads etc. —Repair Splices.
Source Optical Power Level	SL	—Variations in drive conditions. —Variations in slope efficiency. —Variations in threshold current (for laser). —Variations in coupling efficiency. —All these can be production spreads, and possibly also temperature sensitive.
Receiver Sensitivity	RS	—Variations between device parameters, bias conditions, operating wavelength.
Receiver Overload Level	RO	—In addition, any signal impairment (such as dispersion) will affect RS and RO.

arises because the highest optical intensity, available in relatively recent devices, are not yet sufficiently tested for long-term reliability. Similarly, reliability requirements restrict the receiver type (e.g. planar structures preferred to mesa). Also the receiver sensitivity is affected by the restrictions in bias voltage, which may arise from reliability considerations on the silicon integrated circuits. Connectors are avoided and very low loss splices used to conserve the power budget. The terminals will encode the data (e.g. 280 Mbit/s may be encoded to a line rate of, say 300 Mbaud/s), and the extra pulses can be used to ensure adequate timing information and to avoid large excursions from

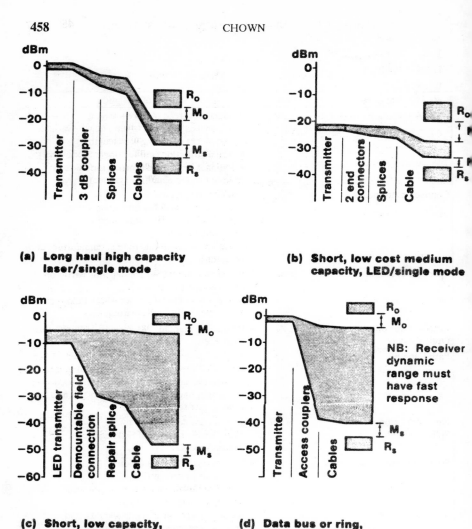

Fig. 6. Some power budget examples.

an equal number of 1's and 0's, (known as the digital sum variation, DSV). The choice of code runs right through the repeater design, and for example, DSV affects the trade-off between receiver sensitivity, dynamic range and bias voltage requirement, where an integrating front-end is used. Finally, the diagram illustrates a warning about interpreting laboratory optimized results. Impressively large, but unrealistic repeater spacings can be demonstrated by selecting best available sources and detectors, by neglecting the reliability considerations just mentioned, and by eliminating the margins. This explains the large gap that always exists between best published results and installed systems.

Turning to short-haul, cost sensitive systems which may be used for a relatively high volume application, it is often attractive to use an LED rather than a laser, and this has previously been considered only suitable for multimode fibres. However, the advantages and general availability of very low loss single-mode fibres make this an attractive alternative achieving adequate power margins over the required distances in spite of the low coupling efficiency. A possible power budget is shown in Fig. 6(b). Here, the loss budget between transmitter and receiver is very limited and careful consideration of tolerances and temperature effects is essential. An attraction is the ability to upgrade to higher bit-rates, using a laser to give extra power to compensate for lower receiver sensitivity, while the single-mode cable already installed could handle these high rates without dispersion.

A tactical military application is represented in Fig. 6(c) based on an LED source and multimode fibre. This example envisages a system that is few kilometers long with several demountable connectors between fibre lengths. Such a system has to be deployed quickly and secretly, then packed up and moved and redeployed on a rapid timescale and under battle conditions. A single cable length may be needed on one occasion and, say, four in series the next, and so a wide dynamic range is required. The demountable connector is the most vulnerable component to rough handling, and a good design concept is (i) to use rugged expanded beam connectors [4, 5] and (ii) to make the greatest possible loss allowance for the connectors. (The second point is subtle: it is not that the connectors have a high loss, but that the higher the failure criterion, the more rugged they will be against that criterion). The use of multimode fibre allows greater tolerance in the connectors, and is appropriate where bandwidth considerations allow it (see next section).

The power budget example reflects these considerations, and shows the bulk of the loss against connectors. Although the normal fibre loss would be quite low, the worst case allowance is for temporary-loss increases induced by ionising radiation. (Fibres with good immunity against radiation have been successfully developed in close conjunction with system design [6]).

The receiver sensitivity is better than for the first example because of a much lower bit-rate. The dynamic range requirement in this application is severe, because the receiver has to cope with the extremes of (i) minimum cable length deployed with no intermediate connectors and conditions of maximum efficiency, and (ii) maximum deployment, with connectors and other components at the minimum specified efficiency, under conditions of radiation. In the final example, Fig. 6(d), the dynamic range requirement on the receiver is even more severe. This power budget refers to one of many possible multiterminal systems-in this case, a fibre ring configuration and the received power ranges between that from the nearest neighbour transmitter and the most distant one [7, 8]. The additional problem for the receiver design is that the change has to be accommodated in the very short time between transmitter bursts, without losing the beginning of the next message. The power budget for a multiterminal bus or ring system is typically dominated by the access couplers, with cable loss being negligible, and this is the area that needs attention if a large number of stations is required.

3. DISPERSION

Whereas attenuation sets length limits governed by the power budget, fibre dispersion can set another limit where very high bit-rates are required, and once again many component parameters will interact. The key to the effect is the spread in the propagating velocities, which transforms into a spread of arrival times, depending on the fibre length. It may result in interaction between neighbouring pulses at high bit-rates and long distances. It is useful to separate the discussion into multimode and single-mode fibres.

3.1 Multimode fibre dispersion

In a geometric description of propagation through a step-index fibre, the range of possible velocities through the fibre can be seen as a path length difference between the most oblique and the most direct [see Fig. 7(a)]. The electromagnetic wave description will, of course, show these as discrete modes, rather than a continuum of possible rays. Each mode (or ray) has its own characteristic propagating velocity (group velocity, V_g), so the two descriptions give agreement on the multimode dispersion resulting from this spread of velocities. This dispersion which could amount to tens of nanoseconds per kilometre, is unacceptable for long-haul high capacity systems, so graded-index multimode fibre has found widespread application in recent years. Here, the rays are curved, as in Fig. 7(b), but the geometrically longer paths, such as B in the figure, occupy lower refractive index regions away from the axis where the velocity is greater. With an optimum refractive index profile, the two factors almost compensate for each other, and all rays have similar average velocities, resulting in an order of magnitude lower dispersion, of perhaps 1-2 ns/km. This performance is acceptable for many long-haul landline systems, but still sets the limit on system length for high bit-rates, now that fibre attenuation is becoming very low.

Modal dispersion is a most difficult parameter to specify, because any measurement depends on the distribution of modes actually propagating [9]. This, in turn, depends on the launch conditions, and can vary through the fibre due to preferential attenuation of the higher-order modes. In addition, there can be mode-conversion which can be thought of as being coupling of energy to other modes from a given mode as it propagates through the fibre. As the length increases, the averaging effect becomes more favourable, resulting in a rather unpredictable sub-linear length dependence.

As the attenuation per unit length is a weighted average over all the propagating modes, it is necessary to measure both dispersion and attenuation under the same realistic conditions to avoid the pitfall of finding that good

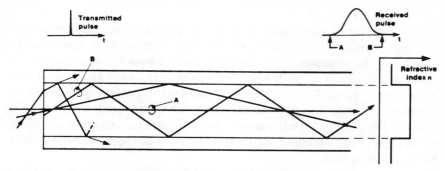

(a) Step-Index — Multipath Pulse Spreading

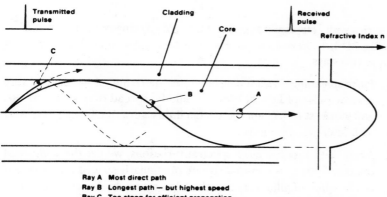

Ray A Most direct path
Ray B Longest path — but highest speed
Ray C Too steep for efficient propagation

(b) Graded Index for Reduced Pulse-Spreading

Fig. 7. Pulse spreading in multimode fibres.

results on both parameters cannot in practice be realised together. For these reasons, system performance is not easily predicted from fibre measurements, and end-to-end connection of fibre lengths may not follow a simple rule.

Although, there is still research on optimizing profiles for very low multi-mode dispersion, it is now widely accepted that multimode fibre should only be chosen where dispersion is not highly critical, and that single-mode fibre is the optimum solution for applications requiring low dispersion and hence high bit-rate transmission. In applications where multimode dispersion is acceptable, this type of fibre is sometimes preferable, because sources and connectors are cheaper generally, although the system designer must then take care to avoid modal noise problems

3.2 Single-mode Dispersion

For a sufficiently small core dispersion and refractive index difference, the conditions can be met to ensure that only a single-mode can propagate efficiently, all others being lost (normally by radiation) after a very short distance.

At the expense of more difficult launching and coupling, this completely eliminates modal dispersion, and the previously insignificant chromatic dispersion is now the only source of velocity variation. Chromatic dispersion is the variation of velocity V_g with optical wavelength λ, and is the sum of two components, waveguide dispersion and material dispersion both of which essentially originate due to finite spectral width of practically available light sources. Fibre design for minimum chromatic dispersion is based on arranging for waveguide and material components to be equal and opposite, as proposed by Dyott [10] in 1970.

It is convenient to think in terms of propagation time per unit length,

$$\tau = 1/V_g \tag{7}$$

and this varies with λ, typically as in Fig. 8(a). Now taking the derivative, $M = d\tau/d\lambda$ in Fig. 8(b), we have a measure of chromatic dispersion. This can affect the signal in two ways:

1. *Pulse spreading*: With a constant laser spectrum, of width $\Delta\lambda$ a very narrow pulse of light at the Tx would be spread over a time $M \cdot \Delta\lambda$, and would start to limit the system if this were significant compared with inter-pulse periods.

2. *Jitter*: With random laser spectral variations, we get the additional effects of jitter. A constant stream of regularly spaced pulses would not be quite regular as received. A laser jump by $\delta\lambda$ between pulses would alter the inter-pulse spacing by $M \cdot \delta\lambda$.

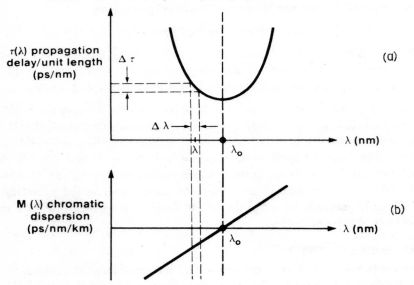

Fig. 8. Single mode chromatic dispersion: Relationship between $\tau(\lambda)$ and $M(\lambda)$ in region of λ_0.

Whether the dispersion is acceptable depends on

(i) Length, speed and modulation method of the system.

(ii) Spectral performance of the laser.

Referring again to Fig. 8(b), there is a cross-over wavelength λ_0 at which $M(\lambda)$ becomes zero, and a laser with a finite $\Delta\lambda$ centered on λ_0 would suffer an extremely small (though not zero) degree of pulse spreading. This is the design philosophy being adopted for the highest capacity long-haul systems (e.g. some trunk landline systems, or submarine systems), usually based around 1300 nm wavelength. The key factor here is the *tolerances*, since actual dispersion will depend on the mismatch between λ and λ_0. Thus, the fibre designer must be asked what production spread he can offer on λ_0. Similarly, for the laser, we need to know the production spread and temperature dependence of λ in addition to spectral width $\Delta\lambda$ and spectral instability $\delta\lambda$ mentioned earlier. The cross-over λ_0 falls naturally in the region of 1300 nm and for systems at 1500 nanometers (with lower fibre loss), there are generally two ways to go. Either the fibre is specially designed to move λ_0 to around 1500 nm (dispersion shifted fibre, DSF) or else a very narrow band laser is used-the distributed feedback DFB devices being an almost universal choice at present. It is of course possible to use a DFB laser and a DSF, where the highest length × capacity product is required, and this allows an increased realiability, through tolerance to changes such as laser spectral behaviour.

4. SYSTEM TRADE-OFFS

To see how the effects of fibre dispersion, laser spectral stability and optical power budget are inter-related in system design, it is necessary to review the operation of a digital regenerator [11].

A stream of binary optical pulses is received and amplified, and fed into a decision circuit, following which a fresh pulse is generated for a "1" decision, and no pulse for a "0" decision. The key parameter defining transmission quality is the bit-error-rate (BER) which is the probability of a wrong decision. (A typical requirement for a high-quality system is BER $< 10^{-9}$). The decision process is, therefore, the heart of the system, and it can be most clearly understood with the help of the eye diagram. This can be obtained experimentally on an oscilloscope, synchronised to the bit-period, as a long random (or pseudo-random) stream of data is received (see Fig. 9). The result is an overlay, seen from the point of view of the decision circuit, and typically has the shape of an eye. Whereas Fig. 8(a) shows an open eye, where all decisions can be made without ambiguity, Fig. 9(b) shows a much more degraded signal with a nearly closed eye. The latter would give rise to some marginal decisions and an unacceptable error rate.

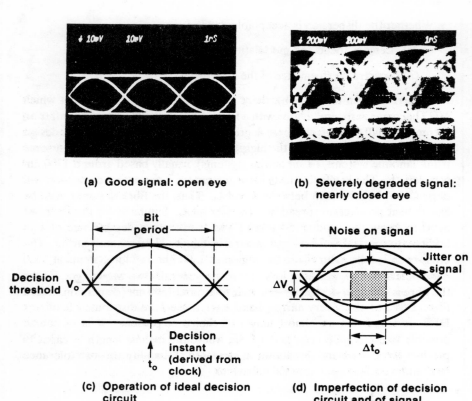

Fig. 9. The eye diagram: (a) good signal: open eye, (b) severally degraded signal: nearly closed eye, (c) operation of ideal decision circuit, (d) imperfection of decision circuit and of signal.

With reference to the eye diagram, we can see that the decision process requires establishment of a regular clock which defines the instant t_0 at which the decision is made, and a threshold V_0 above which a "1" is chosen, as represented in Fig. 8(c). In practice, there are finite tolerances, ΔV_0 and Δt_0, so that a fairer representation is Fig. 8(d). These tolerances arise from finite response times, clock irregularities which depend on the choice of coding, and other random and noise-based effects. For these reasons, the eye must not only be clearly open, but must provide sufficient margins.

This explanation is common to purely electrical digital systems, and we must now consider the additional effects related to the fibre.

1. *Attenuation*: As the received power is reduced, more amplification is required before the decision circuit, which increases the noise on the signal in the eye diagram. (This is equivalent to a reduced S/N, and is the basis for determining the receiver sensitivity RS).

2. *Noise on optical signal*: This can come from the source itself (though not usually significant in a practical digital system), and from modal noise [12, 13] (generally only in a multimode system). A fundamental limit of quantum noise (shot noise in the photon stream) is not generally reached but can be approached. In addition, the spectral instabilities of the source are translated into time-of-arrival instabilities of the received signal after chromatic dispersion. An example is partition noise, where laser power is randomly distributed among modes.
3. *Optical pulse spreading*: This effect, explained above, results in a severe penalty if it approaches the bit period. So long as the effect is constant and predictable, the post-detection electronic filters can compensate to give a well-shaped eye. However, there will be a penalty in that the filter will allow more noise to pass.
4. *Jitter in received optical signal*: The effect, as explained above, arises from the combination of rapid spectral variations in the source, together with chromatic dispersion in the fibre. A change of wavelength between pulses will in effect cause the eye to move horizontally about the decision point, in effect adding to the timing uncertainty Δt_0 [Fig. 9(d)]. A change in wavelength during a single pulse known as chirp will cause a distortion of the received pulse, which may be an expansion or a compression (depending on whether λ is greater or less than λ_0).

System trade-offs are possible between these effects. Consider what happens as the fibre length is increased for a high bit-rate system. Some of the effects mentioned above will increase, with the following two categories usually dominant:

(1) Increased noise at the eye due to reduced received optical power, i.e. loss effects.

(2) Increased eye distortion arising from jitter, i.e. dispersion effects.

If limitation (2) is approached first — i.e. jitter occurs while there is still power margin in hand — it is possible to optimise system length by a trade-off. By increasing the specified received power RS, one increases the S/N, i.e. making a cleaner and more open eye, and this allows more margin for the jitter effects. (In such a trade-off, the BER requirement is understood to be held constant). The amount by which RS must be increased as compensation is known as the 'spectral penalty'. This is such an important parameter, illustrating as it does the component and system design interactions, that its measurement will be explained in more detail.

5. SPECTRAL PENALTY

A method of characterising laser spectral effects is illustrated in Fig. 10. The essential concept is simply that the modulated laser output is first passed

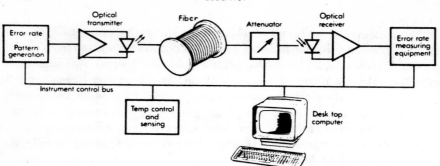

Fig. 10. Schematic diagram of the fibre optic automatic system test facility.

through a short length of fibre (i.e. suffering negligible chromatic dispersion) and then through a long fibre of known dispersion, and a comparison is made to determine the effect of the dispersion. The comparison is based on attenuating the optical signal in each case until a predetermined BER is achieved, which is equivalent to determining RS in each case. For the second case, RS will in effect be higher: this increase is the spectral penalty expressed in dBs.

In such a test, it is easy to get optimistic results unless care is taken. The fibre dispersion must be representative in terms of $d\tau/d\lambda \cdot L$ (in ps/nm) rather than in λ_0, otherwise the result will fail to reflect the worst case mismatch between λ_0 and λ. This is achieved by choosing a fibre with λ_0 well away from λ, so that it is highly dispersive and independent of the exact value of λ. The length can be tailored to give the required total dispersion. It is also necessary to recognise that the laser spectral effects can be highly temperature sensitive, with sharply peaked effects at critical temperatures (in a cyclical manner), and that these critical temperatures themselves may vary with the age of the device. It is, therefore, necessary to cycle the temperature during the experiment, to ensure that the worst-case condition is met.

The results will depend on the operation of the decision circuit used, although every attempt is made to keep this constant and representative. Also, the code is important, a long pseudo-random code being a suitable choice. Lasers have been characterized in this way to give good confidence of reliable operation in a long-haul 280 Mbit/s undersea system [14, 15], and spectral penalties for such systems over 40 km are typically under 1 dB. (However, without this characterisation, lasers which are potentially poor can nevertheless give very good laboratory results, under fortuitous conditions of working near λ_0 and working well clear of the worst-case temperature).

As an extension to this work, Anslow and Goddard [16] have studied modulation-induced effects in conjunction with time-resolved spectral measurements of the laser. They have studied the shift in wavelength through the modulation pulse, and have shown that the resultant spectral penalty is strongly dependent on the sign of the chromatic dispersion, i.e. whether λ is above or below λ_0, as well as its magnitude.

The spectral penalty effect has been used to illustrate component design interaction through fibre dispersion (λ_0 and its tolerance), laser wavelength and spectral stability, regenerator design (especially the decision circuit), and the optical power budget.

6. OTHER COMPONENT CONSIDERATIONS

The key component parameters, such as laser power, fibre attenuation, splice efficiency etc., are well-know. Table 2 brings out a selection of the less obvious requirements to be considered by the system designer.

TABLE 2
Some key parameters for fibre optic components

Parameter	Remarks
OPTICAL SOURCE	
Optical Power	Mean or peak power limited? Launched power into given fibre? Required current drive?
Modulation	Speed of response based on pseudo-random pattern? Optical feedback required?
Spectral performance	Wavelength and temperature coefficient? Spectral width and spectral stability e.g. chirp characteristics? Modal noise parameters?
OPTICAL RECEIVER	
Receiver Sensitivity	At what bit-rate and BER? Spectral sensitivity? Sensitive area (adequate for particular fibre?)
Overload Level	Bias requirements and tolerances? For integrating receivers, overload level depends on digital sum variation of code and on supply rail voltage.
FIBRE	
Spectral Attenuation	Including cabling increment over the temperature range, and with bands associated with practical installation cut-off wavelength in single-mode fibre for good suppression of higher order modes.
Dispersion	Multimode: Addition law for tandem lengths? Adequate for system upgrading? Single-mode: Chromatic dispersion $M(\lambda)$ over working wavelength range. For operation near λ_0, value and tolerance of λ_0 and slope of $M(\lambda)$ around λ_0.
Fibre Geometry	Tolerances and eccentricity determine splice and connector efficiency.
Fibre Strength	Maximum permitted strain? If proof-strained, what is average length between breaks?
Static Fatigue	Weakening of fibre through crack growth under strain. Long term tests of fibre (and splices) under strain. Assume worst case moist environment.
CABLE	
Suitability for Environment and Application	Tensile strength, flexibility, crush resistance, weight, environmental resistance, etc. Widely differing criteria. Optical parameters after installation. Jointing and repair aspects: stripping, termination and jointing.

Parameter	Remarks
CONNECTORS	'Best results' means nothing where tolerances dominate. Environment and handling conditions—What operator training required? Tolerance to dust and dirt? Low losses achieved with randomly selected pairs, or only for matched pairs? Need for power connection in parallel? Reflection characteristics?
Mode Conversion	In measurement, eliminate output coupled into non-propagating modes.
SPLICES	Ignore 'best results'; also 'worst results' if rejected by routine inspection; percentage rework? (problem of multifibre cable). Results apply to different fibres (of same type) or only to resplicing? Strength and static fatigue? Depends on coating reinstatement.
WAVELENGTH MUX DEVICES	Resolution in wavelength and repeatability. What demands are made on source line-width, and wavelength stability? Overall insertion loss between source and detector, via short length of fibre, for each wavelength?
COUPLERS	Uniformity among ports, dependence on modes: in bus or ring, insertion loss of paramount importance (taken as fibre-to-fibre).

Variability of parameters is always important, and the method of measurement and test must be considered at an early stage. There are also many demands special to the particular system of interest which are not listed. These may be related to packaging of components, which should be taken as an integral part of development rather than an afterthought.

Recently, coherent systems, high-density wavelength modulation and optical amplification are becoming feasible. These technologies bring with them other component characteristics which require control, including polarization stability, reflection effects, optical cross-talk and very demanding spectral performance.

7. CONCLUSION

The development of optimized optical communication systems has been most successful when the user environment is taken into account from the earliest stages, including component development. Only by considering overall system implications can the optical technologist see the relative importance of, say, laser power and spectral stability. As the technology matures, systems are more readily designed around off-the-shelf general purpose components and modules.

It seems likely, however, that at the frontier of advanced systems, this type of interaction will continue to be essential. Although this chapter has attempted to illustrate this from experience on a variety of systems that are now becoming conventional, the same pattern is being seen on emerging systems technology, such as coherent optical communications, and optically amplified systems.

REFERENCES

1. G.A. Heath and Chown M., "The UK-Belgium No. 5 optical fibre submarine system", IEEE Journal on Selected Areas in Communication, Vol. SAC-2, pp. 819-826 (1984).
2. P. Worthington, "Cable Design for Optical Submarine Systems", IEEE Journal on Selected Areas in Communication, Vol. SAC-2, pp. 833-838 (1984).
3. J. Irven, Cannell G.J., Byron K.C., Harrison A.P., Worthington R. and Lamb J.G., "Single-mode fibres for submarine cable systems", IEEE Journal of Quantum Electronics, Vol. QE-17, pp. 907-910 (1981).
4. P.R. Cooper, Leach J.S., Harding A.B. and Matthews M.A., "A fibre optic connector suitable for use in a rugged environment", Optics and Laser Technology, Apr, pp. 87-91 (1982).
5. P.H. Bourne and Chown D.P.M., "The Ptarmigan optical fibre subsystem", IERE Conference on Fibre Optics, London, 1-2 Mar., pp. 129-146 (1982).
6. A. Robinson, "Large core high numerical aperture fibres", SIRA conference, Fibre Optics'83, London, Apr., 1/2/1-4 (1983).
7. J.G. Farrington and Chown M., "An optical fibre, multiterminal data system for aircraft", AGARD Conference on Optical Fibres, Integrated Optics and their Military Applications, London, May (1977).
8. D.P.M. Chown, "Dynamic range extension for pin-FET optical receivers", 7th European Conference on Optical Communication, Copenhagen, Sep. 14.5/1-3 (1981).
9. K.C. Byron and Chown M., "Impulse response of optical fibres, 2nd European Conference on Optical Communications", Paris (1976).
10. R.B. Dyott and Stern J.R., "Group delay in glass fibre waveguide", Proceedings of Conference on Trunk Telecommunications by Guided Waves, London, Sep., pp. 176-181 (1970).
11. F.F.E. Owen, "PCM and digital transmission systems", McGraw-Hill, New York, pp. 200-201 (1982).
12. R.E. Epworth, "The phenomenon of modal noise in analogue and digital optical fibre systems", 4th European Conference on Optical Communication, Geneva, Sep. pp. 492-501 (1978).
13. R.E. Epworth, "Modal Noise-causes and cures", Laser Focus, Sep., pp. 109-115 (1981).
14. A. Rosiewicz, Butler B.R. and Hinton R.E.P., "Temperature-dependent lifetests of IRW lasers operating at 1.3 microns", IEE Proceedings, Vol. 132, pp. 97-100 (1985).
15. P.J. Anslow, Farrington J.G., Goddard I.J. and Throssell W.R., "System penalty effects caused by spectral variations and chromatic dispersion in single-mode fibre optic systems", IEEE Journal on Selected Areas in Communication, Vol. SAC-2, pp. 1000-1007 (1984).
16. P.J. Anslow and Goddard I.J., "Modulation-induced spectral penalties in high bit-rate single-mode systems", 11th European Conference on Optical Communication, Venice, Oct. (1985).

20

Coherent Optical Fibre Transmission Systems

T.G. HODGKINSON, D.W. SMITH, R. WYATT AND D.J. MALYON[*]

This article was originally published in the British Telecom Technology Journal (Vol. 3, pp. 5-18) in July 1985, and because of its tutorial nature, it was reprinted in its original form in the Journal of the Institution of Electronics and Telecommunication Engineers in August 1986 by permission of the Editor of BTTJ. My decision to reprint it again in this book demanded that references to coherent systems research work carried out after 1985 be included. However, to modify the main body of the original article was impractical, to briefly refer to this work in a preface would have been inappropriate, so an Appendix entitled 'Overview of coherent systems research 1985-88' has been added to the original article by the authors — *Editor*.

1. INTRODUCTION

Currently all operational optical fibre transmission systems worldwide are of the intensity modulation/direct detection type; in other words, no use is made of the coherence or spectral purity of the optical signal. However, it has been widely appreciated that coherent optical transmission, featuring either optical heterodyne or homodyne detection, could offer improvement in system performance of between 5 to 20 dB. This could lead to increased repeater separations for undersea and inland transmission systems or alternatively, it could allow the transmission capacity of long haul systems to be increased without reducing operating margins; for distribution networks, the improved performance increases the power budget available for optical multiplexing/demultiplexing. In addition to the performance advantage, the inherent selectivity of coherent systems could be used to access the vast optical bandwidth available

---

[*] The authors are with the BT Laboratories, Martlesham Heath, Ipswich, Suffolk IP5 7RE, England.

from single-mode fibre (50,000 GHz in the low-loss fibre window from 1.3-1.6 μm); one possible future application for this is wideband signal distribution.

Although the principles of coherent detection are not new (they are in fact well established at radio frequencies) there has been speculation since the development of the laser in the 1960s concerning the potential application of these techniques at optical frequencies [1, 2]. However, it is only comparatively recently that coherent techniques have found application in optical fibre systems. The reasons are mainly threefold.

- The several hundred GHz wide, multimode spectrum characteristic of a conventional semiconductor laser must be reduced to a narrow linewidth ($\leq$ MHz) single-mode spectrum. Recently this has been achieved using techniques such as external cavities [3] and injection locking [4]; for the future, the DFB laser structure [5] may also satisfy this requirement.

- Research in the areas of single-mode fibre devices and lithium niobate integrated optic components has only recently produced components such as external modulators [6] and directional coupler combiners [7]; these devices have been essential for serious experimental work.

- The polarisation stability of the transmission media is an important consideration for coherent systems and early studies implied that polarisation holding fibre [8] would be necessary. However, recent polarisation measurements of conventional single-mode fibre have shown that this is not the case [9].

As a result of the advances in optical component and semiconductor laser technologies, a period of intense experimental activity took place during 1982-85, especially in Europe and Japan, to demonstrate coherent optical fibre transmission. This chapter briefly reviews the experimental work reported and discusses the potential of coherent detection, along with the constraints for present direct detection receivers. The problems associated with achieving close to ideal performance with coherent detection are also considered and some of BT Laboratories experiments are described.

2. DIRECT DETECTION PERFORMANCE

The sensitivity of an ideal direct detection receiver (ideal assumes that the electronic preamplifier following the photodetector is noiseless) is determined only by the statistical distribution of the detected photons. For a binary ASK transmission system (considering Poisson statistics for the photocurrent) the required average number of received photons for a bit-error-rate (BER) of 10^{-9} is 21 [10]; this represents a receiver sensitivity of -63 dBm for a 1.5 μm wavelength 140 Mbit/s system. In practice it is only possible to meet this quantum limit if the detector has noiseless internal gain. Photomultipliers and silicon avalanche photodiodes come closest to achieving this but their response declines rapidly beyond 1 μm. For PIN diode receivers, operating at wavelengths in the range 1.2 to 1.7 μm, the preamplifier noise dominates over the signal shot noise and more than 400 photons per bit are required in a practical 140 Mbit/s system. At higher data rates the performance of the practical direct detection receiver deviates further from the quantum limited case since the electronic preamplifier usually has a rising noise versus frequency characteristic. Therefore, coherent detection will be of most benefit for high capacity systems working at the longer wavelengths.

3. COHERENT DETECTION PRINCIPLES

With coherent detection the low level received optical signal (e_s), see Fig. 1, is combined, prior to detection, with a second much larger optical signal (e_L) from a local oscillator laser; the angle ϕ is the phase relationship between these two fields defined at some arbitary point in time. Because of the square law photodetection process the signal photocurrent (i_T) is proportional to:

$$i_T \alpha (e_s + e_L)^2 \qquad (1)$$

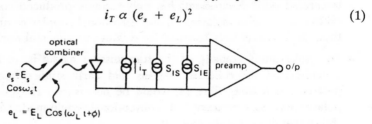

Fig. 1. Basic coherent receiver model.

For ideal coherent detection the wavefronts of the received signal and local oscillator fields must be perfectly matched at the surface of the photodiode; to achieve this polarisation control and single-mode optical combiners are required in the receiver. Given perfect optical mixing the resulting photocurrent, from Eq. (1), is:

$$i_T = \frac{\eta q \lambda}{hc} (2 \sqrt{P_s P_L} \cos(\omega_L t + \phi - \omega_s t) + P_s + P_L), \qquad (2)$$

where P_s is the input signal power in the absence of modulation,
P_L is the local oscillator power,
η is the detector quantum efficiency,
h is Planck's constant,
q is the electronic charge,
λ is the optical wavelength,
c is the velocity of light,
ω_s is the signal frequency
ω_L is the local oscillator frequency,

ϕ is the phase angle between the two optical fields at an arbitarily defined fixed point in time.

If the local oscillator signal is much larger than the input signal the expression for i_T can be replaced by the approximation (i_s) below:

$$i_s = \frac{\eta q \lambda}{hc} \left[2\sqrt{P_s P_L} \cos(\omega_L t + \phi - \omega_s t) \right]. \tag{3}$$

There are two types of coherent detection and they are homodyne and heterodyne. For homodyne detection ω_L and ω_s are equal and Eq. (3) reduces to:

$$i_s = 2\sqrt{P_s P_L} \, \frac{\eta q \lambda}{hc} \cos \phi. \tag{4}$$

When homodyne detection is used the output from the photodiode is a baseband signal and the local oscillator needs to be optically phase locked to the incoming optical signal.

For heterodyne detection ω_L and ω_s are not equal and Eq. (3) can be expressed as:

$$i_s = 2\sqrt{P_s P_L} \, \frac{\eta q \lambda}{hc} \cos(\omega_{IF} t + \phi), \tag{5}$$

where ω_{IF} is the difference between ω_L and ω_s.

With heterodyne detection the output from the photodiode is centred about an intermediate frequency (IF) and this IF is stabilised by incorporating the local oscillator laser in an automatic frequency control (AFC) loop.

From the above expressions it can be seen that the signal photocurrent is proportional to $\sqrt{P_s}$, rather than P_s as in the case of direct detection, and that it is effectively amplified by a factor proportional to $\sqrt{P_L}$. This gain factor increases the optical signal without changing the preamplifier noise term (S_{IE}) and this is the reason why coherent detection gives better receiver sensitivities than direct detection.

Figure 1 shows two receiver noise sources, S_{IS} and S_{IE}. S_{IS} is the shot noise spectral density and S_{IE} is the equivalent current noise spectral density for the preamplifier. Ideally the shot noise term should be the dominant noise source and provided that the local oscillator power is large enough this condition can always be satisfied. The shot noise power in a coherent receiver is dependent on the resultant optical intensity at the surface of the photodiode, but when the local oscillator is much larger than the received signal the shot noise spectral density can be taken as being equal to that generated solely by the local oscillator. Given this condition, the double sided shot noise spectral density is:

$$S_{IS} = P_L \frac{\eta q^2 \lambda}{hc}. \tag{6}$$

If the previous photocurrent terms and the above noise expression are used to derive the shot noise limited signal-to-noise ratio (SNR) the result below will be obtained:

$$\text{SNR} = \frac{\langle P_s \rangle}{KB} \frac{\eta \lambda}{hc}, \tag{7}$$

where *K* has a value determined by the modulation/demodulation combination used, (see Table 1),

B is the double sided noise bandwidth for the receiver,

$\langle P_s \rangle$ is the mean value of P_s with the modulation present.

TABLE 1

Table of *K* values to be used in the SNR expression for the various modulation/demodulation combinations

MOD	SYNC DEMOD	
	HOM	HET
PSK	0.25	0.5
FSK	0.5	1.0
ASK	0.5	1.0

The performance of a digital system is usually expressed as the value of $\langle P_s \rangle$ needed to give a specific BER, 10^{-9} is a typical value used for this purpose. To derive the BER the normalised Gaussian variable *Q* is used [11]; this has the following relationship with Eq. (7):

$$Q_{SNR}^2 = \frac{\langle P_s \rangle}{KB} \cdot \frac{\eta \lambda}{hc}. \tag{8}$$

Transposing this expression for $\langle P_s \rangle$, using the following substitutions and the assumption that the equalisation used gives a raised cosine output pulse spectrum, the shot noise limited receiver sensitivity expression is as given below [in Eq. (9)]:

Q_{SNR} = 6, this corresponds to a BER of 10^{-9},

h = 6.626×10^{-34},

c = 2.998×10^8,

η = 100%,

λ = 1.5×10^{-6},

B = $1.128/T$ where *T* is the bit time,

$$\langle P_s \rangle = 5.4 \times 10^{-18} \frac{K}{T}. \tag{9}$$

4. COHERENT SYSTEM PERFORMANCE

The expression for $\langle P_s \rangle$ shows that the shot noise limited receiver sensitivity is determined by the modulation/demodulation combination used. Table 2 shows the advantage of the various combinations when compared with an ASK heterodyne system [12]. The system performance given by binary PSK is better than that given by binary ASK because optical power is not wasted in a continuous wave carrier component. Furthermore with PSK and FSK systems, for peak-power limited sources, there is an additional 3 dB of transmitter power available. M-ary transmission is also possible, either for spectral conservation (M-ary ASK, M-ary PSK) or to improve receiver sensitivity by spectral expansion (M-ary FSK). In radio there is often pressure to achieve the former whereas the rationale in optical systems may lean to the latter and for some applications the use of M-ary FSK to improve sensitivity could have merit.

Of the demodulation schemes synchronous detection gives the best performance but is more demanding on laser linewidth requirements than the non-synchronous schemes. Homodyne detection by definition is a synchronous detection scheme but heterodyne detection can employ either synchronous or non-synchronous IF demodulation schemes. Any variants of homodyne detection where the local oscillator laser is not phase locked to the incoming signal should be considered as a form of heterodyne detection, an example of which is multiport detection [13, 14].

TABLE 2

Performance advantage of the various modulation/demodulation schemes referenced to the performance of an ASK heterodyne receiver

Synchronous demodulation	Modulation	Improvement on ASK HET (dB)
Homodyne	ASK	3
	PSK	6
Heterodyne	FSK	0
	PSK	3
	DPSK	2

5. COMPARISON OF DIRECT AND COHERENT DETECTION PERFORMANCE

In Fig. 2, the performance of ideal coherent detection is compared with the best practical PINFET and APD direct detection receivers over a range of bit-rates. In this comparison it should be noted that because the best low noise preamplifier designs have a rising noise versus frequency characteristic, the performance of the direct detection receiver degrades by 4.5 dB for each doubling in bit rate; whereas for ideal coherent detection it is only 3 dB. In practice the improvements in sensitivity from using coherent detection may not, ultimately, be so large because improvements in direct detection receivers and/or practical preamplifiers are still possible, and perfect coherent detection may not always be achievable.

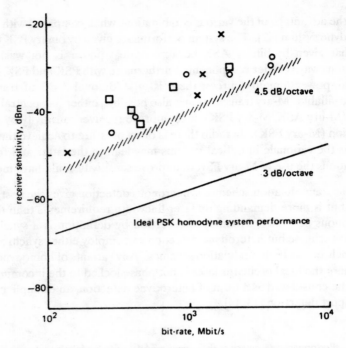

Fig. 2. Performance comparison of ideal PSK homodyne with practical direct detection at a wavelength of 1.5 μm; X : Europe [29, 30, 56, 57], O : USA [58-63], □ : Japan [64-68]

6. PRACTICAL COHERENT SYSTEM CONSTRAINTS

6.1 Effects of Laser Phase Noise

The significance of laser phase noise on system performance will depend on the modulation and demodulation scheme used, with homodyne detection and synchronous IF demodulation placing the greatest demands on laser phase stability. To meet the phase stability required for synchronous detection, the local oscillator laser must be phase locked to the incoming signal for homodyne detection; for heterodyne detection the electrical oscillator in the IF demodulator must be phased locked to the IF. Table 3 compares the laser linewidth requirements for the various modulation/demodulation schemes and indicates which laser sources meet these requirements at present.

TABLE 3
Coherent system linewidth requirements

Modulation	Demodulation		Linewidth to bit rate ratio	Suitable lasers
	Het	Hom		
ASK, FSK, PSK	SYNC	YES	<0.1% [39]	Gas and Ext Cav
DPSK	DELAY	NO	<0.3% [40]	Gas and Ext Cav
ASK, FSK	NON-SYNC	NO	<20% [41,42]	Gas, Ext Cav and DFB

With PSK transmission, since the carrier is normally suppressed, it is necessary to either transmit a low level pilot carrier or locally generate a carrier from the modulation sidebands by either a squaring circuit or Costas loop [15]. To achieve a given BER target without incurring a significant power penalty it is necessary to limit the maximum phase error between the recovered carrier and the signal to be demodulated. Clearly the bandwidth of the phase lock loop, irrespective of whether it is optical or electrical, must be sufficiently broad to track the phase noise (linewidth) of the beat spectrum generated by combining the transmitter and local oscillator signals. However, if the tracking circuit is made excessively wideband the additional phase noise associated with the quantum phase fluctuations of the local oscillator shot noise will dominate. Clearly, for a given required performance there is a maximum laser linewidth which can be tolerated by the scheme. Similar constraints also apply for synchronous demodulation of ASK and FSK modulation, but for these two cases the received carrier component is much larger.

Non-synchronous demodulations can, in principle, be made insensitive to phase noise by ensuring that the IF bandwidth of the heterodyne receiver is large enough. However, practical constraints restrict the maximum bandwidth that can be used; as the bandwidth is increased the extra shot noise power reduces the signal-to-noise ratio at the input to the demodulator and at some threshold value the demodulation process can no longer be assumed to be linear. If the bandwidth is increased beyond this point the system performance is rapidly degraded.

For ASK modulation, square-law or envelope detection are the demodulation choices; they can also be used with FSK modulation provided that a bandpass filter and associated IF demodulator are used to detect each of the transmitted frequencies. Alternatively, a limiter-discriminator can be used to demodulate an FSK signal but the conversion of phase noise to amplitude noise limits its applications.

In practice the amount of local oscillator power incident on the photodetector may be insufficient to achieve shot noise limited detection. Apart from the power available from the laser being limited this situation may arise as a result of other receiver design consideration.

- To ensure there is low loss in the signal path it is often necessary to use an optical combiner with a low coupling coefficient; this results in a high loss in the local oscillator path. With a coupling ratio of 10 : 1 the losses are 0.5 and 10 dB for the signal and local oscillator paths, respectively.

- It excess intensity noise [16] is present at the output of the local oscillator better performance can be achieved by attenuating the output of the local oscillator laser. In the presence of local oscillator excess intensity noise, optimum performance is achieved by operating the laser at a high output

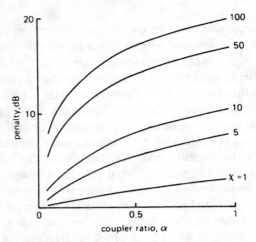

Fig. 3. System performance penalty as a function of the parameters χ and α.
NB The above curves assume that the local oscillator power incident on the photodiode is large enough to give shot noise limited performance irrespective of the value of α.

power to reduce it's excess noise factor (χ), then attenuating it's output by employing a combiner with a low coupling ratio (α) (see Fig. 3). An alternative approach is to use a balanced receiver [17].

When the local oscillator power is limited, irrespective of the reason for this, a low noise preamplifier is a significant asset if performances close to the shot noise limit are to be achieved. This is illustrated by Fig. 4 which shows, theoretically, that the performance of a coherent system will be within 1 dB of the shot noise limit if the local oscillator shot noise spectral density (S_{IS}) is at least four times larger than the equivalent current noise spectral density (S_{IE}) of the receiver [12]. The curves in Fig. 4 are slightly dependent on the value of S_{IE} so they were derived using a value of 10^{-26}; this value is typical for a PINFET preamplifier designed for operation at 140 Mbit/s. If this type of preamplifier is used in a 140 Mbit/s homodyne receiver 1 μW of local oscillator power will give a performance within 1 dB of the shot noise limit

6.2 Polarisation penalty

With heterodyne or homodyne detection the polarisation states of the local oscillator and input signal must be well matched for efficient optical mixing, otherwise larger performance penalties can be incurred. These penalties, along with possible polarisation control techniques, have been reported in detail in an earlier publication [9].

7. COHERENT OPTICAL COMMUNICATION EXPERIMENTS

Coherent optical communication is now being given serious consideration by numerous research laboratories worldwide and work published during the

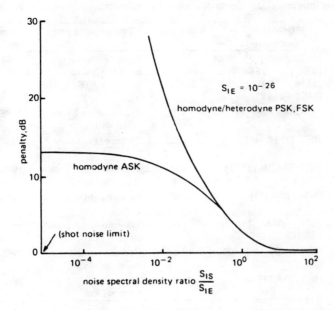

Fig. 4. Effect of preamplifier noise on system performance when the shot noise is not dominant.

period 1981 to 1983 [18], along with that reported up until mid-1985 [19], is summarised in Table 4. This tabulation includes: heterodyne systems using external cavity semiconductor lasers and/or DFB lasers at both the transmitter and receiver [20, 21, 22]; heterodyne detection used in conjunction with optical frequency division multiplexing [23, 24]; the feasibility of optical phase locked loops [25, 26, 27]; and the use of multiport detection [13, 14] as an alternative to homodyne phase lock loop detection. To illustrate some of the differences between the different types of system the following four sections describe some of the 1.5 μm coherent system experiments carried out at BT Laboratories [20, 21, 23, 25].

8. HOMODYNE SYSTEM EXPERIMENT USING GAS LASERS

The experimental arrangement for this system is shown in Fig. 5. The output from the transmitter laser was modulated at a data rate of 140 Mbit/s by an external LiNbO$_3$ phase modulator [6]; the polarisation required at the input to the phase modulator was set by the manual fibre polarisation controller [28]. To satisfy the beat linewidth to bit rate ratio requirement HeNe gas lasers were used at both the transmitter and receiver. The modulated signal was transmitted over 30 km of conventional monomode fibre prior to being combined

TABLE 4
Summary of recently reported coherent detection experiments that quote BER measurements

Research Labs	MOD/DEMOD Combination	Wavelength μm	Transmission path	Year
BTRL UK	HET: ASK [55], FSK [20,21], PSK/DPSK [23,56] HOM: ASK [54], PSK [6,25]	1.5	Operated over distances up to 200 km	1982-85
STL (UK)	Multiport: ASK [14]	1.5	Transmission distance <1 km	1985
CNET (France)	HET: DPSK [46]	0.8	3.9 km of fibre used	1982
HHI* (W Germany)	HET: FM/PM [24,51]	0.8	Transmission distance <1 km	1983-85
TUW (Austria)	HOM: PSK [53]	10.6	Line-of-sight system	1984
UNIV Tokyo (Japan)	HET: ASK [44,45], PSK [43]	0.8	Transmission distance <1 km	1981-83
NEC (Japan)	HET: ASK [49], FSK [22], DSPK [50]	1.3/1.5	Operated over distances up to 105 km	1983-84
NTT (Japan)	HET: FSK [47,48]	0.8	Transmission distance <1 km	1982-83

* This experiment used polarisation holding fibre; all of the others used conventional non-polaristion holding single mode fibre.

with the local oscillator signal in the 1:1 fused fibre coupler [7] at the receiver. To align the polarisation of the local oscillator with that of the received signal a manual fibre polarisation controller was inserted in the local oscillator path. A balanced receiver [17], constructed from two standard direct detection type PINFET receiver modules [29], was used to detect the signal at each output port of the coupler. With homodyne detection any phase variations on the received optical carrier need to be tracked so the local oscillator laser is incorporated in an optical phase lock loop. The balanced receiver design ensures that the phase lock loop is insensitive to intensity fluctuations, especially those associated with the local oscillator. The phase lock loop design used in this experiment maintains a 90° phase difference between the received carrier and local oscillator fields. When the phase difference between these two fields is 90° the two outputs from the balanced receiver are equal, but if the phase difference deviates from 90° one output from the balanced receiver increases and the other decreases. These level variations cause the output from the comparator to vary and this controls the frequency of the local oscillator laser to bring it back into phase quadrature with the received signal. To control the frequency of the local oscillator a PZT mount was used for one of the laser mirrors. To generate the carrier component needed for the control loop, the level of the modulation signal was set to give less than 180° peak-to-

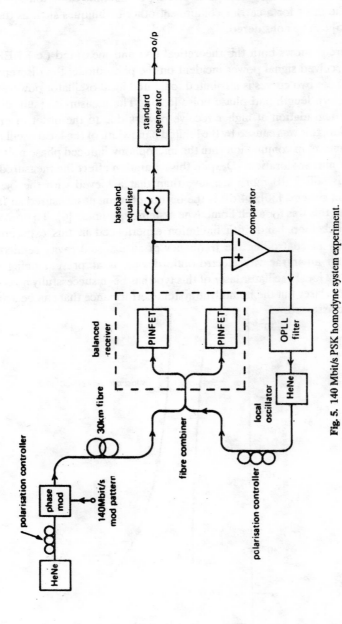

Fig. 5. 140 Mbit/s PSK homodyne system experiment.

peak phase deviation; in this particular experiment the deviation was 170°. To avoid the need for a carrier component other techniques such as the Costas loop [15] can be considered.

Figure 6 shows both the theoretical (A) and measured (●) BER versus mean received signal power incident on the photodiode; the discrepancy between these two curves is attributed to limited local oscillator power, modulation pattern length and phase noise [6,25]. The measured result also shows further degradation at higher receive powers due to the onset of error rate saturation; this was caused by the limited bandwidth of the local oscillator PZT mirror mount in conjunction with the microphony-induced phase noise caused by laser mirror vibration. Despite this saturation effect the measured performance is still 7 dB more sensitive than that achieved with the best direct detection systems [30]; 14 dB is the best improvement measured so far (mid-1985) but this was for a self homodyne system experiment [6]. To overcome the phase lock loop bandwidth limitation experienced in this experiment, the possibility of controlling the frequency of an external cavity semiconductor laser with an intracavity electro-optic device, is at present being studied. Recently a local oscillator laser of this type has been successfully phase locked to a HeNe laser but the homodyne system performance that can be achieved is yet to be assessed.

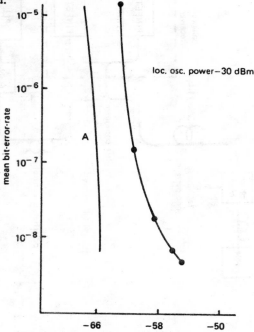

Fig. 6. 140 Mbit/s PSK homodyne system performance.

The advantages of using homodyne detection, compared to heterodyne detection, are that it gives a better receiver sensitivity and requires less receiver bandwidth, which is a significant advantage of homodyne detection especially at Gbit/s data rates. At present, however, these advantages are offset by the need for an optical phase lock loop. Multiport detection [13,14], a technique that detects the in-phase and quadrature components of the received signal, removes the need for an optical phase lock loop without increasing the receiver bandwidth requirement; however, the choice of modulation is restricted and, at best, the performance is only equivalent to that of heterodyne detection.

9. HETERODYNE SYSTEM EXPERIMENT USING EXTERNAL CAVITY LASERS

The experimental arrangement for a 140 Mbit/s FSK system is shown in Fig. 7. The significant differences between this and the previous homodyne experiment are that long external cavity semiconductor lasers [3] were used, there is no need for an optical phase lock loop, and the FSK modulation was obtained by directly current modulating the transmitting laser. The unwanted amplitude modulation produced by direct modulation was minimised by biasing the transmitter laser well above threshold at a current where the output power began to saturate. The residual amplitude modulation depth with 32 mA peak-to-peak modulation current was 20%; the associated frequency deviation was 84 MHz. The output from the transmitter was coupled directly into the transmission fibre which was conventional non-polarisation holding monomode fibre [9]. At the receiver the received signal was combined with the local oscillator field in a 10:1 fused fibre coupler. Only one output port from the coupler is used and this was connected to a standard direct detection type PINFET receiver [29]. The output from the PINFET receiver was bandpass filtered and then demodulated using a delay line discriminator. The baseband signal from the discriminator was filtered and then regenerated as in the homodyne experiment. To maintain a stable 210 MHz IF the local oscillator laser, which could be tuned electromechanically, was incorporated in an AFC loop.

The measured BER for fibre path lengths of 10 km (□), 186 km (X) and 199.7 km (O) are shown in Fig. 8 for a 2^4-1 PRBS modulation pattern; the second 186 km result (Δ) is for a $2^{10}-1$ PRBS modulation pattern.

The solid curves in Fig. 8 are the theoretical plots for frequency deviations to bit rate ratios (h) of 0.7 and 0.6. It can be seen from the plotted results that the system performance was unaffected by the length of the transmission path but it was affected by the length of the modulation pattern; changing the pattern length from 2^4-1 to $2^{10}-1$ degraded the system performance by 1.2 dB. The performance of this system, when using the short modulation pattern, is 6 dB better than has been achieved by direct detection systems [30]; also it is only

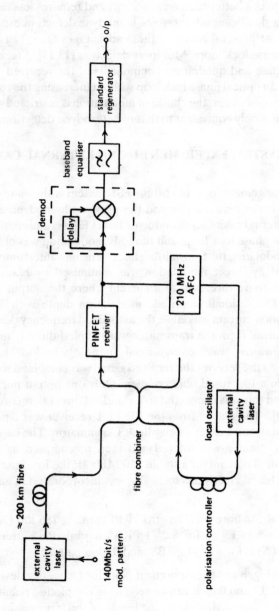

Fig. 7. 140 Mbit/s FSK heterodyne system experiment.

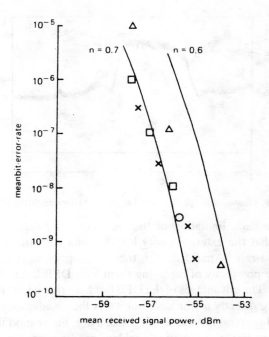

Fig. 8. 140 Mbit/s FSK heterodyne system performance.

1 dB worse, at a BER of 10^{-8}, than the previously discussed homodyne system result. Table 5 summarises the loss budget for this experimental system.

TABLE 5

System budget for the 140 Mbit/s FSK heterodyne system

Transmit power	−2.5 dBm
Fibre loss (199.7 km)	45.5 dB
Joint loss	4.0 dB
Signal loss through coupler	1.0 dB
Receiver sensitivity (10^{-9} BER)	−56 dBm
System margin	2.3 dB

The best binary heterodyne system performance is given by PSK modulation, but when FSK modulation is used the transmit laser can be directly modulated; this avoids the insertion loss of the external modulator needed for PSK systems. However, if the appropriate IF demodulation is used in conjunction with MSK modulation [31] (i.e. FSK modulation with a frequency deviation equal to half the bit rate), a performance similar to that given by PSK modulation should be possible. A further advantage of narrow deviation FSK modulation is its compact IF spectrum (see Fig. 9); this could be of significant importance for Gbit/s heterodyne systems where receiver bandwidth is at a

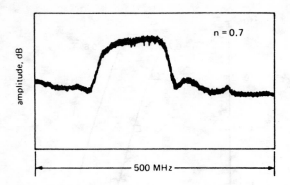

Fig. 9. Measured IF spectrum for 140 Mbit/s narrow deviation FSK.

premium. The main limitation of the external cavity semiconductor laser transmitter is that the external cavity length limits both the FSK frequency deviation and bit-rate. Apart from this they are, at present, relatively complex devices and the possibility of replacing them with DFB lasers has been considered [21,22]. The advantages of the DFB laser are that it is a compact device capable of being directly modulated, but it has the disadvantages of having a relatively broad spectral linewidth along with a non-linear modulation characteristic. The non-linear modulation problem can be overcome using techniques already reported [22,32,33] and to overcome the linewidth problem non-synchronous IF demodulation must be used, but at the expense of system performance.

10. HETERODYNE SYSTEM EXPERIMENT USING A DFB LASER TRANSMITTER

To assess the effect of laser linewidth on system performance the experimental arrangement shown in Fig. 10 was used. The transmitter laser, a 1.5 μm ridge waveguide distributed feedback laser diode [5], was biased above threshold and the current was adjusted to give either a 30 MHz or a 60 MHz FWHM linewidth. FSK modulation of the DFB laser was achieved by directly modulating the bias current; a modulation current of 1 mA was used and this gave a FSK frequency deviation of 1.4 GHz. Due to the sensitivity of the DFB laser to optical reflections more than 40 dB of optical isolation was needed at the laser output. The local oscillator laser was a 10 kHz linewidth tunable external cavity semiconductor laser and this gave 20 μW of power at the photodiode. The highpass filter, PINFET receiver combination gave an IF bandwidth of 300 MHz centred about 420 MHz. Due to limited receiver bandwidth it was necessary to use single filter detection in this experiment; this technique degrades the system performance by 3 dB because half of the received signal is ignored. When using single filter detection the IF spectrum is equivalent to

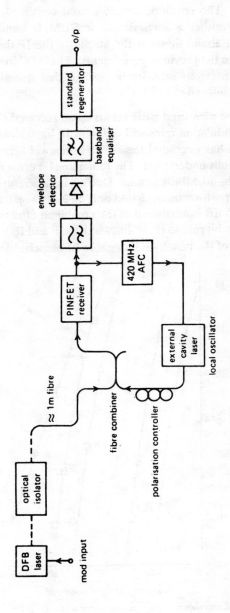

Fig. 10. 140 Mbit/s FSK heterodyne system experiment using a DFB laser transmitter.

that for ASK modulation and this is why envelope detection can be used to demodulate the IF. The envelope detector used consisted of a half wave rectifier followed by either a lowpass 100 or 50 MHz bandwidth filter as appropriate. The baseband signal at the output of the IF demodulator was then regenerated as in the previous experiments. The AFC loop was identical to that used in the previous experiments except that it maintained the IF constant at 420 MHz instead of 210 MHz.

Figure 11 shows the measured BER versus mean received signal power for 140 and 70 Mbit/s modulation rates and different IF linewidths. The onset of error rate saturation has degraded the 60 MHz linewidth result: as yet the reason for this is not fully understood. The remaining bit error rate curve is the 30 MHz IF linewidth, 70 Mbit/s result. Due to insufficient local oscillator power the measured performances should only be 2 dB worse than the theory lines; the remaining 5 dB is attributed to the combined effects introduced by the value of the ratios: bit rate to IF, IF linewidth to IF and IF bandwidth to bit rate. However, most of the penalty is thought to be associated with the first of these three ratios.

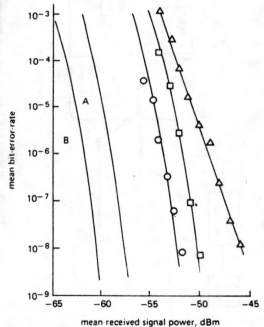

Fig. 11. 140 Mbit/s FSK heterodyne system performance with a DFB laser transmitter.
Theory: 140 Mbit/s (A); 70 Mbit/s (B)
Measured: 30 MHz linewidth
140 Mbit/s (O); 70 Mbit/s (□)
60 MHz linewidth
140 Mbit/s (Δ).

Penalties associated with the non-linear modulation characteristic of the DFB laser were minimised by using a 2^3-1 PRBS modulation pattern; had this effect been significant the demodulated waveforms would have exhibited noticeable patterning effects.

Although the measured performances for this experiment are no better than can be achieved by good direct detection systems [30] they do have the potential of giving a 7 dB improvement. Most of the performance degradation observed is thought to be due to the envelope detector being sensitive to phase noise; a condition that does not exist with ideal envelope detection. Further practical improvements should be possible when receiver bandwidths large enough for dual filter detection become available.

Now that coherent optical systems have been shown to be feasible, future research will be concerned with applications such as multi-Gbit/s long span unrepeatered systems, local area networks and frequency multiplexing. Some preliminary frequency multiplexing experiments have already been carried out at BT Labs [23] and these are dealt with in the following section.

11. FREQUENCY MULTIPLEXED COHERENT SYSTEM EXPERIMENT

Using early research components the possibility of utilising heterodyne detection in conjunction with optical FDM transmission has been demonstrated using the experimental arrangement shown in Fig. 12. Transmitter TX 1 was a packaged external cavity semiconductor laser [34] and TX 2 was a He-Ne laser; both transmitters were modulated by external $LiNbO_3$ phase modulators. Subsequent to modulation the two optical channels were combined in a fused fibre coupler which had one of its outputs connected to the heterodyne receiver by a short length of conventional non-polarisation holding fibre. At the receiver the local oscillator was a packaged external cavity semiconductor laser, the IF was 566 MHz, and synchronous IF demodulation was used. A modulation depth smaller than $\pm 90°$ was used so that there was a carrier component present in the IF spectrum; this was extracted using a bandpass filter and then used as the phase reference for the synchronous IF demodulator. L 60 MHz sub-carrier PFM [35] was used for transmitting video over the link because the lack of modulation components in the vicinity of the IF allowed the design requirements for the IF extraction filter to be relaxed. This is illustrated by Fig. 13 which shows the structure of the IF spectrum when using this type of modulation in conjunction with PSK modulation of the transmitter.

To demonstrate the tunability of the receiver the optical frequency of TX 1 was adjusted to be approximately 0.01 nm away from TX 2, this gives a separation of 1 GHz between the two optical carriers. By tuning the local oscillator in the receiver to be 566 MHz away from TX 1, channel 1 was selected, demodulated and a colour picture displayed on a video monitor;

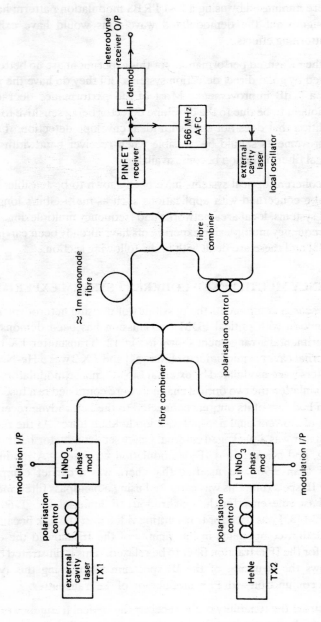

Fig. 12. Experimental 2-channel frequency multiplexed transmission system.

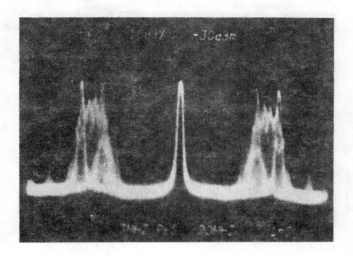

Fig. 13. Intermediate frequency spectrum when using a PFM system modulated by baseband video.

stable operation was achieved and the AFC loop maintained the IF frequency at 566 MHz. Next, the AFC loop was temporarily disconnected and the local oscillator tuning was adjusted to be 566 MHz away from TX 2. The second channel now appeared on the video monitor and stable operation was achieved with the AFC loop again maintaining the intermediate frequency at 566 MHz.

This two channel experiment has shown that stable demodulation of one PFM video channel spaced just 1 GHz away from an adjacent channel is possible. One channel of this system was operated for 12 hours a day over a period of a week and after resetting the control loops each morning the system operated throughout the rest of the day without any need for further adjustment. The stable operation observed from this system is a direct result of the packaged lasers having a better frequency stability [36] than early laboratory versions of the external cavity semiconductor laser.

To assess inter-channel interference the transmitters were modulated by two independent 70 Mbit/s PRBS to give a DPSK signal, and a DPSK IF demodulator was used in place of the synchronous design. Figure 14 shows these two channels at the output of the PINFET receiver. The receiver was tuned to channel 2 and the BER was monitored as the channel spacing was varied; this was achieved by altering the frequency of TX 1. Figure 15 shows the measured BER degradation as a function of channel separation.

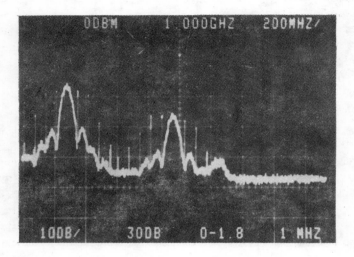

Fig. 14. Simultaneous detection of two frequency multiplexed 70 Mbit/s phase modulated channels.

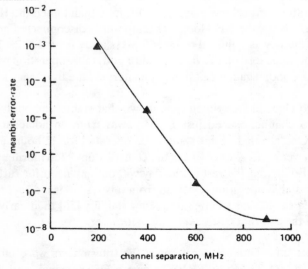

Fig. 15. Effect of channel separation on BER.

For multi-channel transmission, the channel spacings should be large enough to place intermodulation products and image band signals outside the receiver bandwidth; recent theoretical work [37] has reported that the minimum spacing should not be less than $5 \times$ BW (BW is the bandwidth of the modulated carrier [37]). However, for some specific cases, provided the local oscillator output is placed in the optimum position channel spacings as small as 3 × BW should be possible; further experimentation is needed to verify this. Transmission of frequency multiplexed channels over long fibre lengths is still to be demonstrated and an area that needs further study is the variation in the received states of polarisation for the different channels. In addition if a large number of optical channels are to be combined for transmission over a single fibre the effect of fibre non-linearities will need to be considered of which Raman crosstalk [38] could be one of the more troublesome.

12. CONCLUSION

The predicted benefits of coherent optical detection, in terms of improved receiver sensitivity, have been verified in preliminary 140 Mbit/s systems experiments at a wavelength of 1.5 μm. It is expected that the concepts used in these experiments can be extended to Gbit/s operation provided that suitable receiver designs are available. To develop coherent transmission systems for field deployment a key area that will require further attention is line narrowed lasers and at present there are two main lines of approach: the external cavity laser and the DFB laser. For the time being the former approach is offering the best initial gains in terms of system performance but the more compact DFB laser could prove attractive in the longer term.

Interest is now increasing in the application of coherent transmission in future wideband optical networks; this is an application that would make particular use of the wavelength selectivity of coherent transmission. The basic principle of using wavelength multiplexing and coherent detection was demonstrated in the 2 channel experiment. The future prospects of using coherent techniques in this area will depend on cost reduction, where it is hoped that the wider application of integrated optics will play a role.

The ultimate transmission capacity of a network based on coherent optical principles may be restricted to less than the potential 50,000 GHz, by the onset of channel crosstalk introduced by non-linear propagation effects; this is yet to be verified in practice. However, it is already clear that the transmission capacity is likely to be larger than achievable with any other closed transmission medium.

13. APPENDIX: OVERVIEW OF COHERENT SYSTEMS RESEARCH (1985-88)

13.1 Coherent Optical Fibre Transmission System Studies

Theoretical

Published theoretical studies have been mainly concerned with deriving the probability of error expressions for the various types of coherent system and the performance degradation associated with non-ideal operating conditions

[69-90]. A detailed analysis of synchronous coherent optical fibre transmission systems has been published [69] which takes account of polarisation misalignment, reduced modulation depth, preamplifier thermal noise, power coupling ratio of the optical coupler, local oscillator excess intensity noise, reference phase errors, and both balanced and unbalanced receiver designs. The performance degradation caused by the combined effects of additive Gaussian noise and Lorentzian linewidth (quantum phase noise limited) laser sources, is considered for both pilot carrier and Costas phase-locked-loop designs. The outcome of this study is that for all but one of the synchronous coherent receiver designs, laser linewidths of the order 0.01% of the bit-rate are needed for the degradation due to phase noise to be ≤ 1 dB; for the heterodyne Costas loop receiver linewidths an order of magnitude larger can be tolereated.

The non-synchronous coherent system studies have shown that for DPSK and CPFSK modulations, the maximum source linewidths that can be used are of the order 0.2% of the bit-rate provided the maximum IF bandwidth is used. When IF bandwidths wider than necessary for the data-rate are used, narrower linewidths are needed to avoid bit-error-rate saturation [84,89]. For ASK and wide deviation FSK modulations, a trade off exists between IF bandwidth and saturation bit-error-rate provided post detection noise filtering is used [85] (e.g. in the absence of phase noise an IF bandwidth of 10 times the bit-rate only degrades the receiver performance by approximately 1 dB). Therefore, for a small penalty the IF bandwidth can be increased to the extent where linewidths of the order of half the bit-rate can be tolerated [86].

Experimental: Heterodyne

As the various heterodyne system design options have been studied, FSK [91-101] and DPSK [102-106] modulated systems have emerged as the most practicable at present. These systems have been operated at data-rates upto 2 Gbit/s over transmission distances upto 300 km, and performances in the range 5 to 14 dB from the shot noise limit have been measured. The advantages of DPSK modulation are that it gives a performance which is almost identical to that of PSK, without the need for a phase-locked-loop IF demodulator [55, 107]. Its disadvantages are the need for an external modulator and the demand for laser linewidths $\leq 0.2\%$ of the bit-rate. At the expense of reduced performance, FSK modulation overcomes these disadvantages because lasers can be directly FSK modulated, and linewidths which are a significant percentage of the bit-rate can be tolerated provided the two FSK tones are sufficiently far apart (wide deviation FSK).

At present, an important heterodyne systems design aim is to minimise the receiver bandwidth, and one technique which has been proposed is single filter detection [22, 91], but this degrades the performance by 3 dB. A better solution is to use CPFSK [96-98] with a tone separation equal to half the bit-rate (MSK) because this gives the most compact IF spectrum and when used with synchronous IF demodulation the best performance [31]; unfortunately, the laser linewidths required are similar to those needed for DPSK modulation [87,93]. However, the linewidth requirement relaxes as the tone separation increases, and it has been found that for typical DFB laser linewidths, good performance and a fairly compact IF spectrum can be achieved using delay IF demodulation and a tone separation of the order of 0.7 times the bit-rate [97].

An early problem with DFB lasers was their non-linear frequency modulation characteristic, but it is now well known that this can be overcome using either passive equalisation of the modulation signal [108], Manchester [22] or bipolar modulation coding [100,101], multi-contact DFB laser structures [33,91,96] or feedback equalisation [32,109]. DFB lasers have now been successfully directly FSK modulated at data-rates upto 5 Gbit/s [110]. Another early disadvantage with DFB lasers was their broad linewidth, but device research and development have progressed to the stage where sub-megahertz linewidths have now been achieved using an integrated passive cavity structure [193].

For future systems, the balanced receiver will almost certainly be the preferred choice because it makes efficient use of the available local oscillator power and suppresses its excess intensity noise. Balanced designs with bandwidths upto 2 GHz [111,112] have been reported, but it is expected that much larger bandwidths will be achievable, given that unbalanced designs have been reported with bandwidths upto 20 GHz [113,114]. Some receiver designs, such as the high impedance PINFET front end, need larger local oscillator power as the IF is increased, making them unsuitable designs for high bit-rate systems. It is thought that this is one reason why the high bit-rate heterodyne system experiments show a gradual performance degradation with respect to the shot noise limit for increasing bit-rate [115]. To avoid this problem receivers tuned to the IF have been suggested [12,92], and suitable designs are now being assessed in greater detail than ever before [116-129].

Most experimental work has concentrated on ASK, FSK, DPSK and PSK modulation formats, but the possibility of using binary polarisation modulation has been addressed [120]. Using two orthogonal polarisation states to represent the binary modulation levels, polarisation modulation has been shown that to give a 3 dB performance improvement over ASK modulation. The main drawback of this technique is the need for the received polarisation states to be controlled and maintained linear at an orientation determined by the polarisation selective coupler used in the receiver.

Now that advanced optical components are becoming available, the engineering aspects of practical heterodyne transmission systems are being addressed and two field trials have recently been demonstrated. One was a land

565 Mbit/s DPSK system using continuous polarisation control and packaged external cavity lasers [140], the other was an undersea 560 Mbit/s FSK system using polarisation diversity and external grating DFB lasers [94].

Experimental: Homodyne

The interest in homodyne detection is still relatively low key as a result of the significant implementation problems associated with practical optical phase-locked-loops, and the fact that very narrow laser linewidths are required. However, some additional experimental work has been reported since 1985: a semiconductor laser, pilot carrier phase-locked-loop has been demonstrated [121], and a 700 Mbit/s PSK pilot carrier 1.52 μm HeNe phase-locked-loop homodyne system has been operated within 5 dB of the shot noise limit [122].

Homodyne detection without the need for an optical-phase-locked loop can in principle also be achieved by selectively amplifying a residual carrier component, and then recombining this with the modulation sidebands prior to photodetection [115]. To achieve the selective gain Brillioun amplification has been proposed and it has been shown that a gain of 26 dB can be achieved [123]. However, the bit-error-rate performance given by a homodyne receiver using a Brillioun amplified carrier has yet to be assessed.

13.2 Single-Mode Fibre/Cable Studies

Because standard coherent detection systems are sensitive to the received state of polarisation, simple measurements to assess the polarisation fluctuations at the output of standard single-mode fibre cable were carried out in 1982/83. These early results showed that the polarisation fluctuated slowly and unpredictably with time [9]. Since then more detailed measurements on both land [124] and undersea cables [125, 126] have shown these fluctuations to have a low pass characteristic with a corner frequency of the order 0.03 Hz; even during cable laying the corner frequency increased by only an order of magnitude.

The polarisation behaviour of single-mode fibre has also been studied theoretically and a polarisation dispersion model has been proposed [127], which led to a prediction of $\approx$ 10 ps polarisation dispersion pulse broadening over a 100 km transmission distance [128, 129] and analytic confirmation that polarisation orthogonality is maintained over at least a 50 km transmission distance [130] (this fact is made use of by some polarisation insensitive coherent detection schemes [153]). Comparing the polarisation dispersion pulse broadening with that expected as a result of chromatic dispersion [131] shows the latter to be an order of magnitude larger. Therefore, unless chromatic dispersion compensating receivers are used [132], the performance degradation due to polarisation dispersion will almost certainly be a second order effect for most types of coherent transmission system.

13.3 Polarisation Insensitive Coherent Detection

Polarisation Tracking Receivers

Once it was realised that the polarisation at the output of the transmission fibre only changed slowly with time, it appeared feasible that these changes could be tracked out by inserting a suitable number of polarisation controllers in either the local oscillator or received signal path and incorporating them in a feedback control loop. This led to a study of polarisation controllers and the designs suited to automatic control applications [71, 133] have been based on either piezo-electric fibre squeezers, Faraday rotators, electro-optic waveguide devices, wave plates or mechanical fibre cranks. In principle, the minimum number of controllers needed is two [133], but if both the signal and local oscillator polarisation states are time varying and endless control is required from finite range controlling elements, a minimum of four are needed [134, 135]. However, there have been reports of being able to reduce this further: three can be used if the local oscillator polarisation is both stable and an eigenstate of the second controller [136], or two can be used if some passive polarisation selective optics are incorporated into the receiver design along with a second opto-electronic receiver and associated IF processing electronics [137].

Experimental systems using automatic polarisation control have been demonstrated at data-rates up to 565 Mbit/s [71, 107, 133, 138-140] and three different types of feedback control loop have emerged, the difference being the technique used to derive the feedback signal. A microprocessor has been used to make small step changes to the controlling elements drive voltages in order to 'peak search' the IF amplitude [71, 138], a fraction of the received optical power has been tapped off and analysed using bulk optic polarisation sensitive elements [139] and the local oscillator polarisation has been dithered to produce low level, polarisation offset dependent, IF amplitude modulation [133, 140].

The advantage of polarisation tracking receivers is that in principle the full performance potential of coherent detection can be achieved, but at the expense of a more complex receiver design.

Polarisation Diversity Receivers

To avoid using polarisation tracking receivers, but still retain polarisation independent performance, the alternative approach of using polarisation diversity detection can be considered [94,141-149]. Basically, this type of receiver consists of the usual input fibre coupler followed by some form of polarisation selective coupler, each output of which is connected to an opto-electronic receiver and associated IF demodulation stage (the actual component configuration and specification varies slightly depending on whether the receiver is a single ended [142] or a balanced [143] design).

The performance potential of a polarisation diversity receiver is equal to that of a polarisation tracking receiver, but this may not be the case in practice. When using synchronous IF demodulation, the gain of each arm of the polarisation diversity receiver needs to be continuously adjusted (weighted) in inverse proportion to the modulus of the signal in the other arm [144], otherwise the performance will fluctuate by 3 dB. When square-law or delay IF demodulation is used, this weighting occurs naturally, but because of the limitations of practical devices it is expected that FSK [94, 144-146] and DPSK [147, 148] systems will suffer a 0.4 dB performance degradation, and that for ASK systems this could be as large as 2 dB [142].

Polarisation diversity systems experiments have been operated at bit-rates upto 1.2 Gbit/s using all the usual modulation formats, and the best performance to date has been given by a FSK receiver. A performance within 0.3 dB of that given by the same receiver when configured as a standard heterodyne receiver was measured, and its dependence on the received state of polarisation was negligible [146]. The disadvantage of polarisation diversity detection schemes is that they require two opto-electronic receivers and associated IF processing electronics: this may inhibit their use where cost is of prime importance.

Polarisation Insensitive Receivers

Polarisation diversity detection can also be achieved by using orthogonally polarised optical signals [150-153]. It is not immediately obvious but by considering Poincare's sphere it becomes clear that orthogonally polarised fields propagating through a fibre transmission medium may undergo absolute polarisation changes but their orthogonality is unaffected. In practice there will eventually be some deterioration of orthogonality if the transmission medium exhibits polarisation selective loss and/or polarisation dispersion.

The first proposal and demonstration of using orthogonally polarised signals to achieve polarisation insensitive detection was in the guise of a polarisation scrambled ASK system experiment [150] (this technique switches (scrambles) the polarisation of the transmitter output field between orthogonal states at a rate which is a multiple of the data-rate). It has also been demonstrated that polarisation insensitive detection can be achieved using orthogonally polarised FSK modulation [151] (this technique transmits the two FSK tones in orthogonal polarisation states) and the advantage of this over scrambling is that it increases the maximum tolerable transmission path loss by 3 dB. A dual frequency, orthogonally polarised local oscillator has also been shown to be capable of giving polarisation insensitive detection [152] and recent work has shown that polarisation scrambling and orthogonally polarised FSK modulation are in fact special cases of using a dual frequency, orthogonally polarised optical source at the transmitter [153].

The advantage of using a dual frequency, orthogonally polarised source to achieve polarisation insensitive detection is that it removes the need for the complete extra receiver needed when polarisation diversity detection is used. Furthermore, if the dual frequency source is used at the transmitter the receiver complexity is kept to a minimum, and for the special case of ASK systems a standard non-synchronous heterodyne receiver can be used provided the two orthogonally polarised optical carriers are separated in frequency by an amount $\geq$ 3 times the bit-rate [153]. The disadvantages of using a dual frequency source are that the performance is 3 dB worse than given by a polarisation tracking receiver and wider IF bandwidths are needed; typically this is twice the bit-rate plus the frequency separation between the two optical carriers. However, for applications such as medium bit-rate, local loop networks, this wider bandwidth requirement may not be of concern.

13.4 Phase Diversity Systems

Since the publication of references 13 and 14, there has considerable interest in phase diversity coherent optical detection [141,154-169] (this is also known as multiport detection and in-phase and quadrature (I & Q) detection) as an alternative to homodyne detection. The reasons for this is that it offers the reduced receiver bandwidth advantage of homodyne detection and overcomes the need for an optical phase-locked-loop. However, its potential performance is at best only equivalent to that of standard heterodyne detection. The basic operating principle is that two quadrature phase optical signals are generated and processed in such a way that a zero hertz intermediate frequency can be used; hence the reduced bandwidth requirement when compared with that needed for standard heterodyne detection. Phase diversity detection has been demonstrated using a multiport coupler [154-157], a 90 degree optical hybrid [158-161] and a multisegment photodetector illuminated by an overmoded fibre [162].

The multiport coupler receiver would ideally be a 4 × 4 design, but for practical reasons a 3 × 3 design is most common at present, mainly as a result of fabrication difficulties [154]. The 3 × 3 multiport [163-166] and the optical hybrid [159] receiver designs have been studied in detail and both have been shown to have the same sensitivity to laser linewidth and IF centre frequency stability as the equivalent standard heterodyne receiver. However, because the optical hybrid uses the properties of linearly and circularly polarised light to generate the quadrature phase optical signals it is an order of magnitude more sensitive to polarisation fluctuations than both multiport coupler receivers and standard coherent receivers.

Practical studies have reported bit-error-rate measurements in the range 140 Mbit/s to 5 Gbit/s for ASK [154,156,159], FSK [155,161] and DPSK [154,157,159] modulation formats, but they are still in the range 7 to 11 dB away from the shot noise limit. These experiments have also demonstrated that 1.5 μm wavelength He-Ne, external cavity [34, 36, 170, 171] and DFB lasers can be used.

It is interesting to note that optical hybrid phase diversity and polarisation diversity receiver designs are identical (compare [142] and [143] with [159]) and that the type of diversity achieved is determined purely by the polarisation settings used. For phase diversity a circularly polarised local oscillator is used with a linearly polarised signal oriented such that the received power is divided equally between the two arms of the receiver, whereas for polarisation diversity it is only necessary for the local oscillator polarisation to be set such that its output power is divided equally between the two arms of the receiver. If two phase diversity receivers are connected to the outputs of a polarisation selective coupler it is possible to simultaneously achieve polarisation and phase diversity coherent detection [160].

The disadvantage of the phase diversity detection schemes discussed above is that they require two or more receivers depending on which implementation is used, and this may inhibit their use where cost is of prime importance. However, the alternative phase diversity technique of superimposing periodic phase modulation on the transmitted signal at a rate equal to a multiple of the bit-rate, has recently been proposed. This has the advantage of removing the need for extra receivers [167,168], but some reduction in potential performance may have to be tolerated and for some modulation schemes narrower linewidth optical sources may be necessary. Also, it is expected that the receiver bandwidth will have to be increased, so this technique may not have any advantage over heterodyne detection.

13.5 Multichannel Systems

The successful demonstration of coherent optical detection techniques has led to considering their application potential in future wideband communication networks [172 173]. Various new network structures are being studied [172], and common aspects now beginning to be addressed are: laser frequency stabilisation and optical frequency comb generation [174-185], interchannel interference [186-188] and optically generated non-linearities [172, 173].

Numerous schemes have been proposed for producing optical frequency reference combs, but despite the number reported only three basic techniques appear to be involved. One technique is to modulate a single laser source to produce a multiplicity of sidebands; this can be achieved by using either a nonlinear modulation scheme (e.g. phase modulation) or a harmonically rich modulation waveform [174, 175]. The other two techniques, which can be used individually or combined, use heterodyne offset frequency locking loops [176-178] or Fabry Perot interferometer resonance peaks [179-183] to lock several laser sources to a single master laser. If absolute stability is required the master laser can be locked to an atomic frequency standard (absorption line) using previously reported techniques [184, 185].

When using coherent transmission techniques it should in principle be possible to use channel separations equal to the modulation bandwidth; for heterodyne systems this would require the use of an optical image band rejection filter [189, 190]. To achieve such a narrow channel spacing in practice will almost certainly require the use of very narrow linewidth sources and pre-transmission filtering to minimise interchannel interference [187]. The use of channel separations as small as the ideal value have not yet been shown to be feasible, but an experimental heterodyne system has been successfully operated with a channel separation equal to four times the data-rate [188].

When large multiplexes are assembled the power build up in the fibre may reach levels such that Brillioun, Raman and Kerr effects generate intermodulation products, but the upper limit that this will place on the number of channels is at present unknown [172]. If an optical multiplex is amplified using a semiconductor laser, the resulting carrier density modulation caused by the interchannel beat power (4-wave mixing) generates intermodulation products, but recent work has shown that this is not likely to be a problem for channel separations > several GHz, and that at worst it will probably be a second order effect [191].

REFERENCES

1. L.H. Enloe and Rodda J.L., "Laser phase locked loop", Proc. IEEE, Vol. 53, pp. 165-166 (1965).
2. O.E. DeLange and Dietrich A.F., "Optical heterodyne experiments with enclosed transmission paths", Bell Syst. Tech. J. Vol. 47, pp. 161-178 (1968).
3. R. Wyatt and Devlin W.J., "10 kHz linewidth 1.5 μm InGaAsP external cavity laser with 55 nm tuning range", Electron Lett., Vol. 19, pp. 110-112 (1983).
4. R. Wyatt, Smith D.W. and Cameron K.H., "Megahertz linewidth from a 1.5 μm semiconductor laser with HeNe laser injection", Electron Lett., Vol. 18, pp. 292-293 (1982).
5. L.D. Westbrook, Nelson A.W., Fiddyment P.J. and Evans J.S., "Continuous-wave operation of 1.5 μm distributed-feedback-ridge-waveguide lasers", Electron Lett., Vol. 20, pp. 225-226 (1984).
6. D.J. Malyon, Hodgkinson T.G., Smith D.W., Booth R.C. and DaymondJohn B.E., "PSK homodyne receiver sensitivity measurements at 1.5 μm", Electron Lett., Vol. 19, pp. 144-146 (1983).
7. A.P. McDonna, McCartney D.J. and Mortimore D.B., "1.3 μm bi-directional optical transmission over 30 km of installed single-mode fibre using optical couplers", Electron Lett., Vol. 20, pp. 722-723 (1984).
8. T. Okoshi, "Single polarisation single-mode optical fibres", IEEE J. Quantum Electron, Vol. QE-17, pp. 879-884 (1981).
9. D.W. Smith, Harmon R.A. and Hodgkinson T.G., "Polarisation stability requirements for coherent optical fibre transmission systems", Br Telecom Technol. J., Vol. 1, No. 2, pp. 12-16 (1983).

10. S.D. Personick "Fundamental limits in optical communications", Proc. IEEE, Vol. 69, pp. 262-266 (1981).
11. S.D. Personick, "Receiver design for digital fibre optic communication systems I and II", Bell Syst. Tech. J., Vol. 52, pp. 843-887 (1973).
12. T.G. Hodgkinson, Wyatt R., Smith D.W., Malyon D.J. and Harmon R.A., "Studies of 1.5 μm coherent transmission systems operating over installed cable links", IEEE Globecom' 83, San Diego, USA, pp. 21.3.1-21.3.5 (1983).
13. N.G. Walker and Carroll J.E., "Simultaneous phase and amplitude measurements on optical signals using a multiport junction", Electron Lett., Vol. 20, pp. 981-983 (1984).
14. A.W. Davis and Wright S., "A phase insensitive homodyne optical receiver", IEE Colloquim: "Advances in coherent optic devices and technologies", Digest No. 1985/30, London England, pp. 11/1-11/5 (1985).
15. H.K. Phillip Scholtz A.L., Bonek E. and Leeb W.R., "Costas loop experiments for a 10.6 μm communications receiver", IEEE Trans Comm., Vol. 31, pp. 1000-1002 (198).
16. Y. Yamamoto, "AM and FM quantum noise in semiconductor lasers", IEEE J. Quantum Electron, Vol. QE-19, pp. 34-46 (1983).
17. G.L. Abbas, Chan V.W.S. and Yee T.K., "Local oscillator excess noise suppression for homodyne and heterodyne detection", Optics Lett., Vol. 8, pp. 419-421 (1983).
18. T. Okoshi, "Recent progress in heterodyne/coherent optical fibre communications", J. Lightwave Tech., Vol. LT-2, pp. 341-346 (1984).
19. T.G. Hodgkinson, Smith D.W, Wyatt R. and Malyon D.J., "Coherent optical communications", Tech. Dig. OFC/OFS'85, San Diego, USA, pp. 22-23 (1985).
20. R. Wyatt, Smith D.W., Hodgkinson T.G., Harmon R.A. and Devlin W.J., "140 Mbit/s optical FSK fibre heterodyne experiment at 1.54 μm", Electron Lett., Vol. 20, pp. 912-913 (1984).
21. F. Mogensen, Hodgkinson T.G. and Smith D.W., "FSK heterodyne system experiments at 1.5 μm using a DFB laser transmitter", Electron Lett., Vol. 21, pp. 518-519 (1985).
22. K. Emura, Shikada M., Fujita S., Mitu I., Homnou H. and Minemura K., "Novel optical FSK heterodyne single filter detection system using a directly modulated DFB-laser diode", Electron Lett., Vol. 20, pp. 1022-1023 (1984).
23. D.W. Smith, Hodgkinson T.G., Malyon D.J. and Healey P., "Demonstration of a tunable heterodyne receiver in a two channel optical FDM experiment", IEE Colloquium: 'Advances in coherent optic devices and technologies', Digest No. 1985/30, London, England, pp. 13/1-13/5 (1985).
24. E.J. Bachus, Bohnke F., Braun R.P., Eutin W., Foisel H., Heimes K. and Strebel B., "Two channel heterodyne type transmission experiment", Electron Lett., Vol. 21, pp. 35-36 (1985).
25. D.J. Malyon, "Digital fibre transmission using optical homodyne detection", Electron Lett., Vol. 20, pp. 281-283 (1984).
26. A.L. Scholtz, Leeb W.R., Philipp H.K. and Bonek E., "Infrared homodyne receiver with acousto-optically controlled local oscillator", Electron Lett., Vol. 19, pp. 234-235 (1983).
27. W.J. Liddell, "Optical phase locked loop (OPLL)", IEE Colloquium: "Advances in coherent optic devices and technologies", Digest No. 1985/30, London, England, pp. 9/1-9/3 (1985).
28. H.C. Lefevre, "Single-mode fibre fractional wave devices and polarisation controllers", Electron Lett., Vol. 18, pp. 778-780 (1980).
29. M.C. Brain, Smyth P.P., Smith D.R., White B.R. and Chidgey P.J., "PIN-FET hybrid optical receivers for 1.2 Gbit/s transmission systems operating at 1.3 and 1.55 μm wavelength", Electron Lett., Vol. 20, pp. 894-896 (1984).

30. S.D. Walker and Blank L.C., "Ge APD/GaAs FET/op-amp transimpedance optical receiver design having minimum noise and intersymbol interference characteristics", Electron Lett., Vol. 20 pp. 808-809 (1984).
31. S. Haykin, "Communication systems", John Wiley & Sons, second edition, pp. 561-572 (1983).
32. S. Saito, Nilsson O. and Yamamoto Y., "Coherent FSK transmitter using a negative feedback stabilised semiconductor laser", Electron Lett., Vol. 20, pp. 703-704 (1984).
33. S. Yamazaki, Emura K., Shikada M., Yamaguchi M. and Mito I., "Realisation of flat FM response by directly modulating a phase tunable DFB laser diode", Electron Lett., Vol. 21, pp. 283-285 (1985).
34. M.R. Matthews, Cameron K.H., Wyatt R. and Devlin W.J., "Packaged frequency-stable tunable 20 kHz linewidth $1.5\,\mu$m InGaAsP external cavity lasers", Electron Lett., Vol. 21, pp. 113-115 (1985).
35. D.J.T. Heatley and Hodgkinson T.G., "Video transmission over cabled monomode fibre at $1.523\,\mu$m using PFM with 2-PSK heterodyne detection", Electron Lett., Vol 20, pp. 110-112 (1984).
36. K.H. Cameron, Matthews M.R., Wyatt R. and Devlin W.J., "A packaged frequency-stable tunable 20 kHz linewidth $1.5\,\mu$m external cavity laser for use in coherent optical fibre transmission systems", IEE Colloquium: "Advances in coherent optic devices and technologies", Digest No. 1985/30, London, England, pp. 8/1-8/3 (1985).
37. P. Healey, "Effects of intermodulation in multichannel optical heterodyne systems", Electron Lett., Vol. 21, pp. 101-102 (1985).
38. J.F. Mahlein, "Cross-talk due to stimulated Raman scattering in single-mode fibres for optical communication in wavelength division multiplex systems", Opt and Quantum Electron, Vol. 16, pp. 409-425 (1984).
39. I.W. Stanley and Smith D.W., "Progress in coherent optical fibre transmission techniques in the United Kingdon", Proc. SPIE, Fibre Optics: "Short-haul and long-haul measurements and applications", No. 500, San Diego, USA, pp. 13-16 (1984).
40. G. Nicholson, "Probability of error for optical heterodyne DPSK system with Quantum noise", Electron Lett., Vol. 20, pp. 1005-1007 (1984).
41. G. Jacobsen and Garrett I., "Error-rate floor in optical ASK heterodyne systems caused by non-zero (semiconductor) laser linewidth", Electron Lett., Vol. 21, pp. 268-270 (1985).
42. G. Jacobsen and Garrett I., "Influence of (semiconductor) laser linewidth on the error-rate floor in dual-filter optical FSK receivers", Electron Lett., Vol. 21, pp. 280-282 (1985).
43. K. Kikuchi, Okoshi T., Nagamatsu M. and Henmi N., "Bit-error rate of PSK heterodyne optical communication system and its degradation due to spectral spread of transmitter and local oscillator", Electron Lett., Vol. 19, pp. 417-418 (1983).
44. K. Kikuchi, Okoshi t. and Kitano J., "Measurement of bit-error rate of heterodyne type optical communication system-a simulation experiment", IEEE J. Quantum Electron, Vol. QE-17, pp. 2266-2267 (1981).
45. K. Kikuchi, Okoshi T. and Emura K., "Achievement of nearly shot noise limited operation in a heterodyne type PCM ASK optical communication system", Conf. Proc 8th ECOC'82, Cannes France, pp. 419-424 (1982).
46. F. Favre and Leguen D., "Effect of semiconductor laser phase noise on BER performance in an optical DPSK heterodyne type experiment", Electron Lett., Vol. 18, pp. 964-965 (1982).
47. S. Saito, Yamomoto Y. and Kimura T., "S/N and error-rate evaluation for an optical FSK heterodyne detection system using semiconductor lasers", IEEE J. Quantum Electron, Vol. QE-19, pp. 180-193 (1982).

48. S. Saito, Yamamoto Y. and Kimura T., "Optical FSK signal detection in a heterodyne system using semiconductor lasers", Electron Lett., Vol. 18, pp. 470-471 (1982).
49. M. Shikada, Emura K., Fujita S., Kitamura M., Arai M., Kondo M. and Minemura K., "100 Mbit/s ASK heterodyne experiment using 1.3 μm DFB laser diodes", Electron Lett., Vol. 20, pp. 164-165 (1984).
50. M. Shikada, Emura K. and Minemura K., "High sensitivity optical PSK heterodyne differential detection simulation experiment", Proc. IOOC'83, Tokyo, Japan, Paper No. 30C3-4 (1983).
51. E.J. Bachus, Bohnke F., Eutin W. and Strebel B., "Fibre optic digital transmission experiment with heterodyne detection", Electron Lett., Vol. 19, pp. 671-672 (1983).
52. H.K. Philipp, Scholtz A.L. and Leeb W.R., "Homodyning of subnanowatt 140 Mbit/s data at $\lambda = 10.6\,\mu$m", Conf. Proc. 10th ECOC'84, Stuttgart, Germany, pp. 230-231 (1984).
53. T.G. Hodgkinson, Wyatt R. and Smith D.W., "Experimental assessment of a 140 Mbit/s coherent optical receiver at 1.52 μm", Electron Lett., Vol. 18, pp. 523-525 (1982).
54. T.G. Hodgkinson, Smith D.W. and Wyatt R., "1.5 μm optical heterodyne system operating over 30 km of monomode fibre", Electron Lett., Vol. 18, pp. 929-930 (1982).
55. R. Wyatt, Hodgkinson T.G. and Smith D.W., "1.52 μm PSK heterodyne experiment featuring an external cavity diode laser local oscillator", Electron Lett., Vol. 19, pp. 550-552 (1983).
56. R. Goodfellow, Plastow R., Monham K., Carter A., Ritter J.E., Croft T.D. and Gibson M., "Practical demonstration of 1.3 Gbit/s over 107 km of dispersion shifted fibre using a 1.55 μm multimode laser", Postdeadline Papers OFC/OFS'85, San Diego, USA, pp. PD5/1-PD5/4 (1985).
57. S.D. Walker, Blank L.C. and Bickers L., "1.8 Gbit/s 65 km optical transmission experiment using 1.478 μm DFB laser and Ge APD receiver", Electron Lett., Vol. 20, pp. 717-719 (1984).
58. N.S. Bergano, Wagner R.E., Shang H. and Glodis P.F., "150 km 296 Mbit/s dispersion-shifted fibre system experiment", Tech. Dig. OFC/OFS'85, San Diego, USA, pp. 50-51 (1985).
59. V.J. Mazurczyk, Bergano N.S., Wagner R.E., Walker K.L., Olsson N.A., Cohen L.G., Logan R.A. and Campbell J.C., "420 Mbit/s transmission through 203 km using silica-core fibre and a DFB laser", Conf. Proc. 10th ECOC'84, Stuttgart, Germany, Post Deadline paper No. 7 (1984).
60. R.A. Linke, Kasper B.L., Campbell J.C., Dentai A.G. and Kaminow I.P., "120 km lightwave transmission experiment at 1 Gbit/s using a new long-wavelength avalanche photodetector", Electron Lett., Vol. 20, pp. 498-499 (1984).
61. B.L. Kasper, Linke R.A., Walker K.L., Cohen L.G., Koch T.L., Bridges T.J., Burkhardt E.G., Logan R.A., Dawson R.W. and Campbell J.C., "A 130 km transmission experiment at 2 Gbit/s using silica-core fibre and a vapour phase transported DFB laser", Conf. Proc. 10th ECOC'84, Stuttgart, Germany, Post Deadline Paper No. 6 (1984).
62. A.H. Gnauck, Kasper B.L., Linke R.A., Dawson R.W., Koch T.L., Bridges T.J., Burkhardt E.G., Yen R.T., Wilt D.P., Campbell J.C., Ciemiecki Nelson K. and Cohen L.G., "4 Gbit/s transmission over 103 km of optical fibre using a novel electronic multiplexer/demultiplexer", Post deadline Papers OFC/OFS'85, San Diego, USA, pp. PD2/1-PD2/4 (1985).
63. S.K. Korotky, Eisenstein G., Gnauck A.H., Kasper B.L., Veselka J.J., Alferness R.C., Buhl L.L., Burrus C.A., Huo T.C.D., Stulz L.W., Ciemiecki Nelson K., Cohen L.G., Dawson R.W. and Campbell J.C., "4 Gbit/s transmission experiment over 117 km of

optical fibre using a Ti:LiNbO external modulator", Post deadline Papers OFC/OFS'85, San Diego, USA, pp. PD1/1-1/4 (1985).

64. S. Yamamoto, Utaka K., Akiba S., Sakai K., Matsushima Y., Sakaguchi S. and Seki N., "280 Mbit/s single-mode fibre transmission with a DFB laser diode emitting at 1.53 μm", Electron Lett., Vol. 18, pp. 239-240 (1982).

65. K. Iwashita, Nakagawa K., Matsuoka T. and Nakahara M., "400 Mbit/s transmission test using a 1.53 μm DFB laser diode and 104 km single-mode fibre", Electron Lett., Vol. 18, pp. 937-938 (1982).

66. T. Torikai, Sugimoto Y., Taguchi K., Makita K., Ishihara H., Minemura K., Iwakami T. and Kobayashi K., "Low noise and high spped InP/InGaAsP/InGaAs avalanche photodiodes with planar structure grown by vapour phase epitaxy", Conf. Proc. 10th ECOC'84, Stuttgart, Germany, pp. 220-221 (1984).

67. J. Yamada, Machida A., Mukai T. and Kimura T., "800 Mbit/s optical transmission experiment with dispersion-free fibres at 1.5 μm", Electron Lett., Vol. 16, pp. 115-117 (1980).

68. J. Yamada, Kawana A., Nagai H. and Kimura T., "1.55 μm optical transmission experiments at 2 Gbit/s using 51.5 km dispersion-free fibre", Electron Lett., Vol. 18, pp. 98-100 (1982).

69. T.G. Hodgkinson, "Receiver analysis for synchronous coherent optical fibre transmission systems", IEEE J. Lightwave Technol., Vol. LT-5, pp. 573-586, (1987).

70. J. Salz, "Coherent Lightwave Communications", AT&T Tech. J., Vol. 64, pp. 2153-2209 (1985).

71. T. Okoshi and Kikuchi K., "Coherent optical fibre communications", ADOP Advances in Optoelectronics, Pub. KTK Scientific Publishers, (1988).

72. B.S. Glance, "Performance of homodyne detection of binary PSK optical signals", IEEE J. Lightwave Technol., Vol. LT-4, pp. 228-235 (1986).

73. B.S. Glance, "Minimum required power for carrier recovery at optical frequencies", IEEE J. Lightwave Technol., Vol. LT-4, pp. 249-255 (1986).

74. L.G. Kazovsky, "Optical heterodyning versus optical homodyning: a comparison", J. Opt. Commun., Vol. 6, pp. 18-24 (1985).

75. L.G. Kazovsky, "Coherent optical receivers: performance analysis and laser linewidth requirements", Optical Engineering, Vol. 25, pp. 575-579 (1986).

76. L.G. Kazovsky, "Impact of laser phase noise on optical heterodyne communications systems", J. Opt. Commun., Vol. 7, pp. 66-78 (1986).

77. L.G. Kazovsky and Jacobsen G., "Bit error ratio of CPFSK coherent optical receivers", Electron. Lett., Vol. 24, pp. 69-70 (1988).

78. G. Nicholson and Campbell J.C., "Performance estimates for heterodyne and homodyne optical fibre communication system", A.T.R., Vol. 19, pp. 3-13 (1985).

79. G. Nicholson, "Optical source linewidth criteria for heterodyne communication systems with PSK modulation", Optical and Quantum Electron, Vol. 17, pp. 399-410 (1985).

80. G. Nicholson, "Transmission performance of an optical FSK heterodyne system with a single filter envelope detection receiver", IEEE J. Lightwave Technol., Vol. LT-5, pp. 502-509 (1987).

81. J. Franz, "Evaluation of the probability density function and bit error rate in coherent optical transmission systems including laser phase noise and additive Gaussian noise", J. Opt. Commun., Vol. 6, pp. 51-57 (1985).

82. J. Franz, Rapp C. and Soder G., "Influence of baseband filtering on laser phase noise in coherent optical transmission systems", J. Opt. Commun., Vol. 7, pp. 15-20 (1986).

83. J. Franz, "Receiver analysis for incoherent optical ASK heterodyne systems", J. Opt. Commun., Vol. 8, pp. 57-66 (1987).

84. G. Jacobsen and Garrett I., "Theory for optical heterodyne DPSK receivers with post-detection filtering": J. Lightwave Technol., Vol. LT-5, pp. 478-484 (1987).
85. I. Garrett and Jacobsen G., "The effect of laser linewidth on coherent optical receivers with non-synchronous demodulation", J. Lightwave Technol., Vol. LT-5, pp. 551-560 (1987).
86. I. Garrett and Jacobsen G., "Theory for heterodyne optical ASK receivers using square-law detection and post-detection filtering", IEE Proc. J. Vol. 134, pp. 303-312 (1987).
87. I. Garrett and Jacobsen G., "Theory for optical heterodyne narrow deviation FSK receivers with delay demodulation", IEEE J. Lightwave Technol., Vol. 6, pp. 1415-1423 (1988).
88. S. Betti, De Marchis G. and Iannone E. and Martellucci A., "Effect of the non-Lorentzian lineshape of semiconductor laser on a PSK coherent heterodyne optical receiver", Electron. Lett., Vol. 23, pp. 1366-1367 (1987).
89. E. Patzak and Meissner P., "Influence of IF filtering on bit-error-rate floor in coherent optical DPSK systems", IEE Proc. J. Vol. 135, pp. 355-357 (1988).
90. K. Kikuchi, "Impact of 1/f type FM noise on coherent optical communications", Electron Lett., Vol. 23, pp. 885-887 (1987).
91. K. Emura, Yamazaki S., Shikada M., Fujita S., Yamaguchi M., Mito I. and Minemura K., "System design and long-span optical FSK transmission experiment on an optical FSK heterodyne single-filter detection system", IEEE J. Lightwave Technol., Vol. LT-5, pp. 469-477 (1987).
92. K. Iwashita, Imai T. and Matsumoto T., "400 Mbit/s optical FSK transmission experiment over 270 km of single-mode fibre", Electron Lett., Vol. 22, pp. 164-165 (1986).
93. K. Iwashita and Matsumoto T., "Linewidth requirement evaluation and 209 km transmission experiment for optical CPFSK differential detection", Electron Lett., Vol. 22, pp. 791-792 (1986).
94. S. Ryu, Yamamoto S., Namihira Y., Mochizuki K. and Wakabayashi H., "First sea trial of FSK heterodyne optical transmission system using polarisation diversity", Electron Lett., Vol. 24, pp. 399-400 (1988).
95. K. Emura, Yamazaki S., Fujita S., Shikada M., Mito I. and Minemura K., "Over 300 km transmission experiment on an optical FSK heterodyne dual filter detection system", Electron Lett., Vol. 22, pp. 1096-1097 (1986).
96. K. Iwashita and Takachio N., "Optical CPFSK 2 Gbit/s 202 km transmission experiment using a narrow-linewidth multielectrode DFB LD", Electron Lett., Vol. 23, pp. 1022-1023 (1987).
97. R.S. Vodhanel, Gimlett J.L., Cheung N.K. and Tsuji S., "FSK heterodyne transmission experiments at 560 Mbit/s and 1 Gbit/s", IEEE J. Lightwave Technol., Vol. LT-5, pp. 461-468 (1987).
98. K. Manome, Emura K., Yamazaki S., Fujita S., Takano S., Shikada M. and Minemura K., "A 1.2 Gbit/s CPFSK heterodyne detection transmission experiment with optimum system configuration for solitary laser diodes", Conf. Proc. ECOC'87, Vol. 1, pp. 333-336 (1987).
99. J.L. Gimlett, Vodhanel R.S, Choy M.M., Elrefaie A.F., Cheung N.K. and Wagner R.E., "A 2 Gbit/s optical FSK heterodyne transmission experiment using a DFB laser transmitter", J. Lightwave Technol., Vol. LT-5, pp. 1315-1324 (1987).
100. R.S. Vodhanel and Enning B., "1 Gbit/s bipolar optical FSK transmission experiment over 121 km of fibre", Electron Lett., Vol. 24, pp. 163-165 (1988).
101. R. Noe, Maeda M.W., Menocal S.G. and Zah C.E., "AMI signal format for pattern independent FSK heterodyne transmission and two channel cross-talk measurements", Conf. Proc. ECOC'88, Vol. 1, pp. 175-178 (1988).

102. R.A. Linke, Kasper B.L., Olsson N.A. and Alferness R.C., "Coherent lightwave transmission over 150 km fibre lengths at 400 Mbit/s and 1 Gbit/s data-rates using phase modulation", Electron Lett., Vol. 22, pp. 30-31 (1986).
103. A.H. Gnauck, Linke R.A., Kasper B.L., Pollock K.J., Reichmann K.C., Valenzuela R. and Alferness R.C., "Coherent lightwave transmission at 2 Gbit/s over 170 km of optical fibre using phase modulation", Electron Lett., Vol. 23, pp. 286-287 (1987).
104. S. Yamazaki, Murata S., Komatsu K., Koizumi Y., Fujita S., and Emura K., "1.2 Gbit/s optical DPSK heterodyne detection transmission system using monolithic external-cavity DFB LDs", Electron Lett., Vol. 23, pp. 860-862 (1987).
105. J.M.P. Delavaux, Tzeng L.D. and Dixon M., "1.4 Gbit/s optical DPSK heterodyne transmission system experiment", Electron Lett., Vol. 24, pp. 941-942 (1988).
106. R.C. Steele, Creaner M.J., Walker G.R., Walker N.G., Mellis J., Al-Chalabi S.A., Davidson J., Sturgress I., Rutherford M. and Brain M.C., "Field trail of 565 Mbits DPSK heterodyne system over 108 km", Post-Deadline Conf. Proc. ECOC'88, Vol. 2, pp. 61-64 (1988).
107. M.J. Creaner, Steele R.C., Walker G.R. and Walker N.G., "565 Mbit/s PSK transmission system with endless polarisation control", Electron Lett., Vol. 24, pp. 270-271 (1988).
108. S.B. Alexander and Welford D., "Equalisation of semiconductor diode laser frequency modulation with a passive network", Electron Lett., Vol. 21, pp. 361-362 (1985).
109. B. Enning and Vodhanel R.S., "Adaptive quantised feedback equalisation for FSK heterodyne transmission at 150 Mbit/s and 1Gbit/s", Post-Deadline Conf. Proc. OFC'88, pp. PD23/1-PD23/4 (1988).
110. R.S. Vodhanel, Lee T.P. and Tsuji S., "5 Gbit/s optical FSK modulation of a 1530 nm DFB laser", Conf. Proc. ECOC'88, Vol. 1, pp. 171-174 (1988).
111. B.L. Kasper, Burrus C.A., Talman J.R. and Hall K.L., "Balanced dual detector receiver for optical heterodyne communication at Gbit/s rates", Electron Lett., Vol. 22, pp. 413-415 (1986).
112. S.B. Alexander, "Design of wideband optical heterodyne balanced mixer receivers", IEEE J. Lightwave Technol., Vol. LT-5, pp. 523-537 (1987).
113. J.L. Gimlett, "A new low noise 16 GHz PIN/HEMT optical receiver", Post-Deadline Conf. Proc. ECOC'88, Vol. 2, pp. 13-16 (1988).
114. N. Ohkawa, "20 GHz low-noise HEMT preamplifier for optical receivers", Conf. Proc. ECOC'88. Vol. 1, pp. 404-407 (1988).
115. D.W. Smith, "Techniques for multigigabit coherent optical transmission", IEEE J. Lightwave Technol., Vol. LT-5, pp. 1466-1478 (1987).
116. J.X. Kan, Jacobsen G. and Bodtker E., "Noise performance of Gbit/s tuned optical receivers", Electron Lett., Vol. 23, pp. 434-436 (1987).
117. J.X. Kan, Garrett I. and Jacobsen G., "Transformer tuned front ends for heterodyne optical receivers", Electron Lett., Vol. 23, pp. 785-786 (1987).
118. G. Jacobsen, Kan J.X., and Garrett I.,, "Improved design of tuned optical receivers", Electron Lett., Vol. 23, pp. 787-788 (1987).
119. I. Garrett and Jacobsen G.,, "Theoretical analysis of ASK heterodyne optical receivers with tuned front ends", IEE Proc. J., Vol. 135, pp. 255-259 (1988).
120. E. Dietrich. Enning B., Gross R. and Knupke H., "Heterodyne transmission of a 560 Mbit/s optical signal by means of polarisation shift keying", Electron Lett., Vol. 23, pp. 421-422 (1987).
121. D.J. Malyon, Smith D.W. and Wyatt R., "Semiconductor laser homodyne optical phase-locked-loop", Electron Lett., Vol. 22, pp. 421-423 (1986).
122. G. Fischer, "A 700 Mbit/s PSK optical homodyne system with balanced phase-locked-loop", J. Optical Commun., Vol. 9, pp. 27-28 (1988).

123. C.J. Atkins, Cotter D., Smith D.W. and Wyatt R., "Application of Brillouin amplification in coherent optical transmission", Electron Lett., Vol. 22, pp. 556-558 (1987).
124. L. Giehmann and Rocks M., "Measurement of polarisation fluctuations in installed single-mode fibre cables", Optical and Quantum Electron, Vol. 19, pp. 109-113 (1987).
125. Y. Namihira, Ryu S., Mochizuki K., Furusawa K. and Iwamoto Y., "Polarisation fluctuation in optical-fibre submarine cable under 8000m deep sea environmental conditions", Electron Lett., Vol. 23, pp. 100-101 (1987).
126. C.D. Poole, Bergano N.S., Schulte H.J., Wagner R.E., Nathu V.P., Amon J.M. and Rosenberg, "Polarisation fluctuations in a 147 km undersea lightwave cable during installation", Electron Lett., Vol. 23, pp. 1113-1115 (1987).
127. C.D. Poole and Wagner R.E., "Phenomenological approach to polarisation dispersion in long single-mode fibres", Electron Lett., Vol. 22, pp. 1029-1030 (1986).
128. M. Tsubokawa and Sasaki Y., "Limitation of transmission distance and capacity due to polarisation dispersion in a lightwave system", Electron Lett., Vol. 24, pp. 350-352 (1988).
129. C.D. Poole, Bergano N.S., Wagner R.E. and Schulte H.J., "Polarisation dispersion and principle states in a 147 km undersea lightwave cable", IEEE J. Lightwave Technol., Vol. 6, pp. 1185-1190 (1988).
130. L.J. Cimini, Habbab I.M.I., John R.K. and Saleh A.A.M., "Preservation of polarisation orthogonality through a linear optical system", Electron Lett., Vol. 23, pp. 1365-1366 (1987).
131. A.F. Elrefaie, Wagner R.E., Atlas D.A. and Daut D.G., "Chromatic dispersion limitations in coherent optical fibre transmission systems", Electron Lett., Vol. 23, pp. 756-758 (1987).
132. K. Iwashita and Takachio N., "Compensation of 202 km single-mode fibre chromatic dispersion in 4 Gbit/s optical CPFSK transmission experiment", Electron Lett., Vol. 24, pp. 759-760 (1988).
133. R.A. Harmon, Walker G.R. and White T.K., "Polarisation control in a coherent optical fibre system using a dither technique", Proc. SPIE Fibre Optics '87: Fifth International Conference on Fibre Optics and Opto-electronics, No. 734, pp. 63-67 (1987).
134. N.G. Walker and Walker G.R., "Endless polarisation control using four fibre squeezers", Electron Lett., Vol. 23, pp. 290-292 (1987).
135. N.G. Walker and Walker G.R., "Polarisation control for coherent optical fibre systems", Br. Telecom. Technol. J., Vol. 5, pp. 63-76 (1987).
136. R. Noe, "Endless polarisation control experiment with three elements of limited birefringence range", Electron Lett., Vol. 22, pp. 1341-1343 (1986).
137. C.J. Mahon and Khoe G.D., "Endless polarisation state matching using two controllers of finite control range", Electron Lett., Vol. 23, pp. 1234-1235 (1987).
138. T. Okoshi, Cheng Y.H. and Kikuchi K., "New Polarisation control scheme for optical heterodyne receiver using two Faraday rotators", Electron Lett., Vol. 21, pp. 787-788 (1985).
139. H. Honmou, Yamazaki S., Emura K., Ishikawa R., Mito I., Shikada M. and Minemura K., "Stabilisation of heterodyne receiver sensitivity with automatic polarisation control system", Electron Lett., Vol. 22, pp. 1181-1182 (1986).
140. M. Creaner, Steele R., Walker G.R., Walker N.G, Mellis J., Al-Chalabi S., Davidson J. and Brain M.C., "565 Mbit/s DPSK heterodyne system transmission experiment using packaged external cavity lasers and a lithium niobate polarisation controller", Conf. Proc. ECOC'88, Vol. 1, pp. 179-182 (1988).
141. L.G. Kazovsky, "Recent progress in phase and polarisation diversity coherent optical techniques", Invited Papers Conf. Proc. ECOC'87, pp. 83-87 (1987).

142. T.G. Hodgkinson, Harmon R.A. and Smith D.W., "Performance comparison of ASK polarisation diversity and standard coherent optical heterodyne receivers", Electron Lett., Vol. 24, pp. 58-59 (1988).
143. L.D. Tzeng, Emkey W.L. and Jack C.A., "Polarisation insensitive coherent receiver using a double balanced optical hybrid system", Electron Lett., Vol. 23, pp. 1195-1196 (1987).
144. T. Imai, "Polarisation diversity receiver using a simple weight controller for coherent FSK communications", Electron Lett., Vol. 24, pp. 979-980 (1988).
145. T.E. Darcie, Glance B., Gayliard K., Talman J.R., Kasper B.L. and Burrus C.A., "Polarisation diversity receiver for coherent FSK communications", Electron Lett., Vol. 23, pp. 1369-1370 (1987).
146. S. Ryu, Yamamoto S. and Mochizuki K., "Polarisation insensitive operation of coherent FSK transmission system using polarisation diversity", Electron Lett., Vol. 23, pp. 1382-1384 (1987).
147. B. Glance, "Polarisation independent coherent optical receiver", IEEE J. Lightwave Technol., Vol. LT-5, pp. 274-276 (1987).
148. S. Watanabe, Naito T., Chikama T., Kiyonaga T., Onoda Y. and Kuwahara H., "Polarisation-insensitive 1.2 Gbit/s optical DPSK heterodyne transmission experiment using polarisation diversity", Conf. Proc. ECOC'88, Vol. 1, pp. 90-93 (1988).
149. M. Kavehrad and Glance B.S., "Polarisation insensitive frequency shift keying optical heterodyne receiver using discriminator demodulation", IEEE J. Lightwave Technol., Vol. 6, pp. 1386-1394 (1988).
150. T.G. Hodgkinson, Harmon R.A. and Smith D.W., "Polarisation-insensitive heterodyne detection using polarisation scrambling", Electron Lett., Vol. 23, pp. 513-514 (1987).
151. L.J. Cimini, Habbab I.M.I., Yang S., Rustako A.J., Liou K.Y. and Burrus C.A., "Polarisation-insensitive coherent lightwave system using wide deviation FSK and data-induced polarisation switching", Electron Lett., Vol. 24, pp. 358-360 (1988).
152. A.D. Kersey, Yurek A.M., Dandridge A. and Weller J.F., "New polaristion-insensitive detection technique for coherent optical fibre heterodyne communications", Electron Lett., Vol. 23, pp. 924-925 (1987).
153. T.G. Hodgkinson and Cook A.R.J., "Polarisation insensitive coherent detection using orthogonally polarised optical fields", Microwave and Optical Technol. Lett., Vol. 1, pp. 246-249 (1988).
154. A.W. Davis, Pettitt M.J., King J.P. and Wright S., "Phase diversity techniques for coherent optical receivers", IEEE J. Lightwave Technol. Vol. LT-5, pp. 561-572 (1987).
155. M.J. Pettitt, Remedios D., Davis A.W., Hadjifotiou A. and Wright S., "Optical FSK transmission system using a phase diversity receiver", Electron Lett., Vol. 23, pp. 1075-1076 (1987).
156. K. Emura, Vodhanel R.S, Welter R. and Sessa W., "4 to 5 Gbit/s phase diversity homodyne detection experiment", Post-Deadline Conf. Proc. ECOC'88, pp. 57-60 (1988).
157. R. Schneider and Pietzsch J., "Coherent 565 Mbit/s DPSK transmission experiment with a phase diversity receiver", Post-Deadline Papers Conf. Proc. ECOC'87, pp. 5-8 (1987).
158. T.G. Hodgkinson, Harmon R.A. and Smith D.W., "Demodulation of DPSK using in-phase and quadrature detection", Electron Lett., Vol. 21, pp. 867-868 (1985).
159. T.G. Hodgkinson, Harmon R.A., Smith D.W. and Chidgey P.J., "In-phase and quadrature detection using $90°$ optical hybrid receiver: Experiments and design considerations", IEE Proc. J., Vol. 135, pp. 260-267 (1988).

160. T. Okoshi and Cheng Y.H., "Four-port homodyne receiver for optical fibre communications comprising phase and polarisation diversities", Electron Lett., Vol. 23, pp. 377-378 (1987).
161. R. Noe, Sessa W.B., Welter R. and Kazovsky L.G., "New FSK phase diversity receiver in a 150 Mbit/s coherent optical transmission system", Electron Lett., Vol. 24, pp. 567-568 (1988).
162. A.R.L. Travis, Carroll J.E., Epworth R.E. and Bricheno T., "Passive quadrature detection using speckle rotation on a multisegment photodetector", Conf. Proc. OFC/IOOC'87, Paper WF2 (1987).
163. L.G. Kazovsky, Meissner P. and Patzak E., "ASK multiport optical homodyne receivers", IEEE J. Lightwave Technol., Vol. LT-5, pp. 770-791 (1987).
164. L.G. Kazovsky, Elrefaie A.F., Meissner P., Welter R., Crespo P., Gimlett J. and Smith R.W., "Impact of laser intensity noise on ASK two-port optical homodyne detection", Electron Lett., Vol. 23, pp. 871-873 (1987).
165. A.F. Elrefaie, Atlas D.A., Kazovsky L.G. and Wagner R.E., "Intensity noise in ASK coherent lightwave receivers", Electron Lett., Vol. 24, pp. 158-159 (1988).
166. G. Nicholson, "ASK homodyne system receiver using a 6 port fibre coupler", J. Opt. Commun., Vol. 9, pp. 13-16 (1988).
167. I.M.I. Habbab, Kahn J.M. and Greenstein L.J., "Phase insensitive zero IF coherent optical systems using phase switching", Electron Lett., Vol. 24, pp. 974-976 (1988).
168. I.M.I. Habbab and Greenstein L.J., "Phase insensitive zero IF coherent optical detection using sinusoidal phase modulation instead of phase switching", Post-Deadline Conf. Proc. ECOC'88, pp. 65-68 (1988).
169. D. Hoffman, Heidrich, Wenke G. and Langenhorst, "Integrated optical 90° hybrid on $LiNbO_3$ for phase diversity receivers", Post-Deadline Conf. Proc. ECOC'88, pp. 33-36 (1988).
170. R. Wyatt. Cameron K.H. and Matthews M.R., "Tunable narrow line external cavity lasers for coherent optical systems", British Telecom Technol. J., Vol. 3, pp. 5-12 (1985).
171. J. Mellis, Al-Chalabi S., Cameron K.H., Wyatt R., Regnault J.C., Devlin W.J. and Brain M.C., "Miniature packaged external-cavity semiconductor lasers for coherent communications", Conf. Proc. ECOC'88, Vol. 1, pp. 219-222 (1988).
172. I.W. Stanley, Hill G.R. and Smith D.W., "The application of coherent optical techniques to wide-band networks", IEEE J. Lightwave Technol., Vol. LT-5, pp. 439-451 (1987).
173. C. Baack, Bachus E.J. and Heydt G., "Coherent multicarrier techniques in future broadband communication networks", Conf. Proc. ECOC'87, Vol. 2, pp. 79-87 (1987).
174. D.J. Hunkin, Hill G.R. and Stallard W.A., "Frequency-locking of external cavity semiconductor laser using an optical comb generator", Electron Lett., Vol. 22, pp. 388-390 (1986).
175. G.R. Hill, Smith D.W., Lobbet R.A., Hodgkinson T.G. and Webb R.P., "Evolutionary wavelength division multiplexed schemes for broadband networks", Conf. Proc. OFC/IOOC'87, Paper MI2 (1987).
176. G.R. Hill and Stanley I.W., "Application of coherent optical techniques to broadband networks", Optical Engineering, Vol. 26, pp. 349-353 (1987).
177. E.J. Bachus, Braun R.P., Eutin W., Grossman E., Foiselh, Heimes K. and Strebel B., "Coherent optical fibre subscriber line", Electron Lett., Vol. 21, pp. 1203-1205 (1985).
178. D.A. Humphreys and Lobbet R.A., "Investigation of an optoelectronic non-linear effect in a GaInAs photodiode, and its application in a coherent optical communication system", IEE Proc., Vol. 135, pp. 45-51 (1988).

179. M. Ohtsu, "Frequency stabilisation in semiconductor lasers", Optical and Quantum Electron., Vol. 20, pp. 283-300 (1988).
180. B. Glance, Fitzgerald P.J, Pollack K.J., Stone J., Burrus C.A., Eisenstein G. and Stulz L.W., "Frequency stabilisation of FDM optical signals", Electron Lett., Vol. 23, pp. 750-752 (1987).
181. H. Toba, Inoue K., Nosu K. and Motosugi G., "A Multichannel laser diode frequency stabiliser for narrowly spaced optical frequency division multiplexing transmission", J. Optical Commun., Vol. 9, pp. 50-54 (1988).
182. H. Shimosaka, Kaede K. and Murata S., "Frequency locking of FDM optical sources using widely tunable DBR LD's", Conf. Proc. OFC/IOOC'88, Paper THG3 (1988).
183. P. Gambini, Puleo M. and Vezzoni E., "Laser frequency stabilisation for multichannel coherent systems", Conf. Proc. ECOC'88, Vol. 1, pp. 78-81 (1988).
184. B. Villeneuve, Cyr N. and Tetu M., "Use of laser diodes locked to atomic transitions in multiwavelength coherent communications", Electron Lett., Vol. 24, pp. 734-737 (1988).
185. Y.C. Chung and Roxlo C.B., "Frequency locking of a 1.5 μm DFB laser to an atomic krypton line using optogalvanic effect", Electron Lett., Vol. 24, pp. 1048-1049 (1988).
186. L.G. Kazovsky, "Multichannel coherent optical communications systems", IEEE J. Lightwave Technol. Vol. LT-5, pp. 1095-1102 (1987).
187. Y.K. Park, Bergstein S.S, Tench R.E., Smith R.W., Korotky S.K., Burns K.J. and Granlund S.W., "Crosstalk and prefiltering in a two channel ASK heterodyne detection system without the effect of laser phase noise", IEEE J. Lightwave Technol., Vol. 6, pp. 1312-1320 (1988).
188. L.G. Kazovsky and Gimlett J.L., "Sensitivity penalty in multichannel coherent optical communications", IEEE J. Lightwave Technol., Vol. 6, pp. 1353-1365 (1988).
189. B.S. Glance, "An optical heterodyne mixer providing image frequency rejection", IEEE J. Lightwave Technol., Vol. LT-4, pp. 1722-1725 (1986).
190. T.E. Darcie and Glance B.S., "Optical heterodyne image rejection mixer", Electron Lett., Vol. 22, pp. 825-826 (1986).
191. T.G. Hodgkinson and Webb R.P., "Application of communications theory to analyse carrier density modulation effects in travelling wave semiconductor laser amplifiers", Electron Lett, Vol. 24, pp. 1550-1552 (1988).
192. G.T. Forrest, "Commerical telecom diode lasers push below megahertz linewidths", Laser Focus (magazine), October, pp. 112-114 (1988).

21

Integrated Optic Devices

K. Thyagarajan[*] and Ajoy Ghatak[*]

---

1. INTRODUCTION

In Ch. 3 a discussion of the waveguidance properties of planar and strip waveguides was given. In the present chapter we discuss some important integrated optical devices based mainly on the electrooptic effect. Most of the discussion will be based on waveguides fabricated on $LiNbO_3$ although similar principles could be easily extended to other substrate materials.

We briefly discuss the electrooptic effect in Sec. 2 and a phase modulator based on the electrooptic effect in Sec 3. Section 4 deals with polarization modulators and wavelength filters based on periodic coupling. Section 5 discusses the Mach-Zehnder interferometric modulator and some of its application to the building of logic operations.

The optical directional coupler and its application to optical switching, wavelength filtering and polarization filtering is discussed in Sec. 6. Finally Sec. 7 discusses various integrated optic polarizer structures.

2. THE ELECTROOPTIC EFFECT

The electrooptic effect refers to the change in refractive index of a material due to an applied electric field. If the change is proportional to the applied field then it is referred to as linear electrooptic effect or Pockels effect. Pockels effect is one of the most widely used effects in integrated optical devices. In this section we give a brief description of Pockels effect. For more details readers may look up references [1] and [2].

---

[*] Department of Physics, Indian Institute of Technology Delhi, New Delhi-110 016, India.

In the principal axis system (x_1, x_2, x_3), in the absence of an external electric field, the index ellipsoid of a crystal is described by [1]

$$\frac{x_1^2}{n_{11}^2} + \frac{x_2^2}{n_{22}^2} + \frac{x_3^2}{n_{33}^2} = 1, \tag{1}$$

where n_{11}, n_{22} and n_{33} are the principal indices of refraction. For isotropic media $n_{11} = n_{22} = n_{33}$; for uniaxial media (LiNbO$_3$, LiTaO$_3$) $n_{11} = n_{22} \neq n_{33}$ and for biaxial media $n_{11} \neq n_{22} \neq n_{33}$. On application of an electric field the index ellipsoid becomes

$$\frac{x_1^2}{n_{11}'^2} + \frac{x_2^2}{n_{22}'^2} + \frac{x_3^2}{n_{33}'^2} + \frac{2x_2 x_3}{n_{23}'^2} + \frac{2x_1 x_3}{n_{13}'^2} + \frac{2x_1 x_2}{n_{12}'^2} = 1, \tag{2}$$

where the new coefficients are given by

$$\frac{1}{n_{ij}'^2} = \frac{1}{n_{ij}^2} + \Delta\left(\frac{1}{n^2}\right)_{ij} \quad ; \quad i,j = 1, 2, 3, \tag{3}$$

$$\frac{1}{n_{ij}^2} = 0 \quad \text{for} \quad i \neq j, \tag{4}$$

and

$$\Delta\left(\frac{1}{n^2}\right)_{ij} = \sum_{k=1}^{3} r_{ijk} E_k \quad ; \quad i,j = 1, 2, 3. \tag{5}$$

Here E_k represents the component of the applied electric field and r_{ijk} are the components of the electro optic tensor. The electro optic tensor $[r_{ijk}]$ is usually written in a contracted notation of the first two subscripts i.e. as $[r_{ik}]$, $i = 1, 2, ...6$ with the following convention

11 → 1; 22 → 2; 33 → 3; 23, 32 → 4; 13, 31 → 5; 12, 21 → 6.

Comparison of Eqs. (1) and (2) shows that the applied electric field in general may lead to the rotation of the principal axis directions as well as to changes in the principal refractive indices. The orientation of the new axes and their corresponding lengths can be calculated by a rotation of axes such that Eq. (2) does not contain any cross product term.

We consider an example.

Electrooptic effect in Lithium Niobate

Lithium Niobate is a uniaxial medium with $n_{11} = n_{22} = n_0$ and $n_{33} = n_e$ and $n_0 \simeq 2.29$, $n_e \simeq 2.20$ at $\lambda_0 = 0.6\,\mu$m. Thus in the absence of any applied field, the index ellipsoid is

$$\frac{x_1^2 + x_2^2}{n_0^2} + \frac{x_3^2}{n_e^2} = 1. \tag{6}$$

The corresponding electrooptic tensor is described by [1]

$$[r] = \begin{pmatrix} 0 & -r_{22} & r_{13} \\ 0 & r_{22} & r_{13} \\ 0 & 0 & r_{33} \\ 0 & r_{51} & 0 \\ r_{51} & 0 & 0 \\ -r_{22} & 0 & 0 \end{pmatrix} \quad (7)$$

with

$$r_{222} = r_{22} \simeq 3.4 \times 10^{-12} \text{ m/V}; \; r_{113} = r_{13} \simeq 8.6 \times 10^{-12} \text{ m/V};$$
$$r_{333} = r_{33} \simeq 30.8 \times 10^{-12} \text{ m/V}; \; r_{131} = r_{51} \simeq 28 \times 10^{-12} \text{ m/V}. \quad (8)$$

We first consider an applied field along the x_3 direction which is by convention chosen as the direction of the optic axis. Thus we write

$$\vec{E} = \begin{pmatrix} 0 \\ 0 \\ E_3 \end{pmatrix} \quad (9)$$

and the changes in $(1/n^2)$ are described by

$$\Delta\left(\frac{1}{n^2}\right) = \begin{pmatrix} r_{13}E_3 & 0 & 0 \\ 0 & r_{13}E_3 & 0 \\ 0 & 0 & r_{33}E_3 \end{pmatrix} \quad (10)$$

Thus the index ellipsoid in the presence of the electric field along the optic axis is

$$x_1^2\left(\frac{1}{n_0^2} + r_{13}E_3\right) + x_2^2\left(\frac{1}{n_0^2} + r_{13}E_3\right) + x_3^2\left(\frac{1}{n_e^2} + r_{33}E_3\right) = 1. \quad (11)$$

In the above equation, there are no cross terms implying that the applied field does not change the orientation of the principal axes. Since the coefficient of the terms have changed, this implies that the principal refractive indices have changed. Assuming $r_{13}E_3, r_{33}E_3 \ll n_0, n_e$, we may write for the new principal refractive indices

$$n_1' = n_2' \simeq n_0 - \frac{n_0^3 r_{13} E_3}{2}, \quad (12)$$

$$n_3' \simeq n_e - \frac{n_e^3 r_{33} E_3}{2}. \quad (13)$$

Thus by applying an electric field, the refractive indices can be changed.

To get an estimate of the change in index, we assume that a voltage of 1 kV is applied across a 1 cm thick crystal of LiNbO$_3$. Thus the electric field generated is

$$E_3 \simeq \frac{V}{d} = 10^5 \text{ V/m}.$$

For LiNbO$_3$, $n_e = 2.2$, $r_{33} = 30 \times 10^{-12}$ m/V and the change in refractive index is

$$\Delta n_3 = |n_3 - n_e| \simeq \frac{n_e^3 r_{33} E_3}{2} \simeq 10^{-5}.$$

Although the change in index is very small, the accumulated phase change on a propagating light beam could be significant. For example, for a path length of $l = 1$ cm and $\lambda_0 = 1\,\mu$m, the corresponding phase change is

$$\Delta\phi = \frac{2\pi}{\lambda_0} \Delta n \cdot l \simeq 0.2\pi,$$

which is indeed a large phase change.

Electric field along an arbitrary direction

If the electric field has all the three components, then the corresponding change $\Delta\left(\frac{1}{n^2}\right)$ is given by

$$\Delta\left(\frac{1}{n^2}\right) = \begin{pmatrix} 0 & -r_{22} & r_{13} \\ 0 & r_{22} & r_{13} \\ 0 & 0 & r_{33} \\ 0 & r_{51} & 0 \\ r_{51} & 0 & 0 \\ -r_{22} & 0 & 0 \end{pmatrix} \begin{pmatrix} E_1 \\ E_2 \\ E_3 \end{pmatrix}$$

$$= \begin{pmatrix} -r_{22} E_2 + r_{13} E_3 \\ r_{22} E_2 + r_{13} E_3 \\ r_{33} E_{33} \\ r_{51} E_2 \\ r_{51} E_1 \\ -r_{22} E_1 \end{pmatrix} \quad (14)$$

In the decontracted notation we have

$$\Delta\left(\frac{1}{n^2}\right) = \begin{pmatrix} -r_{22} E_2 + r_{13} E_3 & -r_{22} E_1 & r_{51} E_1 \\ -r_{22} E_1 & r_{22} E_2 + r_{13} E_3 & r_{51} E_2 \\ r_{51} E_1 & r_{51} E_2 & r_{33} E_3 \end{pmatrix} \quad (15)$$

Notice that unlike the case when the field was applied along the c-axis, here $\Delta\left(\frac{1}{n^2}\right)$ has off diagonal terms also which from Eq. (2) can be seen to lead to an equation of the index ellipsoid possessing cross product terms. This implies a rotation of the eigen-axes or equivalently to a coupling among old eigen-axes directions. This coupling can be used in the fabrication of TE ↔ TM polarization converters, wavelength filters etc. (see Sec. 4). We can also obtain the change in dielectric constant tensor elements [1]:

$$\Delta \varepsilon_{ij} = - \frac{1}{\varepsilon_0} \varepsilon_{ik} \Delta \left(\frac{1}{n^2}\right)_{kl} \varepsilon_{lj}, \qquad (16)$$

where ε_{ik} are the elements of the dielectric constant tensor of $LiNbO_3$. Substituting

$$\overline{\overline{\varepsilon}} = \varepsilon_0 \begin{pmatrix} n_0^2 & 0 & 0 \\ 0 & n_0^2 & 0 \\ 0 & 0 & n_e^2 \end{pmatrix} \qquad (17)$$

and Eq. (15) in Eq. (16) we have

$$\overline{\overline{\Delta \varepsilon}} = -\varepsilon_0 \begin{pmatrix} n_0^4(-r_{22}E_2 + r_{13}E_3) & -n_0^4 r_{22} E_1 & n_0^2 n_e^2 r_{51} E_1 \\ -n_0^4 r_{22} E_1 & n_0^4(r_{22}E_2 + r_{13}E_3) & n_0^2 n_e^2 r_{51} E_2 \\ n_0^2 n_e^2 r_{51} E_1 & n_0^2 n_e^2 r_{51} E_2 & n_e^4 r_{33} E_3 \end{pmatrix}, \qquad (18)$$

which describes the change in the dielectric permitivity tensor in the presence of an applied electric field.

3. PHASE MODULATOR

We first discuss a simple phase modulator based on $LiNbO_3$. Figure 1 shows two configurations for applying an electric field on a strip waveguide with the help of coplanar electrodes. In Fig. 1 (a), the electrodes are placed on either side of the waveguide and the electric field lines are predominantly horizontal in the region of the waveguide. In Fig. 1 (b), one electrode is placed on the waveguide so that the electric field in the waveguide is predominantly vertical. We have seen in Sec. 2 that in $LiNbO_3$, the largest electrooptic coefficient is r_{33}. In order to have minimum drive voltages, devices are so designed as to use the strongest electrooptic coefficient namely r_{33}. To make use of this coefficient, the mode should be polarized along the c-axis of the crystal and the electric field should also be applied along the c-axis. Hence for the electrode configuration of Fig. 1 (a), the $LiNbO_3$ crystal should be either x cut y-propagating or y cut x-propagating and the mode should be quasi TE polarization as shown. Corresponding to Fig. 1 (b), the crystal should be z cut and the mode should be quasi TM polarization. For the latter case, since the metal electrodes are placed directly on the waveguide and quasi TM polarization is used, there could be substantial losses in the propagating mode (see Sec. 6 of Chapter 3). Hence for this configuration it is usual to put a dielectric buffer layer (typically SiO_2) between the waveguide and the electrode to reduce loading due to metal; this would of course reduce the effective electric field E since the mode is farther away from the electrode which results in a subsequent increase in the voltage requirements; this increase is typically in the range of 20%.

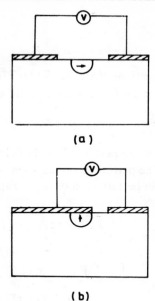

Fig. 1. Two different configurations for applying an electric field on a strip waveguide fabricated in LiNbO$_3$.

In Sec. 2, we had seen that the change in refractive index n_3 due to the applied electric field E_3 was given by

$$\Delta n_3 = - \frac{n_e^3 r_{33} E_3}{2}. \tag{19}$$

The above equation is valid when the applied electric field is uniform over the cross-section of the propagating light wave. In the case of guided waves, since the mode field has a transverse variation and the applied electric field is, in general, inhomogeneous, the change in effective index would be different [3]. As an example, we consider the change in the effective index of the extraordinary mode due to an applied field along the c-axis. In order to obtain an approximate expression for the change in effective index; we first note that if ψ_0 represents the modal field profile in the absence of the applied electric field and β_0^2 the corresponding propagation constant, then ψ_0 satisfies the following equation:

$$\frac{\partial^2 \psi_0}{\partial x^2} + \frac{\partial^2 \psi_0}{\partial y^2} + \left[k_0^2 n_e^2(x, y) - \beta_0^2 \right] \psi_0 = 0, \tag{20}$$

where $n_e^2(x, y)$ is the transverse refractive index profile corresponding to the extraordinary mode. From Eq. (11) we note that the change in $1/n_e^2$ is given by

$$\Delta \left(\frac{1}{n_e^2} \right) = r_{33} E_3.$$

Thus
$$\Delta n_e^2 \simeq -n_e^4 r_{33} E_3 . \tag{21}$$

Since n_e and E_3 are functions of x and y, Δn_e^2 is a function of x and y. Hence in the presence of the applied field, the mode satisfies the following equation:

$$\frac{\partial^2 \psi}{\partial x^2} + \frac{\partial^2 \psi}{\partial y^2} + \left(k_0^2 n_e^2 + k_0^2 \Delta n_e^2 - \beta^2\right) \psi = 0, \tag{22}$$

where ψ and β represent the perturbed mode field profile and propagation constant respectively. The perturbed propagation constant β can be obtained by using simple first order perturbation theory in Eq. (22). Using the standard result we obtain

$$\begin{aligned}\beta^2 &= \beta_0^2 + k_0^2 \frac{\iint \psi_0^*(x,y) \Delta n_e^2 \psi_0(x,y) \, dx \, dy}{\iint \psi_0^*(x,y) \psi_0(x,y) \, dx \, dy} \\ &= \beta_0^2 - k_0^2 r_{33} \frac{\iint n_e^4 \psi_0^*(x,y) E_3(x,y) \psi_0(x,y) \, dx \, dy}{\iint \psi_0^*(x,y) \psi_0(x,y) \, dx \, dy} .\end{aligned} \tag{23}$$

If n_{eff} and $n_{\text{eff}} + \Delta n_{\text{eff}}$ are the mode effective indices in the absence of the applied field and in the presence of the field, then

$$\beta_0 = k_0 n_{\text{eff}} , \quad \beta = k_0 (n_{\text{eff}} + \Delta n_{\text{eff}}) .$$

Thus
$$\begin{aligned}\beta^2 - \beta_0^2 &= (\beta - \beta_0)(\beta + \beta_0) \\ &\simeq 2 k_0 n_{\text{eff}} \cdot k_0 \Delta n_{\text{eff}} \\ &\simeq 2 k_0^2 n_e \Delta n_{\text{eff}},\end{aligned} \tag{24}$$

where we have assumed $\Delta n_{\text{eff}} \ll n_{\text{eff}}$ and for practical waveguides which are weakly guiding $n_{\text{eff}} \simeq n_e$, the extraordinary index of the substrate. Thus from Eq. (23) we obtain

$$2 n_e \Delta n_{\text{eff}} \simeq -r_{33} \frac{\iint n_e^4(x,y) \psi_0^*(x,y) E_3(x,y) \psi_0(x,y) \, dx \, dy}{\iint \psi_0^*(x,y) \psi_0(x,y) \, dx \, dy} . \tag{25}$$

Again for weakly guiding waveguides, we may neglect the transverse dependence of $n_e(x,y)$ in the integral in Eq. (25) and assume $n_e(x,y) = n_e$, the substrate extraordinary index. Thus we obtain

$$\begin{aligned}\Delta n_{\text{eff}} &\simeq \frac{-n_e^3 r_{33}}{2} \frac{\iint \psi_0^*(x,y) E_3(x,y) \psi_0(x,y) \, dx \, dy}{\iint \psi_0^*(x,y) \psi_0(x,y) \, dx \, dy} \\ &= \frac{-n_e^3 r_{33} \overline{E}_3}{2},\end{aligned} \tag{26}$$

where $\overline{E}_3$ is an equivalent electric field defined in terms of the overlap with the mode field distribution as

$$\overline{E}_3 = \frac{\int\int \psi_0^*(x,y) E_3(x,y) \psi_0(x,y)\, dx\, dy}{\int\int \psi_0^*(x,y) \psi_0(x,y)\, dx\, dy}. \tag{27}$$

If $E_3(x, y)$ is uniform, i.e., if $E_3(x, y) = E_3$ independent of x and y, then $\overline{E}_3 = E_3$. Hence to minimise the voltage required for a given phase change, it is necessary to maximize the overlap. The overlap integral can be maximized by using well confined modes. This can be achieved by controlling the fabrication parameters such as width of titanium strip and time and temperature of diffusion.

Another important consideration in the phase modulator design is the insertion loss. Typically, the devices are pigtailed with single-mode fibres and the coupling loss between the fibre and the strip waveguide would depend on the overlap between the fibre mode and the strip waveguide mode. Since fibre modes are usually large (mode field diameter $\simeq 10\,\mu$m at $\lambda_0 = 1.3\,\mu$m), this requires that the strip waveguide mode should also have a transverse dimension $\simeq 10\,\mu$m. This requirement is just the opposite of the requirement on drive voltages. Thus a compromise has to be made between minimization of insertion loss and drive voltages. As an example, Fig. 2 shows the variation of insertion loss and the voltage V_π required to change the phase by π in a waveguide of length 1 mm as a function of the width of Ti strip before diffusion in z-cut $LiNbO_3$. The other parameters correspond to

Ti thickness = 700 Å,

Time of diffusion = 6 hr,

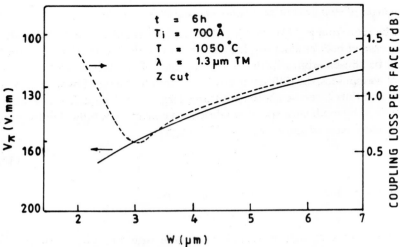

Fig. 2. Variation of insertion loss and the voltages V_π required to change the phase by π in a waveguide of length 1 mm fabricated in z-cut $LiNbO_3$ with the width of Ti strip before diffusion (Reprinted from M. Papuchon, Integrated optical modulation and switching Proc. SPIE, 1985 with permission).

Temperature of diffusion = 1050°C,

Electrode spacing = $6\,\mu$m.

Equation (26) can also be written in a slightly different form. If V is the voltage applied across the electrodes and d is the electrode spacing, then we may write Eq. (26) as

$$\Delta n_{eff} = -\frac{n_e^3 r_{33} V}{2d} \cdot \Gamma, \tag{28}$$

where

$$\Gamma = \bar{E}_3 \frac{d}{V} \tag{29}$$

represents the factor by which the electrooptic change is reduced due to a non-uniform electric field. Now, for a waveguide of length L the phase change is

$$\Delta\phi = \frac{2\pi}{\lambda_0}\Delta n_{eff} \cdot L = \frac{\pi n_e^3 r_{33} V L}{\lambda_0 d}\Gamma. \tag{30}$$

Thus to produce a phase change of π the required voltage-length product is given by

$$V \cdot L = \frac{\lambda_0 d}{n_e^3 r_{33} \Gamma}. \tag{31}$$

Using $\lambda_0 = 1.3\,\mu$m, $d = 6\,\mu$m, $n_e = 2.2$, $r_{33} = 30 \times 10^{-12}$ m/V, $L = 1$ mm, we obtain $V = 25$ V for $\Gamma = 1$. Comparing with Fig. 2 we see that Γ lies typically in the range 0.25 to 0.15.

Advantages of strip waveguide modulators over bulk modulators

It can be seen from Eq. (31) that low drive voltages require a small (d/L) ratio. In the case of bulk modulators, for a given d value the length of interaction L cannot be chosen arbitrarily due to the presence of diffraction effects. In the case of waveguides, the diffraction effects are counterbalanced by waveguiding effects and thus L can be chosen arbitrarily large. In this case, L is essentially determined by modulator speed or fabrication convenience only. For the case of bulk modulators d and L are related through [1]

$$\frac{d^2}{L} \geq \frac{4\lambda_0}{\pi n}. \tag{32}$$

Thus for $d \simeq 6\,\mu$m, $\lambda_0 = 1.3\,\mu$m, $n = 2.2$ we have

$$L \leq 50\,\mu\text{m}.$$

Hence a 1 mm long strip waveguide modulator would require 20 times less voltage for the same phase shift (assuming $\Gamma = 1$).

Another advantage is due to the very small separation between the electrodes. Thus even moderate voltages can create extremely high electric fields. The power per unit bandwidth (which is an important figure of merit of a modulator-see e.g., Ref. [1]) is proportional to d^2/L and thus reductions by a factor of about 1000 from those of bulk devices is feasible in integrated optic form.

At the same time, it must be kept in mind that since the cross-sectional dimensions of the mode are small, there are limitations on the input optical power due to optical damage or onset of non-linear processes etc.

4. POLARIZATION MODULATORS AND WAVELENGTH FILTER

As mentioned in Sec. 2, the electrooptic coefficient r_{51} can be used to couple light between quasi TE and quasi TM polarization. Such a device is shown in Fig. 3 [5]. Interdigital electrodes are fabricated on a strip waveguide formed in an x-cut-y propagating $LiNbO_3$ substrate. Now, the quasi TE polarization (polarized parallel to z) is the extraordinary wave with an effective index $\simeq n_e$ and the quasi TM polarization (polarized parallel to x) is the ordinary polarization with an effective index $\simeq n_0$. For efficient coupling between two modes, two conditions must be fulfilled:

1. There should exist a finite coupling coefficient between the two modes and
2. The two modes should be phase-matched i.e., should have the same propagation constant.

In case they do not have identical propagation constants, spatially periodic coupling can be used to couple the two modes, provided the spatial frequency of the periodic coupling is equal to the difference in propagation constants [2] (see Ch. 3).

In the present case since we wish to couple TE and TM polarizations, the finite coupling coefficient is provided by the proper orientation shown in Fig. 3. Thus the electric field has predominantly only E_x and E_y components and if we neglect contribution from r_{22} as compared to that from r_{51} which is much larger, Eq. (18) becomes

$$\Delta \varepsilon \simeq - \varepsilon_0 \begin{pmatrix} 0 & 0 & n_0^2 n_e^2 r_{51} E_x \\ 0 & 0 & n_0^2 n_e^2 r_{51} E_y \\ n_0^2 n_e^2 r_{51} E_x & n_0^2 n_e^2 r_{51} E_y & 0 \end{pmatrix} \quad (33)$$

Thus $\quad \Delta\varepsilon_{13} = \Delta\varepsilon_{31} \simeq - \varepsilon_0 n_0^2 n_e^2 r_{51} E_x$;

$$\Delta\varepsilon_{23} = \Delta\varepsilon_{32} \simeq - \varepsilon_0 n_0^2 n_e^2 r_{51} E_y. \quad (34)$$

The term $\Delta\varepsilon_{13}$ is responsible for coupling between the polarizations along x and z which in the present case corresponds to coupling between quasi TM and quasi TE polarizations.

Since $LiNbO_3$ is uniaxial with a relatively large birefringence $(n_0 - n_e \simeq 0.086$ at $\lambda_0 \simeq 0.6 \mu m)$, for efficient coupling, periodic coupling as

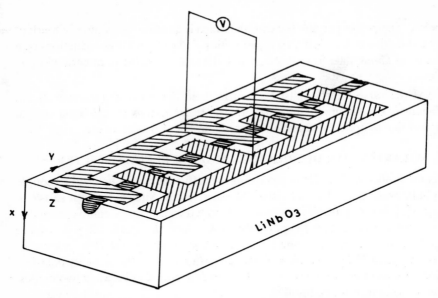

Fig. 3. TE↔TM polarization converter in x-cut LiNbO$_3$ using the electrooptic effect. Periodic electrodes are required to couple the modes efficiently since β(TE) $\neq$ β(TM).

shown in Fig. 3 is necessary. If Λ represents the spatial period of the electrodes, then for efficient coupling

$$\Lambda = \frac{\lambda_0}{\left| n_{eff}^{TE} - n_{eff}^{TM} \right|}, \tag{35}$$

where n_{eff}^{TE} and n_{eff}^{TM} represent the effective indices of the quasi TE and quasi TM polarizations. Assuming $n_{eff}^{TE} \simeq n_e$ and $n_{eff}^{TM} \simeq n_0$, we obtain the required spatial period corresponding to a free space wavelength of 0.6 μm,

$$\Lambda = \frac{\lambda_0}{(n_0 - n_e)} = \frac{0.6}{0.086} \simeq 7 \mu m. \tag{36}$$

Conversion efficiencies of > 99% have been achieved with only 2.5 V for a 6 mm long device.

Such polarization converters can be used for making intensity modulators if they are followed by polarizers or polarization selective couplers.

It is interesting to note from Eq. (35) that for a given electrode period Λ, phase matching is satisfied only at one free space wavelength satisfying

$$\lambda_0 = \Lambda \left| n_{eff}^{TE} - n_{eff}^{TM} \right| \tag{37}$$

Also for such periodic coupling, the bandwidth over which significant conversion takes place is approximately given by [1]

$$\frac{\Delta\lambda_0}{\lambda_0} \simeq \frac{\Lambda}{L}, \tag{38}$$

where L is the length of interaction. For a device length of 6 mm, with $\lambda_0 \simeq 0.6$ μm and $\Lambda = 7\,\mu$m, we have

$$\Delta\lambda_0 \simeq 7\text{Å}.$$

Larger bandwidths can be obtained by choosing smaller L. Fig. 4 shows the filter response for three independently controlled mode converters with slightly different periods of 7 μm, 7.12 μm and 7.24 μm placed on a single strip waveguide in LiNbO$_3$ [6]. Interchannel spacing is approximately 80Å and filter bandwidth is about is 15Å. The measured cross talk between different channels was $\simeq$ 20 dB.

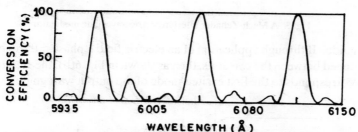

Fig. 4. Filter response corresponding to three different TE↔TM mode converters using periodic electrodes having slightly different periods which are placed on a single strip waveguide in LiNbO$_3$ (Reprinted from R.C. Alferness, Guided wave devices for optical communication, IEE J. Quant. Electron. **QE-17**, 946 (1981) — © IEEE).

In Sec. 6.1 we shall discuss how a directional coupler can be used for fabricating wavelength filters with broad bandwidths.

5. THE MACH ZEHNDER INTERFEROMETRIC MODULATOR

In Sec. 3 we had seen how the electrooptic effect can be used to build a phase modulator. A phase modulator can easily be converted to an intensity modulator by using interference effects. Figure 5 shows the Mach Zehnder interferometric modulator consisting of a single mode strip waveguide splitting into two single mode waveguides through a Y-branch and recombining to form an output single mode guide. The two Y-branches serve as a beam splitter and a beam combiner. If the branch is perfectly symmetric, it serves as a 3 dB splitter. The shaded regions in Fig. 5 correspond to electrodes laid for applying an electric field.

In order to understand the operation, we first note that light coupled into the input waveguide splits equally among the two waveguides at branch B_1. If the two propagating modes arrive at B_2 in phase, then as shown in Fig. 6(a), they excite the fundamental mode of the output waveguide which propagates in the

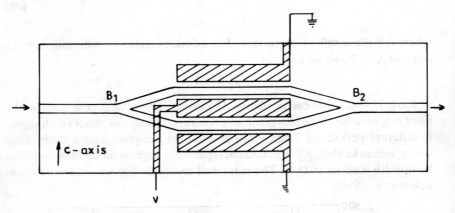

Fig. 5. A Mach-Zehnder interferometric waveguide modulator.

waveguide. If through application of an electric field a phase difference of π is introduced between the two arms, then as shown in Fig. 6(b), the superposition at B_2 corresponds to the first excited mode of the output waveguide. Since the

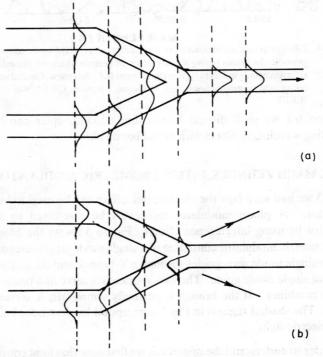

Fig. 6. (a) Modes arriving in phase at the branch excite the fundamental mode in the output waveguide which propagates in the waveguide.
(b) Modes arriving with a π phase difference excite the first excited mode in the output waveguide. If the output waveguide is single moded, light escapes into the substrate.

output waveguide is single moded, the resultant light is radiated into the substrate as a radiation mode and there is no output in the waveguide. For phase differences intermediate between 0 and π, the intensity of the output varies sinusoidally. If the two branches are perfectly symmetric and if we neglect attenuation, then we can write for the power output as

$$P_0 = \frac{P_{in}}{2}(1 + \cos\phi), \qquad (39)$$

where P_{in} is the input power and ϕ is the electrooptically induced phase difference between the two arms.

Figure 5 corresponds to a Mach-Zehnder interferometric modulator in $LiNbO_3$ with the c-axis as shown and the electrode pattern uses the r_{33} coefficient and also the push-pull effect by applying oppositely directed electric fields in the two waveguides. With such a push pull effect, the voltage required to modulate the intensity from a maximum to minimum will be

$$V_\pi = \frac{\lambda_0 d}{2 n_e^3 \, r_{33} \, \Gamma L}, \qquad (40)$$

where L is the length of the electrode. As a typical example, for $\lambda_0 = 0.633\,\mu m$, $d = 4\,\mu m$, $L = 3$ mm and assuming $\Gamma = 0.25$, $V_\pi \approx 5.2$ V.

Figure 7 shows a typical variation of the output intensity when a triangular waveform drive voltage is applied. If V_π represents the voltage required to create a π phase difference between the two arms, the output intensity completes one cycle for every change in voltage of $2V_\pi$. Thus if the drive voltage

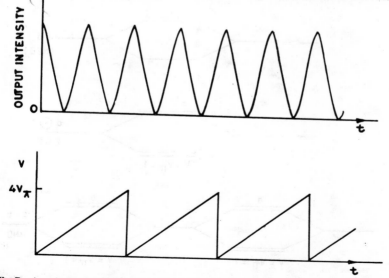

Fig. 7. A typical variation of output intensity for a triangular waveform voltage applied on the electrodes.

exceeds $2 V_\pi$, the output intensity completes more than one cycle. Using such a technique, one can obtain sinusoidal intensity modulation at very high rates by using a lower frequency drive. Using such a technique, Leonberger [7] has demonstrated modulation rates of 1.4 GHz using a 140 MHz electrical drive signal of such an amplitude as to induce a phase difference of 6π.

The Mach-Zehnder interferometric modulator forms a basic building block for many complex integrated optic devices such as for performing logic operations, A to D conversion, bistable operation etc. Here we shall briefly discuss logic operations; for details on other operations readers may look up Refs. [8] and [9].

5.1 Logic Operations

Using the basic Mach-Zehnder interferometer, various logic operations can be performed. Some of these are depicted in Fig. 8. The electrodes are shown as lines adjacent to the waveguides and the electrical input signals are shown as a and b. Let V_π represent the voltage required to change the phase of light propagating in an arm by π. We represent the voltage V_π by '1' and assume that the electrical signals a and b are either 0 or 1 (i.e., the voltage applied is either

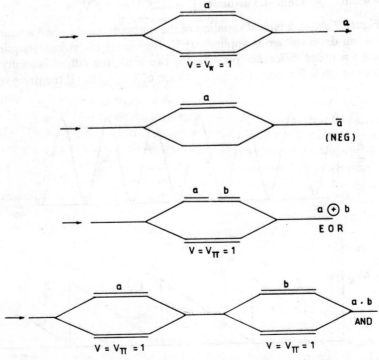

Fig. 8. Typical configurations for performing electrooptic logic operations using Mach-Zehnder waveguide interferometers.

0 or V_π). A continuous wave light beam is incident on the input of the interferometer. In Fig. 8(a), the electrical pulse sequence a is transferred to the same optical pulse sequence. In Fig. 8(b), whenever $a = 0$, the phase difference between the arms is zero and the output is finite. When $a = 1$ (i.e., applied voltage $= V_\pi$) then the phase difference is π and the output is zero. Thus the operation corresponds to NOT. Similarly Figs. 8(c) and (d) correspond to AND and Exclusive OR operations.

6. THE OPTICAL DIRECTIONAL COUPLER

The optical directional coupler consists of two single mode waveguides which are in close proximity over a certain length L as shown in Fig. 9. In Ch. 3, it was shown that the field of a guided mode has an evanescent tail even outside the guiding region. Thus when two waveguides are brought in close proximity, due to interaction between the two wavegudies, energy exchange between the waveguides can take place. The situation is very similar to the case of a pair of coupled pendulums.

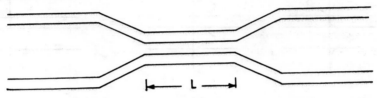

Fig. 9. The optical directional coupler which consists of a pair of single mode waveguides lying close to each other and interacting through the evanescent field in the cladding.

The energy exchange between the two waveguides can be analysed in terms of two equivalent pictures, namely, as a coupling between the modes of individual waveguides or as a beating between the supermodes of the complete structure analysed as a single waveguide. In fact, if the two waveguides are identical and single moded, the coupled structure has two modes the fundamental symmetric and the antisymmetric mode (see Fig. 10). When light is incident in one waveguide, it excites the two modes with almost equal amplitudes and in phase (see Fig. 11). If β_s and β_a represent the corresponding propagation constants, then as the modes travel through the structure, the accumulated phase difference is π at a distance $L = L_c$ satisfying the following equation:

$$L_c = \frac{\pi}{(\beta_s - \beta_a)}. \qquad (41)$$

Thus at this position in the coupler, the two modes are out of phase and the superposition leads to constructive interference in the second waveguide and destructive interference in the first waveguide. Thus after travelling a distance L_c, the energy appears in the second waveguide. This length is called the

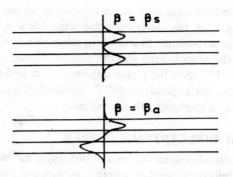

Fig. 10. The two supermodes of a directional coupler consisting of two identical single mode waveguides.

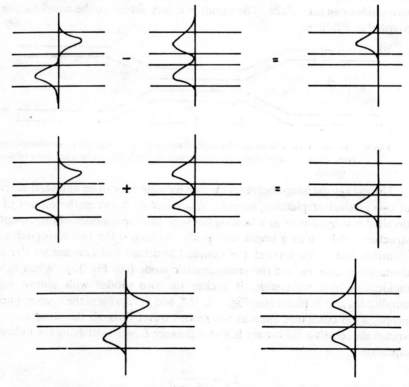

Fig. 11. Light incident on one waveguide excites the two supermodes equally and in phase. As the two supermodes propagate in the directional coupler, they suffer different phase shifts since they have different propagation constants. When the phase difference is π, the superposition of the two modes results in constructive interference in the second waveguide and destructive interference in the first waveguide.

coupling length of the directional coupler. If the modes are allowed to propagate further, energy comes back to the first waveguide after another distance L_c and so on.

Although the above supermode picture clearly shows the energy exchange, we shall use the coupled mode picture to describe the directional coupler as it can be used to obtain an analytic representation and also gives a direct dependence on the various waveguide parameters. Such an analysis also brings out clearly the operation of a directional coupler switch, directional coupler wavelength filter etc.

In the coupled mode analysis, let β_1 and β_2 represent the propagation constants of modes in waveguides 1 and 2 when the two waveguides are non interacting. When the two waveguides are non interacting, the amplitudes of modes excited in each waveguide remain the same and only a change of phase occurs as the mode propagates. In the presence of coupling between the two waveguides, the complex amplitudes $a(z)$ and $b(z)$ of the modes in waveguides 1 and 2 satisfy a set of coupled equations [1]:

$$\frac{da}{dz} = -i\beta_1 a - i\kappa_{12} b, \qquad (42)$$

$$\frac{db}{dz} = -i\beta_2 b - i\kappa_{21} a. \qquad (43)$$

Here κ_{12} and κ_{21} are called coupling coefficients and depend on the waveguide parameters of the two waveguides, the wavelength of operation as well as the separation between the waveguides. For example, for two identical symmetric planar waveguides with film and substrate indices n_f and n_s and thickness d and which are separated by a distance t (see Fig. 12) we have

$$\kappa_{12} = \kappa_{21} = \frac{2\kappa_f^2 \gamma_s e^{-\gamma_s t}}{\beta \left(d + \frac{2}{\gamma_s}\right) \left(\kappa_f^2 + \gamma_s^2\right)}, \qquad (44)$$

where β represents the propagation constant of the mode of each waveguide when they are non-interacting and

$$\kappa_f^2 = k_0^2 n_f^2 - \beta^2 \quad , \quad \gamma_s^2 = \beta^2 - k_0^2 n_s^2. \qquad (45)$$

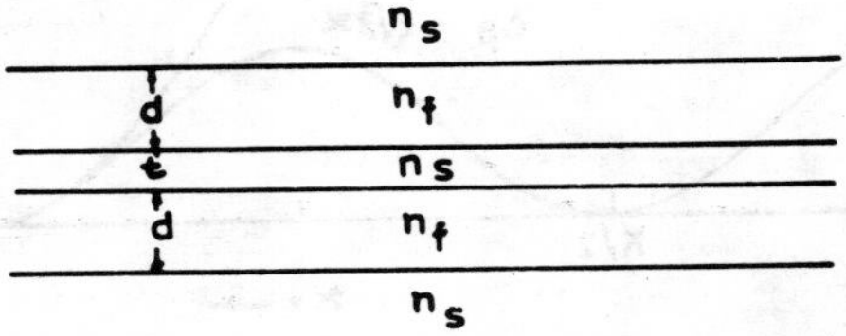

Fig. 12. A directional coupler formed by two identical planar waveguides separated by a distance t.

When the waveguides are non-interacting, we have $\kappa_{12} = \kappa_{21} = 0$ and the solution of Eqs. (42) and (43) are simply

$$a(z) = a(0) e^{-i\beta_1 z} \quad ; \quad b(z) = b(0) e^{-i\beta_2 z}. \tag{46}$$

Assuming unit power to be incident on waveguide 1 at $z = 0$, the solution of Eqs. (42) and (43) can be shown to be [1]

$$P_1(z) = 1 - \frac{1}{1 + \Delta\beta^2/4\kappa^2} \sin^2\left[\sqrt{1 + \frac{\Delta\beta^2}{4\kappa^2}} \kappa z\right], \tag{47}$$

$$P_2(z) = \frac{1}{1 + \Delta\beta^2/4\kappa^2} \sin^2\left[\sqrt{1 + \frac{\Delta\beta^2}{4\kappa^2}} \kappa z\right]. \tag{48}$$

Here $\quad \Delta\beta = \beta_1 - \beta_2 \quad , \quad \kappa^2 = \kappa_{12} \kappa_{21}. \tag{49}$

$\Delta\beta$ represents the phase mismatch between the two waveguides. The above equations clearly show a periodic exchange of energy between the two waveguides as discussed earlier. Plots of $P_2(z)$ vs. z for $\Delta\beta = 0$ and $\Delta\beta = 2\sqrt{3}\,\kappa$ are shown in Fig. 13. From Eq. (48), it follows that for complete transfer of energy from waveguide 1 to 2, a necessary condition is that $\Delta\beta = 0$, i.e., $\beta_1 = \beta_2$. This is equivalent to the phase-matching condition. When $\Delta\beta = 0$, the minimum length for complete transfer of energy is

$$L_c = \frac{\pi}{2\kappa}, \tag{50}$$

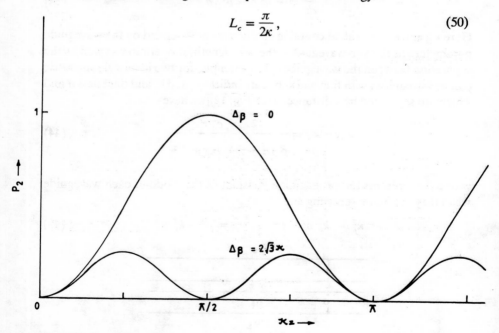

Fig. 13. Variation of P_2 in a directional coupler with identical waveguides ($\Delta\beta = 0$) and non-identical waveguides with $\Delta\beta = 2\sqrt{3}\,\kappa$.

which is called the coupling length. The value $\Delta\beta = 2\sqrt{3}\,\kappa$ is such that at $z = L_c = \pi/2\kappa$, $P_2(z) = 0$ and $P_1(z) = 1$, i.e., with this value of $\Delta\beta$, the power exits from waveguide 1. This is used in the construction of an optical switch. Thus if we consider a directional coupler made of identical waveguides and with a length of interaction equal to L_c, then power incident in waveguide 1 exits from waveguide 2. This corresponds to the cross $\otimes$ state of the switch. Now if through the electrooptic effect one creates a $\Delta\beta = 2\sqrt{3}\,\kappa$, then the power would exit from waveguide 1 which would correspond to the parallel $\ominus$ state. Thus through application of electric field (see Fig. 14) one can switch light energy from one waveguide to another. This is the basic principle behind the directional coupler switch.

Figure 14 shows a directional coupler in z-cut LiNbO$_3$ with COBRA electrodes [10]. For a given voltage, the effective index of one waveguide increases while that of the other reduces leading to a push-pull effect. We can approximately estimate the voltage required as follows. The required $\Delta\beta$ to switch is given by

$$\Delta\beta = 2\sqrt{3}\,\kappa = \frac{\sqrt{3}\,\pi}{L_c} \tag{51}$$

where we have used Eq. (50). Now using push-pull effect

$$\Delta\beta = 2\,\frac{2\pi}{\lambda_0}\,\Delta n \approx 2 \times \frac{2\pi}{\lambda_0}\,\frac{n_e^3 r_{33} V}{2d}\,\Gamma \tag{52}$$

Thus substituting in Eq. (51) we get

$$V \cdot L_c = \frac{\sqrt{3}}{2}\,\frac{\lambda_0 d}{n_e^3 r_{33} \Gamma} \tag{53}$$

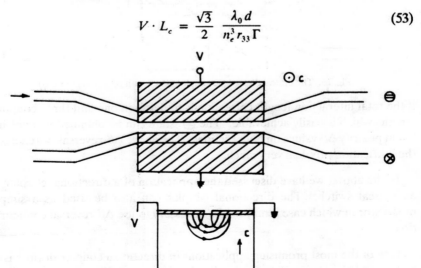

Fig. 14. The optical directional coupler switch.

For $\lambda_0 = 1.3\,\mu\text{m}, d = 6\,\mu\text{m}, n = 2.2, r_{33} = 30 \times 10^{-12}$ m/V, $\Gamma = 0.25$ we get

$$V \cdot L_c \simeq 8.5 \times 10^{-2} \text{ V} \cdot \text{m}$$

Thus for a coupler of length 1 cm, the required switching voltage would be about 8.5 V. Experimentally realized devices correspond very well with the above estimated values.

It is interesting to calculate the corresponding index change required to go from the ⊗ state to the ⊖ state:

$$\Delta n = \frac{n_e^3 r_{33} V}{2d} \Gamma \simeq 5.7 \times 10^{-5},$$

which is indeed very small.

For the directional coupler design shown in Fig. 14, we have seen that the device should have an interaction length exactly equal to L_c to obtain the cross state since the ⊗ state cannot be adjusted electrically. On the other hand, the ⊖ state can always be obtained by applying an external electric field. In fact, for obtaining a ⊗ state with a cross talk of better than −25 dB, κL must be controlled to within ± 3.5%. This is in general difficult to control reproducibly. In order to overcome this, Kogelnik and Schmidt [11] have proposed a $\Delta\beta$ reversed switch in which a pair of split electrodes with equal but opposite polarity values of $\Delta\beta$ are applied on the coupler (see Fig. 15). It can be shown that for such a configuration even the cross state can be controlled electrically

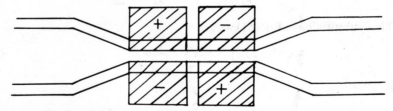

Fig. 15. The optical directional coupler switch with $\Delta\beta$ reversal electrodes.

if the total interaction length corresponds to between one and three coupling lengths which is easily achievable. The ⊖ state can be obtained by applying equal polarity $\Delta\beta$ values on both sections. The above $\Delta\beta$-reversal switch eases the fabrication problem very much.

In the above, we have discussed the application of a directional coupler as an optical switch. The directional coupler can also be used as a simple modulator in which case it may not be necessary to use $\Delta\beta$-reversal configuration.

One of the most promising applications of directional coupler devices is in high speed signal encoding for application in optical fibre communications. Such an external modulator has several advantages vis a vis direct current modulation of semiconductor lasers. The foremost advantage is in terms of spectral purity. Thus even if one uses a single frequency injection laser, for a directly modulated laser, the optical frequency is chirped during a bit time due to coupling of optical refractive index of the laser cavity with the injected electrons. Such a chirp is detrimental to fibre communication systems at bit rates above a few Gbits/sec. The second advantage is in terms of high modulation depth attainable with external modulators. In addition, the laser can be optimized for high spectral purity and power output without considerations to modulation bandwidth. Of course the disadvantage of using an external modulator is in terms of additional insertion loss and the complexity of an additional component in the communication system.

6.1 Directional Coupler Wavelength Filter

In Sec. 6, we had seen that for efficient transfer of energy from one waveguide to another, the propagation constants of the modes of the waveguides should be equal. It can be seen from Eq. (48) that the maximum fractional energy transferred is

$$\eta_m = P_2(z)\Big|_{max} = \frac{1}{1 + \frac{\Delta\beta^2}{4\kappa^2}}. \tag{54}$$

Thus for $\eta_m < 0.01$, $\Delta\beta \gtrsim 20\kappa$. This effect has been used in the construction of wavelength filters using directional couplers.

A directional coupler wavelength filter consists of two non-identical waveguides having different refractive indices and thicknesses. The wider guide has a lower film index and the narrower guide a higher film index. Since the two guides are non-identical, they have different dispersion characteristics (Fig. 16). By adjusting the various parameters of the two waveguides, they can be made to have the same propagation constant at a desired wavelength. Since the dispersions are different, as one moves away from this wavelength, $\Delta\beta$ is no more zero and the energy exchange reduces. Thus two input wavelengths-one corresponding to the phase-matching wavelength and the other sufficiently far away to have $\Delta\beta \gg \kappa$ would thus exit from the two waveguides and be demultiplexed.

In order to get an approximate expression for the bandwidth of the filter, we first consider a non-identical waveguide directional coupler for which at λ_0 $\Delta\beta = 0$. If κ represents the corresponding coupling coefficient, we choose the directional coupler to be of a length $L = \pi/2\kappa$ so that light of wavelength λ_0 gets coupled completely to the second waveguide. Now for $\lambda \neq \lambda_0$, $\Delta\beta \neq 0$ and the power in the second waveguide follows Eq. (48). Thus the value of $\Delta\beta$ for which the fractional coupled power is 1/2 should satisfy the following equation

$$\frac{1}{1 + \frac{\Delta\beta^2}{4\kappa^2}} \sin^2\left[\sqrt{1 + \frac{\Delta\beta^2}{4\kappa^2}} \kappa L\right] = \frac{1}{2}, \tag{55}$$

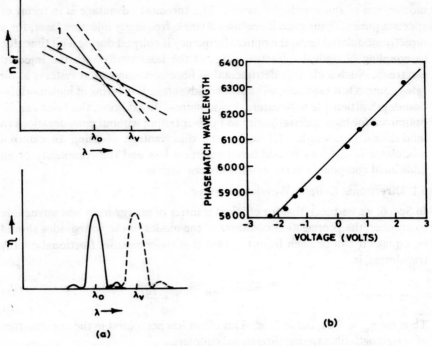

Fig. 16. (a) A schematic of the dispersion characteristics of two non identical waveguides and the corresponding variation of cross over efficiency η with wavelength. Solid curves correspond to the case when no voltage is applied and dashed when a voltage is applied. (b) Measured variation of center wavelength of filter with applied voltage (Adapted from R.C. Alferness and R.V. Schmidt, Tunable Optical waveguide directional coupler filter, App. Phys. Letts. *33* (1978) 161, with permission).

where we have neglected the λ variation of κ in comparison to that of $\Delta\beta$. Since $\kappa L = \pi/2$, Eq. (55) can be solved to obtain

$$\Delta\beta \simeq 1.6\,\kappa = \frac{0.8\,\pi}{L}. \tag{56}$$

Now

$$\Delta\beta = \frac{2\pi}{\lambda_0}\delta n_{eff} = \frac{2\pi}{\lambda_0}\frac{d(\delta n_{eff})}{d\lambda} \cdot \Delta\lambda, \tag{57}$$

where δn_{eff} is the difference in effective indices between the modes of the two waveguides. Hence from Eqs. (56) and (57) we obtain for the full width at half maximum of the filter response

$$\Delta\lambda \simeq \frac{0.8\,\lambda_0}{L\left[\dfrac{d(\delta n_{eff})}{d\lambda}\right]} \tag{58}$$

Since the differential dispersion between two non-identical waveguides is very small, the bandwidth $\Delta\lambda$ is usually large. Experimental values for $\Delta\lambda$ are around 200Å at 0.6 μm [12] for an interaction length of 1.5 cm and peak cross over efficiency was obtained to be ~ 100%. Indeed by using the electrooptic effect, the dispersion curves of the two waveguides can be tuned which leads to a tuning of the center wavelength. Experimentally tuning over 600Å with a 110Å/V shift of center wavelength has been demonstrated (see Fig. 16).

6.2 Polarization Problem

Since $LiNbO_3$ is an anisotropic material and orthogonal polarizations see unequal electrooptic coefficients, devices made in this substrate are in general highly polarization dependent. For example, the phase modulator discussed in Sec. 3 has different V_π values for TE and TM polarization due to a difference of a factor 3 between r_{13} and r_{33}. Hence such integrated optical devices cannot be used with conventional single-mode fibres which do not maintain the state of polarization of the propagating light beam. In order to make them compatible with standard single mode fibres, the devices should be made polarization insensitive.

The simplest devices are the phase modulator and the Mach-Zehnder interferometric modulator. If they are to be polarization insensitive, it is necessary that the phase shift suffered by both polarizations be the same for a given applied voltage. One can achieve this either by using separate electrode patterns along the same waveguide which almost act independently on TE and TM polarizations [13]. A different technique makes use of the inhomogeneous electric field created by coplanar electrodes. Thus by properly positioning the waveguide with respect to the electrodes, one could adjust such that the induced changes in phase for both polarizations are the same [14].

Compared to a simple phase modulator, the directional coupler switch shown in Fig. 14 is polarization dependent due to two reasons:

1. The coupling coefficients for quasi TE and quasi TM polarization are, in general, not the same. This is due to the anisotropic nature of $LiNbO_3$ and also the refractive index difference between the guiding and substrate regions are in general not the same for both polarizations.

2. When an electric field is applied (for example in a z-cut crystal), the quasi TM polarization (which is parallel to c-axis) is affected by the coefficient r_{33} while the quasi TE polarization is affected by r_{13} which is about 3 times smaller as compared to r_{33}. Thus the same applied voltage induces different $\Delta\beta$ values for TE and TM polarizations.

This difference in κ and the induced $\Delta\beta$ between TE and TM polarization leads to the polarization dependence of the devices. There have been various approaches to making the device polarization insensitive. For example, by a proper choice of Ti thickness, waveguide separation, Ti strip width etc. one can make $\kappa(TE) \simeq \kappa(TM)$ [15]. Polarization insensitivity to switching is obtained using tapered coupling between the two waveguides [16].

6.3 Polarization Splitting Directional Coupler

There are certain applications of integrated optic circuits in which one would like to separate spatially the TE and TM polarized light. A directional coupler is usually polarization selective but the difference in coupling lengths of TE and TM modes is usually not large enough to separate the two within one or two coupling lengths. It is possible to separate the TE and TM polarization in a directional coupler by using any of the following schemes:

(a) Design a directional coupler such that the interaction between the waveguides is very strong for one polarization while it is very weak for the other polarization i.e. the coupling length corresponding to one polarization is very small compared to the other polarization.

(b) Design a directional coupler such that the ratio of coupling lengths corresponding to the two polarizations is exactly equal to 2 so that when one polarization is in the cross state, the other is in the parallel state.

(c) Design a directional coupler such that the two waveguides are phase-matched for one polarization but have a $\Delta\beta \gg \kappa$ for the other polarization.

Directional couplers based on the principles given under (a) and (b) can be designed using a thin metal layer sandwiched between the two waveguides [17]. Figure 17 shows a directional coupler structure with polarization splitting characteristics. Since the TM polarization can couple more strongly to the second waveguide via the surface plasmon mode in the metal, the TM mode sees a directional coupler with a much stronger coupling than a directional coupler with the same parameters but without the metal. The TE mode, however, sees a much smaller coupling because the metal layer reduces the interaction between the waveguides. Thus the various parameters can be adjusted so that $\kappa(TM) \gg \kappa(TE)$. If the coupler length equals $\pi/2\kappa(TM)$, one would obtain purely TE polarization in the input waveguide and TM polarization in the other. The main sources leading to finite cross-talk are (i) unequal losses of the symmetric and antisymmetric TM supermodes give rise to unequal

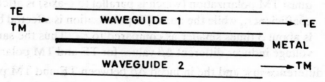

Fig. 17. A polarization splitting directional coupler structure consisting of two interacting waveguides separated by a very thin metal layer.

cancellation of power in the TM mode after one coupling length. (ii) In a practical situation, the coupling length of TE mode, although very large can never be infinite. Therefore, the power of TE polarization in waveguide 2 is always non-zero.

Theoretical estimations using planar waveguides [17] give extinction ratios better than −31 dB with a throughput loss less than 0.1 dB. Experimental devices based on directional couplers formed by a pair of polished fibres [18] show extinction ratios better than −30 dB and throughput loss less than about 1 dB.

Polarization splitting directional couplers based on polarization dependent loading of waveguides by a metal layer have also been demonstrated. In the structure demonstrated by Mikami [19], a directional coupler consisting of identical waveguides is first formed in $LiNbO_3$. One of the waveguides is coated by a thick layer of Aluminium. It is well known that loading by a metal creates a larger change in propagation constant of the TM polarization while creating a much smaller change in the propagation constant of TE polarization. Thus for the directional coupler, $\Delta\beta(TM)$ can be made to have a value very large compared to $\kappa(TM)$ while the small value of $\Delta\beta(TE)$ can be compensated by using a split electrode configuration on the other waveguide and use the principle of the '$\Delta\beta$-reversal' switch. Mikami [19] has demonstrated such a polarization splitter in $LiNbO_3$ with cross-talk less than −20 dB and insertion loss of about 1.7 dB at 1.15 μm.

In the structure discussed above, the small $\Delta\beta$ of the TE mode is compensated by using the electrooptic effect of the substrate in a '$\Delta\beta$-reversal' configuration. Thus such couplers cannot be fabricated on substrates such as glass which do not possess the Pockels effect and they also require an external voltage for obtaining cross state for TE polarization. Recently another configuration has been proposed by Thyagarajan, Diggavi and Ghatak [20]. Here thin metal layers (of thickness $\simeq$ 50-250 Å) are used on both waveguides. The thicknesses of the metal layers and the superstrate refractive index values are so chosen that the directional coupler has $\Delta\beta(TM) \gg \kappa(TM)$ and $\Delta\beta(TE) \simeq 0$. Theoretical analysis has shown that devices with extinction ratios better than −20 dB and throughput loss less than 1 dB should be possible [20].

7. POLARIZERS

Integrated optic polarizers are needed in various applications such as in the chip used in the fibre optic gyroscope [21] or in other interferometric sensors and in coherent communication systems. Also since most integrated optic devices are polarization sensitive, polarizers form an important component. Various techniques have been used for fabrication of such integrated optic polarizers.

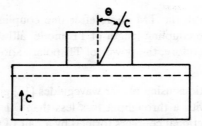

Fig. 18. A typical leaky mode polarizer structure. θ is the angle between the c-axes [22].

7.1 Leaky Mode Polarizers

In leaky mode polarizers a birefringent crystal is placed in close contact with the waveguide, such that the crystal presents a leaky structure for one polarization and a guided structure for the other polarization [22]. Figure 18 shows a typical leaky mode polarizer structure in which a LiNbO$_3$ crystal is used to decouple one of the polarizations while retaining the other polarization. LiNbO$_3$ is a negative uniaxial crystal with $n_0 > n_e$. If Δn_0 and Δn_e are the

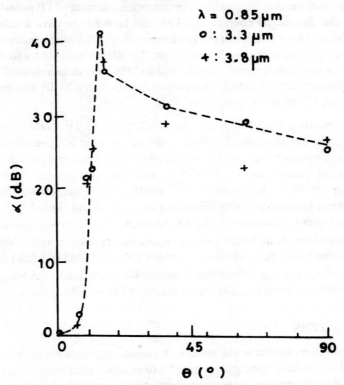

Fig. 19. A typical experimentally observed variation of differential attenuation between TE and TM modes versus θ (Reprinted from M. Papuchon, A. Enard, K. Thyagarajan and S. Vatoux, Anisotropic polarizers for Ti: LiNbO$_3$ strip waveguides, Digest of Topical meeting on Integrated and guided wave optics, Paper WC5 (1984) with permission from Optical Society of America).

changes in ordinary and extraordinary refractive indices brought about by Ti in diffusion, then for Ti diffused waveguide in z-cut crystal $k_0 n_0 < \beta(TE) < k_0(n_0 + \Delta n_0)$ and $k_0 n_e < \beta(TM) < k_0(n_e + \Delta n_e)$. Thus by placing a LiNbO$_3$ crystal with the correct orientation in close contact with the waveguide surface, the TM wave can be coupled out of the waveguide; the TE mode would of course remain guided. Figure 19 shows a typical experimentally observed variation of the differential attenuation between TE and TM modes versus θ (see Fig. 18) at $\lambda_0 = 0.85 \mu m$ for two different waveguides. The length of interaction was 5 mm. The additional loss introduced in the TE mode was too small to be measurable. Design of such leaky mode polarizers can be carried out using a leaky mode analysis [23,24].

7.2 Metal Clad Polarizers

It is well known that a metal-dielectric interface can support a TM guided wave which is also known as the surface plasmon mode (see Ch. 3). Since a metal has an imaginary component of the dielectric constant, such a surface plasmon mode is lossy. At the same time, the metal-dielectric interface cannot support any TE guided mode since one cannot satisfy the boundary conditions. Thus if a waveguide is covered by a metal with an intermediate buffer layer (see Fig. 20), then by a proper choice of buffer refractive index and thickness, one can obtain an efficient coupling of energy from the TM mode of the waveguide to

Fig. 20. A thick metal clad waveguide polarizer structure.

the surface plasmon mode. Since the surface plasmon mode is lossy such a coupling phenomenon results in an attenuation of the TM mode of the waveguide. Since there is no corresponding surface mode for TE polarization, the TE mode of the waveguide would suffer much smaller losses. Hence such a structure can act as an efficient TE pass polarizer. The existence of a critical buffer thickness for resonant absorption of the TM mode was experimentally demonstrated by Thyagarajan et al. [25]. Using mercury as the metal cladding, TM mode attenuations as high as 20 dB over 1 mm interaction length was demonstrated. Figure 21 shows a typical result. The peak attenuation of the TM mode and the width of the resonant peak are dependent on the buffer index [26-28]. For LiNbO$_3$ waveguides, typical buffers are Y_2O_3 ($n \simeq 1.85$), SiO$_2$-TiO$_2$ mixtures ($n \simeq 1.7$-1.9), ZnO etc.

The attenuation introduced by a metal layer has also been used in the characterization of optical directional couplers [29]. Indeed, using mercury as

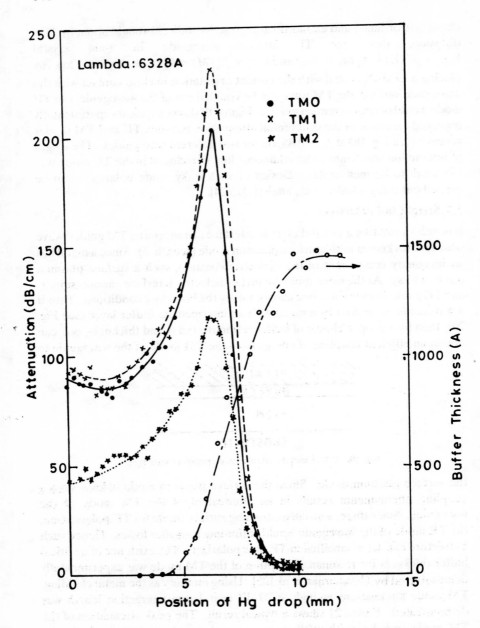

Fig. 21. Variation of TM mode attenuation as a function of the position of the mercury drop on a waveguide covered with a buffer with varying thickness. (Reprinted from K. Thyagarajan, Y. Bourbin, A. Enard, S. Vatoux, and M. Papuchon, Experimental demonstration of TM mode attenuation resonance in planar metal clad waveguides, Optics Letts. **10** (1985) 288, with permission from Optical Society of America.)

the cladding and displacing the mercury drop along a directional coupler in which one waveguide is covered with a sufficiently thick coating of dielectric, it was shown in Bourbin *et al.* [29] that one can non-destructively measure the coupling length of the directional coupler both for TM and TE polarizations. The advantage of the technique is that it works for any wavelength and does not rely on any specific material properties.

7.3 Thin Metal Clad Polarizers

In the thin metal clad polarizer design, the waveguide is covered with a thin layer of metal (thickness in the range 50-200 Å) which in turn is covered by a dielectric overlay (see Fig. 22). When the metal thickness is small, through interaction with the plasmon mode of the thin metal film, it is possible to induce a large leakage loss for the TM mode by choosing an appropriate dielectric

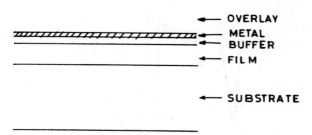

Fig. 22. A thin metal clad waveguide polarizer.

overlay refractive index which is larger than the mode effective index. At the same time, the leakage loss of the TE mode can be maintained very small [18, 30, 31]. Figure 23 shows typical TM mode loss that are obtainable using such a device configuration. If the metal thickness becomes large ($\gtrsim 200$ Å), the polarizer works on the principle of absorption loss introduced by the thin metal layer.

In LiNbO$_3$, one can make polarizers using polarization dependent changes in refractive index using different fabrication processes. Thus the usual Ti-indiffusion technique increases both the ordinary and the extraordinary refractive indices while the proton exchange technique leads to an increase in the extraordinary refractive index and a small decrease in the ordinary refractive index. In order to fabricate a polarizer, first a Ti indiffused waveguide is fabricated with short discontinuities. These discontinuities are later filled by proton exchanged wavegudies. The extraordinary mode is guided along the entire length of the waveguide and hence experiences only a small loss. On the other hand, the ordinary mode is not guided in the proton exchanged region and is radiated into the substrate. It has been shown that by a proper choice of various parameters polarizers with extinction ratio > 50 dB and throughput loss < 0.5 dB are possible [32].

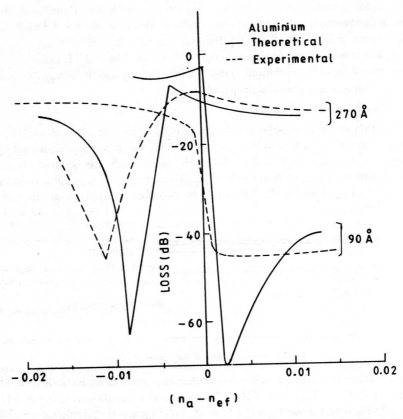

Fig. 23 Variation of TM mode loss with the refractive index of the overlay. The solid curve corresponds to the theoretical estimate and the dashed lines correspond to experimental values of Ref. [30]. Here n_a and n_{ef} correspond to refractive index of overlay and the mode effective index respectively. Metal thicknesses were (a) 90 Å and (b) 270 Å. (Reprinted from K. Thyagarajan, S. Diggavi, A.K. Ghatak, W. Johnstone, G. Stewart and B. Culshaw, Thin metal clad waveguide polarizers analysis and comparison with experiment, Optics Letts. **15** (1990) 1041 with permission from Optical Society of America.)

8. CONCLUSIONS

In this chapter we have discussed a cross-section of integrated optic devices. Although most devices discussed are based on $LiNbO_3$, the principles apply equally to other substrate materials. Since the electro optic effect is the most commonly used effect in making devices, the devices discussed in this chapter are all based on this effect. The Mach Zehnder interferometer and the directional coupler are two generic configurations which form the basis of many devices in integrated optics. In the present chapter some applications of these in logic operations, wavelength filtering, polarization splitting etc. were given.

REFERENCES

1. Ajoy Ghatak and Thyagarajan K., "Optical Electronics", Cambridge University Press, UK, (1989). Reprinted by Foundation Books, India, 1991.
2. A. Yariv and Yeh P., "Optical Waves in Crystals", John Wiley, New York, (1983).
3. M. Papuchon, "Integrated optical modulation and switching" in New Directions in Guided Wave and Coherent Optics, (Eds: D.B. Ostrowsky and E. Spitz) Martinus Nijhoff, The Hague, (1984), Vol. II, p. 371.
4. M. Papuchon, "Integrated Optical Modulation and Switching", Proc. SPIE, 1985.
5. R. Alferness and Buhl L.L., "Electro-optic waveguide TE ↔ TM mode converter with low drive voltage", Optics Letts., Vol. 5, p. 473 (1980).
6. R.C. Alferness, "Guided wave devices for optical communication", IEEE J. Quantum Electron., Vol. QE-17, p. 946 (1981).
7. F.J. Leonberger, "High speed operation of $LiNbO_3$ electrooptic interferometric waveguide modulators", Optics Letts., Vol. 5, p. 312 (1980).
8. H.F. Taylor, "Guided wave electro optic device for logic and computation", Appl. Opt., Vol. 17, p. 1493 (1978).
9. A. Schnapper, Papuchon M. and Puech C, "Optical bistability using an integrated two arm interferometer", *Optics Comm.* 29, p. 364-368 (1979).
10. M. Papuchon, Combemale Y., Mathieu X., Ostrowsky D.B., Reiber L., Roy A.M., Sejourne B. and Werner M., 'Electrically switched optical directional coupler: COBRA', Appl. Phys. Letts., Vol. 27, p. 289 (1975).
11. H. Kogelnik and Schmidt R.V., "Switched directional couplers with alternating $\Delta\beta$", IEEE J. Quantum Electron, Vol. QE-12, p. 396 (1976).
12. R.C. Alferness and Schmidt R.V., "Tunable optical waveguide directional coupler filter", Appl. Phys. Letts., Vol. 33, p. 161 (1978).
13. W.K. Burns, Giallorenzi T.G., Moeller R.P. and West E.J., "Interferometric waveguide modulator with polarization independent operation", Appl. Phys. Letts., Vol. 33, p. 944 (1978).
14. Y. Bourbin, Papuchon M., Vatoux S., Arnoux J.M. and Werner M., "Polarization independent modulators with Ti: $LiNbO_3$ strip wavegudies", Electron. Letts., Vol. 20, p. 495-497 (1984).
15. R.C. Alferness, Schmidt R.V. and Turner E.H., "Characteristics of Ti diffused $LiNbO_3$ optical directional couplers", Appl. Opt., Vol. 18, p. 4012 (1979).
16. R.C. Alferness, "Polarization independent optical directional coupler switch using weighted coupling", Appl. Phys. Letts., Vol. 35, p. 748 (1979).
17. K. Thyagarajan, Diggavi S. and Ghatak A.K., "Design and analysis of a novel polarization splitting directional coupler", Electronics Letts., Vol. 24, p. 869 (1988).
18. W. Johnstone, Stewart G., Culshaw B. and Hart T., "Fibre Optic polarizers and polarizing couplers", Electronics Letts., Vol. 24, p. 866, (1988).
19. S. Mikami, "$LiNbO_3$ coupled-waveguide TE/TM mode splitter", Appl. Phys. Letts., Vol. 36, p. 491 (1980).
20. K. Thyagarajan, Diggavi S. and Ghatak A.K., "Integrated Optic polarization splitting directional coupler", Opt. Letts., Vol. 14, p. 1333 (1989).
21. S. Vatoux, Lefevre H. and Papuchon M., "Integrated Optics, A practical solution for the fibre optic gyroscope", Proc. SPIE (1988) p. 42.
22. M. Papuchon, Enard A., Thyagarajan K. and Vatoux S., "Anisotropic polarizers for Ti: $LiNbO_3$ strip waveguides", Digest of Topical Meeting on Integrated and Guided Wave Optics (Optical Society of America) Paper WC5, (1984).

23. A.K. Ghatak, "Leaky modes in Optical Waveguides" Optical and Quantum Electron, Vol. 17, p. 311 (1985).
24. K. Thyagarajan, Diggavi S. and Ghatak A.K., "Analytical investigations of leaky and absorbing planar structures", Optical and Quantum Electron, Vol. 19, p. 131 (1987).
25. K. Thyagarajan, Bourbin Y., Enard A., Vatoux S. and Papuchon M., "Experimental demonstration of TM mode attenuation resonance in planar metal clad waveguides", Optics Letts., Vol. 10, p. 288 (1985).
26. Y. Yamamoto, Kamiya T. and Yanai H., "Characteristics of Optical guided modes in multilayer metal clad planar optical guide with low index dielectric buffer layer", IEEE J. Quant. Electron., QE-11, p. 729-736, (1975).
27. M. Masuda and Koyama J., "Effects of a buffer layer on TM modes in a metal clad Optical waveguide using Ti diffused $LiNbO_3$", Appl. Opt., 16, p. 2994-3000, (1977).
28. K. Thyagarajan, Kaul A.N. and Hosain S.I., "Attenuation characteristics of single mode metal clad graded index waveguide with a dielectric buffer: A simple and accurate numerical method", Opt. Letts., 11, p. 479-481, (1986).
29. Y. Bourbin, Enard A., Papuchon M. and Thyagarajan K., "The local absorption technique: A straightforward characterization method for many optical devices", J. Lightwave Tech., Vol. LT-5, p. 684 (1987).
30. W. Johnstone, Stewart G., Hart T., and Culsaw B., "Surface plasmon polaritons in thin metal films and their role in fibre optic polarizing devices", J. Lightwave Tech. Vol. 8, p. 538 (1990)
31. K. Thyagarajan, Diggavi S., Ghatak A.K., Johnstone W., Stewart G. and Culshaw B., "Thin metal clad waveguide polarizers: analysis and comparison with experiment", Opt. Letts., Vol. 18, p. 1041 (1990).
32. P.G. Suchoski, Findakly T.K. and Leonberger F.J., "Monolithic low loss, high extinction integrated optical polarizers", Proc. SPIE, Vol. 835, p. 40 (1987).

Part III
Optical Fibre Sensors and Devices

Part III

22

Optical Fibre Sensors and Devices

B.P. PAL[*]

Optical techniques have long played an important role in instrumentation and sensors especially for non-contact measurements. "A sensor can be defined as a device which has the role of converting a change in the magnitude of one physical parameter into a change in magnitude of a second, different parameter, which can be measured more conveniently and perhaps more accurately" [1]. In recent years, the scope of optical techniques in the area of instrumentation and sensors has made a quantum jump with the ready availability of low-loss optical fibres and associated optoelectronic components. The advantages and potentials offered by Fibre Optics in sensing and instrumentation are many as we shall see in the following chapters. During early seventies, when the technology of optical fibres for telecommunication was evolving, the high sensitivity of these fibres in terms of their transmission characteristics to certain external perturbations like bends, microbends, pressure etc. became evident. A great deal of effort was spent at that time to reduce the sensitivity of signal carrying optical fibres to such external effects through suitable fibre and cable designs.

Capitalising on this observation of exceptional sensitivity of optical fibres to external perturbations, an alternate school of thought began to exploit this sensitivity of optical fibres to construct or interrogate a large variety of sensors and instruments. This off-shoot of optical fibre telecommunication soon saw a flurry of R and D activity around the world which led to the emergence of the field of Fibre Optic Sensors and Devices [2]. Some of the key features of this

---

[*] Physics Department, Indian Institute of Technology Delhi, New Delhi-110016, India.

new technology which offer substantial benefits as compared to conventional electric sensors are:

1. Sensed signal is immune to electromagnetic interference (EMI) and radio frequency interference (RFI)
2. Intrinsically safe in explosive environments
3. Highly reliable and secure with no risk of fire/sparks
4. High voltage insulation and absence of ground loops and hence obviate any necessity of isolation devices like optocouplers
5. Low volume and weight, e.g., one kilometer of 200 μm silica fibre weighs only 70 gm and occupies a volume of ~30 cm^3 [1]
6. As a point sensor, they can be used to sense normally inaccessible regions without perturbation of the transmitted signals
7. Potentially resistant to nuclear or ionising radiations
8. Can be easily interfaced with low-loss optical fibre telemetry and hence affords remote sensing by locating the control electronics for LED/lasers and detectors far away from the sensor head
9. Large bandwidth and hence offers possibility of multiplexing a large number of individually addressed point sensors in a fibre network or distributed sensing i.e. continuous sensing along the fibre length
10. Chemically inert and they can be readily employed in chemical, process and biomedical instrumentation due to their small size and mechanical flexibility.

These advantages were sufficient to attract intensive R and D effort around the world to develop a new class of sensors based on fibre optics [3]. This has eventually led to emergence of a variety of fibre optic sensors for accurate sensing and measurement of physical parameters and fields, e.g., pressure, temperature, liquid level, liquid refractive index, liquid pH, antibodies, electric current, rotation, displacement, acceleration, acoustic, electric and magnetic fields and so on. Initial developmental work had concentrated predominantly on military applications like fibre optic hydrophones for submarine and undersea applications and gyroscopes for applications in ships, missiles and aircrafts. Gradually, a large number of civilian applications have also picked up. Fibre Optic Sensors are expected to play a major role in future industrial, medical, aerospace, and consumer applications. According to conservative estimates made by Arthur D. Little, a well known business/market projection Institute the sale of optical fibre sensors will expand from approximately $30 millions in 1989 to $560 millions by 1999. In some markets, annual growth rates will reach 35% or more during the next ten years [4].

TABLE 1

Optical Modulation Schemes by the Mesurand

Type of information	Physical Mechanism	Detection Circuitry	Typical examples
Intensity	Modulation of transmitted light by emission, absorption or r.i. change	Analog/Digital	Pressure, displacement, r.i., temperature, liquid level
Phase	Interference between signal and reference in an interferometer	Fringe counting	Hydrophone, Gyroscope, Magnetometer
Polarization	Changes in gyratory optical tensor	Polarization analyser and amplitude comparison	Magnetic field, transducer for current measurement of high voltage transmission lines
Wavelength	Spectral dependent variation of absorption and emission	Amplitude comparison at two fixed λ's	Temperature measurement

Broadly, a fibre optic sensor may be classified as either intrinsic or extrinsic. In the intrinsic sensor, the physical parameter/effect to be sensed modulates the transmission properties of the sensing fibre whereas in an extrinsic sensor, the modulation takes place outside the fibre. In the former, one or more of the physical properties of the guided light, e.g., intensity, phase, polarisation and wavelength/colour is modulated by the measurand while in the latter case, the fibre merely acts as a conduit to transport the light signal to and from the sensor head. Four of the most common fibre optic sensing techniques are based on interferometry, intensity modulation, fluorescence and spectral modulation of the light used in the sensing process (see Table 1). However, out of these four, the intensity and phase modulated ones offer the most wide spectrum of optical fibre sensors. The advantage of intensity modulated sensors lies in their simplicity of construction and their being compatible to the multimode fibre technology. The phase modulated fibre optic sensors necessarily require an interferometric measurement set-up with associated complexity in construction, although as we shall see in the following chapters, they theoretically offer orders of magnitude higher sensitivity as compared to intensity modulated sensors.

This part of the book begins with a chapter on intensity modulated fibre optic sensors. This is followed by the chapter on interferometric sensors. The subsequent chapters are concerned with multiplexed sensors, application of fibre optics in electrical industry, signal processing schemes for quantitative retrieval of measurands and two chapters on several fibre optic components which have found important applications as in-line fibre components in recent years.

REFERENCES

1. J.P. Dakin, "Optical Fibre Sensors", Summer School on "Principles of Optical Systems" held at Rose Priori, organised by Strathclyde University, Glasgow, May (1987).
2. B. Culshaw, "Fibre Optic Sensing and Signal Processing", Peter Peregrinus, Stevenage (1984).
3. T.G. Giallorenzi, Bucaro J.A., Dandridge A., Cole J.H., Rashley S.C. and Priest R.G., "Optical Fibre Sensor Technology", J. Quant. Electron., Vol. QE-18, pp. 626-665 (1982).
4. "The Outlook for Optical Fibre Sensors", published by Decision Resources, Arthur D. Little, Burlington, MA (1989).

23

Intensity Modulated Optical Fibre Sensors

B.P. Pal[*]

---

1. INTRODUCTION

Intensity modulated fibre optic sensors are the simplest and most widely studied in the field of optical fibre sensors [1-3]. Since early days, a large variety of transduction mecahnisms/techniques have been proposed in the unclassified and patent literature to produce an intensity modulation by the measurand of the guided light that is injected into the sensing fibre. Alternatively, there could be another category of intensity modulated sensors in which the light, after having been modulated by the measurand outside the lead fibre, is returned via the same fibre for detection and demodulation. There exists a huge number of patents and inventions on intensity modulated optical fibre sensors. In the following sections, we outline principles behind several members of the family of intensity modulated fibre optic sensors.

2. GENERAL FEATURE

The general configuration of an intensity modulated sensor is shown in Fig. 1, in which the baseband signal (the measurand) in the form of a sinusoidal varying quantity is seen to modulate the intensity of the light transmitted through the sensor [4]. The modulation envelope is reflected in the voltage output of the detector, which upon calibration, can be used to retrieve measure of the measurand. Intensity modulation can be achieved through a variety of schemes, e.g., displacement of one fibre relative to the other, shutter type, i.e., variable coupling of light between two sets of aligned fibres, collection of modulated light reflected from a target exposed to the measurand through use

---

[*] Physics Department, Indian Institute of Technology Delhi, New Delhi-110016, India.

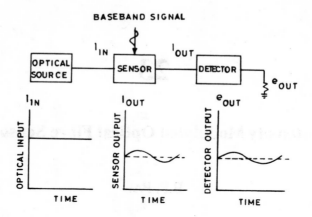

Fig. 1. Schematic of the general configuration of an intensity modulated sensor (after [4]).

of Y-guide or bifurcated fibre probe and loss modulation of light in the core or in the cladding through bending, microbending or evanescent coupling to another fibre/medium.

2.1 Intensity modulation through light interruption

It is known that simple optical sensors can be configured by interrupting direct optical interconnection between an optical source and a detector/detector array without incorporating any fibre optics. However, it is well known from fibre splice studies that coupled light intensity across a fibre joint can be varied through either transverse, longitudinal or angular misalignments or through a combination of them between the axes of two fibres. Figure 2 is a representation of transmission loss vs misalignments for misalignments along any three orientation between two $\sim 50\,\mu$m quasi-parabolic graded core multi-mode fibres of N.A. 0.21 [5]. Here the loss on dB-scale is calculated relative to intensity transmitted across the joint for zero misalignment between the fibre axes. The longitudinal displacement of one fibre relative to the other can be used to measure axial displacement. On the other hand, it is apparent from this figure that joint loss is very sensitive to transverse misalignments between their axes. This phenomenon can be exploited to configure an a.c. pressure or an acoustic sensor in which two fibre end faces are kept close to each other (with $\sim$2-3 μm separation between them) whose axes are otherwise well aligned by maximising optical power throughput across the joint. If a length of the fibre is free to move relative to the input fibre which is kept fixed and an external force field is applied in a direction transverse to their axes, it will induce motion to the free end of the output fibre relative to the fixed fibre, thereby creating an intensity modulation of the light coupled across the joint. It is obvious that the sensitivity of such a device will be inversely proportional to the fibre core diameter, e.g., a displacement between the fibres of one core diameter will

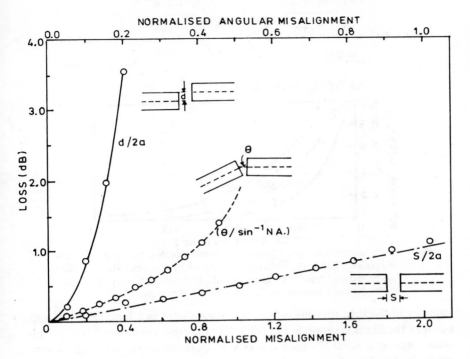

Fig. 2. Transmission loss vs. misalignments between the axes of two multimode fibres ($\phi \simeq 50\,\mu$m) ([5]).

result in approximately 100% light intensity modulation. Approximately the first 20% of displacement yields a linear output [3]. It may be mentioned that in actual experiment, care should be taken to employ cladding mode strippers in order to prevent light from the cladding. In the original experiment [6], the device was found to detect deep sea noise levels in the frequency range 100 Hz to 1 kHz and transverse static displacements down to a few armstrongs. In one such experiment [7], one of the fibres was attached to the diaphragm of the mouthpiece of a telephone set. An electric signal from a signal generator was fed to the diaphragm to make it oscillate and hence oscillate the attached fibre alongwith it, thereby leading to corresponding oscillation in the power coupled across the joint. These electrically driven oscillations amounted to simulation of for example, acoustic waves or a.c. pressures. The response of this simulated acoustic sensor was quite linear upto ~5-8 kHz. Sensitivity of such variable optical power transmissive fibre sensors can be increased by cleaving and polishing the two fibres at a slant angle so as to produce total internal reflection of the guided light from the slant fibre-air interface as shown in Fig. 3(a) [8]. If the slant face of the second fibre is brought sufficiently close to mate with the slant face of the first fibre, due to frustrated total internal reflection, power will leak from the first fibre to the receiving fibre through the slant face. Relative

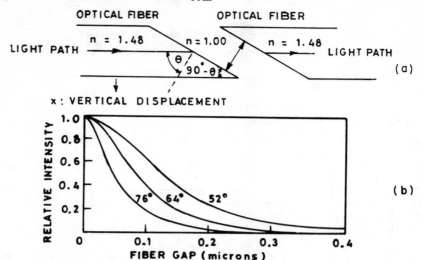

Fig. 3. (a) Frustrated total internal reflection geometry between two angularly cleaved and polished fibres; (b) response curve of such FTIR sensors (both (a) and (b) are reproduced from W.B. Spillman and McMohan, D.H., App. Opt. **19**, 113 (1980) [8] with permission of Optical Society of America).

intensity of the light transmitted through such a joint as a function of air gap between the fibres is shown in Fig. 3(b). The different curves are labelled with slant angle as a variable. Such a sensor is supposed to provide highest sensitivity in this class of transmissive sensors.

2.2 Shutter/Schlieren Multimode Fibre Optic Sensors

Concept of such a shutter device is shown in Fig. 4(a) in which a shutter being actuated by the measurand results in an interruption of the optical beam between two opposed multimode fibres. In the case of dirty environments, the opposed fibre ends are required to be enclosed within a close environment. Several optical fibre microswitch devices are now commercially offered to detect displacement (e.g. with pressure release valves) in hazardous environments [2]. The Plessey Co. in UK has made a pressure operated switch in which a diaphragm is attached to either of the opposed fibres so that only when a correct amount of pressure is applied to the diaphragm will the optical coupling between the fibres take place. The concept of shutter induced interruption of light between opposed fibres can also form the basis of an optical microswitch as shown in Fig. 4(b). In one such shutter mode actual device (as shown in Fig. 5(a)), acoustic waves were made to be incident on a flexible diaphragm (made of rubber 1.5 mm thick and 2 cm in diameter) to which one of the gratings was attached; the other was mounted on the rigid base plate of the housing [9]. The gratings consisted of two $9 \times 3 \times 0.7$ mm cover strip glass substrates on which a 1.16 mm square grating pattern was produced from a 5 μm strip mask by means of photoresist lift off technique through 1200 Å evaporation of chromium [9]. An index matching liquid was inserted between the gratings which were so aligned under a microscope that they were parallel and displaced relative to each other by one half strip width to ensure that the sensor works at the maximum sensitivity region. The overlap area of two glass slides was sealed with a soft epoxy like RTV which enabled displacement of one grating relative to the other and also provided an elastic restoring force. The fibres consisted of two 200 μm plastic core plastic clad fibres. The output laser light (He-Ne) from the input fibre was collimated by means of a GRIN Selfoc lens bonded to it. This collimated beam after transmission through the grating assembly was refocussed by means of another GRIN lens onto the input end of the receiving fibre; this procedure led to isolation of the input and output coupling optics from the gratings. Transmitted light intensity as a function of relative displacement between the gratings is shown in Fig. 5(b). It is apparent from this figure that the sensitivity will be greatest when the bias point is set at either 2.5 μm or 7.5 μm or 12.5 μm and so on because the quantity: dI/dx is largest at these displacements. Such a device was shown to be sensitive to acoustic pressures less than 60 dB (relative to 1 micropascal) over the frequency range 100 Hz to 3 kHz and it could resolve relative displacements as small as few armstrong with a dynamic range of 125 dB. Sensitivity of the sensor can be

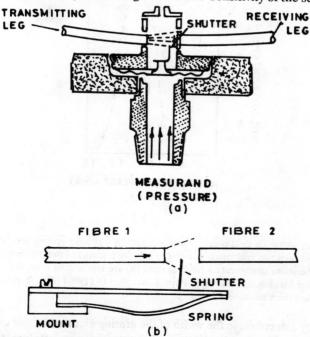

Fig. 4. (a) Pressure sensor based on intensity modulation of light by a shutter across the joint between two multimode fibres (Reprinted by permission. Copyright Instrumentation Society of America 1988. From Fibre Optics Sensors: Fundamentals and Applications, D.A. Krohn [3]); (b) Concept of an optical microswitch (Reprinted from [1] by permission of Kluwer Academic Publishers, Dordrecht).

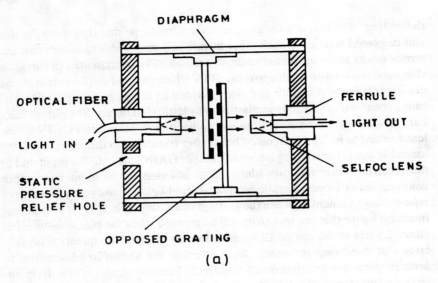

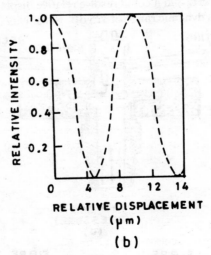

Fig. 5. (a) Acoustic field/pressure sensor based on a moving grating across the joint between two multimode fibres; (b) relative intensity of the transmitted light in the above configuration (both (a) and (b) are reproduced from W.B. Spillman and McMohan, D.H., App. Phys. Letts. **37**, 145 (1980) [9] with permission of American Institute of Physics).

increased by a decrease in the width of the grating elements at the expense of a corresponding decrease in the overall dynamic range. Before testing, the interior of the sensor assembly was filled with distilled water through the pressure relief hole. Further, it was relatively insensitive to static pressure head and responded well to variation in a.c. pressure.

2.3 Reflective fibre optic sensors

In these sensors, in contrast to the transmissive concept of intensity modulation described in the previous sections, measurand is used to induce modulation of the light reflected from a reflecting surface. In its simplest form, an Y-coupler fibre optic probe consisting of two multimode fibres cemented/fused along some portion of their length (two bundles of fibres may also be substituted in their place) to form a power divider would constitute a reflective fibre optic sensor. For example, if the cemented end of the Y-power divider is made to face a light reflecting diaphragm and light is injected through port 1, the intensity of back reflected light that will exit through port 3 will depend on the distance of the reflecting target from the fibre probe (cf. Fig. 6(a)). A typical response curve of such a sensor is shown in Fig. 6 (b), which exhibits a maximum with a steep front slope while the back slope follows an almost inverse square law relationship (i.e. d^{-2} dependence) for the reflected light intensity versus distance (d) of the reflecting target from the fibre probe [3]. As explained in [3], this can be easily understood from geometrical optics argument. If θ represents the maximum light exit angle from the fibre probe corresponding to its N.A., diameter of the illuminating spot on the reflecting target will be approximately $2d \tan \theta \, (=D)$ and hence the diameter of the reflected spot on the probe will be $2D$. As the target is moved away from the probe, the area of the illuminated spot increases in proportion to d^2 while interception of the reflected light by the fibre probe will correspondingly decrease as $1/d^2$. On the other hand, if the target is too close to the fibre probe, a situation would arise in which the reflected light rays would miss the receiving fibre [cf. Fig. 6(c)]. The front slope would disappear if the same fibre is used to collect the reflected light, e.g., in the case of a four port fibre power divider in which the port 2 itself is used both for illuminating the target and collecting the reflected light. Dynamic range of such sensors can be increased by the use of a lens between the fibre probe and the target. The reflective sensors can be used to detect displacement, pressure or even the position of a float in a variable area flow meter [1]. Figure 7 represents the response curve generated by a reflective fibre optic sensor by detecting light reflected from a pressure sensitive diaphragm [3]. In another experiment with an Y-area fibre coupler, in which two 50 μm core fibres were fused against a 200 μm core fibre to form an Y-junction, a.c. pressure variations upto $\sim$ 5 kHz could be measured by detecting the light reflected from a mirrored diaphragm of an audio speaker [10]. Such Y-coupler fibre probes as reflective sensors have reportedly been used to determine surface texture [11], flow of pulp suspension in tube (by employing two Y-fibre probes spaced apart by a preset distance and correlation of the reflected signals picked up by the two probes) in the range $\sim$1-10 m/sec [12], pressure over a range of $\sim$100 psi [13], in medical catheters as intercardiac pressure transducer with a sensitivity $\sim$1 mm of Hg and linearity

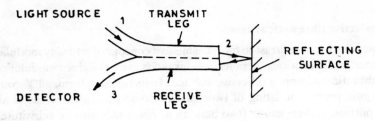
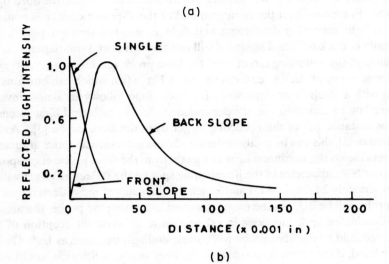
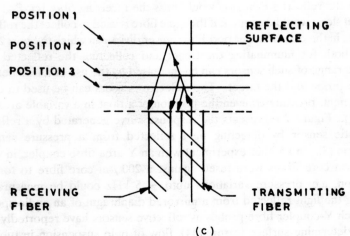

Fig. 6. (a) Y-fibre configuration as a reflective fibre optic sensor; (b) Response of such a reflective sensor in terms of reflected light intensity as a function of distance of the reflecting surface; (c) Physical explanation to the shape of the response curve shown in Fig. 6(b) (all the three figures are reprinted with permission. Copyright Instrument Society of America 1988. From Fibre Optics Sensors: Fundamentals and Applications, by D.A. Krohn [3]).

INTENSITY MODULATED OPTICAL FIBRE SENSORS

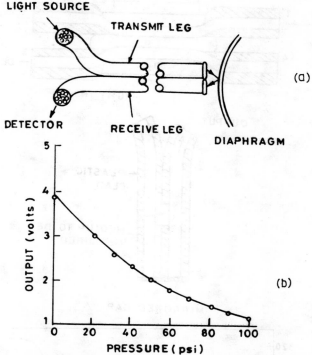

Fig. 7. (a) A pressure sensor working on reflective mode; (b) Response of a reflective fibre optic sensor shown above (both (a) and (b) are reprinted by permission. Copyright Instrument Society of America 1988. From Fibre Optics Sensors: Fundamentals and application, by D.A. Krohn [3]).

in the range 0-200 mm of Hg [14-15], vibrations [16-18], and also in fibre laser doppler anemometry (fibre LDA) [19]. For more references in this area, readers may refer to [20], which contains a huge list of very useful references in the area of fibre optics based sensors.

2.4 Evanescent-wave Fibre Sensors

It is well-known that the guided light in a fibre waveguide penetrates into the cladding to a distance ~ a few wavelengths as the evanescent tail of the waveguide mode. This evanescent tail is precisely responsible for the working of a fibre optic directional coupler. Such a device is formed if two fibres are laid side by side along some portion of their length in which cladding has been depleted to almost negligible thickness. If light is injected into port 1, it will split into ports 2 and 3 with negligible light appearing on port 4 (cf. Fig. 8(a)). The amount of light that will be distributed between ports 2 and 3 is a function of coupling length (L), core-to-core spacing (d) and refractive index n_2. The light exiting through port 3, for example, can be varied by potting the fibre interaction region with a flexible elastomer of index n_2 close to that of the cladding. If appli-

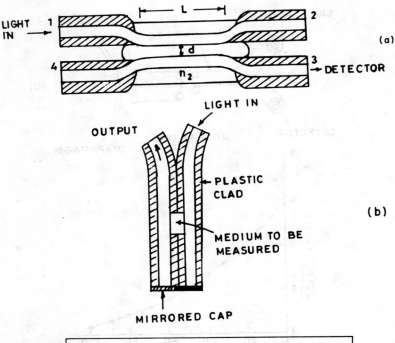

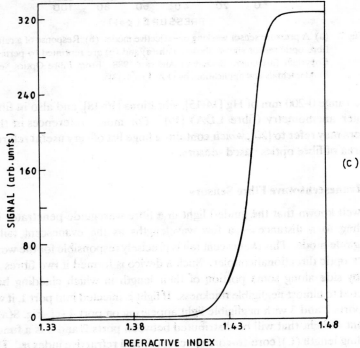

Fig. 8. (a) Geometry of an evanescent field coupled directional coupler/power divider (after [4]) (b) Basic configuration of a cross-talk fibre sensor (c) Cross-talk fibre refractometer (both (b) and (c) are reproduced from [S. Ramakrishnan, J.I.E.T.E. **32**, 307 (1986) 22])

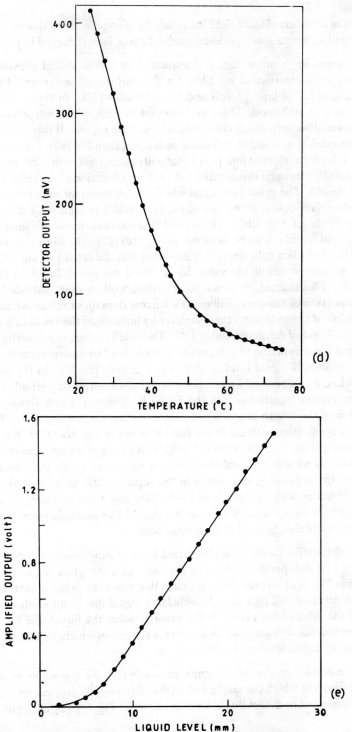

Fig. 8. (d) Cross-talk fibre temperature sensor (e) Cross-talk fibre liquid level sensor (both (d) and (e) are reproduced from S. Ramakrishnan, J.I.E.T.E. **32**, 307 (1986) 221).

cation of an external force field introduced a change in either d or n_2 (and hence L), it will induce a corresponding modulation of light detected in port 3 [21].

A somewhat similar idea was exploited in the so called cross-talk fibre sensors for detection of oil films on the surface of water and for analog measurements of liquid levels and temperatures [22]. In this configuration, cores of two multimode fibres are exposed to each other over a small length with a small air gap, which constitutes the sensing region. If this sensing region is immersed in a liquid of refractive index less than the core refractive index, part of the light injected into port 1 that will get coupled to the second fibre due to cross-talks through this sensing zone will be a function of the refractive index of this liquid. The cross-talk signal which travels down through fibre 2 is made to undergo reflection at the end of the fibre which is eventually detected by a photo-diode (cf. Fig. 8(b)). The typical characteristic curve of signal level as a function of liquid refractive is shown in Fig. 8(c) [22]. The fibre in question was a plastic clad silica core fibre. It is apparent that the cross-talk signal saturates from an almost negligible value for a liquid refractive index in the range 1.4-1.46. This characteristic was used to detect oil on water surface [23]. The linear portion of the curve (albeit in a narrow domain of refractive index) can be exploited to construct a thermometer by immersing the cross-talk zone in a suitable liquid of refractive index 1.45. The small change in refractive index of the liquid as a function of temperature would lead to a corresponding change in the cross-talk signal level as shown in Fig. 8(d) [22-24]. In the same configuration, it can be used as an analog liquid level sensor by partially filling the sensing region lengthwise with the liquid because cross-talk signal will be a function of the length of the liquid column within the sensing zone. Result of such an application with car brake fluid is shown in Fig. 8(e) [22]. By reducing the sensing zone, i.e., the cross talk region to a bare minimum length, it can be alternatively used as a digital on-off sensor for detecting the threshold level of a liquid by suspending the sensor in the liquid container so that the sensing zone coincides with the threshold level. This way it can be used, for example, to trigger an alarm signal if a hazardous liquid like petrochemical is found to ever cross the danger level in its storage tank.

An alternative mode of digital liquid level sensing involves optimally connecting two independent fibres to the base of a 90° glass micro-prism. As shown in Fig. 9(a), the ray paths are such that when the prism is surrounded by air, the detector will pick up relatively large signal due to total internal reflection at the sides of the prism. If the prism touches the liquid, the light suffers attenuation due to frustrated total internal reflection which, in fact, results in a drop in the signal level at the detector end.

In a modified version of the above principle [22], an Y-coupler made out of plastic fibres in which the single end of the fibre coupler was cut into a conical shape was used to detect liquid level as shown in Fig. 9(b) or configure a fibre

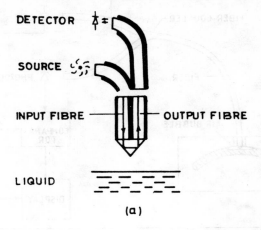

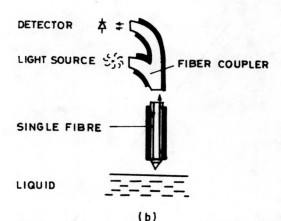

Fig. 9. (a) Liquid threshold level sensor based on two independent fibres and a microprism (see text) (b) Liquid level sensor with a Y-fibre coupler and an in-line microprism (see text) (reproduced from S. Ramakrishnan, J.I.E.T.E. **32**, 307 (1986) [22])

optic spirit level [25] (cf. Fig. 9(c)) by introducing the sensing tip at the centre of a spirit level. The detected signal level would obviously be higher if the sensing tip faces the air bubble in contrast to the signal when it will face the liquid. This configuration of an Y-fibre coupler tip can also be exploited to construct a fibre optic refractometer [26]. Typical characteristic curve of such a fibre refractometer is shown in Fig. 9(d) [22].

2.5 Microbend optical fibre sensors

A widely studied intensity modulated sensor is based on microbend induced excess transmission loss of an optical fibre to detect/measure displacement, acoustic pressure (hydrophone), strain and temperature [7, 27-31]. It has long been known that if a portion of a fibre lay is forced by a deformer to go through

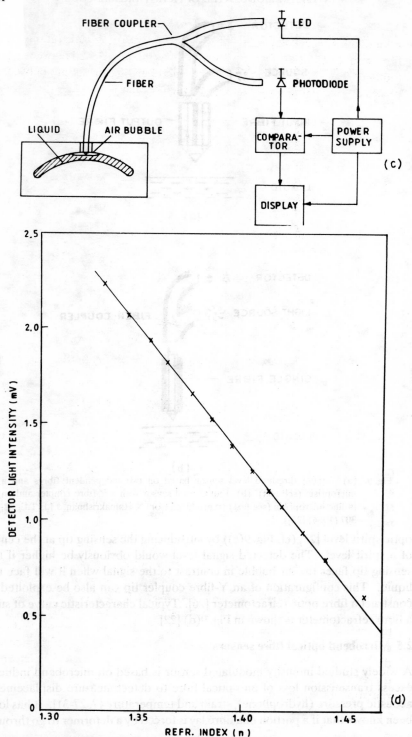

Fig. 9. (c) Surface level indicator with a fibre probe coupled to a spirit level (d) Y-fibre coupler refractometer (both (c) and (d) are reproduced from S. Ramakrishnan, J.I.E.T.E. **32**, 307 (1986) [22]).

a continuous succession of small bends on a microscopic scale (cf. Fig. 10), the fibre would exhibit excess loss. The fibre may be sandwiched between a pair of toothed or serrated plates to induce microbending or simply between two pieces of sand paper under pressure. Such regular or random meandering of

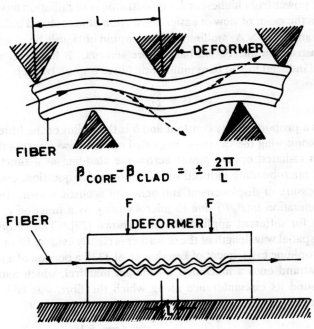

Fig. 10. Principle of microbend induced attenuation in an optical fibre (see text) (after [4]).

fibre axis results in redistribution of guided power between modes of the fibre and also coupling of power from one mode/mode group to another. If the spatial wavelength (Λ) of periodic deformation satisfies the so called phase-matching condition:

$$\beta_p - \beta_q = \frac{2\pi}{\Lambda}, \qquad (1)$$

where β_p and β_q represent modal propagation constants of two modes, power transfer will take place from p-th to q-th mode. If the q-th mode happens to be a radiation mode, this transfer of power will result in a net transmission loss of the guided modes. From a coupled mode analysis, it can be shown that in the case of a step index fibre of core radius a, core index n_1 and relative core-cladding index difference Δ, critical spatial wavelength (Λ_{cr}) of a deformer required to induce heavy transfer of power from highest order guided mode to radiation modes will be

$$\Lambda_{cr} = \frac{\pi a}{\sqrt{\Delta}} = \frac{\sqrt{2}\,\pi.a.n_1}{N.A.}. \qquad (2)$$

Typically, in a 62.5 μm core multimode fibre of the step-index type of N.A. 0.30, the spatial deformation periodicity required to induce heavy microbending loss would be ~0.69 mm whereas for a 200 μm core similar fibre of N.A. 0.18 (e.g. plastic clad silica core fibre), the required periodicity to induce large transfer of power from highest order guided modes to radiation modes is ~3.6 mm. From the point of view of easier fabrication of microbend inducer, use of large core and large N.A. multimode fibres would obviously be of advantage to use in constructing microbend based fibre sensors. It has been shown that microbend induced loss in bare multimode fibres is given by [32]

$$\alpha = \frac{K}{\Delta^3} \frac{a^4}{b^6}, \tag{3}$$

where K is a proportionality constant and b is the radius of the fibre cladding. Thus, by monitoring the decrease in guided optical power across the core or increase in radiated optical power across the cladding as a function of the amount of microbending induced on the fibre under question, one can construct a pressure or displacement and hence an acoustic sensor. In Fig. 11, a plot of attenuation $\ln(P_0/P)$ due to microbending as a function of deformer periodicity for different applied forces is shown [29]. The resonance at a particular spatial wavelength of the deformer is clearly evident from the figure. In the hydrophone experiment of Lagakos et. al. [33], a portion of a multimode fibre was wound onto a machined aluminium mandrel, which had a helical groove around its circumference along which the fibre was laid to induce

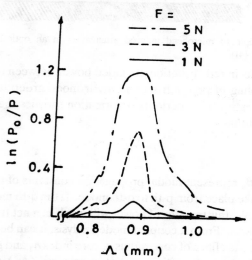

Fig. 11. Micro-bend induced attenuation as a function of spatial wavelength of the microbend inducer (reproduced from W.H.G. Horsthuis and Fluitman, J.H.J., Sensors and Actuators 3, 99 (1982/83) with permission. Copyright Elsevier Sequoia S.A., Lausanne); spatial wavelength can be easily varied for a given pair of microbend inducers through $\cos \phi$ factor by sandwiching the fibre at an angle (ϕ) to the toothed ridges.

microbending. The mandrel was covered with an acoustically sensitive compliant rubber boot of diameter 36 mm for acoustic coupling to the fibre.

In the case of a parabolic index core fibre, adjacent guided modes are separated from each other by a constant difference in wavenumber space [34] given by

$$\Delta\beta = \frac{\sqrt{2\Delta}}{a} \Rightarrow \Lambda = \frac{2\pi}{\Delta\beta} = \frac{2\pi a n_1}{N.A.} \qquad (4)$$

as the spatial wavelength of a deformer required to couple power between adjacent modes. Since all modes are equally spaced, above deformer periodicity will lead to a continuous transfer of power to adjacent modes and hence eventually to radiation modes thereby exhibiting a detectable drop in guided mode light intensity at the end of the fibre. If T represents the guided mode transmission coefficient of the fibre and W the optical power at the fibre input, the power received by the detector will be $W.T$, which will result in a detector output current as

$$i = \left(\frac{W.T}{h\nu}\right) \eta e . \qquad (5)$$

Here η is quantum efficiency of the detector, e the electronic charge while h and ν respectively, represent Planck's constant and frequency of the light wave injected into the fibre. Let P represent a change in pressure applied to the deformer. This will lead to an effective force ΔF on the microbent fibre causing the amplitude X of the bent fibre to change by ΔX. As a result of this change in amplitude of fibre deformation, the fibre will suffer an incremental loss. Hence, transmission coefficient T of the fibre will change by ΔT and one can relate ΔT to ΔX through [35]

$$\Delta T = \left(\frac{\Delta T}{\Delta X}\right) D . \Delta P , \qquad (6)$$

where

$$D \cdot \Delta P = \Delta X . \qquad (7)$$

D being a constant that depends on the pressure change ΔP. A typical plot for drop in guided light transmission as a function of static force applied to a step index optical fibre microbend sensor is shown in Fig. 12 [7]. The curves are labelled with light launch angle as a variable. It is apparent that for higher angle launching which amounts to preferential excitation of higher order modes, leads to a greater drop in transmission for the same amount of applied force.

The transduction coefficient $\Delta T/\Delta P$ can be written as [33]

$$\frac{\Delta T}{\Delta P} = \left(\frac{\Delta T}{\Delta X}\right) \left(\frac{\Delta X}{\Delta P}\right) . \qquad (8)$$

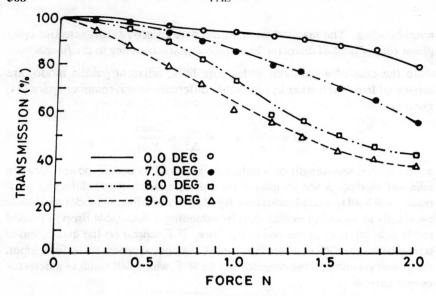

Fig. 12. Transmitted power as a function of weight in a microbend sensor; different curves correspond to different light launch angles (adapted with permission of American Institute of Physics copyright 1980, from, "Fibre Optic Pressure Sensor", by J.N. Fields, C.K. Asawa, O.G. Romer and M.K. Barnoski, J. Acoustic Society of America **67**, 816-818 (1980); very similar results were obtained by us also [7]).

Comparing Eq. (6) with (8), we find that $D = \Delta X/\Delta P$.

The detector output (the signal) (cf. Eq. (5)) will consequently change to

$$i_s = \frac{\eta e W}{h\nu} \left(\frac{\Delta T}{\Delta X}\right) \cdot D \cdot \Delta P. \tag{9}$$

However, a noise current (i_N) will also be over-riding the signal. Assuming a shot-noise limited detection system, the mean square detector noise current is given by [36]

$$i_N^2 = 2e\,(\eta e\,W\,T/h\nu).\Delta f, \tag{10}$$

where Δf is detector bandwidth. Minimum detectable pressure is calculated for the condition: S/N power ratio (i_S^2/i_N^2) as unity, which yields from Eqs. (8-10) [35]

$$\Delta P_{min} = D^{-1} \left(\frac{\Delta T}{\Delta X}\right)^{-1} \left[\frac{2T \cdot h \cdot \nu \cdot \Delta f}{\eta W}\right]^{1/2}. \tag{11}$$

It can be seen from Eq. (11) that minimum detectable pressure will be lower if ($\Delta T/\Delta X$) as well as D are relatively large. In order to achieve this (i) the fibre should be so designed that it is highly sensitive to microbending and (ii) D should be large through appropriate design of the deformer [33].

Obviously, one can accomplish the former by reversing the design procedures commonly employed to minimise mode-coupling losses in telecommunication grade fibres. Even the fibre coating material could be used as a design parameter to enhance the acoustic sensitivity of the sensor as shown in Fig. 13 [4]. Initially, microbend fibre optic sensors were mainly used to detect acoustic fields as hydrophones and displacement sensors [31,33]. However,

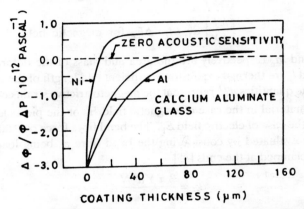

Fig. 13. Acoustic field response for different coating materials (after [4]).

depending on the construction of the deformer, the same concept can be exploited as a generic sensor to determine several other environmental changes like temperature, acceleration and electric and magnetic fields [35]. Denoting the generic environmental change as ΔE (in place of ΔP in Eq. (6)) one can rewrite Eqs. (6) and (7) as [35]

$$\Delta T = \left(\frac{\Delta T}{\Delta X}\right) \cdot D \cdot \Delta E \qquad [12(a)]$$

with

$$D \cdot \Delta E = \Delta X. \qquad [12(b)]$$

For a generic microbend sensor which measures ΔX as defined by Eq (12), the minimum ΔX will be (cf. Eq. (11))

$$\Delta X_{min} = \left(\frac{\Delta T}{\Delta X}\right)^{-1} \left[\frac{2 T h \nu \Delta f}{\eta e}\right]^{1/2}, \qquad (13)$$

which on combining with Eq. (11) can be written as

$$\Delta E_{min} = D^{-1} \Delta X_{min} \qquad (14)$$

Following detailed discussions provided by Logakos et al [35] with regard to design optimisation of the deformer (in terms of parameter D) for various environments, change in transmission coefficient can be expressed as follows [35]:

$$\Delta T = \left(\frac{\Delta T}{\Delta X}\right) \cdot A_p \cdot k_f^{-1} \cdot \Delta P \text{ for pressure}, \qquad [15(a)]$$

$$= \left(\frac{\Delta T}{\Delta X}\right) \cdot \alpha_s \cdot l_s \cdot \Delta\theta \quad \text{for temperature,} \qquad [15(b)]$$

$$= \left(\frac{\Delta T}{\Delta X}\right) m_p \cdot k_f^{-1} \cdot \Delta a \quad \text{for acceleration,} \qquad [15(c)]$$

$$= \left(\frac{\Delta T}{\Delta X}\right) d_{33}^H \cdot l_s \cdot \Delta H_F \quad \text{for magnetic field,} \qquad [15(d)]$$

$$= \left(\frac{\Delta T}{\Delta X}\right) d_{33}^E \cdot l_s \cdot \Delta E_F \quad \text{for magnetic field,} \qquad [15(d)]$$

where A_p and m_p respectively are the area and mass of the deformer plates while α_s and l_s are thermal expansion coefficient and length of the spacers (cf. Fig. 14). The quantities $d_{33}^{H,E}$ represent the magnetostrictive strain coefficient of the spacer material in the case of magnetic field H_F or the piezoelectric stain constant in the case of electric field E_F. The bend fibre force constant k_f in Eq. (15) can be evaluated by considering the bend fibre as being loaded at the centre and clamped at the ends [35]

$$k_f = \frac{3\pi Y d^4 N}{\Lambda^3}, \qquad (16)$$

with Y and d representing the Young's modulus and diameter of the fibre, respectively and N the number of bent intervals. Assuming ΔX_{min} as 0.01 Å, the minimum detectable changes for a variety of environments are listed in Table 1. Thus, a microbend fibre optic sensor is a versatile sensor which could be suitably configured to measure a variety of environmental changes.

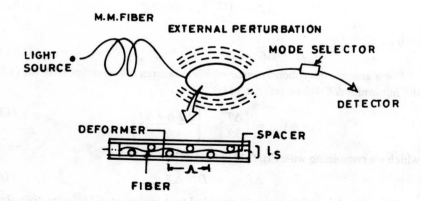

Fig. 14. Generic configuration of microbend sensor (reproduced from N. Lagakos, Cole, J.H. and Bucaro, J.A., App. Opt. 26, 2171 (1987) [35] with permission of Optical Society of America).

TABLE 1 [35]

Environment	D	Minimum detectability	Remarks
Pressure	$A_p k_f^{-1}$	N. 3×10^{-4} dyne/cm²	$A_p = 1$ cm² $k_f^{-1} = 33 \times 10^{-8}$/dyne/cm for silica fibre
Temperature	$\alpha_s l_s$	4×10^{-6}°C	$l_s = 1$ cm, $\alpha_s = 24\times10^{-6}$/°C for Aluminium
Acceleration	$m_p k_f^{-1}$	N. 3×10^{-4} cm/sec²	1 cm² and 1 mm thick lead plate
Magnetic field	$d_{33}^H l_s$	9.6×10^{-5} Oe	$l_s = 1$ cm and $d_{33}^H = 10.4\times10^{-7}$/G (for Permalloy)
Electric field	$d_{33}^E l_s$	1.7×10^{-1} V/m	$l_s = 1$ cm and $d_{33}^E = 593\times10^{-12}$ m/V (for PZT-5H piezoceramic)

2.6 Fibre optic refractometers

In a number of ways, optical fibres can be used to measure refractive index of a liquid through intensity modulation of light by the measurand. In Fig. 15, light is seen to get partially coupled to fibre 2 from fibre 1 through a fixed small air gap [1]. Due to the air gap, only a fraction of the light emitted (into a cone characteristic of its N.A.) will be intersected and collected by the fibre 2. If, however, a liquid of refractive index $> n_{air}$ fills up the gap between the fibres, a greater fraction of the light emitted by fibre 1 will be picked up and guided by fibre 2 as shown in the figure due to an effective decrease in fibre N.A.. The ratio of the power collected by fibre 2 through the liquid to that through air is a linear function of the refractive index of the liquid [1]. This configuration has been also used to detect acid level in batteries [37] and for that matter as a

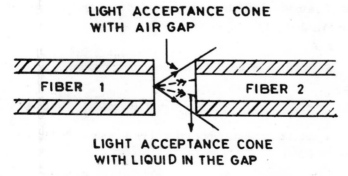

Fig. 15. A fibre refractometer configuration, which can also be used as a threshold liquid level sensor (reproduced from [1] with permission of Kluwer Academic Publishers, Dordrecht).

threshold liquid level sensor because, with and without the liquid in the gap between the two fibres, the signal level at the output of fibre 2 will be different.

In another configuration, a bare tapered multimode fibre has been used as a refractometer [38]. Figure 16 represents geometry of such a tapered fibre.

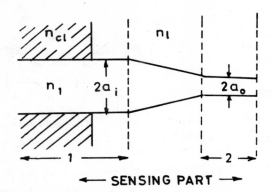

Fig. 16. Geometry of a tapered multimode fibre refractometer (reproduced from A. Kumar, Subrahmonium T.V.B., Sharma A.D., Thyagarajan K., Pal B.P. and Goyal I.C., "A Noval Refractometer using Tapered Optical Fibres", Electron. Letts. 20, 534 (1984) [38] with permission. Copyright 1984 IEE, Stevenage).

The tapered portion can be though of as interconnecting two fibres: #1 of core dia $2a_i$; and #2 of core dia $2a_o (a_o < a_i)$. Fibres 1, 2 and tapered interconnecting zone all having the same core and cladding refractive indices n_1 and n_2, respectively except for the initial section of fibre 1 in which the cladding index is n_{cl}. A guided mode of effective index $\widetilde{\beta}_1$ ($= n_1 \cos \theta_1$, θ_1 being the characteristic mode propagation angle) in fibre 1 gets transformed to a corresponding characteristic propagation angle $\theta(Z)$ as it propagates down the taper as [38]

$$a(Z) \sin \theta(Z) = a_i \sin \theta_1, \tag{17}$$

where $a(Z)$ represents radius of the taper at a distance Z from its thick end. Accordingly, a mode of effective index $\widetilde{\beta}_1$ in fibre 1 will get transformed through the taper to a mode of $\widetilde{\beta} = \widetilde{\beta}_2$ in fibre 2 with $\widetilde{\beta}_2$ as [38]

$$\widetilde{\beta}_2 = n_1 \cos \theta_2 = n_1 \left[1 - R^2 \frac{(n_1^2 - \widetilde{\beta}_1^2)}{n_1^2}\right]^{1/2}$$

$$= \left[n_1^2 - R^2 (n_1^2 - \widetilde{\beta}_1^2)\right]^{1/2}, \tag{18}$$

where $R (= a_i/a_o)$ represents taper ratio. For a mode to be guided in fibre 2, one must have [38]

$$\widetilde{\beta}_1 \geq \left[n_1^2 - \frac{n_1^2 - n_l^2}{R^2}\right]^{1/2} \equiv \widetilde{\beta}_{\min}^2. \tag{19}$$

If P_o represents the total power injected into the guided modes of fibre 1, then the power in the modes with $\widetilde{\beta}_1 > \widetilde{\beta}_{\min}$ will be

$$P_b = P_o \left[\frac{n_1^2 - \widetilde{\beta}_{\min}^2}{n_1^2 - n_{cl}^2}\right]^{1/2}, \tag{20}$$

which on substitution of $\tilde{\beta}_{min}$ from Eq. (19) becomes

$$P_b = P_o \frac{n_1^2 - n_l^2}{R^2 \left(n_1^2 - n_{cl}^2\right)}. \tag{21}$$

It is evident from this equation that power coupled to fibre 2 through the taper increases linearly with proportional decrease in n_l^2. This result can be exploited to construct a fibre refractometer by starting with a plastic clad silica core fibre from a small portion of which plastic has been removed [38]. This bare portion of the fibre is then converted into a taper by heating and pulling and immersing the taper zone in a liquid of refractive index $n_l(<n_1)$. Immersing the taper subsequently in a number of other liquids while taking care to clean tapered zone each time appropriately, and monitoring the corresponding power reaching the fibre 2, one can generate a calibration curve for a given fibre taper. Thus by measuring the power exiting fibre 2 when the taper is immersed in a liquid of unknown refractive index, one can use this calibration curve to determine the refractive index of the unknown liquid. Experimental results alongwith theoretically expected results for a taper (of ratio 3.2) are depicted in Fig. 17. In principle, the same technique can be used to construct a temperature sensor by encapsulating the taper with a metallic encapsulation filled with a liquid whose refractive index is temperature sensitive.

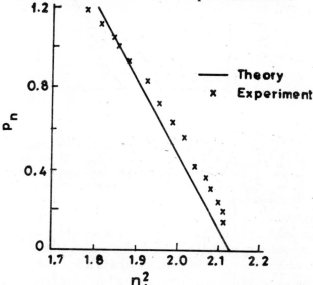

Fig. 17. A comparison between theory and experiment of a tapered fibre refractometer (reproduced from A. Kumar, Subrahmonium T.V.B., Sharma A.D., Thyagarajan K., Pal B.P. and Goyal I.C., "A Novel Refractometer using Tapered Optical Fibres", Electron. Letts. 20, 534 (1984) [38] with permission. Copyright 1984 IEE, Stevenage).

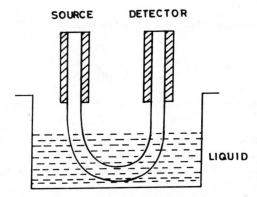

Fig. 18. Schematic of a U-shaped fibre refractometer.

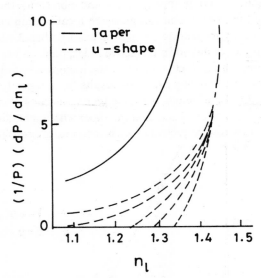

Fig. 19. Comparison in sensitivity between tapered and bent fibre refractometers (reproduced from A. Kumar, Subrahmonium T.V.B., Sharma A.D., Thyagarajan K., Pal B.P. and Goyal I.C., "A Novel Refractometer using Tapered Optical Fibres", Electron. Letts. **20**, 534 (1984) [38] with permission. Copyright 1984 IEE, Stevenage).

In a companion study, Takeo et al. [39] have reported a fibre optic refractometer in which a plastic clad silica fibre has been bent into the form of an U-shape by removing plastic cladding from a section of the fibre (cf. Fig. 18). The bent section of the fibre, when immersed in a liquid of refractive index n_l ($> n_{air}$) yields a signal different from that in a bare fibre in the same manner as in a tapered fibre refractometer. However, as shown in Fig. 19 a comparison of the sensitivity of the measured liquid refractive index in these two configurations has shown that the tapered fibre refractometer is more sensitive than that of the bent fibre refractometer at any value of the refractive index [38].

2.7 Intensity modulated fibre optic thermometers

It is well known that metal clad optical fibres exhibit high attenuation at optical frequencies due to the metal having an imaginary part for its refractive index [40]; longer the length of the metal cladding, larger will be the attenuation. This fact has been exploited in [41] to construct a fibre optic thermometer. It essentially worked like an ordinary mercury thermometer. A plastic clad silica fibre was used as the fibre from which a portion (~4 cm) of the plastic cladding was removed. The fibre was run through a glass reservoir (a capillary tube 1 mm in diameter having a bulb at one end) of mercury taking care that the lower end of the unclad portion of the fibre barely touches the upper level of mercury at the ambient temperature as shown in Fig. 20. As the temperature surrounding the mercury reservoir is raised, the mercury level rises in accordance with conventional thermometer principle and thereby becomes the cladding for the bare portion of the fibre. As the temperature rises, more and more length of the fibre becomes metal clad which leads to more and more attenuation of the guided light due to absorption by the metallic mercury. The variation of normalised power ($P_N = P/P_o$, with P_o as the initial power) detected at the fibre end with length (l) of the mercury cladding in the capillary is shown in Fig. 21(a). A functional dependence for P_N with l as $P_N = b/(l+l_o)$ was found to yield a good fit to the experimental data; this is also evident from the linear nature of the curve for P_N^{-1} vs l (curve II). Since l will be proportional to

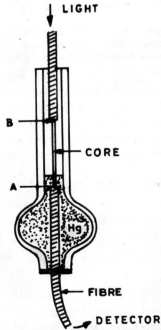

Fig. 20. Metal clad fibre temperature sensor (reproduced from A. Kumar, Shenoy M.R., Pal B.P. and Goyal I.C., J.I.E.T.E. **32**, 347 (1986) [41]).

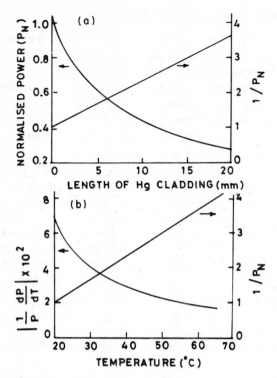

Fig. 21. (a) Transmitted power as a function of length of the mercury column; (b) Temperature response of the mercury clad fibre thermometer. (both (a) and (b) are reproduced from A. Kumar, Shenoy M.R., Pal B.P. and Goyal I.C., J.I.E.T.E. **32**, 347 (1986) [41]).

temperature (T), P_N^{-1} vs. T will also yield a linear curve as is evident from Fig. 21(b) where abcissa representing 'T' is replaced with corresponding temperature scale. The temperature sensitivity of this thermometer can be increased by suitably modifying the design of the mercury reservoir so as to affect longer increase in length of the mercury column per unit degree rise in temperature. Another interesting aspect of this device is that the high sensitivity range of operation of the thermometer can be shifted to other temperature ranges by simply shifting the threshold level of the bare portion of the fibre such that it starts touching the mercury level at the lower limit of the temperature range to be measured. Although the time response of this thermometer is of the same order as that of an ordinary mercury thermometer, it offers the advantage of remote measurement.

In another version [42], a fibre optic thermometer especially for medical applications was constructed by exploiting light intensity modulation induced by a thermosensitive cladding surrounding a short bare portion of an otherwise plastic clad silica fibre (cf. Fig. 22). The sensor works on a reflective mode, i.e.,

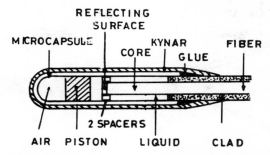

Fig. 22. Fibre optic thermometer for medical applications (after M. Brenci, Conforti G., Falciai R. and Scheggi A.M., "Opitcal Fibre Thermometer for Medical Use", SPIE Vol. 494, Novel Optical Fibre Techniques for Medical Applications, pp. 13-17 (1984) [42] with permission.)

light injected into the PCS fibre is reflected back from one end of the fibre and the returned light intensity is monitored. Since the refractive index of the liquid (a mineral oil) cladding around the sensing zone is temperature dependent, effective N.A. of the fibre for collecting the reflected signal gets modulated in accordance with variations in temperature used to construct the thermometer (cf. Fig. 22). The reported sensitivity of such a thermometer was 0.1°C with accuracy $\sim \pm 0.3°C$ and response time ~ 1 sec [43].

Another approach of measuring temperature involves the use of a fibre optic probe tipped with rare earth phosphors which fluoresce under excitation with UV light. The fluorescent radiation appears in the visible range (hence, does not interfere with the exciting UV light) and is collected by the same fibre for detection. In practice, relative intensities of two sharp characteristic fluorescent emission lines are measured, the ratio of which is a measure of the temperature of the phosphor. The emitted lines are a direct function of temperature and the above ratio is independent of extraneous noise sources. A phosphor which has been successfully employed in such sensors is known as europium-activated gadolinium oxysulfide [44]. The sensor can work over the temperature range −50°C to 250°C with an accuracy $\sim 1°C$, which can be increased to $\sim 0.1°C$ in narrower temperature range [3]. Though commercial thermometers based on this fluorescence technique tend to be more expensive than their optomechanical counterparts, they have a faster response due to their smaller size [45]. Fluorescent temperature sensors based on fibre optics have been successfully used in the so called Hypothermia for cancer therapy where affected tissues are locally heated with rf field requiring temperature to be accurately monitored during the treatment [3].

2.8 Chemical Analysis

Optical fibres have been used in a variety of ways for chemical analysis [46]. Fluorescent techniques have been extensively exploited in this domain of optical fibre sensors. For example, in remote fibre fluorimetry (RFF), as in the

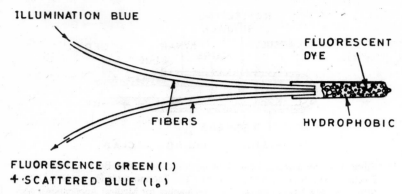

Fig. 23. A fibre optic fluorescence probe (reprinted with permission from J. Peterson, Fitzgerald R. and Buckhold D., "Fibre Optic Probe for in Vivo Measurement of Oxygen Partial Pressure", Analytical Chemistry **62**, 56 (1984), copyright 1984, American Chemical Society [48]).

case of fluorescence based fibre optic thermometers, a high intensity light beam is injected into a typically large core quartz fibre. The output end of the fibre is immersed in the sample to stimulate fluorescence, which is collected by the same fibre (often referred to as an "optrode") for detection and processing perhaps by an on-line computer [3]. This technique has been reported to have been used to monitor contamination of ground water with pollutants like toluene, xylene associated with gasoline spills and leakage [3,47]. In certain situations, a target of known fluorescence may be employed, whose fluorescence may be modulated by the sample when brought in contact with the target [3,48]. RFF technique has been also used to measure partial pressure of oxygen in a given environment by measuring fluorescence quenching due to chemical reaction of oxygen with a fluorescent dye attached to a fibre probe (see Fig. 23). Remote fibre fluorescence quenching technique has been extended to also measure concentration of glucose [49], halides [50] and pH over the range 3 to 8 [3, 51].

Plastic clad silica (PCS) fibres having a dye-doped uv-curable silicone acrylate as the cladding have been recently employed as a chemical sensor for

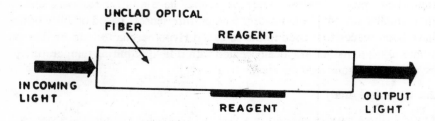

Fig. 24. A fibre optic chemical sensor (reproduced with permission from W. Seitz, "Chemical Sensors based on Fibre Optics", Analytical Chemistry, **56**, 667 (1984), copyright 1984, American Chemical Society [49]).

ammonia [52]. In the actual experiment, spectral loss characteristics of such a fibre was measured once without and then in the presence of ammonia; the dye related absorption peak at 665 nm (450 dB/km in air) was found to decrease to 90 dB/km in 100% ammonia. Ammonia vapours were also detected by bonding a suitable immobilised dye directly as cladding on an otherwise bare core fibre (cf. Fig. 24). On reaction with ammonia, the dye changed colour which resulted in a loss in transmission due to enhanced absorption of the evanescent wave associated with the guided modes of the fibre [53].

2.9 Distributed sensing with fibre optics

Lately there has been a great deal of interest for distributed sensing, in particular, of temperature to determine thermal distribution and location of hot spots using fibre optics. The technique is essentially based on the well known time-domain reflectometry (OTDR) concept [54] extensively used in optical telecommunication to obtain non-destructive measurement of fibre loss and locations of any discontinuity e.g. break, splice along a fibre length. In the first implementation, a liquid core fibre was employed to detect the Rayleigh backscatter signature from a long length of the fibre; any localised temperature rise led to an increase in backscatter signal which could be detected and processed at the input end of the fibre to locate the measurand as well as obtain its temperature [55]. There were, however, some doubts about the life and stability of liquid core fibres. This technique was later on modified in which

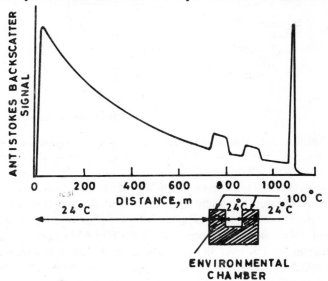

Fig. 25. Fibre optic distributed temperature sensor based on backscattered light measurement (reproduced from A.H. Hartog, Leach A.P. and Gold M.P., "Distributed Temperature Sensing in Solid Core Fibres", Electron, Letts. *21*, 1061 (1985) [56] with permission. Copyright 1985 IEE, Stevenage).

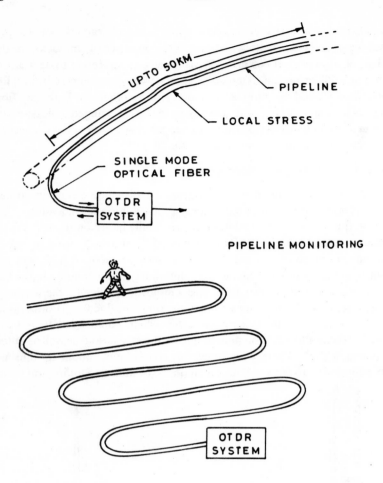

Fig. 26. Distributed fibre optic sensor based on OTDR principle (reprinted with permission. Copyright Instrument Society of America 1984. From "Distributed fibre optic sensors", by S. Kinsley [58]).

solid core conventional telecommunication grade germano-silicate multimode fibres were substituted for liquid core fibres and the signal processing was based on detection of anti-stokes Raman line of the backscattered light, which was picked up with an interference filter of FWHM ($\Delta\lambda$) 11 nm. The equipment could resolve 1°K over fibre lengths greater than 1 km with a spatial resolution of 7.5 m over the temperature range −50 to 100°C. A typical anti-stokes Raman OTDR signature as a function of location in the fibre is reproduced in Fig. 25 from [56]. The ambient temperature was 24°C, the return of the signal to ambient level between the two hot measurement zones is evident from the figure. A commercial version is now available which can measure temperature (accuracy ± 1°C) and locations at 1000 points (spatial resolution 7.5 metre) over 4 loops, each loop being upto 2 km long. Measurement time per loop is 12s. Spatial resolution could be enhanced to ~4 cm in the point probe mode of operation by tightly coiling the fibre on a former. Measurement range is –100°C to 125°C, which can be extended to 600°C with special fibre coating to preserve its strength [57].

The conventional OTDR can also be employed to locate local stresses along a pipeline by bonding a continuous length of fibre to the walls of the pipe. The stressed locations would yield alarm signals through excess loss in the backscatter signature [3] due to actuation of microbending. In an analogous manner, by operating as a pressure sensor, stepping in of an intruder in a security fence can be detected and located by laying fibres underneath sensitive areas for defense as shown in Fig. 26. The OTDR concept has also been reported to have been used to locate hot spots, e.g., in power cables and high voltage transformers by exploiting the change in refractive index of the fibre cladding induced by local heating at the region of hot spots [3].

3. CONCLUSION

A survey of several fibre optic sensor concepts that have been reported in the literature has been presented. It is seen how very simple concepts could be exploited to configure a variety of optical sensors based on optical fibres. The list is never-ending; only a few have been described here which to our opinion, has good market potential and requires investments in terms of engineering studies.

REFERENCES

1. R.S. Medlock, "Fibre Optic Intensity Modulated Sensors" in "Optical Fibre Sensors", Ed. By A.N. Chester, S. Martelluci and A.M. Verga Scheggi, Martinus Nijhoff, Dordrecht (1987).
2. G.D. Pitt, Extance P., Neat R.C., Batchelder D.N., Jones R.E., Barnett J.A. and Pratt R.H., "Optical Fibre Sensors", Proc. IEE **132**, pt. J., 214-218 (1985).
3. D.A. Krohn, "Fibre Optic Sensors-Fundamentals and Applications", Instrument Society of America (1988).
4. C.M. Davis, Carome E.E., Weik M.H., Ezekiel S. and Einzig R.E., "Fibre Optic Sensors Technology Handbook", Dynamic Systems Inc., Virginia (1982).
5. R.N. Chakraborty and Pal B.P., Internal Study on "Development of a Fibre Optic Fusion Splicing Machine", IIT Delhi (1985).
6. W.B. Spillman and Gravel R.L., "Moving Fibre Optic Hydrophone", Opt. Letts. **5**, 30-33 (1980).
7. N.R. Paramanthan and Pal B.P., Internal Study on "Fibre Optic Acoustic Sensors", IIT Delhi (1984).
8. W.B. Spillman and McMohan D.H., "Frustrated Total Internal Reflection Multimode Fibre Optic Hydrophone", App. Opt. **19**, 113-117 (1980).
9. W.B. Spillman and McMohan D.H., "Schlieren Multimode Fibre Optic Hydrophone", App. Phys. Letts. **37**, 145-147 (1980).

10. M. Priyamvade and Pal B.P., Internal Study on "Fabrication of Fused Y-Coupler", IIT Delhi (1989).
11. S. Uena, "New Method of Detecting Surface Texture by Fibre Optics", Bull. Japan Soc. of Prec. Engg. **7**, 87-90 (1973).
12. K. Oki, Akehata T. and Shirai T., "A New Method of Evaluating the Size of Moving Particles with a Fibre Optic Probe", Powder Tech. **11**, 51-54 (1975).
13. C.R. Tallman, Wingate F.P. and Ballard E.O., "Fibre Optic Coupled Pressure Transducer", ISA Trans. **19**, 49-51 (1980).
14. L.H. Lindstroem, "Miniaturised Pressure Transducer Intended for Intravascular Use", IEEE Trans. on Bio-medic. Engg. **BME-17**, 207-219 (1970).
15. H. Matsmoto, Saegusa M., Saito K. and Mizoi K., "The Development of a Fibre-Optic Catheter-tip Pressure Transducer", J. Med. Engg. Tech. **2**, 239-242 (1978).
16. M.A. Nokes, Hill B.C. and Barelli A.E., "Fibre Optic Heterodyne Interferometer for Vibration Measurements in Biological Systems", Rev. Sc. Instrum. **49**, 722-728 (1978).
17. F. Parmigiani, "A High Sensitivity Laser Vibration Meter Using a Fibre-Optic Probe", Opt. Quant. Electron. **10**, 533-537 (1978).
18. S. Uena, Shibata N. and Tsujiuchi J., "Flexible Coherent Optical Probe for Vibration Measurements", Opt. Commn. **23**, 407-410 (1977).
19. K. Kyuma, Tai S., Hamanaka K. and Nuhoshita M., "Laser Doppler Velocimeter with a Novel Fibre Optic Probe", App. Opt. **20**, 2424-2428 (1981).
20. A.L. Harmer, "Optical Fibre Sensors"-A Survey Report prepared by Battele-Geneva Research Centre, Switzerland.
21. S. Sheem and Cole J., "Acoustic sensitivity of single-mode optical fibre power divider", Opt. Letts. **4**, 322-326 (1979).
22. S. Ramakrishnan, "Multimode Optical Fibre Sensors", J.I.E.T.E. (India) **32**, 307-310 (1986).
23. S. Ramakrishnan and R. Th. Kersten, "A multipurposes cross-talk sensor using multimode fibres", Proc. 2nd Opt. Fibre Sensor Conf., Stuttgart, 105-110 (1984).
24. S. Ramakrishnan and R. Th. Kersten, "Faseroptic Sensor mit Uberwachungsmoglichkeiten", German Patent P. 3415242 (1984).
25. S. Ramakrishnan and R. Th. Kersten, "Faseroptische Wasserwage", German patent P. 3236436 (1982).
26. S. Ramakrishnan and R. Th. Kersten, "Faseroptische Flussigkeitsbrechzahl-Messvorichtung", German Patent P. 3302089 (1983).
27. J.N. Fields and Cole J.H., "Fibre Microbend Acoustic Sensor", App. Opt. **19**, 3265-3267 (1980).
28. J.N. Fields, Asawa C.K., Smith C.P. and Morrison R.J., "Fibre Optic Hydrophone", Advances in Ceramics **2**, Ed. B. Bendow and S.S. Mitra, Am. Ceram. Soc., 529-539 (1981).
29. W.H.G. Horsthuis and Fluitman J.H.J., "The Development of Fibre Optic Microbend Sensors", Sensors and Actuators **3**, 99-110 (1982/83).
30. N. Lagakos, Litovitz T., Macedo P., Mohr R. and Meister R., "Multimode Optical Fibre Displacement Sensor", App. Opt. **20**, 167-170 (1981).
31. N. Lagakos, Macedo P., Litovitz T., Mohr R. and Meister R., "Fibre Optic Displacement Sensor", Advnaces in Ceramics **2**, Ed. B. Bendow and S.S. Mitra, Am. Ceram. Soc. Bull. 539-545 (1981).
32. R. Olshansky, "Distortion Losses in Cabled Optical Fibres", App. Opt. **14**, 20-21 (1975).
33. N. Lagakos, Trott W.J., Hickman T.R., Cole J.H. and Bucaro J.A., "Microbend Fibre Optic Acoustic Sensor as an Extended Hydrophone", IEEE J. Quant. Electron. **QE-18**, 1633-1638 (1982).

34. C.N. Kurtz and Streifer W., "Scalar Analysis of Radially Inhomogeneous Guiding Media", J. Opt. Soc. Am. **57**, 779-786 (1972).
35. N. Lagakos, Cole J.H. and Bucaro J.A., "Microbend Fibre Optic Sensor", App. Opt. **26**, 2171-2176 (1987).
36. A. Yariv, "Introduction to Optical Electronics", Holt Reinehart and Winston, New York (1971).
37. K. Spenner, Singh M.D., Schulte H. and Boehnel, H.J., "Experimental Investigation on Fibre Optic Liquid Level Sensors and Refractometers". Proc. 1st Int. Conf. on Opt. Fib. Sensors, London, 96-99 (1983).
38. A. Kumar, Subrahmoniam T.V.B., Sharma A.D., Thyagarajan K., Pal B.P. and Goyal I.C., "A Novel Refractometer using Tapered Optical Fibres", Electron. Letts. **20**, 534-535 (1984).
39. A. Ankiewicz, Pask C. and Snyder A., "Slowly Varying Tapers", J. Opt. Soc. Am. **72**, 198-203 (1982).
40. M.J. Adams, "An Introduction to Optical Waveguides", John Wiley and Sons, New York (1981).
41. A. Kumar, Shenoy M.R., Pal B.P. and Goyal I.C., "A Novel Temperature Sensor using Mercury Cladded Optical Fibres", J.I.E.T.E. **32**, 347-348 (1986).
42. M. Brenci, Conforti G., Falciai R. and Scheggi A.M., "Optical Fibre Thermometer for Medical Use", Proc. SPIE **497**, 13-17 (1984).
43. A.M. Scheggi, "Optical Fibre Sensors in Medicine" in Ref. [1].
44. K.A. Wickersheim and Alves R.V., "Fluoroptic Thermometry: A New Immune Technology", in Biomedical Thermology, Alan R. Liss, New York, 547-554 (1982).
45. D.A. Knone and Vinarub E.I., "Fibre Optics Invade Process Control", Photonics Spectra, 51-57 (Feb., 1984).
46. I. Chabay, "Optical Waveguides", Analytical Chemistry **54**, 1071A-1080A (1982).
47. W. Chudky, Kenhy K., Jarvis G. and Pohlig K., "Monitoring of Ground Water Contaminants using Laser Fluorescence and Fibre Optics", Proc. Instrument Soc. of America, Houston, TX, 1237-1243 (1986).
48. J. Peterson, Fitzgerald R. and Buckhold D., "Fibre Optic Probe for in Vivo Measurement of Oxygen Partial Pressure", Analytical Chemistry **56**, 62-67 (1984).
49. W. Seitz, "Chemical Sensors based on Fibre Optics", Analytical Chemistry **56**, 667-670 (1984).
50. E. Urbans, Offenbacker H. and Wolfbeis O., "Optical Sensor for Continuous Determination of Halides", Ibid **56**, 427-429 (1984).
51. L. Saari and Seitz W., "pH Sensor based on Immobilised Fluoresceinamine", Ibid **54**, 821-823 (1982).
52. L.L. Blyler, Jr. Ferrara J.A. and MacChesney J.B., "A Plastic Clad Silica Fibre-Chemical Sensor for Ammonia", Optical Fibre Sensor Conference, New Orleance, 369-371 (February, 1988).
53. J. Giuliani, Wohltjen W. and Jarvis N., "Reversible Optical Waveguide Sensor for Ammonia Vapours", Optics Letters **8**, 54-56 (1983).
54. M.K. Barnoski and Jensen S.M., "Fibre Waveguides: A Novel Technique for Investigating Attenuation Characteristics", App. Opt. **15**, 2112-2115 (1976).
55. A.H. Hartog, "A Distributed Temperature Sensor Based on Liquid Core Optical Fibres", IEEE J. Lightwave Tech. **LT-3**, 498-509 (1983).
56. A.H. Hartog, Leach A.P. and Gold M.P., "Distributed Temperature Sensing in Solid-Core Fibres", Electron. Letts. **21**, 1061-1062 (1985).
57. York Technology, Chandlers Ford, U.K.
58. S. Kinsley, "Distributed Fibre Optic Sensors", Proc. of the Instrument Society of America, Houston, Texas, 315-330 (1984).

24

Interferometric Optical Fibre Sensors[*]

B. CULSHAW[**]

1. INTRODUCTION

Optical fibre sensors offer significant potential in applications where intrinsic safety, high levels of electromagnetic interference or high sensitivity is of paramount importance [1]. Of the numerous technologies which have been demonstrated, interferometric sensors have shown the highest levels of sensitivity and the greatest technical challenge in their implementation [2]. Indeed interferometric optical fibre sensors have, for some measurands, demonstrated the highest sensitivity yet achieved.

This chapter concentrates on interferometric sensors and presents a review of their basic principles of operation and limitations in achievable performance levels. The chapter also presents a brief review of the interaction mechanisms available between the measurand and the phase delay in the fibre and examines the practical implications of realistically available components upon sensor performance. The technology of fibre optic interferometric sensors is, in itself, now well understood and characterised. However, there is still considerable scope for future development at the system level and in the final section of this chapter we shall examine some of the emerging ideas which may enable a multitude of sensors to be assembled in a network and/or may finally result in sensors exhibiting the ultimate in sensitivity.

It will become apparent that interferometric sensors offer considerable potential in a wide range of engineering applications. It will also become

---

[*] Updated reprint from J.I.E.T.E., Vol. 32 (Special issue on Optoelectronics and Optical Communication), July-August (1986).

[**] Department of Electronic & Electrical Engineering, University of Strathclyde, Royal College Building, 204 George Street, Glasgow G1 1XW

apparent that the successful realisation of a practical sensor presents a number of technological challenges. Many of these remain to be solved at the engineering level. The extent of the impact of the technology upon the science of measurement will only become evident in the fullness of time. However, of the interferometric sensors which have been demonstrated, the optical fibre gyroscope [3] will almost certainly become a production device and magnetometers [4] and hydrophones [5] offer significant promise.

2. BASIC PRINCIPLES OF INTERFEROMETRIC OPTICAL FIBRE SENSORS

An interferometric optical fibre sensor operates by measuring an environmentally induced change in the delay along a fibre as a change in optical phase. The change in the delay may be introduced by changes in ambient temperature or pressure, in mechanical longitudinal and/or radial strain or through non-reciprocal effects in a loop as experienced in the optical fibre gyroscope. In all cases, the phase change is detected using one or other form of an optical fibre interferometer (see Fig. 1). The output from one port of such an interferometer in which the splitting ratios of the coupler are exactly 50% is related to the phase difference ϕ between the two paths through:

$$P = \frac{P_0}{2} \{1 + \cos\phi\} \quad (1)$$

with a complementary relationship for the other output. The measurand introduces a change in the value of ϕ and so we can determine the sensitivity of the

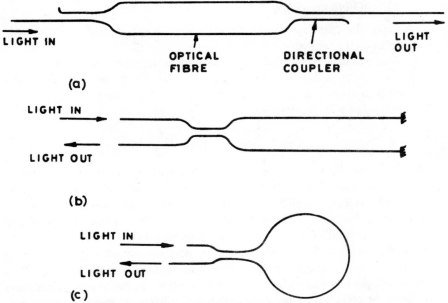

Fig. 1. Optical Fibre interferometers (a) Mach-Zehnder, (b) Michelson (c) Sagnac.

interferometer by differentiating the above expression. This sensitivity is clearly a maximum when the phase angle ϕ is 90° or odd multiples thereof. At the maximum sensitivity bias point, the minimum detectable phase change is dictated by shot noise. In broad terms, this limit is approximately the reciprocal of the square root of the number of photons incident upon the photodetector in the integration time of the receiver. For optical power P in milliwatts and time T in seconds, this corresponds approximately to $10^{-8}\,(TP)^{1/2}$ radians. In practice, achieving this detection limit may be frustrated by intensity noise in the optical source, phase noise in the optical source for other than zero fringe detection and thermal noise in the receiver. With due attention to these points, detection sensitivities within a factor of two of shot noise may be relatively easily obtained.

There are five main techniques whereby the phase changes induced by the environmental parameter change may be detected. These are:

- Maintaining a static phase difference of 90° using mechanical feedback network.
- Using a frequency tunable source to compensate for phase drift between the two arms of the interferometer.
- Using cross-multiply and add circuitry from both outputs of a two beam interferometer [6].
- Using synthetic heterodyne and related systems [7].
- Using phase sweeping about the zero order fringe as, for example, in the optical fibre gyroscope.

A detailed explanation of all these techniques may be found in appropriate references in the bibliography. With the exception of the last named, all the detection schemes mentioned are designed to operate in interferometers in which the static path difference is non-zero and is also liable to substantial drifting with environmental changes, especially temperature. The perceived phase change $\delta\phi$ in this situation may be expressed as:

$$\delta\phi = \frac{2\pi \Lambda L}{\lambda}\left\{\frac{\delta L}{L} + \frac{\delta f}{f}\right\}, \qquad (2)$$

where L is the physical path difference between the two interferometer arms and f is the operating frequency of the optical source. If we use a tunable source (for instance, a single mode semiconductor laser diode) [8] then it is clearly possible to compensate for phase changes by applying an equal and opposite apparent phase through the frequency shift. This may often be preferable to applying the same change by mechanically altering the length of one of the arms using a piezoelectrically driven feedback element [9]. The cross-multiply and add technique allow for drift in the static phase difference between the two arms so that neither of these feedback techniques is necessary.

However, the circuitry is complex and the technique appears to be losing favour. Heterodyne detection has the advantage that the received signal is put on to an electrical intermediate frequency which may be readily demodulated using standard radio techniques. The various synthetic heterodyne approaches which have been described in the literature have the inherent disadvantage that they require a delay between the two interfering arms during which time the bias on the laser is modified [10]. There are substantial noise penalties to pay for this procedure and a directly frequency shifted conventional heterodyne approach would be preferable. Only the lack of the suitable shifting element prevents this from being a standard detection technique.

The effects of source coherence and laser phase noise on the performance of interferometers operating away from the zero fringe cannot be underestimated [11]. With increasing path difference between the two arms of the interferometer, two phenomena conspire to reduce the signal to noise ratio which may be achieved. The first is the reduction in the signal level due to the fact that the source is becoming less coherent with the delayed version of itself; the second which is usually much more important is the increase in noise level due to phase to amplitude conversion induced by the delays in the interferometer. This, in fact, is one of the principal limitations of interferometric fibre optic sensors. As a rule of thumb, the noise penalties become substantial when the path difference is of the order of 0.1% of the source coherence length. This really implies that for operation with a standard semiconductor laser, the two arms of the interferometer should be balanced to within less than 1 cm. Sometimes this is extremely difficult to realise in practice. Of course, in the fibre optic gyroscope where the whole system operates on the zero fringe, these considerations do not apply. The level of the signal to noise penalty as a function of the path difference is shown in Fig. 2.

The sensitivity of the interferometer to the desired measurand (and, of course, to undesired measurands) depends upon two basic criteria. The first is the efficiency of the interface between the measurand and the optical delay in the fibre and the ability of this interface to reject interfering measurands. The second is the relative spectral occupancy of the desired and undesired signals, e.g., separating slowly varying temperature changes from acoustic fluctuation, is a relatively straightforward filtering operation.

The phase sensitivities of fibre optics interferometers to external measurands are readily evaluated. Typically, these are in the region of 10 radians per microstrain per metre, 10 radians per bar per metre and 100 radians per degree centigrade per metre. The exact value depends upon the detailed constitution of the fibre and also upon the wavelength, the figures quoted being taken at 850 nanometers. These figures indicate that the fibre optic interferometer is relatively sensitive to variations in strain and in environmental temperature and less so as to pressure changes. This qualitative

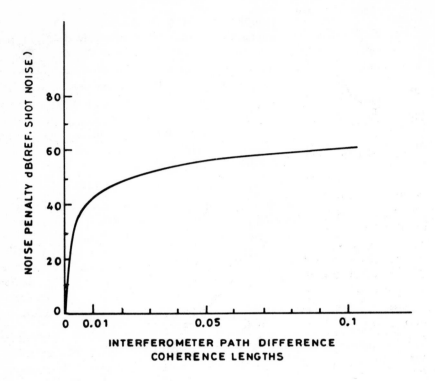

Fig. 2. Noise penalty vs. interferometer path difference.

hypothesis may be confirmed by comparing the energy required to produce a unit phase change when applied in each of the proceding formats.[1] It then transpires that the strain and temperature sensitivities are comparable whilst the pressure sensitivity is orders of magnitude less. Consequently, there is a considerable insentive to introduce measurand transformers into the design of interferometric sensors. These are carefully designed to change, for instance, a pressure field into a strain field in hydrophones or a magnetic field, again into a strain field, in magnetometers. The former may be implemented using a compliant plastic coating whilst the latter will require the use of magnetostrictive material. These sensitivity figures also emphasise the effects of temperature change on the performance of an interferometer. For instance, if the interferometer arms are only 1 cm different in path length, a change of 10°C in the temperature of the total assembly will sweep the relative phase difference between the two arms through 10 radians. This is about 100 times more than can be tolerated to maintain a stable bias point for quadrature operation. The need for a detection system which compensates for bias point drift is then immediately apparent.

The interaction process in the Sagnac interferometer is slightly different and the phase difference between the two paths is given by

$$\phi = \frac{8\pi R L}{\lambda_0 C} \cdot \Omega . \tag{3}$$

In terms of normalised sensitivity, this corresponds to a phase deviation of the order of 0.1 radians per (radian per second) per metre squared. The other non-reciprocally modulated fibre optic sensor is the magnetic field sensor which depends upon Faraday rotation which is, of course, an induced circular birefringence. The level of this, again, depends upon the glasses used and the wavelength, but for silica, the Verdet constant is in the order of 3.3×10^{-40} per ampere turn implying a relatively low sensitivity to magnetic fields unless a Sagnac interferometer [13] could be configured with circularly polarised light propagating in the two directions around the loop. Sensitivity of such a device would be in the region of approximately 10^{-2} Ampere turns in a one second integration time. The principal difficulty here lies in ensuring that circularly polarised light propagates in the two directions around the loop. For comparison, the earth's magnetic field is approximately 27 Amperes per metre.

The discussion so far has concentrated on two beam interferometers in which either each path is a separate fibre connected to the other using directional couplers in either a Mach-Zehnder or Michelson configuration or the paths are counterpropagating within the same fibre as in the Sagnac interferometer. There are many alternative approaches to designing a two path interferometer and perhaps the most promising is that based upon launching equal amounts of power into the two principal modes of a birefringent fibre. The sensitivity of such an interferometer is considerably reduced since now we are looking at the differential effect imposed upon these two modes [14]. However, this reduction — typically about two orders of magnitude — causes no significant difficulties in practice, especially when the resulting simplification in the sensor (Fig. 3) is taken into account. There is another advantage to the birefringence sensor in that a configuration such as that shown in Fig. 4 may be used to ensure that the interferometer operates on very near the zero-fringe [15]. There are considerable advantages in the noise performance of such a device, effectively relaxing the specification of the optical source.

It will have become apparent that one of the principal sensitivity criteria in fibre optic interferometric sensors involves the interaction length of the light passing through the measurand field. This may be increased by increasing the length of fibre or alternatively by ensuring that the light passes through the same length of fibre a number of times. We are then led on to the subject of multipath interferometry — that is the use of Fabry-Perot interferometers (see Fig. 5) [16]. The enhanced sensitivity of the multipath interferometer manifests itself in the Airey function response shown in Fig. 5. For optimum sensitivity,

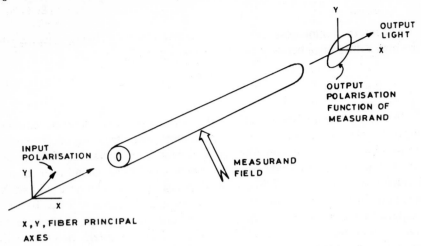

Fig. 3. Birefringent fibre optic sensor.

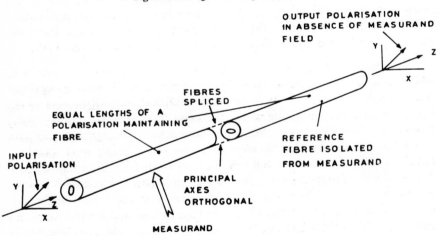

Fig. 4. Balanced birefringent sensor.

the bias point must be defined at the point of maximum slope and the tolerance on this is much more stringent than on the two path interferometer case — though the detailed comparison is complex since an equivalent two path interferometer will be much longer. The increase in sensitivity may also be deceptive since the frequency selectivity of a Fabry-Perot is also enhanced. In practice, it is often the phase noise of the source which limits the sensitivity which may be achieved. In the ideal case where both phase and intensity noise on the optical source are at the shot noise limit, the intrinsic sensitivities of both Fabry-Perot and two path interferometers are virtually identical. The Fabry-Perot is, however, beginning to make some impact in the form of the recirculating ring as applied to fibre optic gyroscopes [17].

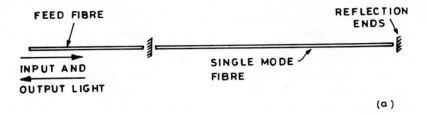

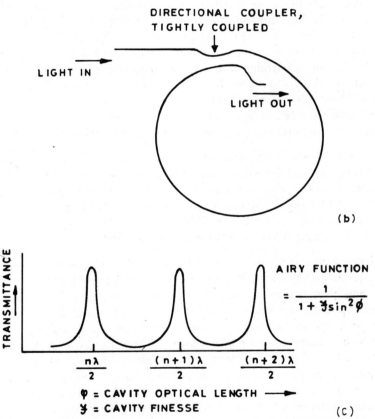

Fig. 5. Fabry-Perot fibre interferometers (a) reflecting (b) recirculating and (c) the Airy function.

The interferometer is then, in all its forms, an extremely sensitive means of measuring distance and in interferometric sensors, we are concerned with the measurement of environmentally induced path difference. The "ruler" is the wavelength of the light used in the interferometer and so our measurement can only be as accurate as our knowledge of this quantity. This can have important implications especially in the design of accurate measuring systems based on fibre interferometers and solid state sources.

3. APPLICATIONS OF INTERFEROMETRIC OPTICAL FIBRE SENSORS

There are to date no truly commercial versions of interferometric sensors though there are a number of very promising laboratory prototypes which have been demonstrated. These fall into three principal groups — hydrophones, magnetometers and gyroscopes. Much has been published on all of these three systems [18-20].

Many forms of the optical fibre magnetometer have been described in literature. The basic principle is identical in all of these. A single-mode fibre is attached to (or coated with) a magnetostrictive material. The length of this material is proportional to the magnetic field which is applied and so a longitudinal strain is introduced in the fibre path which in turn depends upon magnetic field. In its simplest from the magnetically sensible fibre forms one arm of a Mach-Zehnder or Michelson interferometer. For signals at relatively high frequencies (above approximately 100 Hertz), a field sensitivity of the order of 10^{-7} amperes per metre on a 1 metre length of fibre is quite feasible.

For very slowly varying magnetic fields, the difficulties alluded to earlier concerning thermal drifts dominate the detection process. However, by using a modified version of the flux gate magnetometer, good performance can also be achieved for slowly varying signals [21]. The system is, however, still relatively complex though it is unprecedented in sensitivity — exceeded only by the SQUID[*] — continues to stimulate further development.

Hydrophones fabricated using fibre optic interferometers also offer very high sensitivity but here it is, perhaps, the flexibility in the achievement of a variety of polar responses that is the most advantageous feature. (See Fig. 6). Furthermore, the availability in the future of an all optical array of receiving elements is also particularly attractive. The principal design problem in fibre optic hydrophones lies in controlling the sensitivity of signal and reference coils to the acoustic field [22]. Considerable literature has evolved on the use of compliant coating on the sensor coils and on the stiffening of the reference coil using metallic coatings. A detailed account of this may be found in the appropriate references. The final result is that fibre optic interferometric sensors with sensitivities and frequency responses more than adequate for underwater applications can now be readily demonstrated. The status of this technology when compared to the well established ceramic sensors and the more recently developed piezoelectric film transducers has still to be resolved. A more than adequate technical performance is available from the fibre optic sensors. However, its comparative value in terms of overall system cost is still uncertain, both from the sensor and system compatibility points of view.

---

[*] Super cooled quantum interference device.

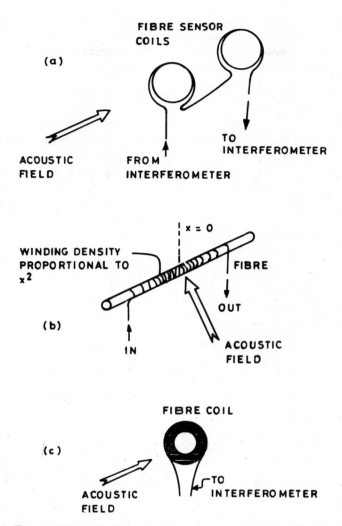

Fig. 6. Forms of fibre optic hydrophone for (a) gradient polar response, (b) shaded Gaussian polar response and (c) toroidal receiver polar response.

Fibre optic gyroscopes continue to receive considerable attention both in research laboratories and in development organisations. The principal advantages here are concerned with the lack of moving parts, the relatively few precision mechanical assembly steps in the sensor coil and the very inexpensive final construction. When compared with ring lasers and mechanical sensors, this lack of precision assembly steps becomes very apparent. The major reason why fibre gyroscopes have not yet made a much greater impact is that there remain some fundamental limitations on the achievable performance. These limitations are concerned not with mechanical assembly but with the properties of the optical transmission medium and of the source used to energise the sensor. There is also an important trade off, at least in systems used to date, between the availability of a low noise detection system on the one hand and of a linear and well defined scale factor performance on the other [23].

A consequence of this is that the major developments of fibre optic gyroscopes have been targeted towards the relatively modest performance end of the spectrum. These are gyroscopes used for short range missile guidance systems, altitude heading and reference and related applications where sensitivity and drift of 1-100° per hour can be tolerated with scale factors of the order 0.1%. These application areas are, of course, very cost sensitive and mechanical sensors with adequate performance (apart from acceleration sensitivity) are already available. In the last few of years, there has been a shift in gyroscope research towards seeking "inertial" grade or better performance. Here noise and drift levels of the order of 10^{-20} per hour are required together with scale factor stability of parts per million. The currently available mechanical and ring laser sensors are all inherently extremely expensive, largely due to the amount of precision machining involved in their fabrication. The fibre gyroscope may be able to considerably simplify this, though of course, it must be acknowledged that the necessary test procedures and calibration will be very similar for all precision instruments. For the future, there is a school of thought which subscribes to the concept that extremely delicate geophysical measurements (looking towards 10^{-50} per hour or less) may be achieved using fibre optic sensors. Even the most optimistic projection of current technology falls short of this performance level, though, of course, it is theoretically possible to achieve this sensitivity.

The future of the fibre optic gyroscope and, indeed, the other principal interferometric sensors mentioned here will be determined in the relatively near future and in the next decade, the technology will have matured and production systems will be available for appropriate applications. It seems likely that these will lie in areas where extremely high sensitivity is required (especially for the magnetometer and the hydrophone) or for the gyroscope in areas where the relative mechanical simplicity and shape flexibility of the instrument will be particularly attractive.

4. COMPONENTS FOR INTERFEROMETRIC SENSORS

It is inherent within the design of any interferometric fibre optic sensor that the entire interferometer must be configured within a fibre optic path. There are a few exceptions to this—most notably in gyroscopes—in which some components are either in bulk optics or, more commonly, in integrated optics. In this type of sensor, there are inherent instabilities and losses stemming from reflections at the interface between the fibre optic and the bulk or integrated optic device and from modal mismatches between the propagating regions between the two media. In most cases, therefore, it is desirable to stay within the fibre path throughout. This then leads to a requirement for fibre optic directional couplers and for fibre optic modulators.

Coupler technology is now available on the production line. High quality low cost couplers are now offered by a variety of manufacturers using the fused process and most frequently optimised for 1.3 micron wavelength systems – a development stimulated, in the main, by interest in fibre optic local area networks. These couplers could be suitable for sensor use though in the main the environmental specification for communications use is less stringent than that for sensor use. The other principal coupler technology involves polishing a fibre mounted in a silica block until the propagating optical field is available at the surface. These polished couplers are labour intensive and so inherently more expensive that the fused equivalent. They do, however, offer substantially greater flexibility since they are readily adjusted for a variety of coupling ratios and the optical field at the surface of the polished block may interact with numerous other media including crystals or metal films for polarisers and non-linear substances for use in amplifiers and switches. The surface may also be used with advantage with a periodic structure to stimulate resonant responses. Again the polished technology has now entered production phase though its inherent expense makes it suitable only for more specialist applications than situations requiring a simple power splitter. (Fig. 7) [24, 25].

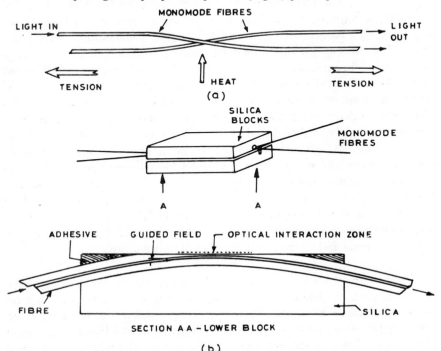

Fig. 7. Fused (a) and polished (b) single-mode fibre optic directional couplers.

All fibre light modulators have invariably utilised the sensitivity of the optical phase to an applied acoustic strain [26]. This is most suited to the application of a sinusoidally varying phase modulation at a frequency which corresponds to the resonant frequency of the driving element. In some sensor systems—for instance, the gyroscope—this is exactly what is required for detection at around the zero-fringe. However, there are also situations in which other modulating functions are desirable. Most usually this involves some form of frequency shifter. In bulk optics, this would be realised using a Bragg cell, but in the all-fibre form, the spatially extended interaction region is not available apart, that is, from *along* the fibre. One approach to the frequency shifter, has, then, been to use a longitudinal wave interaction rather than a transverse wave interaction [27]. In common with the Bragg cell, this longitudinal wave interaction device requires two outputs which are spaced in wave number by the acoustic wave number involved. In an all fibre system, this can only be realised by coupling between modes of different wave numbers. In the experimental versions of this system which have been reported, this has been effected in birefringent fibres. An alternative approach is to use the so-called "Serrodyne" which is a means of converting phase modulation into a frequency shift by applying a phase saw tooth [28]. Using acoustic devices this is rendered difficult by mechanical ringing on the flyback, though some considerable progress has been made along this path (Fig. 8). There may also be other interaction mechanisms which can be exploited but as yet have not been. Of these, perhaps the most relevant is the use of non-linear interactions involving inserting a modulating wave through one port of a modified coupler element. The other major important component in a sensor system is the optical source itself. In particular, the phase stability of the source has a critical impact upon the noise performance of the resulting interferometer. In some systems—most notably the gyroscope—the form of the source spectrum is also important in determining non-linear interactions which may in turn influence the apparent phase in the arms of the interferometer. The situation is difficult since, in effect, the sensor community relies upon the availability of semi-conductor sources which have been developed for communications. Even in coherent communications, the required phase stability is considerably less stringent than in sensors. The basic criterion is that the phase should be "stable" for a period corresponding to a quantity of the order of the inverse of the system bandwidth [29]. The exact criteria for stability depends on the type of measurement and on the detailed design of the interferometer.

There is relatively little published information on the importance of the source in sensors. The need for such information will become much greater with the advent of more complex systems using multiple interferometers and inevitably involving a wide range of time delays within the different elements of the system. This is discussed in more detail below. The other area in which

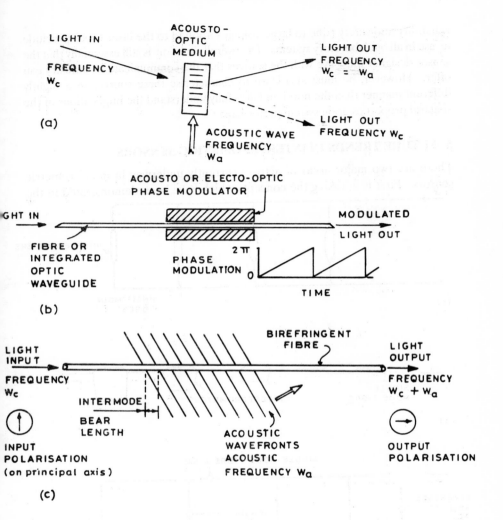

Fig. 8. Optical frequency shifters (a) Bragg cell (b) serrodyne and (c) travelling wave.

source parameters are important is in the interferometer's scale factor. In all cases, variation in optical delay is being measured as a variation in optical phase and the multiplying factor between the two is, in effect, the optical wavelength. The wavelength of a semiconductor diode varies by typically of the order of 0.1% per degree C change in junction temperature. Therefore, measurements to be made with scale factors significantly better than 1% require the use of increasingly sophisticated temperature control equipment or the facility to perform on-line measurement of the source wavelength. In many ways, a gas laser is the ideal source since it is easily made to operate in a single longitudinal mode with a very long coherence length. However, considerations of bulk, reliability and safety (due to large voltages applied to the laser tube) preclude its use in all but laboratory systems. The overall situation is still essentially that the sensor designer must live with the sources that the communications industry can offer. However, he must also learn to characterise these sources in a slightly different manner than the norm and also fully understand the implications of the spectral properties upon system performance.

5. FUTURE TRENDS IN INTERFEROMETRIC SENSORS

There are two major areas of activity currently underway in interferometric sensors. First is in taking the concepts which have been demonstrated in the

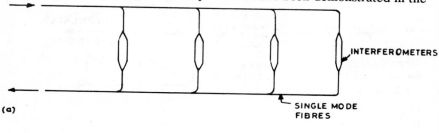

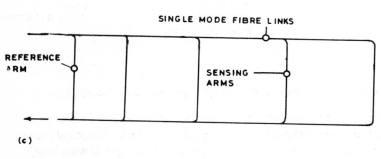

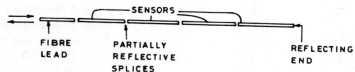

Fig. 9. Forms of interferometric multiplexer (a) time delay, matched parallel interferometers (b) differential time delay, series interferometers, (c) differential delays, parallel arms and common reference (d) reflective splice architecture.

laboratory into production engineering. There are innumerable practical difficulties to be overcome, some of which have already been mentioned, but which are largely concerned with environmental and temporal stability.

The second major development lies in the research laboratory where there is considerable interest in a multiplexing system whereby several sensors may be interrogated simultaneously along a single linking fibre. These systems all relay upon a differential path to encode the location of each particular sensor. The path difference may be within the sensor itself (via static path encoding) or it may be in terms of the physical distance from the source to the receiver via the sensor element. Some examples of these formats are shown in Fig. 9. In order to identify each of the sensors within the network, it is fundamentally necessary to modulate the waveform applied to the network [30]. This operation may be implemented either by bias current modulation of, for instance, a laser diode or by an external (perhaps integrated optic) modulator. The system imposes stringent conditions on the spectrum of the waveform introduced into the network. In this respect, interferometric sensors are more demanding than fibre optic sensors based on modulation of other optical parameters in that the coherence function must be maintained for all sensing elements. There have been a number of suggested multiplexing systems for interferometers including one which postulates a correlation detection system thereby overcoming some of the coherence constraints by effectively always working on the zero-fringe [31, 32]. The experimental demonstrations which have been reported in the open literature have achieved limited success and only one has reported a reasonable number of sensing elements [33]. There is clearly considerable scope for further development.

6. CONCLUDING COMMENTS

This chapter has very briefly reviewed the principles and applications of interferometric fibre optic sensors. The principal advantage of these sensors lies in their remarkably high sensitivity. Against this there are stability difficulties with off-sets and scale factors which depend critically on the source characteristics.

The engineering of some sensors, particularly AC magnetometers, hydrophones and gyroscopes is at the prototype stage and adequate environmental and technical characteristics have been demonstrated. Initial evaluations of multiplexed interferometrics have shown promise, though techniques whereby the optical power may be equitably balanced among the sensors in an array need to be evolved.

ACKNOWLEDGEMENTS

The author would like to acknowledge stimulating discussions with numerous colleagues in industry and academic institutions. Particular thanks are due to other members of the Opto-electronics Research Group at the University of Strathclyde, and to John Nuttall, John Dakin and Alan Rogers.

REFERENCES

1. B. Culshaw, "Fibre Optic Sensors and Signal Processing", Peter Peregrunus, Stevenage (1984).
2. T.G. Giallorenzi, Bucaro J.A., Dandridge A., Siegal G.H., Cole J.H., Rashleigh S.C. and Priest R.G., "Optical Fibre Sensor Technology" IEEE *J. Quantum Electronics*, **18**, 4, p. 626 (1982).
3. Bergh R.A., Lefevre H.C. and Shaw H.J., "Fibre Optic Gyroscopes", IEEE *J. Lightwave Technology* **LT-2**, April (1984).
    or
    B. Culshaw and Giles I.P., "Optical Fibre Gyroscopes", *J. Phys. E.* **16**, p. 5 January 1983.
4. R.H. Pratt, Jones R.E., Extance P., Pitt G.D. and Foulds K.W., "Optical fibre magnetometer using a stabilised semiconductor laser source", Proceedings Second Intnl. Conf. on Optical Fibre Sensors, Stuttgart, p. 45 (Proceedings Published by VDE-Verlag, Berlin) September 1984.
5. N. Lagakos, Ku G., Jarzunski J., Cole J.H. and Bucaro J.A., "Desensitization of the ultrasonic response of single mode fibres", IEEE *J. Lightwave Technology* **LT-3**, 5, p. 1036, October (1985).
6. A.D. Kersey, Corke M., Jones J.D.C. and Jackson D.A., "Signal recovery techniques for unbalanced fibre interferometric sensors illuminated by laser diodes". *Proc. First Intnl. Conf. on Optical Fibre Sensors*, London p. 43 (Published by IEE, London) (1983).
7. J.H. Cole, Danver B.A. and Bucaro J.A., "Synthetic heterodyne interferometric demodulation", IEEE *J. Quantum Electronics* **QE.18** p. 694 (1982).
8. A. Dandrindge and Goldberg L., "Current induced frequency modulation in laser diodes", Electronics Letters, **18** p. 302 (1982).
9. D.A. Jackson, Priest R., Dandrindge A. and Tveten A.B., "Elimination of drift in a single mode optical fibre interferometer using a piezoelectrically stretched coiled fibre", Applied Optics, **19** p. 2926 (1980).
10. D. Uttam, Giles I.P., Culshaw B., Davies D.E.N., "Remote interferometric sensors using frequency modulated laser source", *Proc. First Intnl. Conf. on Optical Fibre Sensors*, London, p. 182 (1983).
11. D. Uttam and Culshaw B., "Semiconductor Lasers in advanced interferometric optical fibre sensors", IEE *Proceedings J.* **132**, 3, p. 184, June (1985).
12. A.J. Rogers, "Optical measurements of current and voltage on power systems", IEE *Journal Electrical Power Applications*, 2, 4, p. 120, (1979).
13. H.J. Arditty, Lefevre H.C., Graindorge P., "Current sensor using state of the art fibre optic interferometric techniques", Proceedings IOOC p. 128 (IEEE New York) (1981).
14. W. Eickoff, "Temperature sensing by mode-mode interference in birefringent optical fibres", Optical Letters, 6, 4, p. 204 (1981).
15. J.P. Dakin, Broderick S., Carless D.C., Wade C.A., "Operation of a compensated polarimetric sensor with a semiconductor light source", Proc. 2nd Intnl. Conf. on Optical Fibre Sensors, Stuttgart, p. 241 1984.
16. D.D. Atherton, Reay N.K., Ring J. and Hicks T.R.., "Febry-Perot Interferometry", Optical Engineering **20**, 6, p. 806 (1981).

17. R.E. Meyer, Ezekial S., Stowe D.W. and Tekippe V., "Passive fibre optic ring resonator for rotation sensing", Optics Letters **8** p. 644, (1983).
18. Proceeding International Conferences on Optical Fibre Sensors, London (1983) (IEE), Stuttgart (1984) (VDE-Verlag), San Diego (1985) (IEEE/OSA), Tokyo (1986).
19. Proceedings SPIE Conferences on Optical Fibre Sensors and related topics, SPIE, Bellingham, Wash. U.S.A.
20. Proceeding International Annual Meeting on Gyroscope Technology, Stuttgart (DGON).
21. A.D. Kersey, Jackson D.A. and Corke M., "Single mode fibre optic magnetometer with D.C. bias field stabilization", IEEE *Journal Lightwave Tech.* LT-3, 4, p. 836 August (1985).
22. N. Lagakos and Bucaro J.A., "Pressure desensitization of optical fibres", Applied optics, **20** p. 3276 (1981).
23. B. Culshaw and Nuttall J.D., "Fibre Optic Gyroscopes for Inertial Navigation", Proceedings AGARD meeting on Military Applications of Guided Wave Optics, Istanbul, September (1985).
24. G. Georgiou and Boucouvalas A.C., "Low-loss single mode optical couplers", Proc IEE - J **132**, 5 p. 297, October (1985).
25. R.A. Bergh, Kotler G. and Shaw H.J., "Single mode fibre optic directional coupler", Electronics Letters, **16** p. 260, (1980).
26. D.E.N. Davies and Kingsley S.A., "Method of phase modulating signals in optical fibres; application to optical telemetry systems", Electronics Letters, **10** p. 21, (1974).
27. W.P. Risk, Youngquist R.C., Kino G.S. and Shaw H.J. "Acoustic-optic frequency shifting in birefringent fibre", Optics Letters, **9** p. 309 (1984).
28. J.P. Dakin, Wade G.A. and Haji-Michael C., "A fibre optic serrodyne frequency translator based on a piezoelectrically strained fibre phase shifter", IEE Proc. J. **132**, 5, p. 287, October (1985).
29. I.P. Giles, Uttam D., Culshaw B. and Davies D.E.N., "Coherent optical fibre sensors with modulated laser sources", Electronics Letters, **19**, 1, p. 14 January (1983).
30. S.A. Al-Chalabi, Culshaw B., Davies D.E.N., Giles I.P. and Uttam D., "Multiplexed optical fibre interferometers; an analysis based on radar systems;, IEE. Proc. J. **132**, 2 p. 150, April (1985).
31. S.A. Al-Chalabi, Culshaw B. and Davies D.E.N., "Partially coherent sources in interferometric sensors", Proc. First Intnl. Conf. on Optical Fibre Sensors, London, p. 132 (1983).
32. J.L. Brooks, Wentworth R.H., Youngquist R.C., Tur M., Kim P.Y. and Shaw H.J., "Coherence multiplexing of fibre optic interferometric sensors", IEEE Journal Lighwave Technology **LT-3**, 5, p. 1062 October (1985).
33. J.P. Dakin, Wade C.A. and Withers P.B., "Engineering improvements to multiplexed interferometric optical fibre sensors", Proc. SPIE **522** p. 226 (FO'85 London May 1985).

25

Fused Single-Mode Optical Fibre Couplers[*]

F.P. PAYNE[**]

1. INTRODUCTION

Single-Mode optical fibre directional couplers are important components in optical fibre systems. Their applications include power division [1,2], wavelength division multiplexing [2,3], optical fibres [2,4], local area networks [5], Mach-Zehnder interferometery [6], optical fibre gyroscopes [7], polarisation control devices [8,10] and ring resonators [11].

The two most common methods of fabricating single-mode couplers are by polishing [12] and fusion [1,2]. The main purpose of this chapter will be to introduce the reader to the most important concepts and ideas associated with fused single-mode couplers.

Perhaps it will be useful to begin by looking briefly at how the polished and fused couplers are made because it is here that the main advantages of the fusion method become apparent. The polished coupler is made by polishing away the cladding of a single-mode fibre to within about one micron of the core. The fibre is usually supported during this operation by mounting it in a curved 'V' groove cut in a silica block. The coupler is formed by placing two such polished half-couplers one over the other together with either an index matching oil or UV curable epoxy between them (Fig. 1). The power coupling in the resulting coupler can be tuned by sliding the two polished halves relative to each other, and then fixing them permanently if required by curing the epoxy.

---

[*] Reprint from J.I.E.T.E., Vol. 32 (Special issue on Optoelectronics and Optical Communication), July-August (1986).

[**] Department of Electronics and Information Engineering, The University of Southampton, Highfield, Southampton SO5 3DG, U.K. Present address: Department of Engineering, University of Cambridge, Cambridge CB2 1PZ, U.K.

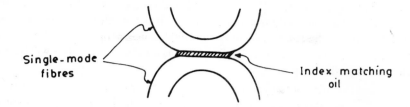

Fig. 1. The polished single-mode fibre coupler.

The insertion losses of the resulting coupler will largely be determined by the accuracy of the polishing process. It is essential to minimise any surface roughness on the polished faces. Fracture of the fibre during polishing is a common problem, and great care is needed when the neighbourhood of the core is reached. Another problem with this method is the difficulty to measure accurately how far the polished face is from the core, as this will determine the coupling characteristics when the coupler is assembled.

The fused single-mode coupler is made by a quite different technique [1,2]. Two single-mode fibres are twisted together and held by two movable supports. A small oxy-butane flame is applied to the fibres so that they melt together. Simultaneously, the two fibre supports are moved apart so that a fused tapered region is formed. To ensure a low-loss coupler, it is essential that the fibre supports move apart in a straight line with no sideways motion or vibration. The speed of separation will control the shape of the resulting taper and this also has a significant influence on the resulting loss. The movement of the fibre supports is most conveniently controlled by small d.c. motors. The oxy-butane flame also requires careful consideration. It is necessary to strike a compromise between gas flow and flame temperature. A high gas flow may result in a higher flame temperature, but it will also disturb the taper and cause an increase in loss. Constant gas pressure must also be accurately maintained by means of suitable control valves. The quality of the single-mode fibre used in the fusion process is extremely important. The core and cladding must be highly circular and concentric with one another. Inferior quality fibres can result in very high losses in the resulting coupler. The polished cladding coupler is much less sensitive to fibre imperfections (the reasons for this difference will be discussed later).

Both polished and fused couplers can be made with very low loss (less than 0.05 dB). However, the fused coupler is quicker and cheaper to make. It is also possible to monitor its power splitting ratio during fabrication and to make a coupler with any required splitting ratio at a given wavelength. This is impossible with the polished coupler, whose characteristics can only be measured after it is assembled. The polished coupler does have certain advantages however. Once assembled, it is tunable by sliding the two polished halves over each other. This can be useful when constructing high-Q ring resonators for use in optical fibre lasers [13]. The polished coupler can also be made with depressed cladding and birefringent fibres. It is very difficult to make fused couplers using these types of fibre, although some progress has been made recently along this direction [21]. The reasons for these difficulties will become clear later in the chapter.

One of the aims of the engineer is to relate the measured characteristics of a device to the theoretical understanding of the physical principles involved. At first sight, the fused tapered coupler appears a formidable problem to analyse. However, as any microwave engineer will know, it is possible to say a great deal about the operation of a multi-port device, (such as a directional coupler) from a knowledge of the external symmetry of that device. Surprisingly, this approach is only rarely applied to optical fibre components. This is a pity, as the microwave engineer's approach provides a great deal of insight into the expected characteristics of many devices. It can also tell us, on the basis of fundamental arguments, whether a device with certain characteristics can, in principle, be constructed. The next section of this chapter, describes those features of optical fibre couplers that are independent of the method of fabrication. Some of the more interesting measured characteristics of fused couplers are presented next. This is followed by a description of the extent of our current theoretical knowledge to show that all the measured characteristics of fused couplers can be adequately explained. A detailed comparison of theory and experiment is presented next and followed by a discussion on the problems that are presently considered important.

2. PHYSICAL PRINCIPLES

2.1 Coupling co-efficient

We can describe an optical fibre directional coupler as a four-port device in which the input and output ports consist of single-mode fibres. The internal construction of the coupler is not specified but represented as a 'black box' (Fig. 2). The ports are numbered 1 to 4 and are assumed to be identical single-mode fibres. All properties of the coupler will be described in terms of the power flow along these four ports. It might appear that very little can be said about the characteristics of this device without any knowledge of its internal construction. However, as we shall see, the opposite is true. By making a few very general assumptions, all the main characteristics of the coupler can be determined. We assume the following properties:-

1. The coupler is loss less and passive
2. The coupler is symmetric about a plane midway between ports 1 to 3, and ports 2 and 4, as shown in Fig. 2.
3. All four ports are free from reflections.
4. There is no electromagnetic interaction between the input and output ports.

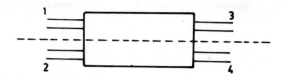

Fig. 2. Four port representation of an optical fibre coupler. The input and output ports are single-mode fibres and the box contains some unspecified interaction between them. The device has a reflection symmetry about the dotted line.

We shall now show that any four-port device satisfying these conditions automatically becomes a directional coupler, with power input to port 1 divided between ports 3 to 4 and no power emerging from port 2.

It is convenient to denote a signal on a port by one of the column vectors:

$$\vec{\Psi}_{1,3} \equiv \begin{pmatrix} 1 \\ 0 \end{pmatrix}, \quad \vec{\Psi}_{2,4} \equiv \begin{pmatrix} 0 \\ 1 \end{pmatrix}. \tag{1}$$

Let us suppose that a signal of unit amplitude is incident on port 1,

$$\vec{\Psi}_{in} = \begin{pmatrix} 1 \\ 0 \end{pmatrix}. \tag{2}$$

To describe the propagation of this signal through the coupler, we must express it in terms of the eigenstates of the pairs of ports (1,2) and (3,4). From the assumed symmetry of the coupler, these eigenstates are nothing but the even and odd modes:

$$\vec{\Psi}_e = \frac{1}{\sqrt{2}} \begin{pmatrix} 1 \\ 1 \end{pmatrix}, \quad \vec{\Psi}_o = \frac{1}{\sqrt{2}} \begin{pmatrix} 1 \\ -1 \end{pmatrix}. \tag{3}$$

They satisfy the normalisation conditions:

$$\vec{\Psi}_e^T \cdot \vec{\Psi}_e = 1, \quad \vec{\Psi}_o^T \cdot \vec{\Psi}_o = 1, \quad \vec{\Psi}_e^T \cdot \vec{\Psi}_o = 0. \tag{4}$$

The input state can now be expressed as

$$\vec{\Psi}_{in} = \left(\vec{\Psi}_e + \vec{\Psi}_o \right) / \sqrt{2}. \tag{5}$$

Since we have assumed the coupler to be lossless and the connecting ports to be single-mode fibres, the only states that can emerge from ports 3 and 4 are $\vec{\Psi}_e$ and $\vec{\Psi}_o$. The assumed symmetry prevents any coupling between these states and because the ports are reflectionless, no power will emerge from port 2. The output state emerging from ports 3 and 4 must be related to the input state by a linear transformation.

$$\Psi_{out} = U \cdot \vec{\Psi}_{in}. \tag{6}$$

Power conservation forces the matrix U to be unitary.

$$U^+ \cdot U = 1. \tag{7}$$

It is also clear that $\vec{\Psi}_e$ and $\vec{\Psi}_0$ must be eigenvectors of U with unit modulus eigenvalues:

$$U \cdot \vec{\Psi}_{e,0} = \vec{\Psi}_{e,0} \exp(i\phi_{e,0}). \tag{8}$$

The phases $\phi_{e,0}$ represent the phase shifts of the even and odd modes resulting from the internal construction of the coupler. After a little algebra, the output state can be expressed as

$$\Psi_{out} = (A + B) \begin{pmatrix} 1 \\ 0 \end{pmatrix} + (A - B) \begin{pmatrix} 0 \\ 1 \end{pmatrix}, \tag{9}$$

where $\quad A = 0.5 \exp(i\phi_e), \; B = 0.5 \exp(i\phi_0)$.

The two terms in this equation give the amplitudes of the signals emerging from ports 3 and 4 respectively, whilst the U matrix is given by

$$U = \begin{pmatrix} A + B & A - B \\ A - B & A + B \end{pmatrix}. \tag{10}$$

Normally only the power emerging from the output ports are of interest, and these are given by

$$P_3 = \cos^2 C, \quad P_4 = \sin^2 C, \tag{11}$$

where the coupling co-efficient C is given by

$$C = 0.5 |\varphi_e - \varphi_0|. \tag{12}$$

We see that the coupler is described by just one parameter, the coupling co-efficient, whose value can only be determined from a knowledge of the internal construction of the coupler. To evaluate it, we need to know the total phase shifts of the even and odd modes after passing through the coupler. In

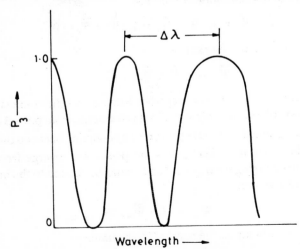

Fig. 3. Schematic representation of the spectral dependence of a coupler.

general, C is wavelength dependent and Eq. (11) tells us that the output powers must vary periodically with wavelength, as shown schematically in Fig. 3. The period $\Delta\lambda$ will, in general, be wavelength dependent and is given by

$$\Delta\lambda = \pi / |\partial C / \partial \lambda|. \tag{13}$$

A further interesting result can be deduced from our analysis concerning the 3-dB power splitter, for which $C = \pi/4$. From Eq. (10), we can see that the amplitudes of the signals emerging from ports 3 and 4 will be 90 degrees out of phase. This phase shift is quite independent of how the directional coupler is constructed internally.

2.2 Polarisation effects

In the above discussion, we have not mentioned the polarisation properties of light. The symmetry of the coupler indicates that the even and odd modes can each have one of two orthogonal polarisations, which we shall call X and Y. We must specify coupling co-efficients C_x and C_y corresponding to light input with either X or Y polarisations. The output powers will then be

$$P_3^x = \cos^2 C_x \quad P_4^x = \sin^2 C_x$$
$$P_3^y = \cos^2 C_y \quad P_4^y = \sin^2 C_y \tag{14}$$

The coupling co-efficients are still given by Eq. 12, but with the phase shifts referred to appropriate polarisations. In general, C_x and C_y will be different and this has some very interesting consequences. Let us assume that a coupler can be made so that C_x and C_y satisfy the following relations:

$$C_x = n\pi, \quad C_y = \left(m + \frac{1}{2}\right)\pi \tag{15}$$

with n and m as integers. Then for unit power input to port 1, the output powers, corresponding to the two possible polarisations, are

$$P_3^x = 1 \qquad P_3^y = 0$$
$$P_4^x = 0 \qquad P_4^y = 1. \tag{16}$$

This means that X-polarized light emerges completely from port 3 and Y-polarized light emerges from port 4. This property will cause the coupler to act as a polarisation beam splitter, with unpolarized light in port 1 being split equally into X and Y components at ports 3 and 4 (Fig. 4). The power from port 3 is then given by

$$P_3 = 0.5 (\cos^2 C_x + \cos^2 C_y) \tag{17}$$

This can be rewritten as

$$P_3 = \frac{\{1 + \cos(C_x + C_y)\cos(C_x - C_y)\}}{2} \tag{18}$$

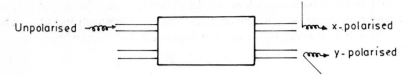

Fig. 4. The polarising beam splitter coupler separates unpolarised light into its polarised components.

If C_x and C_y are wavelength dependent, then P_3 must have a *modulated* periodic response, as indicated by Fig. 5. The *modulation* minima correspond to the polarisation beam splitting points. The two periods $\Delta\lambda$ and $\delta\lambda$ indicated in Fig. 5 are given by

$$\Delta\lambda = 2\pi / |\partial (C_x + C_y)/\partial\lambda|, \qquad (19)$$

$$\partial\lambda = \pi / |\partial (C_x - C_y)/\partial\lambda|. \qquad (20)$$

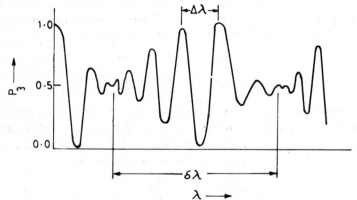

Fig. 5. Schematic representation of the spectral dependence of a polarising beam splitter coupler.

This is as far as general principles can take us. To proceed further with the analysis, we must consider the details of the internal construction of the coupler. It will help us to obtain some physical insight if we now consider some of the experimental measurements that have been made on fused couplers.

3. EXPERIMENTAL PROPERTIES

A very large number of experimental measurements have been made on fused single-mode couplers at our laboratory during the previous three years. Only a few of those results are presented here. For further details, the readers are referred to [2].

3.1 Wavelength dependence [2]

The wavelength dependence of the output power shows a periodic variation with an almost constant period. This is illustrated in Fig. 6 for a coupler pulled

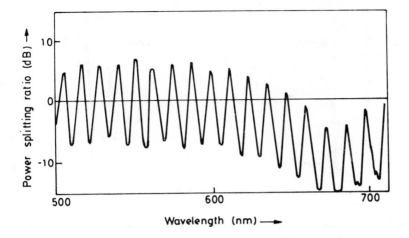

Fig. 6. Measured spectral behaviour of a fused coupler with a taper ratio of 40 [2].

through about 15 mm to a final taper ratio of 40 (ratio of initial to final fibre diameters, as measured at the taper waist). The initial fibre diameter was 105 μm. The coupler was potted in a silicone resin of refractive index 1.42. The constant period $\Delta\lambda$ tells us much about the coupling co-efficient. Referring to Eq. (13), this implies that the coupling co-efficient is proportional to the wavelength,

$$C = \alpha\lambda. \tag{21}$$

This result is important, as a naive argument would suggest that each phase shift ϕ_e and ϕ_o would be proportional to the wavenumber, and hence imply a coupling co-efficient proportional to $1/\lambda$. This is clearly not the case. The constant period $\Delta\lambda$ is obtained for all fused couplers measured in our laboratory and is an important result for any detailed theory to describe.

3.2 Polarisation effects [8-10]

The wavelength response for unpolarized light of a coupler with a 200 mm long tapered region is shown in Fig. 7 [9]. The periodic behaviour characteristics of the shorter coupler described above is now modulated. This is precisely what we expect from our general analysis (see Sec. 4) for polarisation effects. To check that this conclusion is indeed correct, the coupler was excited separately with X and Y polarized light. The result is shown in Fig. 8. The modulated response has now completely disappeared. The numerical summation of the individual responses shown in Fig. 8 almost exactly reproduces the modulated behaviour shown in Fig. 7. In a long coupler, the X and Y polarized states have slightly different coupling strengths, resulting in slightly different beat lengths. Eventually, the two polarisations become completely dephased, with each

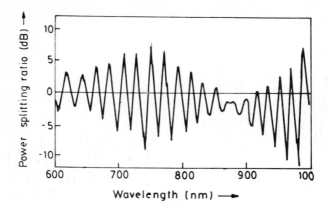

Fig. 7. Measured spectral dependence of a 200 mm long fused coupler, showing modulations due to polarisation effects [9].

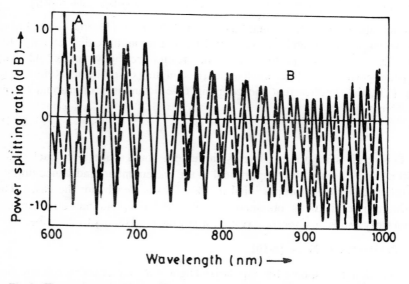

Fig. 8. The measured response to X and Y polarised light of the coupler shown in Fig. 7. The points marked A and B correspond to the modulation minima of Fig. 7 [9].

emerging from a different output port. This corresponds to the points A and B in Fig. 8. The spectral behaviour of a 300 mm long fused coupler is shown in Fig. 9 [4]. A large number of modulations are shown, and it is interesting to not that the modulation period decreases with increasing wavelength. This observation is important, since if the difference in coupling strengths C_x and C_y was due to stress birefringence in the coupler cross-section, an increase in the modulation period with wavelength would be expected [17,18].

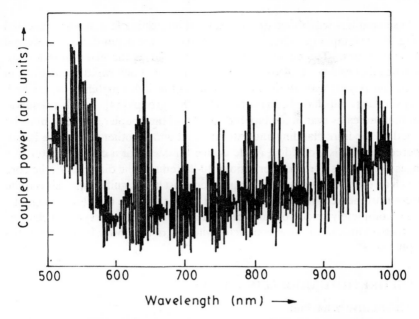

Fig. 9. Measured spectral response of a 300 mm long fused coupler to unpolarised light [4].

3.3 Dependence on external refractive index [2,14]

The variation of the output characteristics with external refractive index for two couplers with taper ratios of 10 and 20 is shown in Fig. 10 [2]. The taper lengths

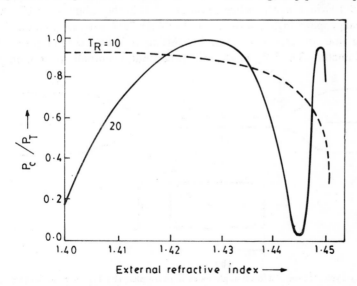

Fig. 10. Experimental variation of splitting ratio as a function of external refractive index. The couplers had taper ratios of 10 and 20 [2].

of these couplers was about 10 mm. Several interesting features are revealed in Fig. 10. The taper ratio has a profound effect on the dependence on external refractive index. The curves in Fig. 10 show that as the external index approaches that of silica (1.458), one of the couplers shows rapid oscillations in the output power, with all the power being lost as index matching is reached. Very similar results have been reported by other groups [14]. It is this observation that indicates that in the tapered section of the coupler, the light must be guided by the fibre cladding and external medium. In other words, the fused-tapered coupler is a cladding mode device. Indeed, for a core diameter of 5 μm, as used in this fibre, a taper ratio of 10-20 reduces the diameter of the core to such an extent that its waveguiding action must be insignificant. This is quite different to the behaviour in the polished coupler, where the electromagnetic field is completely guided by the core. In Fig 10, we see that very small changes in refractive index can cause complete switching of the power between the two output ports.

4. THEORETICAL MODELLING [15,19]

4.1 Qualitative behaviour

The fused tapered coupler made in our laboratory have a structure similar to that of Fig. 11(a). The tapered section CD is approximately parallel and typically has a length of 10-20 mm, although much longer fused lengths are possible. A cross-section XX' through the tapered section has the 'dumb-bell' shape shown schematically in Fig. 11(b). The dimension a lies in the range 2.5 μm to about 10 μm, corresponding to taper ratios in the range 40 to 10. An optical signal propagating along an input fibre AB will be in an LP_{01} mode. On propagating along the tapering section BC, this mode 'sees' a core of gradually reducing diameter. The LP_{01} mode will spread out, eventually becoming an

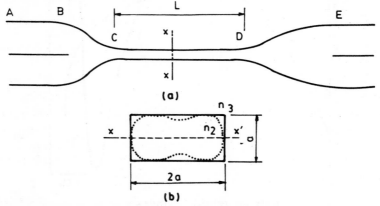

Fig 11. (a) Typical shape of a fused single-mode fibre coupler; (b) Typical cross-section through the neck of the tapered coupler. The dotted line is a rectangular approximation discussed in the text.

LP_{01} cladding mode. It is very important that this occurs adiabatically, otherwise coupling to higher order cladding modes will occur, resulting in a power loss in the coupler. In the tapered section, the core is so small that it has no significant effect on the mode guidance, which is now by an effective cladding waveguide consisting of the whole coupler cross-section (see Fig. 11(b)). On leaving the taper, the electromagnetic field adiabatically evolves into core modes along the tapered section DE. This type of coupler closely fulfills all the general conditions assumed in previous sections. The coupling co-efficient C is given by the *total* phase shift of the even and odd modes propagating through the whole coupler, BCDE. This is given by

$$C = \frac{1}{2} \int_{BCDE} |\beta_e - \beta_o| \, dz. \quad (22)$$

In Eq (22), $\beta_{e,o}$ are the propagation constants of the coupler's even and odd modes. As it stands, Eq. (22) is the exact expression for the coupling co-efficient. However, for any practical device, it is impossible to evaluate Eq. (22) analytically without making some approximations. To proceed further, we make three approximations:

1. The integral in Eq. (22) is dominated by the cladding mode region CD shown in Fig. 11(a).
2. The tapered section CD is assumed to have a constant cross-section of length L, and constant refractive index n_2.
3. The 'dumb-bell' cross-section shown in Fig. 11(b) is appoximated by a rectangle with sides a and $2a$.

The coupling co-efficient can now be expressed by

$$C \approx \frac{1}{2} |\beta_e - \beta_o| \cdot L, \quad (23)$$

where $\beta_{e,o}$ are the propagation constants of the lowest order even and odd modes of the rectangular guide.

4.2 A first approximation

In the cladding waveguide of Fig. 11(b), the dimensions are large compared to one wavelength, and the index difference n_2-n_3 is *not* weakly guiding. The electromagnetic field will be tightly confined by this waveguide with only a very small fraction of the filed penetrating into the surrounding index n_2. As a first approximation, we will assume the field to vanish exactly on the boundary of this waveguide. Neglecting polarisation effects for the moment, we then need to solve the scalar wave equation

$$\nabla^2 \Psi + (n_2^2 k^2 - \beta^2) \Psi = 0 \quad (24)$$

with $\Psi = 0$ on the waveguide boundary. Under these conditions β is given, *independently* of the shape of the cross-section, by an equation of the form

$$n_2^2 k^2 - \beta^2 = \text{constant}. \tag{25}$$

The 'constant' in Eq. 25 is independent of wavelength and has the dimensions of an inverse-length squared. Thus, we can say quite generally that

$$n_2^2 k^2 - \beta^2 = \alpha^2 / a^2. \tag{26}$$

In Eq. 26 the parameter α is a dimensionless number which depends only on detailed shape of the cross-section. It depends on no other parameter. In general, it will be different for the even and odd modes, which we will denote by $\alpha_{e,0}$. We can now expand β in powers of $1/k$, giving to first-order, the following expression

$$\beta_{e,0} = n_2 k - \alpha_{e,0}^2 / (2n_2 k a^2). \tag{27}$$

The coupling co-efficient is then approximated by

$$C = \frac{|\alpha_e^2 - \alpha_0^2|}{8\pi} \cdot \frac{\lambda}{n_2 a^2} \cdot L. \tag{28}$$

It should be emphasized that Eq. (28) is valid for any coupler cross-section, and to first order, it displays explicitly the dependence of C on wavelength, refractive index, and taper ratio through the dimension a. We see that the coupling co-efficient is proportional to wavelength, precisely as required by the spectral measurements discussed earlier.

The constant spectral period is seen to be a direct consequence of the cladding mode operation of the coupler, and strongly suggests that our neglect of the fibre cores is a valid assumption. We also note that the spectral period is expected to be inversely proportional to both the (taper ratio)^2 and the length L. Again, this is well in keeping with the experimental measurements. To make more quantitative comparisons, we now turn to the rectangular approximation to the coupler cross-section.

4.3 A second approximation

It will be convenient to define a normalized frequency variable V through

$$V = \frac{2\pi a}{\lambda} (n_2^2 - n_3^2)^{1/2}. \tag{29}$$

For the fused coupler cross-section, V is much larger than 1 and this allows an accurate asymptotic analysis of the rectangular approximation.

The rectangular dielectric guide is conveniently analysed by the effective index method [22] and we give here only the main results of that analysis [15-19]. The coupling co-efficients corresponding to X and Y polarized modes are:

$$C_x + C_y = \frac{3\pi\lambda}{32 n_2 a^2} \left[\frac{1}{\left(1 + \frac{1}{V}\right)} + \frac{1}{\left\{\frac{1 + n_3^2}{(V n_2)}\right\}^2} \right], \quad (30)$$

$$C_x - C_y = \frac{3\pi\lambda}{16 n_2 a^2} \left[1 - \left(\frac{n_3^2}{n_2^2}\right) \right] \cdot \frac{1}{V} \cdot L. \quad (31)$$

For large V we see that Eq. (30) reduces in form to Eq. (28) whilst the polarisation effects, as described by Eq. (31), disappear. The spectral period and polarisation modulation, given by Eqs. (19) and (20) are now easily determined. In particular, the following very simple relation can be derived

$$\frac{\partial \lambda}{\Delta \lambda} = \frac{\text{constant}}{\lambda}. \quad (32)$$

The constant in Eq. (32) is independent of wavelength. The dependence on external refractive index comes through the V dependence of Eq. (30). We now examine how well this analysis describes the measurements made on real couplers.

5. COMPARISON WITH EXPERIMENT

5.1 Wavelength dependence

For a coupler potted in a medium of index n_3, slightly lower than n_2, the spectral period is given by Eqs. (19) and (30) as

$$\Delta\lambda = \frac{32 n_2 a^2}{3L} \cdot \frac{(1 + 1/V)^3}{(1 - 1/V)}. \quad (33)$$

For the coupler described earlier (Fig. 11(a)) [2], $a = 2.63\,\mu\text{m}$, $L = 15$ mm, $n_2 = 1.458$ and $n_3 = 1.42$. Since V depends on λ, the period is not quite constant but increases with wavelength. At two different wavelengths, the predicted and measured $\Delta\lambda$ corresponding to Fig. 6 are

λ (nm)	$\Delta\lambda$ (theory) nm	$\Delta\lambda$ (experiment) nm
500	10.2	11.3
700	11.8	13.2

Considering all the approximations that have been made, this agreement is surprisingly good, and probably much better than we deserve. Even the small increase of $\Delta\lambda$ with wavelength is accounted for.

5.2 Polarisation effects

The polarisation properties are mainly determined by Eqs. (20), (31) and (32). Equation (32) predicts that the modulation period $\delta\lambda$ decreases with increasing wavelength. This agrees qualitatively with the measurements shown in Fig. 9 [4]. To check the quantitative prediction of Eq. (32), the ratio $\delta\lambda/\Delta\lambda$ determined from Fig. 9 was plotted against $1/\lambda$ in Fig. 12. The experimental points lie close to a straight line, in complete agreement with out model.

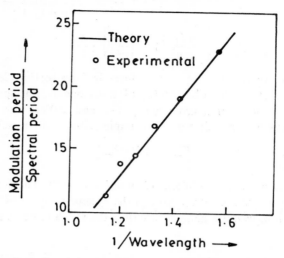

Fig. 12. Experimental variation of the ratio of modulation period to spectral period against $1/\lambda$ for the coupler shown in Fig. 9. Theory predicts a straight line dependence.

5.3 Dependence on external refractive index

The sensitivity to external refractive index comes from the V-dependence in Eq. (30). This sensitivity increases with the taper ratio through the $1/a^2$ dependence and also as n_3 approaches n_2. Both these conclusions are in general agreement with the measurements shown in Fig. 10.

For both these couplers, the initial fibre diameter was $105\,\mu$m and the taper length was about 10 mm. The measurement wavelength was 633 nm. Using these parameters in Eq. (30), the predicted variation of output power with refractive index is shown in Fig. 13. Comparing this with Fig. 10, and bearing in mind, the simplicity of the model, the theory provides a very good description of the measured results. In particular, the position of the maxima and minima are very well reproduced.

6. CONCLUSION

The fused single-mode coupler has emerged as a versatile fibre optic component which is quick, simple and inexpensive to fabricate. As a result of much experimental and theoretical work, the basic principles of the device are well understood. It is now possible to understand some of the problems encountered when depressed cladding or highly birefringent fibres are used in the fabrication process. Referring to Fig. 11, the problems arise from the requirement of an adiabatic expansion of the LP_{01} mode at it propagates down the

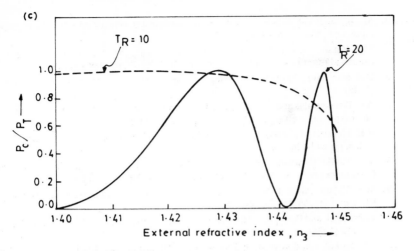

Fig. 13. Theoretical prediction of the variation of splitting ratio with external refractive index for the couplers of Fig. 10.

tapering section BC. Any coupling to higher order modes will result in losses. It is easy to achieve the adiabatic condition in matched clad fibres, however, in depressed clad fibres, the adiabatic limit places very severe constraints on the rate of tapering [20]. In practice, the rate of tapering required is so slow that the coupler would be impractically long [20]. Since highly birefringent fibres have depressed cladding regions, they suffer from the same problems. However, some progress has been made with this problem [21]. Nevertheless, this still remains the outstanding problem in fused-coupler technology.

ACKNOWLEDGEMENT

This chapter arose from some informal lectures given to the Optical Waveguide Group at IIT Delhi in November 1985. I would like to thank Professors Ghatak, Pal, Goyal and Kumar for their very kind hospitality during my stay in Delhi. I would also like to thank Prof. Pal for his infinite-patience whilst waiting for the final version of the manuscript. I have learnt much about this subject from my colleagues at Southampton. In particular, I thank Sam Yataki, whose superb experimental work has revealed much about fused couples, Con Hussey who introduced me to many of the ideas discussed here and Robert Mears for many helpful discussions.

REFERENCES

1. B.S. Kawasaki, Hill K.O. and Lamont R.G., "Biconical taper single-mode fibre coupler", Appl. Opt., Vol. 6, pp. 327-328 (1981).
2. F. De Fornel, Ragdale C.M. and Mears R.J., "Analysis of single-mode fused tapered couplers", IEE Proc. J. Optoelectron, Vol. 131, pp. 221-228 (1984).

3. M. Digonnet and Shaw H.S., "Wavelength multiplexing in single-mode fibre couplers", Appl. Opt., Vol. 22, pp. 484-491 (1983).
4. M.S. Yataki, Payne D.N. and Varnham M.P., "All-fibre wavelength filters using concatenated fused-taper couplers", Electron Lett., Vol. 21, pp. 248-249 (1985).
5. D.B. Mortimore, "Low-loss 8×8 single-mode star coupler", *ibid.*, Vol. 21, pp. 502-504 (1985).
6. A.C. Boucouvalas and Georgiou G., "Fibre-optic interferometric tunable switch using the thermo-optic effect", *ibid*, Vol. 21, pp. 512-514 (1985).
7. R.E. Meyer et al., "Passive fibre-optic ring resonator for rotation sensing", Opt. Lett. Vol. 8, pp. 644-646 (1983).
8. M.S. Yataki, Varnham M.P. and Payne D.N., "Fabrication and properties of very long fused taper couplers", Proc. 8th Optical Fibre Conf., San Diego (1985).
9. M.S. Yataki, Payne D.N. and Varnham M.P., "All-fibre polarising beamsplitter", Electron Lett., Vol. 21, pp. 249-251 (1985).
10. T. Bricheno and Baker V., "All-fibre polarisation splitter/combiner", *ibid.*, Vol. 21, pp. 251-252 (1985).
11. L.F. Stokes, Chodorow M. and Shaw H.J., "All single-mode fibre resonator", Opt. Lett., Vol. 7, pp. 288-290 (1982).
12. R.A. Bergh, Kotler G. and Shaw H.J., "Single-mode fibre optic directional coupler", Electron Lett., Vol. 16, pp. 260-261 (1980).
13. W.V. Sorin and Yu M.H., "Single-mode fibre ring dye laser", Opt. Lett., Vol. 10, pp. 550-552 (1985).
14. R.G. Lamont, Johnson D.C. and Hill K.O., "Power transfer in fused biconical-taper single-mode fibre couplers: dependence on external refractive index", Apl. Opt., Vol. 24, pp. 327-332 (1985).
15. J.D. Love and Hall M., "Polarisation modulation in long couplers", Electron Lett., Vol. 21, pp. 519-521 (1985).
16. F.P. Payne, Hussey C.D. and Yataki M.S., "Modelling fused single-mode fibre couplers", Vol. 21, pp. 461-462 (1985).
17. F.P. Payne, Hussey C.D. and Yataki M.S., "Polarisation analysis of strongly fused and weakly fused tapered couplers", Vol. 21, pp. 561-563 (1985).
18. F.P. Payne, Hussey C.D. and Yataki M.S., "Polarisation in fused single-mode fibre couplers", Proc. 11th European Conference on Optical Communication, pp. 571-574 (1985).
19. A.W. Snyder, "Polarising beamsplitter from fused-taper couplers", Electron Lett., Vol. 21, pp. 623-625 (1985).
20. W.J. Stewart and Love J.D., "Design limitation on tapers and couplers in single-mode fibres", Proc. 11th European Conference on Optical Communication, 1985, pp. 559-562.
21. M. Abebe, Villaruel C.A. and Burns W.K., "Reproducible process for the fabrication of polarisation-preserving couplers", Technical Digest of OFC'86, Atlanta, Georgia, pp. 82-84 (1986).
22. E.A.J. Marcatili, "Dielectric rectangular waveguide and directional coupler for integrated optics", Bell. Sys. Tech. J., Vol. 48, pp. 2071-2102 (1969).

26

Single-Mode All Fibre Components

D. Uttamchandani[*]

---

1. INTRODUCTION

The operation of most current and future optical fibre communication, sensing and signal processing systems will rely heavily on the availability of both passive and active in-line optical fibre components examples of which include polarizers, multiplexers, modulators and more recently, fibre lasers and amplifiers.

For fibre optic systems, three classes of components are generally available. These are all-fibre components, integrated optic components and finally micro-optic components. In the latter, miniature versions of bulk components such as lenses, beam splitters, polarizers, gratings and so on are assembled in small packages to perform the required function. Micro-optic components are, in general, more suitable for multimode fibre systems and will not be discussed further in this chapter.

Integrated optic technology can be used to fabricate most of the components required in single-mode optical communication systems. However, for practical use, the components must have fibre pigtails and there are problems encountered in efficiently matching the field profiles of a circularly symmetric single-mode fibre pigtail to that of a rectangular cross-section integrated optic component. Also there is the not insignificant problem of reliably attaching the fibre end to the integrated optic chip whilst maintaining a low insertion loss and low reflection from the connections. Nevertheless, integrated optic components are gradually becoming more widely available.

---

[*] Department of Electronic and Electrical Engineering, University of Strathclyde, 204 George Street, GLASGOW G1 1XW, Scotland (UK)

An all-fibre optical component is simply defined as one in which the necessary function is performed on the light whilst it is still guided by the fibre. Thus, all-fibre components are of great importance since they help in avoiding many of the problems encountered with integrated optics. This chapter will review some of the components which are available or have been reported for use in communications, sensing and signal processing. Both passive devices such as directional couplers, wavelength multiplexers, wavelength demultiplexers, polarizers, isolators and active devices such as phase, intensity and frequency modulators will be considered.

2. DIRECTIONAL COUPLERS

Directional couplers are used for dividing optical power from one input fibre to usually two or sometimes three output fibres. Alternatively, the device can be used to combine power from two or more input fibres into one output fibre. In fibre optic systems, division or recombination of optical power is required in areas such as coherent communications for mixing an incoming optical signal with a local oscillator [1], fibre interferometric sensors [2] such as Michelson and Mach Zehnder interferometers and signal processing systems [3,4] such as filters and matrix multipliers. The directional coupler is, therefore, an extremely versatile device and is now widely available commercially from a number of companies. The two most common forms of directional couplers, namely the 2×2 fused coupler and the polished single-mode fibre coupler are described below.

2.1 Fused Single-Mode Couplers

Fused directional couplers, or strictly speaking fused biconical taper couplers have been reported by a number of research groups [5,6,7]. During the manufacture of 2×2 couplers, two fibres are stripped of their primary and secondary coatings and held together, usually by a twist. Light of the desired wavelength is coupled into one of the fibres and the intensity at the two outputs is monitored. The fibres are then fused together using a source of heat such as a methane/oxygen flame or an electrical resistive heater. The fused fibres are then tapered by heating and pulling apart the two ends of the fused region whilst monitoring the intensity at the fibre outputs. Figure 1 shows the power variation in the two arms of a typical coupler during elongation. When the desired coupling ratio is achieved, the heating and pulling is stopped and the coupler is attached to a thin silica rod in order to prevent bends and strain on the fused region. The final diameter of the fused region depends on the coupling ratio achieved but, typically, after starting with two 125 μm cladded fibres, the dimensions of the fused region is around 5 μm × 10μm. Insertion losses of less than 0.1 dB are usually achieved.

The analysis of fused tapered couplers by the theory of evanescent wave coupling does not adequately explain their behaviour. A simple and more

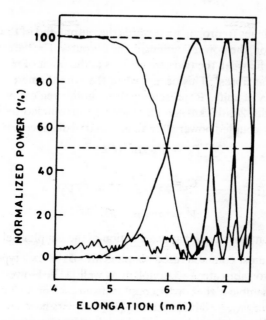

Fig. 1. Coupled power vs. elongation in a fused biconical taper coupler [10] (© 1987 OSA. Reproduced with permission).

appropriate model has been developed based on interference between the propagating modes of the composite waveguide formed in the tapered coupling region [8,9,10]. For analytical purposes, the dumb-bell shaped coupling region is approximated by a rectangular waveguide [cf. companion chapter on fused couplers by F.P. Payne]. As the light from the input fibre propagates in the down tapered region where the core diameter is continually decreasing, the mode spot size increases until the field is no longer guided by the core. This occurs when [10]

$$\frac{\beta^2 - n_2^2 k^2}{(n_1^2 - n_2^2) k^2} = \frac{2\alpha_c n_2}{(n_1^2 - n_2^2)^{1/2}}, \tag{1}$$

where β = propagation constant of fibre mode,
n_1 = core index,
n_2 = cladding index,
k = $2\pi/\lambda_0$,
α_c = core taper angle.

The light now propagates in the composite waveguide comprising the cladding material of index n_2 and the surrounding material of index n_3 (which can be air) and exciting the modes of this structure. The functioning of the coupler can be explained in terms of the interference between the lowest order even (symmetric) and odd (antisymmetric) supermodes of the composite structure.

As the light continues to propagate towards the output end of the coupler, i.e. the up-taper region, the power coupling at the output will be determined by the phase difference between the two supermodes at the point of recapture of the light by the output fibres. This occurs when the core diameter yields a normalised frequency of unity. Representing the coupling region by a rectangular waveguide of width $2a$, thickness a and index n_2 surrounded by a medium of index n_3 the cross-coupled power after an interaction distance L is given by [10]

$$P_2 = P_0 \sin^2(CL), \qquad (2)$$

where
$$C = \frac{3\pi\lambda_0}{32n_2a^2} \frac{1}{(1+1/V)^2} \qquad (3)$$

and
$$V = ak(n_2^2 - n_3^2)^{1/2} \qquad (4)$$

This expression gives a good approximation of the coupler behaviour.

Although the most common fused couplers are of the 2×2 type, it has been shown possible to fabricate 3×3 couplers as well [11]. However, it is more convenient to assemble higher order couplers from 2×2 or 3×3 coupler building blocks. In this way, 8×8 [12] and 9×9 [13] couplers have been assembled. Such single-mode star couplers will find applications in future single-mode optical fibre LAN's and high-speed optical data buses. The 8×8 device shown schematically in Fig. 2 is fabricated without the use of fusion splices and

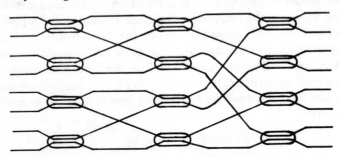

Fig. 2. Structure of an 8×8 star coupler [12] (© 1985 O.S.A. Reproduced with permission).

comprises twelve 3 dB couplers each with an excess loss of around 0.13 dB (average). The 9×9 coupler is assembled from six 3×3 couplers with their tails spliced together as shown in Fig. 3.

2.2 Polished Single-Mode Couplers

The evanescent field of light in a fibre can be accessed by laterally polishing the fibre. This involves cementing the unjacketed fibre in a groove cut in the shape of an arc of radius 25 cm or so in a quartz block, and then laterally lapping and polishing the mounted fibre and block to a distance of a few microns from the core/cladding interface. The approximate depth of the polishing can be

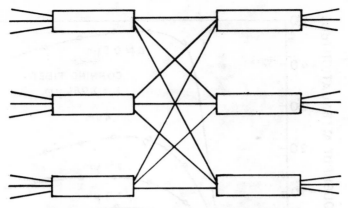

Fig. 3. Structure of a 9×9 star coupler [13] (© 1987 OSA. Reproduced with permission).

deduced from simple calculations [14] involving the radius of the arc shaped groove and the length of the major axis of the ellipse formed on the side of the fibre by the polishing process (cf. Fig. 4). The proximity of the core may then be more accurately deduced by monitoring the transmitted intensity of light through the polished region. This is achieved by placing an oil whose refractive index is close to that of the effective index of the fibre on top of the polished region and measuring the attenuation of the transmitted light [15]. Figure 5 shows the variation in transmission through the fibre as a function of oil index and core proximity of a polished block. Although polishing is a lengthy

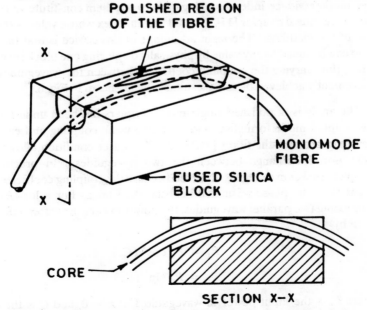

Fig. 4. A half coupler block [16] (© 1985 IEE, Reproduced with permission).

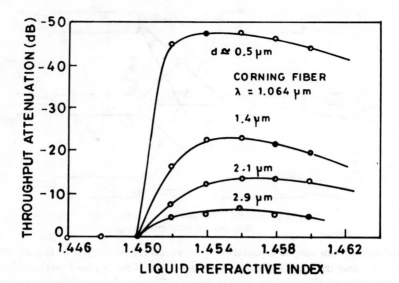

Fig. 5 Throughput attenuation vs. oil refractive index in a half-coupler block for different values of core/flat distance d [15] (© 1985 OSA. Reproduced with permission).

process, it does not change the core dimension or the strength of the fibre. An automated polishing/monitoring process has been reported in [16]. Two such polished blocks (henceforth referred to as half-coupler blocks) laid on top of one another with an index matching oil between them constitute an evanescent wave directional coupler [17]. The oil has an index whose value is the same as that of the cladding. The main advantage of this device is that the coupling ratio can be tuned to any value by physically displacing one block relative to the other, thus varying the coupling coefficient between the waveguides. This is conveniently achieved by using a suitable jig as shown in Fig. 6.

The analysis of polished single-mode couplers has been undertaken using the coupled mode formalism and assuming weak coupling and exact phase-matching between the fibres [14,18,19]. The latter condition allows for complete power exchange between the two waveguides. In practice, near complete power exchange is observed because the coupling coefficient is much larger than the phase mismatch between the fibres. For a lossless coupler comprising two parallel waveguides, the power in each arm after a distance z is given by

$$P_1 = P_0 \cos^2 (Cz)$$
$$P_2 = P_0 \sin^2 (Cz) \qquad (5)$$

where P_0 is the input power in waveguide 1 at $z = 0$, and C is the coupling coefficient between the modes of the waveguides. For a pair of identical

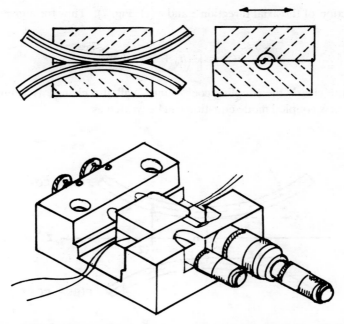

Fig. 6. Polished directional coupler and an assembled coupler in an adjustable jig [18] (© 1982 IEEE. Reproduced with permission).

parallel step index fibres whose cores are separated by a distance h, C is given by

$$C = \frac{\lambda_0}{2\pi n_1} \cdot \frac{U^2}{a^2 V^2} \cdot \frac{K_0\left[\frac{Wh}{a}\right]}{K_1^2[W]} \tag{6}$$

where

$$U = a\left[\left(\frac{2\pi n_1}{\lambda_0}\right)^2 - \beta^2\right]^{1/2},$$

$$W = a\left[\beta^2 - \left(\frac{2\pi n_2}{\lambda_0}\right)^2\right]^{1/2},$$

$$V = \frac{2\pi a}{\lambda_0}(n_1^2 - n_2^2)^{1/2},$$

K_0, K_1 = modified Bessel functions of the second kind.

In the polished coupler however, the waveguides are not parallel and will have a lateral offset y. The separation of the waveguide cores h now becomes a function of the axial direction z and y (cf. Fig. 7). Thus for a given value of y [18]

$$h(z) = \left[\left(h_0 + \frac{z^2}{R} \right)^2 + y^2 \right]^{1/2}, \quad (7)$$

Quite clearly the coupling coefficient C will also change as a function of z and a new coupled mode equation can be written as

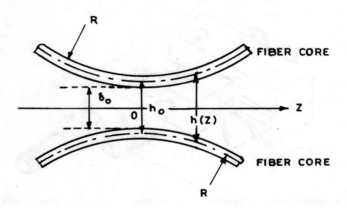

Fig. 7. Relative positions of fibre cores in a polished coupler [19] (© 1982 IEEE. Reproduced with permission).

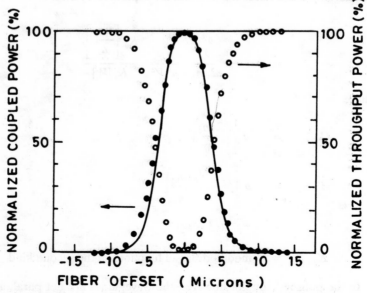

Fig. 8 Experimental curve and theoretical curve (solid line) of a directional coupler response at $\lambda = 0.6328\mu m$, $h_0 = 5.4\mu m$, $R = 25$ cm and $n_2 = 1.4567$ [19] (© 1982 IEEE. Reproduced with permission).

$$P_1 = P_0 \cos^2 \left\{ \int_{-\infty}^{\infty} C(z)dz \right\},$$

$$P_2 = P_0 \sin^2 \left\{ \int_{-\infty}^{\infty} C(z)dz \right\}, \tag{8}$$

The integrals can be numerically integrated by replacing the two curved fibres by a series of elemental sections of incremental length Δz. Typical theoretical and experimental results are shown in Fig. 8.

3. POLARIZERS

Fibre optic polarizers are often required for coherent communications and in sensor systems such as the Faraday effect current sensor and, in particular, the optical fibre gyroscope. Many designs have been proposed for fibre optic polarizers and some of these are discussed below.

3.1 Fibre Polarizers Based on Polished Coupler Blocks

The polished half coupler block is a convenient way of accessing the evanescent field of an optical fibre thereby enabling interaction with the guided mode of the fibre. Though such an interaction, fibre polarizers have been fabricated in a number of ways.

3.1.1 Birefringent crystal polarizer

Fibre polarizers have been constructed by placing an appropriate birefringent crystal on the polished surface of the fibre [20]. Potassium pentaborate has been used and oriented over the polished half-block such that for TM polarized light (for which the electric field is perpendicular to the polished surface) the index of this crystal is lower than that of the original fibre cladding, while for TE polarized light (whose electric field is parallel to the polished surface) the index of the crystal is slightly larger than that of the fibre cladding. Consequently, TM polarized light in the fibre experiences little loss during its propagation through the polished region whilst TE polarized light is radiated out of the fibre into the bulk crystal. The extinction ratio of the polarizer can be varied by rotating the crystal about the axis normal to the fibre surface since this tunes the index seen by the TE mode. Extinction ratios greater than 60 dB have been measured with low (< 0.1 dB) insertion loss.

3.1.2 Metal clad waveguide polarizers

Metal clad waveguide polarizers can be fabricated by polishing a fibre close to its core and depositing a metal film on the polished surface. Such a multilayer metal clad waveguide system is shown in its planar form in Fig. 9, and has been analysed in detail [21]. Two interesting properties of such waveguides emerge. The first property is that the TM mode has no cut-off while the TE polarized mode has a certain cut-off thickness. The second is that when both modes are above cut-off the attenuation of the TE mode is lower than that of the TM

METAL CLADDING	n_1
LOW INDEX LAYER	n_2
CORE	n_3
SUBSTRATE	n_4

Fig. 9. Schematic of a multi-layer metal clad waveguide.

mode. This arises because the ohmic losses of the TE modes are lower since the corresponding fields are smaller at the metal boundary. These properties of metal clad waveguides have been used to fabricate a polarizer by depositing around 1500 Å of aluminium on a polished fibre [22]. At short wavelengths, i.e. when both modes are above cut off, a polarization extinction ratio of 14 dB was observed, with the TM mode exhibiting the greater loss. At longer wavelengths, the TE mode is cut off leaving the TM mode guided in the metal clad region. For both cases, the insertion loss due to the absolute attenuation of each polarization state is high (around 11 dB/cm for the transmitted TE polarization). For lower insertion loss devices, a thin and short metal film is required.

An alternative polarizer that has been demonstrated using metal cladding is based on the resonant coupling of a guided mode from a waveguide to a surface plasmon wave. Such a device has been reported first as a polarizer for integrated optics applications [23]. Surface plasmon waves[*] are perpendicularly polarized electro-magnetic waves which exist at boundaries between metals and dielectric materials. The fundamental polarization selectivity of such waves has aroused a great deal of interest in their use in fibre optic polarizers. Figure 10 shows the approach used in the realisation of a surface plasmon fibre

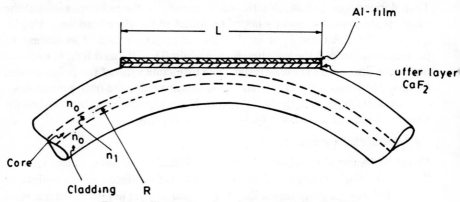

Fig. 10. A fibre optic surface plasmon wave polarizer [25].

---

[*] Physics of surface plasmon waves is discussed in Ch. 3 by A. Ghatak and K. Thyagarajan.

optic polarizer [24]. A plasmon wave of effective index n_s is supported at the interface between the low index buffer and the metal layer in contact with it. The value of n_s is given by

$$n_s = \left(\frac{\varepsilon_m n_b^2}{\varepsilon_m + n_b^2} \right)^{1/2}, \qquad (9)$$

where ε_m is the complex dielectric constant of the metal and n_b the refractive index of the buffer layer. For a metal such as aluminium $\varepsilon_m = -40$ (ignoring the optical losses) at a wavelength of 0.63 μm. With a buffer layer of index 1.42, n_s is approximately 1.46 which is typically the effective index of a single-mode fibre. Thus TM light guided in the fibre will couple to the surface plasmon wave and escape leaving the TE polarization still guided. Polarizers operating on this principle have been measured to have an extinction ratio greater than 45 dB and an insertion loss less than 1 dB [25].

3.1.3 Fibre polarizers using thin metal films

Recently there has been great interest shown in the use of thin metal films (i.e. metal films of thicknesses less than the skin depth at optical frequencies) in fibre optic polarizers. Such thin film polarizers have already been investigated for integrated optics applications [26]. The use of thin films ensures that ohmic losses which increase the attenuation of metal clad waveguide polarizers described earlier can be reduced.

A thin film polarizer has been demonstrated which cuts off the TE mode in the interaction region of a polished half-block [27]. The device is fabricated by laterally polishing a single-mode fibre into the core region so that both polarization modes are initially cut off. A thin metal film is then sandwiched between the fibre core and dielectric material of the same index as the fibre core (Fig.11). Such a waveguide supports as the fundamental mode the ω^+

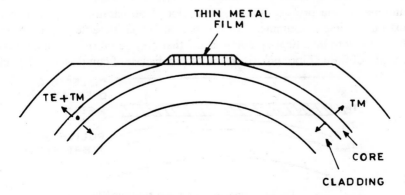

Fig. 11. A fibre optic surface plasmon wave cut-off polarizer [27]
(© 1986 OSA. Reproduced with permission).

surface plasmon wave which is polarized in the TM direction and which exhibits no cut off thickness. The fundamental TE mode has a finite cut off thickness. Consequently, as light propagates through the interaction region, the TE mode is cut off since the remaining core thickness is arranged to be less than the cut off thickness for the TE mode. The TM mode propagates through the interaction region as the symmetric surface plasmon oscillation and couples back into the fibre. The extinction ratio reported at 0.8 micron wavelength is 47 dB with an insertion loss of around 1 dB. Similar work has recently been reported in [28].

An alternative scheme utilising thin metal films for fibre polarizers has been reported [29] where polarization dependent synchronous coupling between two adjacent waveguides is used to devise a polarizer. A schematic diagram of this arrangement is shown in Fig. 12 where one of the waveguides is a fibre waveguide polished down to near its core while the second waveguide is formed by depositing a thin metal film on the polished fibre and then coating the metal with a dielectric overlay. Thus, the thin metal film bound by two dielectric regions constitutes the second waveguide. This guide can only support the TM surface plasmon mode. The symmetric mode is of interest since its effective index can be made to match that of the optical fibre. For example, by choosing a 270 Å thickness of aluminium and an overlay index of 1.44, synchronous coupling has been achieved from the TM mode of the fibre waveguide of effective index 1.451 to the surface waveguide. The extinction ratio measured was 46 dB while the insertion loss of the TE polarization remaining in the fibre waveguide was around 0.5 dB.

3.2 Polarisers Fabricated in Continuous Fibre Length

The fabrication of fibre optic polarizers using continuous lengths of optical fibre is far more desirable than using the polished block techniques that have been described above. Although the lateral polishing of fibres can be automated, the process still requires a lot of preparatory steps and can be extremely time consuming. For this reason, a considerable amount of effort has gone into investigating components that may be fabricated in continuous lengths of fibre. Fibre polarizers are one such class of components. For metal

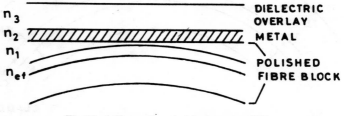

Fig. 12. A fibre optic polarising structure [29]
(© 1988 IEE. Reproduced with permission).

clad waveguide polarizers, the principal requirement is for access to the evanescent field of the guided mode by a metal film and this has led to the fabrication of special optical fibres. The fabrication of an eccentric core optical fibre, Fig. 13, has been described in [30]. In this fibre, the core is fabricated close to the cladding/outer coating surface. A part of this coating is then stripped. The bare fibre is slowly etched in a solution of hydrofluoric acid

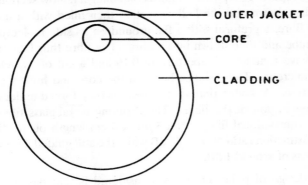

Fig. 13. Cross-section of an eccentric core fibre (after [30]).

in order to vary the thickness of the cladding which serves as a buffer layer. Coatings of aluminium are then deposited on the fibre which still has a large diameter. Experimental results have shown that at $\lambda = 1.15\,\mu$m the TM mode is attenuated by 22 dB (polarizer length = 21.4 mm) and the TE mode suffers a loss of around 0.66 dB. The polariser performance is critically dependent on the buffer thickness.

A second technique of devising a fibre with an accessible evanescent field involves fabricating a fibre using two cladding materials which have different etch rates [31]. The core is concentric with the cladding. One side of the core has a pure silica cladding while the other side has a borosilicate glass cladding which has a nine times faster etching rate in hydrofluoric acid than pure silica. Thus, after etching, a fibre of D shaped cross-section is left and a metal film is deposited on the flat. Measurements on a 4 cm length of fibre show a maximum extinction ratio of 37 dB of the TM mode but with large insertion losses of several dB for the TE mode. This is a result of the surface roughness remaining on the fibre, a consequence of the long etching times used.

An alternative technique used to fabricate fibres of D shaped cross-section is by cutting and polishing a flat on the fibre preform before drawing the preform into a fibre. Long lengths of such fibre have been fabricated and the fibre loss measured at 1.52 μm was 20 dB/km. Such high attenuations are not necessarily a limitation in the use of these fibres since most optical fibre components are expected to have short lengths [32]. The core/flat separation in these fibres can be controlled to within a few microns. A fibre polarizer has been reported [33] where a film of indium metal is deposited on the flat of an elliptical core D-shaped fibre. For a polarizer length of 40 mm, the extinction ratio is 39 dB with an insertion loss of 0.2 dB.

An interesting approach to producing low cost polarizers in optical fibre form has been reported [34] where a fibre containing a hollow section close to the core is first drawn and the hollow section then filled with a metal. To fabricate the fibre, a preform with a flat ground on its side is sleeved with a close fitting tube and then drawn into a fibre. The fibre thus fabricated was reported to have a numerical aperture of 0.16 and a cut off wavelength of around 1.25 microns. The distance between the core and hollow section is around 3 microns. A molten tin/indium alloy was then forced under pressure into the hollow region of the fibre. The resulting metal/glass fibre can be handled as a conventional fibre. At 1.3 μm a 6 cm length of this fibre had exhibited an extinction ratio of over 45 dB while the still guided TE mode had an insertion loss of around 1 dB.

The final design of polarizers to be considered is that based on coiled birefringent fibres. Birefringent fibres maintain the polarization of light transmitted along one of the principal axes. However, high birefringent fibres based on stress induced birefringence can also polarize light because, as a result of large internal stress, one polarization mode has stronger guidance than the other [35]. Consequently, if the fibre is bent with an appropriate radius of curvature, a differential attenuation occurs between the two polarization modes as the mode closer to cut-off leaks (power) at a faster rate. Typical results from a polarizing device using birefringent fibre [36] wound into a coil of radius 15 mm are shown in Fig. 14. Typically at 0.82 μm, the polarizer has an insertion loss of around 2 dB for the guided mode and an extinction ratio between the two polarization modes greater than 25 dB.

4. POLARIZATION SPLITTERS

Polarizing splitters are four port devices with the property of splitting an input randomly polarized beam of light into two orthogonal polarization components. Each polarization component appears at one of the output ports. Such devices can be used in optical fibre communication systems employing polarization diversity or in certain optical fibre sensors. Fibre optic polarization splitters can be fabricated either by using fused tapered fibre couplers or polished fibre couplers.

4.1 Polarization Effects in Fused Tapered Couplers

The polarization behaviour of fused tapered couplers fabricated from non-birefringent fibres has been reported in [37,38]. It has been shown that these devices can behave as polarization splitters when the couplers have long cou-

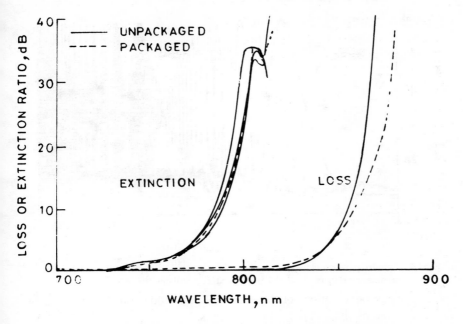

Fig. 14. Extinction ratio and loss for coiled birefringent fibre polarizers [36] (© 1984 OSA. Reproduced with permission).

pling lengths of between 100 and 200 mm. Extinction ratios of light in the output arms lie between 12 and 17 dB. Analysis of these devices can again be carried out by approximating the coupling region by a rectangular waveguide and deriving the propagation constants of the two polarizations of the fundamental symmetric and antisymmetric modes [39–41].

Let the propagation constants be denoted as $\beta_0^{\|}$ and $\beta_0^{\perp}$ for the symmetric mode and $\beta_1^{\|}$ and $\beta_1^{\perp}$ for the antisymmetric mode. For polarization splitting action at a given wavelength, the two parallel polarizations, for example, must be in phase after propagating through the coupling region of length L, while the two perpendicular polarizations must simultaneously be out of phase after travelling through the same length. This requirement can be expressed as

$$(\beta_0^{\|} - \beta_1^{\|}) L = 2m\pi \qquad (10)$$

$$(\beta_0^{\perp} - \beta_1^{\perp}) L = (2n + 1)\pi \ ; \quad m, n \text{ being integers.} \qquad (11)$$

Figure 15 shows the coupled power versus elongation of a fused taper coupler made with non-birefringent fibres. The slow envelope modulation is due to the polarization properties just described. The minima in the envelope occur when the even and odd modes are in phase for one of the two polarization eigen states and out of phase for the other polarization eigen-states. This corresponds to polarization splitting. The maxima in the envelope occur when

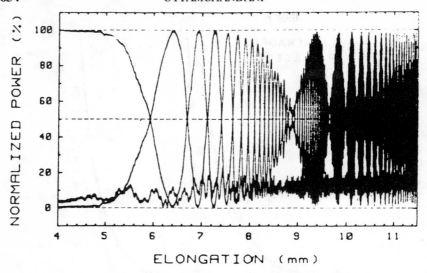

Fig. 15. Coupled power vs elongation in a fused biconical taper coupler showing envelope modulation [10] (© 1987 OSA. Reproduced with permission).

both the eigen polarizations of the even and odd modes are simultaneously in-phase or out of phase at the coupler outputs.

Polarization splitters can also be fabricated from fused high birefringent fibres[*] by aligning the stress applying parts parallel to one another before fusion and tapering [42]. The tapered region again supports four modes (two transverse modes each with two polarization states) and operates in exactly the same way as described for the non-birefringent fibre polarizing couplers. An extinction ratio of 17 dB has been measured on the light emerging from each arm of the coupler after polarization splitting has occurred. Since the polarization maintaining fibre has stress birefringence built into it, the beat lengths for the polarization eigen modes in the tapered region are much shorter than in ordinary non-birefringent fibres. Consequently, the overall coupling lengths to obtain polarization splitting in "hi-bi" (high-birefringent) fibre couplers are upto an order of magnitude shorter than in non-birefringent fibre couplers. Typical coupling region lengths are around 20 mm compared to 200 mm in non-birefringent fibres [43].

4.2 Polarization Effects in Polished Couplers

It has already been shown that fused biconical taper directional couplers can operate as polarizing beam splitters by exploiting birefringence (stress or form) in the coupling region. However, because these devices are many beat lengths long, they act as polarization splitters over a fairly narrow wavelength range. Polarization splitters based on polished couplers can be made physically

---

[*] Often referred in the literature as hi-bi fibres.

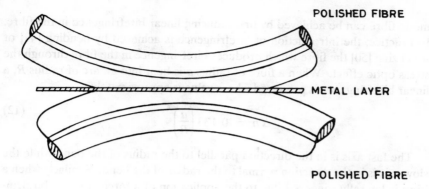

Fig. 16. A polarizing beamsplitter based on coupling through surface plasmon waves [47] (© 1988 IEE. Reprdouced with permission).

shorter and broadband in their operation provided high birefringent fibre is used [44]. An experimental device has been reported based on the principle of matching the propagation constants for one polarization and simultaneously maximising the mismatch for the orthogonal polarization. Only a 9 dB extinction ratio was obtained in the output arm of the coupler [45]. More recent experiments based on the same principle have yielded upto 25 dB extinction on both ports [46].

A more interesting polarization splitting directional coupler has been proposed using two identical parallel waveguides which interact through a thin metal layer [47] (Fig. 16). The presence of the metal layer causes strong coupling from the TM mode of the input guide to the adjacent guide via the surface plasmon mode of the metal film which is polarization selective. The TE mode of the input guide however couples less strongly to the adjacent guide because, for this polarization, the presence of the metal reduces the interaction between the guides. The theoretical extinction ratio predicted is 30 dB with better than 0.1 dB insertion loss. These results apply to a single-mode guide at 0.633 μm with an aluminium film of 80 Å thickness. An experimental device comprising two polished coupler blocks of non-birefringent fibres with a thin metal film between them has been reported as a polarization splitting coupler [48].

5. POLARIZATION CONTROLLERS

Polarization control is required in coherent optical communication systems which use normal (non-birefringent) single-mode fibres. Without control, the variation in state of polarization of an incoming optical signal would give rise to deep fades if the signal and local oscillator polarizations were not the same. There are similar requirements for polarization control in interferometric fibre sensors. The control of the state of polarization of light guided in a single-mode fibre can be achieved by first inducing linear birefringence in the fibre. In practice, the introduction of birefringence is achieved by bending [49] or squeezing [50] the fibre which produces birefringence in the fibre through the stress optic effect. When a fibre of radius a is bent into an arc of radius R, a linear birefringence is produced whose value is approximately

$$\Delta n = 0.133 \left(\frac{a}{R}\right)^2 . \tag{12}$$

The fast axis is in the direction parallel to the radius of the bend while the slow axis is in the direction normal to the radius of the bend. Similarly when a fibre is laterally squeezed due to the application of a force, a linear birefringence is produced in the fibre with a slow axis parallel to the direction of the force and a fast axis perpendicular to the direction of the force.

One of the most commonly used and easily assembled polarization controllers is based on a design first described in [51] and illustrated in Fig. 17. This device can transform any input state of polarization to any desired output state and consists of three fibre loops. The first and third loops have a single turn and produce $\lambda/4$ retardation (i.e. a quarter wave plate) whilst the middle loop has two turns and produces $\lambda/2$ retardation (i.e. a half wave plate). An angular rotation of the plane of any loop corresponds to a rotation of the equivalent wave plate. Light of an arbitrary elliptical polarization enters the first $\lambda/4$ loop whose orientation can be adjusted so that on exit the light is linearly polarized. This light then enters the middle $\lambda/2$ loop and an adjustment of the orientation of this loop corresponds to a rotation of the state of polarization of the linear light. Thus only two stages are necessary to transform any elliptical input state to any linear output state. When the linear light travels through the third $\lambda/4$ loop, it again emerges elliptical. Thus all possible input polarization states can be transformed to all possible output states. Note that in the type of device just described, the loops have a fixed birefringence but of variable orientation. In

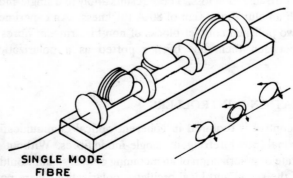

SINGLE MODE
FIBRE

Fig. 17. Polarisation controller for single-mode fibre [51]
(© IEEE. Reproduced with permission).

general, this device is only used for manual polarization control of light in a fibre since it tends to be rather bulky in design.

An alternative control scheme which is suitable for automatic polarization control is now described. This uses fibre squeezers which introduce a variable amount of birefringence on the fibre but in a fixed direction. For this reason, these devices operate in a different manner from the loop polarization controllers described above. To provide endless polarization control between two time varying states, four fibre squeezers are required to be placed alternately at 0° and 45° to the vertical [52] as shown in Fig. 18. The general transform of polarization from an input state A to an output state E is illustrated in Fig. 19 which shows the side elevation of a Poincare sphere. Four transformation steps are shown corresponding to the four fibre squeezers. Controlling action is transferred smoothly between alternate squeezers in order to allow individual squeezers to reset from their birefringence limits. This enables maintenance of continuous polarization control and avoids potential signal loss which would otherwise occur during the reset of individual squeezers.

6. OPTICAL ISOLATORS

Optical isolators are devices which permit low loss transmission of optical power propagating in one direction through the device but have high attenuation for power propagating in the opposite direction. Such devices are required, e.g., to isolate semiconductor lasers from reflections generated in an optical communication or sensor system. Spurious reflections can broaden the output spectrum of the semiconductor laser or cause fluctuations in its output frequency.

Fibre optical isolators based on the Faraday effect have been demonstrated using conventional single-mode fibres [53] and high birefringent fibres [54, 55]. In this section, an example of a Faraday effect isolator designed from conven-

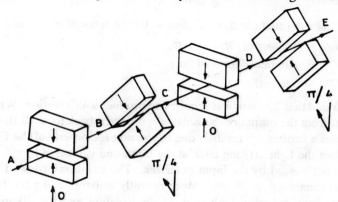

Fig. 18. Polarisation controller based on fibre squeezers (reproduced with permission from British Telecom Tech. Journ. Vol 5 No. 2 April 1987).

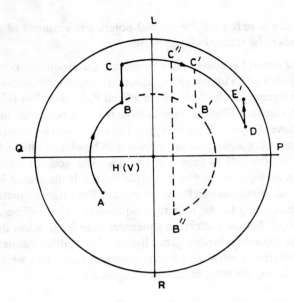

Fig. 19. Polarisation transformation of four fibre squeezers of Fig. 18 shown on a Poincaré sphere [52] (© 1987 IEE. Reproduced with permission).

tional fibre is described. A 15 metre length of fibre was first twisted along its axes at a twist rate of around 20 turns per metre in order to introduce a circular birefringence in the fibre. Circular birefringence was introduced in order to overcome any linear birefringence in the fibre which would otherwise overwhelm the required Faraday effect. The fibre was wound around a former, around which copper wire was then wound as shown in Fig. 20. The toroid consisted of 55 turns of fibre and 1250 turns of wire. The passage of an electric current through the wire rotated the direction of the linear polarization of the input by an amount ϕ where

$$\phi = V N_i N_f I. \tag{13}$$

Here V = Verdet constant of silica = 0.0156 seconds of arc/amp turn,
 N_i = number of turns in wire,
 N_f = number of turns of fibre,
 I = current.

It was found that 2.5 amps was required to produce a 45° rotation. When the linear light from the output of the isolator is reflected back towards the input, it undergoes a further 45° rotation due to the non-reciprocity of the Faraday effect. Thus the light arriving back at the input end is rotated by 90° and is, therefore, not passed by the input polarizer. The isolation provided by this device was measured as 18 dB. More recently isolators using birefringent fibres have been reported with over 44 dB isolation and 0.4 dB insertion loss [55].

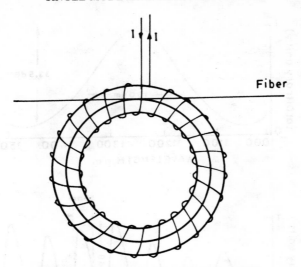

Fig. 20. Single-mode fibre optical isolater [53]
(© 1981 OSA. Reproduced with permission).

7. SINGLE-MODE FIBRE FILTERS

A single-mode optical fibre filter is considered as a two port device with the characteristic of rejecting or transmitting specific optical wavelengths from the input port to the output port. The two technologies that have been most commonly utilised in fibre filter fabrication are single fibre tapering and fibre polishing.

Tapered fibres have been observed to have a near sinusoidal output intensity variation as a function of wavelength [56] (See also the Ch. 25 by F.P. Payne). In the tapered region, coupling occurs between the core modes and higher order cladding modes and this is observed as power oscillations at the fibre output as it is drawn into a taper. Since the coupling coefficients and the effective indices of the relevant modes are wavelength dependent, the tapered fibre possesses a filtering characteristic. For short tapers fabricated with a few power oscillations, the filter response shows a broad pass band. Fig. 21(a) shows the response from a short taper. It is observed that the pass band is around 190 nm width. For tapers fabricated with 20 oscillations, the wavelength response is clearly periodic with narrower passbands as shown in Fig. 21(b). In order to select a specific wavelength in the taper filter response, the use of concatenated tapers has been proposed and demonstrated [57]. The overall response of the filter is a product of the responses of individual tapers. The spectral transmittance of a single taper is given by the expression

$$T(\lambda) = \cos^2\left[\frac{\pi(\lambda - \lambda_0)}{2\Delta\lambda}\right], \tag{14}$$

where λ_0 is the central wavelength and $\Delta\lambda$ is the spectral width of the half period in the wavelength response. For N concatenated tapers,

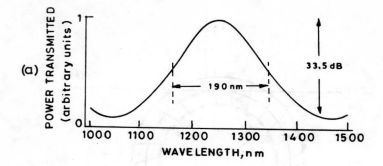

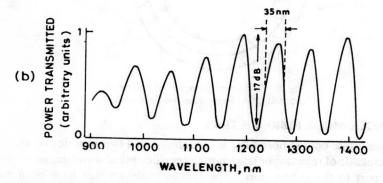

Fig. 21 (a) Wavelength transmission of a tapered fibre with 3 power oscillations. (b) Wavelength transmission of a tapered fibre with 20 power oscillations [56] (© 1985 IEE. Reproduced with permission).

$$T_N(\lambda) = \prod_{i=1}^{N} \cos^2\left[\frac{\pi(\lambda - \lambda_0)}{2\Delta\lambda_i}\right]. \tag{15}$$

Now, if $\Delta\lambda_i = 2^{i-1}\Delta\lambda_1$ and $N \to \infty$ then

$$T_N(\lambda) = \text{sinc}^2\left[\frac{\pi(\lambda - \lambda_c)}{\Delta\lambda_1}\right]. \tag{16}$$

This corresponds to a narrow band filter centred at λ_0 and with a half power width of approximately $0.89\,\Delta\lambda_1$. Such a filter has been realised where four concatenated tapers were used which were fabricated with 32, 16, 8 and 4 power transfer cycles (cf. Fig. 22). The measured response of this filter is shown in Fig. 23, where λ_0 is 766 nm and the FWHM of the filter response is 5.8 nm.

Fibre filters can also be fabricated from single or concatenated fused taper couplers [58]. Although such devices are strictly four port devices, they are considered in this section as filters when only a single input and single output ports are used. Fused taper couplers act as filters since the propagation constants of the super modes and consequently the beat length of the coupler

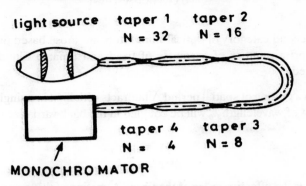

Fig. 22. Fibre filter comprising 4 concatenated tapers [57]
(© 1986 OSA. Reproduced with permission).

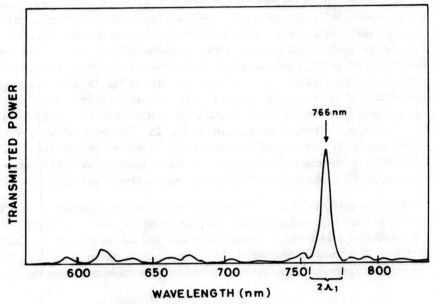

Fig. 23. Wavelength transmission of filter of Fig. 22 [57]
(© 1986 OSA. Reproduced with permission).

change with wavelength. The wavelength response of fused couplers is also periodic. The spectral width of a half period decreases with increasing length of the coupling region. Narrow band filter operation can be achieved by concatenating fused biconical taper couplers. Thus the overall wavelength response of N concatenated couplers is given by the product of the wavelength response of the individual couplers

$$T_N(\lambda) = \prod_{i=1}^{N} \frac{1}{2^N} \left[1 + \sin\left(\frac{2\pi(\lambda - \lambda_0)}{2\Delta\lambda}\right) \right], \qquad (17)$$

where the symbols have been defined earlier.

The second category of optical fibre filters are those based on half coupler blocks and a number of examples of fixed and tunable filters will now be considered.

When a grating of spatial period Λ interacts with light in a single-mode fibre, then light of wavelength λ_0 will be coupled in the backward direction when

$$\Lambda = \frac{\lambda_0}{2n_e}, \qquad (18)$$

where n_e is the effective index of the fibre. A grating of this periodicity is a first order Bragg grating. Such devices can be considered as band reject filters during transmission in the forward direction or band pass filters if the contra-propagating light is collected. First order Bragg filters have been fabricated by placing a diffraction grating on a polished half coupler so that the evanescent field of the guided light interacts with the grating [59] or, alternatively, by fabricating the grating on the core itself [60] as shown in Fig. 24. The measured reflectivity of the surface relief grating is illustrated in Fig. 25. An interesting approach that has been demonstrated to realise tunability in grating based optical fibre filters makes use of a special fan shaped first order grating [61]. The grating is illustrated schematically in Fig. 26. The pitch of the grating varies across its surface and by moving the grating relative to the fibre, tunability can be achieved. Measurements have shown a FWHM reflection bandwidth of 0.8 nanometers with a tuning range varying from 1.46 to 1.54 μm.

An alternative approach to designing filters based on using side polished fibres has been reported [62] where coupling occurs between the fibre mode and modes of a high index overlay waveguide formed on the polished half coupler. A schematic diagram of the arrangement is shown in Fig. 27. By solving the eigenvalue equation of a symmetric waveguide and using the fact

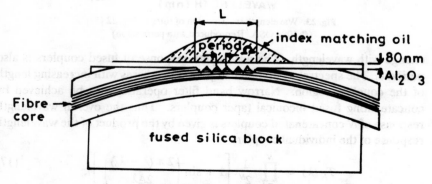

Fig. 24. Fibre grating filter [60]
(© 1986 IEE. Reproduced with permission).

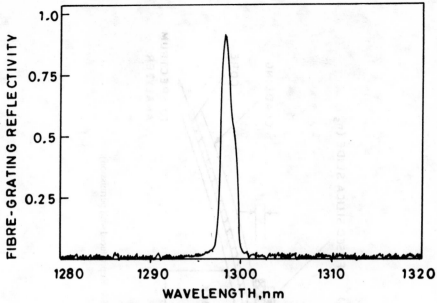

Fig. 25. Wavelength reflectivity of grating filter of Fig. 24 [60]
(© 1986 IEE. Reproduced with permission).

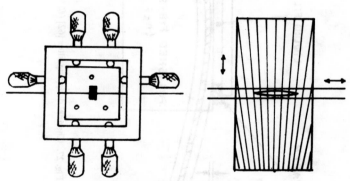

Fig. 26. Wavelength tunable filter [61]
(© IEE. Reproduced with permission).

that for efficient coupling, the effective indices of the fibre waveguide and overlay waveguides must be equal, the following equation can be obtained

$$\lambda_m \approx \frac{2t}{m} \left(n_g^2 - n_e^2 \right)^{1/2} \tag{19}$$

Here n_e is the effective index of the fibre mode, n_g the film index and λ_m represents the wavelength at which coupling occurs to the mth mode of the overlay waveguide. Figures 28(a) and 28(b) show the transmission response of the filter for a thin film ($t \approx 1.5\ \mu$m) and a thick film ($t \approx 80\ \mu$m) overlay

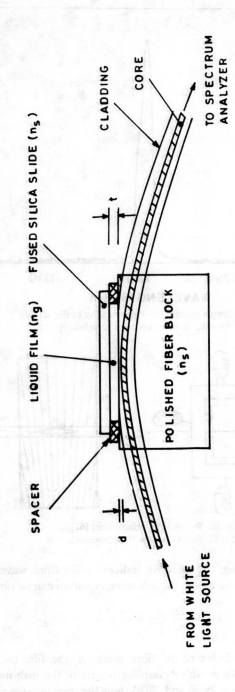

Fig. 27. Channel dropping filter [62] (© 1987 OSA. Reproduced with permission).

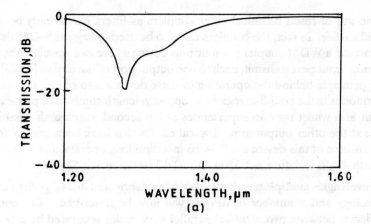

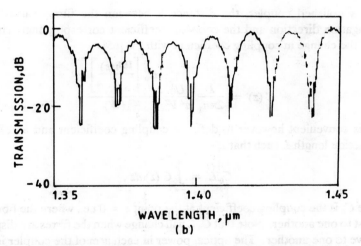

Fig. 28. (a) Wavelength transmission of channel dropping filter using a thin overlay of oil [62], (b) Wavelength transmission of channel dropping filter using a thick overlay of oil [62] (© 1987 OSA. Reproduced with permission).

waveguide. For thin films, the insertion loss of the device is around 0.5 dB with 20 dB rejection while for a thick film, the insertion loss is between 1 and 4 dB for the same rejection.

8. WAVELENGTH MULTIPLEXERS AND DEMULTIPLEXERS

Wavelength division multiplexing (WDM) offers a means of increasing the capacity of an optical fibre communication system by introducing more optical carrier wavelengths down the same transmission channel. Consequently a great deal of research has gone into developing practical low loss single-mode WDM components.

The use of fused biconical taper couplers as filters has already been discussed earlier. In fact, the couplers can also be used as rugged WDM devices. In practice, a WDM coupler is a four port device where two wavelengths at the common input port transmit, each to one output port of the device [63,64]. The basic principle behind the operation of these devices is to ensure that the two supermodes in the coupling regions at one wavelength combine in phase at one output arm whilst the two supermodes at the second wavelength combine in phase at the other output arm. Typical results that have been quoted for the performance of this device are 0.04 dB insertion loss, operation at 1.3 and 1.52 μm with 43 dB isolation at 1.3 μm and 30 dB isolation at 1.52 μm.

Wavelength multiplexers have also been fabricated using polished fibre technology and a number of devices will now be described. The coupling coefficient between two identical parallel waveguides separated by a constant distance h has been already described in Eq. (6).

In a polished coupler, the distance h between the fibres varies in the propagation direction and the coupling coefficient correspondingly changes. Thus, the change in coupling coefficient with z is [65]

$$C(z) = \frac{\lambda}{2\pi n_1} \frac{U^2}{a^2 V^2} \frac{K_0\left[\frac{Wh(z)}{a}\right]}{K_1^2(W)}. \tag{20}$$

It is convenient however to define a coupling coefficient and an effective interaction length L such that

$$C_0 L = \int_{-\infty}^{\infty} C(z)\, dz, \tag{21}$$

where C_0 is the coupling coefficient at the point $z = 0$ i.e., where the fibres are closest to one another. Note that $C_0 L$ will change when the fibres are displaced relative to one another. The optical power in each arm of the coupler is given by

$$P_1 = P_0 \cos^2(C_0 L)$$

and
$$P_2 = P_0 - P_1 = P_0 \sin^2(C_0 L). \tag{22}$$

For ideal wavelength multiplexing, one input wavelength λ_1 should become fully cross-coupled while the second wavelength λ_2 should remain uncoupled. In mathematical terms, this requires that

$$(c_0 L)_{\lambda_1} = \left(m + \frac{1}{2}\right)\pi$$

and
$$(c_0 L)_{\lambda_2} = n\pi, \tag{23}$$

where m and n are integers.

In practice, this condition may be achieved by tuning the coupler, i.e., by varying the separation of the fibres on the polished blocks using a suitable jig. A typical response of the wavelength multiplexer is shown in Fig. 29 where the power output from one port of the coupler is shown against wavelength. The power output from the second port would simply be opposite to that shown.

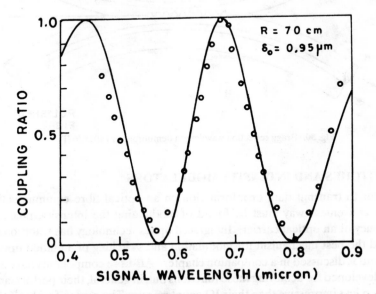

Fig. 29. Experimental and theoretical (solid line) transmission response of wavelength demultiplexer [65] (© 1983 OSA. Reproduced with permission).

Directional couplers have also been fabricated from two dissimilar polished fibres. In practice, this ensures that the coupling coefficients between the two fibres are matched only over a narrow spectral range and consequently power exchange will take place only at a specific wavelength [66].

Finally a WDM device based on a first order Bragg grating is described. The device consists of a polished directional coupler formed between fibres of slightly differing propagation constants with a Bragg grating formed between them [67]. If the grating has a periodicity Λ, the following equations describe the behaviour of the device [68]

$$\lambda_1 = (n_{e1} + n_{e2}) \Lambda$$
$$\lambda_2 = 2n_{e1} \Lambda, \qquad (24)$$

where n_e is the effective index of each fibre. Figure 30 shows a schematic diagram of the device. Wavelengths differing by as little as 12 Å have been separated by this device.

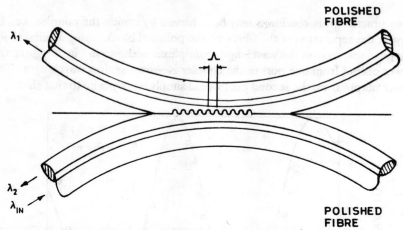

Fig. 30. Bragg reflection wavelength demultiplexer (after [67]).

9. SWITCHES AND INTENSITY MODULATORS

In order to transmit data or information on an optical fibre communication link, a convenient way must be found of modulating the intensity, phase or frequency of an optical carrier. Integrated optics technology has traditionally offered the best mechanism for modulation and switching of a guided optical wave and is discussed in a companion chapter. All-fibre components have also been developed for achieving these functions but, in general, their performance has been less impressive than their IO counterparts. The reason for this is that silica is a "passive" material demonstrating no intrinsic electro-optic effect. Nevertheless, in-line active components have been developed. In the following sections, a few examples of all-fibre modulators mentioned above are given.

A number of fibre optic intensity modulators have been reported of which probably the most interesting are those based on the use of liquid crystalline materials. An intensity modulator has been demonstrated where a liquid crystal with birefringent axes of refractive indices 1.46 and 1.52 was used [69]. The liquid crystal molecules are long and narrow and when the molecules are lined up (nematic phase), the high refractive index lies parallel to the long axis of the molecules. If an electric field is applied to the liquid crystal, the long axis of the molecules tend to align with the field. An intensity modulator arrangement is shown in Fig. 31 using a polished half coupler block. The molecules are initially aligned vertically. When the input polarization of light is TM, the light leaks out of the fibre since it experiences a higher overlay index. On application of a horizontal electric field, the molecules are reoriented horizontally and the TM polarization now sees the lower overlay index and is consequently transmitted. Thus the intensity modulator is polarization dependent. The intensity modulator was measured to have a 3 dB cut off frequency of 1 kHz. A similar

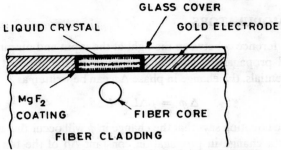

Fig. 31. Liquid crystal based intensity modulator [69]
(© 1986 OSA. Reproduced with permission).

kind of device has been reported in which a tapered fibre is located in a liquid crystal film embedded between two electrodes [70].

More recently, a polarization desensitised intensity modulator has been reported [71] that also uses a liquid crystal overlay. The device is fabricated using a half coupler block chemically treated in such a manner that the refractive index of the material seen by both polarizations of light in the fibre is above the cladding index in one switching state and below the cladding index in the other state. Measurements made on this device show that for TM polarized light, the insertion loss is 2.3 dB and the on/off ratio is 16.9 dB, while the same quantities for TE polarized light are 0.48 dB and 15.4 dB respectively. The switch rise time for both polarizations is 2.5 milliseconds while the fall time is 7 milliseconds and 35 milliseconds for TM and TE polarizations, respectively.

An interferometric fibre optical switching device has been reported [72] where a compact all fibre interferometer was fabricated with each arm coated with an electrically resistive materials (Fig. 32). The passage of an electric current through the resistive region of one arm generates heat, causing a change in the effective index of the mode in the fibre. Depending on the relative phase difference between the modes in the two arms, the light appears either at port 1 or port 2 of the output. The power consumption of this device was around 25 milliwatts with a switching speed of around 40 kHz.

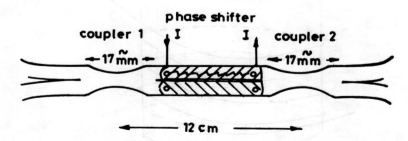

Fig. 32. Compact Mach-Zehnder intensity modulator [72]
(© 1985 IEE. Reproduced with permission).

10. PHASE MODULATORS

The phase difference ϕ between the light at the input and output of a fibre of length L and propagation constant β is given by the product βL radians. Taking differentials, the change in phase $\Delta\phi$ can be written as

$$\Delta\phi = \beta\Delta L + L\Delta\beta. \tag{25}$$

This simple equation says that the change in ϕ will occur due to a change in length ΔL or a change in propagation constant $\Delta\beta$ of the fibre. The most common form of fibre optic phase shifter is one which uses a piezo electric material to stretch the fibre. The piezo electric material usually used is PZT (lead zirconate titanate) in the form of a cylinder. To realise a phase modulator, a few turns of a single-mode fibre are wound around the cylinder and held under a small tension whilst a voltage is applied between the inside and outside walls of the cylinder as shown in Fig. 33. The corresponding change in circumference of the cylinder generates a change in length of the fibre [73]. For a PZT cylinder of 1" diameter, the phase modulation index is around 0.04 rad/turn/volt at 0.633 μm optical wavelength when the cylinder is operated away from its resonance frequency. It is usually desirable to operate away from resonance in order to obtain a uniform response over a large bandwidth. Typically these devices are used upto operating frequencies of a few hundred kHz. At higher frequencies, the phase deviations from phase modulators become small because of the reduction in the physical size of the transducers and the acoustic wavelengths. At these frequencies, it becomes more convenient to generate phase modulation by inducing radial strain in the fibre [74] or by launching a transverse acoustic wave across the fibre cross-section. A high frequency phase modulator operating at frequencies between 400 and 500 MHz has been reported [75] using the latter technique. The device

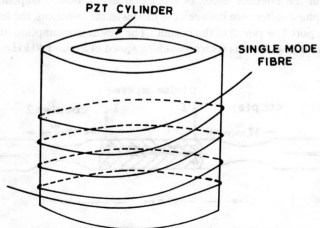

Fig. 33. Phase modulator using PZT cylinder [73].

works by launching an acoustic wave into the centre of a single-mode fibre using a piezo electric transducer fabricated on the outer surface of the fibre (Fig. 34). The 2 mm long device that was reported produces a peak phase deviation of about 2 radians for an acoustic drive power of 2 watts. Care was taken to provide acoustic damping (to prevent unwanted reflections) and heat sinking in the device. For optimum operation, the core diameter of the fibre should be less than half an acoustic wavelength in order to prevent the cancellation of phase modulation. For a fibre with a 5 μm core, the minimum acoustic wavelength would thus have to be 10 μm. The acoustic velocity in silica is around 6,000 metres per second yielding a maximum operating frequency of 600 MHz for the modulator. Finally, we describe phase modulators based on the use of piezo-electric polymers. One approach that has been reported involves the bonding of an optical fibre to a PVDF (polyvinylidene difluoride) film [76] and, more interestingly, the development of a fibre with a PVDF jacket [77,78]. More recently [79] a fibre optic phase modulator has been reported using a radially poled piezo-electric co-polymer jacket which generates a uniform radial strain in the fibre. This avoids inducing birefringence in the fibre core which would otherwise make the modulator polarization sensitive. The phase sensitivity of this device to electric field was 1.24×10^{-5} radian/volt m^{-1} for a 1 metre length of the fibre, whilst it operated over a frequency range from 20 Hz to 20 kHz.

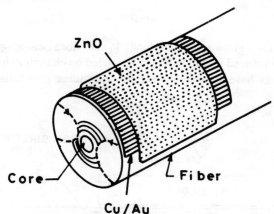

Fig. 34. High frequency phase modulator [75]
(© 1988 IEEE. Reproduced with permission).

11. FREQUENCY MODULATORS

Single-mode fibre optic frequency shifters are often required in applications such as the fibre optic gyroscope and heterodyne interferometers in fibre form. Two approaches to frequency shifting are generally used—the first being the fibre optic equivalent of an acousto optic cell and the second a fibre optic equivalent of a serrodyne frequency translator. Only the first technique will be considered in this section. A frequency shifter has been developed in fibre optic form using a travelling acoustic wave to couple light between the two eigen polarizations of a high birefringence fibre [80]. Two equations must be satisfied for this device to work

$$\beta_2 = \beta_1 \pm \beta_a \sin\theta$$
$$\omega_2 = \omega_1 \pm \omega_a. \qquad (26)$$

Here ω_1 and β_1 represent the angular frequency and propagation constant of light in one polarization, ω_2 and β_2 represent the same quantities in the other polarization and ω_a and β_a are the acoustic frequency and propagation constant. The quantity θ represents the angle between the optical fibre axis and the acoustic wave front as shown in Fig. 35. When the acoustic wave propagates in the same direction as light in the fibre, light in the fast mode of the fibre is coupled to the slow mode with a frequency upshift while light in the slow mode couples to the fast mode with a corresponding frequency downshift. The reverse situation holds if the acoustic wave travels against the direction of light propagation. The phase-match conditions of Eq. (27) yield the following equation for the frequency shift, f_a as

$$f_a = \frac{V_L}{L_b \sin\theta}. \qquad (27)$$

This shows that for a given acoustic velocity V_L and fibre beat length L_b, the frequency shift f_a produced on the light can be varied by changing the angle θ. Experimental results have shown that a PZT transducer generating a bulk

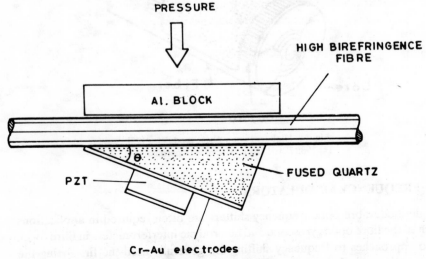

Fig. 35. Bulk acoustic wave frequency shifter (after [80]).

acoustic wave of 15 MHz requires 2.25 watts of continuous electrical power for the original optical carrier and unwanted sidebands to be suppressed by better than 20 dB.

12. CONCLUSIONS

In-line optical fibre components have generated a great deal of research interest and over the last few years, a number of single-mode fibre components have become commercially available. Virtually all of these have been passive components such as fibre couplers, wavelength multiplexers and filters. Active components such as all-fibre high-speed switches, intensity and phase modulators are undergoing extensive research but there is no indication of imminent commercial availability of such products. When they are eventually developed, such devices will find use in areas other than optical communications such as guided wave signal processing and optical sensor multiplexing systems.

REFERENCES

1. Y. Yamamoto and Kimura T., "Coherent Optical Fibre Transmission Systems", IEEE *J. Quant. Electron.*, Vol. QE-17, pp. 919-935 (1981).
2. T.G. Giallorenzi, Bucaro J.A., Dandridge A., Siegel G.H., Cole J.H., Rashleigh S.C. and Priest R.G., "Optical Fibre Sensor Technology", IEEE *J. Quant Elect.*, Vol. QE-18, pp. 626-665 (1982).
3. K.P. Jackson, Bowers J.E., Newton S.A. and Cutler C.C., "Microbend Optical Fibre Tapped Delay Line for Gigahertz Signal Processing", *Appl. Phys. Lett.*, 41, pp. 139-141 (1982).
4. M. Shabeer, Andonovic I. and Culshaw B., "Fibre Optic Bipolar Tap Implementation Using An Incoherent Optical Source", *Opt. Lett.*, 12, pp. 726-728 (1987).
5. B.S. Kawasaki, Hill K.O. and Lamont R.G., "Biconical Taper Single Mode Fibre Coupler", *Opt. Lett.*, 6, pp. 327-328 (1981).
6. C.M. Ragdale, Payne D.N., De Fornel F. and Mears R.J., "Single Mode Fused Biconical Taper Fibre Couplers", Proc. 1st Int. Conf. on Opt. Fibre Sensors, London, pp. 75-78 (1983).
7. T. Bricheno and Fielding A., "Stable Low Loss Single Mode Couplers", *Elect. Lett.*, 20, pp 230-232 (1984).
8. F.P. Payne, Hussey C.D. and Yataki M.S., "Modelling Fused Single Mode Fibre Couplers", *Elect. Lett.*, 21, pp. 461-462 (1985).
9. J. Bures, Lacroix S. and Lapierre J., "Analyse d'un Coupleur bidirectionnel a fibres optiques monomodes fusionnees", *App. Opt.*, 22, pp. 1918-1922 (1983).
10. F. Bilodeau, Hill K.O., Johnson D.C. and Faucher S., "Compact, Low Loss, Fused Biconical Taper Couplers: Overcoupled Operation and Antisymmetric Supermode Cut off", *Opt. Lett.*, 12, pp. 634-636 (1987).
11. W.K. Burns, Moeller R.P. and Villarruel C.A., "Observation of Low Noise In a Passive Fibre Gyroscope", *Elect. Lett.*, 18, pp. 648-649 (1982).
12. D.B. Mortimore, "Low Loss 8×8 Single Mode Star Coupler", *Elect. Lett.*, 21, pp. 502-504 (1985).

13. C.C. Wang, Burns W.K. and Villarruel C.A., "9×9 Single Mode Fibre Optic Star Couplers", *Opt. Lett.*, **10**, pp. 49-51 (1985).
14. O. Parriaux, Gidon S. and Kuznetsov A.A., "Distributed Coupling on Polished Single Mode Optical Fibres", *App. Opt.* **20**, pp. 2420-2423 (1981).
15. M.J.F. Digonnet, Feth J.R., Stokes L.F. and Shaw H.J., "Measurement of the Core Proximity in Polished Fibre Substrates and Couplers", *Opt. Lett.* **10**, pp. 463-465 (1985).
16. S.T. Nicholls, "Automatic Manufacture of Polished Single Mode Fibre Directional Coupler", *Elect. Lett.* **21**, pp. 825-826 (1985).
17. R.A. Bergh, Kotler G. and Shaw H.J., "Single Mode Fibre Optic Directional Coupler", *Elect. Lett.*, **16**, pp. 260-261 (1980).
18. T. Findakly and Chen C.L., "Optical Directional Couplers with Variable Spacing", *App. Opt.*, **17**, pp. 769-773 (1978).
19. M.J.F. Digonnet and Shaw H.J., "Analysis of a Tunable Single Mode Optical Fibre Coupler", *J. Quant. Elect.*, Vol. QE-18, pp. 746-754 (1982).
20. R.A. Bergh, Lefevre H.C. and Shaw H.J., "Single Mode Fibre Optic Polarizer", *Opt. Lett.*, **5**, pp. 479-481 (1980).
21. Y. Yamamoto, Kamiya T. and Yanai H., "Characteristics of Optical Guided Modes in Multilayer Metal Clad Optical Guide with Low-Index Dielectric Buffer Layer", *J. Quant. Elect.*, Vol. QE-11, pp. 729-736 (1975).
22. W. Eickhoff, "In Line Fibre Optic Polarizer", *Elect. Lett.* **16**, pp. 762-764 (1980).
23. S. Wright, De Oliveira A.D. and Wilson M.G.F., "Optical Waveguide Polarizer with Synchronous Absorption", *Elect. Lett.*, **15**, pp. 510-512 (1979).
24. O. Parriaux, Gidon S. and Cochet F., "Fibre Optic Polarizer Using Plasmon Guided Wave Resonance", *Proc. 7th Europ. Conf. on Opt. Comm.*, Copenhagen, Paper P6 (1981).
25. D. Gruchmann, Petermann K., Staudigel L. and Weidel E., "Fibre Optic Polarizers with High Extinction Ratio", *Proc. 9th Europ. Conf. on Opt. Comm.*, Geneva, pp. 305-308 (1983).
26. K.H. Rollke and Sohler W., "Metal Clad Waveguide as Cut off Polarizer for Integrated Optics", *J. Quant. Elect.*, Vol. QE-13, pp. 141-145 (1977).
27. J.R. Feth and Chang C.L., "Metal Clad Fibre Optic Cut off Polarizer", *Opt. Lett.*, **11**, pp. 386-388 (1986).
28. S. Markatos, Zervas M.N. and Giles I.P., "Optical Fibre Surface Plasmon Wave Devices", *Elect. Lett.*, **24**, pp. 287-288 (1988).
29. W. Johnstone, Stewart G., Culshaw B. and Hart T., "Fibre Optic Polarizers and Polarizing Couplers", *Elect. Lett.*, **24**, pp. 866-868 (1988).
30. T. Hosaka, Okamoto K. and Noda J., "Single Mode Fibre Type Polarizer", *J. Quant. Elect.*, Vol. QE-18, pp. 1569-1572 (1982).
31. T. Hosaka, Okamoto K. and Edahiro T., "Fabrication of Single Mode Fibre Type Polarizer", *Opt. Lett.*, **8**, pp. 124-126 (1983).
32. C.A. Millar, Ainslie B.J., Brierley M.C. and Craig S.P., "Fabrication and Characterisation of D-fibres with a Range of Accurately Controlled Core/Flat Distances", *Elect. Lett.*, **22**, pp. 322-324 (1986).
33. R.B. Dyott, Bello J. and Handerek V.A., "Indium Coated D-shaped Fibre Polarizer", *Opt. Lett.*, **12**, pp. 287-289 (1987).
34. L. Li, Wylangoswski G., Payne D. and Birch R., "Low Cost Metal/Glass Fibre Polarizers Produced in Continous Lengths", *Proc. 4th Int. Conf. on Opt. Fib. Sensors*, Tokyo, pp. 163-166 (1986).

35. A.W. Snyder and Ruhl F., "New Single Mode Single Polarization Optical Fibre", *Elect. Lett.*, **19**, pp. 185-186 (1983).
36. M.P. Varnham, Payne D.N., Barlow A.J. and Tarbox E.J., "Coiled Birefringent Fibre Polarizers", *Opt. Lett.*, **9**, pp. 306-308 (1984).
37. M.S. Yataki, Payne D.N. and Varnham M.P., "All-Fibre Polarizing Beamsplitter", *Elect. Lett.*, **21**, pp. 249-251 (1985).
38. T. Bricheno and Baker V., "All Fibre Polarizing Splitter/Combiner", *Elect. Lett.*, **21**, pp. 251-252 (1985).
39. J.D. Love and Hall M., "Polarization Modulation in Long Couplers", *Elect. Lett.*, **21**, pp. 519-521 (1985).
40. F.P. Payne, Hussey C.D. and Yataki M.S., "Polarization Analysis of Strongly Fused and Weakly Fused Tapered Couplers", *Elect. Lett.*, **21**, pp. 561-563 (1985).
41. A.W. Snyder, "Polarizing Beamsplitters from Fused Tapered Couplers", *Elect. Lett.*, **21**, pp. 623-625 (1985).
42. I. Yokohama, Okamoto K. and Noda J., "Fibre Optic Polarizing Beamsplitter Employing Birefringent Fibre Coupler", *Elect. Lett.*, **21**, pp. 415-416 (1985).
43. I. Yokohama Okamoto K. and Noda J., "Analysis of Fibre Optic Polarizing Beamsplitters Consisting of Fused Taper Couplers", *J. Light. Tech.*, Vol. LT-4, pp. 1352-1359 (1986).
44. A.W. Snyder and Stevenson A.J., "Polished Type Couplers Acting as Polarizing Beamsplitters", *Opt. Lett.*, **11**, pp. 254-256 (1986).
45. R.H. Stolen, Ashkin A., Bowers J.E., Dziedzic J.M. and Pleibel W., "Polarization Selctive Fibre Directional Coupler", *J. Light. Tech.*, Vol. LT-3, pp. 1125-1129 (1985).
46. H.C. Lefevre, Simonpietri P. and Graindorge P., "High Performance Polarization Splitting Fibre Coupler", *Elect. Lett.*, **24**, pp. 1304-1305 (1988).
47. K. Thyagarajan, Diggavi S. and Ghatak A.K., "Design and Analysis of a Novel Polarisation Splitting Directional Coupler", *Elect. Lett.*, **24**, pp. 869-870 (1988).
48. M.N. Zervas Markatos S. and Giles I.P., "Polarising beamsplitter of Polished Type", SPIE Proceedings, Vol. 988, paper 988-10.
49. R. Ulrich, Rashleigh S.C. and Eickhoff W., "Bending Induced Birefringence in Single Mode Fibres", *Opt. Lett.*, **5**, pp. 273-275 (1980).
50. M. Johnson, "In-Line Fibre Optical Polarization Transformer", *App. Opt.*, **18**, pp. 1288-1289 (1979).
51. H.C. Lefevre, "Single Mode Fibre Fractional Wave Devices and Polarization Controllers", *Elect. Lett.*, **16**, pp. 778-780 (1980).
52. N.G. Walker and Walker G.R., "Endless Polarization Control Using Four Fibre Squeezers", *Elect. Lett.*, **23**, pp. 290-292 (1987).
53. T. Findakly, "Single Mode Fibre Isolator in Toroidal Configuration", *App. Opt.*, **20**, pp. 3989-3990 (1981).
54. E.H. Turner and Stolen R.H., "Fibre Faraday Circulator or Isolator", *Opt. Lett.*, **6**, pp. 322-323 (1981).
55. G.W. Day, Payne D.N., Barlow A.J. and Ramskov-Hansen J.J., "Design and Performance of Tuned Fibre Coil Isolators", *J. Light. Tech.*, Vol. LT-2, pp. 65-60 (1984).
56. A.C. Boucouvalas and Georgiou G., "Biconical Taper Coaxial Coupler Filter", *Elect. Lett.*, **21**, pp. 1033-1034 (1985).
57. S. Lacroix, Gonthier F. and Bures J., "All Fibre Wavelength Filter from Successive Biconical Tapers", *Opt. Lett.*, **11**, pp. 671-673 (1986).
58. M.S. Yataki, Payne D.N. and Varnham M.P., "All Fibre Wavelength Filters Using Concatenated Fused Taper Couplers", *Elect. Lett.*, **21**, pp. 248-249 (1985).

59. W.V. Sorin and Shaw H.J., "A Single Mode Fibre Evanescent Grating Reflector", *J. Light. Tech.*, Vol. LT-3, pp. 1041-1043 (1985).
60. I. Bennion, Reid D.C.J., Rowe C.J. and Stewart W.J., "High Reflectivity Monomode Fibre Grating Filters", *Elect. Lett.*, 22, pp. 341-343 (1986).
61. M.S. Whalen, Tennant D.M., Alferness R.C., Koren U. and Bosworth R., "Wavelength Tunable Single Mode Fibre Grating Reflector", *Elect. Lett.*, 22, pp. 1307-1308 (1986).
62. C.A. Millar, Brierley M.C. and Mallinson S.R., "Exposed Core Single Mode Fibre Channel Dropping Filter Using a High Index Overlay Waveguide", *Opt. Lett.*, 12, pp. 282-284 (1987).
63. C.M. Lawson, Kopera P.M., Hsu T.Y. and Tekippe V.J., "In-Line Single Mode Wavelength Division Multiplexer/Demultiplexer", *Elect. Lett.*, 20, pp. 963-964 (1984).
64. G. Georgiou and Boucouvalas A.C., "High Isolation Single Mode Wavelength Division Multiplexer/Demultiplexer", *Elect. Lett.*, 22, pp. 62-63 (1986).
65. M. Digonnet and Shaw H.J., "Wavelength Multiplexing in Single Mode Fibre Couplers", *App. Opt.*, 22, pp. 484-491 (1983).
66. M.S. Whalen and Walker K.L., "In-Line Optical Fibre Filter for Wavelength Multiplexing", *Elect. Lett.*, 21, pp. 724-725 (1985).
67. M.S. Whalen, Divino M.D. and Alferness R.C., "Demonstration of a Narrowband Bragg Reflection Filter in a Single Mode Fibre Directional Coupler", *Elect. Lett.*, 22, pp. 681-682 (1986).
68. P. Yeh and Taylor H.F., "Contradirectional Frequency Selective Couplers for Guided Wave Optics", *App. Opt.*, 19, pp. 2848-2855 (1980).
69. K. Liu, Sorin W.V. and Shaw H.J., "Single Mode Fibre Evanescent Polarizer/Amplitude Modulator Using Liquid Crystals", *Opt. Lett.*, 11, pp. 180-182 (1986).
70. C. Veilleux, Lapierre J. and Bures J., "Liquid Crystal Clad Tapered Fibres", *Opt. Lett.*, 11, pp 733-735 (1986).
71. R. Kashyap, Winter C.S. and Nayar B.K., "Polarization Desensitized Liquid Crystal Overlay Optical Fibre Modulator", *Opt. Lett.*, 13, pp. 401-403 (1988).
72. A.C. Boucouvalas and Georgiou G., "Fibre Optic Interferometric Tunable Switch Using the Thermo-Optic Effect", *Elect. Lett.*, 21, pp. 512-514 (1985).
73. S.A. Kingsley and Davies D.E.N., "Method of Phase Modulating Signals in Optical Fibres: Application to Optical Telemetry Systems", *Elect. Lett.*, 10, pp. 21-22 (1974).
74. S.A. Kingsley, "Optical Fibre Phase Modulator", *Elect. Lett.*, 11, pp. 453-454 (1975).
75. A.A. Godil, Patterson D.B., Heffner B.L., Kino G.S. and Khuri-Yakub B.T., "All Fibre Acousto-Optic Phase Modulators Using Zinc Oxide Films on Glass Fibre", *J. Light. Tech.*, 6, pp. 1586-1590 (1988).
76. K.P. Koo and Sigel G.H., "An Electric Field Sensor Utilizing a Piezoelectric Polyvinylidene Fluoride (PVF_2) Film in a Single Mode Fibre Interferometer", *J. Quant. Elect.*, Vol. **QE-18**, PP. 670-675 (1982).
77. L.J. Donalds, French W.G., Mitchell W.C., Swinehart R.M. and Wei T., "Electric Field Sensitive Optical Fibre Using Piezoelectric Polymer Coating", *Elect. Lett.*, 18, pp. 327-328 (1982).
78. P.D. DeSouza and Mermelstein M.D., "Electric Field Detection with A Piezoelectric Polymer Jacketed Single Mode Optical Fibre", *App. Opt.*, 21, pp. 4214-4218 (1982).
79. M. Imai, Shimizu T., Ohtsuka Y. and Odajima A., "An Electric Field Sensitive Fibre with Coaxial Electrodes for Optical Phase Modulation", *J. Light. Tech.*, Vol. **LT-5**, pp. 926-931 (1987).
80. W.P. Risk, Youngquist R.C., Kino G.S. and Shaw H.J., "Acousto-Optic Frequency Shifting in Birefringent Fibre", *Opt. Lett.*, 9, pp. 309-311 (1984).

27

Signal Processing in Monomode Fibre Optic Sensor Systems

J.D.C. JONES[*]

---

1. INTRODUCTION

An optical sensor may be formally defined as a device in which an optical signal is changed in some reproducible way by an external stimulus, such as temperature or strain. This definition covers a very wide range of devices, because an optical beam is characterised by a number of independent variables such as intensity, wavelength spectrum, phase and state of polarisation. In an optical sensor, any one or a combination of these may be modulated by the measurand (that is, by the parameter which is to be measured).

A fibre optic sensor is, of course, an optical sensor which makes use of optical fibres. The fibre may actually be used as the sensing element, or simply as a flexible waveguide which conveys light to and from the region of measurement. Those devices in which the beam is guided by a fibre at the measurement region are called intrinsic, the other type is called extrinsic. As will be described further below, fibre components may be used to control or modulate the guided beam, thus facilitating signal processing.

This chapter is solely concerned with the use of monomode optical fibre, which has the very important property of preserving the spatial coherence of the guided beam [1]. Such fibre is therefore, most suitable for sensor applications which exploit coherent optical techniques [2] — that is, those sensors in which the transduction mechanism is the modulation of optical phase,

---

[*] Optoelectronics and Laser Engineering, Department of Physics, Heriot-Watt University, Riccarton, Edinburgh EH14 4AS, UK.

measured by interferometry (interferometric sensors), or those based on the modulation of the state of polarisation (polarimetric sensors).

Interferometry is a well established classical optical technique, and has been used for making very high resolution measurements [3]. However, classical interferometry demands the use of precision optical components mounted with great stability, and is thus generally impractical for applications outside the research laboratory. The use of fibre optic techniques has greatly extended the range of applications of interferometry. It should also be pointed out that monomode fibre sensors share the advantages of other optical sensors, such as intrinsic safety, freedom from electromagnetic interference, and potential compatibility with optical communication systems.

It is the purpose of this chapter to describe the operating principles of monomode fibre optic sensors, based either on phase or polarisation state modulation. It will then be shown how the signal from the sensing element may be processed using a combination of optical and electronic techniques to produce an output which is linearly related to the measurand.

2. TRANSDUCTION MECHANISMS

2.1 Sensor Transfer Function

It is possible to completely describe an optical beam by its electric field vector, E. It is therefore, possible to describe the propagation of a beam through a sensing element by the equation:

$$E_{out} = T(X) E_{in} \qquad (1)$$

as shown in Fig. 1, where E_{in} and E_{out} are the electric field vectors of the beam, before and after passing through the sensing element, respectively. The matrix T describes the optical properties of the sensing element, which are a function of the measurand field X. The measurand may be a scalar quantity, such as temperature, or a vector quantity, such as an electromagnetic field.

The purpose of the sensor system is to recover the value of the measurand. This is achieved by first measuring the fields E_{in} and E_{out}, and hence deriving the matrix T. From a knowledge of the manner in which T depends on X, the

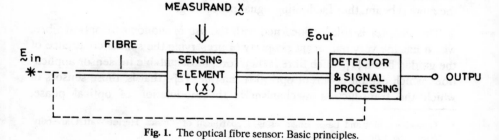

Fig. 1. The optical fibre sensor: Basic principles.

measurand may be determined. The aim of this chapter is to describe the practical techniques which may be employed to carry out these steps.

It has previously been noted that the principal transduction mechanisms which are employed with monomode fibre sensors are those based on phase or polarisation modulation, and this chapter will be restricted to those mechanisms. For simplicity, we shall assume that the optical source is perfectly monochromatic, and that it is fully polarised. We therefore describe the electric field using a two element vector, such that

$$E = (E_x, E_y) e^{i\omega t}, \qquad (2)$$

where E_x and E_y are the electric field amplitudes in the orthogonal (x and y) directions, transverse to the directions of propagation of the beam; ω is the optical frequency, and t is time. The state of polarisation is described by the relative magnitudes of E_x and E_y, and the phase difference between them; E_x and E_y are thus complex scalars. The polarisation properties of the sensing element may then be described by a 2×2 complex matrix (the Jones matrix). With these simplifications, Eq. (1) becomes

$$E_{out} = aB(X) \exp[i\varphi_1(X)] E_{in}, \qquad (3)$$

where B is the Jones Matrix of the sensing element, and φ_1 is the mean phase retardance experienced by the beam as a result of propagating through the sensing element; a is a scalar describing the transmittance of the system, which we shall assume to be constant.

2.2 Phase Modulated Sensors

For simplicity, we shall first consider non-birefringent systems; that is, those in which the phase velocity of the light is independent of state of polarisation. In this case, $B = I$, the identity matrix, and Eq. (3) becomes

$$E_{out} = aE_{in} \exp[i\varphi_1(X)]. \qquad (4)$$

We shall now examine the way in which φ_1 depends on X. For any optical system, the phase retardance may be written

$$\varphi_1 = \frac{2\pi n l}{\lambda}, \qquad (5)$$

where n is the refractive index, l is the physical path length and λ is the vacuum wavelength of the light. Generally speaking, when the measurand is applied to the sensing element (such as a length of monomode optical fibre) it will act to change both the physical length and the refractive index. For example, we may find the sensitivity of φ_1 to temperature, T, by differentiating Eq. (5) to give

$$\frac{\delta \varphi_1}{\delta T} = \frac{2\pi}{\lambda} \left[n \frac{\delta l}{\delta T} + l \frac{\delta n}{\delta T} \right], \qquad (6)$$

where the first term in the bracket corresponds to the thermal expansion of the fibre, and the second to the dependence of refractive index on temperature (the thermo-optic effect). We may determine the sensitivity of the fibre to other measurands by analogy. For example, for strain measurement, the sensitivity arises from contributions due to physical extension, and also the strain-optic effect. Some practical values for fibre sensitivities are given in Table 1 [4, 5].

TABLE 1

Temperature and strain sensitivity coefficients measured using 0.1 m of highly birefringent monomode optical fibre (York Technology 'bow tie' fibre, 3 mm beat length) at a wavelength of 633 nm

X	$(1/l)(\delta\varphi_1/\delta X)$	$(1/l)(\delta\varphi_2/\delta X)$
T	100	5 rad K^{-1} m^{-1}
$\Delta l/l$	6.5×10^6	6.5×10^4 rad m^{-1}

2.3 Polarisation Modulated Sensors

(a) Linear birefringence

We shall now consider the effects of birefringence. Birefringence is induced by anisotropies in the sensing element. For example, if it is stressed, then the phase velocities for linear states of polarisation in the directions parallel and perpendicular to the direction of stress become unequal. A good example of this is those fibres which are designed to have a high degree of intrinsic linear birefringence. This is achieved by manufacturing the fibres with a high degree of internal stress, as described in [6]. In common with all linearly birefringent materials [7], there exist two specific linear states of polarisation (the eigenmodes) which propagate without change, and without coupling; however, the eigenmodes propagate at different phase velocities, and hence experience different phase retardances. For such fibres, Eq. (3) becomes

$$E_{out} = aE_{in} \begin{bmatrix} e^{i\varphi_2/2} & 0 \\ 0 & e^{-i\varphi_2/2} \end{bmatrix} \exp[i\varphi_1], \qquad (7)$$

where the coordinate axes have been chosen so that they are aligned with the eigenmodes.

These special linearly birefringent fibres are very suitable for use in sensor applications, because their modal retardance (that is, φ_2) is a function of such parameters as temperature and strain [4, 5]. We can calculate the sensitivity by analogy with the phase modulated sensor. The modal retardance may be written

$$\varphi_2 = 2\pi(n_s - n_f)l/\lambda, \qquad (8)$$

where n_s and n_f are the effective refractive indices corresponding to the polarisation eigenmodes. To consider the effect of a temperature change (for example), we differentiate as before to find

$$\frac{\delta\varphi_2}{\delta T} = \frac{2\pi}{\lambda}\left[\Delta n\,\frac{\delta l}{\delta T} + l\,\frac{\delta n}{\delta T}\,\Delta n\right], \tag{9}$$

where $\Delta n = n_s - n_f$. The first term in the brackets corresponds to the thermal expansion of the fibre, and the second to the change in relative refractive index with temperature — that is, to the temperature dependence of the birefringence. Sensitivities to other measurands, such as strain and pressure, may be determined by analogy.

The measurand dependence of linear birefringence is a strong function of the structure of the particular fibre used. For example, a common technique by which fibres can be made birefringent is to build-in some thermal stress during manufacture. It is not surprising that such fibres show an enhanced temperature sensitivity. An example of typical sensitivities for a particular fibre is shown in Table 1.

(b) Circular birefringence

In a sensor based on the modulation of linear birefringence, such as those discussed above, the measurand produces a change in the ellipticity of the state of polarisation. It is also possible to construct sensors based on measurand induced changes in polarisation azimuth [8]. We could describe the propagation of light through such a sensor by the expression

$$E_{out} = aE_{in}\begin{pmatrix}\cos\varphi_3 & -\sin\varphi_3 \\ \sin\varphi_3 & \cos\varphi_3\end{pmatrix}\exp(i\varphi_1), \tag{10}$$

where the matrix denotes a rotation through an angle φ_3. It is then possible to calculate the sensitivity of φ_3 to a measurand by differentiating, as we have already done for φ_1 and φ_2. A linear state of polarisation may be regarded as the sum of two circularly polarised components of equal intensity, one right-handed, the other left. It is simple to show that the azimuth of the linear state depends on the phase difference between the left and right components. A rotation of polarisation azimuth therefore, occurs as a result of propagation through a circularly birefringent material. A circularly birefringent material is one in which the polarisation eigenmodes are the right and left circularly polarised states, which propagate with different phase velocities.

Circularly birefringent fibres are not readily available, so that sensors based on this effect are much less common than those which exploit linear birefringence. Nevertheless, circular birefringence-based sensors have been demonstrated. A very important example is the measurement of magnetic field based on the Faraday effect [7, 9, 10]. This effect may be considered as magnetically induced circular birefringence: when a beam of polarised light propagates through a medium in a magnetic field then the polarisation azimuth is rotated. This technique may be employed in a low birefringence fibre (as is usual) or in a special fibre with high circular birefrigence.

3. OPTICAL PROCESSING

We saw in Sec. 2 that an important transduction mechanism is the modulation of the phase, φ_1, of an optical signal. However, it is clearly impossible to measure this phase directly, because the optical frequency is extremely high (~ 500 THz for visible light) and thus beyond the bandwidth of practical photodetectors. It is therefore, necessary to coherently mix the phase modulated optical signal with one or more reference beams of closely similar frequency to produce a low frequency difference signal within the detector bandwidth. This process is, of course, optical interferometry.

Many different types of fibre interferometer have been exploited for sensor applications, and a representative range of these are described below. We shall also describe how the concepts of interferometry can be extended for use in polarisation state measurement.

3.1 Two Beam Interferometers

3.1.1 The Mach-Zehnder interferometer

The simplest interferometers are those based on the mixing of two optical beams. As an example, consider the fibre optic Mach Zehnder interferometer shown in Fig. 2. Light is coupled from a coherent (laser) source into a monomode optical fibre, and is amplitude divided into two paths, denoted a and b, by means of a directional coupler (DC_1). A directional coupler is the fibre optic analogue of the conventional beamsplitter. In a sensor, we could consider path a to represent the signal beam, and b to represent the reference beam. The signal beam is phase modulated by the measurand, whereas the phase of the reference beam remains constant. The signal and reference

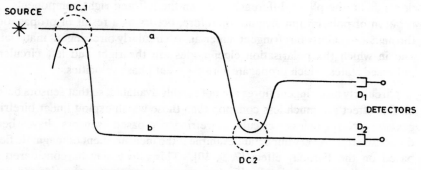

Fig. 2. The fibre optic Mach-Zehnder interferometer; DC: Directional coupler.

beams are recombined at a second directional coupler (DC$_2$), giving two optical outputs at the detectors, D$_1$ and D$_2$.

To analyse the behaviour of the interferometer, we shall make several simplifying assumptions:

1. The laser has a very long coherence length;
2. The directional couplers divide the optical power equally between their two outputs;
3. Polarisation effects are small enough to be neglected; and
4. There are no power losses in the system

Let us denote the phase retardance of the signal and reference arms by φ_a and φ_b respectively as calculated from Eq. (1), and hence determine the electric field at detector D_1, thus

$$E_1 = \frac{1}{2} E_0 \left[\exp(i\varphi_a) + \exp(i\varphi_b + \pi) \right], \tag{11}$$

where E_0 is the source electric field. Note that this equation incorporates the fact that a beam coupled across a directional coupler experiences a $\pi/2$ phase shift in comparison with the transmitted beam. Similarly, we may show that the field at detector D_2 is

$$E_2 = \frac{1}{2} E_0 \left[\exp\left(i\varphi_a + \frac{\pi}{2}\right) + \exp\left(i\varphi_b + \frac{\pi}{2}\right) \right]. \tag{12}$$

We can now calculate the intensities, I, at the detectors using

$$I \propto \langle E.E^* \rangle \tag{13}$$

so that
$$I_1 = \frac{1}{2} I_0 \left[1 - \cos(\varphi_a - \varphi_b) \right] \tag{14}$$

and
$$I_2 = \frac{1}{2} I_0 \left[1 + \cos(\varphi_a - \varphi_b) \right], \tag{15}$$

where $I_0 = E_0 \cdot E_0^*$. We have thus obtained two outputs, each of which depends on the phase difference between the arms of the interferometer. It may be seen that the two outputs vary in antiphase, so that $I_1 + I_2 = I_0$ for all values of $\varphi_a - \varphi_b$, thus ensuring that energy is conserved.

Equations (14) and (15) suggest that at either output, the intensity will range from a maximum value $I_{max} = I_0$ to a minimum $I_{min} = 0$, depending on the value of $\varphi_a - \varphi_b$. In practice, this complete modulation will not be observed, and Eqs. (14) and (15) become

$$I_{2,1} = \frac{1}{2} I_0 \left[1 \pm V \cos(\varphi_a - \varphi_b) \right], \tag{16}$$

where V is the *visibility* of the interference, defined by

$$V = \frac{I_{max} - I_{min}}{I_{max} + I_{min}}, \tag{17}$$

which can take any value between zero and unity.

The visibility will be less than unity for one or a combination of the following reasons:

1. The coherence length of the source is not large enough in comparison with the optical path imbalance between the interferometer arms;
2. The components of signal and reference beams reaching a specific detector are unequal in intensity;
3. The states of polarisation of the signal and reference beams are unequal.

Source coherence effects are discussed further below. In practice, single longitudinal mode diode lasers [12] are often used as sources and have a coherence length of several meters, which is generally much greater than the path imbalance of the interferometer. Equality of signal and reference beam intensities can be ensured by selecting directional couplers with suitable power splitting ratios.

It is difficult to ensure that the signal and reference beams have equal states of polarisation. This is because normal circular core optical fibre is slightly birefringent, with much of this birefringence arising from extrinsic sources, such as bends and twists in the fibre [13]. The resulting output polarisation is therefore, variable and unpredictable. The directional couplers may also introduce some birefringence. One solution is to construct the entire interferometer using highly birefringent fibre [6], and to illuminate only one polarisation eigenmode. Because the eignmodes do not couple, the state of polarisation of the guided beam is preserved. However, this approach also demands that polarisation preserving couplers are used. A more straightforward solution is therefore, to use normal circular core fibre, and to incorporate polarisation controllers, described in Sec. 4, into the interferometer. It is then possible to adjust the states of polarisation in the two beams until they are equal, indicated by maximum observed visibility. Such polarisation controllers may operate by introducing an adjustable amount of bending or twisting into the fibre [14].

3.1.2 *The Michelson interferometer*

Another commonly used form of two beam interferometer is the Michelson, shown in Fig. 3. It is similar to the Mach-Zehnder, except that it is a reflective configuration, in that the far ends of fibres *a* and *b* behave as mirrors, so that the guided beam returns and recombines at the same coupler as was used for beam division. These mirrors may be formed by depositing a reflective coating directly onto the fibre ends (either a metallic, or a dielectric multi-layer coating). Alternatively, conventional mirrors may be used, either butted against the fibres, or with suitable lenses interposed. However, the simplest approach is to leave the fibre ends uncoated, and to rely on the small (about 4%) Fresnel reflectivity occurring at the fibre (quartz)-air interface. In this case, the low reflectivity leads to a much reduced detected power. A special type of fibre reflector is described in Sec. 3.1.3 below.

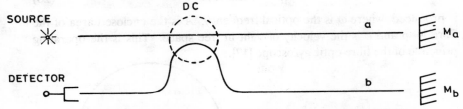

Fig. 3. Fibre optic Michelson interferometer; DC: Directional coupler; Ma, Mb: Mirrors.

The transfer function may be derived for the Michelson by analogy with the Mach-Zehnder, and it may be shown that

$$I_1 = \frac{1}{2} I_0 \left[1 + V \cos (\varphi_a - \varphi_b)\right] \tag{18}$$

and

$$I_2 = \frac{1}{2} I_0 \left[1 - V \cos (\varphi_a - \varphi_b)\right]. \tag{19}$$

In calculating φ_a and φ_b, allowance must be made for the fact that in this reflective configuration, the beams traverse each fibre twice, once in the outward and once in the return direction. For the visibility to be unity, assumptions 1, 3 and 4 in Sec. 3.1.1 above must be satisfied, and the reflectivities of the two mirrors must be equal. However, it may be shown that the visibility is independent of the coupler splitting ratio.

In some applications, the Michelson is preferable to the Mach-Zehnder because the sensing fibre is connected only at one end, so that it may be used as a probe. Also, only one coupler is required, and its splitting ratio is unimportant. Conversely, although the interferometer has two complementary outputs, one of these is directed towards the source, and so it is not readily accessible. The availability of complementary outputs facilitates signal processing, as will be discussed further in Sec. 4. It is clear that considerable optical power is fed back to the source, which may therefore, become unstable. Diode lasers are particularly susceptible to optical feedback [15], and an isolator (for example, based on the Faraday effect [16]) should be used.

3.1.3 The Sagnac interferometer

The Sagnac interferometer is a specialised form of two beam interferometer, which is said to be reciprocal, and is shown in Fig. 4. We see that the two beams each propagate around the same closed loop, but in opposite directions. At first sight, it would appear that the two beams have identical optical path lengths, so that no phase shift would ever occur. However, certain stimuli do produce non-reciprocal (that is, unequal) phase shifts. The most important of these is angular velocity. If the closed loop of fibre is rotated in its own plane with an angular velocity Ω, then the effective optical paths of the two beams become unequal, and a phase shift

$$\varphi_\Omega = 4\Omega\omega A/c^2 \tag{20}$$

is produced, where ω is the optical frequency, A is the enclosed area of the fibre loop and c is the velocity of light in free space. This is the operating principle of the fibre optic gyroscope [17].

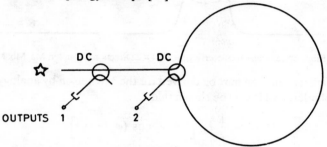

Fig. 4. Fibre optic Sagnac interferometer; DC: Directional coupler.

A great advantage of the Sagnac configuration is that it shows a high degree of common mode rejection to those stimuli which are reciprocal. For example, a change in temperature affects both arms equally and so does not produce an erroneous phase shift. However, to maximise this essential common mode rejection, it is important that the two counter propagating paths should be as equivalent as possible. For example, for output 2 shown in Fig. 4, the two beams do not experience equal numbers of transmissions and couplings as they pass through the directional couplers. For this reason, it is usual to take the output from a Sagnac interferometer at output 1 shown in Fig. 4, even though introducing the second coupler reduces the optical power level observed at the detector.

Another application of the Sagnac interferometer is in the measurement of magnetic fields using the Faraday effect [18, 19]. Because the phase shift (or polarisation rotation) produced by the Faraday effect depends on the direction of the magnetic field, it is non-reciprocal, and hence produces a change in the observed phase shift. Although this effect is valuable in realising magnetometers, it is a disadvantage in gyroscopes, where stray magnetic fields (such as those produced by the earth) may perturb the output [20].

The Sagnac interferometer may also be used as a loop reflector [21, 22]. That is, it can be made to act as a mirror. For example, when using a coupler with equal splitting ratio, in the absence of non-reciprocal phase shifts, and neglecting polarisation effects and attenuation, all the optical power is coupled back into the input lead so that the interferometer acts as a fully reflective mirror. By varying the power splitting ratio of the coupler, any desired power fraction can be reflected.

3.2 Multiple Beam Interferometers

3.2.1 Fabry-Perot

Multiple beam interferometers are used in situations where very high resolution is required. That is, it is possible to construct multiple beam interferometers in which the intensity varies with phase more rapidly than is possible with two beam designs. Probably the best known multiple beam interferometer is the Fabry-Perot [7], and a fibre optic version is shown in Fig. 5 [23, 24]. The Fabry-Perot may be considered as an optically resonant cavity, in which the beam is multiply reflected between the two highly reflective mirrors. In a fibre optic version, the mirrors may be formed by using the same techniques as for the Michelson, described in Sec. 3.1.

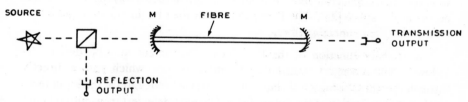

Fig. 5. Fibre Fabry-Perot Interferometer: M = mirror.

The derivation of the transfer function for a Fabry-Perot interferometer is given in many standard text books [7], and it may be shown that the transmitted intensity is

$$I = I_0 / [1 + F \sin^2 (\varphi/2)], \qquad (21)$$

where φ is the round-trip phase retardance, and the finesse, F, is

$$F = 4R (1 - R)^2, \qquad (22)$$

where R is the mirror reflectivity, neglecting attenuation. The form of this transfer function is shown in Fig. 6 and comprises a set of sharp peaks, periodically spaced. As the finesse is increased (by increasing the mirror reflectivity), the peaks become sharper. It may be seen that if the interferometer is operated on the side of one of the peaks, the intensity varies more quickly with phase than for the two beam interferometer (see, for example, Eq. (14)). In the absence of attenuation, the reflection output is complementary. That is, the sum of the transmitted and reflected intensities is constant.

Because of the requirement for high reflectivity mirrors, it is fairly difficult to fabricate a fibre Fabry-Perot having a large finesse. Conversely, it is particularly easy to make a low finesse Fabry-Perot, in which the mirrors are simply

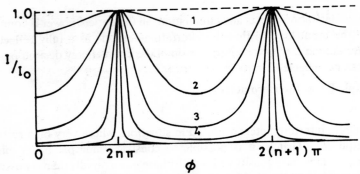

Fig. 6. Transfer function of a Fabry-Perot interferometer 1 → 5: Finesse increasing.

the cleaved end faces of the fibre, which show a small Fresnel reflection [23]. This very simple interferometer is useful for many applications where a small sensing element is required. It may be deployed remotely by taking a short length of cleaved fibre (which acts as the Fabry-Perot sensing element) and splicing it to a connecting fibre of arbitrary length in such a way that a reflection occurs at the splice [25]. The Fabry-Perot may then be illuminated and interrogated via the connecting fibre.

The transfer function of the low finesse instrument may be derived as follows. The strongest transmitted beam is the one which passes directly through the fibre, losing a little intensity by external reflection at the input face, and internal reflection at the output face. The next strongest transmitted beam is transmitted to the output face, and undergoes internal reflection at both the output and input faces before finally emerging in the transmission direction. Beams undergoing a larger number of reflections have such a small intensity (because of the low reflectivity of the fibre ends) that they may be neglected. The transfer function is therefore, that of a two beam interferometer, and may be written as

$$I_T/I_o = (1 - R)^2 + R^2(1 - R)^2 + 2R(1 - R)^2 \cos\varphi, \quad (23)$$

where the visibility is thus

$$V_T = 2R/(1 + R^2). \quad (24)$$

An obvious disadvantage of this arrangement is that the visibility obtained is very low, ~ 0.08 with $R \approx 0.04$. This result is unsurprising, given that the two beams are of such unequal intensities. We have not considered polarisation or coherence effects, so that the actual visibility will be smaller still.

A much more useful device is obtained if the interferometer is operated in reflection. The transfer function is then

$$I_R/I_o = R + R(1 - R)^2 - 2R(1 - R)^2 \cos\varphi \quad (25)$$

and the visibility is

$$V_R = 2R(1 - R)^2/[R + R(1 - R)^2] \quad (26)$$

so that for $R \simeq 0.04$, $V_R \simeq 0.99$. The price paid for the enhanced visibility is that the mean intensity is much reduced, in this case to $I \simeq 0.08\ I_0$. The transfer functions for the cases of both transmission and reflection are shown in Fig. 7, and it may be seen that the outputs are approximately complementary.

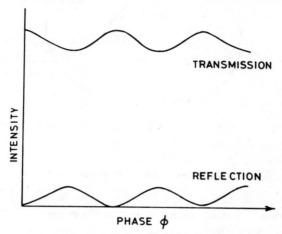

Fig. 7. Transfer function of a low finesse Fabry-Perot interferometer.

3.2.2 Ring resonator

A special form of multiple interferometer can be conveniently implemented in fibre form. This is the ring resonator, shown in Fig. 8, which has the considerable advantage that it requires no mirrors [26]. Light from the source is launched into the input fibre, and some light is transmitted by the coupler to the detector. However, most of the light is coupled into the ring where it circulates. On each circulation, some light is coupled out of the ring and is transmitted to the detector. Multiple beams are therefore incident on the detector, where they interfere, yielding the transfer function

$$\frac{I}{I_0} = (1 - \gamma_0) \left[\frac{(1-k)(1-k_r)}{\left(1 + \sqrt{[kk_r]}\right)^2 - 4\sqrt{[kk_r]}\sin^2\frac{1}{2}\left(\frac{\varphi}{2} - \pi\right)} \right], \quad (27)$$

where k and γ_0 are the power coupling coefficint and excess loss of the directional coupler, respectively, and ϕ is the phase retardance for one circulation of the ring; k_r is the resonant coupling ratio and is given by

$$k_r = (1 - \gamma_0)\ e^{-2\alpha_0 L}, \quad (28)$$

where α_0 is the attenuation coefficient and L is the length of the fibre ring. This analysis does not include coherence or polarisation effects.

Equation (27) shows that the transfer function of the ring resonator is very similar to that of a Fabry-Perot operated in reflection. It is also possible to

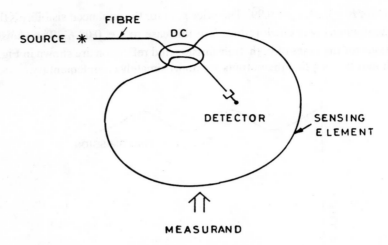

Fig. 8. Fibre optic ring resonator.

produce a complementary transmission-like output by inserting a second coupler in the ring. The ring resonator finds applications not only in sensing, but also as a resonant cavity for fibre lasers.

3.3 Polarimetric Techniques

In Sec. 2, it was explained that an important class of monomode fibre optic sensors is formed by those in which the transduction mechanism is the modulation of the state of polarisation of the guided beam by the measurand—that is, those sensors based on changes in birefringence. In this section, we discuss optical techniques by which this change in state of polarisation may be converted to a change in intensity. Polarimetry may be considered as analogous to two beam interferometry. In an interferometer, the two beams (ideally of equivalent state of polarisation) follow spatially separate paths, and mix coherently on a detector. In polarimetry, the two beams may follow physically identical paths, but are orthogonal states of polarisation. The orthogonal states are then forced to interfere by resolving them both in the same direction using a polarisation analyser.

We shall first consider devices in which the change in linear birefringence of a sensing element is measured. The argument will then be extended to include the measurement of changes in circular birefringence. It will also be shown that it is possible to combine polarisation and phase measurement in a single device.

3.3.1 Linear birefringence

An arrangement for the measurement of changes in linear birefringence is shown in Fig. 9. The sensing element could conveniently be a length of monomode fibre, with a high degree of intrinsic linear birefringence, of the type

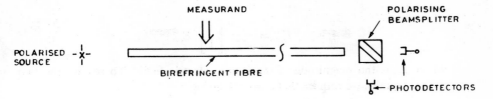

Fig. 9. Basic polarimetric sensor.

described in Sec. 2. The measurement principle is interference between the polarisation eigenmodes. It is, therefore, necessary to use a polarised source which couple to both of the fibre eigenmodes. Some form of polarisation analyser is, therefore, required to resolve the eigenmodes along a common direction. For example, as shown in Fig. 9, a polarising beamsplitter may be used. A half wave plate is useful for rotating the polarisation azimuth of the beam emerging from the fibre, so that the angle between the eigenaxes of the fibre and of the polarising beamsplitter, α, can be adjusted.

To derive the transfer function of the system, we note that the electric field at either of the detectors may be written as

$$\mathbf{E_i} = \mathbf{P}_i(\alpha) \, \mathbf{B_L} \, E_{in} \, e^{i\varphi_1} \quad ; \quad i = 1, 2 \tag{29}$$

in the absence of attenuation and coherence effects, where E_{in} is the electric field at the source, and φ_1 is the mean phase retardance for the optical system; $P_i(\alpha)$ and B_L are the Jones matrices for the polarising beamsplitter and the fibre, respectively. We see from Eq. (7) that

$$\mathbf{B_L} = \begin{bmatrix} e^{i\varphi_2/2} & 0 \\ 0 & e^{-i\varphi_2/2} \end{bmatrix}, \tag{30}$$

where φ_2 is the modal retardance. For the beamsplitter

$$\mathbf{P}_i(\alpha) = \mathbf{R}(-\alpha) \, \mathbf{P}_i(0) \, \mathbf{R}(+\alpha), \tag{31}$$

where $\mathbf{R}(\alpha)$ is the rotation matrix

$$\mathbf{R}(\alpha) = \begin{bmatrix} \cos\alpha & -\sin\alpha \\ \sin\alpha & \cos\alpha \end{bmatrix}, \tag{32}$$

where $\mathbf{P}_i(0)$ are the Jones matrices for the beamsplitter with $\alpha = 0$, such that

$$\mathbf{P}_1(0) = \begin{bmatrix} 1 & 0 \\ 0 & 0 \end{bmatrix} \quad ; \quad \mathbf{P}_2(0) = \begin{bmatrix} 0 & 0 \\ 0 & 1 \end{bmatrix}. \tag{33}$$

For simplicity, we shall consider the optimum case, which gives unit visibility polarisation interference, with maximum intensity. To achieve this, it is necessary to populate equally the polarisation modes of the fibre, and to resolve them equally with the polarising beamsplitter. For example, we may choose as an input a linear state with an azimuth of $\pi/4$ rads relative the fibre eigenaxes, so that

$$E_{in} = \frac{1}{\sqrt{2}} \begin{bmatrix} 1 \\ 1 \end{bmatrix} E_0, \qquad (34)$$

where E_0 is the magnitude of the source electric field. To resolve the two modes equally, we require that $\alpha = \pi/4$, giving

$$\mathbf{P}_1\left(\frac{\pi}{4}\right) = \frac{1}{2}\begin{bmatrix} 1 & 1 \\ 1 & 1 \end{bmatrix} \quad ; \quad \mathbf{P}_2\left(\frac{\pi}{4}\right) = \frac{1}{2}\begin{bmatrix} 1 & -1 \\ -1 & 1 \end{bmatrix}. \qquad (35)$$

We may thus substitute Eqs. (34) and (35) into Eq. (29), and calculate the output intensities

$$I_1 = \frac{1}{2} I_0 \left(1 + \cos\varphi_2\right), \qquad (36)$$

$$I_2 = \frac{1}{2} I_0 \left(1 - \cos\varphi_2\right), \qquad (37)$$

where I_0 is the source intensity. We thus see that the two outputs are complementary.

It would be possible to repeat the analysis for different input states of polarisation by modifying Eq. (34), and for different values of α by substitution in Eq. (31). We recall that a condition for optimum performance was that the eigenmodes were equally populated. To achieve this, it is necessary only that the azimuth is $\pm \pi/4$; the state need not be linear. Indeed, it may be advantageous to use a circular input state, which will populate the eigenmodes equally, but has the advantage that it is unnecessary to explicitly locate the positions of the eigenaxes. If either the input state or the angle α deviate from their optimum values, then either the visibility or mean intensity (or both) will be reduced. However, the two outputs will remain complementary.

3.3.2 Circular birefringence

We shall now extend the arguments of Sec. 3.3.1 to apply to a system based on measurand induced modulation of circular birefringence. That is, those devices in which the transduction mechanism is the rotation of the azimuth of the state of polarisation. The experimental arrangement is basically the same as that of Fig. 9, except that the sensing element is now circularly birefringent, so that the eigenmodes are the left and right circularly polarised states. The transfer function may be derived in the same way as for Sec. 3.3.1, except that the Jones matrix for the sensing element now takes the form

$$\mathbf{B}_c = \begin{bmatrix} \cos\varphi_3 & -\sin\varphi_3 \\ \sin\varphi_3 & \cos\varphi_3 \end{bmatrix} \qquad (38)$$

as shown in Eq. (10)

The change in polarisation state is measured by interfering the two eigenmodes. It is therefore, necessary to use a polarised source which populates both eigenmodes. In order to obtain unit visibility interference, the modes must be populated equally, so that the input state should be linear, for example we may choose

$$E_{in} = E_0 \begin{bmatrix} 1 \\ 0 \end{bmatrix}. \tag{39}$$

It is still convenient to use a linear polarisation analyser. For example, we may use a polarising beamsplitter as in Fig. 9. However, the azimuth of the beamsplitter eigenaxes is now unimportant and we may choose, for example, $\alpha = 0$, so that the $\mathbf{P}_i$ take the form shown in Eq. (33).

Under these conditions, the field at the first detector may be found by substituting Eqs. (33), (38) and (39) into Eq. (29) yielding

$$E_1 = \begin{bmatrix} 1 & 0 \\ 0 & 0 \end{bmatrix} \begin{bmatrix} \cos\varphi_3 & -\sin\varphi_3 \\ \sin\varphi_3 & \cos\varphi_3 \end{bmatrix} E_0 e^{i\varphi_1} \begin{bmatrix} 1 \\ 0 \end{bmatrix} \tag{40}$$

leading to the output intensity

$$I_1 = \frac{1}{2} I_0 \left(1 + \cos 2\varphi_3\right) \tag{41}$$

and, by analogy, for the other output

$$I_2 = \frac{1}{2} I_0 \left(1 - \cos 2\varphi_3\right). \tag{42}$$

The outputs are therefore complementary, as before. The derivation may be extended to include different input states of polarisation, by modifying Eq. (39). It will be found that for non-linear input states the visibility is reduced.

3.3.3 Combined interferometry and polarimetry

It should be evident that a fundamental problem with sensors based either on interferometry or polarimetry is that their transfer functions are periodic, in all cases. This means that the output is ambiguous, in the sense that the transfer functions all contain a term of the form $\cos\varphi$, and that the inverse function $\cos^{-1}$ is multi-valued. The unambiguous measurement range, therefore, corresponds to a phase change of, at most, 2π radians, and is thus very restricted. Some electronic techniques are available which allow the measurement range to be extended, and these are discussed in Sec. 5. An alternative approach, which we shall explore here, is to extend the measurement range optically. This method involves simultaneously measuring phase and state of polarisation [27, 28]. A related approach, based on spectral techniqes, is described in Sec. 3.4.

Table 1 shows that polarimetric sensors are generally less sensitive than their interferometric counterparts. For example, for a given length of fibre, the polarisation fringe (i.e., a phase change in φ_2 of 2π radians) corresponds to a much larger change in the measurand than would be required to produce one interference fringe. In the case of temperature, the same temperature change would produce a phase shift of 20 interference fringes or one polarisation fringe, for the conditions of Table 1. We may therefore envisage an instrument in which phase and state of polarisation were measured simultaneously. The unambiguous operating range would be within one polarisation fringe, whereas the resolution would be set by the much more sensitive interferometer.

Another application of simultaneous phase, φ_1, and polarisation (e.g., φ_2) measurement is for multi-measurand sensing. For example, consider a sensing element made from birefringent fibre which is exposed to simultaneously changing temperature, T, and strain, Δl. By measuring φ_1 and φ_2, and solving a simultaneous equation in the two unknowns T and Δl, it is possible to explicitly determine changes in temperature and strain [29]. Of course, if this approach is adopted, then the extended range available in the single measurand measurement is sacrificed.

A number of optical arrangements are available for simultaneous determination of phase and polarisation. For illustrative purposes, a technique based on a modified Michelson interferometer will be described. The experimental arrangement is shown in Fig. 10, and the sensing element is a length of fibre with high linear birefringence. To analyse the transfer function of the arrangement, we may write for the electric field at the ith detector

$$E_i = \frac{1}{2} \mathbf{P}_i \left(\mathbf{I} e^{i\varphi_r} + \mathbf{B}_L e^{i\varphi_1} \right) E_{in}, \qquad (43)$$

where we assume that there are no coherence effects or attenuation, and that the beamsplitter gives equal power division. The first term in the bracket corresponds to reflection from the mirror with associated phase retardance φ_r, and the second to the beam which has passed through the sensing element, with mean phase retardance φ_1, as in Sec. 3.1.2; $\mathbf{I}$ is the identity matrix, and $\mathbf{B}_L$ the Jones matrix for the fibre as in Eq. (30). We assume that there are no polarisa-

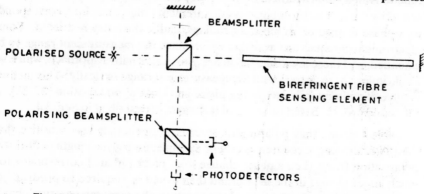

Fig. 10. Interferometer for simultaneous phase and polarisation measurements.

tion changes either in the beamsplitter or in reflection from the mirror, and that the eigenaxes of the beamsplitter are aligned with those of the fibre, so that the Jones matrices for the polarising beamsplitter, P_j, are given by Eq. (33).

To obtain unit visibility interference, it is necessary to populate equally the fibre eigenmodes, as Sec. 3.3.1 above. We may therefore choose a linear input state of azimuth $\pi/4$ as in Eq. (34). We may, therefore, substitute into Eq. (43) and calculate the output intensities

$$I_1 = \frac{1}{4} I_0 \left[1 + \cos\left(\varphi_1 + \frac{1}{2}\varphi_2 - \varphi_r\right)\right], \tag{44}$$

$$I_2 = \frac{1}{4} I_0 \left[1 + \cos\left(\varphi_1 - \frac{1}{2}\varphi_2 - \varphi_r\right)\right]. \tag{45}$$

It is therefore possible to obtain φ_1 and φ_2 explicitly, as required. As for any Michelson interferometer, the complementary outputs are directed back towards the source.

This illustrative example has shown how it is possible to measure simultaneously linear birefringence, represented by φ_2, and optical path length, represented by φ_1. However, it should be evident that analogous arrangements may be devised for the simultaneous measurement of circular birefringence and path length.

3.4 Spectral Techniques

3.4.1 Two wavelength

The basis of the combined interferometric-polarimetric technique described in Sec. 3.3.3 above was that two optical outputs were derived, containing $\varphi_1 + \frac{1}{2}\varphi_2$ and $\varphi_1 - \frac{1}{2}\varphi_2$, which had slightly different periods, and that it was therefore, possible to extend the unambiguous measurement range. A disadvantage of the technique is that it is strongly dependent on the birefringence properties of the fibre used. An alternative approach which has a long history in classical interferometry is to use two different sources of slightly different wavelength in the interferometer.

A suitable experimental arrangement is shown in Fig. 11. Two sources are used, with wavelengths λ_1 and λ_2. Let us consider that only the first source is illuminated then, with the simplifying assumptions 1 to 4 of Sec. 3.1.1, the transfer function is

$$I(\lambda_1) = I_0(\lambda_1) \left[1 + \cos \varphi(\lambda_1)\right] \quad ; \quad \varphi(\lambda_1) = 2\pi n l / \lambda_1 \tag{46}$$

whereas, with only the second source illuminated

$$I(\lambda_2) = I_0(\lambda_2) \left[1 + \cos \varphi(\lambda_2)\right] \quad ; \quad \varphi(\lambda_2) = 2\pi n l / \lambda_2 \tag{47}$$

with the assumption that λ_1 is close to λ_2, so that we can neglect dispersion (i.e., treat n as a constant with respect to wavelength). With both sources illuminated, and where

$$I_0(\lambda_1) = I_0(\lambda_2) = I_0$$

$$I = I(\lambda_1) + I(\lambda_2) = 2I_0 \left[1 + V \cos\left\{\pi n l \left(\frac{\lambda_1 + \lambda_2}{\lambda_1 \lambda_2}\right)\right\}\right] \tag{48}$$

with
$$V = \cos\left[\pi n l \left(\frac{\lambda_2 - \lambda_1}{\lambda_1 \lambda_2}\right)\right]. \tag{49}$$

Therefore, by making a simultaneous measurement of both the visibility and the phase of the interference, the unambiguous measurement range has been extended by a factor of $\lambda_2/(\lambda_2 - \lambda_1)$ in comparison with the interferometer illuminated by λ_1 alone.

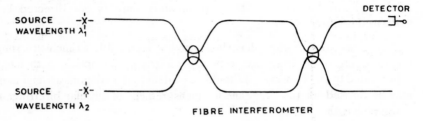

Fig. 11. Extended measurement range interferometer using two source wavelengths.

An alternative approach would be to illuminate the interferometer using the wavelengths λ_1 and λ_2 sequentially, or to use two wavelength selective detectors. In this way, it would be possible to obtain the phases $\varphi(\lambda_1)$ and $\varphi(\lambda_2)$ directly. We may hence calculate the phase difference between the outputs

$$\varphi(\lambda_1) - \varphi(\lambda_2) = 2\pi n l \left(\frac{1}{\lambda_1} - \frac{1}{\lambda_2}\right) = 2\pi n l \left(\frac{\lambda_2 - \lambda_1}{\lambda_1 \lambda_2}\right), \tag{50}$$

which we see is equivalent to the phase which would be produced if the interferometer was illuminated by a source wavelength $\lambda_1 \lambda_2 / (\lambda_2 - \lambda_1)$.

The basis of operation of all two wavelength devices is to use the differential phase (or visibility) information in order to identify the order of interference, within the unambiguous range of the technique. It is therefore, essential that λ_1 and λ_2 are accurately known, and remain stable during the measurement. Otherwise, the order of interference may be mis-identified, thus leading to a serious error in the recovered value of the measurand.

3.4.2 Low coherence source

It should be clear that it would be possible to extend the measurement range further by using three, or even more wavelengths. One may generalise this argument to an infinite number of wavelengths, corresponding to a single source of finite spectral width [31]. Such a source is obviously not monochromatic, and the spectral width, $\Delta\lambda$, may be characterised in terms of the coherence length of the source, where

$$l_c \sim \lambda^2 / \Delta\lambda. \tag{51}$$

If the path imbalance in the interferometer, Δl, exceeds the coherence length, then the visibility of the resulting interference will be small. For example, for a two beam interferometer in which conditions 2 to 4 of Sec. 3.1.1 are satisfied, the transfer function is given by Eq. (16), and the visibility becomes

$$V = \exp - \left(|\Delta l| / l_c \right). \tag{52}$$

We therefore, see that the position of maximum visibility corresponds to $\Delta l = 0$, thus defining a unique point in the interferometer transfer function.

A useful practical arrangement is shown in Fig. 12. This comprises a remote sensing interferometer of nominal path imbalance Δl_s ($\neq 0$) connected by a fibre to a local receiving interferometer of adjustable path imbalance Δl_R ($\neq 0$). In operation, Δl_R is adjusted until maximum visibility is obtained, at which point

$$\left| \Delta l_R \right| - \left| \Delta l_S \right| = 0 \tag{53}$$

and thus Δl_s may be determined.

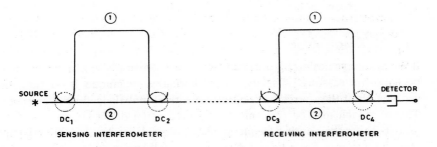

Fig. 12. Tandem 'white light' interferometer; DC-directional coupler.

4. MODULATORS AND COMPONENTS

Signal processing in fibre interferometers is largely concerned with controlling the guided beam. For example, the optical arrangements of Sec. 3 have involved the use of passive components for amplitude division of the beam, and for various polarisation operations. We shall see in Sec. 5 that it is often also necessary to employ various active components for the modulation of phase, polarisation or optical frequency. Devices which carry out these operations are described below.

4.1 Phase Modulation

4.1.1 Path length tuning

A direct means of controlling phase retardance of the guided beam is to vary the optical path length. The first technique reported, and one which is in general use, is to employ a piezo-electric transducer to strain the fibre in its axial direction [32]. The amplitude of the induced phase change can be enhanced by wrapping a length of the fibre round a piezo-electric cylinder. For example, in a typical arrangement using 100 fibre turns on a 50 mm diameter cylinder made from PZT-5, a phase change of about 4.5 rad can be produced per unit applied voltage. However, when the fibre is wrapped in a coil, some birefringence is introduced (see Sec. 4.4), which may cause problems. Another disadvantage of the piezo-electric transducer is that the modulation frequency is limited by mechanical resonances to a maximum in the order of 20 kHz. As an alternative, it is possible to use piezo-electric fibre coatings, such as PVDF.

A related approach is to make use of thermal modulation, by depositing a metallic coating on the fibre, and passing an electric current through the coating [33]. In this case, only a low voltage (but high current) power supply is required. This approach is advantageous because the thermo-optic coefficient of the fibre is fairly high, and because the coating can be applied over a substantial length of fibre, allowing phase modulations of large amplitude. The magnitude of the effect, when thin coatings are used, can be calculated from the data of Table 1. When thick coatings are used, it is necessary to allow for the additional phase change which is produced by the thermal expansion of the coating.

4.1.2 Wavelength tuning

It is not always desirable to use an electrically active phase shifting element, perhaps for reasons of safety. An alternative technique is to tune the wavelength of the source, in conjunction with an unbalanced interferometer. This approach may be conveniently adopted when diode laser sources are used, because their optical frequency shows some dependence on the injection (supply) current. For a typical diode laser, a frequency shift of about 3 GHz $(mA)^{-1}$ is produced. The maximum amplitude of frequency modulation that can be applied is limited by the onset of mode hopping; that is, by the abrupt switching of the laser output to an adjacent longitudinal cavity mode. For most lasers, the maximum practical frequency deviation is about 50 GHz.

To calculate the phase shift, we recall from Eq. (5) that

$$\varphi_1 = 2\pi n \frac{l}{\lambda} = 2\pi n l \frac{\nu}{c}, \tag{54}$$

where ν is the optical frequency of the source. Hence an optical frequency shift $\Delta\nu$ will produce a phase change

$$\Delta\varphi_1 = 2\pi n l \frac{\Delta \nu}{c} \qquad (55)$$

and we note that the phase change is thus proportional to the path imbalance. However, for reasons that will be explained in Sec. 5.3, it is desirable to minimise this path imbalance in a practical design. An obvious disadvantage of the technique is that the current modulation inevitably introduces an unwanted intensity modulation of the laser.

4.2 Polarisation Modulation

4.2.1 Polarisation ellipticity

State of polarisation can be controlled either by inducing birefringence into fibres which naturally have a low birefringence (that is, fibres with nominal cylindrical symmetry), or by exploiting the properties of highly birefringent fibre. For example, many polarisation controllers are based on introducing controlled bends or twists into fibres using various types of transducers (for a summary of such techniques, see reference [35]). It was shown in Sec. 2 that the modal retardance, φ_2, of a fibre with high linear birefringence controls the ellipticity of the stated of polarisation. It is therefore possible to make a polarisation controller by, for example, straining or heating a birefringent fibre. For example, a piezo-electric transducer or a metallic coating can be used, in a manner analogous to that described in Sec. 4.1.1. However, it is generally undesirable to coil a birefringent fibre onto a piezo-electric cylinder, because excessive bending will cause power coupling between the eigenmodes.

An alternative approach is to use a frequency modulated source, as in Sec. 4.1.2. The optical path lengths corresponding to the eigenmodes are unequal, so that a change in optical frequency will modulate φ_2. However, the difference in path length is small, being one wavelength per beat length of fibre (a typical beat length is a few mm), so that long fibres are needed to produce useful modulations. To overcome this, a two-beam interferometer (fibre or otherwise) may be constructed, in which polarisation components are used to produce orthogonally polarised beams in the two arms. The phase difference between the two arms is thus equivalent to φ_2. The path imbalance can be designed as any convenient value, so that any desired polarisation control can be achieved by laser frequency modulation [36]. Alternatively, any phase modulation technique can be used in one of the interferometer arms to produce polarisation modulation. It is difficult to construct such a polarisation interferometer in all fibre form, because the polarisation components used must be of high quality, and stable in their properties [37].

4.2.2 Polarisation azimuth

It is straightforward to convert a polarisation state of controlled ellipticity to one of controlled azimuth, simply by using a quarter waveplate whose eigenaxis is at $\pi/4$ rads to the polarisation eigenmodes of the incident beam [38]. An experimental arrangement is shown in Fig. 13. A linearly polarised beam of azimuth $\pi/4$ is guided by a birefringent fibre of controllable modal retardance, φ_2. The output beam thus has controllable ellipticity, and then passes through a quarter wave plate. The electric field at the output is thus given by

$$E_{out} = \mathbf{Q}\ \mathbf{B}_L\ E_{in}\ ;\ E_{in} = \frac{1}{\sqrt{2}} \begin{bmatrix} 1 \\ 1 \end{bmatrix}, \qquad (56)$$

where $\mathbf{Q}$ is the Jones matrix of the quarter wave plate, such that

$$\mathbf{Q} = \frac{1}{2} \begin{bmatrix} 1 & i \\ i & 1 \end{bmatrix}, \qquad (57)$$

therefore

$$E_{out} \propto \begin{bmatrix} \sin(\pi/4 + \varphi_2/2) \\ \cos(\pi/4 + \varphi_2/2) \end{bmatrix}, \qquad (58)$$

thus producing a linearly polarised beam whose azimuth depends on $\varphi_2/2$, as required. The quarter waveplate could conveniently be made from fibre, as described in Sec. 4.2.1.

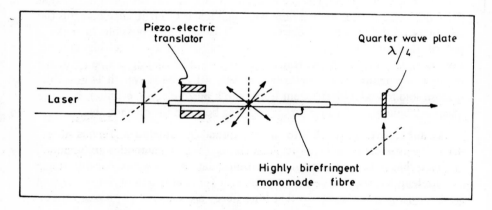

Fig. 13. Single fibre polarisation azimuth modulator. Polarisation azimuths and eigenaxes shown thus ←→.

4.2.3 Complete polarisation control

In many situations, it is important to have complete control over the state of polarisation, both in ellipticity and in azimuth. This basically demands two separate modulators which are concatenated. An example is shown in Fig. 14, and comprises two sections of linearly birefringent fibre, each with separately controllable modal retardance, φ_2, perhaps by using a piezo-electric stretcher [39]. In one implementation, the input state of polarisation is linear, and is of azimuth $\pi/4$ with respect to the eigenaxes of the first fibre section. In this case, through control of the modal retardances, say α and β for the first and second fibre sections respectively, it is possible to produce any desired output state. To see this, we may calculate the output electric field from

$$E_{\text{out}} = \mathbf{B}_\beta \, \mathbf{R} \left[\frac{-\pi}{4} \right] \mathbf{B}_\alpha \, \mathbf{R} \left[+\frac{\pi}{4} \right] E_{\text{in}} \quad ; \quad E_{\text{in}} = \begin{bmatrix} 1 \\ 0 \end{bmatrix} \quad (59)$$

where $\mathbf{B}_\alpha$ and $\mathbf{B}_\beta$ are the Jones matrices for the first and second fibre sections respectively. We thus find that

$$E_{\text{out}} \propto \begin{bmatrix} \cos(\alpha/2) \\ \sin(\alpha/2) \exp - i \, (\beta + \pi/2) \end{bmatrix} \quad (60)$$

which is an arbitrary state of polarisation which may take any value depending on the values of α and β. It is obviously possible to use this controller to perform the inverse function; that is, to convert an arbitrary state of polarisation to a known linear one. To carry out the more general task of converting an arbitrary state to any other arbitrary state requires a controller incorporating additional sections of fibre.

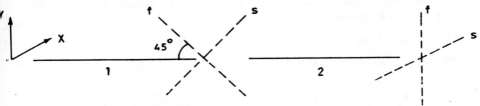

Fig. 14. Schematic of SOP control scheme; 1, 2: Highly linearly birefringent fibre; eigenmodes shown thus: f: fast; s: slow. The coordinate system is shown by X and Y.

4.2.4 Passive components

The most important passive component is the polarisation analyser, and devices have been reported based on both normal and highly birefringent fibre. The evanescent field polariser exploits the fact that a metal clad waveguide shows a strong polarisation dependence for attenuation. A device based on this principle may be fabricated by locally polishing the cladding of a fibre to expose the evanescent field, and then depositing a metal coating. Extinction ratios of $\geq$ 45dB have been reported for this type of component [40]. An alternative approach involves depositing some birefringent material onto the polished fibre [41]. If the refractive indices for the eigenstates of the birefringent material are greater and less than the effective refractive index of the guided wave, respectively, then one of the states will be very strongly attenuated with respect to the other. By using a birefringent crystal, extinction ratios of > 60 dB have been obtained [42].

Polarisers may readily be fabricated using birefringent fibre. For a sufficiently long wavelength, the attentuation coefficients for the eigenmodes differ significantly. Therefore, at this wavelength, if the beam propagates through a sufficient length of the fibre, then it will become polarised. The effect may be enhanced by exploiting the fact that for a given wavelength, the increased attenuation coefficients produced by bending the fibre are polarisation dependent [43]. These coiled fibre polarisers have been demonstrated with extinction ratios ~ 60 dB [44]. An alternative means of polarisation analysis is to use a polarisation sensitive directional coupler, and such components are described in Sec. 4.4.

We noted above that polarisation controllers could be made by bending or squeezing the fibre, thus inducing birefringence to form fibre equivalents of wave plates or rotators. For example, a coiled fibre exhibits linear birefringence, where the retardance is adjustable via the bend radius and length, and the tension. A full polarisation state controller may be realised by forming two coils on a single fibre, each with a linear retardance of $\lambda/4$ [45]. By rotating the coils relative to the input state, and to each other, any output state may be generated. This component is useful in laboratory experiments, where environmentally produced changes in polarisation are usually slow, and can therefore, be compensated manually.

4.3 Frequency Modulation

One of the most important methods of signal processing in fibre interferometers is the heterodyne technique, which is described in Sec. 5.3. To implement this technique requires that a shift in optical frequency is produced in one of the arms of the interferometer. In a classical optical system, a very useful frequency shifting component is the Bragg cell, this device operates by coupling ultrasonic travelling waves into an optical medium such as a quartz block. The acoustic pressure waves effectively produce a moving diffraction grating, where the grating is produced by the changes in refractive index corresponding to the pressure differences caused by the acoustic wave. Light which is diffracted by this moving grating is consequently frequency shifted. Commercially available Bragg cells typically provide a frequency shift of 40 or 80 MHz.

4.3.1 Fibre optic frequency shifters

The bulk-optic Bragg cell is clearly not easily compatible with fibre optic systems. For this reason, research into fibre optic frequency shifters is very active at present. One development, which has strong analogies with the Bragg cell, is shown in Fig. 15 [46]. It uses a birefringent optical fibre which is clamped between a quartz block and a backing plate. Travelling acoustic waves, propagating with a component in the direction of the fibre axis, are coupled into the quartz block. The acoustic wavelength and the direction of the fibre are chosen so that the component of the acoustic wavelength in the direction of the fibre axis is resonant with the polarisation beat length of the fibre. These travelling waves periodically squeeze the fibre, which is orientated so that the axis of the squeezing bisects the fibre eigenaxes. Light is initially launched into only one of the fibre eigenmodes. However, the squeezing causes the eigenmodes to be coupled, thus transferring optical power to the

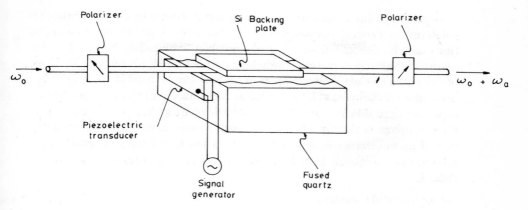

Fig. 15. Frequency shifter using travelling acoustic waves in birefringent fibre.

other mode. Because this coupling is caused by a moving disturbance, the coupled power (which is all in the previously un-illuminated mode) is frequency shifted. The frequency shifted beam may be selected at the output using a polariser.

In an alternative design, the fibre is not supported by a quartz block, but is held at its ends in free space [47]. Near to one end of the fibre, ultrasonic flexure waves are coupled into it. That is, one end of the fibre is made to vibrate up and down. This causes bends to travel along the fibre, with the fibre supports designed so that the bends move unidirectionally. The bending causes mode coupling, and therefore, a frequency shift is produced, as before. Another possibility is to use two-moded low birefringence fibre [48]; that is to say, a fibre illuminated with a sufficiently short optical wavelength that the two lowest order transverse modes propagate, rather than just the fundamental mode. Travelling flexure waves may then be used to produce mode coupling and frequency shifting, as before.

4.3.2 Pseudo frequency shifting by phase modulation

The techniques described above are progressing rapidly, but cannot yet be described as mature. Therefore, many currently used heterodyne processing techniques are based on using periodic phase modulation in order to produce a pseudo frequency shift. The simplest example is the use of serrodyne (sawtooth ramp) phase modulation [49]. During the linearly rising part of the ramp, a constant rate of change of phase is produced, which is equivalent to a frequency shift. During the flyback, the phase returns to its original value, and the process repeats. Because of the flyback, the frequency shift is impure: it is not a true single sideband frequency shift in the sense that the flyback corresponds to a frequency shift of opposite sign. However, a satisfactory heterodyne carrier (see Sec. 5.3) can be produced by band-pass filtering, with appropriate adjustment of the phase modulation amplitude.

In principle, any asymmetric modulating waveform can be used to synthesise a heterodyne carrier. Alternatively, even symmetrical modulating waveforms (for example, sinusoidal) can be made asymmetric by gating the signal at the output of the interferometer, by switching the photodiode on and off [50]. Any of the phase modulation techniques of Sec. 4.1 can be used in these pseudo-heterodyne techniques. The use of harmonic (sinusoidal) phase modulation is especially desirable for use with piezo-electric phase shifters, because they can then be driven at their mechanical resonance frequency. Serrodyne modulation of piezo-electrics is only satisfactory at low frequencies, otherwise they tend to ring excessively at their resonant frequencies as a result of the abrupt flyback.

4.4 Directional Couplers

The practical realisation of the optical systems described in Sec. 3 requires a range of passive components. For example, components for beam division are essential in interferometers, and various types of polarisation components described in Sec. 4.2.4 are required in polarimetric instruments. It is obviously possible to construct hybrid systems which are part fibre, and part bulk optic components. However, it is clearly desirable, where possible, to construct all fibre systems. We describe some of the most useful components below.

The directional coupler is the fibre optic equivalent of the beamsplitter, and thus lies at the heart of most interferometer designs. The basic arrangement is shown in Fig. 16, it is a 2 × 2 coupler, which is the simplest possible kind, having two input fibres and two outputs, and is bi-directional. There are two fabrication techniques which are commonly used: polished and fused. These are described below.

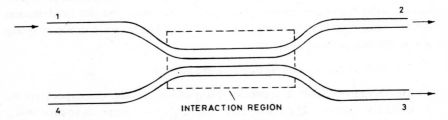

Fig. 16. Schematic representation of a directional coupler.

4.4.1 Fabrication techniques

A polished coupler [51, 52] is fabricated by setting the fibre in a suitable substrate, such as a groove in a quartz block. The block is then polished to remove locally much of the cladding material, so that the evanescent field of the guided wave becomes exposed. The polished region is then placed in contact with the polished region of a second similar fibre-block assembly, and the two are coupled by a thin film of index matching fluid. Micropositioners may be used to adjust laterally the relative positions of the two blocks. In this way, the power coupling ratio (the fraction of the power coupled from one fibre core to the other) can be adjusted. This facility can be very useful in laboratory applications, although polished couplers are rather expensive for general use.

In the fused coupler [53, 54], the two fibres are slightly twisted together, and heated until they soften. They are then drawn axially, so that the diameter in the coupling region is greatly reduced. In the tapered region, the core diameters are vanishingly small, and the beam is guided in the cladding-core composite structure. The power is carried by the two lowest order modes, symmetric and antisymmetric. As described below, power exchange between the two fibres then occurs as a result of beating between the modes. The power splitting ratio may be controlled during the fabrication process, whilst the fibres are being drawn. Following manufacture, the splitting ratio is fixed. Fused couplers are now mass-produced, and their use is widespread. They are robust, and reasonably inexpensive.

4.4.2 Specifications

Couplers can be characterised in terms of the following parameters: insertion loss, backscatter and extinction ratio. The insertion loss is defined by

$$\Lambda = P_1 / (P_2 + P_3), \tag{61}$$

the backscatter by

$$b = P_1 / P_4, \tag{62}$$

where P_i is the optical power in the ith arm of the coupler, as shown in Fig. 16. The extinction ratio relates to a coupler designed to couple as much power as possible from the first fibre into the second, and in this case is given by

$$\eta = P_3 / P_2. \tag{63}$$

It is usual to express Λ, b and η in decibels, and Table 2 shows typical values for the best that can currently be achieved for both the polished and fused types.

TABLE 2
Demonstrated performance parameters for polished and fused type directional couplers

	Polished	Fused
Insertion loss Λ	0.005 [55]	0.05 [57]
Backscatter b	70 [56]	60
Extinction ratio η	50 [56]	30 [58]

4.4.3 Polarisation effects

All couplers show polarisation effects. The coupling ratio depends on polarisation state, and the coupler itself is birefringent so that the guided beam changes its state of polarisation. The polishing process makes a fibre asymmetrical, and therefore, birefringent [56, 59]. The polarisation eignemodes are in the plane of polishing, and perpendicular to it. Therefore, the state of polarisation can be preserved provided that only a single eigenmode is excited. Fused couplers show considerably greater birefringence due to anisotropic stresses in the fused region [60, 61].

4.4.4 Theory

It is possible to physically represent the coupling between fibres by considering the two adjacent cores as a single waveguide, which supports the two lowest order modes, being the symmetrical and anti-symmetrical modes referred to above [62, 63]; these two modes have propagation constants β_s and β_a respectively. The two modes are excited approximately equally by the incoming beam and are initially in phase, so that all the power is in the core of the first fibre. As the modes propagate, they slip in relative phase, given that $\beta_s \neq \beta_a$. Eventually the condition

$$\left|\beta_s - \beta_a\right| l_0 = \pi \tag{64}$$

will be satisfied, where l_0 is the coupling length. At this point, the modes are in anti-phase, and all the power will be in the second fibre. Clearly, any desired coupling ratio can be produced by controlling the interaction length to produce the desired modal retardance (i.e., the phase difference between the modes). When the modal retardance is $\pi/2$, then power is divided equally between the two output ports. It is then easily shown that there is a phase difference of $\pi/2$ between the two outputs, as noted previously in Sec. 3.1.

As the interaction length is increased, then power will be coupled backwards and forwards between the two output ports. For a typical fused coupler, the interaction length can be as short as 1 mm [56].

4.4.5 Special couplers

The propagation constants, β_s and β_a, depend both on the state of polarisation and the wavelength. In this way it is possible to make couplers which are the equivalent of polarising beamsplitters [64, 67], or for wavelength division [68, 69]. For example, for some value of the interaction length, the following conditions will hold:

$$\left|\beta_s^{(1)} - \beta_a^{(1)}\right| l = 2N\pi \tag{65}$$

$$\left|\beta_s^{(2)} - \beta_a^{(2)}\right| l = (2N + 1)\pi \tag{66}$$

where the superscripts (1) and (2) refer to the two polarisation eigenstates. In this case, state (1) will pass through the coupler to port 2, whereas state (2) will be completely coupled to port (3). Alternatively, (1) and (2) may represent two wavelengths, λ_0 and $\lambda_0 + \Delta\lambda$, which would then be transmitted and coupled respectively. For wavelength division couplers, for given values of β_s and β_a, the channel spacing, $\Delta\lambda$, decreases with interaction length, l.

SIGNAL PROCESSING IN MONOMODE FIBRE OPTIC SENSOR SYSTEMS   687

The construction of true polarisation preserving couplers demands the use of highly birefringent fibre. The technical difficulties are then increased considerably. One serious problem is that the eigenaxes of the two fibres must be made accurately parallel.

It is also possible to construct couplers with a large number of ports. One approach is to splice together a number of 2×2 devices, or to fabricate a set of 2×2 couplers simultaneously to form a network (sometimes called a 'tree-coupler'). One type of coupler which has special signal processing applications, noted in Sec. 5.2, is the 3×3 [70]. In such a device, the waveguiding properties of the composite interaction region are complicated, and lead to phase differences between the outputs; it is these phase differences which are exploited in the signal processing applications.

5. ELECTRONIC PROCESSING

A fundamental problem with all monomode fibre optic sensors, whether based on interferometry or polarimetry, is that their transfer functions are periodic. We have shown that the complementary outputs from a two beam system take the form

$$I_1 = \frac{1}{2} I_0 (1 - V \cos\varphi) \tag{67}$$

$$I_2 = \frac{1}{2} I_0 (1 + V \cos\varphi) \tag{68}$$

from Eq. (16) (for example). The effect of this periodicity is that the value of the recovered measurand is ambiguous, and the sensitivity is variable.

We may write

$$\varphi = \varphi_m + \varphi_d, \tag{69}$$

where φ the phase difference between the two beams is made up of the measurand induced phase φ_m and an unwanted term, φ_d, arising from environmental drift and noise. We may calculate the sensitivity from

$$\frac{\delta I_1}{\delta \varphi_m} = \frac{1}{2} V I_0 [\sin\varphi_m \cos\varphi_d + \cos\varphi_m \sin\varphi_d]. \tag{70}$$

It is instructive to consider the case where the measurand induced phase, φ_m is small, so that we can approximate Eq. (70) to

$$\frac{\delta I_1}{\delta \varphi_m} \simeq \frac{1}{2} I_0 V \sin\varphi_d. \tag{71}$$

We therefore, see that the sensitivity is dependent on the drift term φ_d, and therefore, is certainly not constant. Furthermore, when $\varphi_d = 0$ (modulo π), the sensitivity is zero, and the signal fades completely. Ideally, we would wish to operate with $\varphi_d = \pi/2$ (modulo π), where the sensitivity is maximum; this is called the quadrature condition.

The objective of any signal processing scheme is to produce an output of constant sensitivity, which is free from fading, and is linear with respect to the measurand. There exist two basic classifications of processing scheme: homodyne and heterodyne. In the homodyne technique, phase control is used within the interferometer (or polarisation control in a polarimeter) in order to hold it at the quadrature point, thus maintaining constant sensitivity. In the heterodyne technique, a frequency shift is imposed on one of the beams to generate a heterodyne carrier signal which is modulated by the measurand. A number of electronic techniques exist for the demodulation of such a carrier. We shall describe a representative range of both types of techniques below, and conclude with a discussion of the noise present in the output signal.

5.1 Active Homodyne Processing

The basic principle of active homodyne signal processing is as follows. The intensity output from the interfrometer (or polarimeter) is measured relative to some reference level, where the reference level corresponds to the output at quadrature. The relative measurement is therefore, zero when the interferometer is at quadrature. Otherwise, the relative measurement serves as an error signal in a servo loop feeding back to a phase control element in the interferometer, which is hence driven to quadrature.

As an example, Fig. 17(a) shows an arrangement based on a Mach-Zehnder interferometer with a piezo-electric element for phase modulation [32]. However, the method is general, and can be adapted to cover the various optical arrangements of Sec. 3, using the modulators described in Sec. 4.

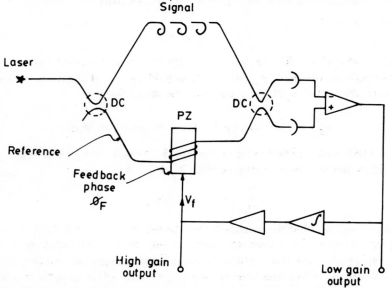

Fig. 17. (a) Active homodyne signal processing using a piezoelectric phase modulator.

An advantage of using the Mach-Zehnder interferometer is that the complementary outputs are readily available, so that when they are combined differentially a signal of the form

$$i \propto I_1 - I_2 = -I_0 V \cos \varphi \qquad (72)$$

is produced, following from Eqs. (67) and (68). We see that the output, i, is thus zero at the quadrature condition. Therefore, under these circumstances, no separate reference level is required.

Let us suppose that the servo loop is initially in equilibrium, and that the equilibrium is disturbed by a phase change in the signal arm produced by the measurand. The output, i, will then no longer be zero, and is fed back via an integrator thus applying a voltage change to the piezo-electric element, and so producing a change in reference phase exactly equal to the change in signal phase, thus restoring the interferometer to quadrature. The piezo-electric modulator is a linear device, so that the change in phase is directly proportional to the applied voltage. Therefore, the change in measurand is linear with the modulator voltage, and the coefficient of proportionality may be found by calibration.

Two modes of operation are possible with this processing scheme depending on whether the frequency of the measurand is within or above the bandwidth of the servo loop. We assumed above that the response of the loop was sufficient to follow the measurand; this is the high gain bandwidth product mode (HGBWP). Let us now consider that the measurand frequency is high, that its amplitude is small, and that we wish to recover the measurand in the presence of large amplitude, low frequency phase drifts. This is a situation which often occurs in practice. For example, in the fibre optic hydrophone [71], one seeks to recover a small acoustic signal in the presence of large, but slow, temperature changes.

In the above circumstances, we may write

$$\varphi = \varphi_d + \varphi_{om} \sin \omega_m t + \varphi_f, \qquad (73)$$

where φ_{om} is the amplitude of the measurand, ω_m is its circular frequency, and φ_f is the phase change produced by the modulator. If the servo loop is too slow to follow the measurand, then its function will be to maintain

$$\varphi_d + \varphi_f = \pi / 2 \text{ (modulo } 2\pi) . \qquad (74)$$

The output from the differential amplifier now becomes

$$i \propto \cos (\pi/2 + \varphi_{om} \sin \omega_m t) , \qquad (75)$$

which in the small signal limit approximates to

$$i \propto - \varphi_{om} \sin \omega_m t . \qquad (76)$$

Thus the output, i, is linear with the measurand. This is the low gain bandwidth product mode (LGBWP) of operation.

For many applications, the use of a piezo-electric phase modulator is undesirable, because it is a high voltage component which stores energy and may, therefore, be a risk to safety. A suitable alternative approach is to control the phase by using a frequency modulated laser diode in conjunction with an unbalanced interferometer. This technique was discussed in Sec. 4.1.2. A disadvantage of all schemes based on unbalanced interferometers is that they increase the noise floor of the system, for reasons which will be discussed in Sec. 5.4.

Active homodyne processing is also appropriate for use in polarimetric sensors. However, in this case, the phase modulator must be replaced with a polarisation modulator, using the techniques described in Sec. 4.2.

Active homodyne processing has the considerable advantage of simplicity. However, the measurement range is always restricted by the finite range of phase (or polarisation) compensation which can be provided by the modulator. For example, consider the use of the piezo-electric phase modulator in a situation where the measurand is increasing steadily. In consequence, the feed back phase compensation must increase concomitantly. Eventually the modulator will reach the limit of its range, set either by the maximum available voltage, or the maximum safe fibre strain. At this point, the zero must be reset, and the value of the measurand will be lost. Similar arguments apply to all modulator types. For example, if one attempts to modulate the frequency of a laser diode too far, it will discontinuously switch operation to an adjacent longitudinal mode; the maximum practical amplitude of frequency modulation is ~ 30 GHz.

Two approaches exist which allow phase measurement over a wide range: passive homodyne processing, discussed in Sec. 5.2, and heterodyne processing in Sec. 5.3.

5.2 Passive Homodyne Processing

The objective of any passive homodyne processing scheme is to derive two (or more) outputs from the interferometer (or polarimeter) which bear a constant phase relationship to each other; that is, we wish to produce outputs of the form

$$I_A = \frac{1}{2} I_0 \left[1 + V \cos \varphi \right], \tag{77}$$

$$I_B = \frac{1}{2} I_0 \left[1 + V \cos (\varphi + \varphi_B) \right], \tag{78}$$

where φ_B is a bias phase introduced by some passive component. From Eqs. (77) and (78), we see that the two outputs cannot fade simultaneously, provided that $\varphi_B \neq 0$ (modulo π). Ideally, we require that $\varphi_B = \pi/2$ (modulo π), so that when one output has faded, the other is at quadrature, and vice versa. We shall first describe some of the techniques that can be used to generate these quadrature outputs, and then explain methods by which the quadrature outputs can be processed to give a signal which is linear with phase.

5.2.1 3 × 3 directional coupler

One way in which quadrature outputs can be derived is to use a 3 × 3 directional coupler [70] as the recombiner in a Mach-Zehnder interferometer, as shown in Fig. 17(b). We recall that in a 2 × 2 directional coupler, a phase shift of $\pi/2$ is produced between the transmitted and coupled beams. In the 3 × 3 coupler, phase shifts are also produced between the outputs, with the magnitude of the phase changes dependent on the properties of the coupler. Thus, for the three outputs of the interferometer in the general, we can write

$$I_1 = 2B_2 [1 + \cos\varphi], \qquad (79)$$

$$I_2 = B_1 + B_2 \cos\varphi + B_3 \sin\varphi_3, \qquad (80)$$

$$I_3 = B_1 + B_2 \cos\varphi - B_3 \sin\varphi_3, \qquad (81)$$

where for simplicity, we have assumed that the visibility is unity. We see that any phase changes between the outputs can be represented by choosing appropriate values of B_1, B_2 and B_3. The three outputs may then be combined electronically in the following manner

$$I_2 + I_3 - 2B_1 = 2B_2 \cos\varphi, \qquad (82)$$

$$I_2 - I_3 = 2B_3 \sin\varphi, \qquad (83)$$

thus giving the quadrature outputs required.

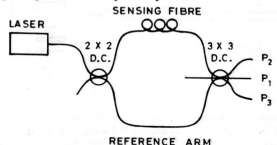

Fig. 17. (*b*) Passive homodyne interferometer utilising a 3 × 3 directional coupler.

5.2.2 Phase switched technique

An alternative technique for providing the bias φ_B is to use a phase modulator. For example, we may drive the modulator with a square wave giving a phase modulation amplitude of $\pi/2$, so that when the square wave is low

$$\varphi_A = \frac{1}{2} I_0 [1 + V \cos\varphi] \qquad (84)$$

and when it is high

$$\varphi_B = \frac{1}{2} I_0 \left[1 + V \cos\left(\varphi + \frac{\pi}{2}\right)\right] \qquad (85)$$

thus yielding the desired quadrature outputs. Any of the phase modulation techniques of Sec. 4.1 can be used, but in a passive arrangement, the frequency modulated diode laser in conjunction with an unbalanced interferometer is the most suitable approach [72].

5.2.3 Polarimetric techniques

It would be possible to adapt the phase switched technique to a polarimetric sensor by using one of the polarisation modulators of Sec. 4.2. However, there are more straightforward passive techniques for producing the bias. An example is shown in Fig. 18 which illustrates a simple polarimetric sensor, employing a fibre with high linear birefringence as the sensing element [73]. The output is divided into two; one beam passes through an analyser to a detector, producing the output given by Eq. (36):

$$I_A = \frac{1}{2} I_0 \left[1 + V \cos\varphi_2\right], \tag{86}$$

the other output passes through a quarter wave plate, aligned with the eigenaxes parallel to those of the fibre, thus introducing a bias of $\pi/2$ so that

$$I_B = \frac{1}{2} I_0 \left[1 + V \cos\left(\varphi_2 + \frac{\pi}{2}\right)\right] \tag{87}$$

as required.

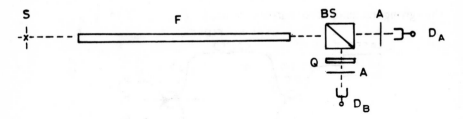

Fig. 18. Simple polarimetric sensor utilising passive homodyne signal processing
S — Source, linearly polarised with azimuth of $\pi/4$ relative to the fibre polarisation eigenaxes
F — Highly birefringent fibre sensing element
BS — Beamsplitter
A — Polarisation analyser aligned at $\pi/4$ relative to the fibre eigenaxes
Q — Quarter wave plate with eigenaxes coincident with those of the fibre
D_A, D_B — Photodetectors.

5.2.4 Electronic processing

Once the quadrature outputs have been produced, a number of electronic techniques exist to produce a signal linear with phase. We shall first describe a method which is applicable to the demodulation of small amplitude, high frequency measurands. In this case, from Eqs. (77) and (78) with

$$\varphi_B = \pi/2, \quad \varphi = \varphi_d + \varphi_{om} \sin \omega_m t \tag{88}$$

we see that

$$I_A = \frac{1}{2} k I_0 \left[1 + V \cos \varphi_d - V \sin \varphi_d \cdot \varphi_{om} \sin \omega_m t \right], \quad (89)$$

and
$$I_B = \frac{1}{2} k I_0 \left[1 + V \sin \varphi_d - V \cos \varphi_d \cdot \varphi_{om} \sin \omega_m t \right]. \quad (90)$$

We now high pass filter these signals to produce

$$I'_A = -\frac{1}{2} k I_0 V \varphi_{om} \sin \varphi_d \cdot \sin \omega_m t, \quad (91)$$

and
$$I'_B = -\frac{1}{2} k I_0 V \varphi_{om} \cos \varphi_d \cdot \sin \omega_m t \quad (92)$$

so that by squaring and adding we obtain

$$\sqrt{I'^2_A + I'^2_B} = \frac{1}{2} k I_0 V \varphi_{om} \sin \omega_m t \quad (93)$$

which is linear with the measurand phase, as required.

A more sophisticated technique which is applicable to measurands of larger magnitude is shown in Fig. 19 [74]. We first subtract the d.c. pedestals from I_A and I_B, with $\varphi_B = \pi/2$, to produce signals of the form

$$i_A = a \cos (\varphi_d + \varphi_{om} \sin \omega_m t), \quad (94)$$
$$i_B = a \sin (\varphi_d + \varphi_{om} \sin \omega_m t), \quad (95)$$

where a is a constant, and then evaluate the following function

$$i_s = \int \left[i_A \frac{d}{dt} i_B - i_B \frac{d}{dt} i_A \right] dt. \quad (96)$$

We must assume that the measurand frequency is much higher than that of φ_d, so that we can assume $d\varphi_d/dt = 0$. Hence, after some manipulation:

$$i_s = a^2 \varphi_m, \quad (97)$$

which is of the form required.

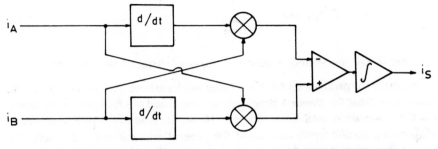

Fig. 19. Passive homodyne demodulation by differentiating and cross-multiplying.

5.2.5 Conclusions

The advantages of the passive homodyne scheme over active homodyne are that the measurement range is increased, and that no electrically active elements are required in the region of measurement. However, it has the disadvantage of requiring optical elements (such as the 3 × 3 coupler), or using an unbalanced interferometer (thus increasing noise — see Sec. 5.4). Also, the electronic processing involved is quite complex. An alternative approach which also yields a wide measurement range is heterodyne processing, and this is discussed below.

5.3 Heterodyne Processing

Figure 20(a) shows one possible implementation of a true heterodyne processing scheme. The basic requirement is that a frequency offset is produced between the two beams, so that at one of the outputs

$$I_1 = \frac{1}{2} k I_0 \left[1 + V \cos (\varphi + \omega t) \right], \qquad (98)$$

where ω is the circular offset frequency. We see that Eq. (98) has the form of a phase modulated heterodyne carrier which can be demodulated using one of a number of established electronic techniques.

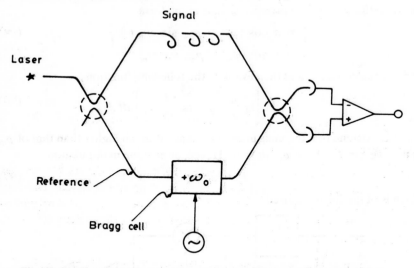

Fig. 20. (*a*) True heterodyne signal processing using a Bragg cell frequency shifter.

Methods for producing the offset frequency have been discussed in Sec. 4.3. It was noted that the classical Bragg cell accomplished this function, but it is not readily compatible with fibre systems. However, true fibre optic frequency shifters are insufficiently developed for general applications. It is for this reason that many heterodyne processing schemes for the fibre interferometer make use of the pseudo frequency shifters, based on phase modulation, discussed in Sec. 4.3.

It is possible to generate an electronic heterodyne carrier without using an optical frequency shifting element, by developing the passive homodyne technique. A block diagram is shown in Fig. 20(b). The starting point is quadrature signals of the form shown in Eqs. (94) and (95).

$$i_A = a \cos\varphi, \tag{99}$$

and
$$i_B = a \sin\varphi. \tag{100}$$

We then cross-multiply with quadrature components of a reference oscillator of circular frequency ω and add, giving

$$i_H = i_A \sin \omega t + i_B \cos \omega t, \tag{101}$$

so that from Eqs. (99) and (100)

$$i_H = a \sin(\varphi + \omega t), \tag{102}$$

which has the form of a phase modulated heterodyne carrier.

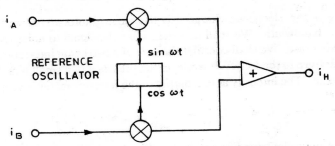

Fig. 20. (b) Quadrature recombination heterodyne demodulation.

5.4 Noise Considerations

The resolution of a sensor is defined as the minimum detectable change in measurand. The resolution is set by the noise floor of the system, and for the monomode fibre sensor the principal sources of noise are: (i) environmental effects on the sensing element and on other parts of the system; (ii) receiver noise; and (iii) source noise.

The most serious environmental perturbations arise from changes in ambient temperature and vibration, which acting on the sensing element cause phase changes indistinguishable from those produced by the measurand. Temperature changes are less serious, because they are usually of a much lower frequency than the measurand, and hence their effect may be removed by filtering.

Minimisation of environmentally produced noise is an important aspect of the sensor design. However, the actual noise levels in practical systems are very much dependent on the specific application, and it is not possible to state a typical value.

Receiver noise arises from two different sources. A fundamental limit is set by the photodetector shot noise, and is imposed by the statistical (photon) nature of light. This matter is discussed in Sec. 5.4.1 below. Noise is also produced within the signal processing electronics.

The laser source itself introduces two types of excess noise into the system: intensity noise and frequency noise [75, 76]. Intensity noise is simply fluctuation of the power of the laser, and is discussed in Sec. 5.4.2. Frequency noise is a fluctuation of the optical frequency of the source, described in Sec. 5.4.3. Diode laser sources are preferred for many applications, and our discussion will be restricted to these.

5.4.1 Detector noise

The shot noise current produced by any photodetector is given by

$$i_{SN} = (2ei\Delta f)^{1/2}, \qquad (103)$$

where e is the electronic charge, i is the mean photocurrent and Δf is the detector bandwidth. We shall now determine the signal to noise ratio arising from shot noise. We shall consider the case of a two beam interferometer held at quadrature in the LGBWP mode, in the presence of a small high frequency measurand. The intensity at one output of the interferometer is thus

$$I_1 = \frac{1}{2} I_0 \left[1 - V \sin\left(\varphi_{om} \sin \omega_m t\right)\right]. \qquad (104)$$

If the photodetector has an efficiency η, then in the small signal limit the photocurrent at the frequency ω_m is

$$i_s = \frac{1}{2} \eta I_0 V \varphi_m \qquad (105)$$

whereas the shot noise current is

$$i_{SN} = (2e i_1 \Delta f)^{1/2}, \qquad (106)$$

where

$$i_1 \simeq \frac{1}{2} \eta I_0 \qquad (107)$$

so that from Eq. (103)

$$i_{SN} = (e\eta I_0 \Delta f)^{1/2}. \qquad (108)$$

We can therefore, calculate the signal to noise ratio (SNR) from Eqs. (105) and (108) giving

$$SNR = \frac{1}{2} \left(\frac{\eta I_0}{e \Delta f}\right) V \varphi_m. \qquad (109)$$

It is instructive to consider the minimum phase change which can be detected, which we shall assume to correspond to a SNR of unity. In the optimum case of unity visibility

$$\varphi_m \text{ (min)} = 2 \, (e\Delta f / \eta I_0)^{1/2} \, . \tag{110}$$

Some numerical examples are given in Table 3.

It may be seen that the resolution depends only on the optical power, and improves as the power is increased.

TABLE 3

Minimum detectable phase, expressed in μ rads and for a signal to noise ratio of unity, as limited by various noise sources. The results are based on the laser noise data of Figs. 21 and 22, and assume a detector efficiency of 0.5 AW^{-1}, with a (Hz)$^{1/2}$ detector bandwidth

Noise Source	Measurand frequency	Optical path difference			
		0.01 m		1.0 m	
		Optical Power			
		1 μW	1 mW	1 μW	1 mW
Detector shot noise	10 Hz	1.1	0.04	1.1	0.04
	1 kHz	1.1	0.04	1.1	0.04
Intensity noise	10 Hz	8.4	8.4	8.4	8.4
	1 kHz	0.6	0.6	0.6	0.6
Phase noise	10 Hz	53	53	5300	5300
	1 Hz	3.0	3.0	300	300

5.4.2 Intensity noise

Intensity noise is power fluctuations of the source, and is described by the excess noise parameter

$$\xi \, (f, \, \Delta f) = P(f, \, \Delta f)/P_0 \, , \tag{111}$$

where $P(f, \Delta f)$ is the rms optical power present as fluctuations with frequencies between f and $f + \Delta f$ and P_0 is the mean power. Some data for a typical diode laser are shown in Fig. 21[2], plotted in decibels where

$$\xi \, (db) = 20 \, log \, \xi \, . \tag{112}$$

It may be seen that the resolution depends only on the optical power, and improves as the power is increased.

It is now possible to calculate the SNR, which we shall do for the case of the two beam interferometer held at quadrature in the LGBWP mode, for a small high frequency measurand. The photocurrent for one output of the interferometer is thus, by analogy with Eq. (104) and using Eq. (111)

$$i_1 = \frac{1}{2} \eta I_0 \, (1 + \xi) \left[1 - V \sin \left(\varphi_{om} \sin \omega_m t \right) \right] , \tag{113}$$

the signal current at frequency ω_m is given by Eq. (105), and the corresponding intensity noise current is

$$i_{IN} = \frac{1}{2} \eta I_0 \xi \tag{114}$$

giving
$$SNR = V\varphi_m/\xi, \tag{115}$$

so that the minimum detectable phase (corresponding to SNR = 1) when the visibility is unity is

$$\varphi_m \text{ (min)} = 1/\xi. \tag{116}$$

Some typical values for the intensity noise limited resolution are given in Table 3.

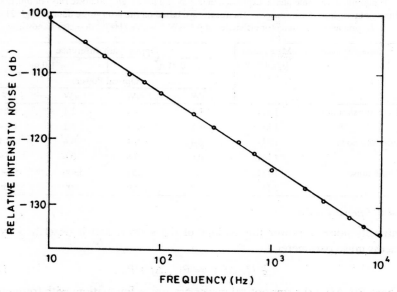

Fig. 21. Typical diode laser intensity noise spectrum.

In practice, it is fairly straightforward to compensate for the effects of intensity noise [76]. It is only necessary to derive an intensity reference from the source (for example, by using an additional directional coupler or beamsplitter), and then to electronically divide the intensity output of the interferometer by this reference signal. For some signal processing schemes, the compensation is even easier. For example, when the active homodyne scheme is implemented with an interferometer for which the complementary outputs are accessible (as in Sec. 5.1), then the quadrature reference level is zero. This method of signal processing is, therefore, unaffected by source intensity noise.

5.4.3 Frequency noise

A serious source of noise is created by frequency jitter in the output of the laser. Some data for a typical diode laser are shown in Fig. 22, which shows that the amplitude of the optical frequency variations, Δv, diminishes as the observation frequency is increased. Therefore, frequency noise worsens the resolution for low frequency measurands. We see from Eq. (55) that the effect of frequency noise in an interferometer with an optical path imbalance of nl will cause phase changes indistinguishable from those produced by the measurand. The corresponding resolution is thus

$$\varphi_m (\min) = (2\pi n\lambda / c)\Delta v . \tag{117}$$

It is evident that the effect of frequency noise is reduced by minimising the path imbalance, and for a perfectly balanced interferometer, frequency noise has no effect. However, many signal processing schemes demand a path imbalance in the interferometer, in order that phase modulation may be produced by controlling the frequency of the source, as explained in Sec. 5.1. Furthermore, any processing scheme based on source modulation demands some minimum range of phase modulation amplitude, so that given the finite possible frequency modulation amplitude of the source, some minimum path imbalance requirement is imposed.

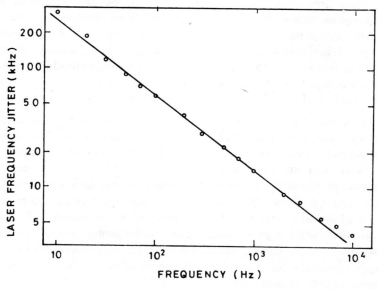

Fig. 22. Typical diode laser frequency noise spectrum.

It is possible to compensate for frequency noise, in a manner analogous to that used for the compensation of intensity noise. However, the compensation is practically more difficult, because it is necessary to derive a frequency reference from the source. A technique which has been successfully demonstrated [77] is to use a conventional high finesse Fabry-Perot interferometer to monitor the source frequency, and hence provide an error signal used in a feedback system to stabilise the source. In this way frequency noise reductions in the range 10 to 30 dB have been demonstrated.

6. CONCLUSIONS

We have reviewed a wide range of techniques for signal processing in monomode fibre optic sensor system. We may enumerate the elements which make up the complete system:

1. The sensing element;
2. Optical processing;
3. Electronic processing;
4. The optical/electronic interface;
5. Data communication.

The operation of the system may be summarised as follows. In the sensing element, the measurand interacts with the fibre to produce changes in its waveguiding properties. For monomode fibres, we are concerned with coherent optical techniques so that the transduction mechanisms involve changes in the optical path length of the sensing element, or in its polarisation properties. It is necessary to illuminate the sensing element, in almost all cases, using a coherent source, and then use optical processing to produce an intensity which is dependent on the measurand. The optical processing schemes used are passive, and are based on classical interferometry and polarimetry.

The purpose of the electronic processing is to transduce the optical intensity signal to an electronic signal which is linear with the measurand. The essential first step is to use a photodetector to convert the optical intensity to a photocurrent. Most electronic processing schemes demand the use of some active component as an interface between the electronic and optical systems. For example, one may control the phase in an interferometer by current modulation of a diode laser. However, fibre modulators are of considerable importance, for the control of phase, polarisation and frequency. Finally, the linear output signal is in electronic form, and may, therefore, conveniently be transmitted electronically. However, the signal could, of course, be reconverted to an optical form for transmission.

Despite the considerable practical success of the techniques described in this chapter, it is worth enumerating the limitations which are imposed on fibre sensor systems by current technology, and to indicate the areas of research which may lead to a relaxation of these limitations. A basic difficulty arises with the design of the sensing element itself, where the principal problem is that of cross-sensitivity. For example, the great strain resolution of the fibre interferometer can rarely be exploited in practice unless the strain frequency is large, because of cross-sensitivity to temperature. One reason for this restriction is that almost all fibre sensors are based on the use of fibres similar to those used for telecommunications, so that the designer has little control over the properties of the sensing element. However, fibres designed specifically for sensor applications are beginning to become available. It is also possible for the designer to modify the properties of the fibre — for example, through the use of coatings — to enhance or suppress the sensitivity to a specific variable.

Passive fibre component technology is now well advanced, and it has, therefore, been possible to demonstrate almost all classical interferometer (and polarisation interferometer) configurations in a fibre form. Design options for the optical system have been greatly enhanced through the availability of high quality components such as directional couplers, and components which show useful wavelength or polarisation selectivity.

Active component technology is less well advanced. For example, most modulators operate either mechanically or thermally, and consequently have a relatively slow response. The availability of a true fibre optic frequency shifter would be very valuable in many signal processing schemes. Because of the limited performance and range of types of available active fibre components, a good deal of signal processing in fibre sensors is achieved electronically. However, this situation may be expected to change with the rapid expansion in the research area of active fibre devices (such as fibre lasers and amplifiers constructed using doped fibres). Also, by using active integrated optical devices in conjunction with fibre optics, the range of design options is further enhanced.

The two areas of future development which may be seen as the most important for the advance of the subject are firstly the development of specific sensing elements, and secondly the transfer of signal processing from the electronic to the optical domain. Indeed, it is only through the development of optical signal processing that true compatibility with optical data communication can be achieved.

REFERENCES

1. M.J. Adams 'An introduction to optical waveguides' (Wiley, 1981).
2. D.A. Jackson and Jones J.D.C., 'Fibre optic sensors', *Optica Acta* Vol. 33, 1469 (1986).
3. W.H. Steel 'Interferometry' (Cambridge University Press, 1983).
4. J.D.C. Jones, Akhavan Leilabady P., and Jackson D.A., 'Monomode fibre optic sensors: optical processing schemes for recovery of phase and polarisation state information', *Int J. Optical Sensors* Vol. 1, 123 (1986).
5. S.C. Rashleigh, 'Polarimetric sensors: Exploiting the axial stress in high birefringence fibres', IEE Conf. Publ. 221, 210 (London, 1983).
6. D.N. Payne, Barlow A.J. and Ramskov Hansen J., 'Development of low and high birefringence optical fibres', IEEE *J. Quantum Electron*, Vol. 18, 477 (1982).
7. G.R. Fowles 'Introduction to modern optics' (Holt, Rinehart and Wilson, 1975).
8. D. Langeac, 'Temperature sensing in twisted single mode fibres', *Electron. Letts.*, Vol. 18, 1022 (1982).

9. A.M. Smith, 'Polarisation and magneto-optic properties of single mode optical fibre', *Appl. Opt.*, Vol. 17, 52 (1978).
10. M. Berwick, Jones J.D.C. and Jackson D.A., 'Alternating current measurement and non-invasive data ring utilising the Faraday effect in a closed loop fibre magnetometer', *Opt. Letts.*, Vol. 12, 293 (1987).
11. P.D. McIntyre and Snyder A.W., 'Power transfer between optical fibres', *J. Opt. Soc. Am.*, Vol. 653, 1518 (1973).
12. J. Wilson and Hawkes J.F.B. 'Optoelectronics: an introduction' (Prentice Hall, 1983).
13. I.P. Kaminow, 'Polarisation in optical fibres', IEEE *J. Quantum. Electron.*, Vol. QE 17, 15 (1981).
14. H.C. Lefevre, 'Single mode fibre fractional wave devices and polarisation controllers', *Electron. Letts.*, Vol. 16, 778 (1980).
15. L. Goldberg, Taylor H.F., Dandridge A., Weller J.F. and Miles R.O., 'Spectral characteristics of semiconductor lasers with optical feedback', IEEE *J. Quantum Electron.*, Vol. QE 18, 555 (1982).
16. R.H. Stolen and Turner E.H., 'Faraday rotation in highly birefringent optical fibres', *Appl. Opt.*, Vol. 19, 842 (1980).
17. B. Culshaw and Giles I.P., 'Fibre optic gyroscopes', *J. Phys. E.*, Vol. 16, 5 (1983).
18. P. Akhavan Leilabady, Wayte A.P., Berwick M., Jones J.D.C. and Jackson D.A., 'A pseudo-reciprocal fibre optic Faraday rotation sensor: current measurement and data communication applications', *Opt. Comm.*, Vol. 59, 173 (1986).
19. P.A. Nicati and Robert P., 'Stabilised current sensor using a Sagnac interferometer', *J. Phys. E.*, Vol. 21, 791 (1988).
20. K. Bohm, Petermann K. and Weidel E., 'Sensitivity of a fibre optic gyroscope to environmental magnetic fields', *Opt. Letts.*, Vol. 7, 180 (1982).
21. I.D. Miller, Mortimore D.B., Urquhart P., Ainslie B.J., Craig S.P., Millar C.A. and Payne D.B., 'A Nd^{3+}—doped cw fibre laser using all-fibre reflectors', *Appl. Opt.*, Vol. 26, 2197 (1987).
22. C.A. Millar, Miller I.D., Mortimore D.B., Ainslie B.J. and Urquhart P., 'Fibre laser with adjustable fibre reflector for wavelength tuning and variable output coupling', IEE *Proc. J. (Optoelectron)* Vol. 135, 303 (1988).
23. A.D. Kersey, Jackson D.A. and Corke M, 'A simple fibre Fabry-Perot sensor', *Opt. Comm.*, Vol. 45, 71 (1983).
24. J. Stone, 'Optical fibre Fabry-Perot interferometer with finesse of 300', *Electron. Letts.*, Vol. 21, 504 (1985).
25. P. Akhavan Leilabady, Jones J.D.C. and Jackson D.A., 'Combined interferometric-polarimetric fibre optic sensor capable of remote operation', *Opt. Comm.*, Vol. 57, 77 (1986).
26. S. Tai, Kyuma K., Hamanaka K. and Nakayama T., 'Applications of fibre optic ring resonators using laser diodes', *Optica Acta.*, Vol. 33, 1539 (1986).
27. M. Corke, Jones J.D.C., Kersey A.D. and Jackson D.A., 'Combined Michelson and polarimetric fibre optic interferometric sensor', *Electron. Letts.*, Vol. 21, 148 (1985).
28. P. Akhavan Leilabady, Jones J.D.C., Corke M. and Jackson D.A., 'A dual interferometer implemented in parallel on a single birefringent optical fibre', *J. Phys. E.*, Vol. 19, 143 (1986).
29. P. Akhavan Leilabady, Jones J.D.C. and Jackson D.A., 'Interferometric strain measurement using optical fibres', SPIE Proc., Vol. 586, 230 (1985).
30. W.H. Steel, 'Interferometry' (Cambridge University Press, 1983).

31. S.A. Al-Chalabi, Culshaw B. and Davies D.E.N., 'Partially incoherent sources in interferometric sensors', IEE Conf. Publ. 221, 132 (1983).
32. D.A. Jackson, Priest R., Dandridge A. and Tveten A.B., 'Elimination of drift in a single-mode optical fibre interferometer using a piezo-electrically stretched coiled fibre', *Appl. Opt.*, Vol. 19, 2926 (1980).
33. S.J. Petuchowski, Siegel G.H. and Giallorenzi T.G., 'A sensitive fibre optic Fabry-Perot interferometer', IEEE *J. Quantum. Electron.*, Vol. QE 18, 4 (1982).
34. A. Dandridge and Goldberg L., 'Current induced frequency modulation in diode lasers', *Electron. Letts.*, Vol. 18, 302 (1982).
35. R.P. Tatam, Pannell C.N., Jones J.D.C. and Jackson D.A., 'Full polarisation state control utilising linearly birefringent monomode optical fibre', *J. Lightwave Tech.*, Vol. LT 5, 980 (1987).
36. D.A. Jackson, Kersey A.D., Akhavan Leilabady P. and Jones J.D.C., 'High frequency non-mechanical optical linear polarisation state rotator', *J. Phys. E.*, Vol. 19, 146 (1986).
37. R.P. Tatam, Hill D.C., Jones J.D.C. and Jackson D.A., 'All-fibre-optic polarisation state azimuth control: application to Faraday rotation', *J. Lightwave Tech.*, Vol. 6, 1171 (1988).
38. D.A. Jackson and Jones J.D.C., 'Extrinsic fibre optic sensors for remote measurement', *Opt. and Laser Tech.*, Vol. 18, 243 (1986).
39. R.P. Tatam, Pannell C.N., Jones J.D.C. and Jackson D.A., 'Full polarisation state control utilising linearly birefringent monomode optical fibre', *J. Lightwave Tech.*, Vol. LT 5, 980 (1987).
40. D. Gruchmann, Petermann K., Staudigel L. and Weidel E., 'Fibre optic polarisers with high extinction ratio', Proc. 9th European Conference on Opt. Comm. 305 (Elsevier, Amsterdam, 1983).
41. K. Liu, Sorin W.V. and Shaw H.J., 'Single mode fibre evanescent polariser/amplitude modulator using liquid crystals' *Opt. Letts.*, Vol. 11, 180 (1986).
42. R.A. Bergh, Lefevre H.C. and Shaw H.J., 'Single mode fibre optic polariser', *Opt. Lett.*, Vol. 5, 479 (1980).
43. M.P. Varnham, Payne D.N., Birch R.D. and Tarbox E.J., 'Single polarisation operation of highly birefringent fibres', *Electron. Letts.*, Vol. 19, 246 (1983).
44. M.P. Varnham, Payne D.N., Barlow A.J. and Tarbox E.J., 'Coiled birefringent fibre polarisers', *Opt. Lett.*, Vol. 9, 306 (1984).
45. H.C. Lefevre, 'Single mode fibre fractional wave devices and polarisation controllers', *Electron. Letts.*, Vol. 16, 778 (1980).
46. W.P. Risk, Youngquist R.C., Kino G.S. and Shaw H.J., 'Acousto-optic frequency shifting in birefringent fibre', *Opt. Lett.*, Vol. 9, 309 (1984).
47. C.N. Pannell, Tatam R.P., Jones J.D.C. and Jackson D.A., 'A fibre optic frequency shifter utilising travelling flexure waves in birefringent fibre', *J. IERE.*, Vol. 58, S 92 (1988).
48. B.Y. Kim, Blake J.N., Engan H.E. and Shaw H.J., 'All fibre optic acousto-optic frequency shifter', *Opt. Letts.*, Vol. 11, 389 (1986).
49. D.A. Jackson, Kersey A.D., Corke M. and Jones J.D.C., 'Pseudo-heterodyne detection scheme for optical interferometers', *Electron. Letts.*, Vol. 18, 1081, (1982).
50. J.H. Cole, Danver B.A. and Bucaro J.A., 'Synthetic heterodyne interferometric demodulation', IEEE *J. Quantum Electron.*, Vol. QE 18, 694 (1982).
51. R.A. Bergh, Kotler G. and Shaw H.J., 'Single mode fibre optic directional coupler', *Electron. Letts.*, Vol. 16, 260 (1980).

52. O. Parriaux, Gidon S. and Kuznetsov A.A., 'Distributed coupling on polished single mode optical fibres', *Appl. Opt.*, Vol. 20, 2420 (1981).
53. B.S. Kawasaki, Hill K.O. and Lamont R.G., 'Biconical taper single mode fibre coupler', *Opt. Lett.*, Vol. 6, 327 (1981).
54. C.A. Villarruel and Moeller R.P., 'Fused single mode fibre access coupler', *Electron. Letts.*, Vol. 17, 243 (1981).
55. M.H. Yu and Hall D.B., 'Low loss fibre ring resonator', SPIE Proc. — Fibre Optic and Laser Sensors II, 104 (May 1984).
56. M.J.F. Digonnet and Shaw H.J., 'Analysis of a tunable single mode optical fibre coupler', IEEE *J. Quantum. Electron.*, Vol. QE 18, 746 (1982).
57. J.D. Beasley, Moore D.R. and Stowe D.W., 'Evanescent wave fibre optic couplers: three methods' Proc. OFC'83, ML 5 (February 1983).
58. F. Bilodeau, Hill K.O., Johnson D.C. and Faucher S., 'Compact, low loss, fused biconical taper couplers: overcoupled operation and antisymmetric supermode cut off', *Opt. Lett.*, Vol. 12, 634 (1987).
59. R.H. Stolen, 'Polishing induced birefringence in single mode fibres', *Appl. Opt.*, Vol. 25, 344 (1986).
60. T. Bricheno and Fielding A., 'Stable low loss single mode couplers', *Electron. Letts.*, Vol. 20, 230 (1984).
61. C.A. Villaruel, Abebe M. and Burns W.K., 'Polarisation preserving single mode fibre coupler', Electron. Letts., Vol. 19, 17 (1983).
62. L. Eyges and Wintersteiner P., 'Modes of an array of dielectric waveguides', *J. Opt. Soc. Am.*, Vol. 71, 1351 (1981).
63. M.D. Feit and Fleck J.A., 'Propagating beam theory of optical fibre cross-coupling', *J. Opt. Soc. Am.*, Vol. 71, 1361 (1981).
64. R.B. Dyott and Bello J., 'Polarisation-holding directional coupler made from elliptically cored fibre having a D section', *Electron. Letts.*, Vol. 19, 601 (1983).
65. W. Pleibel, Stolen R.H. and Rashleigh S.C., 'Polarisation preserving couplers with self-aligning birefringent fibres', *Electron. Letts.*, Vol. 19, 825 (1983).
66. I. Yokohama, Kawachi M., Okamoto K. and Noda J., 'Polarisation-maintaining fibre couplers with low excess loss', *Electron. Letts.*, Vol. 22, 929 (1986).
67. M. Kawachi, Kawasaki B.S., Hill K.O. and Edahiro T., 'Fabrication of single-polarisation single-mode-fibre couplers', Electron. Letts., Vol. 18, 962 (1982).
68. M. Digonnet and Shaw H.J., 'Wavelength multiplexing in single-mode fibre couplers', Appl. Opt., Vol. 22, 484 (1983).
69. G. Meltz, Dunphy J.R., Morey W.W. and Snitzer E., 'Cross-talk fibre optic temperature sensor', *Appl. Opt.*, Vol. 22, 464 (1983).
70. K.P. Koo, Tveten A.B. and Dandridge A., 'Passive stabilisation scheme for fibre interferometers using 3 × 3 fibre directional couplers', *Appl. Phys. Letts.*, Vol. 41, 616 (1982).
71. T.G. Giallorenzi, Bucaro J.A., Dandridge A., Sigel G.H., Cole J.H., Rashleigh S.C. and Priest R.G., 'Optical fibre sensor technology', IEEE *J. Quantum Electron.* Vol. QE 18, 626 (1982).
72. A.D. Kersey, Jackson D.A. and Corke M., 'Demodulation scheme for interferometric sensors employing laser frequency switching', *Electron. Letts.*, Vol. 19, 102 (1983).
73. S.C. Rashleigh, 'Polarimetric sensors exploiting the axial stress in high birefringence fibres', IEE Conf. Publ. 221, 210 (1983).

74. K.P. Koo, Tveten A.B. and Dandridge A., 'Passive stabilisation scheme for fibre interferometers using 3 × 3 directional couplers', *Appl. Phys. Letts.*, Vol. 41, 616 (1982).
75. A. Dandridge, Tveten A.B., Miles R.O. and Giallorenzi T.G., 'Laser noise in fibre optic interferometric systems', *Appl. Phys. Letts.*, Vol. 37, 526 (1980).
76. A. Dandridge and Tveten A.B., 'Noise reduction in fibre optic interferometer systems', *Appl. Opt.*, Vol. 20, 2237 (1981).
77. F. Favre and LeGuen D., 'High frequency stability of laser diode for heterodyne communication systems', *Electron. Letts.*, Vol. 16, 709 (1980).

28

Fibre Optic Sensor Multiplexing

KJELL BLØTEKJÆR[*]

1. INTRODUCTION

Many sensor systems require more than one sensor. There may be several sensors of the same kind, in close proximity or distributed over a wide area, say in an industrial plant. Also, in most systems, measurement of more than one physical parameter is required. Typically, temperature and pressure are two such parameters. In order for fibre optic sensors to be competitive, means must be found to link several sensors together in some kind of a network, and still be able to read the status of each sensor individually. These methods will be the subject of this article. Emphasis will be on basic principles, not on technical details. We shall establish a systematic classification of multiplexing techniques, which will enable us to see the similarities between various published concepts.

Some general knowledge of fibre optic sensors is assumed, such as the distinction between amplitude modulated and phase modulated sensors, interferometric detection techniques, etc. No specific sensors will be considered. It is irrelevant for the present discussion whether temperature, pressure, or any other physical parameter is to be measured.

Fibre optic sensor multiplexing has been reviewed at recent conferences [1-10].

2. GENERIC TOPOLOGICAL CONFIGURATION

We shall start by making some assumptions. These are not essential, but they will simplify the exposition. We assume that the sensors are identical, except

---

[*] Division of Physical Electronics, Norwegian Institute of Technology, N-7034 Trondheim-NTH, Norway.

for differences which are deliberately introduced in order to address the sensors individually. We also assume that the length of fibre used to connect two nearest neighbour sensors is the same for all sensors in the network. The fibres may be longer than the physical distance between sensors. Hence, the sensors need not necessarily be regularly spaced. A third assumption is that the sensors are interconnected via one, two, or at least a small number of optical fibres, independent of the number of sensors. This excludes star configurations, two-dimensional arrays, and similar networks.

A typical example of a sensor system with the described properties is a so-called "streamer" used in seismic profiling. A streamer is a linear array of a thousand or so hydrophones spaced equidistantly a few meters apart.

There are two important aspects of fibre optic sensor multiplexing. One is the network configuration, also referred to as the topology of the network. The topology is a description of how the sensors are interconnected to form a network. The second aspect is the addressing technique, which is the modulation, detection, and signal processing used to read the status of each sensor individually, with little or no interference (crosstalk) from other sensors of the network.

We shall discuss various addressing techniques with reference to a generic topological configuration, which is shown in Fig. 1. Later on we shall show that

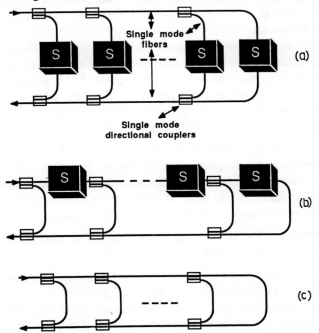

Fig. 1. (a) Generic topological configuration. (b) Modified version of the generic configuration, used with some addressing techniques. (c) Intrinsic sensor version of the generic configuration.

several other topologies have properties which are very close to our generic configuration, but we shall also discuss topologies which are basically different.

Figure 1(a) shows what is called a ladder configuration. The sensors are connected to an input and an output fibre optic bus like the rungs of a ladder. The coupling between the rungs and the buses takes place in directional couplers, also referred to as power splitters. These split the power in an input fibre between two output fibres, in a specified ratio. The same component is used to recombine the light in two fibres into one. With few exceptions we shall assume that single mode fibres are used, although multimode fibres may be used for amplitude modulated sensors.

Some of the addressing techniques to be discussed measure essentially the difference between the amplitudes or phases of light propagated through two rungs in the ladder. For such schemes, the modified configuration of Fig. 1(b) is preferable, because the status of each sensor is read directly.

At this point, we shall make no assumptions about the sensors. To emphasize their generality, we depict them in the figures literally as black boxes. They may be designed to measure any physical quantity. They may be extrinsic sensors, in which the light is brought to and from the sensors in fibres, but the actual modulation of light by the measurand (physical parameter to be measured) takes place outside the fibre. Or they may be intrinsic sensors, in which the measurand changes the properties of the optical fibre itself, thereby modulating the amplitude or phase of the light propagating through the fibre. The modulation imposed on the lightwave by the measurand may be either amplitude or phase modulation. Other modulation forms are also found in the literature, such as polarization, frequency, or colour modulation. With a sufficiently wide interpretation, these modulation forms may be considered as special cases of amplitude or phase modulation.

Figure 1(c) shows a special realization of a network of intrinsic sensors. The figure emphasizes the simplicity of such sensors. It consists entirely of single mode fibres interconnected by directional couplers. Phase modulation is imposed on the fibres by variations of temperature, pressure, etc. Amplitude modulation may be imposed by means of periodic bends applied to the fibre.

The network of Fig. 1(c) belongs to the class of Figs. 1(a) or (b), depending on which parts of the network are exposed to the measurand.

There are some fundamental problems associated with fibre optic sensors. For amplitude modulated sensors, the main problem is one of reference. Changes in attenuation of fibre optic components due to temperature, ageing, etc., may interfere with the desired sensor signal. For phase modulated sensors, the basic problem is to convert the fundamentally non-linear response of an interferometer to a linear one. These problems are not specific to sensor networks, they are just as important for single sensors. Several techniques for solving the problems have been suggested and realized. Therefore, we shall make the tacit assumption that a suitable technique is being used to solve these problems. With the same arrogance, we refuse to bother about polarization fluctuations and other essential or trivial problems associated with fibre optic sensors.

3 INCOHERENT DETECTION

The detection and addressing schemes can be classified as being either incoherent or coherent. In the former, which we shall now discuss, no optical interference effects are used. No optical reference is carried from the source to the detector, and a coherent source is not required. We shall see that in some cases, there is actually an upper limit to the coherence length. Since an optical detector responds only to amplitude (power) modulation, incoherent detection can be used only for amplitude modulated sensors. It should be noted that this does not exclude sensors where the basic effect of the measurand is to change the phase of the light wave. Figure 2 shows the basic principle of the archetype phase sensitive fibre optic sensor, a fibre optic version of the classical Mach-Zehnder interferometer. The input light is divided into two fibres, one of which is the sensing arm. The measurand changes the phase of the light in this arm. When the phase modulated light is recombined with the light in the reference arm, the phase modulation is converted into amplitude modulation. Viewed from the two terminals, the device acts as an amplitude modulated sensor, although the primary effect of the measurand is to change the phase of a lightwave.

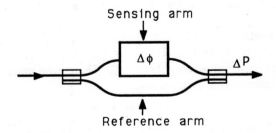

Fig. 2. Fibre optic version of Mach-Zehnder interferometer. In this version, the two arms are assumed to have (almost) the same optical path lengths.

3.1 Time Division Multiplexing

The problem of addressing individual sensors has many analogies in the radar problem of separating individual targets, and the multiplexing schemes are largely adopted from radar technology. It is natural, therefore, to start with the classical radar scheme, a time multiplexed, pulsed system. Historically, this was also the first multiplexing technique discussed for fibre optic sensors [11]. As shown schematically in Fig. 3, a single pulse input results in a multiple pulse output. Each pulse is amplitude modulated by one sensor only. The pulse width τ must be shorter than the time delay τ_d between neighbouring sensors, which is

$$\tau_d = 2L/N_1 c, \tag{1}$$

where c is the velocity of light in vacuum, N_1 is the group index of the fibre mode, and L is the length of fibre connecting two nearest neighbour sensors (it is assumed here that the input and output buses have the same length, but this is not essential). Note that L is not necessarily equal to the physical distance between sensors. By coiling the fibre, L can be made arbitrarily large.

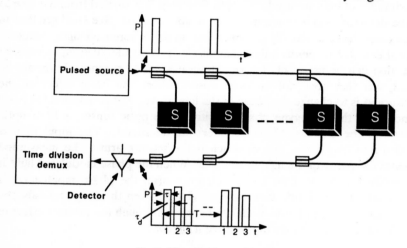

Fig. 3. Time division multiplexing.

The pulse repetition frequency must be such that one pulse from the last sensor arrives before the next pulse from the first sensor. This means that the time T between pulses must be at least $N\tau_d$, N being the number of sensors. Thus, the maximum duty cycle η becomes

$$\eta = \tau_d/T = 1/N. \tag{2}$$

We note that the duty cycle is reduced with increasing number of sensors. This is not the only reason why the sensitivity decreases with increasing number of sensors. The power transmitted down the input bus is shared among the sensors, thus reducing the power through each sensor. In addition, nature does not allow us to couple all power from a rung to the return bus and at the same time transmit power in the bus through the coupler without loss. In fact, if we couple a certain fraction of the power from a single mode rung to a single mode bus, the same fraction will be coupled from the bus to the open end of the coupler, and will be lost. This is a consequence of reciprocity. We shall now analyze the effect of this property on the transmitted power [4].

If we are restricted to using directional couplers with the same coupling coefficient, a little consideration shows that sensor number $N-1$ transmits the lowest power. This power is maximized by choosing the coupling coefficient K (fraction of power coupled from the bus to each sensor) equal to

$$K = 1/(N-1) . \qquad (3)$$

The power transmitted from source to detector through this critical sensor is

$$P_{N-1} = P_0 (N-2)^{2(N-2)} / (N-1)^{2(N-1)}, \qquad (4)$$

where P_0 is the power transmitted from the source to the input fibre. This result is valid in the ideal case of lossless elements. For large N, Eq. (4) approaches

$$P_{N-1} \approx P_0 / e^2 \, (N-1)^2 , \qquad (5)$$

which shows that the power is approximately inversely proportional to the number of sensors squared, rather than the number of sensors directly, which would be the case if no power was lost.

If we allow individual coupling coefficients for each directional coupler, we can improve somewhat on the power budget. The optimum choice of coupling coefficient for coupler number k is

$$K_k = 1/(N-k+1), \qquad (6)$$

and the power transmitted through each sensor is

$$P_k = P_0/N^2 , \qquad (7)$$

which for large N is an improvement by a factor of e^2 as compared to the case of equal coupling coefficients.

The above discussion shows that the average power transmitted through a sensor is inversely proportional to the number of sensors squared. Therefore, the number of sensors which can be multiplexed with reasonable sensitivity is rather limited. The ideal system would transmit an average power inversely proportional to N. In order to approach this situation, one must improve on the duty cycle as well as the loss in coupling the light from the sensors into the return bus.

3.2 Improving the Duty Cycle

The radar community has developed modifications of the simple pulse modulation, in order to increase the duty cycle. The idea is to use some kind of modulation on a continuous wave. Such schemes can also be used in sensor arrays. They have advantages only if the source is peak power limited. If the source is average power limited, the peak power must be decreased in the same rate as the duty cycle is increased, and nothing is gained. Semiconductor lasers are used as sources in most fibre optic sensor systems. Since these lasers are to

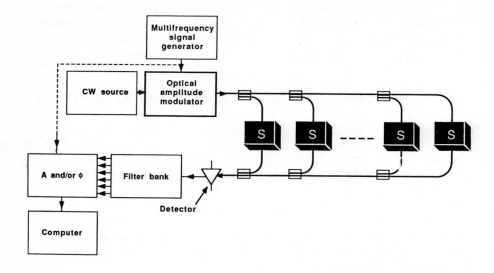

Fig. 4. Sinusoidal modulation. The dashed line indicates an optional reference used for phase measurement.

a large extent peak power limited, the duty cycle can be increased by suitable choice of modulation format. The penalty to be paid is an increase in complexity of the electronics.

It is well known that the properties of a network can be determined in either time or frequency domain. Therefore, an obvious alternative to pulse modulation is amplitude modulation with a set of sinusoids [12]. Figure 4 shows the principle. The source is amplitude modulated by a sum of M sinusoids with different frequencies $\Omega_m, m = 1, 2...M$. In Fig. 4, a separate optical modulator is shown for clarity, but a semiconductor laser source could be modulated directly through the drive current.

After detection, the frequencies are separated, and the amplitude and/or phase of each component is measured.

The complex amplitude at frequency Ω_m is

$$P_{om} e^{j\phi_m} = \sum_k P_k e^{-jk\Omega_m \tau_d}, \qquad m = 1, 2, ...M \qquad (8)$$

where P_k is the power transmitted through sensor number k. There are $2M$ real equations in the complex set (8), and N unknowns, P_k. If we measure both amplitude P_{om} and phase ϕ_m of all frequency components, we need $M = N/2$ modulation frequencies. If we measure amplitude or phase, but not both, we need $M = N$ frequencies. A computer may be used to solve Eq. (8) with respect to P_k.

The system shown in Fig. 4 uses parallel processing of all the modulation frequencies. Series processing is also possible. The modulating signal can be frequency switched, and the frequencies processed in sequence.

Binary codes, such as Barker codes or pseudorandom codes, can also be used for modulation [13,14]. They consist of specific bit sequences of specified length, with the property that their autocorrelation functions are strongly peaked at zero delay. As shown in Fig. 5, the output signal is correlated with delayed versions of the bit sequences to address each sensor separately. This processing can also be done in series rather than in parallel as shown in the figure.

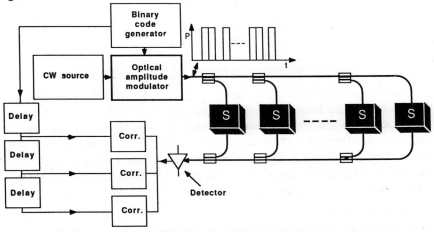

Fig. 5. Binary coded modulation.

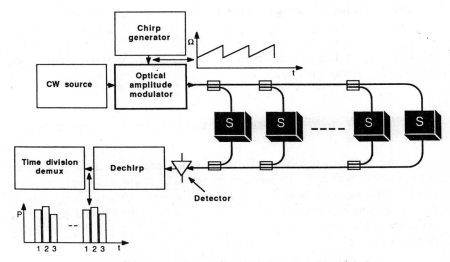

Fig. 6. Chirp modulation. (a) Dechirping and time demultiplexing,

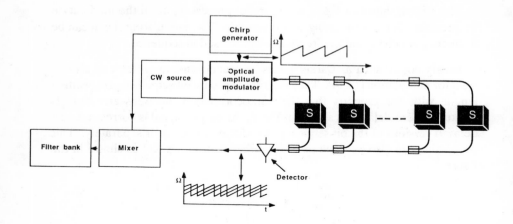

Fig. 6 (b) Difference frequency generation and frequency demultiplexing.

Still another modulation is frequency chirp, i.e., a modulation signal whose frequency changes linearly with time. The optical source is amplitude modulated, but the modulation frequency varies as a ramp function. Two entirely different signal processing techniques can be used. The time domain approach shown in Fig. 6(a) uses a dispersive element, such as a surface acoustic wave delay line to dechirp the signal, i.e., to convert the chirp to a short pulse [15,16]. The individual sensor signals can then be recovered by time division demultiplexing, as discussed above.

In the frequency domain approach, shown in Fig. 6(b), the output signal is mixed with the modulating signal, producing a difference frequency [15]. If the modulation frequency during the chirp is given by

$$\Omega = \Omega_0 + a \cdot t, \qquad (9)$$

where a is a constant that describes the chirp rate, the difference frequency generated by sensor number k is

$$\Delta \Omega_k = ak\tau_d. \qquad (10)$$

The signals from individual sensor signals can be separated using bandpass filters.

All the modulation formats discussed in the present section require an optical source with limited coherence length. The coherence length must be short compared to the time delay τ_d between sensors. Otherwise, the signals from neighbouring sensors would interfere, and the output would depend on arbitrary phase shifts along the fibre. Later on, we shall see that this interference can be used with success in coherent detection schemes.

3.3 Wavelength Division Multiplexing

One way of avoiding both the duty cycle problem and the loss in the output couplers is to use wavelength division multiplexing [17,18]. Figure 7 shows the principle. A cw multiwavelength source transmits light into the input fibre bus. The couplers are wavelength selective. Each coupler couples one wavelength

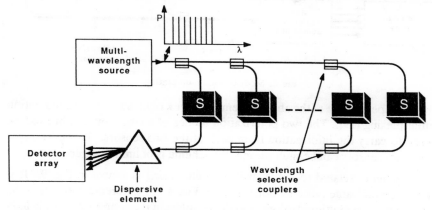

Fig. 7. Wavelength division multiplexing.

completely to the sensor, transmitting all other wavelengths further along the bus. The same kind of couplers are used to couple light from the sensors into the return bus. In principle, this can be done without loss. At the output, the light is sent through some kind of wavelength demultiplexing device. This may be a dispersive element, such as a prism, which separates the wavelengths spatially, and directs each wavelength to an element of a detector array.

The source can be a multiwavelength laser or several quasimonochromatic sources coupled to the same bus. Wavelength selective couplers have been made, but they are presently in a state of development which makes it difficult to assess the possible future performance of this multiplexing scheme.

3.4 Mode Division Multiplexing

Another scheme with properties similar to wavelength division multiplexing is mode division multiplexing, shown schematically in Fig. 8. A continuous wave is launched into the input fibre bus, which may be an ordinary single mode fibre. The same applies to the fibre in the rungs. However, the return bus is a multimode fibre, and the couplers are made to couple light from the single-mode fibre selectively into one mode of the multimode fibre. Each sensor occupies one mode of the multimode fibre. At the output, the modes are separated and detected by a detector array. This scheme is mentioned here for completeness, but it is very far from being realised. Mode selective couplers and filters would have to be developed, and mode coupling in the multimode fibre would pose problems.

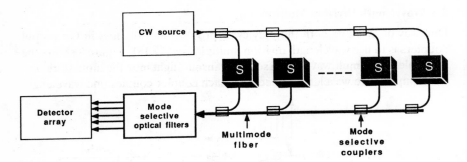

Fig. 8. Mode division multiplexing.

One special case of the above scheme has been realized, namely polarization multiplexing [19]. The two polarization modes of a single-mode fibre can be used to carry the information from each of the two sensors. However, since only two sensors can be multiplexed, the scheme is of limited value.

A scheme related to mode division multiplexing is shown in Fig. 9. It is based on the same pulse modulation as in Fig. 3, the difference being that the return bus is a multimode fibre, and the couplers in the return bus couple light from the single-mode fibre into all modes of the multimode fibre, ideally with even power distribution among the modes. Such a coupler can be realised as shown in Fig. 10 [20]. A small mirror is placed in the core of the multimode fibre, at 45 degrees to the fibre axis. The light from the single-mode fibre is focused onto the mirror. The focusing element may be a graded-index lens. By proper design, this coupler can distribute the power evenly between the modes, with low loss. The loss experienced by the power transmitted through the coupler in the multimode bus is inversely proportional to the number of modes.

Figure 11 shows the power returned from one sensor, relative to the ideal equal division of power between sensors without loss. The abscissa is the number of sensors. The solid lines refer to the case when the coupling coeffi-

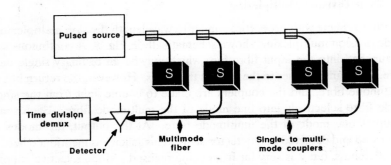

Fig. 9. Time division multiplexing with multimode return bus.

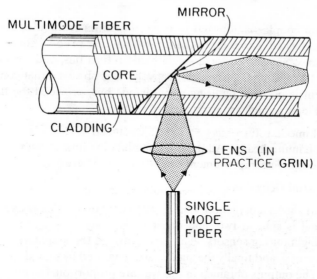

Fig. 10. Single mode to multimode coupler with mode mixing.

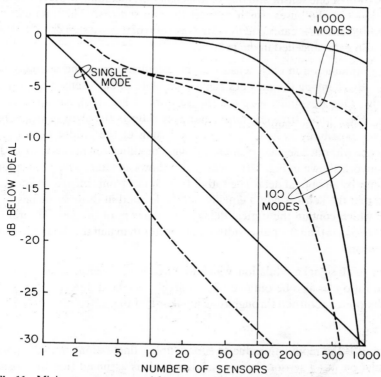

Fig. 11. Minimum power returned from one sensor, provided all components are ideal. Power is shown in decibels relative to a completely lossless system, plotted versus number of sensors. The solid lines refer to couplers adjusted for equal power from all sensors, whereas the dashed lines are for equal coupling coefficients. The curves shown are for a single mode return bus, and for multimode buses with 100 and 1000 modes.

cient of each coupler is adjusted to make the power from all sensors equal, whereas the dashed lines refer to the suboptimal case of equal coupling coefficients. Curves are shown for a single-mode return bus, and for multimode buses with 100 and 1000 modes, respectively. It is observed that considerable improvement is possible by using a multimode bus, provided the number of modes is at least as high as the number of sensors.

The multimode return bus system was discussed here in combination with time division multiplexing. All the other modulation formats mentioned in the section on duty cycle improvement can also be used here.

3.5 Differential Detection

Figure 12(a) shows a detection scheme with some interesting properties. The output signal is split in two, and one part is delayed by τ_d, the time delay between neighbouring sensors. The ratio between the amplitudes of the two signals is formed, and finally the signals are separated by time division demultiplexing. The outputs obtained in this way are proportional to the ratio of the attenuation of one sensor and that of its neighbour, including the attenuation in those parts of the buses which connect the two sensors. The attenuation in preceding sensors is cancelled by taking the ratio of the two signals. The source noise also gets cancelled in the process.

The attenuation in each sensor can be calculated from the demultiplexed ratios. By changing the network slightly the state of each sensor can be read directly. One possibility is to use the network of Fig. 1(b), as shown in Fig. 12(b). Each pulse from the ratiometer now reflects the state of one sensor. Another possibility is to let every second rung of the ladder be a passive reference without a sensing element. A slight modification of this idea leads to the network shown in Fig. 12(c), which also shows an alternative realization of the delay by an optical delay line rather than by electronic means. It should be noted that the sensors could equally well be located in those branches of the rungs which contain the optical delay. The delay τ_p of the parallel branches must be longer than the pulse width τ and shorter than half the delay τ_d between the sensors.

For small signal modulation, which we often have in sensors, the equivalent of the ratio can also be obtained in a suitably weighted difference amplifier. Let the power returned through rung number k of Fig. 12(b)

$$P_k = P(1 + m), \qquad (11)$$

where m is the modulation due to source noise, fluctuations in the leads and modulation in all sensors before number k. It is assumed that the absolute value of m is small compared to unity. The power returned through branch number $k + 1$ will then be

$$P_{k+1} = CP(1 + m)(1 + m_k) \approx CP(1 + m + m_k), \qquad (12)$$

where C is a constant depending on the couplers, and m_k is the modulation imposed by the sensor number k. By forming the expression $P_{k+1}/C-P_k$, we see

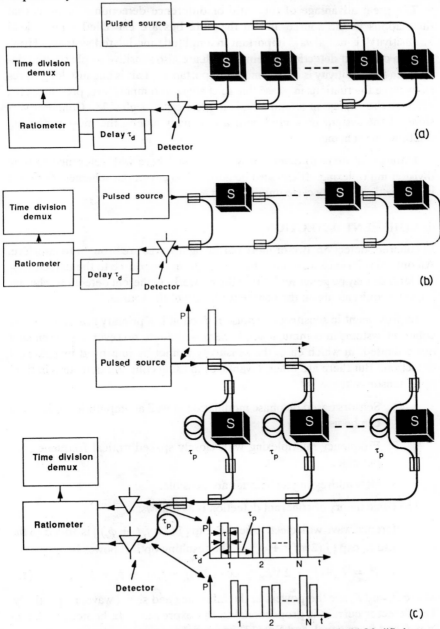

Fig. 12. Ratio detection. (a) Original network. (b) Modified network. (c) Modified network with individual, internal references. The delay is introduced electronically in (a) and (b), optically in (c).

that the modulation m is eliminated. A problem is that the factor C may be different for different sensors.

The great advantage of this ratio or difference detection scheme is that fluctuations in the source and in the fibre bus are cancelled. This "lead insensitivity" is not always important in amplitude modulated systems. However, in coherent detection systems, which are also sensitive to phase modulation, lead insensitivity is of paramount importance. This is because long fibres show large fluctuations in phase due to changes in temperature, pressure, etc., whereas the change in attenuation is in most cases negligible. Coherent versions of the system presented here are actually among the most promising multiplexing schemes.

Differential or ratio detection was discussed here with reference to time division multiplexing. It can also be adopted to any of the schemes discussed in the section on duty cycle improvement.

4. COHERENT DETECTION

In coherent detection, two or more partial light waves are brought to interfere. An optical reference wave must be carried from the source to the detector, or a reference may be generated within the network. Coherent detection schemes impose requirements on the coherence length of the source.

Improvement in sensitivity by noise reduction is a primary reason for using coherent systems in communication. Sensor systems, however, are often shot noise limited, in which case, the sensitivity cannot be improved by coherent detection. But there are other advantages in using coherent detection in fibre optic sensor systems:

1. Sensors based on phase modulation as well as amplitude modulation can be used.

2. Frequency multiplexing with closely spaced optical frequencies is possible.

3. New addressing techniques are possible.

The basic theory of coherent detection is as follows:

A reference wave with electric field $E_r \exp[j(2\pi f_r t + \phi_r)]$ is added to the signal field $E_s \exp[j(2\pi f_s t + \phi_s)]$. The resulting optical power is:

$$P = P_r + P_s + 2\sqrt{P_r P_s} \cos\left[2\pi(f_s - f_r)t + \phi_s - \phi_r\right], \quad (13)$$

where P_r and P_s are the powers in the reference and signal wave, respectively. The detector current is proportional to this expression. In homodyne detection, f_s and f_r are equal, and Eq. (13) becomes

$$P_{\text{homo}} = P_r + P_s + 2\sqrt{P_r P_s} \cos\left[\phi_s - \phi_r\right]. \quad (14)$$

We see that the power depends on the phase ϕ_s as well as the amplitude P_s of the signal. Hence, both amplitude and phase modulation can be used. In practice, however, coherent detection is used primarily with phase modulating sensors.

In heterodyne detection, f_s and f_r differ by an amount Δf, and Eq. (13) becomes

$$P_{\text{hetero}} = P_r + P_s + 2\sqrt{P_r P_s}\,\cos\left[2\pi\Delta f t + \phi_s - \phi_r\right]. \qquad (15)$$

Thus, the detector current varies sinusoidally with time. The amplitude of the sinusoid depends on the signal amplitude P_s. The phase ϕ_s can be recovered by measuring the phase between the detector output and the sinusoid at frequency Δf which is used to shift the optical frequency of the reference with respect to the signal.

4.1 Homodyne Detection

Introducing homodyne detection into the time multiplexed system of Fig. 3 results in the system shown in Fig. 13. Part of the light from a continuous wave source is split off, and added as a reference to the signal before detection. The reference and the signal from the sensors interfere, resulting in an output which reflects both amplitude and phase changes in the sensor signals. The individual sensor signals are recovered using time division demultiplexing.

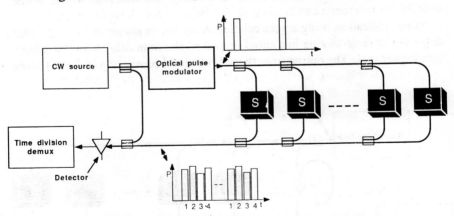

Fig. 13. Homodyne detection using a continuous source.

4.2 Heterodyne Detection

To provide heterodyne detection an optical frequency shifter may be introduced either in the reference arm, as shown in Fig. 14, or in the signal arm. As described above, the detector output contains a sinusoid at a frequency equal to the difference between the two optical frequencies. The amplitude of this sinusoid reflects the modulation of the signal amplitude. The phase of the lightwave through a sensor can be obtained by measuring the phase between

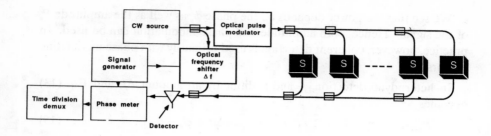

Fig. 14. Heterodyne detection.

the detector output and the sinusoid used to frequency shift the reference, as indicated in the figure.

In general, optical frequency shifters are acoustooptic devices. Till now, they have been available only as bulk component, so-called Bragg cells. Work is going on to develop all-fibre frequency shifters [21,22].

4.3 Coherence Requirement

For coherent detection, the source must have a coherence length longer than the total delay $N\tau_d$ of the network. Otherwise, the reference would not interfere with the signal from all sensors. If a sufficiently coherent source is not available, a reference can be constructed by repeated delay of one pulse. A possible realization using a recirculating delay line is shown in Fig. 15. The delay of the recirculating line should equal τ_d, the time delay between neighbouring sensors. The output from the delay line will consist of a series of pulses

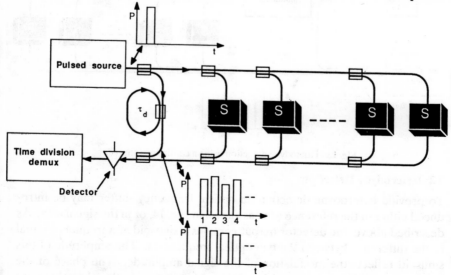

Fig. 15. Homodyne detection using a pulsed source with delayed references.

with decreasing amplitude. Each pulse is coherent with the corresponding signal pulse.

The train of reference pulses can be generated by several other fibre optic networks. One possibility is to use a ladder network identical to the sensing network, except that the sensors are removed. A simpler solution involves a reflective delay line which we shall discuss in a later section (Fig. 23(d)) [7].

An alternative is to read the sensors consecutively by switching the delay, as shown in Fig. 16. The switch does not operate at the pulse repetition frequency, but at a lower rate, determined by the number of sensors and the actual bandwidth of the sensor system.

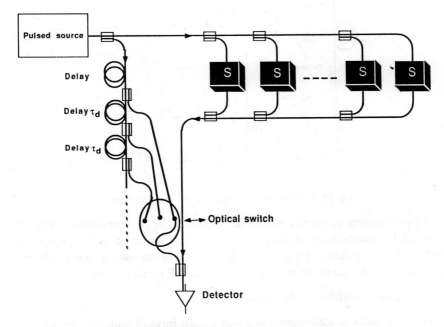

Fig. 16. Sequential processing in homodyne system.

Another method for meeting the coherence requirement is shown in Fig. 17. Here, separate references are provided for each sensor, and the signal from different sensors are detected by separate detectors. Essentially, this is a series to parallel conversion. By introducing appropriate delays, the reference will appear at an output coupler coincident with the pulse from one particular sensor. Pulses from other sensors will appear at the detector at times when there is no reference, and hence no interference signal will be generated. We shall see that this system can be modified to increase the duty cycle by using a continuous wave source. This will be discussed in the section on coherence multiplexing.

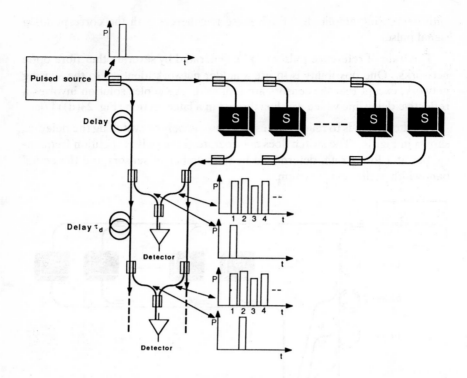

Fig. 17. Space division multiplexing with pulsed source.

By introducing frequency shift in the reference, the systems shown in Fig. 15 through 17 become heterodyne systems. Amplitude modulation does not work well with the systems of Figs. 16 and 17, because an output signal will be generated even if no reference is present. This will result in cross-talk.

4.4 Frequency Division Multiplexing

Frequency division multiplexing is a well known form of multiplexing used in communication (cf. companion chapter in this book). Figure 18(a) shows a sensor system based on frequency division multiplexing[*]. Optical frequency shifters (Bragg cells or all-fibre acoustooptic devices) are inserted in the input (or output) bus, to shift the optical frequency by an amount Δf between each pair of sensors. The signal through sensor number k will be at a frequency shifted by $k\Delta f$ relative to the reference. By interference with the reference,

---

[*] It is a common practice to distinguish between wavelength and frequency division multiplexing. The term *wavelength* is used when the optical frequency differences are too large to be accessible to electronic processing. Demultiplexing of a wavelength division multiplexed system must be performed by optical means. The term *frequency* is used when the optical frequency differences are in the microwave region or below.

frequencies $k\Delta f$ will be generated at the detector output, and these can be separated by conventional frequency demultiplexing techniques, such as a set of bandpass filters. In order to suppress interference between signals from different sensors, the reference must be strong compared to the signal. The ratio of reference to signal power must be chosen to reduce cross-talk to an acceptable level.

Another frequency division multiplexing scheme is shown in Fig. 18(b) [23]. A pulsed source is used, and the method for generating a reference is the same as for the homodyne system of Fig. 15, except that a frequency shifter is inserted in the recirculating delay line. A series of reference pulses are generated, separated in frequency as well as in time. Interference between signal and reference pulses gives rise to separate frequency outputs for each sensor. In principle, there is no cross-talk, but the pulsed source represents a reduction of duty cycle compared to the system of Fig. 18(a).

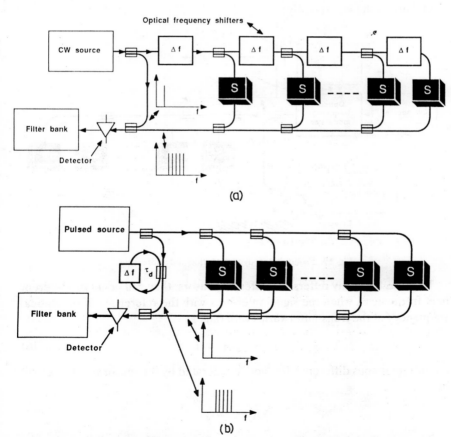

Fig. 18. Frequency division multiplexing. (a) Continuous wave, coherent reference. (b) Pulsed, delayed reference.

The systems shown in Figs. 18(a) and 18(b) are applicable to amplitude modulated sensors. The status of phase modulated sensors can be read by comparing the phase of the output at $k\Delta f$ with the corresponding harmonic of the sinusoid used to drive the frequency shifters. The electronics required for this function is not shown in the figure.

4.5 Pseudoheterodyne

The technique shown schematically in Fig. 19 is referred to as pseudoheterodyne, serrodyne, or FMCW (frequency modulated continuous wave) [24,25]. It has some similarity with the incoherent chirp technique described previously. There, the light was *amplitude* modulated with a chirp, i.e., the modulation signal had a linearly increasing frequency. In the present system, the light is *frequency* modulated, the optical frequency varies as a ramp function of time. Semiconductor lasers can be frequency modulated through the drive current, and no separate modulator is required, although in Fig. 19 the modulator is shown explicitly.

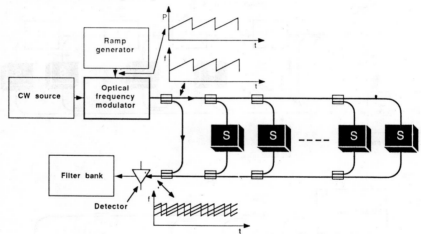

Fig. 19. Pseudoheterodyne detection and addressing.

Because the delay differs for different sensors, they will generate different beat frequencies when the signal interferes with the reference. If the optical frequency f at the transmitter varies with time as

$$f = f_0 + a \cdot t \tag{16}$$

the instantaneous difference frequency generated by the sensor number k will be

$$\Delta f_k = ka\tau_d. \tag{17}$$

These frequencies can be separated by filtering. Phase modulation in the sensors results in a modulation of the instantaneous beat frequency. Thus, the status of the sensors can be obtained by FM demodulation. The frequency steps occurring at the end of each chirp will generate spurious frequency components, and care must be taken in the design to avoid interference with these frequency components.

Cross-talk due to interference between sensor signals occur in this system, and the discussion of this problem in the previous section applies to the present system as well.

4.6 Two-pulse Systems

All the coherent systems described above suffer from one common draw-back: they are lead sensitive. Any phase shift in the fibre optic buses will result in a signal which will mix with the desired signal from the sensor. Since the phase of a lightwave in a fibre is extremely sensitive to environmental perturbations, this lead sensitivity is much more serious for coherent than for incoherent systems, because the latter are only sensitive to amplitude fluctuations.

It was shown above that incoherent systems can be made lead insensitive by using ratio or difference detection. Similar techniques are applicable to coherent systems. The idea is to generate references internally in the ladder network, by relying on the interference between the light from two rungs of the ladder. This is shown in Fig. 20 [26]*. The system in Fig. 20(a) is the coherent analog of the incoherent system in Fig. 12(b). The source generates two pulses, which are delayed by τ_d relative to each other. Alternatively, one pulse of length $2\tau_d$ can be used. Each rung of the ladder returns two pulses. The last pulse from rung number k-1 appears at the detector simultaneously with the first pulse from rung number k. The two signals interfere, producing a detector output depending on the phase shift in sensor number k, but independent of phase changes in those parts of the buses which are common to the two rungs.

Since this multiplexing scheme is based on interference between two pulses, the two must be coherent. If a sufficiently coherent source is not available, a double pulse can be generated by delaying one pulse and adding the delayed pulse to the original one [27]. Figure 20(b) shows such a system, applied to the alternative network of Fig. 12(c) [28,29]. The system shown here incorporates an optional frequency shifter in one of the arms in order to provide heterodyne detection [26]. Of course, the various elements of Figs. 20(a) and (b) can be interchanged. Again, we emphasize that the sensors may just as well be in the long arms of the interferometers which constitute the rungs of the ladder.

---

* The work reported in reference [26] was actually carried out with a reflective array, but the principle may equally well be applied to a ladder network.

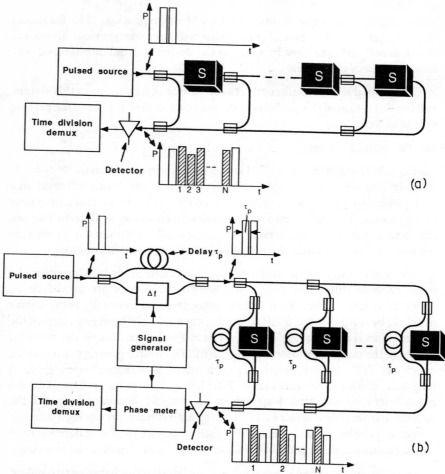

Fig. 20. Double pulse technique for lead insensitive addressing. In the diagrams showing pulse sequences, pulse overlap is indicated by cross-hatching. (a) and (b) show two different methods for generating the double pulse. The topologies in (a) and (b) can be interchanged. Also, heterodyne detection, shown in (b), can be applied to the system in (a).

The unbalanced, Mach-Zehnder interferometer used to produce the double pulse in Fig. 20(b) may equally well be inserted in the return bus, i.e., after the sensors rather than before. The single pulse source then generates a series of pulses, one from each rung of the ladder. The extra interferometer creates a delayed version of the pulse train, adding it to the original one to produce interference between signals from two neighbouring rungs. This version is usually found in the literature. It also opens a possibility of spatially separating the sensor outputs in a manner analogous to the scheme shown in Fig. 17. A system based on the network of Fig. 20(b) is shown in Fig. 21(a). The source emits a single pulse, and a delayed reference pulse is generated internally in

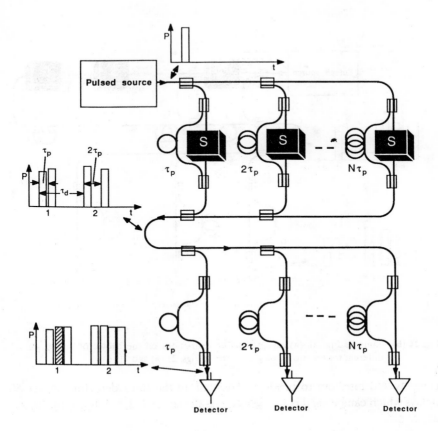

Fig. 21 (a) Space division multiplexing derived from the double pulse technique. In the diagrams showing pulse sequences, pulse overlap is indicated by cross-hatching.

each rung. The delay is different for each sensor. In the receiver, the incoming power is divided into N parts, and there is one detector for each sensor. The sensor signal and its reference is brought to interfere by again splitting the signal and introducing a delay matching the delay between sensor and reference. In order to avoid cross-talk the longest delay between a sensor and its reference must be smaller than the delay between two sensors. As an alternative, one interferometer and one detector may be used, by switching the delay in the interferometer with an arrangement similar to that shown in Fig. 16.

The network in Fig. 20(a) can replace the network in Fig. 21(a), provided different delays are inserted in one of the buses between the rungs, as shown in Fig. 21(b). Each pulse from the source will produce a series of N pulses with different time intervals. There are restrictions on the choice of delays, in order to avoid cross-talk. For example, if the first and second delays are τ_s and $2\tau_s$, respectively, the third delay cannot be $3\tau_s$, because the signals from the first two

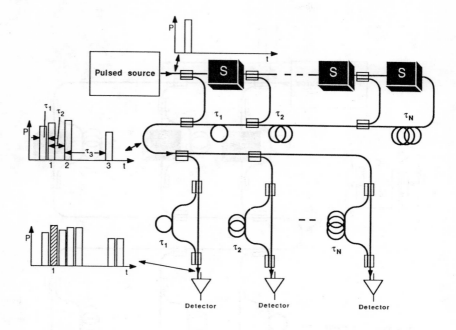

Fig. 21 (b) Same as Fig. 21 (a) except that the restrictions on the choice of delays are different for the two topologies shown in Figs. 21 (a) and 21 (b).

rungs would interfere to produce cross-talk in the third detector. A set of delays which can be used is $1_k \tau_s$, where 1_k is the series 1, 2, 4, 5, 8, 10, 14, 15, 16, 21, 22,...

The networks in Fig. 21 can also be used with frequency division multiplexing, either the real heterodyne or the pseudoheterodyne schemes described above.

4.7 Coherence Multiplexing

Coherence multiplexing is a name given to schemes which utilize the finite coherence length of a continuous wave source to address individual sensors. The basic principle underlying coherence multiplexing is that two waves interfere constructively or destructively, depending on their relative phase, if they are derived from the same source, and are delayed relative to each other by a time interval short compared to the source coherence time. On the other hand, if the delay between the waves is long compared to the coherence time, there is no interference, and information about their relative phase is lost in the detection process. The advantage of coherence multiplexing compared to time division multiplexing is that the former enables yield of one hundred percent duty cycle.

An example of a coherence multiplexed system is obtained by replacing the pulsed source in the system shown in Fig. 17 by a continuous wave source with coherence length short compared to the delay τ_d between sensors [30]. In each detector interference with the reference will occur only for the sensor signal which has a delay matching the delay of the reference. The other sensors will only contribute to the total power on the detector, and thereby to the shot noise. Compared to the pulsed system of Fig. 17 coherence multiplexing has the advantage of providing one hundred percent duty cycle. In fact, the pulsed system may be modified by using a periodic train of pulses, with period small compared to the total delay $N\tau_d$ of the array. This will work provided the pulses are mutually incoherent. In the limit when there is no spacing between the pulses, the source actually emits a continuous wave with coherence time equal to the pulse width. This viewpoint emphasizes the similarity between a pulsed and a coherence multiplexed system.

The switched reference system of Fig. 16 and the internal reference systems of Fig. 21 can also be used for coherence multiplexing simply by replacing the pulsed source by a continuous one with an appropriate coherence time [30]. In the next section we shall see another topology to which coherence multiplexing was first applied [31]. We can also combine coherence multiplexing with frequency division multiplexing by using a continuous, short coherence length source in the system of Fig. 18(b) [23]. The individual sensors can be addressed by frequency demultiplexing rather than the spatial separation common to most coherence multiplexed schemes.

Coherence multiplexing has one main drawback: even if the waves with different delay to not cause a coherent interference, they do interfere to produce amplitude noise from any phase noise present in the lightwave. The conversion from phase to amplitude noise actually approaches a maximum for delay long compared to the coherence time. Since semiconductor lasers have relatively high phase noise, this conversion imposes severe restrictions on the number of sensors which can be multiplexed with good sensitivity [32].

5. ALTERNATIVE TOPOLOGICAL CONFIGURATION

The above discussion of addressing techniques refers to the ladder topology of Fig. 1(a), although for some of the schemes it is advantageous to use the slightly modified version shown in Fig. 1(b). We shall now discuss some other topologies which show strong resemblance with the ladder structure already discussed. We shall also see an example of a topological configuration with entirely different properties.

5.1 Recursive Array

In the ladder configuration of Fig. 1(b), the rungs of the ladder play no other role than transmitting part of the light from the input to the output bus. It might be tempting, therefore, to avoid the rungs altogether, by combining the functions of the input and output couplers in the buses. The result is the topology shown in Fig. 22. An obvious advantage of this system is that the number of couplers is reduced by fifty percent. More important is the improvement in transmitted power. In Eq. (3) through (7), we gave the results of an analysis of the power transmitted through the configuration of Fig. 1(a). The results for the configuration of Fig. 1(b) are obtained by substituting $N+1$ for N. For equal coupling coefficients of $1/N$ we then obtain:

$$P_{N-1} = P_0 (N-1)^{2(N-1)} / N^{2N}. \tag{18}$$

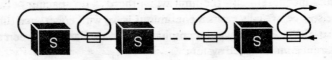

Fig. 22. Recursive network derived from the topology in Fig. 1(b).

For large N, this becomes

$$P_{N-1} \approx P_0 / e^2 N^2. \tag{19}$$

For the recursive array of Fig. 22, we find that the optimum choice of coupling coefficient is $1/(2N-1)$, and the power transmitted through the critical sensor number N-1 is

$$P_{N-1} = P_0 (2N-2)^{2N-2} / (2N-1)^{2N-1}, \tag{20}$$

which approaches

$$P_{N-1} \approx P_0 / e(2N-1) \tag{21}$$

for large N. This is an improvement of approximately $eN/2$. We observe that the power through each sensor is inversely proportional to the number of sensors. This is the desired property which we discussed in Sec. 2.2, and which could otherwise be obtained by using wavelength or mode division multiplexing or a multimode return bus in a time division multiplexed scheme.

The recursive array can be applied to all addressing techniques which we have described for the network of Fig. 1(b), i.e., the differential technique of Fig. 12(b), the double pulse technique of Fig. 20(a), and with different delays it can be applied to the space division multiplexing scheme of Fig. 21(b). As we have mentioned before, the same network can also be used in pseudo-heterodyne and coherence multiplexing schemes.

This topology has drawbacks too. These are associated with the recursive nature of the array. Part of the backward travelling waves is coupled into the forward bus at each coupler. Thus, waves travel back and forth, and give rise to an infinite number of overlapping signals. The amplitude of these spurious signals depends on the coupling coefficients, which must be small enough to keep the cross-talk at an acceptable level.

5.2 Reflective Arrays

Figure 23 shows some configurations which in most respects are equivalent to the ladder structures we have discussed till now. They are all based on reflecting the light back into the input bus. The output light must be separated from the input by means of a directional coupler.

The configuration in Fig. 23(a) is equivalent to that in Fig. 1(a), and all addressing schemes which can be used for the latter can also be used for the former. Similarly, Fig. 23(b) is equivalent to Fig. 1(b), and Fig. 23(c), is equivalent to Fig. 22. The recursive nature of this configuration is due to waves being reflected back and forth within the bus. The changes required to generate equivalents of the modified ladder structures, such as Figs. 18(a), 20(b) or 21(a) and (b), should be evident. We have seen that the recursive array has a higher power efficiency than the other ladder networks. The same

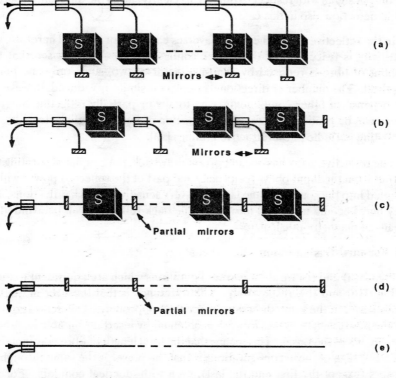

Fig. 23. Reflective networks. (a), (b) and (c) correspond to the ladder networks of Figs. 1(a), 1(b) and 22, respectively, (d) shows an intrinsic sensor version of the network in (c). The distributed sensor in (e) is also a realization of the same network.

applies to the reflective network of Fig. 23(c). For this reason, this network has been studied more extensively than any other network.

Figure 23(d) shows a realization of the configuration in Fig. 23(c) by means of intrinsic phase modulation sensors. These are simply pieces of fibre, connected by partially reflecting mirrors [26]. An even simpler configuration is shown in Fig. 23(e). There, the reflection is provided by internal scattering in the fibre (Rayleigh or Raman scattering). This system is referred to as a distributed sensor rather than multiplexed. Since the distributed sensor will always have a finite spatial resolution, due to the finite pulse width, there is really not much difference between the two systems of Figs. 23(d) and (e). We note that since they are equivalent to the recursive array of Fig. 22, which in turn is almost equivalent to the configuration of Fig. 1(b), the addressing techniques applicable to these systems are those described as differential incoherent detection and two-pulse coherent detection schemes. For a distributed sensor, the status at a specific position is obtained by differentiating the output signal with respect to time. This clearly shows the differential nature of the detection also in this case.

In the reflective arrays the wave traverses each sensor twice. Therefore, the sensitivity is twice that of the ladder configurations. We also see that the amount of fibre is reduced by almost a factor of two, since only one bus is required. The number of directional couplers is similarly reduced, however at the expense of other components, e.g., totally or partially reflecting mirrors. These can be produced by vacuum deposition of multilayer dielectric films or by writing periodic photorefractive gratings [33].

The reflective array has one important drawback, the coupler separating the output from the input obeys reciprocity and part of the reflected power will be coupled into the source. Some lasers are very sensitive to back reflections, and they may become noisy and unstable. The back reflection can be avoided by means of magnetooptic isolators.

5.3 Forward Transmission

In Fig. 24(a) and (b) we show two configurations which are equivalent to those of Fig. 1(a) and (b), respectively. The difference is that the light in the two buses travel in the same direction rather than oppositely. The delay required for the addressing schemes to work is obtained by inserting suitable lengths of fibre in one of the buses. This means that the total length of fibre may increase. An advantage of these configurations is that the power is the same through all sensors (except the first and the last), even with identical couplers. For the configuration of Fig. 24(a) the optimum coupling coefficient is $2/(N+1)$, and the transmitted power is

$$P_k = P_0 \cdot 4 (N-1)^{(N-1)} / (N+1)^{(N+1)}. \qquad k = 2, 3,...N-1 \qquad (22)$$

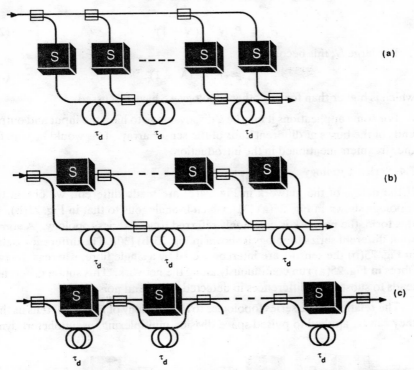

Fig. 24. Forward transmission networks (a), (b) and (c) correspond to the networks in Figs. 1(a), 1(b) and 22, respectively, although the analogy is not as complete as that between reflective and ladder networks.

For large N, this becomes

$$P_k \approx P_0 \cdot 4/e^2 (N+1)^2, \tag{23}$$

which means nearly a factor of four improvement relative to the structure of Fig. 1(a). For the structure of Fig. 24(b), N should be replaced by $N+1$. A slight improvement is possible by choosing the coupling coefficients for the first and the last coupler different from the others.

We derived the recursive array of Fig. 22 from the non-recursive one in Fig. 1 by a logic argument. The same argument can be used to derive the network of Fig. 24(c) from that of Fig. 24(b) [34]. Again, we emphasize that the sensors and the delays may be located in any one of the two buses.

The topologies of Figs. 22 and 24(c) are similar in the sense that light is coupled back and forth between the buses in each coupler, with the possibility of generating numerous cross-talk terms. However, the analogy is not complete, since the network in Fig. 24(c) is not recursive, i.e., its pulse response has finite extent.

The optimum choice of coupling coefficient for this topology is $1/(N+1)$, and the transmitted power is

$$P_k = P_0 N^N / (N+1)^{(N+1)}. \tag{24}$$

For large N, this becomes

$$P_k \approx P_0 / e(N+1), \tag{25}$$

which is higher than for any other topology we have discussed.

For some applications it may be a disadvantage to have the input and output ends of the buses at different ends of the sensor array. This would be true for the streamers mentioned in the introduction.

5.4 Series Topology

If the delays of the network in Fig. 24(c) are made different, we obtain the topology shown in Fig. 25(a) [31], which is analogous to that in Fig. 21(b). In this form, the network is commonly referred to as a series topology. A somewhat different series topology is shown in Fig. 25(b) [30]. The difference is that in Fig. 25(b) the sensors are interconnected by a single fibre, whereas the two fibres in Fig. 25(a) run continuously along the network. This subtle difference leads to substantial differences in detected signal and noise [35].

The relation of the series topologies to the topology of Fig. 21(b) tells us that they can be applied to pulsed space division multiplexing, pseudoheterodyne,

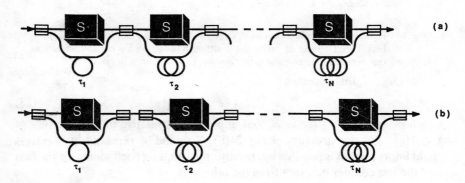

Fig. 25. Series networks. (a) is a modified version of Fig. 24(c), but it is used here in a different mode, (b) is not equivalent to any other network we have discussed.

and coherence multiplexing. Note that in these applications, the networks will be quite different from the network of Fig. 24(c) as it was described in the section above. There, small coupling coefficients of the order of $1/N$ were assumed to optimize power transmission and minimize cross-talk. When the series topologies are applied as described here, the coupling coefficients may be of the order of 0.5*. With this design, cross-talk becomes a serious problem,

---

* For the topology of Fig. 25(b) it is fairly obvious that this is the optimum choice. For the topology of Fig. 25(a) the situation is more complex [35].

and the delays must be carefully chosen in order to eliminate it [36, 37]. It turns out that the delay increases exponentially with the number of sensors. Therefore, these topologies have severe limitations. The situation becomes more acceptable if some small cross-talk can be tolerated [36].

6. BIREFRINGENT AND TWO-MODE FIBRES

Polarimetric sensors have been studied extensively, and recently there has also been some interest in using two-mode fibres for sensor applications [38]. Multiplexing systems based on any of these kinds of sensors fit snugly into our description and classification of networks. Consider the Mach-Zehnder interferometer of Fig. 2. The essential action of this device is to split the light into two separate paths and recombine the two after they have experienced a relative phase change. If we let the two paths be the two polarization modes of a birefringent fibre, the interferometer becomes a polarimeter. The directional couplers are replaced by polarization couplers, which may be simply a splice between two fibres with their principal axes rotated relative to each other. Or even simpler, coupling may be obtained by squeezing the fibre [39].

Similarly, the two transverse (spatial) modes of a two-mode fibre can replace the two arms of a Mach-Zehnder interferometer. Suitable couplers for these modes are made by periodically bending the fibre [40].

The two polarizations of a birefringent fibre and the two spatial modes of a two-mode fibre propagate with slightly different delays. Therefore, interferometers with different pathlengths can be realized in these systems. With this in mind, it is relatively straight forward to see how we can realize analogs of the networks shown in Figs. 12(c), 20(b), 21(a), 24(c), 25(a), and 25(b).

As an example, a sensor system based on distributed [41] or lumped [42] coupling from one mode to the other along a two-mode fibre belongs to the class of topologies shown in Fig. 24(c). The delay difference between the two modes is used to address the individual sensors.

7. CONCLUSION

We have discussed two important aspect of fibre optic sensors: network topologies and addressing techniques. Emphasis has been on a systematic classification. We discussed addressing techniques with reference to two closely related topologies. Through our systematic approach we were able to find analogies to several other topologies, and thereby see which addressing techniques could be applied to which topologies. Also, we were able to fit polarimeteric and two-mode sensors into the same system.

Our exposition is not complete. Some published systems do not fit into our classification. Examples are networks with star configuration, two dimensional arrays [43], and a special recursive ladder network [44].

We have only given rudimentary analyses of power budgets and discussed qualitatively such aspects as noise and cross-talk. Although some studies of these and other important aspects of fibre optic sensor multiplexing are reported in the literature, a lot remains to be done in order to assess the multitude of networks and systems which have been suggested. A careful analysis of sensitivity, noise, cross-talk, stability, linearity, dynamic range, etc., accounting for nonideal components, would have to be undertaken in order to draw definite conclusions about the features of the various multiplexing schemes.

ACKNOWLEDGEMENTS

The author was introduced to the subject of fibre optic sensor multiplexing as a visiting scholar at Stanford University. He is indebted to H. John Shaw and his students, Janet Brooks, Byoung Yoon Kim, and Robert Wentworth for stimulating discussions and valuable contributions. He also acknowledges the cooperation with his colleague Helge Engan and their students Jan Ove Askautrud and Kjell Kråkenes. This work was sponsored by the Royal Norwegian Council for Scientific and Industrial Research (NTNF, Oslo).

REFERENCES

1. D.E.N. Davies, "Signal Processing for Distributed Optical Fibre Sensors", Optical Fibre Sensors, Second International Conference, pp. 285-295 (VDE-Verlag GmbH 1984).
2. D.E.N. Davies, "Optical Fibre Distributed Sensors and Sensor Networks", Fibre Optic Sensors, Proc. SPIE 586, pp. 52-57 (1985).
3. A.J. Rogers, "Distributed Optical-Fibre Sensors", Optical Fibre Sensors, NATO ASI Series E-No. 132, pp. 143-163 (1986).
4. B. Culshaw, "Distributed and Multiplexed Fibre Optic Sensor System", Optical Fibre Sensors, NATO ASI Series E-No. 132, pp. 165-184 (1986).
5. R. Kist, "Fibre-Optic Sensors for Networks", Optical Fibre Sensors, Fourth International Conference, pp. 128-129, (The Institute of Electronics and Communication Engineers of Japan 1986).
6. S.A. Kingsley, "Advances in Distributed FODAR (Fibre Optic Detection and Ranging)", Fibre Optic and Laser Sensors IV, Proc. SPIE 718, pp. 66-77 (1986).
7. R.E. Wagoner, Clark T.E., "Overview of Multiplexing Techniques for All-Fibre Interferometer Sensor Arrays", Fibre Optic and Laser Sensors IV, Proc. SPIE 718, pp. 80-91 (1986).
8. A.J. Rogers, "Distributed Sensors: a Review", Fibre Optic Sensors II, Proc. SPIE 798 pp. 26-35 (1987).
9. A.D. Kersey, Dandridge A., Tveten A.B., "Overview of Multiplexing Techniques for Interferometric Fibre Sensors", Fibre Optic and Laser Sensors V, Proc. SPIE 838, pp. 184-193 (1987).
10. A.D. Kersey, Dandridge A., "Distributed and Multiplexed Fibre Optic Sensors", Optical Fibre Sensors, 1988 Technical Digest Series, Vol. 2, pp. 60-71 (Optical Society of America, Washington D.C. 1988).

11. A.R. Nelson, McMahon D.H., Gravel R.L., "Passive Multiplexing System for Fibre-Optic Sensors", *Appl. Opt.*, **16**, 2917-2920 (1980).
12. J. Mlodzianowski, Uttamchandani D., Culshaw B., "A Frequency Domain Approach to the Multiplexing of Intensity Based Optical Fibre Sensors", *J. Opt. Sensors*, **2**, 17-23 (1987).
13. J.K. A. Everard, "Novel Signal Processing Techniques for Enhanced OTDR Sensors", Fibre Optic Sensors II, Proc. SPIE 798, pp. 42-46 (1987).
14. J.-J. Bernard, Depresles E., "Correlation-Optical-Time-Domain-Reflectometry for High Resolution Distributed Fibre Optic Sensing", Fibre Optic and Laser Sensors V, Proc. SPIE 838, pp. 206-209 (1987).
15. A.R. Nelson, McMahon D.H., "Passive Multiplexing Techniques for Fibre Optic Sensor Systems", Int. Fibre Optical Communications J., Vol. 2, pp. 27-30 (1981).
16. A.R. Nelson, McMahon D.H., van de Vaart H., "Multiplexing System for Fibre Optic Sensors Using Pulse Compression Techniques", *Electron. Lett.*, **17**, 263-264 (1981).
17. G. Winzer, "Wavelength Multiplexing Components—A Review of Single-Mode Devices and their Applications", *J. Lightwave Technol.*, **2**, 369-378 (1984).
18. H. Ishio, Minowa J., Nosu K., "Review and Status of Wavelength-Division Multiplexing Technology and its Application", *J. Lightwave Technol.*, **2**, 448-463 (1984).
19. M. Corke, Jones J.D.C., Kersey A.D., Jackson D.A., "Dual Fabry-Perot Interferometers Implemented in Parallel on a Single Monomode Optical Fibre", Optical Fibre Sensors, Third International Conference, pp. 128-129 (OSA/IEEE 1985).
20. K. Bløtekjær, US Patent No. 4,750,795.
21. W.P. Risk, Kino G.S., Shaw H.J., "Fibre Optic Frequency Shifter Using a Surface Acoustic Wave Incident at an Oblique Angle", *Opt. Lett.*, **11**, 115-117 (1986).
22. B.Y. Kim, Blake J.N., Engan H.E., Shaw H.J., "All-Fibre Acousto-Optic Frequency Shifter", *Opt. Lett.*, **11**, 389-391 (1986).
23. D.-T. Jong, Hotate K., "Frequency Division Multiplication of Optical Fibre Sensors Using an Optical Delay Line with a Frequency Shifter", Optical Fibre Sensors, 1988 Technical Digest Series, Vol. 2, pp. 76-79 (Optical Society of America, Washington D.C. 1988).
24. D.A. Jackson, Kersey A.D., Corke M., Jones J.D.C., "Pseudo-heterodyne Detection Scheme for Optical Interferometers", *Electron. Lett.* **18**, 1081 (1982).
25. I.P. Giles, Uttam D., Culshaw B., Davies D.E.N., "Coherent Optical-Fibre Sensors with Modulated Laser Sources", *Electron. Lett.*, **19**, 14 (1983).
26. J.P. Dakin, Wade C.A., "Novel Optic Fibre Hydrophone Array Using a Single Laser Source and Detector", *Electron. Lett.*, **20**, 53-54 (1984).
27. E.L. Green, Holmberg G.E., Gremillon J.C., Allard F.C., "Remote Passive Phase Sensor", Optical Fibre Sensors, Third International Conference, pp. 130-131 (OSA/IEEE 1985).
28. J.L. Brooks, Tur M., Kim B.Y., Fesler K.A., Shaw H.L., "Fibre Optic Interferometric Sensor Arrays with Freedom from Source Phase-Induced Noise", *Opt. Lett.*, **11**, 473-475 (1986).
29. J.L. Brooks, Moslehi B., Kim B.Y., Shaw H.J., "Time-Domain Addressing of Remote Fibre-Optic Interferometric Sensor Arrays", *J. Lightwave Technol.*, **5**, 1014-1023 (1987).
30. J.L. Brooks, Wentworth R.H., Youngquist R.C., Tur M., Kim B.Y., Shaw H.J., "Coherence Multiplexing of Fibre Optic Interferometric Sensors", *J. Lightwave Technol.*, **3**, 1062-1072 (1985).

31. S.A. Al-Chalabi, Culshaw B., Davies D.E.N., "Partially Coherent Sources in Interferometric Sensors", Optical Fibre Sensors, First International Conference, pp. 132-135 (IEE 1983).
32. R.H. Wentworth, Shaw H.J., "Expected Noise Levels for Interferometric Sensors Multiplexed Using Partially Coherent Light", Fibre Optic and Laser Sensors III, Proc. SPIE 566, pp. 212-217 (1985).
33. W.W. Morey, Meltz G., Glenn W.H., "Fibre Optic Bragg Grating Sensors", Proc. SPIE 1169, 98-107 (1987).
34. A.D. Kersey, Dandridge A., "Tapped Serial Interferometric Fibre Sensor Array with Time Division Multiplexing", Optical Fibre Sensors, 1988 Technical Digest Series, Vol. 2, pp. 80-83 (Optical Society of America, Washington D.C. 1988).
35. R.H. Wentworth, "Theoretical Noise Performance of Coherence-Multiplexed Interferometric Sensors", *J. Lightwave Technol.*, 7, 941-956 (1986).
36. K. Bløtekjær, Wentworth R.H., Shaw H.J., "Choosing Relative Optical Path Delays in Series-Topology Interferometric Sensor Arrays", *J. Lightwave Technol.*, 5, 229-235 (1987).
37. I. Sakai, Parry G., Youngquist R.C., "Multiplexing Fibre-Optic Sensors by Frequency Modulation: Cross-term Considerations", *Opt. Lett.*, 11, 183-186 (1986).
38. B.Y. Kim, "Few-Mode Fibre Devices", Optical Fibre Sensors, 1988 Technical Digest Series, Vol. 2, pp. 146-149 (Optical Society of America, Washington D.C. 1988).
39. R.C. Youngquist, "Brooks J.L., Risk W.P., Kino G.S., Shaw H.J., "All-Fibre Components Using Periodic Coupling", IEE Proceedings, Vol. 132, pp. 277-286 (1985).
40. J.N. Blake, Kim B.Y., Shaw H.J., "Fibre Optic Modal Coupler Using Periodic Microbending", *Opt. Lett.*, 11, 177-179 (1986).
41. K. Kurozawa, Hattori S., "Distributed Fibre-Optic Sensor Using Forward Travelling Light in Polarization Maintaining Fibre", Fibre Optic Sensors II, Proc. SPIE 798, pp. 36-41 (1987).
42. G. Kotrotsios, Denervaud P., Falco L., Parriaux O., "High Dynamic Dual Mode Fibre Transitometry", Optical Sensing and Metrology, International Congress. Hamburg, Germany (1988).
43. A. Dandridge, Tveten A.B., Kersey A.D., Yurek A.M., "Multiplexing of Interferomtric Sensors Using Phase Carrier Techniques", *J. Lightwave Technol.*, 5, 947-952 (1987).
44. W.B. Spillman, Jr., Lord J.R., "Self-Referencing Multiplexing Technique for Fibre-Optic Intensity Sensors", *J. Lightwave Technol.*, 5, 865-869 (1987).

29

Optical Fibres for Power Systems

A.J. ROGERS[*]

1. INTRODUCTION

The subject known as optoelectronics effectively dates from the invention of the laser in 1960. Since then the subject has grown steadily to the point where it is now making its presence felt in most areas of industrial activity, and especially in the communications and measurement functions.

The electricity supply industry (ESI) is no exception to these trends. Optical fibre communications links are already being installed by the UK's Electricity Utilities, and they are under serious consideration by many other electrical power utilities throughout the world.

The organisation, control and maintenance of any electricity supply system present a set of complex measurement problems.

The stable and efficient operation of any electricity supply network requires a variety of reliable communication systems. Data concerning the status of localised units such as generating stations and switching sub-stations must be available to the local operators; data also must be available to the regional and national centres so that an overall system control may be imposed. Such control is necessary both to optimise the network within the constraints acting at any one time and to mitigate the effects of any local plant failure. The communication systems requirements range from the relatively low bandwidth (but highly reliable) telemetry needed to operate circuit breakers in case of major faults, through normal administrative telephone traffic, up to high bandwidth computer input and output data streams. As the centralised control

---

[*] Department of Electronic and Electrical Engineering, King's College London, Strand, London WC2R 2LS, U.K.

of supply systems becomes more and more sophisticated, more reliable automatic acquisition of data from the individual units is needed. This leads to increasing demands on the integrity and bandwidth of the communication systems.

Optical fibres, besides their enormous bandwidth capabilities, offer two singular advantages for communications in the Electricity Supply Industry: they are good electrical insulators and they are immune from electromagnetic interference. With the Central Electricity Generating Board's main transmission voltage presently at 400 kV, and with currents of a few kilo-Amps to be controlled and switched, these advantages obviously make optical fibre communications very attractive for the industry. The variety of measurement functions involved spans almost the entire range of physical parameters. The primary measurement requirement is to provide a continually updated statement of the condition of the power-supply system, and of the demands being made on it, so that it may be configured to operate with maximum efficiency (within its operating constraints) at any given time.

Another important measurement function is that of system protection. Measurements must be made which give rapid indication of any anomalous condition, so as to allow effective (usually automatic), protective action to be taken. For example, a 400 kV overhead transmission line which has been brought down by lightning must not be permitted to damage transformers or generators by passing a 'short-circuit' current into the earth for any length of time; the faulty line must be electrically isolated within a few milliseconds.

Finally, measurement is necessary for a number of diagnostic and 'metering' tasks. The former relates to those measurements which must be made in order to understand and solve special, unforeseen problems which arise during the normal course of operations (e.g., rusty bolts in nuclear reactors, unstable cooling towers, etc). The latter relates to the need for the correct billing of customers.

Power-system operations may be divided into two broad primary functions: generation and transmission. The former is concerned, basically, with the sources of power—the power stations—and measurements required at these sites extend over a large range of physical quantities: the current and the voltage on generators and transformers; the temperature of boilers and condensers; flow rates in cooling systems; strain, displacement, pressure and vibration on all critical vessels and rotating machinery; and, for nuclear stations, radiation levels in many types of environments, personnel-accessible and otherwise.

The transmission function, on the other hand, is concerned with the countrywide distribution of the power emerging from the power stations. This is arranged via a large number of switching substations, which allow active network control. At these substations, we shall again require to measure the current and the voltage being fed on to the various lines, the temperature in the transformers, the displacement in the line isolators (for it is in the substation that line protection is effected), etc. It is clear that the range and types of measurement functions employed will require that a broad range of measurement technology be brought to bear on them.

But why use optical techniques? The primary advantage is that virtually all physically measurable quantities are accessible to optical techniques — the propagation of light in material media is sensitive to a complete range of external physical fields.

Secondly, the use of dielectric, insulating materials and measurement media implies easy high-voltage insulation, immunity from electromagnetic interference (which can be a serious problem in the presence of large currents and voltages) and resistance to environmental attack (e.g., rust). Other advantages worthy of mention are the prospects for measurement devices which are electrically passive, and which thus require no power at inaccessible points (e.g., high voltage), which have very large measurement bandwidths and easily adjustable sensitivity (by varying the optical path length), which do not suffer from hysteresis or saturation effects, which have no moving parts to cause maintenance problems, and which interface easily to optical communication links, for purposes of telemetering the information to convenient indication points. The devices can also be significantly cheaper than many of the conventional equipments for performing power-system measurements, sometimes by an order of magnitude.

Of course, there are also disadvantages. The fact that optical propagation is readily influenced by many external fields means that it is sometimes difficult to separate wanted from unwanted effects. For this and other reasons, the signal processing is often complex. Difficulties which are more specific to the ESI are those of interfacing optical communications and measurement devices to a pre-existing system which has designed without any reference to them, and the fact that their installation and operation are foreign to the established practices and expertise of the traditional power engineer.

Nevertheless, the advantages of using optical techniques in both communications and measurement functions within the ESI are compelling; and it is to optical fibres that we turn for most of the new solutions.

2. ELEMENTS OF PROPAGATION IN OPTICAL FIBRES

An optical fibre is a thin wire of glass. Its overall thickness is about 100 μm, almost the same as that of a human hair.

A typical fibre structure is shown in Fig. 1. Here we see a central core, normally of glass or silica, surrounded by a cladding, also of glass or silica, but which has a slightly lower refractive index (about one per cent less). Consider a ray of light travelling within the core. When it strikes the core/cladding boundary, it will normally pass into the cladding, travelling then at a slightly reduced angle to the boundary, in accordance with Snell's well-known law of refraction. But if the incident angle is very shallow with respect to the boundary, the ray may suffer 'total internal reflection' at the boundary; the boundary now acts like a perfect mirror. Under this condition, the ray will bounce from side to side along the fibre until it emerges at the far end. The light is thus trapped in the fibre core, which acts as an optical waveguide.

However, not all reflection angles are possible for a ray to become guided along the length of the fibre, even when the 'total internal reflection' condition is satisfied; for light is a wave motion and the waves represented by the various rays will interfere with each other to cancel or reinforce, depending on the relative phases of the waves. The result is that a discrete reflection angle is allowed only if it satisfies the condition shown in Fig. 1. The condition requires

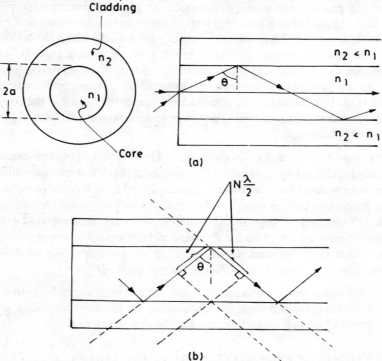

Fig. 1. A typical optic fibre structure consists of a central core of glass or silica, surrounded by a cladding of slightly lower refractive index (n). If $\sin \theta > n_2/n_1$ the ray is totally internal reflected at the core/cladding boundary so that the ray bounces down the core and is trapped within it. The light is thus 'guided' by the core. Only if the path differences $N \lambda/2$ shown correspond to an integral number of half-wavelengths of the light will the angle lead to a self-reproducing wave interference pattern and thus to a stable 'propagation mode' of the waveguide.

that, for two parallel rays striking the boundary at opposite ends of a diameter, the difference in effective path length to the boundary points is a whole number of half-wavelengths of the light in the core. The difference in phase is thus a 'whole number times π' radians. For each of these allowable angles, an interference pattern is formed which is stable and self-reproducing along the fibre. These allowable patterns are the propagation modes of the fibre, and it is clear that the number of these for a given fibre will depend on the geometry of the

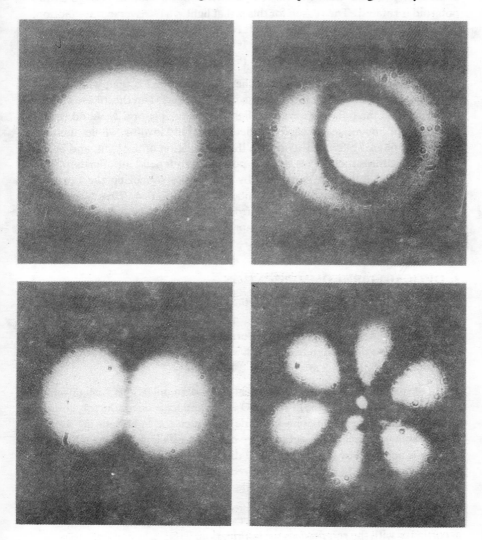

Fig. 2. Light-intensity distributions (or interference patterns) over the cross-section of an optical fibre waveguide for various propagation modes, showing the effect of changing the incident angle of the light.

fibre and the wavelength of the light. A given interference pattern (or mode) will be characterised by a certain distribution of light intensity over the cross-section of the fibre. Some typical patterns for a variety of modes are shown in Fig. 2.

The modes possess another interesting and very valuable property, they show polarization effects. If the light emerging from the fibre is passed through a sheet of polaroid, then the viewed patterns will be seen to change as the polaroid is rotated. The reason for this is that light is a transverse wave motion. The electric and magnetic fields which constitute a light wave vibrate at right angles to the direction of propagation. If the vibrations of each of the fields are confined to one plane then we say that the light is linearly polarized.

Now the really important point in the present context is that, when a light wave is reflected at a boundary, a phase change occurs, and this phase change depends upon the polarization of the light. Thus the previously stated condition that, for any mode, successive reflections should involve 'whole numbers times π' of phase shift will also involve the polarization of the light, and hence the mode patterns themselves become polarization-dependent. Furthermore, if we alter the reflection condition, by bending or squeezing the fibre for example, then the polarization of the mode will change also. Thus we see that the polarization of the light emerging from the fibre may be quite sensitive to conditions external to the fibre. Perhaps this means that we can use fibres to tell us about those external conditions?

3. OPTICAL FIBRE COMMUNICATION

It is fairly clear that the guiding properties of an optical fibre provide us with a convenient method for communicating information between any two points. Provided that we can impress the information on a beam of light, we may connect the two points in question with a length of optical fibre, and pass the light along the fibre to form a communications link.

A convenient source of light for optical fibre communications is a light-emitting diode (LED) (cf. Fig. 3). An LED is a compact semiconductor, source, (typically of gallium aluminium arsenide) in which the light-emitting area is comparable with that of the optical fibre core. An efficient light launch arrangement is achieved by butting the LED directly on to the fibre end. The light output from the LED results from the recombination of positive and negative electric charge carriers in the body of the semiconductor and this is achieved by passing electric current into the device. It is thus particularly easy to impress information on this light output, it is necessary only to vary the input current in accordance with the required signal information.

The question which now naturally arises is, over what distance can effective communication be established using an optical fibre? How long can the fibre

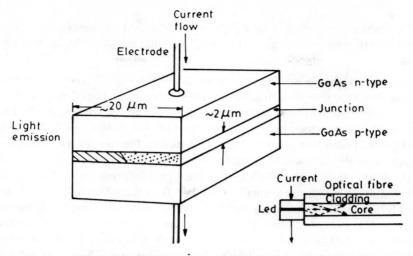

Fig. 3. In a light-emitting diode, the electric current flow causes electrons and holes to recombine in the semiconductor junction regions. The energy of recombination is released as light, whose level can be controlled by the current flow. The light-emitting area has similar dimensions to those of the fibre core, and is thus a convenient source for optical fibre propagation.

be? The first requirement is that there must be sufficient light power at the output end to reconstruct the signal information accurately. This means that the fibre should not attenuate the optical radiation too much. Figure 4 shows the attenuation of a typical optical fibre plotted against wavelength. The initial steady fall from low to higher wavelengths is due to the scattering of the optical beam by small imperfections and impurities in the glass structure. This is known as Rayleigh scattering, and it varies inversely as the fourth power of the optical wavelength. This effect causes the light to be scattered out of the core and thus to be lost. The peaks which occur at the longer wavelengths are due to absorption of the radiation by impurity molecules in the silica, an absorption which causes loss via a slight heating of the fibre. It is clear that good

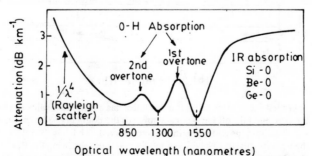

Fig. 4. Rayleigh scattering causes the light attenuation within a fibre to fall steadily as the wavelength increases, until various molecular resonances give rise to absorptions which increase the attenuation again. The best wavelengths for optical fibre propagation will be where the attenuation is lowest, in one of the troughs.

wavelengths to use will be 1300 nanometers and 1550 nanometers, and these are indeed the favoured values for present generation optical fibre communications.

Low attenuation is not the only requirement for a good communications system. We have seen that, in general, a fibre will support a number of modes. A little examination will reveal that each mode will travel at a different velocity. This is easily understood by considering again the ray picture of Fig. 1. Each mode corresponds to a given ray angle and, clearly, the steeper the angle, the greater is the distance which the mode must travel in order to reach the far end of the fibre. If the light which carries the signal information is distributed amongst many modes (as normally it will be), then the information will quite quickly become distorted by the fact that various parts of it travel at different speeds. We cannot tolerate too much of this distortion before the signal becomes unintelligible, and the allowable length of the fibre is limited for a given information rate, or bandwidth. This phenomenon is known as 'modal dispersion' and may be alleviated in two ways.

Consider first the fibre structure shown in Fig. 5(a). Here we have a core with a refractive index varying from a large value at the centre to a smaller one at the core/cladding interface. The result of this is that rays with steeper angles spend more time in regions of low refractive index and hence travel faster than they would in a guide with a uniform index. This speeds up the slower modes and hence reduces the modal dispersion. As already discussed such a fibre is

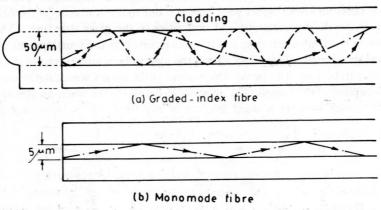

Fig. 5. In graded-index fibre, the refractive index has a maximum value at the centre of the core and falls off parabolically towards the boundary. The core now acts like a continuous lens and ray paths become curved. This has the effect of reducing the velocity differences between modes and thus increasing the fibre bandwidth. For the monomode fibre geometry, only one mode can propagate. There is now no modal dispersion and the bandwidth is very high. However, the core diameter is now very small and it is difficult to launch light into it. It is also difficult to join two fibres together. One mode also means only one polarization state at each point along the fibre.

known as 'graded-index' fibre and this type of fibre is used in the majority of multimode optical fibre communications links now being installed.

Consider now Fig. 5(b). In this case, we have tailored the fibre geometry so that as if only one ray angle is possible. Only one mode can propagate. Obviously there can now be no modal dispersion. This structure is known as monomode fibre and it will clearly allow a very large information bandwidth. It is used for the very high capacity communications links which are now being designed for trunk-telephone and submarine-cable usage.

The monomode fibre possesses another important feature, since there is only one mode there is only one polarization state. We shall return to this point later.

We have considered several ideas: let us now consider some numbers. Attenuations in fibres at 1300 nanometres are now down to only 0.4 decibels per kilometre, and at 1550 nanometre to 0.2 decibels per kilometre. These figures are so low that the distances between repeater stations are presently limited by bandwidth requirements rather than by those of signal power. For trunk intercity communications, a bandwidth of one gigahertz can be achieved with a repeater spacing of 30 kilometres. For sub-oceanic systems, repeater spacings must be larger, around 300 kilometres, but even here bandwidths of 100 megahertz and more are possible.

The available bandwidths are, therefore, enormous and the distances between repeater stations significantly greater than for copper cables. It is these features which make optical fibre communications so attractive.

4. OPTICAL FIBRE COMMUNICATION SYSTEMS IN THE ESI

Figure 6 illustrates a particular arrangement which uses both the insulating and the interference – immune properties of optical fibres in communicating information to solve a particularly difficult measurement problem which typifies one kind of optical fibre communications problem in the ESI.

In order to study the insulating properties of SF_6 gas (an insulating medium widely used in power system plants), it was necessary to record the voltage waveform across a test gap immersed in the gas as the voltage was raised to a level sufficient to cause electrical breakdown in the gap. The voltage necessary was of the order 1 MV, so that, when breakdown occurred, the electrical noise was exceptionally severe, causing large levels of interference on the measuring equipment. This problem was solved by making the measurement within a screened box and then communicating the information to a suitably distant recording point via an optical fibre telemetry link. The recording point was sufficiently distant (~250 m) for the interference levels, and for the step-change of earth potential resulting from breakdown, to be negligible.

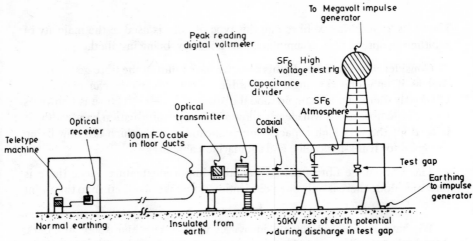

Fig. 6. Schematic of fibre optic link for transmission of data from C.E.R.L. SF$_6$ megavolt test rig to control room.

Electrical interference also can be both severe and difficult to predict in power stations and in switching sub-stations. An interesting case in point is that of the pumped-storage power station operating at Dinorwic in North Wales [1]. Pumped-storage stations enable water to be pumped from a lower to an upper dam when demand for electricity is low (e.g., at night), so that the water is available at short notice (~ 10 s) to drive turbines which provide for sudden large demands. The Dinorwic turbine hall was built inside a cavern within a mountain consisting largely of slate. Slate has anisotropic electrical properties and the flow of earth currents, at times of major electrical faults, is difficult to predict; this leads to a corresponding unpredictability in interference levels for telecommunications.

It was decided, therefore, to install an optical fibre telecommunications test link, to run from the upper dam down to a point close to the lower dam, to demonstrate the advantages of such a link facing this type of difficulty [1]. The optical fibre cable was supplied by the Pirelli General Company.

The total length of the link was 5.3 km. The cable was installed in five sections, the individual fibres being fusion-spliced at the four joint bays. The cable comprised four nylon-coated fibres loosely-laid in a paper-coated steel-laminated tube. The tube was encased in a polyethylene sheath to give an overall diameter of 10 mm. The steel was necessary to provide the strength during duct pulling and was not cross-connected at the joint bays, in order to provide longitudinal electrical is¹ation.

A closed circuit TV system was operated over the link using two of the fibres, one for ppm video transmission, the other (also ppm) for camera control. The transmitters used lasers operating at 850 nm wavelength; the receivers used avalanche photodiodes (APD). The camera was placed at the top dam level with the monitor and the control console at the lower level. The other two fibres were looped at the uppermost joint bay and were used to transmit digital data at 34 Mb/s, with automatic monitoring of the error rate. The terminal equipment for the digital monitoring was supplied by Telettra SpA and it operated over 6.6 km of cable with an 850 nm LED transmitter and again, an APD receiver. The bit-error rate over a period of 12 months remained lower than 10^{-10}. The mean attenuation of the fibres remained substantially constant over this period at 3.5 dB km^{-1} (at 850 nm), and the overall performance of the link was entirely satisfactory.

For central control of the CEGB's distribution and transmission system, a national telecommunications ring network is desirable. A support structure for a wholly CEGB-owned network already exists in its high-voltage transmission towers. Figure 7 shows how this structure might be used for national telecommunications. An optical fibre cable is incorporated within a specially designed

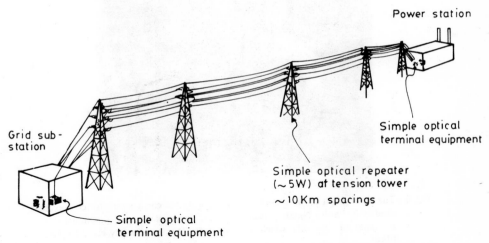

Fig. 7. Optical fibres incorporated within the earth-conductor of the overhead-line transmission system can carry enormous quantities of information between terminals situated in control centres, power stations or substations. If necessary, the signals can be boosted by repeaters routed on conveniently located transmission towers.

earth conductor, at the top of the structure. A conventional earth conductor normally occupies this position and acts to provide a lightning shield for the phase conductors. It also carries fault currents under certain conditions. The optical fibre earth conductor continues to perform these functions and, in addition, allows interference-free high-bandwidth communications.

A section of the new earth conductor is shown in Fig. 8. The optical fibre cable is enclosed at the centre of the conductor and consists of six individual fibres, each lying loosely within a polymer tube [2]. The six tubes are laid

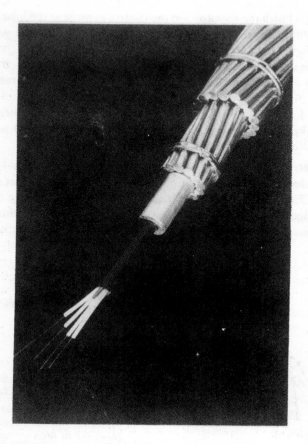

Fig. 8. The photograph shows a section of the optical fibre earth conductor composite used in the Fawley-Nursling installation. Two layers of stranded aluminium alloy surround an alloy tube which loosely houses four protectively coated optical fibres.

helically about a central polymeric strength member to allow the expected 0.5 per cent strain to be taken up by radial movement of the fibres, rather than by a tightening of the fibres within their tube. This latter effect would cause transverse stress which, in turn, would lead to large increases in attenuation as the result of the phenomenon known as 'microbending': small sharp bends are formed in the fibre, and these cause loss due to the violation of the total-internal-reflection condition.

A 24 kilometre length of the optical fibre earth conductor was installed for evaluation on the 400 kilovolt Fawley to Nursling transmission line in October 1981. A communications link was established at 34 megahertz and a close watch was kept for any variations in the basic propagation properties of the fibres [3]. This system led to the identification of a serious fibre attenuation problem due to the diffusion and chemical reaction of atomic hydrogen (from water ingress) within the fibre material.

5. MEASUREMENT APPLICATIONS

We have already noted that the propagation modes in fibres possess polarization properties, and that these properties can be influenced by external conditions. We also hinted that it might thus be possible to use fibres as indications of those external conditions, that is, to use them as measurement sensors.

It is well worth considering whether optical-fibres can help us to make measurements, for this would have many advantages for the Electricity Supply Industry.

The measurements would be made with a medium which is a dielectric rather than a metal, and a good insulator; it carries no electrical current and is, therefore, intrinsically safe. It is easily configured, installed, and maintained, and will interface easily with interference-free optical fibre telemetry links. It is made from a cheap material; silica is simply a pure form of sand. Let us begin looking at these possibilities by considering more carefully the polarization properties of optical fibres.

We have already noted that the propagation modes in fibres possess polarization properties. These properties depend on the relationship between some preferred direction in the fibre and the light-wave's electric and magnetic field directions. In the matter of defining the polarization properties of any given mode, the preferred directions in the fibre are those lying in the plane which is tangential to the core/cladding boundary at the point where reflection takes place. If we impose other forms of directionality on the optical fibre we might expect these to alter the polarization of the propagating light, and this is indeed the case. For example, if we bend the fibre in a particular plane then the fibre geometry is altered more in that plane than in any other and the polarization of the propagating modes is altered as a result of that induced directionality.

We may also impose directionality by subjecting the fibre to external electric and magnetic fields, since we know that each of these acts in a particular direction. Consider the situation shown in Fig. 9. Here we have a beam of linearly polarized light propagating in a block of glass which is under the influence of a magnetic field. The field direction is the same as that of the light propagation. Now we know that light travels in glass more slowly than it does in a vacuum and that this is because it interacts with the glass material—in fact with the atomic electrons in the glass. The electrons rotate in orbits about the atomic nucleus and, when a magnetic field is present, they can rotate more easily in one sense about the field direction than in the other. This is because a rotating electron also creates a magnetic field and the total energy is less

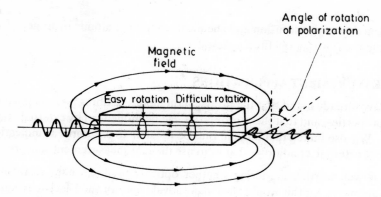

Fig. 9. Electrons rotate more easily in the clockwise direction about a magnetic field than in the anticlockwise direction. This asymmetry causes clockwise rotation of the polarization direction of the linearly polarized light wave, through its interaction with the electrons as it passes through the medium. This provides a means for imposing a preferred polarization direction on an optical fibre.

when that field is in opposition to the imposed one. In the case shown, it will be easier for the electron to rotate anticlockwise than to rotate clockwise, as viewed by an observer looking in the direction of light propagation. If the light is initially linearly polarized, then, as it interacts with the electrons during its passage through the glass, it will come under the influence of the anti-clockwise bias which has been imposed by the magnetic field. Its polarization direction will thus be rotated in the anti-clockwise sense. This effect is known as the Faraday magneto-optic effect after the man who discovered it in 1845, and it can be exploited in optical fibres to measure magnetic fields.

Since magnetic fields are produced by electric currents, it follows that fibres can also be used to measure the currents. Figure 10 shows how this can be

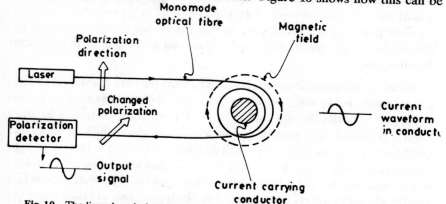

Fig. 10. The linearly polarized light propagating in the monomode fibre has its polarization direction rotated by the action of the magnetic field (due to the current flowing in the conductor) as the fibre loops around it. The total rotation which occurs throughout the fibre length is a measure of the current in the conductor.

done. The circular magnetic field surrounding a current acts always along the axis of a fibre looped around a conductor, so that linearly polarized light propagating in the fibre will have its direction continuously rotated, by an amount proportional to the electric current.

How can we ensure that the light in the fibres is always linearly polarized? We have already seen that there is only one way to ensure that we have only one polarization state—we must have only one mode. Monomode fibre must be used. A monomode fibre current-measurement device is shown in Fig. 11. It

Fig. 11. A prototype optical fibre current sensor with a fibre loop of diameter one metre. This was used to measure current on 22 kV main connection bus-bar at Fawley power station.

was constructed to measure the current in a 22 kilovolt bar carrying power from a 500 megawatt generator, and was capable of measuring currents from 10 Ampères to 15 kiloAmpères with a response time of only 0.2 microseconds [4]. The laser and the detector are to be seen in the box below the fibre loop, which is one metre in diameter. A noteworthy feature of this type of arrangement is that the rotation is proportional to the line integral of the magnetic field along the fibre path, and is thus (in principle, at least) directly proportional to the current (from Ampère's theorem) independently of the shape or the size of the loop (provided only that it encloses the current.)

This idea has been implemented in a number of other practical devices, one of which is illustrated in Fig. 12. Here the fibre loop encloses the base of a high-voltage transmission tower, and it is able to measure the current flowing to earth when fault current flows between a phase conductor and the tower. This arrangement was used successfully by the CEGB during planned fault-

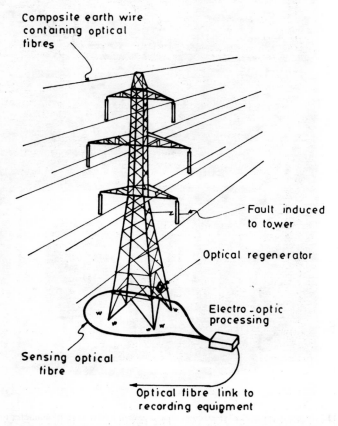

Fig. 12. Optical-fibre measurement of transmission-tower earth current.

throwing tests in September 1985, and provides a simple solution to what would otherwise constitute an expensive measurement operation [5].

Can we measure voltage in a similar way? Yes, we can, but it is more difficult. The extra difficulty arises as a result of the fact that the electric fields, which are produced by the voltage differences, affect the linear, rather than the rotary, motions of the atomic electrons. For a material like glass, without any natural directionality, this means that we can only affect a transverse wave motion, like light, via a transverse electric field. An electron vibrating in the direction of light propagation can have no effect in the direction at right angles. So we cannot stretch a straight fibre between a high voltage and a low voltage point and expect the voltage to affect the polarization properties.

We can, however, arrange the fibre shown in Fig. 13, in a spiral. Now we have the electric field acting transversely to the fibre at each point, so the atomic electrons are able to move less easily in the direction of the field than in

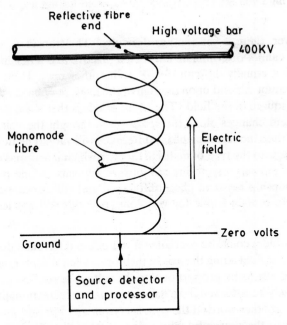

Fig. 13. Voltage can be measured by allowing an electric field to act transversely on the fibre, thus altering the polarization state of the propagating light. The light passes up to the high-voltage point and back again, allowing all the processing to occur at earth potential.

the direction at right angles to it. Consequently, light which is polarized in the same direction as the field travels more slowly than light polarized in the orthogonal direction (cf. Fig. 14). The result is that if light waves in each of these two direction enter the fibre in phase, they will emerge with a phase

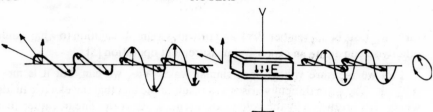

Fig. 14. The effect of a transversely acting electric field on a light wave propagating in a medium is to delay the wave polarized in the direction at right angles to the field with respect to that parallel with the field. A wave linearly polarized at 45° to the field will thus be converted into an elliptically polarized wave, and the form of the ellipse can be used to measure the electric field, and hence the voltage which gave rise to it.

difference between them, and this phase difference will depend upon the integrated effect of the transverse electric field along the length of the fibre. This effect can thus be used to measure the voltage which cause the fields and one way of doing this is also shown in Fig. 13, where the light is back reflected from the high voltage end of the fibre to avoid having any active components there.

However, there is yet another difficulty. The transverse vibrations of the electrons cannot discriminate between a positive and a negative electric field. They find it equally difficult to vibrate in either one. Hence the phase difference cannot depend upon the sign of the field; the phase will in fact depend upon the square of the field. The result of this is that if the distribution of the electric field changes along the spiral (even thought the voltage may remain constant) then the resulting phase difference will vary. The field distribution is very sensitive to the type of material through which it acts and can change for a number of reasons, say, if dirt or moisture deposits on one part of the spiral. The relationship between phase difference and voltage cannot, therefore, be expected to remain stable for very long, and this will not lead to a reliable voltmeter.

The problem could be overcome if the phase difference depended linearly on an electric field along the axis of the fibre. Then the integrated effect of the field would always be proportional to the voltage regardless of its distribution. This can only be achieved if we have some natural directionality in the core of the fibre – in other words if the core is crystalline. For this, and other reasons, research into the successful fabrication of crystalline fibres is being pursued. Despite several attempts, however, suitable crystalline-cored fibres do not yet exist

Whilst awaiting the right fibre there is another way round the voltage measurement problem which also has much wider implications. Consider the arrangement shown in Fig. 15. A short pulse of light is launched into a length of monomode fibre and as the pulse propagates along the fibre, a very small fraction of the light will be continuously scattered back towards the launch end

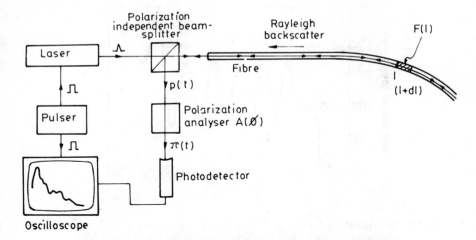

Fig. 15. Polarization-optical time-domain reflectometry (POTDR) allows the spatial distribution of polarization properties of a monomode fibre to be determined, by time resolving the polarization state of the light backscattered from a light pulse propagating down the fibre. This allows the spatial distribution of any external field which affects the polarization properties of the fibre also to be determined.

by small imperfections in the fibre material. Since the velocity of light in the fibre is known (about 2×10^8 meters per second) it is possible, by noting how the returning light level varies with time, to determine from which point in the fibre each portion of the light has been returned. This technique, entirely analogous to radar, is known as Optical Time Domain Reflectrometry (OTDR) and is routinely used in telecommunications to locate fibre breaks, bad joints or very 'lossy' sections of fibre [6]. It can be used for both monomode and multimode fibres.

Suppose, now, that in a monomode fibre we also note how the polarization state of the returning light varies with time. This now tells how the polarization properties of the fibre vary along its length and, if some external agency such as an electric or magnetic field is acting on the fibre to alter the fibre's polarization properties, then we have a method for measuring the distribution of the agency along the length of the fibre. This technique, known as Polarization Optical Time Domain Reflectrometry (POTDR) [7] could clearly be a very powerful probe for determining the way in which various important quantities varied with position, as well as with time.

The technique has already been well explored in the laboratory and has, for example, been used to show the cyclical variation of a monomode fibre's polarization properties when it is wound on a circular drum (see Fig. 16). We see also how the technique could help solve the voltage-measurement problem in the device shown in Fig. 13. Thus we now have the means available to measure the electric field distribution along the axis of the fibre spiral—

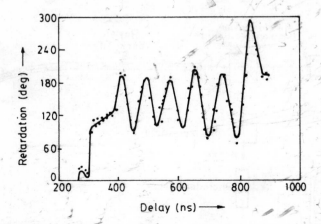

Fig. 16. The cyclical variation of a fibre's polarization properties which results from winding it on a drum is here revealed by means of the POTDR technique.

whatever it is, and however it may change. Armed with POTDR, we can always calculate the true voltage, and it has many other possible uses: the location of faults on high-voltage lines; the monitoring of vibration patterns on large structures such as boilers or generators; the temperature distributions in transformers or oil-filled cables; the location of sources of electromagnetic interference.

However, POTDR is not without its problems. The back-scattered signal is weak and, to get good spatial resolution, processing of the signal must be very fast; with a light speed of 2×10^8 meters per second, only one nanosecond is allowed for a positional accuracy of one meter. Moreover, each of the applications mentioned above does required a specialised form of fibre if it is to be successful. These special fibres do not yet exist. Again we see where the main thrust of research must lie for widely successful measurement applications of optical fibres.

Nevertheless, in the meantime, all is not lost. In the first place we need not rely solely on polarization properties of fibres for measurement functions. Consider the range of devices shown in Fig. 17. Each one uses a multimode fibre and relies on changes in light power to make its measurement. The 'fire-alarm' device is perhaps the simplest of all types of optical fibre sensor. Light is allowed to bridge the gap in the fibre caused by breaking the fibre and bringing the two ends quite closely together again. Any effect (such as temperature acting on a bimetallic strip) which can cause misalignment of the two ends, will vary the light power arriving at the output end of the fibre, and this can thus be used as a measure of the strength of the effect.

The flow sensor uses the fact that the mode structure depends on the fibre geometry (since the geometry fixes the phase condition) and thus, if the fibre

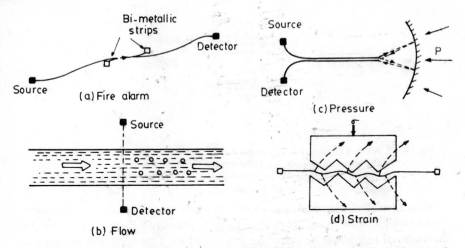

Fig. 17. Four quite simple optical fibre sensors employing light power modulation. Each one uses a multimode fibre and relies solely on changes in light power to make its measurement.

oscillates, the mode structure oscillates with it. If a fibre is stretched across a liquid flow tube, then it will oscillate at a frequency which will depend on the flow rate, due to the turbulence it creates and the frequency at which eddies are produced. Hence a light detector which picks up part of the mode structure will give an output whose frequency will be a measure of the flow rate.

In the pressure sensor, the pressure will vary the amount of light reflected from the first fibre back into the second fibre. In the final application illustrated, we may sense displacement by causing it to vary the bend radii for a corrugation in the fibre. The bending will cause some modes to fall outside the total-internal-reflection condition and hence cause light to be lost from the core. The smaller the bend radius the greater the loss of light power. All these intensity sensors have the advantages of being simple, reliable, insulating, intrinsically safe, and cheap [9].

Figure 18 illustrates the principle on which another broad class of optical fibre measurement sensors is based—the so-called Mach-Zehnder principle [9]. Light propagates in a monomode fibre which forks into two separate fibres and which then come together again. The light splits into two at the fork and, when the light beams are rejoined, they interfere to cancel or reinforce at the second junction, depending on their phase relationship. Thus the light power emerging at the output of the device is very sensitive to any phase differences between the two arms. If one arm is protected from the external world, the other can be exposed to a quantity which is capable of causing a phase change in the light propagating within it. The output light power then gives a measure of the quantity.

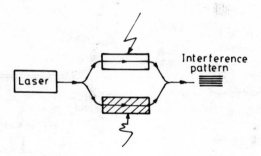

Fig. 18. A 'Mach-Zehnder' interferometric sensor uses the phase difference introduced by the 'measurand' (quantity to be measured) into one arm of an optical bridge. The phase displacement between the arms is observed in the interference pattern when the two light paths are recombined.

Now almost all physical quantities will cause phase changes in the light propagating in the exposed arm. For example, pressure will alter the density of the material and hence its refractive index; and we have seen how an electric field can alter the light velocity in a material. This type of device can thus measure many quantities, and it is extremely sensitive, since it can easily detect changes in phase of one thousandth of a radian [10]. This corresponds to a variation of about one part in one hundred million in the quantity to be measured.

An interesting special case (Sagnac configuration) of a Mach-Zehnder device is the optical fibre gyroscope [11], shown in Fig. 19. Here, the two arms are within the same fibre, which is in the form of a number of loops, the two beams passing along it in opposite directions. If any mechanical rotation of the

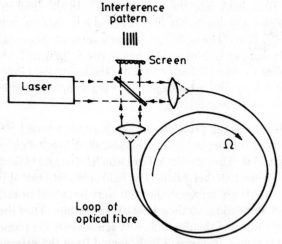

Fig. 19. An optical fibre gyroscope is a form of 'Mach-Zehnder' device where the two arms use the same fibre but propagation occurs in opposite directions. Any rotation of the system introduces a phase difference between the two directions which is sensed via the interference pattern.

device occurs, one beam has further to travel in the fibre than the other, so that a phase displacement results. This shows up in a variation of the interference pattern. Rotations of less than one tenth of a degree per hour have been measured in this way. It is clear that such a device has significant advantages over an internal (rotating disc) gyroscope, since there are no metallic elements or moving parts.

We can also go some way towards making distributed measurements with multimode fibres. Consider the arrangement shown in Fig. 20. Here we have several sections of fibre separated by thin plates of ruby glass. Ruby glass transmits long light wavelengths easily but short wavelengths poorly, and the region of transition between these states is quite narrow. Moreover, the value of the centre wavelength for the transition region depends markedly on

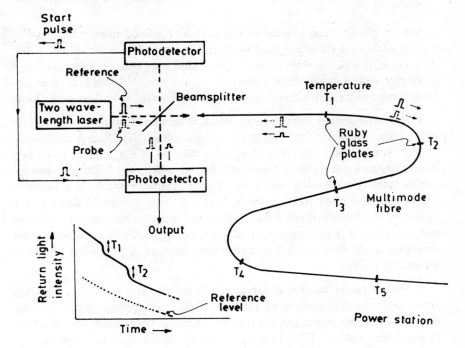

Fig. 20. A multipoint temperature sensor may be constructed in the following way: two optical pulses, of different wavelengths, are launched into a multimode fibre, along which is positioned a series of ruby-glass plates at the locations where temperature measurements are derived. The absorption of the red wavelength by the ruby-glass is dependent on temperature whereas that of the blue wavelength is not. Light backscattered from the fibre is detected with the help of a beam splitter. The point at which backscatter occurs for any part of the returning light is known from the time it takes to reach the photodetector. Thus, the absorption introduced by each individual plate can be determined, and so can the temperature at the plate. The blue light provides a reference level, allowing removal of all other effects which can cause optical power variations.

temperature. Suppose we transmit light of two wavelengths along our sectioned device, one wavelength being in the transmission range close to the transition, the other lying within the transition region. The ratio of the two received intensities will clearly vary with temperature, since one varies with temperature whilst the other does not.

All the effects such as bending of the fibre, junction losses etc., will be closely similar for the two wavelengths, so that the ratio will effectively vary only with temperature. If we now pulse the two light wavelengths and look at the backscattered light, as in OTDR, then it will be possible to determine the differential attenuation due to each piece of ruby glass separately, and thus to determine the temperature at a chosen number of positions along the length of the fibre. Such a device has been constructed in the laboratory and has worked well [12].

Multimode fibre can be used, and up to about ten fixed measurement points may be addressed: the attenuation becomes too severe for many more than this. A multipoint measurement sensor of this kind may best be described as a quasi-distributed device (it is sometimes called a multiplexed sensor as discussed in a companion chapter) to distinguish it from the fully-distributed kind.

Distributed temperature measurement, in particular, has many important applications in electricity supply [8]. High-voltage oil-filled power transformers can develop 'hot spots' due to degradation of the primary/secondary insulation, and the result is sometimes catastrophic. Oil-filled high-voltage cables suffer from similar problems. Power-generator windings sometimes overheat, and a close watch must be kept over long lengths of these. Such applications ideally require a fully distributed temperature measurement arrangement, as it is not usually possible to predict in advance just where the problems might occur, and thus a very large number of points must be kept under observation.

A measurement method capable of fulfilling the above requirements has come under investigation quite recently. Unlike all previous method so far discussed this one makes use of a non-linear, rather than linear, optical effect. A non-linear optical effect, broadly, is one where the light propagating in a material medium is of such large intensity that it is capable of altering its own propagation conditions. A common consequence of this is the generation of optical frequencies other than that of the incident light.

The particular non-linear effect on which the present method relies is the Raman effect, whereby molecular vibration and rotations within a material medium give rise to a modulation of the light propagating within it. The interaction must properly be described at the quantum level, of course, and the effect corresponds to the absorption (by a molecule), of an incident photon, and the re-emission of a photon of different energy (and therefore different wavelength), either greater or smaller than that of the incident photon. In either case, the energy difference is equal to one of the discrete vibrational or rotational energies of the molecule. Now, if the emitted photon is to have greater energy than the original, the molecule must provide the extra energy required, and this can only occur if the molecule is already in an excited state. But the number of molecules in excited states under conditions of thermal equilibrium depends directly on the absolute temperature, and thus the level of radiation scattered at higher energy can be used to measure absolute temperature. This higher-energy radiation is known as anti-Stokes radiation, and that scattered at lower energy is called Stokes radiation, the latter normally having the higher level.

The arrangement which uses this phenomenon in a distributed temperature measurement system is shown in Fig. 21. An optical pulse with high peak power is launched into a multimode optical fibre, and backscattered levels at the Stokes and anti-Stokes frequencies are measured as a function of time. The temperature dependence of the anti-Stokes radiation is normalised on dividing by the Stokes level, since the latter acts as a natural reference (as does the ruby glass device) to remove any effects, other than temperature, which cause loss. An experimental system has provided a temperature resolution around 5 °K with a spatial resolution of a few meters [13], and there is good reason to believe that these figures can be improved.

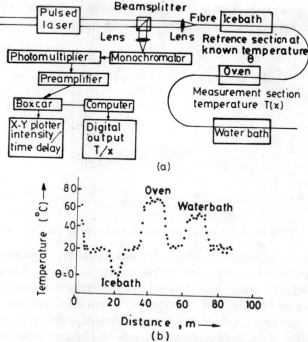

Fig. 21. Differential absorption ratio thermometry.

Two important features of this device should be emphasised: the measurement is independent of the fibre material; and any fibre, via this technique, is potentially a distributed thermometer (over its first 1 km or so), no matter for what purpose it was originally installed.

6. THE FUTURE

What does the future hold for optical fibre measurement sensors? We have noted that much materials research needs to be done to produce fibres with properties better-suited to the measured function. If these were available, more active attention could be turned to devising a range of distributed optical-fibre sensors, for these are so much more versatile than point sensors, especially for large plant applications. POTDR would benefit considerably from the availability of suitable fibres, but it still faces the problem of low-level backscatter.

For better methods of implementing distributed systems, we may look towards an entirely new kind of optics—non-linear optics. A non-linear optical effect is one where the refractive index of a material is actually changed by the light which is passing through it. We noted earlier that a light wave interacts with the atomic electrons in a material as it passes through it.

Suppose there are two light waves passing in opposite directions through a medium, as in Fig. 22. Under normal circumstances the electrons are quite happy to respond to each wave quite independently, so that the two waves are blissfully unaware of each other's existence—each propagates as if the other

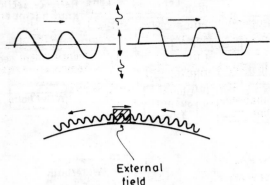

Fig. 22. Pulse/wave nonlinear interaction in an optical fibre.

were not there. However, if one of the waves has a very large amplitude, the electrons are strained within the atomic structure. This strain will also be sensed by the other wave. In this case, the refractive index for the second wave has been altered by the first, and the two waves effectively interact. This interaction may take the form of a change in velocity, absorption, wavelength or polarization state, depending on the precise physical situation. Now we know that the atomic electrons can also be influenced by external fields: electric or magnetic fields, for example. Hence it follows that external fields may themselves influence the optical interaction between two waves.

How can all this help us to provide better distributed optical fibre measurement systems? Consider the arrangement in Fig. 23. Here, a continuous optical wave is passing in one direction down an optical fibre, and an optical

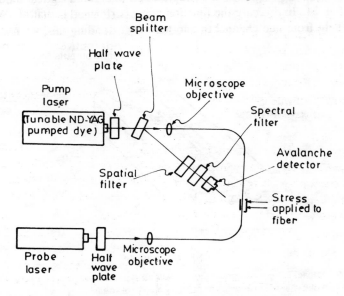

Fig. 23. Set-up for pulse/wave interaction in an optical fibre.

pulse is passing in the opposite direction. We know the speed of the optical pulse so we know exactly where it is at any time. If the interaction between the two waves is being influenced by an external agency, then this influence will be evident in the time variation of the continuous wave as it emerges from the fibre. We shall thus be able to infer the magnitude of the external field at each point along the fibre.

In this case we have not used a low level backscatter signal but a high level forward-transmitted signal. Our signal processing is consequently easier. There are several non-linear effects which might be exploited for this purpose, and research in this area has only just begun [14].

Some signal processing will always be required for both communications and measurement purposes. As bandwidths become greater and as measurements become faster, so more and faster processing will need to be done, preferably on the optical signal itself. This requirement has led to the development of a new set of techniques which combine optics with electronics, and which are grouped under the generic title of 'integrated optics' [15].

Integrated optics (cf. Ch. 21) is concerned with the development of compact, solid packages which consist, basically, of a wafer of suitable material modified and processed in a variety of ways so as to allow a variety of electronic controls over the light which enters the package. In much the same way, an integrated-circuit silicon chip allows control over electric currents, and both packages have the considerable advantage of lending themselves to reliable mass production techniques. An example of an integrated-optical module is shown in Fig. 24. In this case the function being performed is that of a simple switch of light from one channel to another. By extending this, we may look

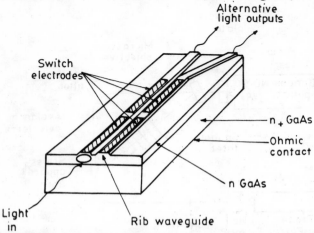

Fig. 24. A compact integrated-optical module will allow rapid switching of a light wave from one channel to another by applying a control voltage.

forward to the prospect of being able to devise optical modules which will amplify light, switch it, attenuate it, change its direction, its colour, its velocity or split it into two or more other beams, all under electronic control. The marriage between such a package and a microprocessor will clearly be a particularly powerful one, adding intelligence to the control.

Non-linear-optical effects in integrated-optical packages can be exploited in a variety of ways but probably the most exciting of these is in the provision of very fast optical switches: we can cause light to switch light, this switching can be performed in about one picosecond (one million millionth of a second) and has obvious implications for the next generation of very fast computers.

It is evident, then, that optical technology can provide us with the means for gathering, communicating and processing enormous quantities of information at much greater speeds than has hitherto been possible. Optical fibres occupy a central role in this technology. They can perform the communications and measurement functions safely and reliably in electrically hostile environments, and, since they are manufactured from one of the world's most abundant substances, sand, they can also perform them cheaply.

Optical fibre communications are fairly well established. Optical fibre measurement and optical signal processing are still in their infancy. There is much to be done.

REFERENCES

1. J.R. Osterfield, Norman S.R., McIntosh D.N., Rogers A.J., Castelle R., Tamburello, "An Optical Fibre Link in a Mountainous Environment", Proc. 29th Int. Wire and Cable Symposium, Cherry Hill, USA, pp. 202-210 (Nov. 1980).
2. P. Gaylard, Dey P., Holden G., "Telecommunications by Optical Fibres in Overhead Power Conductors", Wire Industry, pp. 473-475 (July 1981).
3. B.J. Maddock and Pullen F.D., "Telecommunications Using Optical Fibres on Power Lines", IEE Conf. Publication No. 193, pp. 76-79 (1981).
4. A.M. Smith, "Optical Fibre Current Measurement Device at a Generating Station", European conference on Optical Systems and Applications, Utrecht, SPIE Publication No. 236, pp. 352-357.
5. A.J. Rogers, "Optical-Fibre Current Measurement", Proc. EFOC/LAN'88, Amsterdam, IGI Europe, pp. 203-211 (July 1988).
6. M.K. Barnoski, Rourke, M.D., Jensen, S.M., and Melville R.T., "Optical Time Domain Reflectometry" *Appl. Opt.*, 16, 2375-2379 (1977).
7. A.J. Rogers, "Polarization-Optical Time Domain Reflectrometry", *Appl. Opt.*, 20, 1060-1074 (1981).
8. A.J. Rogers, "Distributed Optical Fibre Sensors for the Measurement of Pressure, Strain and Temperature", *Physics Reports*, 169, No. 2, pp. 99-143 (1988).
9. B. Culshaw, "Optical Fibre Sensors and Signal Processing', Peter Peregrinus (1984).
10. B. Culshaw, "Interferometric Optical Fibre Sensors", *Int. J. of Optical Sensors*, 1, No. 3, pp. 237-254 (1983).
11. V. Vali and Shorthill R.W., "Fibre Ring Interferometer", *Appl. Opt.* 15, No. 5, pp. 1099-1100 (1976).
12. E. Theocharous, "Differential Absorption Distributed Thermometer", Proc. 1st Int. Conf. on Optical Fibre Sensors, London, IEE Conf. Publication No. 221, pp. 10-12 (1983).
13. J.P. Dakin, Pratt D.J., Bibby G.W. and Ross J.N., "Distributed Anti-Stokes Raman Thermometry", Proc. 3rd Int. Conf. on Optical Fibre Sensors, San Diego, post-deadline paper (1985).
14. A.J. Rogers, "Distributed Optical Fibre Sensors Based on Counter-Propagating Beams and Non-Linear Interactions", Proc. Int. Conf. on Optical Science and Engineering, SPIE, Hamburg, paper 1011-01 (1988).
15. R.G. Hunsperger, "Integrated Optics: Theory and Technology", 2nd Edition, Springer-Verlag, Berlin (1984).

Index

Acousto optic modulator, 24
Advantage of strip waveguide over bulk modulator, 520
Aerial fibre cable, 323, 324
Alternate Mark Inversion (AMI), 441, 444, 446
Amplitude shift keying (ASK), 433-435, 437, 438, 471, 474-477, 486, 494, 495, 498, 499
Analog to Digital Conversion (ADC), 7, 424, 426, 433, 439
Anisotropic fibres, 158
APCM, 428
Attenuation in optical fibres, 79
    absorbtive, 81
    connector, 447
    hydrogen induced, 118, 304, 309, 310
    macrobend, 84, 86, 116, 304-306, 308, 314-316, 318
    microbend, 86, 87, 101, 116-120, 304, 307
    scattering, 116
    single-mode fibre, 275
    splice, 89, 90, 101, 117, 250, 447
    radiation, 83, 120
Automatic frequency control (AFC), 473, 483, 488, 491

Beam walk-off, 148
Beat length, 127, 138, 251, 297, 298, 300-302
Bipolar Junction Transistor (BJT), 394, 396, 401, 408
Birefringence, 48, 126-129, 131, 137, 138, 155, 156, 250, 297, 298, 521

    circular, 137, 661, 672
    geometrical, 128, 129
    linear, 635, 660, 670
    modal, 135
    stress-induced, 129, 132
Bit-error-rate (BER), 471, 474, 477, 482, 483, 488, 491
BNZS coding, 445, 446
Bow-tie fibres, 136
BPSK, 434, 436

Cerenkov configuration, 210, 211, 215, 217
Chirping, 17, 18, 347, 533
Chopper, 253
Chromatic dispersion, 13, 68, 149, 150, 156, 162, 163, 168, 496
Circularly birefringent fibres, 137, 138
Cladding mode stripper, 254, 256
Coherent communication systems, 143, 331, 347, 470
Coherent detection, 471, 472, 473, 475, 500, 714, 720
    coherence requirement, 722
Coherence length, 150, 154, 169, 175
Colour centres, 167, 168
Concatenated fibre links, 274
Confocal focusing, 200
Costas loop, 477
Coupled mode analysis, 40
Coupled-wive equation, 353, 354, 529
Coupling length, 42
CPSK, 436, 437, 438

Critical periodicity for microbending in fibres, 86, 406
Curie temperature, 184, 196
Cut-back method, 265, 270, 271, 273
Cut-off wavelength, 62, 102, 103, 112, 113, 117, 119, 120, 121, 251, 252, 466

dBm, 453
Depressed clad fibre, 112, 121, 122
Depth of modulation, 165, 533
Differential mode delay (DMD), 276
Differential mode loss, 251, 269-271
Differential PCM, 430
Digital sum variation, 458
Digital to analog converter (DAC), 7, 515
Direct detection, 470, 471, 472, 475
Dispersion, 294, 296, 460, 467
    Gaussian temporal pulse, 69
    graded core single-mode fibre, 78, 79
    intermodal, 68
    intramodal, 79
    material, 68, 76, 77, 119, 120
    waveguide, 68, 69, 77, 119, 120
Dispersion coefficient, 75, 344, 345, 462
    material, 68, 76, 104, 119
    waveguide, 75, 76, 104, 119
Dispersion flattened fibres, 14, 104, 105, 124
Dispersion shifted fibres, 8, 13, 19, 104, 105, 123
DPCM, 430
DPSK, 436, 438

Effective cut-off wavelength, 62, 102, 104, 113, 114, 288-292
    length dependence, 113, 294
Eigen value equation under WKB, 68
Electric field induced second harmonic (EFISH) generation, 163, 169, 170, 178, 181, 182, 183
Electric supply industry, 741-743, 749, 753
Electrooptic effect, 150, 512, 513
Electro-optic modulator, 24
Electro-optic tensor, 513
Elliptic core fibres, 48, 129, 130, 131, 132, 160, 300

Embedded strip waveguide, 46, 48
Equilibrium mode distribution (EMD), 265, 266, 274
Equivalent step index (ESI) profile, 96, 100, 101, 113, 250, 251, 252, 286, 287
Error rates, 437
Extraordinary mode/wave, 517, 521
Eye diagram, 404, 464

Fabry-Perot cavity/interferometer, 336, 337, 343, 351, 500
Faraday rotator, 497
FDDI, 10
Feedback cascode amplifier, 400, 401
Fibre ageing due to hydrogen migration, 118, 119
Fibre amplifier, 14, 19
Fibre diameter monitor, 226
Fibre directional coupler, 602, 620, 662, 663, 684-687
    coupling coefficient, 606, 609, 622
    dependence on external refractive index, 611, 616
    fused, 602, 603, 617, 620
    physical principle, 604
    polarisation effects, 607, 609, 615, 633-635, 685
    polished, 602, 603, 623, 625
    theoretical modelling, 612
    3×3, 691
    wavelength dependence, 608, 615
Fibre drawing, 224, 225, 226
    double crucible method, 228
    neckdown region of fibre preform, 227
    rod-in-tube method, 227, 228
Fibre far-field, 99, 100, 252, 285, 286
Fibre filter, 639
    channel dropping, 641
    grating, 641, 642
    tapered fibre, 640
    tunable, 643
Fibre isolator, 620, 637, 638
Fibre near field, 95, 99-101, 252
Fibre optic Bragg cell, 597
Fibre optic chemical analysis, 577
Fibre optic frequency modulator, 651

INDEX

Fibre optic frequency shifter, 682, 683, 694
Fibre optic gyroscope, 142, 548, 586, 590, 592, 593, 602
Fibre optic hydrophone, 548, 588, 592
Fibre optic intensity modulator, 648
    liquid crystal, 648
    Mach-Zehnder, 649
Fibre optic loop relector, 666
Fibre optic magnetometer, 592
Fibre optic phase modulator, 649
Fibre optic refractometer, 571-574
Fibre optic serrodyne, 597
Fibre optic switch, 648
Fibre optic thermometer, 574-577, 764, 766
Fibre polariser, 620, 626
    birefringent crystal, 627
    coiled birefringent, 632
    continuous length, 630
    D-fibres, 631
    metal clad, 627
    thin metal clad, 629
Fibre polarisation controller, 479, 480
Fibre polarisation splitter, 632
Fibre preform, 223
Fibre reinforced plastic (FRP), 325
Fibre sensor
    acceleration, 570
    birefringent, 589-590
    cross talk, 561
    current, 755-757
    distributed, 578-580, 765
    electric field, 571
    evanescent wave, 559, 560
    extrinsic, 549
    fire alarm, 761
    flow measurement, 762
    frustrated total internal reflection, 553
    interferometric, 584-600
    intrinsic, 549
    liquid level, 561-562
    magnetic field, 570
    microbend, 563, 565-571
    polarimetric, 670-672
    pressure, 551-555, 569, 762
    reflective, 557
    Shutter/Schlieren, 553, 555
    temperature, 561, 570, 574-577
    transfer function, 658
    voltage, 758
Fibre sensor multiplexing, 706
    binary codes, 713
    chirp modulation, 713
    coherence, 730
    coherent detection, 720
    differential detection, 718
    frequency division, 724, 725
    incoherent detection, 709
    ladder configuration, 708
    mode division, 715, 716
    recursive array configuration, 731
    reflective array, 733
    series topology, 736
    space division, 728-729
    time division, 709-711
    two-mode fibre, 737
    wavelength division, 715
Fibre squeezers, 497

Field Effect Transistor (FET), 390, 391, 393, 399, 406, 407
Frequency division multiplexing (FDM), 415, 417, 420, 448, 489
Frequency doubling, 147, 177, 195
Frequency locking loop, 500
Frequency mixing, 147, 152
    sum and difference, 151, 152
Frequency shift keying (FSK), 347, 433, 434, 435, 437, 474, 475-477, 483-488, 494-496, 498, 499
Fusion splicing machine, 92

Gas lens, 4, 5
Graded index fibres, 64, 65, 224, 229, 268
Group delay, 68, 70
Group index, 70
Group velocity, 56, 149
Guided modes, 56, 57, 64
    number of modes, 68

HD-WDM, 19

Heterodyne detection, 470, 476, 483, 489, 587, 694, 721
    detector noise, 696
    frequency noise, 698
    intensity noise, 697
    noise consideration, 695
    pseudo-heterodyne, 726
High damage threshold, 148, 149
Homodyne detection, 470, 473, 480, 483, 496, 721
    active processing, 688
    passive processing, 690-693, 695

Index ellipsoid, 513-515
Infrared fibres, 8, 10, 107, 124
Interference filter, 253
Interferometric fibre sensor, 142, 658
    basic principle, 585
    components, 594-598
    Fabry-Perot, 589-591, 667-669
    Mach-Zehnder, 589, 592, 662, 762, 763
    Michelson, 589, 592, 664, 665
    noise penalty, 588
    Sagnac, 589, 665
    sensitivity, 587, 589
Intermediate frequency (IF), 473, 476, 477, 485, 486, 488, 489, 491, 494, 495, 497, 498
Intersymbol interference (ISI), 405, 441, 446-448

Jitter, 462, 465

Knife-edge method, 285

Lambertain source, 88
Laplace spot size, 101
Lasers
    AlGaGs, 6, 8, 15, 340
    broad area, 339
    buried heterostructure, 339, 340
    cleaved coupled-cavity (c^3), 346
    DCPBH DFB, 361-363, 366-369
    diode; 19, 75, 117, 249, 253, 416

    distributed Bragg reflector (DBR), 41, 43, 352
    distributed feedback (DFB), 8, 16-19, 39, 43, 103, 346, 348, 352, 353, 355, 356-360, 364-367, 370, 471, 476, 479, 486-489, 493, 495
    EMBH DFB, 361
    Fabry Perot (FP), 356-358, 360
    gain switching, 16
    heterostructure semiconductor, 337, 338, 340
    index guided, 361
    injection, 382
    InGaAsP, 8, 15, 18, 340, 341, 353, 360, 370
    longitudinal modes, 16, 337, 344, 351
    Phase noise, 476, 587
    population inversion, 336
    Q-switching, 18
    relaxation oscillation, 343
    ridge waveguide (RWG) DFB, 361
    Schawlow-Townes formula, 360
    semiconductor, 12-16
    single-frequency semiconductor, 16, 344, 345
    Spontaneous emission, 335-337, 342
    Stimulated emission, 335-337, 342
Leaky modes, 35, 36, 49
Leaky ray correction, 256-257
LED, 6, 12, 75, 89, 249, 746, 751
Light-current (L-I) characteristics of lasers, 341-343
Limited phase space (LPS), 265
Local area network (LAN), 10, 12, 489, 602, 622
Lock-in-amplifier, 253
log-PCM, 428, 429
Loose tube structures, 320, 321
Loss stability, 118
LP-modes, 58, 61

Mach-Zehnder interferometer, 662, 728
    logic operations, 526
    modulator, 512, 523, 524, 525, 535
    power output, 525
    visibility, 663

INDEX

Y-branch, 523
Magneto-optic modulation, 300
Matched clad fibre, 104, 105, 112, 120, 121, 122
Material dispersion, 13, 208
Microcracks on fibre surface, 224, 251
Modal field profile, 94
Modal noise, 12, 16, 113, 120
Mode conversion, 31, 523
Mode coupling parameter, 128
Mode cut-off, 29, 31, 60, 61, 121
Mode field diameter (MFD), 94, 96, 251, 282, 283, 286, 289, 292, 293, 519
Mode field radius (MFR), 94, 114, 115, 118, 121
Mode filter, 265
Mode scramblers, 274
Mode spot size, 94, 95-98, 251
Modified chemical vapour deposition (MCVD), 120, 232-234, 236, 237, 239, 240, 243, 244
Modulation
    amplitude (AM), 17, 347, 415-417, 421, 425
    CPFSK, 494
    delta (DM), 430, 431
    DPSK, 476, 491, 494, 495, 496, 498, 499
    frequency (FM), 17, 347, 415, 417, 420, 421, 425, 682
    intensity, 470, 549, 551, 552
    phase (PM), 415, 420, 421, 549, 659, 678
    polarisation, 521, 549, 679
    PRBS, 483, 489, 491
    pulse amplitude (PAM), 415, 421-423
    pulse code (PCM), 68, 416, 421, 423, 425, 427, 432, 433
    pulse length (PLM), 415, 421-423
    pulse position (PPM), 415, 421-423
    wavelength, 549
Molecular hyperpolarisability, 170, 178
Monochromator, 253

Nonlinear integrated optics (NLIO), 198, 199, 201, 205, 206, 208

Nonlinear polarisation, 202, 203, 215
Nonlinear susceptibility, 148, 152
Normalised propagation constant, 58
NRZ pulses, 18, 441, 444, 445
Numerical aperture, 53, 54, 98, 100, 251, 254, 255, 257-259, 262-264, 267, 311

OEIC, 20, 21
Optic axis, 514
Optical frequency division multiplexing (OFDM), 19
Optical Kerr effect, 152
Optical rectification, 151, 152
Optical waveguide switch, 531, 535
    cross state, 531, 532
    parallel state, 531, 532
    $\Delta\beta$-reversal, 532
Optical Time Domain Reflectometry (OTDR), 32, 271, 272
Optical waveguide
    anisotropic, 158
    anisotropic polarizer, 35
    asymmetric, 31
    cut-off frequency, 29
    leaky mode, 35
    metal clad, 32
    planar, 24-25
    periodic, 39
    quasi-mode, 37
    rectangular, 44, 45
    strip, 24, 25, 46, 48, 519
    symmetric, 28
Optimum profile fibres, 70, 71
Outside vapour-phase deposition (OVD), 122, 230-232, 243, 244
Overlap integral, 205, 285, 519

PANDA fibres, 136, 137
Parametric amplification, 147
Parametric interactions, 198, 202
Parametric oscillation, 147
Partial Response Signalling (PRS), 441
PCM Eye pattern, 447
PCM-MUX, 432
Periodic modulation of nonlinearity, 165

INDEX

Periodic modulation of waveguide dimension, 164
Periodic modulation of waveguide refractive index, 164, 165, 171, 174
Periodic poled fibres, 183, 195
Periodic structures, 146, 147, 149, 150, 153, 162
Periodic waveguides, 38
    coupled-mode analysis, 40
Petermann spot size, 101, 296
Phase conjugation, 149
Phase diversity detection, 499, 500
Phase locked loop (PLL), 438, 439, 473, 477, 479, 480, 482, 494, 496
Phase matching, 146, 148-150, 200, 205, 206, 208, 209, 521, 530
    modal, 153, 156, 157, 160, 161
    non-critical, 148, 160, 195
    periodic, 150, 153, 166
    quasi, 150, 184, 185, 195, 196
    condition/criteria, 40, 153, 522
    Type I, 153
    Type II, 153
    spatially periodic nonlinearity, 162
    double phase matching, 196
Phase mis-match factor, 153, 530
Phase modulator, 516, 535
Phase shift keying (PSK), 347, 433-435, 474-477, 481, 485, 495
Photodetectors
    APD, 249, 271, 273, 331, 332, 375, 378, 379, 383, 386-388, 406, 408, 416, 472, 475, 751
    BER, 377, 379, 383, 463, 466, 467, 471, 477, 482, 483, 488, 491, 492
    dark current, 374, 375, 378, 383
    depletion region, 374, 378
    Johnson noise, 378
    SNR, 376, 378
    dynamic range, 379
    heterodyne detection, 347, 470, 473, 475, 483, 496
    homodyne detection, 347, 470, 473, 475, 479, 480, 496, 499
    InGsAsP system, 383
    internal current gain factor, 377, 380
    noise equivalent power (NEP), 378
    PIN, 249, 281, 331, 332, 373, 375, 378, 383, 386, 388, 406, 408, 416, 472
    PINFET, 475, 480, 483, 486, 491
    quantum efficiency, 375, 385
    responsivity, 376, 385, 386
    speed of response, 378, 387
Photophone experiment, 3, 4, 331
Piezoelectric fibre squeezer, 497
PIN-FET receivers, 406, 478, 480, 483, 491, 495
Planar waveguide, 25
    asymmetric, 27
    symmetric, 28
Plasma activated CVD (PCVD), 239, 243, 244
Plasma enhanced MCVD (PMCVD), 238, 239, 243, 244
Plasma outside deposition (POD), 240
p-n junction, 333-337, 342, 372-374
Pockels effect, 512, 537
Polarisation control, 478, 479, 497, 602, 635, 636, 680
Polarisation converter, 42, 516, 522
Polarisation crosstalk, 128
Polarisation dispersion, 469
Polarisation diversity receiver, 497, 498
Polarisation holding fibre, 480
Polarisation insensitive receiver, 497, 498
Polarisation maintaining fibre, 126-128
Polarisation mode dispersion, 126
Polarisation optical time domain reflectometry (POTDR), 759-760
Polarisation penalty in coherent detection, 478
Polarisation selective coupler, 522
Polarisation tracking receiver, 498
Power budget, 452, 455, 456, 470
Power-law index profile, 65, 66
    number of modes, 68
Power per unit bandwidth, 521
Power spectrum of curvature, 100, 315, 322
Prism-coupling technique, 301
    m-lines, 303
Profile dispersion, 70

INDEX

Proton exchange (PE) waveguide, 201, 208, 209, 216, 218, 220, 542
    phase matching diagram, 209
Pulse dispersion
    multimode fibres, 64, 72, 73
    single-mode fibres, 74
Push-pull effect, 525, 431

Q-switching, 18
QPSK, 436-438
Quantization noise, 427
Quantum phase noise, 494
Quasi-modes, 37-39

Radiation modes, 56, 60
Rayleigh scattering, 117, 120, 123, 269, 297, 299, 300, 747
Recombination
    radiative, 335, 340, 343
    non-radiative, 335, 343
    stimulated, 337, 338
    Auger, 343
Receiver overload level (RO), 455, 457, 467
Receiver sensitivity (RS), 454, 457, 459, 466, 467, 485
Refracted near field, 255, 258, 260, 261, 282
Resonance condition, 38
Ribbon cable, 327, 328
Ring resonator, 602, 669
RMS spot size, 99-102
RZ pulses, 443-445

Sampling theorem, 415, 421
Scalar wave equation, 56
Second harmonic generation (SHG), 147, 148, 151, 153, 161, 166, 198, 200, 204
    conversion efficiency, 185, 195-198, 203, 207
    Cerenkov configuration/radiation, 161, 209, 210, 212, 214, 216, 218
    non-periodic waveguides, 170
    periodic domain reversal, 176, 177
    periodic structures, 171, 175
    quasi-phase matching, 220, 221
Second order nonlinearity, 151, 163, 176, 199

Segmented (SEGCOR) core fibres, 8, 105, 123, 124
Self phase matching, 149
Self phase modulation, 152, 169
Self written gratings in fibres, 152, 166, 167
Sellmeier fit for refractive index, 295
SF_6 gas, 749
Shot noise spectral density, 473
Side-pit fibres, 132, 134, 139
Side-tunnel fibres, 50, 132-134, 139
Single polarisation bandwidth, 134
Single polarization single mode (SPSM) waveguide/fibres, 27, 29, 31, 49, 124, 127, 134, 138, 140, 141
Single side band signal (SSB), 416, 417
Slotted cable structure, 321, 325
Slowly varying approximation, 41
SNR, 417, 420, 428, 429, 439, 474
'Solgel' process, 106
Solitons, 10
SONET, 10, 18
Spatial modulation of refractive index, 174
Spectral penalty, 465
State of polarisation (SOP), 126
Steady-state mode distribution (SMD), 265, 266, 274
Stimulated Raman scattering, 147, 149
Stress applying parts (SAP), 135, 137
Stress corrosion, 319
Stress induced fibres, 134, 135
Submarine cable, 327
Surface plasmon mode/wave, 375, 628, 629

Tapped delay line (TDL), 447
TAT (Trans-atlantic transmission), 13
TDM-MUX, 431
TE-modes, 25-31, 33, 35, 36, 40, 48, 140, 206, 212, 536, 539, 541
Third order nonlinear susceptibility, 152, 168, 180
Three wave mixing, 153
Time division multiplex (TDM), 415, 422
TM-modes, 25, 26, 28-33, 36, 48, 142, 201, 208, 536, 539, 541
Transimpedance Front-end, 397
Transmitted near field, 255, 258, 261, 263

Transverse interferometry, 254, 261

Vapour deposition methods, 229
    chemical vapour deposition method, 233
    collapsing step, 231
    inside-vapour-deposition method, 231
    modified chemical vapour deposition method, 233-235, 237, 238
    outside-vapour-deposition method, 231, 232
    plasma activated CVD method, 240
    plasma enhanced MCVD method, 239, 240
    Plasma outside deposition method, 241
    vapour phase axial deposition method, 242-245

Variable aperture far-field (VAFF), 285
Vector wave equation, 55
V-number, 55, 57
Voltage controlled oscillator (VCO), 438, 439

Waveguide directional coupler, 527
    supermodes, 528
    coupling length, 529-531
    coupling coefficients, 529
    switch, 531, 532, 535
    polarisation splitting, 536, 537
    wavelength filter, 533, 534
Waveguide dispersion, 13, 68
Waveguide polariser, 537
    leaky mode, 537, 538
    metal clad, 539, 540
    thin metal clad, 540
Wavelength division multiplexing (WDM), 14, 16, 17, 19, 451, 602
Wavelength filter, 42, 521, 533
Wavelength multi/demultiplexers, 620, 643, 645
    Bragg grating, 646, 647
Weakly guiding fibres/waveguides, 56, 58, 65, 119, 154, 518

Zero dispersion wavelength, 13, 77, 78, 81, 121, 122, 463, 466, 467
Zero dispersion slope, 78